BRIEF TABLE OF INTEGRALS

1. $\int u^n\,du = \dfrac{u^{n+1}}{n+1} + C,\ n \neq -1$

2. $\int \dfrac{1}{u}\,du = \ln|u| + C$

3. $\int e^u\,du = e^u + C$

4. $\int a^u\,du = \dfrac{1}{\ln a}a^u + C$

5. $\int \sin u\,du = -\cos u + C$

6. $\int \cos u\,du = \sin u + C$

7. $\int \sec^2 u\,du = \tan u + C$

8. $\int \csc^2 u\,du = -\cot u + C$

9. $\int \sec u \tan u\,du = \sec u + C$

10. $\int \csc u \cot u\,du = -\csc u + C$

11. $\int \tan u\,du = -\ln|\cos u| + C$

12. $\int \cot u\,du = \ln|\sin u| + C$

13. $\int \sec u\,du = \ln|\sec u + \tan u| + C$

14. $\int \csc u\,du = \ln|\csc u - \cot u| + C$

15. $\int u \sin u\,du = \sin u - u\cos u + C$

16. $\int u \cos u\,du = \cos u + u\sin u + C$

17. $\int \sin^2 u\,du = \frac{1}{2}u - \frac{1}{4}\sin 2u + C$

18. $\int \cos^2 u\,du = \frac{1}{2}u + \frac{1}{4}\sin 2u + C$

19. $\int \tan^2 u\,du = \tan u - u + C$

20. $\int \cot^2 u\,du = -\cot u - u + C$

21. $\int \sin^3 u\,du = -\frac{1}{3}\left(2 + \sin^2 u\right)\cos u + C$

22. $\int \cos^3 u\,du = \frac{1}{3}\left(2 + \cos^2 u\right)\sin u + C$

23. $\int \tan^3 u\,du = \frac{1}{2}\tan^2 u + \ln|\cos u| + C$

24. $\int \cot^3 u\,du = -\frac{1}{2}\cot^2 u - \ln|\sin u| + C$

25. $\int \sec^3 u\,du = \frac{1}{2}\sec u \tan u + \frac{1}{2}\ln|\sec u + \tan u| + C$

26. $\int \csc^3 u\,du = -\frac{1}{2}\csc u \cot u + \frac{1}{2}\ln|\csc u - \cot u| + C$

27. $\int \sin au \cos bu\,du = \dfrac{\sin(a-b)u}{2(a-b)} - \dfrac{\sin(a+b)u}{2(a+b)} + C$

28. $\int \cos au \cos bu\,du = \dfrac{\sin(a-b)u}{2(a-b)} + \dfrac{\sin(a+b)u}{2(a+b)} + C$

29. $\int e^{au} \sin bu\,du = \dfrac{e^{au}}{a^2+b^2}\left(a\sin bu - b\cos bu\right) + C$

30. $\int e^{au} \cos bu\,du = \dfrac{e^{au}}{a^2+b^2}\left(a\cos bu + b\sin bu\right) + C$

31. $\int \sinh u\,du = \cosh u + C$

32. $\int \cosh u\,du = \sinh u + C$

33. $\int \operatorname{sech}^2 u\,du = \tanh u + C$

34. $\int \operatorname{csch}^2 u\,du = -\coth u + C$

35. $\int \tanh u\,du = \ln(\cosh u) + C$

36. $\int \coth u\,du = \ln|\sinh u| + C$

37. $\int \ln u\,du = u\ln u - u + C$

38. $\int u \ln u\,du = \frac{1}{2}u^2 \ln u - \frac{1}{4}u^2 + C$

39. $\int \dfrac{1}{\sqrt{a^2 - u^2}}\,du = \sin^{-1}\dfrac{u}{a} + C$

40. $\int \dfrac{1}{\sqrt{a^2 + u^2}}\,du = \ln\left|u + \sqrt{a^2 + u^2}\right| + C$

41. $\int \sqrt{a^2 - u^2}\,du = \dfrac{u}{2}\sqrt{a^2 - u^2} + \dfrac{a^2}{2}\sin^{-1}\dfrac{u}{a} + C$

42. $\int \sqrt{a^2 + u^2}\,du = \dfrac{u}{2}\sqrt{a^2 + u^2} + \dfrac{a^2}{2}\ln\left|u + \sqrt{a^2 + u^2}\right| + C$

43. $\int \dfrac{1}{a^2 + u^2}\,du = \dfrac{1}{a}\tan^{-1}\dfrac{u}{a} + C$

44. $\int \dfrac{1}{a^2 - u^2}\,du = \dfrac{1}{2a}\ln\left|\dfrac{a+u}{a-u}\right| + C$

***Note*:** Some techniques of integration, such as integration by parts and partial fractions, are reviewed in the *Student Resource Manual* that accompanies this text.

Tenth Edition

A FIRST COURSE IN DIFFERENTIAL EQUATIONS

with Modeling Applications

Tenth Edition

A FIRST COURSE IN DIFFERENTIAL EQUATIONS

with Modeling Applications

DENNIS G. ZILL
Loyola Marymount University

BROOKS/COLE
CENGAGE Learning

Australia • Brazil • Japan • Korea • Mexico • Singapore • Spain • United Kingdom • United States

A First Course in Differential Equations with Modeling Applications, Tenth Edition
Dennis G. Zill

Publisher:
Richard Stratton

Senior Sponsoring Editor:
Molly Taylor

Development Editor: Leslie Lahr

Assistant Editor:
Shaylin Walsh Hogan

Editorial Assistant: Alex Gontar

Media Editor: Andrew Coppola

Marketing Manager: Jennifer Jones

Marketing Coordinator:
Michael Ledesma

Marketing Communications Manager:
Mary Anne Payumo

Content Project Manager:
Alison Eigel Zade

Senior Art Director: Linda May

Manufacturing Planner: Doug Bertke

Rights Acquisition Specialist:
Shalice Shah-Caldwell

Production Service: MPS Limited,
a Macmillan Company

Text Designer: Diane Beasley

Projects Piece Designer:
Rokusek Design

Cover Designer:
One Good Dog Design

Cover Image:
©Stocktrek Images

Compositor: MPS Limited,
a Macmillan Company

Library of Congress Control Number: 2011944307
ISBN-13: 978-1-111-82705-2
ISBN-10: 1-111-82705-2

Brooks/Cole
20 Channel Center Street
Boston, MA 02210
USA

Printed in the United States of America
1 2 3 4 5 6 7 16 15 14 13 12

Contents

4 HIGHER-ORDER DIFFERENTIAL EQUATIONS 116

5 MODELING WITH HIGHER-ORDER DIFFERENTIAL EQUATIONS 192

6 SERIES SOLUTIONS OF LINEAR EQUATIONS 231

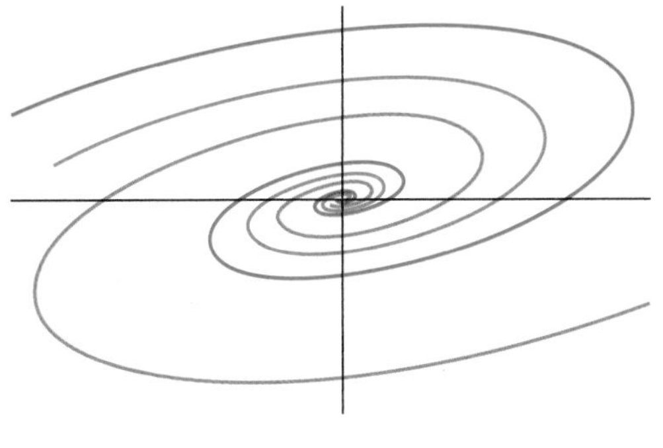

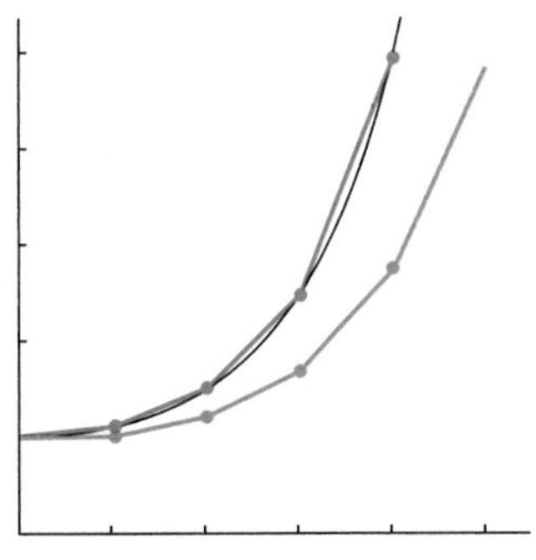

APPENDIXES

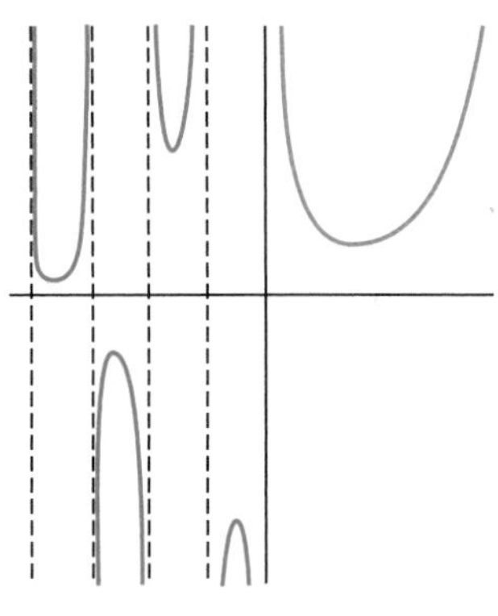

Preface

TO THE STUDENT

Authors of books live with the hope that someone actually *reads* them. Contrary to what you might believe, almost everything in a typical college-level mathematics text is written for you, and not the instructor. True, the topics covered in the text are chosen to appeal to instructors because they make the decision on whether to use it in their classes, but everything written in it is aimed directly at you, the student. So I want to encourage you—no, actually I want to *tell* you—to read this textbook! But do not read this text like you would a novel; you should not read it fast and you should not skip anything. Think of it as a *work*book. By this I mean that mathematics should always be read with pencil and paper at the ready because, most likely, you will have to *work* your way through the examples and the discussion. Before attempting any of the exercises, work *all* the examples in a section; the examples are constructed to illustrate what I consider the most important aspects of the section, and therefore, reflect the procedures necessary to work most of the problems in the exercise sets. I tell my students when reading an example, copy it down on a piece of paper, and do not look at the solution in the book. Try working it, then compare your results against the solution given, and, if necessary, resolve any differences. I have tried to include most of the important steps in each example, but if something is not clear you should always try—and here is where the pencil and paper come in again—to fill in the details or missing steps. This may not be easy, but that is part of the learning process. The accumulation of facts followed by the slow assimilation of understanding simply cannot be achieved without a struggle.

Specifically for you, a *Student Resource Manual* (*SRM*) is available as an optional supplement. In addition to containing worked-out solutions of selected problems from the exercises sets, the *SRM* contains hints for solving problems, extra examples, and a review of those areas of algebra and calculus that I feel are particularly important to the successful study of differential equations. Bear in mind you do not have to purchase the *SRM*; by following my pointers given at the beginning of most sections, you can review the appropriate mathematics from your old precalculus or calculus texts.

In conclusion, I wish you good luck and success. I hope you enjoy the text and the course you are about to embark on—as an undergraduate math major it was one of my favorites because I liked mathematics with a connection to the physical world. If you have any comments, or if you find any errors as you read/work your way through the text, or if you come up with a good idea for improving either it or the *SRM*, please feel free to contact me through my editor at Cengage Learning:

molly.taylor@cengage.com

TO THE INSTRUCTOR

In case you are examining this textbook for the first time, *A First Course in Differential Equations with Modeling Applications, Tenth Edition*, is intended for either a one-semester or a one-quarter course in ordinary differential equations. The longer version of the textbook, *Differential Equations with Boundary-Value Problems, Eighth Edition*, can be used for either a one-semester course, or a two-semester course covering ordinary and partial differential equations. This longer book includes six additional chapters that cover plane autonomous systems of differential equations, stability, Fourier series, Fourier transforms, linear partial differential equations and boundary-value problems, and numerical methods for partial differential equations. For a one semester course, I assume that the students have successfully completed at least two semesters of calculus. Since you are reading this, undoubtedly you have already examined the table of contents for the topics that are covered. You will not find a "suggested syllabus" in this preface; I will not pretend to be so wise as to tell other teachers what to teach. I feel that there is plenty of material here to pick from and to form a course to your liking. The textbook strikes a reasonable balance between the analytical, qualitative, and quantitative approaches to the study of differential equations. As far as my "underlying philosophy" it is this: An undergraduate textbook should be written with the student's understanding kept firmly in mind, which means to me that the material should be presented in a straightforward, readable, and helpful manner, while keeping the level of theory consistent with the notion of a "first course."

For those who are familiar with the previous editions, I would like to mention a few of the improvements made in this edition.

- Eight new projects appear at the beginning of the book. Each project includes a related problem set, and a correlation of the project material with a section in the text.
- Many exercise sets have been updated by the addition of new problems—especially discussion problems—to better test and challenge the students. In like manner, some exercise sets have been improved by sending some problems into retirement.
- Additional examples have been added to many sections.
- Several instructors took the time to e-mail me expressing their concerns about my approach to linear first-order differential equations. In response, Section 2.3, Linear Equations, has been rewritten with the intent to simplify the discussion.
- This edition contains a new section on Green's functions in Chapter 4 for those who have extra time in their course to consider this elegant application of variation of parameters in the solution of initial-value and boundary-value problems. Section 4.8 is optional and its content does not impact any other section.
- Section 5.1 now includes a discussion on how to use both trigonometric forms

  $$y = A\sin(\omega t + \phi) \quad \text{and} \quad y = A\cos(\omega t - \phi)$$

  in describing simple harmonic motion.
- At the request of users of the previous editions, a new section on the review of power series has been added to Chapter 6. Moreover, much of this chapter has been rewritten to improve clarity. In particular, the discussion of the modified Bessel functions and the spherical Bessel functions in Section 6.4 has been greatly expanded.

STUDENT RESOURCES

- ***Student Resource Manual*** (*SRM*), prepared by Warren S. Wright and Carol D. Wright (ISBN 9781133491927 accompanies *A First Course in Differential Equations with Modeling Applications, Tenth Edition* and ISBN 9781133491958 accompanies *Differential Equations with Boundary-Value Problems, Eighth Edition*), provides important review material from algebra and calculus, the solution of every third problem in each exercise set (with the exception of the Discussion Problems and Computer Lab Assignments), relevant command syntax for the computer algebra systems *Mathematica* and *Maple*, lists of important concepts, as well as helpful hints on how to start certain problems.

INSTRUCTOR RESOURCES

- ***Instructor's Solutions Manual*** (*ISM*) prepared by Warren S. Wright and Carol D. Wright (ISBN 9781133602293) provides complete, worked-out solutions for all problems in the text.
- ***Solution Builder*** is an online instructor database that offers complete, worked-out solutions for all exercises in the text, allowing you to create customized, secure solutions printouts (in PDF format) matched exactly to the problems you assign in class. Access is available via

  www.cengage.com/solutionbuilder
- ***ExamView*** testing software allows instructors to quickly create, deliver, and customize tests for class in print and online formats, and features automatic grading. Included is a test bank with hundreds of questions customized directly to the text, with all questions also provided in PDF and Microsoft Word formats for instructors who opt not to use the software component.
- ***Enhanced WebAssign*** is the most widely used homework system in higher education. Available for this title, Enhanced WebAssign allows you to assign, collect, grade, and record assignments via the Web. This proven homework system includes links to textbook sections, video examples, and problem specific tutorials. Enhanced WebAssign is more than a homework system—it is a complete learning system for students.

ACKNOWLEDGMENTS

I would like to single out a few people for special recognition. Many thanks to Molly Taylor (senior sponsoring editor), Shaylin Walsh Hogan (assistant editor), and Alex Gontar (editorial assistant) for orchestrating the development of this edition and its component materials. Alison Eigel Zade (content project manager) offered the resourcefulness, knowledge, and patience necessary to a seamless production process. Ed Dionne (project manager, MPS) worked tirelessly to provide top-notch publishing services. And finally, I thank Scott Brown for his superior skills as accuracy reviewer. Once again an especially heartfelt thank you to Leslie Lahr, developmental editor, for her support, sympathetic ear, willingness to communicate, suggestions, and for obtaining and organizing the excellent projects that appear at the front of the text. I also extend my sincerest appreciation to those individuals who took the time out of their busy schedules to submit a project:

Ivan Kramer, *University of Maryland—Baltimore County*
Tom LaFaro, *Gustavus Adolphus College*
Jo Gascoigne, *Fisheries Consultant*
C. J. Knickerbocker, *Sensis Corporation*
Kevin Cooper, *Washington State University*
Gilbert N. Lewis, *Michigan Technological University*
Michael Olinick, *Middlebury College*

Finally, over the years these textbooks have been improved in a countless number of ways through the suggestions and criticisms of the reviewers. Thus it is fitting to conclude with an acknowledgement of my debt to the following wonderful people for sharing their expertise and experience.

REVIEWERS OF PAST EDITIONS

William Atherton, *Cleveland State University*
Philip Bacon, *University of Florida*
Bruce Bayly, *University of Arizona*
William H. Beyer, *University of Akron*
R. G. Bradshaw, *Clarkson College*

Dean R. Brown, *Youngstown State University*
David Buchthal, *University of Akron*
Nguyen P. Cac, *University of Iowa*
T. Chow, *California State University—Sacramento*
Dominic P. Clemence, *North Carolina Agricultural and Technical State University*
Pasquale Condo, *University of Massachusetts—Lowell*
Vincent Connolly, *Worcester Polytechnic Institute*
Philip S. Crooke, *Vanderbilt University*
Bruce E. Davis, *St. Louis Community College at Florissant Valley*
Paul W. Davis, *Worcester Polytechnic Institute*
Richard A. DiDio, *La Salle University*
James Draper, *University of Florida*
James M. Edmondson, *Santa Barbara City College*
John H. Ellison, *Grove City College*
Raymond Fabec, *Louisiana State University*
Donna Farrior, *University of Tulsa*
Robert E. Fennell, *Clemson University*
W. E. Fitzgibbon, *University of Houston*
Harvey J. Fletcher, *Brigham Young University*
Paul J. Gormley, *Villanova*
Layachi Hadji, *University of Alabama*
Ruben Hayrapetyan, *Kettering University*
Terry Herdman, *Virginia Polytechnic Institute and State University*
Zdzislaw Jackiewicz, *Arizona State University*
S. K. Jain, *Ohio University*
Anthony J. John, *Southeastern Massachusetts University*
David C. Johnson, *University of Kentucky—Lexington*
Harry L. Johnson, *V.P.I & S.U.*
Kenneth R. Johnson, *North Dakota State University*
Joseph Kazimir, *East Los Angeles College*
J. Keener, *University of Arizona*
Steve B. Khlief, *Tennessee Technological University (retired)*
C. J. Knickerbocker, *Sensis Corporation*
Carlon A. Krantz, *Kean College of New Jersey*
Thomas G. Kudzma, *University of Lowell*
Alexandra Kurepa, *North Carolina A&T State University*
G. E. Latta, *University of Virginia*
Cecelia Laurie, *University of Alabama*
James R. McKinney, *California Polytechnic State University*
James L. Meek, *University of Arkansas*
Gary H. Meisters, *University of Nebraska—Lincoln*
Stephen J. Merrill, *Marquette University*
Vivien Miller, *Mississippi State University*
Gerald Mueller, *Columbus State Community College*
Philip S. Mulry, *Colgate University*
C. J. Neugebauer, *Purdue University*
Tyre A. Newton, *Washington State University*
Brian M. O'Connor, *Tennessee Technological University*
J. K. Oddson, *University of California—Riverside*
Carol S. O'Dell, *Ohio Northern University*
A. Peressini, *University of Illinois, Urbana—Champaign*
J. Perryman, *University of Texas at Arlington*
Joseph H. Phillips, *Sacramento City College*
Jacek Polewczak, *California State University Northridge*
Nancy J. Poxon, *California State University—Sacramento*
Robert Pruitt, *San Jose State University*

K. Rager, *Metropolitan State College*
F. B. Reis, *Northeastern University*
Brian Rodrigues, *California State Polytechnic University*
Tom Roe, *South Dakota State University*
Kimmo I. Rosenthal, *Union College*
Barbara Shabell, *California Polytechnic State University*
Seenith Sivasundaram, *Embry-Riddle Aeronautical University*
Don E. Soash, *Hillsborough Community College*
F. W. Stallard, *Georgia Institute of Technology*
Gregory Stein, *The Cooper Union*
M. B. Tamburro, *Georgia Institute of Technology*
Patrick Ward, *Illinois Central College*
Jianping Zhu, *University of Akron*
Jan Zijlstra, *Middle Tennessee State University*
Jay Zimmerman, *Towson University*

REVIEWERS OF THE CURRENT EDITIONS

Bernard Brooks, *Rochester Institute of Technology*
Allen Brown, *Wabash Valley College*
Helmut Knaust, *The University of Texas at El Paso*
Mulatu Lemma, *Savannah State University*
George Moss, *Union University*
Martin Nakashima, *California State Polytechnic University—Pomona*
Bruce O'Neill, *Milwaukee School of Engineering*

Dennis G. Zill
Los Angeles

Tenth Edition

A FIRST COURSE IN DIFFERENTIAL EQUATIONS

with Modeling Applications

Is AIDS an Invariably Fatal Disease?

by **Ivan Kramer**

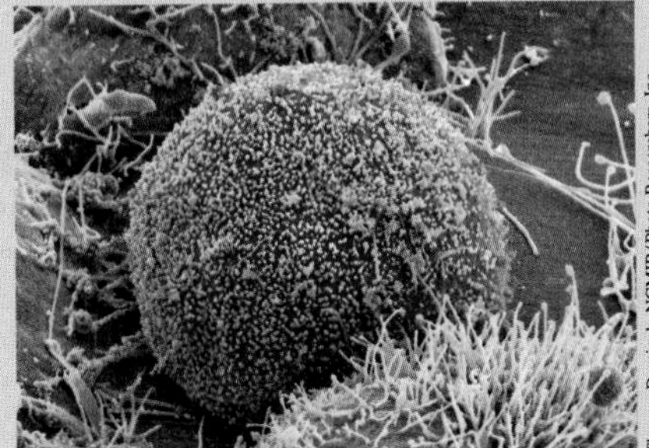
Thomas Deerinck, NCMIR/Photo Researchers, Inc.

Cell infected with HIV

This essay will address and answer the question: Is the acquired immunodeficiency syndrome (AIDS), which is the end stage of the human immunodeficiency virus (HIV) infection, an invariably fatal disease?

Like other viruses, HIV has no metabolism and cannot reproduce itself outside of a living cell. The genetic information of the virus is contained in two identical strands of RNA. To reproduce, HIV must use the reproductive apparatus of the cell it invades and infects to produce exact copies of the viral RNA. Once it penetrates a cell, HIV transcribes its RNA into DNA using an enzyme (reverse transcriptase) contained in the virus. The double-stranded viral DNA migrates into the nucleus of the invaded cell and is inserted into the cell's genome with the aid of another viral enzyme (integrase). The viral DNA and the invaded cell's DNA are then integrated, and the cell is infected. When the infected cell is stimulated to reproduce, the proviral DNA is transcribed into viral DNA, and new viral particles are synthesized. Since anti-retroviral drugs like zidovudine inhibit the HIV enzyme reverse transcriptase and stop proviral DNA chain synthesis in the laboratory, these drugs, usually administered in combination, slow down the progression to AIDS in those that are infected with HIV (hosts).

What makes HIV infection so dangerous is the fact that it fatally weakens a host's immune system by binding to the CD4 molecule on the surface of cells vital for defense against disease, including T-helper cells and a subpopulation of natural killer cells. T-helper cells (CD4 T-cells, or T4 cells) are arguably the most important cells of the immune system since they organize the body's defense against antigens. Modeling suggests that HIV infection of natural killer cells *makes it impossible for even modern antiretroviral therapy to clear the virus* [**1**]. In addition to the CD4 molecule, a virion needs at least one of a handful of co-receptor molecules (e.g., CCR5 and CXCR4) on the surface of the target cell in order to be able to bind to it, penetrate its membrane, and infect it. Indeed, about 1% of Caucasians lack coreceptor molecules, and, therefore, are completely *immune* to becoming HIV infected.

Once infection is established, the disease enters the acute infection stage, lasting a matter of weeks, followed by an *incubation period*, which can last two decades or more! Although the T-helper cell density of a host changes quasi-statically during the incubation period, literally billions of infected T4 cells and HIV particles are destroyed—and replaced—daily. This is clearly a war of attrition, one in which the immune system invariably loses.

A model analysis of the essential dynamics that occur during the incubation period to invariably cause AIDS is as follows [**1**]. Because HIV rapidly mutates, its ability to infect T4 cells on contact (its infectivity) eventually increases and the rate T4 cells become infected increases. Thus, the immune system must increase the destruction rate of infected T4 cells as well as the production rate of new, uninfected ones to replace them. There comes a point, however, when the production rate of T4 cells reaches its maximum possible limit and any further increase in HIV's infectivity must necessarily cause a drop in the T4 density leading to AIDS. Remarkably, about 5% of hosts show no sign of immune system deterioration for the first ten years of the infection; these hosts, called *long-term nonprogressors*, were originally thought to be possibly immune to developing AIDS, but modeling evidence suggests that these hosts will also develop AIDS eventually [**1**].

In over 95% of hosts, the immune system gradually loses its long battle with the virus. The T4 cell density in the peripheral blood of hosts begins to drop from normal levels (between 250 over 2500 cells/mm^3) towards zero, signaling the end of the incubation period. The host reaches the AIDS stage of the infection *either* when one of the more than twenty opportunistic infections characteristic of AIDS develops (clinical AIDS) *or* when the T4 cell density falls below 250 cells/mm^3 (an additional definition of AIDS promulgated by the CDC in 1987). The HIV infection has now reached its potentially fatal stage.

In order to model survivability with AIDS, the time t at which a host develops AIDS will be denoted by $t = 0$. One possible survival model for a cohort of AIDS patients postulates that AIDS is not a fatal condition for a fraction of the cohort, denoted by S_i, to be called the *immortal fraction* here. For the remaining part of the cohort, the probability of dying per unit time at time t will be assumed to be a constant k, where, of course, k must be positive. Thus, the survival fraction $S(t)$ for this model is a solution of the linear first-order differential equation

$$\frac{dS(t)}{dt} = -k[S(t) - S_i]. \tag{1}$$

Using the integrating-factor method discussed in Section 2.3, we see that the solution of equation (1) for the survival fraction is given by

$$S(t) = S_i + [1 - S_i]e^{-kt}. \tag{2}$$

Instead of the parameter k appearing in (2), two new parameters can be defined for a host for whom AIDS is fatal: the *average survival time* T_{aver} given by $T_{\text{aver}} = k^{-1}$ and the *survival half-life* $T_{1/2}$ given by $T_{1/2} = \ln(2)/k$. The survival half-life, defined as the time required for half of the cohort to die, is completely analogous to the half-life in radioactive nuclear decay. See Problem 8 in Exercise 3.1. In terms of these parameters the entire time-dependence in (2) can be written as

$$e^{-kt} = e^{-t/T_{\text{aver}}} = 2^{-t/T_{1/2}} \tag{3}$$

Using a least-squares program to fit the survival fraction function in (2) to the actual survival data for the 159 Marylanders who developed AIDS in 1985 produces an immortal fraction value of $S_i = 0.0665$ and a survival half life value of $T_{1/2} = 0.666$ year, with the average survival time being $T_{\text{aver}} = 0.960$ years [**2**]. See Figure 1. Thus only about 10% of Marylanders who developed AIDS in 1985 survived three years with this condition. The 1985 Maryland AIDS survival curve is virtually identical to those of 1983 and 1984. The first antiretroviral drug found to be effective against HIV was zidovudine (formerly known as AZT). Since zidovudine was not known to have an impact on the HIV infection before 1985 and was not common

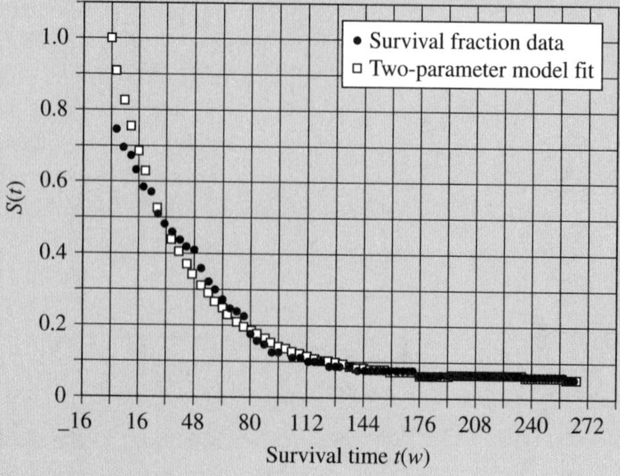

FIGURE 1 Survival fraction curve $S(t)$.

therapy before 1987, it is reasonable to conclude that the survival of the 1985 Maryland AIDS patients was not significantly influenced by zidovudine therapy.

The small but nonzero value of the immortal fraction S_i obtained from the Maryland data is probably an artifact of the method that Maryland and other states use to determine the survivability of their citizens. Residents with AIDS who changed their name and then died or who died abroad would still be counted as alive by the Maryland Department of Health and Mental Hygiene. Thus, the immortal fraction value of $S_i = 0.0665$ (6.65%) obtained from the Maryland data is clearly an upper limit to its true value, which is probably zero.

Detailed data on the survivability of 1,415 zidovudine-treated HIV-infected hosts whose T4 cell densities dropped below normal values were published by Easterbrook et al. in 1993 [**3**]. As their T4 cell densities drop towards zero, these people develop clinical AIDS and begin to die. The longest survivors of this disease live to see their T4 densities fall below 10 cells/mm^3. If the time $t = 0$ is redefined to mean the moment the T4 cell density of a host falls below 10 cells/mm^3, then the survivability of such hosts was determined by Easterbrook to be 0.470, 0.316, and 0.178 at elapsed times of 1 year, 1.5 years, and 2 years, respectively.

A least-squares fit of the survival fraction function in (2) to the Easterbrook data for HIV-infected hosts with T4 cell densities in the 0–10 cells/mm^3 range yields a value of the immortal fraction of $S_i = 0$ and a survival half-life of $T_{1/2} = 0.878$ year [**4**]; equivalently, the average survival time is $T_{\text{aver}} = 1.27$ years. These results clearly show that zidovudine is not effective in halting replication in all strains of HIV, since those who receive this drug eventually die at nearly the same rate as those who do not. In fact, the small difference of 2.5 months between the survival half-life for 1993 hosts with T4 cell densities below 10 cells/mm^3 on zidovudine therapy ($T_{1/2} = 0.878$ year) and that of 1985 infected Marylanders not taking zidovudine ($T_{1/2} = 0.666$ year) may be entirely due to improved hospitalization and improvements in the treatment of the opportunistic infections associated with AIDS over the years. Thus, the initial ability of zidovudine to prolong survivability with HIV disease ultimately wears off, and the infection resumes its progression. Zidovudine therapy has been estimated to extend the survivability of an HIV-infected patient by perhaps 5 or 6 months on the average [**4**].

Finally, putting the above modeling results for both sets of data together, we find that the value of the immortal fraction falls somewhere within the range $0 < S_i < 0.0665$ and the average survival time falls within the range 0.960 years $< T_{\text{aver}} <$ 1.27 years. Thus, the percentage of people for whom AIDS is not a fatal disease is less than 6.65% and may be zero. These results agree with a 1989 study of hemophilia-associated AIDS cases in the USA which found that the median length of survival after AIDS diagnosis was 11.7 months [**5**]. A more recent and comprehensive study of hemophiliacs with clinical AIDS using the model in (2) found that the immortal fraction was $S_i =$ 0, and the mean survival times for those between 16 to 69 years of age varied between 3 to 30 months, depending on the AIDS-defining condition [**6**]. **Although bone marrow transplants using donor stem cells homozygous for CCR5 delta32 deletion *may* lead to cures, to date clinical results consistently show that AIDS is an invariably fatal disease.**

Related Problems

1. Suppose the fraction of a cohort of AIDS patients that survives a time t after AIDS diagnosis is given by $S(t) = \exp(-kt)$. Show that the average survival time T_{aver} after AIDS diagnosis for a member of this cohort is given by $T_{\text{aver}} = 1/k$.
2. The fraction of a cohort of AIDS patients that survives a time t after AIDS diagnosis is given by $S(t) = \exp(-kt)$. Suppose the mean survival for a cohort of hemophiliacs diagnosed with AIDS before 1986 was found to be $T_{\text{aver}} = 6.4$ months. What fraction of the cohort survived 5 years after AIDS diagnosis?

3. The fraction of a cohort of AIDS patients that survives a time t after AIDS diagnosis is given by $S(t) = \exp(-kt)$. The time it takes for $S(t)$ to reach the value of 0.5 is defined as the survival half-life and denoted by $T_{1/2}$.
   (a) Show that $S(t)$ can be written in the form $S(t) = 2^{-t/T_{1/2}}$.
   (b) Show that $T_{1/2} = T_{aver} \ln(2)$, where T_{aver} is the average survival time defined in problem (1). Thus, it is always true that $T_{1/2} < T_{aver}$.
4. About 10% of lung cancer patients are cured of the disease, i.e., they survive 5 years after diagnosis with no evidence that the cancer has returned. Only 14% of lung cancer patients survive 5 years after diagnosis. Assume that the fraction of *incurable* lung cancer patients that survives a time t after diagnosis is given by $\exp(-kt)$. Find an expression for the fraction $S(t)$ of lung cancer patients that survive a time t after being diagnosed with the disease. Be sure to determine the values of all of the constants in your answer. What fraction of lung cancer patients survives two years with the disease?

References

1. Kramer, Ivan. What triggers *transient* AIDS in the acute phase of HIV infection and *chronic* AIDS at the end of the incubation period? *Computational and Mathematical Methods in Medicine,* Vol. 8, No. 2, June 2007: 125–151.
2. Kramer, Ivan. Is AIDS an invariable fatal disease?: A model analysis of AIDS survival curves. *Mathematical and Computer Modelling* 15, no. 9, 1991: 1–19.
3. Easterbrook, Philippa J., Emani Javad, Moyle, Graham, Gazzard, Brian G. Progressive CD4 cell depletion and death in zidovudine-treated patients. *JAIDS,* Aug. 6, 1993, No. 8: 927–929.
4. Kramer, Ivan. The impact of zidovudine (AZT) therapy on the survivability of those with progressive HIV infection. *Mathematical and Computer Modelling,* Vol. 23, No. 3, Feb. 1996: 1–14.
5. Stehr-Green, J. K., Holman, R. C., Mahoney, M. A. Survival analysis of hemophilia-associated AIDS cases in the US. *Am J Public Health,* Jul. 1989, 79 (7): 832–835.
6. Gail, Mitchel H., Tan, Wai-Yuan, Pee, David, Goedert, James J. Survival after AIDS diagnosis in a cohort of hemophilia patients. *JAIDS,* Aug. 15, 1997, Vol. 15, No. 5: 363–369.

ABOUT THE AUTHOR

Courtesy of Ivan Kramer

Ivan Kramer earned a BS in Physics and Mathematics from The City College of New York in 1961 and a PhD from the University of California at Berkeley in theoretical particle physics in 1967. He is currently associate professor of physics at the University of Maryland, Baltimore County. Dr. Kramer was Project Director for AIDS/HIV Case Projections for Maryland, for which he received a grant from the AIDS Administration of the Maryland Department of Health and Hygiene in 1990. In addition to his many published articles on HIV infection and AIDS, his current research interests include mutation models of cancers, Alzheimers disease, and schizophrenia.

Project for Section 3.2

The Allee Effect

by Jo Gascoigne

Dr Jo with Queenie; Queenie is on the left

Courtesy of Jo Gasoigne

The top five most famous Belgians apparently include a cyclist, a punk singer, the inventor of the saxophone, the creator of Tintin, and Audrey Hepburn. Pierre François Verhulst is not on the list, although he should be. He had a fairly short life, dying at the age of 45, but did manage to include some excitement—he was deported from Rome for trying to persuade the Pope that the Papal States needed a written constitution. Perhaps the Pope knew better even then than to take lectures in good governance from a Belgian. . . .

Aside from this episode, **Pierre Verhulst** (1804–1849) was a mathematician who concerned himself, among other things, with the dynamics of natural populations—fish, rabbits, buttercups, bacteria, or whatever. (I am prejudiced in favour of fish, so we will be thinking fish from now on.) Theorizing on the growth of natural populations had up to this point been relatively limited, although scientists had reached the obvious conclusion that the growth rate of a population (dN/dt, where $N(t)$ is the population size at time t) depended on (i) the birth rate b and (ii) the mortality rate m, both of which would vary in direct proportion to the size of the population N:

$$\frac{dN}{dt} = bN - mN. \tag{1}$$

After combining b and m into one parameter r, called the **intrinsic rate of natural increase**—or more usually by biologists without the time to get their tongues around that, just r—equation (1) becomes

$$\frac{dN}{dt} = rN. \tag{2}$$

This model of population growth has a problem, which should be clear to you—if not, plot dN/dt for increasing values of N. It is a straightforward exponential growth curve, suggesting that we will all eventually be drowning in fish. Clearly, something eventually has to step in and slow down dN/dt. Pierre Verhulst's insight was that this *something* was the capacity of the environment, in other words,

How many fish can an ecosystem actually support?

He formulated a differential equation for the population $N(t)$ that included both r and the **carrying capacity** K:

$$\frac{dN}{dt} = rN\left(1 - \frac{N}{K}\right), \quad r > 0. \tag{3}$$

Equation (3) is called the **logistic equation**, and it forms to this day the basis of much of the modern science of population dynamics. Hopefully, it is clear that the term $(1 - N/K)$, which is Verhulst's contribution to equation (2), is $(1 - N/K) \approx 1$ when $N \approx 0$, leading to exponential growth, and $(1 - N/K) \to 0$ as $N \to K$, hence it causes the growth curve of $N(t)$ to approach the horizontal asymptote $N(t) = K$. Thus the size of the population cannot exceed the carrying capacity of the environment.

The logistic equation (3) gives the overall growth rate of the population, but the ecology is easier to conceptualize if we consider *per capita* growth rate—that is, the growth rate of the population per the number of individuals in the population—some measure of how "well" each individual in the population is doing. To get *per capita* growth rate, we just divide each side of equation (3) by N:

$$\frac{1}{N}\frac{dN}{dt} = r\left(1 - \frac{N}{K}\right) = r - \frac{r}{K}N.$$

This second version of (3) immediately shows (or plot it) that this relationship is a straight line with a maximum value of $\frac{1}{N}\frac{dN}{dt}$ at $N = 0$ (assuming that negative population sizes are not relevant) and $dN/dt = 0$ at $N = K$.

Er, hang on a minute . . . "a maximum value of $\frac{1}{N}\frac{dN}{dt}$ at $N = 0$?!" Each shark in the population does best when there are . . . zero sharks? Here is clearly a flaw in the logistic model. (Note that it is now a *model*—when it just presents a relationship between two variables dN/dt and N, it is just an equation. When we use this equation to try and analyze how populations might work, it becomes a model.)

The assumption behind the logistic model is that as population size decreases, individuals do better (as measured by the *per capita* population growth rate). This assumption to some extent underlies all our ideas about sustainable management of natural resources—a fish population cannot be fished indefinitely unless we assume that when a population is reduced in size, it has the ability to grow back to where it was before.

This assumption is more or less reasonable for populations, like many fish populations subject to commercial fisheries, which are maintained at 50% or even 20% of K. But for very depleted or endangered populations, the idea that individuals keep doing better as the population gets smaller is a risky one. The Grand Banks population of cod, which was fished down to 1% or perhaps even 0.1% of K, has been protected since the early 1990s, and has yet to show convincing signs of recovery.

Warder Clyde Allee (1885–1955) was an American ecologist at the University of Chicago in the early 20th century, who experimented on goldfish, brittlestars, flour beetles, and, in fact, almost anything unlucky enough to cross his path. Allee showed that, in fact, individuals in a population can do worse when the population becomes very small or very sparse.* There are numerous ecological reasons why this might be—for example, they may not find a suitable mate or may need large groups to find food or express social behavior, or in the case of goldfish they may alter the water chemistry in their favour. As a result of Allee's work, a population where the *per capita* growth rate declines at low population size is said to show an **Allee effect**. The jury is still out on whether Grand Banks cod are suffering from an Allee effect, but there are some possible mechanisms—females may not be able to find a mate, or a mate of the right size, or maybe the adult cod used to eat the fish that eat the juvenile cod. On the other hand, there is nothing that an adult cod likes more than a snack of baby cod—they are not fish with very picky eating habits—so these arguments may not stack up. For the moment we know very little except that there are still no cod.

Allee effects can be modelled in many ways. One of the simplest mathematical models, a variation of the logistic equation, is:

$$\frac{dN}{dt} = rN\left(1 - \frac{N}{K}\right)\left(\frac{N}{A} - 1\right). \tag{4}$$

where A is called the **Allee threshold**. The value $N(t) = A$ is the population size below which the population growth rate becomes negative due to an Allee effect—situated at

*Population size and population density are mathematically interchangeable, assuming a fixed area in which the population lives (although they may not necessarily be interchangeable for the individuals in question).

a value of N somewhere between $N = 0$ and $N = K$, that is, $0 < A < K$, depending on the species (but for most species a good bit closer to 0 than K, luckily).

Equation (4) is not as straightforward to solve for $N(t)$ as (3), but we don't need to solve it to gain some insights into its dynamics. If you work through Problems 2 and 3, you will see that the consequences of equation (4) can be disastrous for endangered populations.

Related Problems

1. **(a)** The logistic equation (3) can be solved explicitly for $N(t)$ using the technique of partial fractions. Do this, and plot $N(t)$ as a function of t for $0 \le t \le 10$. Appropriate values for r, K, and $N(0)$ are $r = 1$, $K = 1$, $N(0) = 0.01$ (fish per cubic metre of seawater, say). The graph of $N(t)$ is called a **sigmoid growth curve**.

   (b) The value of r can tell us a lot about the ecology of a species—sardines, where females mature in less than one year and have millions of eggs, have a high r, while sharks, where females bear a few live young each year, have a low r. Play with r and see how it affects the shape of the curve. *Question*: If a marine protected area is put in place to stop overfishing, which species will recover quickest—sardines or sharks?

Doug Perrine/Getty Images

Copper sharks and bronze whaler sharks feeding on a bait ball of sardines off the east coast of South Africa

2. Find the population equilibria for the model in (4). [*Hint*: The population is at equilibrium when $dN/dt = 0$, that is, the population is neither growing nor shrinking. You should find three values of N for which the population is at equilibrium.]
3. Population equilibria can be stable or unstable. If, when a population deviates a bit from the equilibrium value (as populations inevitably do), it tends to return to it, this is a stable equilibrium; if, however, when the population deviates from the equilibrium it tends to diverge from it ever further, this is an unstable equilibrium. Think of a ball in the pocket of a snooker table versus a ball balanced on a snooker cue. Unstable equilibria are a feature of Allee effect models such as (4). Use a phase portrait of the autonomous equation (4) to determine whether the nonzero equilibria that you found in Problem 2 are stable or unstable. [*Hint*: See Section 2.1 of the text.]
4. Discuss the consequences of the result above for a population $N(t)$ fluctuating close to the Allee threshold A.

References

1. Courchamp, F., Berec L., and Gascoigne, J. 2008. *Allee Effects in Ecology and Conservation.* Oxford University Press.
2. Hastings, A. 1997. *Population Biology—Concepts and Models.* Springer-Verlag, New York.

ABOUT THE AUTHOR

Courtesy of Jo Gascoigne

After a degree in Zoology, **Jo Gascoigne** thought her first job, on conservation in East Africa, would be about lions and elephants—but it turned out to be about fish. Despite the initial crushing disappointment, she ended up loving them—so much, in fact, that she went on to complete a PhD in marine conservation biology at the College of William and Mary, in Williamsburg, Virginia, where she studied lobster and Caribbean conch, and also spent 10 days living underwater in the Aquarius habitat in Florida. After graduating, she returned to her native Britain and studied the mathematics of mussel beds at Bangor University in Wales, before becoming an independent consultant on fisheries management. She now works to promote environmentally sustainable fisheries. When you buy seafood, make good choices and help the sea!

Wolf Population Dynamics

by C. J. Knickerbocker

A gray wolf in the wild

Early in 1995, after much controversy, public debate, and a 70-year absence, gray wolves were re introduced into Yellowstone National Park and Central Idaho. During this 70-year absence, significant changes were recorded in the populations of other predator and prey animals residing in the park. For instance, the elk and coyote populations had risen in the absence of influence from the larger gray wolf. With the reintroduction of the wolf in 1995, we anticipated changes in both the predator and prey animal populations in the Yellowstone Park ecosystem as the success of the wolf population is dependent upon how it influences and is influenced by the other species in the ecosystem.

For this study, we will examine how the elk (prey) population has been influenced by the wolves (predator). Recent studies have shown that the elk population has been negatively impacted by the reintroduction of the wolves. The elk population fell from approximately 18,000 in 1995 to approximately 7,000 in 2009. This article asks the question of whether the wolves could have such an effect and, if so, could the elk population disappear?

Let's begin with a more detailed look at the changes in the elk population independent of the wolves. In the 10 years prior to the introduction of wolves, from 1985 to 1995, one study suggested that the elk population increased by 40% from 13,000 in 1985 to 18,000 in 1995. Using the simplest differential equation model for population dynamics, we can determine the growth rate for elks (represented by the variable r) prior to the reintroduction of the wolves.

$$\frac{dE}{dt} = rE, \quad E(0) = 13.0, E(10) = 18.0 \tag{1}$$

In this equation, $E(t)$ represents the elk population (in thousands) where t is measured in years since 1985. The solution, which is left as an exercise for the reader, finds the combined birth/death growth rate r to be approximately 0.0325 yielding:

$$E(t) = 13.0\, e^{0.0325t}$$

In 1995, 21 wolves were initially released, and their numbers have risen. In 2007, biologists estimated the number of wolves to be approximately 171.

To study the interaction between the elk and wolf populations, let's consider the following predator-prey model for the interaction between the elk and wolf within the Yellowstone ecosystem:

$$\begin{aligned} \frac{dE}{dt} &= 0.0325E - 0.8EW \\ \frac{dW}{dt} &= -0.6W + 0.05EW \\ E(0) &= 18.0,\ W(0) = 0.021 \end{aligned} \tag{2}$$

where $E(t)$ is the elk population and $W(t)$ is the wolf population. All populations are measured in thousands of animals. The variable t represents time measured in years from 1995. So, from the initial conditions, we have 18,000 elk and 21 wolves in the year 1995. The reader will notice that we estimated the growth rate for the elk to be the same as that estimated above $r = 0.0325$.

Before we attempt to solve the model (2), a qualitative analysis of the system can yield a number of interesting properties of the solutions. The first equation shows that the growth rate of the elk (dE/dt) is positively impacted by the size of the herd ($0.0325E$). This can be interpreted as the probability of breeding increases with the number of elk. On the other hand the nonlinear term ($0.8EW$) has a negative impact on the growth rate of the elk since it measures the interaction between predator and prey. The second equation $dW/dt = -0.6W + 0.05EW$ shows that the wolf population has a negative effect on its own growth which can be interpreted as more wolves create more competition for food. But, the interaction between the elk and wolves ($0.05EW$) has a positive impact since the wolves are finding more food.

Since an analytical solution cannot be found to the initial-value problem (2), we need to rely on technology to find approximate solutions. For example, below is a set of instructions for finding a numerical solution of the initial-value problem using the computer algebra system MAPLE.

```
e1 := diff(e(t),t)- 0.0325 * e(t) + 0.8 * e(t)*w(t) :
e2 := diff(w(t),t)+ 0.6 * w(t) - 0.05 * e(t)*w(t) :
sys := {e1,e2} :
ic := {e(0)=18.0,w(0)=0.021} :
ivp := sys union ic :
H:= dsolve(ivp,{e(t),w(t)},numeric) :
```

The graphs in Figures 1 and 2 show the populations for both species between 1995 and 2009. As predicted by numerous studies, the reintroduction of wolves into Yellowstone had led to a decline in the elk population. In this model, we see the population decline from 18,000 in 1995 to approximately 7,000 in 2009. In contrast, the wolf population rose from an initial count of 21 in 1995 to a high of approximately 180 in 2004.

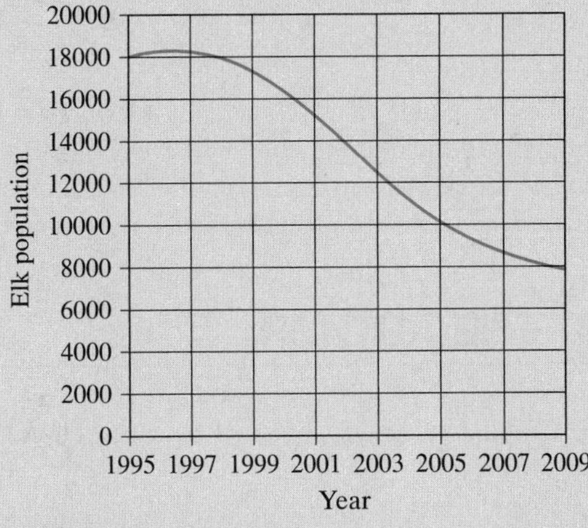

FIGURE 1 Elk population

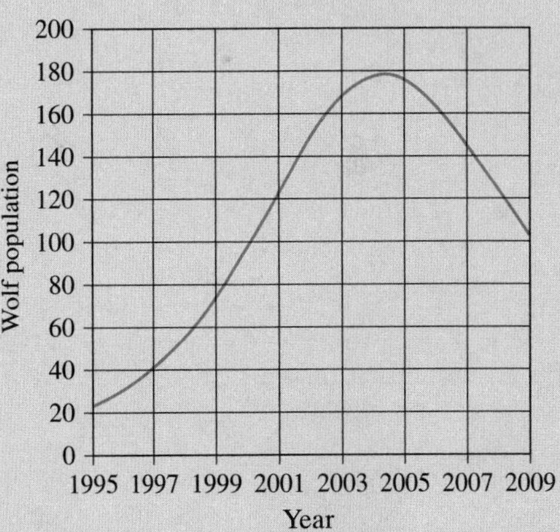

FIGURE 2 Wolf population

The alert reader will note that the model also shows a decline in the wolf population after 2004. How might we interpret this? With the decline in the elk population over the first 10 years, there was less food for the wolves and therefore their population begins to decline.

Figure 3 below shows the long-term behavior of both populations. The interpretation of this graph is left as an exercise for the reader.

Information on the reintroduction of wolves into Yellowstone Park and central Idaho can be found on the Internet. For example, read the U.S. Fish and Wildlife Service news release of November 23, 1994, on the release of wolves into Yellowstone National Park.

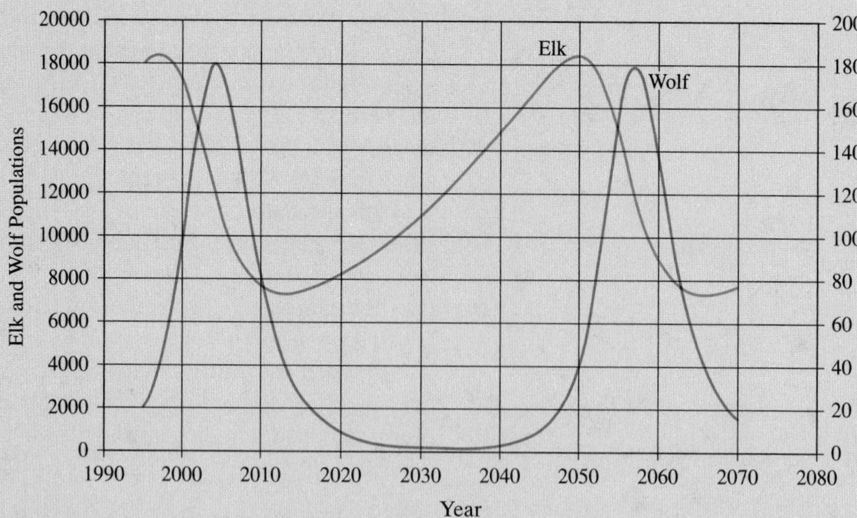

FIGURE 3 Long-term behavior of the populations

Related Problems

1. Solve the pre-wolf initial-value problem (1) by first solving the differential equation and applying the initial condition. Then apply the terminal condition to find the growth rate.
2. Biologists have debated whether the decrease in the elk from 18,000 in 1995 to 7,000 in 2009 is due to the reintroduction of wolves. What other factors might account for the decrease in the elk population?
3. Consider the long-term changes in the elk and wolf populations. Are these cyclic changes reasonable? Why is there a lag between the time when the elk begins to decline and the wolf population begins to decline? Are the minimum values for the wolf population realistic? Plot the elk population versus the wolf population and interpret the results.
4. What does the initial-value problem (1) tell us about the growth of the elk population without the influence of the wolves? Find a similar model for the introduction of rabbits into Australia in 1859 and the impact of introducing a prey population into an environment without a natural predator population.

ABOUT THE AUTHOR

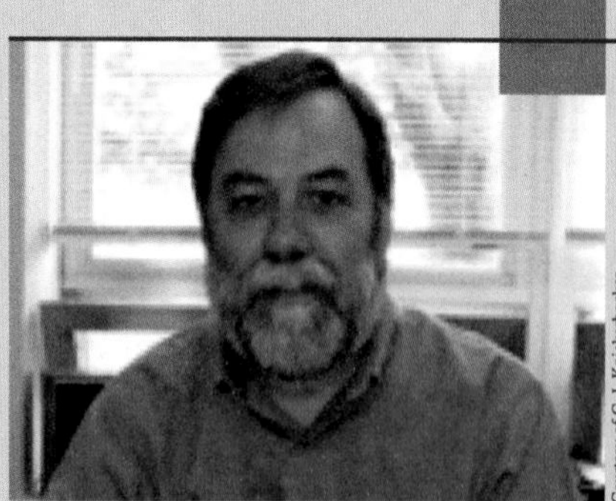

C. J. Knickerbocker
Professor of Mathematics and Computer Science (retired)
St. Lawrence University
Principal Research Engineer
Sensis Corporation

C. J. Knickerbocker received his PhD in mathematics from Clarkson University in 1984. Until 2008 he was a professor of mathematics and computer science at St. Lawrence University, where he authored numerous articles in a variety of topics, including nonlinear partial differential equations, graph theory, applied physics, and psychology. He has also served as a consultant for publishers, software companies, and government agencies. Currently, Dr. Knickerbocker is a principal research engineer for the Sensis Corporation, where he studies airport safety and efficiency.

Bungee Jumping

by Kevin Cooper

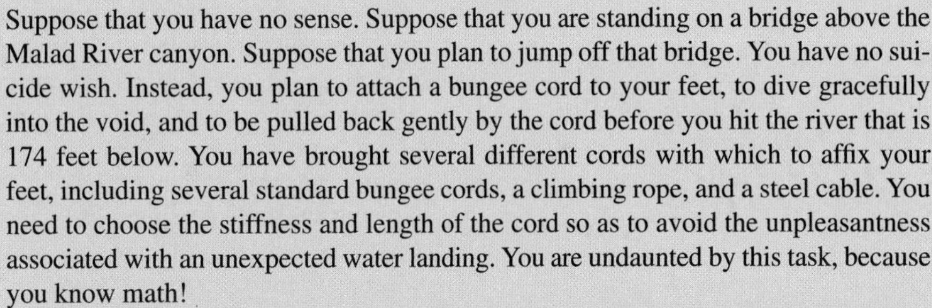

Suppose that you have no sense. Suppose that you are standing on a bridge above the Malad River canyon. Suppose that you plan to jump off that bridge. You have no suicide wish. Instead, you plan to attach a bungee cord to your feet, to dive gracefully into the void, and to be pulled back gently by the cord before you hit the river that is 174 feet below. You have brought several different cords with which to affix your feet, including several standard bungee cords, a climbing rope, and a steel cable. You need to choose the stiffness and length of the cord so as to avoid the unpleasantness associated with an unexpected water landing. You are undaunted by this task, because you know math!

Sean Nel/Shutterstock.com

Bungee jumping from a bridge

Each of the cords you have brought will be tied off so as to be 100 feet long when hanging from the bridge. Call the position at the bottom of the cord 0, and measure the position of your feet below that "natural length" as $x(t)$, where x increases as you go down and is a function of time t. See Figure 1. Then, at the time you jump, $x(0) = -100$, while if your six-foot frame hits the water head first, at that time $x(t) = 174 - 100 - 6 = 68$. Notice that distance increases as you fall, and so your velocity is positive as you fall and negative when you bounce back up. Note also that you plan to dive so your head will be six feet below the end of the chord when it stops you.

You know that the acceleration due to gravity is a constant, called g, so that the force pulling downwards on your body is mg. You know that when you leap from the bridge, air resistance will increase proportionally to your speed, providing a force in the opposite direction to your motion of about βv, where β is a constant and v is your velocity. Finally, you know that Hooke's law describing the action of springs says that the bungee cord will eventually exert a force on you proportional to its distance past its natural length. Thus, you know that the force of the cord pulling you back from destruction may be expressed as

$$b(x) = \begin{cases} 0 & x \le 0 \\ -kx & x > 0 \end{cases}$$

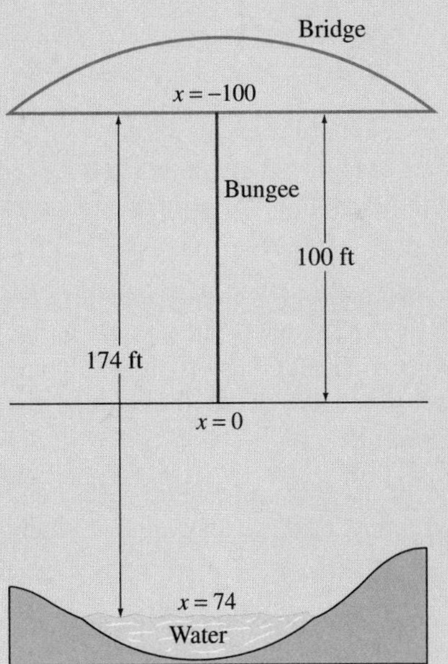

FIGURE 1 The bungee setup

The number k is called the *spring constant*, and it is where the stiffness of the cord you use influences the equation. For example, if you used the steel cable, then k would be very large, giving a tremendous stopping force very suddenly as you passed the natural length of the cable. This could lead to discomfort, injury, or even a Darwin award. You want to choose the cord with a k value large enough to stop you above or just touching the water, but not too suddenly. Consequently, you are interested in finding the distance you fall below the natural length of the cord as a function of the spring constant. To do that, you must solve the differential equation that we have derived in words above: The force mx'' on your body is given by

$$mx'' = mg + b(x) - \beta x'.$$

Here mg is your weight, 160 lb., and x' is the rate of change of your position below the equilibrium with respect to time; i.e., your velocity. The constant β for air resistance depends on a number of things, including whether you wear your skin-tight pink spandex or your skater shorts and XXL T-shirt, but you know that the value today is about 1.0.

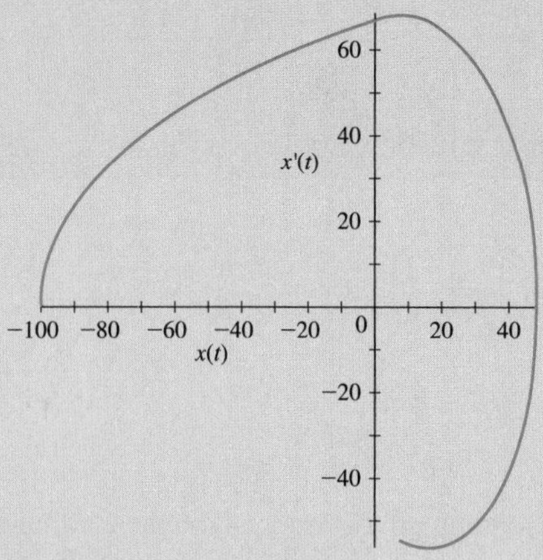

FIGURE 2 An example plot of $x(t)$ against $x'(t)$ for a bungee jump

This is a nonlinear differential equation, but inside it are two linear differential equations, struggling to get out. We will work with such equations more extensively in later chapters, but we already know how to solve such equations from our past experience. When $x < 0$, the equation is $mx'' = mg - \beta x'$, while after you pass the natural length of the cord it is $mx'' = mg - kx - \beta x'$. We will solve these separately, and then piece the solutions together when $x(t) = 0$.

In Problem 1 you find an expression for your position t seconds after you step off the bridge, before the bungee cord starts to pull you back. Notice that it does not depend on the value for k, because the bungee cord is just falling with you when you are above $x(t) = 0$. When you pass the natural length of the bungee cord, it does start to pull back, so the differential equation changes. Let t_1 denote the first time for which $x(t_1) = 0$, and let v_1 denote your speed at that time. We can thus describe the motion for $x(t) > 0$ using the problem $x'' = g - kx - \beta x'$, $x(t_1) = 0$, $x'(t_1) = v_1$. An illustration of a solution to this problem in phase space can be seen in Figure 2.

This will yield an expression for your position as the cord is pulling on you. All we have to do is to find out the time t_2 when you stop going down. When you stop going down, your velocity is zero, i.e., $x'(t_2) = 0$.

As you can see, knowing a little bit of math is a dangerous thing. We remind you that the assumption that the drag due to air resistance is linear applies only for low speeds. By the time you swoop past the natural length of the cord, that approximation is only wishful thinking, so your actual mileage may vary. Moreover, springs behave nonlinearly in large oscillations, so Hooke's law is only an approximation. Do not trust your life to an approximation made by a man who has been dead for 200 years. Leave bungee jumping to the professionals.

Related Problems

1. Solve the equation $mx'' + \beta x' = mg$ for $x(t)$, given that you step off the bridge—no jumping, no diving! Stepping off means $x(0) = -100$, $x'(0) = 0$. You may use $mg = 160$, $\beta = 1$, and $g = 32$.

2. Use the solution from Problem 1 to compute the length of time t_1 that you freefall (the time it takes to go the natural length of the cord: 100 feet).

3. Compute the derivative of the solution you found in Problem 1 and evaluate it at the time you found in Problem 2. Call the result v_1. You have found your downward speed when you pass the point where the cord starts to pull.

4. Solve the initial-value problem

$$mx'' + \beta x' + kx = mg, \quad x(t_1) = 0, \quad x'(t_1) = v_1.$$

For now, you may use the value $k = 14$, but eventually you will need to replace that with the actual values for the cords you brought. The solution $x(t)$ represents the position of your feet below the natural length of the cord after it starts to pull back.

5. Compute the derivative of the expression you found in Problem 4 and solve for the value of t where it is zero. This time is t_2. Be careful that the time you compute is greater than t_1—there are several times when your motion stops at the top and bottom of your bounces! After you find t_2, substitute it back into the solution you found in Problem 4 to find your lowest position.

6. You have brought a soft bungee cord with $k = 8.5$, a stiffer cord with $k = 10.7$, and a climbing rope for which $k = 16.4$. Which, if any, of these may you use safely under the conditions given?

7. You have a bungee cord for which you have not determined the spring constant. To do so, you suspend a weight of 10 lb. from the end of the 100-foot cord, causing the cord to stretch 1.2 feet. What is the k value for this cord? You may neglect the mass of the cord itself.

ABOUT THE AUTHOR

Courtesy of Kevin Cooper

Kevin Cooper, PhD, Colorado State University, is the Computing Coordinator for Mathematics at Washington State University, Pullman, Washington. His main interest is numerical analysis, and he has written papers and one textbook in that area. Dr. Cooper also devotes considerable time to creating mathematical software components, such as *DynaSys*, a program to analyze dynamical systems numerically.

Project for Section 5.3

The Collapse of the Tacoma Narrows Suspension Bridge

by Gilbert N. Lewis

AP Photo

Collapse of the Tacoma Narrows Bridge

Sir Armstrong/Shutterstock.com

The rebuilt Tacoma Narrows bridge (1950) and new parallel bridge (2009)

In the summer of 1940, the Tacoma Narrows Suspension Bridge in the State of Washington was completed and opened to traffic. Almost immediately, observers noticed that the wind blowing across the roadway would sometimes set up large vertical vibrations in the roadbed. The bridge became a tourist attraction as people came to watch, and perhaps ride, the undulating bridge. Finally, on November 7, 1940, during a powerful storm, the oscillations increased beyond any previously observed, and the bridge was evacuated. Soon, the vertical oscillations became rotational, as observed by looking down the roadway. The entire span was eventually shaken apart by the large vibrations, and the bridge collapsed. Figure 1 shows a picture of the bridge during the collapse. See [**1**] and [**2**] for interesting and sometimes humorous anecdotes associated with the bridge. Or, do an Internet search with the key words "Tacoma Bridge Disaster" in order to find and view some interesting videos of the collapse of the bridge.

The noted engineer von Karman was asked to determine the cause of the collapse. He and his coauthors [**3**] claimed that the wind blowing perpendicularly across the roadway separated into vortices (wind swirls) alternately above and below the roadbed, thereby setting up a periodic, vertical force acting on the bridge. It was this force that caused the oscillations. Others further hypothesized that the frequency of this forcing function exactly matched the natural frequency of the bridge, thus leading to resonance, large oscillations, and destruction. For almost fifty years, resonance was blamed as the cause of the collapse of the bridge, although the von Karman group denied this, stating that "it is very improbable that resonance with alternating vortices plays an important role in the oscillations of suspension bridges" [**3**].

As we can see from equation (31) in Section 5.1.3, resonance is a linear phenomenon. In addition, for resonance to occur, there must be an exact match between the frequency of the forcing function and the natural frequency of the bridge. Furthermore, there must be absolutely no damping in the system. It should not be surprising, then, that resonance was not the culprit in the collapse.

If resonance did not cause the collapse of the bridge, what did? Recent research provides an alternative explanation for the collapse of the Tacoma Narrows Bridge. Lazer and McKenna [**4**] contend that nonlinear effects, and not linear resonance, were the main factors leading to the large oscillations of the bridge (see [**5**] for a good review article). The theory involves partial differential equations. However, a simplified model leading to a nonlinear ordinary differential equation can be constructed.

The development of the model below is not exactly the same as that of Lazer and McKenna, but it results in a similar differential equation. This example shows another way that amplitudes of oscillation can increase.

Consider a single vertical cable of the suspension bridge. We assume that it acts like a spring, but with different characteristics in tension and compression, and with no damping. When stretched, the cable acts like a spring with Hooke's constant, b, while, when compressed, it acts like a spring with a different Hooke's constant, a. We assume that the cable in compression exerts a smaller force on the roadway than when stretched the same distance, so that $0 < a < b$. Let the vertical deflection (positive direction downward) of the slice of the roadbed attached to this cable be denoted by $y(t)$, where t represents time, and $y = 0$ represents the equilibrium position of the road. As the roadbed oscillates under the influence of an applied vertical force (due to the von Karman vortices), the cable provides an upward restoring force equal to by when $y > 0$ and a downward restoring force equal to ay when $y < 0$. This change in the Hooke's Law constant at $y = 0$ provides the nonlinearity to the differential equation. We are thus led to consider the differential equation derived from Newton's second law of motion

$$my'' + f(y) = g(t),$$

where $f(y)$ is the nonlinear function given by

$$f(y) = \begin{Bmatrix} by & \text{if } y \geq 0 \\ ay & \text{if } y < 0 \end{Bmatrix},$$

$g(t)$ is the applied force, and m is the mass of the section of the roadway. Note that the differential equation is linear on any interval on which y does not change sign.

Now, let us see what a typical solution of this problem would look like. We will assume that $m = 1$ kg, $b = 4$ N/m, a $= 1$ N/m, and $g(t) = \sin(4t)$ N. Note that the frequency of the forcing function is larger than the natural frequencies of the cable in both tension and compression, so that we do not expect resonance to occur. We also assign the following initial values to y: $y(0) = 0$, $y'(0) = 0.01$, so that the roadbed starts in the equilibrium position with a small downward velocity.

Because of the downward initial velocity and the positive applied force, $y(t)$ will initially increase and become positive. Therefore, we first solve this initial-value problem

$$y'' + 4y = \sin(4t), \quad y(0) = 0, \quad y'(0) = 0.01. \tag{1}$$

The solution of the equation in (1), according to Theorem 4.1.6, is the sum of the complementary solution, $y_c(t)$, and the particular solution, $y_p(t)$. It is easy to see that $y_c(t) = c_1\cos(2t) + c_2\sin(2t)$ (equation (9), Section 4.3), and $y_p(t) = -\frac{1}{12}\sin(4t)$ (Table 4.4.1, Section 4.4). Thus,

$$y(t) = c_1\cos(2t) + c_2\sin(2t) - \frac{1}{12}\sin(4t). \tag{2}$$

The initial conditions give

$$y(0) = 0 = c_1,$$
$$y'(0) = 0.01 = 2c_2 - \frac{1}{3},$$

so that $c_2 = (0.01 + \frac{1}{3})/2$. Therefore, (2) becomes

$$\begin{aligned} y(t) &= \frac{1}{2}\left(0.01 + \frac{1}{3}\right)\sin(2t) - \frac{1}{12}\sin(4t) \\ &= \sin(2t)\left[\frac{1}{2}\left(0.01 + \frac{1}{3}\right) - \frac{1}{6}\cos(2t)\right]. \end{aligned} \tag{3}$$

We note that the first positive value of t for which $y(t)$ is again equal to zero is $t = \frac{\pi}{2}$. At that point, $y'(\frac{\pi}{2}) = -(0.01 + \frac{2}{3})$. Therefore, equation (3) holds on $[0, \pi/2]$.

After $t = \frac{\pi}{2}$, y becomes negative, so we must now solve the new problem

$$y'' + y = \sin(4t), \quad y\left(\frac{\pi}{2}\right) = 0, \quad y'\left(\frac{\pi}{2}\right) = -\left(0.01 + \frac{2}{3}\right). \tag{4}$$

Proceeding as above, the solution of (4) is

$$\begin{aligned} y(t) &= \left(0.01 + \frac{2}{5}\right)\cos t - \frac{1}{15}\sin(4t) \\ &= \cos t\left[\left(0.01 + \frac{2}{5}\right) - \frac{4}{15}\sin t\cos(2t)\right]. \end{aligned} \tag{5}$$

The next positive value of t after $t = \frac{\pi}{2}$ at which $y(t) = 0$ is $t = \frac{3\pi}{2}$, at which point $y'(\frac{3\pi}{2}) = 0.01 + \frac{2}{15}$, so that equation (5) holds on $[\pi/2, 3\pi/2]$.

At this point, the solution has gone through one cycle in the time interval $[0, \frac{3\pi}{2}]$. During this cycle, the section of the roadway started at the equilibrium with positive velocity, became positive, came back to the equilibrium position with negative velocity, became negative, and finally returned to the equilibrium position with positive velocity. This pattern continues indefinitely, with each cycle covering $\frac{3\pi}{2}$ time units. The solution for the next cycle is

$$y(t) = \sin(2t)\left[-\frac{1}{2}\left(0.01 + \frac{7}{15}\right) - \frac{1}{6}\cos(2t)\right] \quad \text{on} \quad [3\pi/2, 2\pi],$$

$$y(t) = \sin t\left[-\left(0.01 + \frac{8}{15}\right) - \frac{4}{15}\cos t\cos(2t)\right] \quad \text{on} \quad [2\pi, 3\pi]. \tag{6}$$

It is instructive to note that the velocity at the beginning of the second cycle is $(0.01 + \frac{2}{15})$, while at the beginning of the third cycle it is $(0.01 + \frac{4}{15})$. In fact, the velocity at the beginning of each cycle is $\frac{2}{15}$ greater than at the beginning of the previous cycle. It is not surprising then that the amplitude of oscillations will increase over time, since the amplitude of (one term in) the solution during any one cycle is directly related to the velocity at the beginning of the cycle. See Figure 2 for a graph of the **deflection function** on the interval $[0, 3\pi]$. Note that the maximum deflection on $[3\pi/2, 2\pi]$ is larger than the maximum deflection on $[0, \pi/2]$, while the maximum deflection on $[2\pi, 3\pi]$ is larger than the maximum deflection on $[\pi/2, 3\pi/2]$.

It must be remembered that the model presented here is a very simplified one-dimensional model that cannot take into account all of the intricate interactions of real bridges. The reader is referred to the account by Lazer and McKenna [**4**] for a more complete model. More recently, McKenna [**6**] has refined that model to provide a different viewpoint of the torsional oscillations observed in the Tacoma Bridge.

Research on the behavior of bridges under forces continues. It is likely that the models will be refined over time, and new insights will be gained from the research. However, it should be clear at this point that the large oscillations causing the destruction of the Tacoma Narrows Suspension Bridge were not the result of resonance.

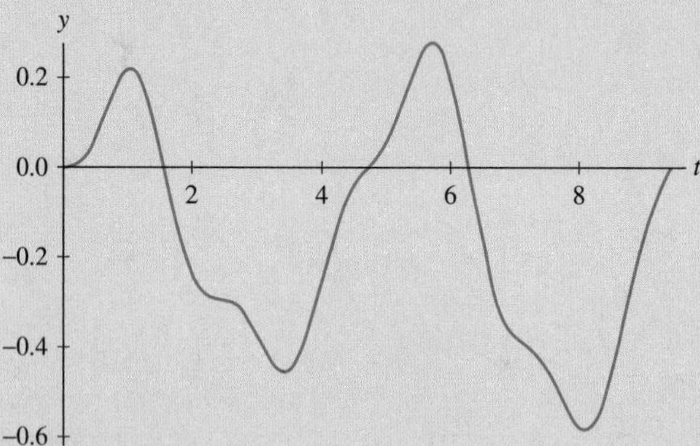

FIGURE 2 Graph of deflection function $y(t)$

Related Problems

1. Solve the following problems and plot the solutions for $0 \leq t \leq 6\pi$. Note that resonance occurs in the first problem but not in the second.

   (a) $y'' + y = -\cos t$, $y(0) = 0$, $y'(0) = 0$.
   (b) $y'' + y = \cos(2t)$, $y(0) = 0$, $y'(0) = 0$.

2. Solve the initial-value problem $y'' + f(y) = \sin(4t)$, $y(0) = 0$, $y'(0) = 1$, where

$$f(y) = \begin{Bmatrix} by & \text{if } y \geq 0 \\ ay & \text{if } y < 0 \end{Bmatrix},$$

and

(a) $b = 1, a = 4$, (Compare your answer with the example in this project.)
(b) $b = 64, a = 4$,
(c) $b = 36, a = 25$.

Note that, in part (a), the condition $b > a$ of the text is not satisfied. Plot the solutions. What happens in each case as t increases? What would happen in each case if the second initial condition were replaced with $y'(0) = 0.01$? Can you make any conclusions similar to those of the text regarding the long-term solution?

3. What would be the effect of adding damping ($+cy'$, where $c > 0$) to the system? How could a bridge design engineer incorporate more damping into the bridge? Solve the problem $y'' + cy' + f(y) = \sin(4t)$, $y(0) = 0$, $y'(0) = 1$, where

$$f(y) = \begin{Bmatrix} 4y & \text{if } y \geq 0 \\ y & \text{if } y < 0 \end{Bmatrix},$$

and

(a) $c = 0.01$
(b) $c = 0.1$
(c) $c = 0.5$

References

1. Lewis, G. N., "Tacoma Narrows Suspension Bridge Collapse" in *A First Course in Differential Equations*, Dennis G. Zill, 253–256. Boston: PWS-Kent, 1993.
2. Braun, M., *Differential Equations and Their Applications*, 167–169. New York: Springer-Verlag, 1978.
3. Amman, O. H., T. von Karman, and G. B. Woodruff, *The Failure of the Tacoma Narrows Bridge*. Washington D.C.: Federal Works Agency, 1941.
4. Lazer, A. C., and P. J. McKenna. Large amplitude periodic oscillations in suspension bridges: Some new connections with nonlinear analysis. *SIAM Review* 32 (December 1990): 537–578.
5. Peterson, I., Rock and roll bridge. *Science News* 137 (1991): 344–346.
6. McKenna, P. J., Large torsional oscillations in suspension bridges revisited: Fixing an old approximation. *American Mathematical Monthly* 106 (1999):1–18.

ABOUT THE AUTHOR

Dr. Gilbert N. Lewis is professor emeritus at Michigan Technological University, where he has taught and done research in Applied Math and Differential Equations for 34 years. He received his BS degree from Brown University and his MS and PhD degrees from the University of Wisconsin-Milwaukee. His hobbies include travel, food and wine, fishing, and birding, activities that he intends to continue in retirement.

Courtesy of Gilbert N. Lewis

Project for Section 7.3

Murder at the Mayfair Diner

by Tom LoFaro

© Ronald C. Saari

The Mayfair diner in Philadelphia, PA

Dawn at the Mayfair Diner. The amber glow of streetlights mixed with the violent red flash of police cruisers begins to fade with the rising of a furnace orange sun. Detective Daphne Marlow exits the diner holding a steaming cup of hot joe in one hand and a summary of the crime scene evidence in the other. Taking a seat on the bumper of her tan LTD, Detective Marlow begins to review the evidence.

At 5:30 a.m. the body of one Joe D. Wood was found in the walk in refrigerator in the diner's basement. At 6:00 a.m. the coroner arrived and determined that the core body temperature of the corpse was 85 degrees Fahrenheit. Thirty minutes later the coroner again measured the core body temperature. This time the reading was 84 degrees Fahrenheit. The thermostat inside the refrigerator reads 50 degrees Fahrenheit.

Daphne takes out a fading yellow legal pad and ketchup-stained calculator from the front seat of her cruiser and begins to compute. She knows that Newton's Law of Cooling says that the rate at which an object cools is proportional to the difference between the temperature T of the body at time t and the temperature T_m of the environment surrounding the body. She jots down the equation

$$\frac{dT}{dt} = k(T - T_m), \quad t > 0, \tag{1}$$

where k is a constant of proportionality, T and T_m are measured in degrees Fahrenheit, and t is time measured in hours. Because Daphne wants to investigate the past using positive values of time, she decides to correspond $t = 0$ with 6:00 a.m., and so, for example, $t = 4$ is 2:00 a.m. After a few scratches on her yellow pad, Daphne realizes that with this time convention the constant k in (1) will turn out to be *positive*. She jots a reminder to herself that 6:30 a.m. is now $t = -1/2$.

As the cool and quiet dawn gives way to the steamy midsummer morning, Daphne begins to sweat and wonders aloud, "But what if the corpse was moved into the fridge in a feeble attempt to hide the body? How does this change my estimate?" She re-enters the restaurant and finds the grease-streaked thermostat above the empty cash register. It reads 70 degrees Fahrenheit.

"But when was the body moved?" Daphne asks. She decides to leave this question unanswered for now, simply letting h denote the number of hours the body has been in the refrigerator prior to 6:00 a.m. For example, if $h = 6$, then the body was moved at midnight.

Daphne flips a page on her legal pad and begins calculating. As the rapidly cooling coffee begins to do its work, she realizes that the way to model the environmental temperature change caused by the move is with the unit step function $\mathcal{U}(t)$. She writes

$$T_m(t) = 50 + 20\mathcal{U}(t - h) \tag{2}$$

and below it the differential equation

$$\frac{dT}{dt} = k(T - T_m(t)). \tag{3}$$

Daphne's mustard-stained polyester blouse begins to drip sweat under the blaze of a midmorning sun. Drained from the heat and the mental exercise, she fires up her cruiser and motors to Boodle's Café for another cup of java and a heaping plate of scrapple and fried eggs. She settles into the faux leather booth. The intense air-conditioning conspires with her sweat-soaked blouse to raise goose flesh on her rapidly cooling skin. The intense chill serves as a gruesome reminder of the tragedy that occurred earlier at the Mayfair.

While Daphne waits for her breakfast, she retrieves her legal pad and quickly reviews her calculations. She then carefully constructs a table that relates refrigeration time h to time of death while eating her scrapple and eggs.

Shoving away the empty platter, Daphne picks up her cell phone to check in with her partner Marie. "Any suspects?" Daphne asks.

"Yeah," she replies, "we got three of 'em. The first is the late Mr. Wood's ex-wife, a dancer by the name of Twinkles. She was seen in the Mayfair between 5 and 6 p.m. in a shouting match with Wood."

"When did she leave?"

"A witness says she left in a hurry a little after six. The second suspect is a South Philly bookie who goes by the name of Slim. Slim was in around 10 last night having a whispered conversation with Joe. Nobody overheard the conversation, but witnesses say there was a lot of hand gesturing, like Slim was upset or something."

"Did anyone see him leave?"

"Yeah. He left quietly around 11. The third suspect is the cook."

"The cook?"

"Yep, the cook. Goes by the name of Shorty. The cashier says he heard Joe and Shorty arguing over the proper way to present a plate of veal scaloppine. She said that Shorty took an unusually long break at 10:30 p.m. He took off in a huff when the restaurant closed at 2:00 a.m. Guess that explains why the place was such a mess."

"Great work, partner. I think I know who to bring in for questioning."

Related Problems

1. Solve equation (1), which models the scenario in which Joe Wood is killed in the refrigerator. Use this solution to estimate the time of death (recall that normal living body temperature is 98.6 degrees Fahrenheit).

2. Solve the differential equation (3) using Laplace transforms. Your solution $T(t)$ will depend on both t and h. (Use the value of k found in Problem 1.)

3. (CAS) Complete Daphne's table. In particular, explain why large values of h give the same time of death.

h	time body moved	time of death
12	6:00 p.m.	
11		
10		
9		
8		
7		
6		
5		
4		
3		
2		

4. Who does Daphne want to question and why?

5. **Still Curious?** The process of temperature change in a dead body is known as *algor mortis* (*rigor mortis* is the process of body stiffening), and although it is not

perfectly described by Newton's Law of Cooling, this topic is covered in most forensic medicine texts. In reality, the cooling of a dead body is determined by more than just Newton's Law. In particular, chemical processes in the body continue for several hours after death. These chemical processes generate heat, and thus a near constant body temperature may be maintained during this time before the exponential decay due to Newton's Law of Cooling begins.

A linear equation, known as the *Glaister equation*, is sometimes used to give a preliminary estimate of the time t since death. The Glaister equation is

$$t = \frac{98.4 - T_0}{1.5} \tag{4}$$

where T_0 is measured body temperature (98.4° F is used here for normal living body temperature instead of 98.6° F). Although we do not have all of the tools to derive this equation exactly (the 1.5 degrees per hour was determined experimentally), we can derive a similar equation via linear approximation.

Use equation (1) with an initial condition of $T(0) = T_0$ to compute the equation of the tangent line to the solution through the point $(0, T_0)$. Do not use the values of T_m or k found in Problem 1. Simply leave these as parameters. Next, let $T = 98.4$ and solve for t to get

$$t = \frac{98.4 - T_0}{k(T_0 - T_m)}. \tag{5}$$

ABOUT THE AUTHOR

Courtesy of Tom LoFaro

Tom LoFaro is a professor and chair of the Mathematics and Computer Science Department at Gustavus Adolphus College in St. Peter, Minnesota. He has been involved in developing differential modeling projects for over 10 years, including being a principal investigator of the NSF-funded IDEA project (http://www.sci.wsu.edu/idea/) and a contributor to CODEE's ODE Architect (Wiley and Sons). Dr. LoFaro's nonacademic interests include fly fishing and coaching little league soccer. His oldest daughter (age 12) aspires to be a forensic anthropologist much like Detective Daphne Marlow.

Earthquake Shaking of Multistory Buildings

by Gilbert N. Lewis

s76/ZUMA Press/Newscom

Collapsed apartment building in San Francisco, October 18, 1989, the day after the massive Loma Prieta earthquake

Large earthquakes typically have a devastating effect on buildings. For example, the famous 1906 San Francisco earthquake destroyed much of that city. More recently, that area was hit by the Loma Prieta earthquake that many people in the United States and elsewhere experienced second-hand while watching on television the Major League Baseball World Series game that was taking place in San Francisco in 1989.

In this project, we attempt to model the effect of an earthquake on a multi-story building and then solve and interpret the mathematics. Let x_i represent the horizontal displacement of the ith floor from equlibrium. Here, the equilibrium position will be a fixed point on the ground, so that $x_0 = 0$. During an earthquake, the ground moves horizontally so that each floor is considered to be displaced relative to the ground. We assume that the ith floor of the building has a mass m_i, and that successive floors are connected by an elastic connector whose effect resembles that of a spring. Typically, the structural elements in large buildings are made of steel, a highly elastic material. Each such connector supplies a restoring force when the floors are displaced relative to each other. We assume that Hooke's Law holds, with proportionality constant k_i between the ith and the $(i + 1)$st floors. That is, the restoring force between those two floors is

$$F = k_i(x_{i+1} - x_i),$$

where $x_{i+1} - x_i$ is the displacement (shift) of the $(i + 1)$st floor relative to the ith floor. We also assume a similar reaction between the first floor and the ground, with proportionality constant k_0 . Figure 1 shows a model of the building, while Figure 2 shows the forces acting on the ith floor.

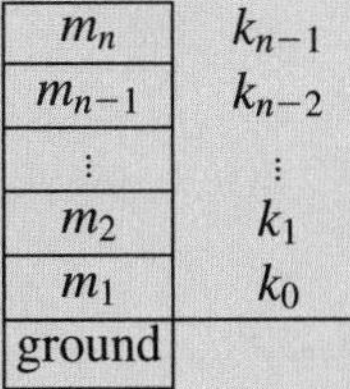

FIGURE 1 Floors of building

	m_{i+1}	$k_i(x_{i+1} - x_i)$
$k_{i-1}(x_i - x_{i-1})$	m_i	
	m_{i-1}	

FIGURE 2 Forces on ith floor

We can apply Newton's second law of motion (Section 5.1), $F = ma$, to each floor of the building to arrive at the following system of linear differential equations.

$$m_1\frac{d^2x_1}{dt^2} = -k_0x_1 + k_1(x_2 - x_1)$$
$$m_2\frac{d^2x_2}{dt^2} = -k_1(x_2 - x_1) + k_2(x_3 - x_2)$$
$$\vdots \qquad \vdots$$
$$m_n\frac{d^2x_n}{dt^2} = -k_{n-1}(x_n - x_{n-1}).$$

As a simple example, consider a two-story building with each floor having mass $m = 5000$ kg and each restoring force constant having a value of $k = 10000$ kg/s^2. Then the differential equations are

$$\frac{d^2x_1}{dt^2} = -4x_1 + 2x_2$$
$$\frac{d^2x_2}{dt^2} = 2x_1 - 2x_2.$$

The solution by the methods of Section 8.2 is

$$x_1(t) = 2c_1 \cos \omega_1 t + 2c_2 \sin \omega_1 t + 2c_3 \cos \omega_2 t + 2c_4 \sin \omega_2 t,$$
$$x_2(t) = (4 - \omega_1^2)c_1 \cos \omega_1 t + (4 - \omega_1^2)c_2 \sin \omega_1 t + (4 - \omega_2^2)c_3 \cos \omega_2 t + (4 - \omega_2^2)c_4 \sin \omega_2 t,$$

where $\omega_1 = \sqrt{3 + \sqrt{5}} = 2.288$, and $\omega_2 = \sqrt{3 - \sqrt{5}} = 0.874$. Now suppose that the following initial conditions are applied: $x_1(0) = 0$, $x_1'(0) = 0.2$, $x_2(0) = 0$, $x_2'(0) = 0$. These correspond to a building in the equilibrium position with the first floor being given a horizontal speed of 0.2 m/s. The solution of the initial value problem is

$$x_1(t) = 2c_2 \sin \omega_1 t + 2c_4 \sin \omega_2 t,$$
$$x_2(t) = (4 - \omega_1^2)c_2 \sin \omega_1 t + (4 - \omega_2^2)c_4 \sin \omega_2 t,$$

where $c_2 = (4 - \omega_2^2)0.1/[(\omega_1^2 - \omega_2^2)\omega_1] = 0.0317 = c_4$. See Figures 3 and 4 for graphs of $x_1(t)$ and $x_2(t)$. Note that initially x_1 moves to the right but is slowed by the drag of x_2, while x_2 is initially at rest, but accelerates, due to the pull of x_1, to overtake x_1 within one second. It continues to the right, eventually pulling x_1 along until the two-second mark. At that point, the drag of x_1 has slowed x_2 to a stop, after which x_2 moves left, passing the equilibrium point at 3.2 seconds and continues moving left, draging x_1 along with it. This back-and-forth motion continues. There is no damping in the system, so that the oscillatory behavior continues forever.

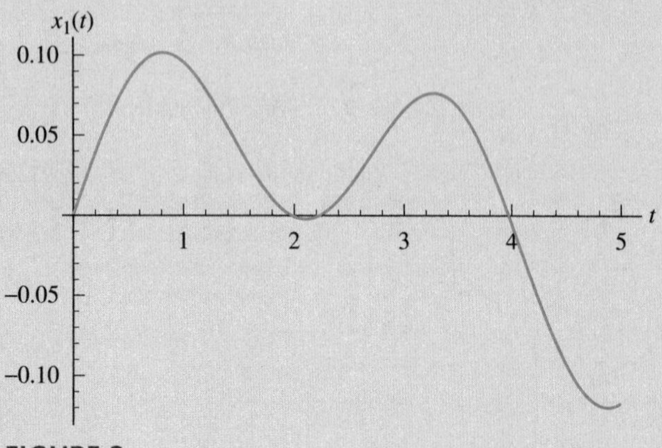

FIGURE 3 Graph of $x_1(t)$

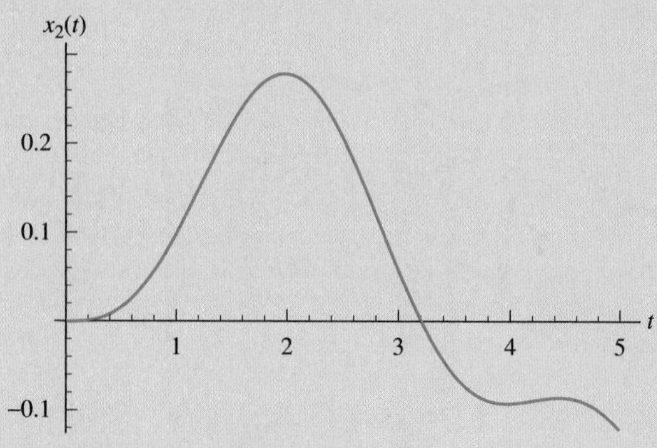

FIGURE 4 Graph of $x_2(t)$

If a horizontal oscillatory force of frequency ω_1 or ω_2 is applied, we have a situation analogous to resonance discussed in Section 5.1.3. In that case, large oscillations of the building would be expected to occur, possibly causing great damage if the earthquake lasted an appreciable length of time.

Let's define the following matrices and vector:

$$\mathbf{M} = \begin{pmatrix} m_1 & 0 & 0 & \cdots & 0 \\ 0 & m_2 & 0 & \cdots & 0 \\ \vdots & & & & \vdots \\ 1 & 0 & 0 & \cdots & m_n \end{pmatrix},$$

$$\mathbf{K} = \begin{pmatrix} -(k_0 + k_1) & k_1 & 0 & 0 \ldots & 0 & 0 & 0 \\ k_1 & -(k_1 + k_2) & k_2 & 0 \ldots & 0 & 0 & 0 \\ 0 & k_2 & -(k_2 + k_3) & k_3 \ldots & 0 & 0 & 0 \\ \vdots & & & & & & \vdots \\ 0 & 0 & 0 & 0 \ldots & k_{n-2} & -(k_{n-2} + k_{n-1}) & k_{n-1} \\ 0 & 0 & 0 & 0 \ldots & 0 & k_{n-1} & -k_{n-1} \end{pmatrix}$$

$$\mathbf{X}(t) = \begin{pmatrix} x_1(t) \\ x_2(t) \\ \vdots \\ x_n(t) \end{pmatrix}$$

Then the system of differential equations can be written in matrix form

$$\mathbf{M}\frac{d^2\mathbf{X}}{dt^2} = \mathbf{KX} \quad \text{or} \quad \mathbf{MX}'' = \mathbf{KX}.$$

Note that the matrix **M** is a diagonal matrix with the mass of the ith floor being the ith diagonal element. Matrix **M** has an inverse given by

$$\mathbf{M}^{-1} = \begin{pmatrix} m_1^{-1} & 0 & 0 & \cdots & 0 \\ 0 & m_2^{-1} & 0 & \cdots & 0 \\ \vdots & & & & \vdots \\ 0 & 0 & 0 & \cdots & m_n^{-1} \end{pmatrix}.$$

We can therefore represent the matrix differential equation by

$$\mathbf{X}'' = (\mathbf{M}^{-1}\mathbf{K})\mathbf{X} \quad \text{or} \quad \mathbf{X}'' = \mathbf{AX}.$$

Where $\mathbf{A} = \mathbf{M}^{-1}\mathbf{K}$, the matrix **M** is called the **mass matrix**, and the matrix **K** is the **stiffness matrix**.

The eigenvalues of the matrix **A** reveal the stability of the building during an earthquake. The eigenvalues of **A** are negative and distinct. In the first example, the eigenvalues are $-3 + \sqrt{5} = -0.764$ and $-3 - \sqrt{5} = -5.236$. The natural frequencies of the building are the square roots of the negatives of the eigenvalues. If λ_i is the ith eigenvalue, then $\omega_i = \sqrt{-\lambda_i}$ is the ith frequency, for $i = 1, 2, \ldots, n$. During an earthquake, a large horizontal force is applied to the first floor. If this is oscillatory in nature, say of the form $\mathbf{F}(t) = \mathbf{G}\cos\gamma t$, then large displacements may develop in the building, especially if the frequency γ of the forcing term is close to one of the natural frequencies of the building. This is reminiscent of the resonance phenomenon studied in Section 5.1.3.

As another example, suppose we have a 10-story building, where each floor has a mass 10000 kg, and each k_i value is 5000 kg/s^2. Then

$$\mathbf{A} = \mathbf{M}^{-1}\mathbf{K} = \begin{pmatrix} -1 & 0.5 & 0 & 0 & 0 & 0 & 0 & 0 & 0 & 0 \\ 0.5 & -1 & 0.5 & 0 & 0 & 0 & 0 & 0 & 0 & 0 \\ 0 & 0.5 & -1 & 0.5 & 0 & 0 & 0 & 0 & 0 & 0 \\ 0 & 0 & 0.5 & -1 & 0.5 & 0 & 0 & 0 & 0 & 0 \\ 0 & 0 & 0 & 0.5 & -1 & 0.5 & 0 & 0 & 0 & 0 \\ 0 & 0 & 0 & 0 & 0.5 & -1 & 0.5 & 0 & 0 & 0 \\ 0 & 0 & 0 & 0 & 0 & 0.5 & -1 & 0.5 & 0 & 0 \\ 0 & 0 & 0 & 0 & 0 & 0 & 0.5 & -1 & 0.5 & 0 \\ 0 & 0 & 0 & 0 & 0 & 0 & 0 & 0.5 & -1 & 0.5 \\ 0 & 0 & 0 & 0 & 0 & 0 & 0 & 0 & 0.5 & -0.5 \end{pmatrix}$$

The eigenvalues of **A** are found easily using *Mathematica* or another similar computer package. These values are -1.956, -1.826, -1.623, -1.365, -1.075, -0.777, -0.5, -0.267, -0.099, and -0.011, with corresponding frequencies 1.399, 1.351, 1.274, 1.168, 1.037, 0.881, 0.707, 0.517, 0.315, and 0.105 and periods of oscillation $(2\pi/\omega)$ 4.491, 4.651, 4.932, 5.379, 6.059, 7.132, 8.887, 12.153, 19.947, and 59.840. During a typical earthquake whose period might be in the range of 2 to 3 seconds, this building does not seem to be in any danger of developing resonance. However, if the k values were 10 times as large (multiply **A** by 10), then, for example, the sixth period would be 2.253 seconds, while the fifth through seventh are all on the order of 2–3 seconds. Such a building is more likely to suffer damage in a typical earthquake of period 2–3 seconds.

Related Problems

1. Consider a three-story building with the same m and k values as in the first example. Write down the corresponding system of differential equations. What are the matrices **M**, **K**, and **A**? Find the eigenvalues for **A**. What range of frequencies of an earthquake would place the building in danger of destruction?
2. Consider a three-story building with the same m and k values as in the second example. Write down the corresponding system of differential equations. What are the matrices **M**, **K**, and **A**? Find the eigenvalues for **A**. What range of frequencies of an earthquake would place the building in danger of destruction?
3. Consider the tallest building on your campus. Assume reasonable values for the mass of each floor and for the proportionality constants between floors. If you have trouble coming up with such values, use the ones in the example problems. Find the matrices **M**, **K**, and **A**, and find the eigenvalues of **A** and the frequencies and periods of oscillation. Is your building safe from a modest-sized period-2 earthquake? What if you multiplied the matrix **K** by 10 (that is, made the building stiffer)? What would you have to multiply the matrix **K** by in order to put your building in the danger zone?
4. Solve the earthquake problem for the three-story building of Problem 1:

$$\mathbf{MX}'' = \mathbf{KX} + \mathbf{F}(t),$$

where $\mathbf{F}(t) = \mathbf{G}\cos\gamma t$, $\mathbf{G} = E\mathbf{B}$, $\mathbf{B} = [1 \quad 0 \quad 0]^T$, $E = 10{,}000$ lbs is the amplitude of the earthquake force acting at ground level, and $\gamma = 3$ is the frequency of the earthquake (a typical earthquake frequency). See Section 8.3 for the method of solving nonhomogeneous matrix differential equations. Use initial conditions for a building at rest.

Project for Section 8.3

Modeling Arms Races

by Michael Olinick

NICOLAS ASFOURI/AFP/Getty Images/Newscom

Weapons and ammunition recovered during military operations against Taliban militants in South Waziristan in October 2009

The last hundred years have seen numerous dangerous, destabilizing, and expensive arms races. The outbreak of World War I climaxed a rapid buildup of armaments among rival European powers. There was a similar mutual accumulation of conventional arms just prior to World War II. The United States and the Soviet Union engaged in a costly nuclear arms race during the forty years of the Cold War. Stockpiling of ever-more deadly weapons is common today in many parts of the world, including the Middle East, the Indian subcontinent, and the Korean peninsula.

British meteorologist and educator Lewis F. Richardson (1881–1953) developed several mathematical models to analyze the dynamics of arms races, the evolution over time of the process of interaction between countries in their acquisition of weapons. Arms race models generally assume that each nation adjusts its accumulation of weapons in some manner dependent on the size of its own stockpile and the armament levels of the other nations.

Richardson's primary model of a two country arms race is based on *mutual fear*: A nation is spurred to increase its arms stockpile at a rate proportional to the level of armament expenditures of its rival. Richardson's model takes into account internal constraints within a nation that slow down arms buildups: The more a nation is spending on arms, the harder it is to make greater increases, because it becomes increasingly difficult to divert society's resources from basic needs such as food and housing to weapons. Richardson also built into his model other factors driving or slowing down an arms race that are independent of levels of arms expenditures.

The mathematical structure of this model is a linked system of two first-order linear differential equations. If x and y represent the amount of wealth being spent on arms by two nations at time t, then the model has the form

$$\frac{dx}{dt} = ay - mx + r$$
$$\frac{dy}{dt} = bx - ny + s$$

where a, b, m, and n are positive constants while r and s are constants which can be positive or negative. The constants a and b measure mutual fear; the constants m and n represent proportionality factors for the "internal brakes" to further arms increases. Positive values for r and s correspond to underlying factors of ill will or distrust that would persist even if arms expenditures dropped to zero. Negative values for r and s indicate a contribution based on goodwill.

The dynamic behavior of this system of differential equations depends on the relative sizes of ab and mn together with the signs of r and s. Although the model is a relatively simple one, it allows us to consider several different long-term outcomes. It's possible that two nations might move simultaneously toward mutual disarmament, with x and y each approaching zero. A vicious cycle of unbounded increases in x and y is another possible scenario. A third eventuality is that the arms expenditures asymptotically approach a stable point (x^*, y^*) regardless of the initial level of arms expenditures. In other cases, the eventual outcome depends on the starting point. Figure 1 shows one possible situation with four different initial

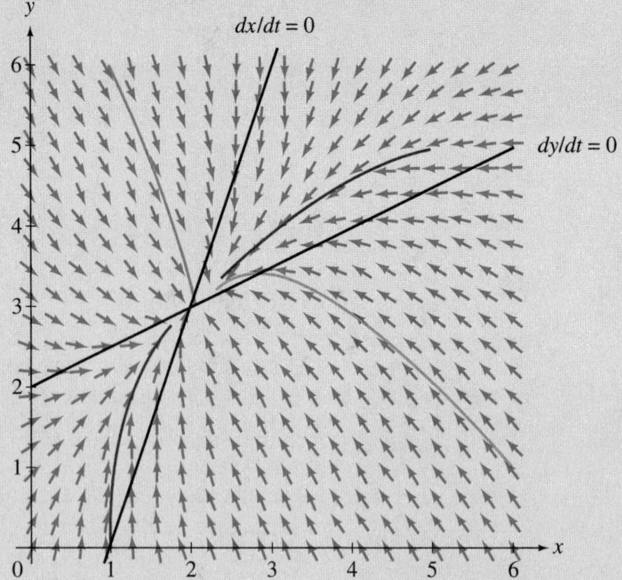

FIGURE 1 Expenditures approaching a stable point

levels, each of which leads to a "stable outcome," the intersection of the nullclines $dx/dt = 0$ and $dy/dt = 0$.

Although "real world" arms races seldom match exactly with Richardson's model, his pioneering work has led to many fruitful applications of differential equation models to problems in international relations and political science. As two leading researchers in the field note in [**3**], "The Richardson arms race model constitutes one of the most important models of arms race phenomena and, at the same time, one of the most influential formal models in all of the international relations literature."

Arms races are not limited to the interaction of nation states. They can take place between a government and a paramilitary terrorist group within its borders as, for example, the Tamil Tigers in Sri Lanka, the Shining Path in Peru, or the Taliban in Afghanistan. Arms phenomena have also been observed between rival urban gangs and between law enforcement agencies and organized crime.

The "arms" need not even be weapons. Colleges have engaged in "amenities arms races," often spending millions of dollars on more luxurious dormitories, state- of-the-art athletic facilities, epicurean dining options, and the like, to be more competitive in attracting student applications. Biologists have identified the possibility of evolutionary arms races between and within species as an adaptation in one lineage may change the selection pressure on another lineage, giving rise to a counter-adaptation. Most generally, the assumptions represented in a Richardson-type model also characterize many competitions in which each side perceives a need to stay ahead of the other in some mutually important measure.

Related Problems

1. (a) By substituting the proposed solutions into the differential equations, show that the solution of the particular Richardson arms model

$$\frac{dx}{dt} = y - 3x + 3$$

$$\frac{dy}{dt} = 2x - 4y + 8$$

with initial condition $x(0) = 12$, $y(0) = 15$ is

$$x(t) = \frac{32}{3}e^{-2t} - \frac{2}{3}e^{-5t} + 2$$
$$y(t) = \frac{32}{3}e^{-2t} + \frac{4}{3}e^{-5t} + 3$$

What is the long-term behavior of this arms race?

(b) For the Richardson arms race model (a) with arbitrary initial conditions $x(0) = A$, $y(0) = B$, show that the solution is given by

$$\begin{aligned} x(t) &= Ce^{-5t} + De^{-2t} + 2 \\ y(t) &= -2Ce^{-5t} + De^{-2t} + 3 \end{aligned} \quad \text{where} \quad \begin{aligned} C &= (A - B + 1)/3 \\ D &= (2A + B - 7)/3 \end{aligned}$$

Show that this result implies that the qualitative long-term behavior of such an arms race is the same ($x(t) \to 2$, $y(t) \to 3$), no matter what the initial values of x and y are.

2. The qualitative long-term behavior of a Richardson arms race model can, in some cases, depend on the initial conditions. Consider, for example, the system

$$\frac{dx}{dt} = 3y - 2x - 10$$
$$\frac{dy}{dt} = 4x - 3y - 10$$

For each of the given initial conditions below, verify that the proposed solution works and discuss the long-term behavior:

(a) $x(0) = 1, y(0) = 1 : x(t) = 10 - 9e^t, y(t) = 10 - 9e^t$
(b) $x(0) = 1, y(0) = 22 : x(t) = 10 - 9e^{-6t}, y(t) = 10 + 12e^{-6t}$
(c) $x(0) = 1, y(0) = 29 : x(t) = -12e^{-6t} + 3e^t + 10, y(t) = 16e^{-6t} + 3e^t + 10$
(d) $x(0) = 10, y(0) = 10 : x(t) = 10, y(t) = 10$ for all t

3. (a) As a possible alternative to the Richardson model, consider a *stock adjustment model* for an arms race. The assumption here is that each country sets a desired level of arms expenditures for itself and then changes its weapons stock proportionally to the gap between its current level and the desired one. Show that this assumption can be represented by the system of differential equations

$$\frac{dx}{dt} = a(x^* - x)$$
$$\frac{dx}{dt} = b(y^* - y)$$

where x^* and y^* are desired constant levels and a, b are positive constants. How will x and y evolve over time under such a model?

(b) Generalize the stock adjustment model of (a) to a more realistic one where the desired level for each country depends on the levels of both countries. In particular, suppose x^* has the form $x^* = c + dy$ where c and d are positive constants and that y^* has a similar format. Show that, under these assumptions, the stock adjustment model is equivalent to a Richardson model.

4. Extend the Richardson model to three nations, deriving a system of linear differential equations if the three are mutually fearful: each one is spurred to arm by the expenditures of the other two. How might the equations change if two of the nations are close allies not threatened by the arms buildup of each other, but fearful of the armaments of the third. Investigate the long-term behavior of such arms races.

5. In the real world, an unbounded runaway arms race is impossible since there is an absolute limit to the amount any country can spend on weapons; *e.g.* gross national product minus some amount for survival. Modify the Richardson model to incorporate this idea and analyze the dynamics of an arms race governed by these new differential equations.

References

1. Richardson, Lewis F., *Arms and Insecurity: A Mathematical Study of the Causes and Origins of War.* Pittsburgh: Boxwood Press, 1960.
2. Olinick, Michael, *An Introduction to Mathematical Models in the Social and Life Sciences.* Reading, MA: Addison-Wesley, 1978.
3. Intriligator, Michael D., and Dagobert L. Brito, "Richardsonian Arms Race Models" in Manus I. Midlarsky, ed., *Handbook of War Studies.* Boston: Unwin Hyman, 1989.

ABOUT THE AUTHOR

Courtesy of Michael Olinick

After earning a BA in mathematics and philosophy at the University of Michigan and an MA and PhD from the University of Wisconsin (Madison), **Michael Olinick** moved from the Midwest to New England where he joined the Middlebury College faculty in 1970 and now serves as Professor of Mathematics. Dr. Olinick has held visiting positions at University College Nairobi, University of California at Berkeley, Wesleyan University, and Lancaster University in Great Britain. He is the author or co-author of a number of books on single and multivariable calculus, mathematical modeling, probability, topology, and principles and practice of mathematics. He is currently developing a new textbook on mathematical models in the humanities, social, and life sciences.

1 Introduction to Differential Equations

The words *differential* and *equations* certainly suggest solving some kind of equation that contains derivatives $y', y'', \ldots$. Analogous to a course in algebra and trigonometry, in which a good amount of time is spent solving equations such as $x^2 + 5x + 4 = 0$ for the unknown number x, in this course *one* of our tasks will be to solve differential equations such as $y'' + 2y' + y = 0$ for an unknown function $y = \phi(x)$.

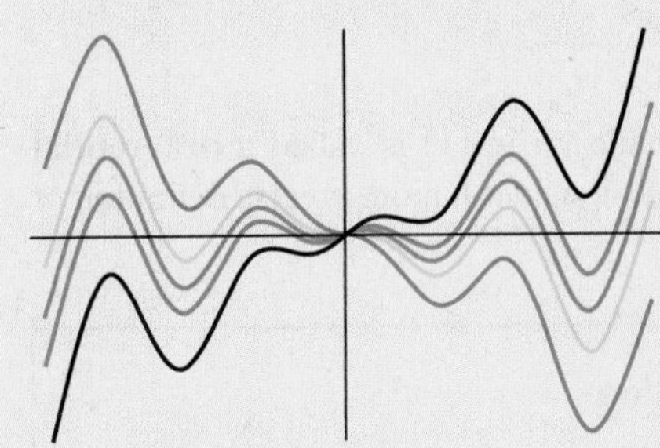

The preceding paragraph tells something, but not the complete story, about the course you are about to begin. As the course unfolds, you will see that there is more to the study of differential equations than just mastering methods that mathematicians over past centuries devised to solve them.

But first things first. In order to read, study, and be conversant in a specialized subject, you have to master some of the terminology of that discipline. This is the thrust of the first two sections of this chapter. In the last section we briefly examine the link between differential equations and the real world. Practical questions such as

How fast does a disease spread? *How fast does a population change*?

involve rates of change or derivatives. And so the mathematical description—or *mathematical model*—of phenomena, experiments, observations, or theories may be a differential equation.

1.1 DEFINITIONS AND TERMINOLOGY

REVIEW MATERIAL

- The definition of the derivative
- Rules of differentiation
- Derivative as a rate of change
- Connection between the first derivative and increasing/decreasing
- Connection between the second derivative and concavity

INTRODUCTION The derivative dy/dx of a function $y = \phi(x)$ is itself another function $\phi'(x)$ found by an appropriate rule. The exponential function $y = e^{0.1x^2}$ is differentiable on the interval $(-\infty, \infty)$, and, by the Chain Rule, its first derivative is $dy/dx = 0.2xe^{0.1x^2}$. If we replace $e^{0.1x^2}$ on the right-hand side of the last equation by the symbol y, the derivative becomes

$$\frac{dy}{dx} = 0.2xy. \tag{1}$$

Now imagine that a friend of yours simply hands you equation (1)—you have no idea how it was constructed—and asks, *What is the function represented by the symbol* y? You are now face to face with one of the basic problems in this course:

How do you solve such an equation for the function $y = \phi(x)$?

A Definition The equation that we made up in (1) is called a **differential equation.** Before proceeding any further, let us consider a more precise definition of this concept.

DEFINITION 1.1.1 Differential Equation

An equation containing the derivatives of one or more unknown functions (or dependent variables), with respect to one or more independent variables, is said to be a **differential equation (DE).**

To talk about them, we shall classify differential equations according to **type, order,** and **linearity.**

Classification by Type If a differential equation contains only ordinary derivatives of one or more unknown functions with respect to a *single* independent variable, it is said to be an **ordinary differential equation (ODE).** An equation involving partial derivatives of one or more unknown functions of two or more independent variables is called a **partial differential equation (PDE).** Our first example illustrates several of each type of differential equation.

EXAMPLE 1 Types of Differential Equations

(a) The equations

an ODE can contain more than one unknown function

$$\frac{dy}{dx} + 5y = e^x, \quad \frac{d^2y}{dx^2} - \frac{dy}{dx} + 6y = 0, \quad \text{and} \quad \frac{dx}{dt} + \frac{dy}{dt} = 2x + y \tag{2}$$

are examples of ordinary differential equations.

(b) The following equations are partial differential equations:*

$$\frac{\partial^2 u}{\partial x^2} + \frac{\partial^2 u}{\partial y^2} = 0, \quad \frac{\partial^2 u}{\partial x^2} = \frac{\partial^2 u}{\partial t^2} - 2\frac{\partial u}{\partial t}, \quad \frac{\partial u}{\partial y} = -\frac{\partial v}{\partial x}. \tag{3}$$

Notice in the third equation that there are two unknown functions and two independent variables in the PDE. This means u and v must be functions of *two or more* independent variables. ≡

≡ Notation

Throughout this text ordinary derivatives will be written by using either the **Leibniz notation** $dy/dx, d^2y/dx^2, d^3y/dx^3, \ldots$ or the **prime notation** $y', y'', y''', \ldots$. By using the latter notation, the first two differential equations in (2) can be written a little more compactly as $y' + 5y = e^x$ and $y'' - y' + 6y = 0$. Actually, the prime notation is used to denote only the first three derivatives; the fourth derivative is written $y^{(4)}$ instead of y''''. In general, the nth derivative of y is written d^ny/dx^n or $y^{(n)}$. Although less convenient to write and to typeset, the Leibniz notation has an advantage over the prime notation in that it clearly displays both the dependent and independent variables. For example, in the equation

unknown function
or dependent variable

$$\frac{d^2x}{dt^2} + 16x = 0$$

independent variable

it is immediately seen that the symbol x now represents a dependent variable, whereas the independent variable is t. You should also be aware that in physical sciences and engineering, Newton's **dot notation** (derogatorily referred to by some as the "flyspeck" notation) is sometimes used to denote derivatives with respect to time t. Thus the differential equation $d^2s/dt^2 = -32$ becomes $\ddot{s} = -32$. Partial derivatives are often denoted by a **subscript notation** indicating the independent variables. For example, with the subscript notation the second equation in (3) becomes $u_{xx} = u_{tt} - 2u_t$.

≡ Classification by Order

The **order of a differential equation** (either ODE or PDE) is the order of the highest derivative in the equation. For example,

second order — first order

$$\frac{d^2y}{dx^2} + 5\left(\frac{dy}{dx}\right)^3 - 4y = e^x$$

is a second-order ordinary differential equation. In Example 1, the first and third equations in (2) are first-order ODEs, whereas in (3) the first two equations are second-order PDEs. First-order ordinary differential equations are occasionally written in differential form $M(x, y)\,dx + N(x, y)\,dy = 0$. For example, if we assume that y denotes the dependent variable in $(y - x)\,dx + 4x\,dy = 0$, then $y' = dy/dx$, so by dividing by the differential dx, we get the alternative form $4xy' + y = x$.

In symbols we can express an nth-order ordinary differential equation in one dependent variable by the general form

$$F(x, y, y', \ldots, y^{(n)}) = 0, \tag{4}$$

where F is a real-valued function of $n + 2$ variables: $x, y, y', \ldots, y^{(n)}$. For both practical and theoretical reasons we shall also make the assumption hereafter that it is possible to solve an ordinary differential equation in the form (4) uniquely for the

*Except for this introductory section, only ordinary differential equations are considered in *A First Course in Differential Equations with Modeling Applications,* Tenth Edition. In that text the word *equation* and the abbreviation DE refer only to ODEs. Partial differential equations or PDEs are considered in the expanded volume *Differential Equations with Boundary-Value Problems,* Eighth Edition.

highest derivative $y^{(n)}$ in terms of the remaining $n + 1$ variables. The differential equation

$$\frac{d^n y}{dx^n} = f(x, y, y', \ldots, y^{(n-1)}), \tag{5}$$

where f is a real-valued continuous function, is referred to as the **normal form** of (4). Thus when it suits our purposes, we shall use the normal forms

$$\frac{dy}{dx} = f(x, y) \quad \text{and} \quad \frac{d^2y}{dx^2} = f(x, y, y')$$

to represent general first- and second-order ordinary differential equations. For example, the normal form of the first-order equation $4xy' + y = x$ is $y' = (x - y)/4x$; the normal form of the second-order equation $y'' - y' + 6y = 0$ is $y'' = y' - 6y$. See (*iv*) in the *Remarks*.

Classification by Linearity An nth-order ordinary differential equation (4) is said to be **linear** if F is linear in $y, y', \ldots, y^{(n)}$. This means that an nth-order ODE is linear when (4) is $a_n(x)y^{(n)} + a_{n-1}(x)y^{(n-1)} + \cdots + a_1(x)y' + a_0(x)y - g(x) = 0$ or

$$a_n(x)\frac{d^n y}{dx^n} + a_{n-1}(x)\frac{d^{n-1}y}{dx^{n-1}} + \cdots + a_1(x)\frac{dy}{dx} + a_0(x)y = g(x). \tag{6}$$

Two important special cases of (6) are linear first-order ($n = 1$) and linear second-order ($n = 2$) DEs:

$$a_1(x)\frac{dy}{dx} + a_0(x)y = g(x) \quad \text{and} \quad a_2(x)\frac{d^2y}{dx^2} + a_1(x)\frac{dy}{dx} + a_0(x)y = g(x). \tag{7}$$

In the additive combination on the left-hand side of equation (6) we see that the characteristic two properties of a linear ODE are as follows:

- The dependent variable y and all its derivatives $y', y'', \ldots, y^{(n)}$ are of the first degree, that is, the power of each term involving y is 1.
- The coefficients $a_0, a_1, \ldots, a_n$ of $y, y', \ldots, y^{(n)}$ depend at most on the independent variable x.

A **nonlinear** ordinary differential equation is simply one that is not linear. Nonlinear functions of the dependent variable or its derivatives, such as $\sin y$ or $e^{y'}$, cannot appear in a linear equation.

EXAMPLE 2 Linear and Nonlinear ODEs

(a) The equations

$$(y - x)dx + 4xy\,dy = 0, \quad y'' - 2y' + y = 0, \quad x^3\frac{d^3y}{dx^3} + x\frac{dy}{dx} - 5y = e^x$$

are, in turn, *linear* first-, second-, and third-order ordinary differential equations. We have just demonstrated that the first equation is linear in the variable y by writing it in the alternative form $4xy' + y = x$.

(b) The equations

nonlinear term: coefficient depends on y ↓ nonlinear term: nonlinear function of y ↓ nonlinear term: power not 1 ↓

$$(1 - y)y' + 2y = e^x, \quad \frac{d^2y}{dx^2} + \sin y = 0, \quad \text{and} \quad \frac{d^4y}{dx^4} + y^2 = 0$$

are examples of *nonlinear* first-, second-, and fourth-order ordinary differential equations, respectively. ≡

Solutions As was stated before, one of the goals in this course is to solve, or find solutions of, differential equations. In the next definition we consider the concept of a solution of an ordinary differential equation.

DEFINITION 1.1.2 Solution of an ODE

Any function ϕ, defined on an interval I and possessing at least n derivatives that are continuous on I, which when substituted into an nth-order ordinary differential equation reduces the equation to an identity, is said to be a **solution** of the equation on the interval.

In other words, a solution of an nth-order ordinary differential equation (4) is a function ϕ that possesses at least n derivatives and for which

$$F(x, \phi(x), \phi'(x), \ldots, \phi^{(n)}(x)) = 0 \qquad \text{for all } x \text{ in } I.$$

We say that ϕ *satisfies* the differential equation on I. For our purposes we shall also assume that a solution ϕ is a real-valued function. In our introductory discussion we saw that $y = e^{0.1x^2}$ is a solution of $dy/dx = 0.2xy$ on the interval $(-\infty, \infty)$.

Occasionally, it will be convenient to denote a solution by the alternative symbol $y(x)$.

Interval of Definition You cannot think *solution* of an ordinary differential equation without simultaneously thinking *interval*. The interval I in Definition 1.1.2 is variously called the **interval of definition,** the **interval of existence,** the **interval of validity,** or the **domain of the solution** and can be an open interval (a, b), a closed interval $[a, b]$, an infinite interval (a, ∞), and so on.

EXAMPLE 3 Verification of a Solution

Verify that the indicated function is a solution of the given differential equation on the interval $(-\infty, \infty)$.

(a) $dy/dx = xy^{1/2}$; $y = \frac{1}{16}x^4$ **(b)** $y'' - 2y' + y = 0$; $y = xe^x$

SOLUTION One way of verifying that the given function is a solution is to see, after substituting, whether each side of the equation is the same for every x in the interval.

(a) From

left-hand side: $$\frac{dy}{dx} = \frac{1}{16}(4 \cdot x^3) = \frac{1}{4}x^3,$$

right-hand side: $$xy^{1/2} = x \cdot \left(\frac{1}{16}x^4\right)^{1/2} = x \cdot \left(\frac{1}{4}x^2\right) = \frac{1}{4}x^3,$$

we see that each side of the equation is the same for every real number x. Note that $y^{1/2} = \frac{1}{4}x^2$ is, by definition, the nonnegative square root of $\frac{1}{16}x^4$.

(b) From the derivatives $y' = xe^x + e^x$ and $y'' = xe^x + 2e^x$ we have, for every real number x,

left-hand side: $$y'' - 2y' + y = (xe^x + 2e^x) - 2(xe^x + e^x) + xe^x = 0,$$

right-hand side: 0.

Note, too, that in Example 3 each differential equation possesses the constant solution $y = 0$, $-\infty < x < \infty$. A solution of a differential equation that is identically zero on an interval I is said to be a **trivial solution.**

Solution Curve The graph of a solution ϕ of an ODE is called a **solution curve.** Since ϕ is a differentiable function, it is continuous on its interval I of definition. Thus there may be a difference between the graph of the *function* ϕ and the graph of the *solution* ϕ. Put another way, the domain of the function ϕ need not be the same as the interval I of definition (or domain) of the solution ϕ. Example 4 illustrates the difference.

EXAMPLE 4 Function versus Solution

The domain of $y = 1/x$, considered simply as a *function*, is the set of all real numbers x except 0. When we graph $y = 1/x$, we plot points in the xy-plane corresponding to a judicious sampling of numbers taken from its domain. The rational function $y = 1/x$ is discontinuous at 0, and its graph, in a neighborhood of the origin, is given in Figure 1.1.1(a). The function $y = 1/x$ is not differentiable at $x = 0$, since the y-axis (whose equation is $x = 0$) is a vertical asymptote of the graph.

Now $y = 1/x$ is also a solution of the linear first-order differential equation $xy' + y = 0$. (Verify.) But when we say that $y = 1/x$ is a *solution* of this DE, we mean that it is a function defined on an interval I on which it is differentiable and satisfies the equation. In other words, $y = 1/x$ is a solution of the DE on *any* interval that does not contain 0, such as $(-3, -1)$, $\left(\frac{1}{2}, 10\right)$, $(-\infty, 0)$, or $(0, \infty)$. Because the solution curves defined by $y = 1/x$ for $-3 < x < -1$ and $\frac{1}{2} < x < 10$ are simply segments, or pieces, of the solution curves defined by $y = 1/x$ for $-\infty < x < 0$ and $0 < x < \infty$, respectively, it makes sense to take the interval I to be as large as possible. Thus we take I to be either $(-\infty, 0)$ or $(0, \infty)$. The solution curve on $(0, \infty)$ is shown in Figure 1.1.1(b).

(a) function $y = 1/x$, $x \neq 0$

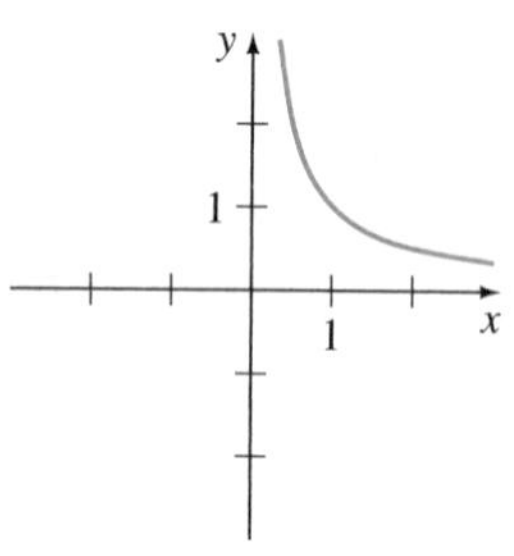

(b) solution $y = 1/x$, $(0, \infty)$

FIGURE 1.1.1 In Example 4 the function $y = 1/x$ is not the same as the solution $y = 1/x$

Explicit and Implicit Solutions You should be familiar with the terms *explicit functions* and *implicit functions* from your study of calculus. A solution in which the dependent variable is expressed solely in terms of the independent variable and constants is said to be an **explicit solution.** For our purposes, let us think of an explicit solution as an explicit formula $y = \phi(x)$ that we can manipulate, evaluate, and differentiate using the standard rules. We have just seen in the last two examples that $y = \frac{1}{16}x^4$, $y = xe^x$, and $y = 1/x$ are, in turn, explicit solutions of $dy/dx = xy^{1/2}$, $y'' - 2y' + y = 0$, and $xy' + y = 0$. Moreover, the trivial solution $y = 0$ is an explicit solution of all three equations. When we get down to the business of actually solving some ordinary differential equations, you will see that methods of solution do not always lead directly to an explicit solution $y = \phi(x)$. This is particularly true when we attempt to solve nonlinear first-order differential equations. Often we have to be content with a relation or expression $G(x, y) = 0$ that defines a solution ϕ implicitly.

DEFINITION 1.1.3 Implicit Solution of an ODE

A relation $G(x, y) = 0$ is said to be an **implicit solution** of an ordinary differential equation (4) on an interval I, provided that there exists at least one function ϕ that satisfies the relation as well as the differential equation on I.

It is beyond the scope of this course to investigate the conditions under which a relation $G(x, y) = 0$ defines a differentiable function ϕ. So we shall assume that if the formal implementation of a method of solution leads to a relation $G(x, y) = 0$, then there exists at least one function ϕ that satisfies both the relation (that is, $G(x, \phi(x)) = 0$) and the differential equation on an interval I. If the implicit solution $G(x, y) = 0$ is fairly simple, we may be able to solve for y in terms of x and obtain one or more explicit solutions. See (*i*) in the *Remarks*.

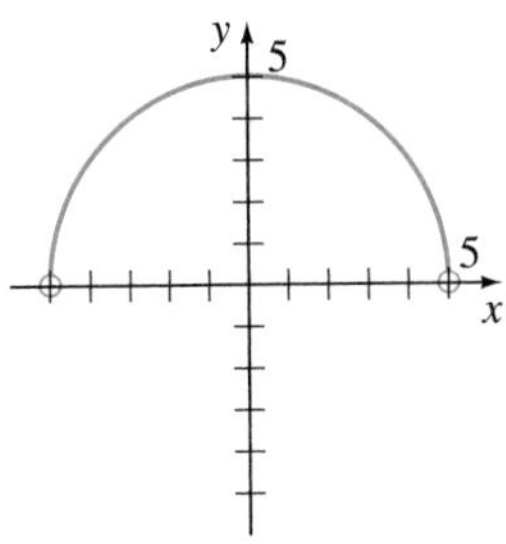

(a) implicit solution

$x^2 + y^2 = 25$

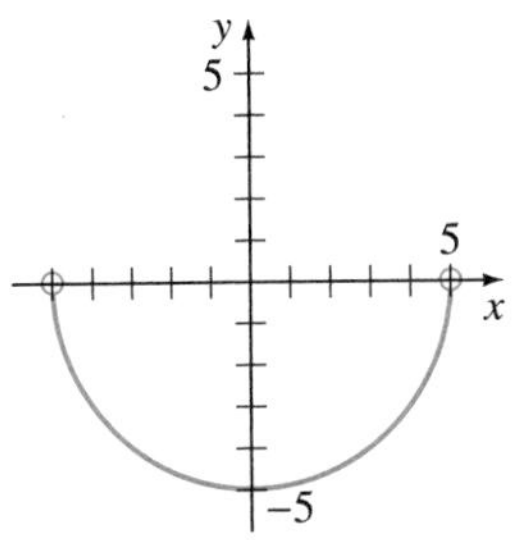

(b) explicit solution

$y_1 = \sqrt{25 - x^2}, \ -5 < x < 5$

(c) explicit solution

$y_2 = -\sqrt{25 - x^2}, \ -5 < x < 5$

FIGURE 1.1.2 An implicit solution and two explicit solutions of (8) in Example 5

EXAMPLE 5 Verification of an Implicit Solution

The relation $x^2 + y^2 = 25$ is an implicit solution of the differential equation

$$\frac{dy}{dx} = -\frac{x}{y} \tag{8}$$

on the open interval $(-5, 5)$. By implicit differentiation we obtain

$$\frac{d}{dx}x^2 + \frac{d}{dx}y^2 = \frac{d}{dx}25 \quad \text{or} \quad 2x + 2y\frac{dy}{dx} = 0.$$

Solving the last equation for the symbol dy/dx gives (8). Moreover, solving $x^2 + y^2 = 25$ for y in terms of x yields $y = \pm\sqrt{25 - x^2}$. The two functions $y = \phi_1(x) = \sqrt{25 - x^2}$ and $y = \phi_2(x) = -\sqrt{25 - x^2}$ satisfy the relation (that is, $x^2 + \phi_1^2 = 25$ and $x^2 + \phi_2^2 = 25$) and are explicit solutions defined on the interval $(-5, 5)$. The solution curves given in Figures 1.1.2(b) and 1.1.2(c) are segments of the graph of the implicit solution in Figure 1.1.2(a). ≡

Any relation of the form $x^2 + y^2 - c = 0$ *formally* satisfies (8) for any constant c. However, it is understood that the relation should always make sense in the real number system; thus, for example, if $c = -25$, we cannot say that $x^2 + y^2 + 25 = 0$ is an implicit solution of the equation. (Why not?)

Because the distinction between an explicit solution and an implicit solution should be intuitively clear, we will not belabor the issue by always saying, "Here is an explicit (implicit) solution."

≡ Families of Solutions The study of differential equations is similar to that of integral calculus. In some texts a solution ϕ is sometimes referred to as an **integral** of the equation, and its graph is called an **integral curve.** When evaluating an antiderivative or indefinite integral in calculus, we use a single constant c of integration. Analogously, when solving a first-order differential equation $F(x, y, y') = 0$, we *usually* obtain a solution containing a single arbitrary constant or parameter c. A solution containing an arbitrary constant represents a set $G(x, y, c) = 0$ of solutions called a **one-parameter family of solutions.** When solving an nth-order differential equation $F(x, y, y', \ldots, y^{(n)}) = 0$, we seek an **$n$-parameter family of solutions** $G(x, y, c_1, c_2, \ldots, c_n) = 0$. This means that *a single differential equation can possess an infinite number of solutions* corresponding to the unlimited number of choices for the parameter(s). A solution of a differential equation that is free of arbitrary parameters is called a **particular solution.**

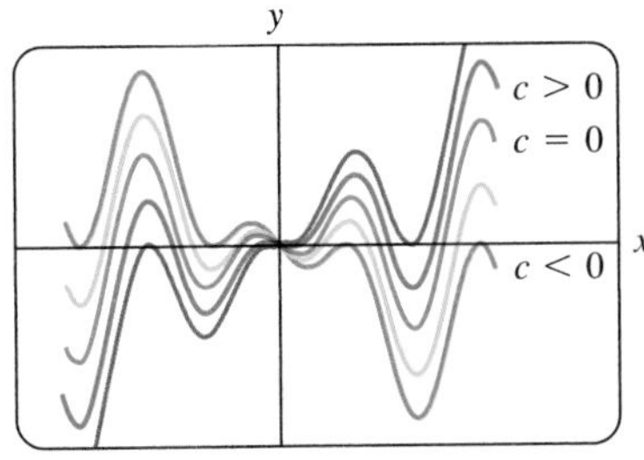

FIGURE 1.1.3 Some solutions of DE in part (a) of Example 6

EXAMPLE 6 Particular Solutions

(a) The one-parameter family $y = cx - x\cos x$ is an explicit solution of the linear first-order equation

$$xy' - y = x^2 \sin x$$

on the interval $(-\infty, \infty)$. (Verify.) Figure 1.1.3 shows the graphs of some particular solutions in this family for various choices of c. The solution $y = -x\cos x$, the blue graph in the figure, is a particular solution corresponding to $c = 0$.

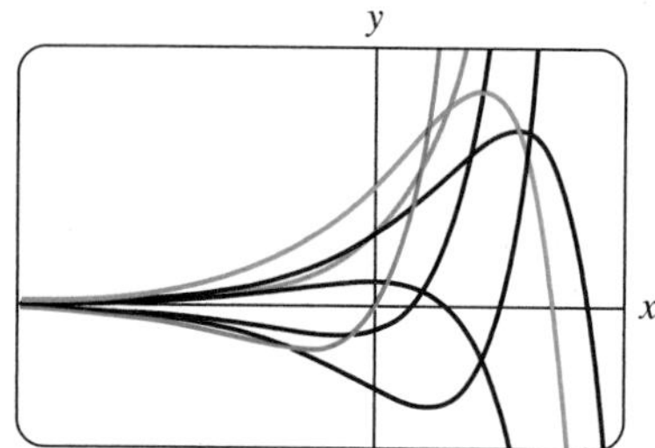

FIGURE 1.1.4 Some solutions of DE in part (b) of Example 6

(b) The two-parameter family $y = c_1e^x + c_2xe^x$ is an explicit solution of the linear second-order equation

$$y'' - 2y' + y = 0$$

in part (b) of Example 3. (Verify.) In Figure 1.1.4 we have shown seven of the "double infinity" of solutions in the family. The solution curves in red, green, and blue are the graphs of the particular solutions $y = 5xe^x$ ($c_1 = 0$, $c_2 = 5$), $y = 3e^x$ ($c_1 = 3$, $c_2 = 0$), and $y = 5e^x - 2xe^x$ ($c_1 = 5$, $c_2 = 2$), respectively. ≡

Sometimes a differential equation possesses a solution that is not a member of a family of solutions of the equation—that is, a solution that cannot be obtained by specializing *any* of the parameters in the family of solutions. Such an extra solution is called a **singular solution.** For example, we have seen that $y = \frac{1}{16}x^4$ and $y = 0$ are solutions of the differential equation $dy/dx = xy^{1/2}$ on $(-\infty, \infty)$. In Section 2.2 we shall demonstrate, by actually solving it, that the differential equation $dy/dx = xy^{1/2}$ possesses the one-parameter family of solutions $y = \left(\frac{1}{4}x^2 + c\right)^2$. When $c = 0$, the resulting particular solution is $y = \frac{1}{16}x^4$. But notice that the trivial solution $y = 0$ is a singular solution, since it is not a member of the family $y = \left(\frac{1}{4}x^2 + c\right)^2$; there is no way of assigning a value to the constant c to obtain $y = 0$.

In all the preceding examples we used x and y to denote the independent and dependent variables, respectively. But you should become accustomed to seeing and working with other symbols to denote these variables. For example, we could denote the independent variable by t and the dependent variable by x.

EXAMPLE 7 Using Different Symbols

The functions $x = c_1 \cos 4t$ and $x = c_2 \sin 4t$, where c_1 and c_2 are arbitrary constants or parameters, are both solutions of the linear differential equation

$$x'' + 16x = 0.$$

For $x = c_1 \cos 4t$ the first two derivatives with respect to t are $x' = -4c_1 \sin 4t$ and $x'' = -16c_1 \cos 4t$. Substituting x'' and x then gives

$$x'' + 16x = -16c_1 \cos 4t + 16(c_1 \cos 4t) = 0.$$

In like manner, for $x = c_2 \sin 4t$ we have $x'' = -16c_2 \sin 4t$, and so

$$x'' + 16x = -16c_2 \sin 4t + 16(c_2 \sin 4t) = 0.$$

Finally, it is straightforward to verify that the linear combination of solutions, or the two-parameter family $x = c_1 \cos 4t + c_2 \sin 4t$, is also a solution of the differential equation. ≡

The next example shows that a solution of a differential equation can be a piecewise-defined function.

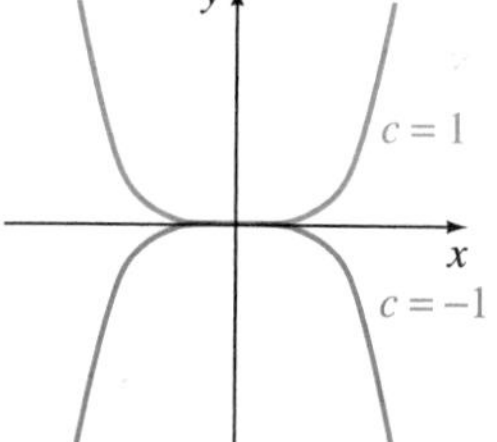

(a) two explicit solutions

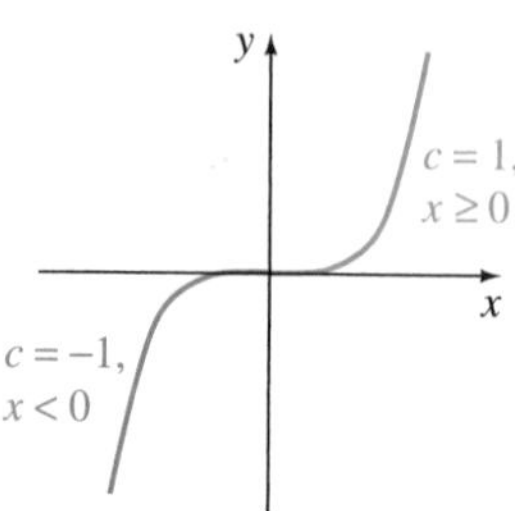

(b) piecewise-defined solution

FIGURE 1.1.5 Some solutions of DE in Example 8

EXAMPLE 8 Piecewise-Defined Solution

The one-parameter family of quartic monomial functions $y = cx^4$ is an explicit solution of the linear first-order equation

$$xy' - 4y = 0$$

on the interval $(-\infty, \infty)$. (Verify.) The blue and red solution curves shown in Figure 1.1.5(a) are the graphs of $y = x^4$ and $y = -x^4$ and correspond to the choices $c = 1$ and $c = -1$, respectively.

The piecewise-defined differentiable function

$$y = \begin{cases} -x^4, & x < 0 \\ x^4, & x \geq 0 \end{cases}$$

is also a solution of the differential equation but cannot be obtained from the family $y = cx^4$ by a single choice of c. As seen in Figure 1.1.5(b) the solution is constructed from the family by choosing $c = -1$ for $x < 0$ and $c = 1$ for $x \geq 0$. ≡

≡ **Systems of Differential Equations** Up to this point we have been discussing single differential equations containing one unknown function. But often in theory, as well as in many applications, we must deal with systems of differential equations. A **system of ordinary differential equations** is two or more equations involving the derivatives of two or more unknown functions of a single independent variable. For example, if x and y denote dependent variables and t denotes the independent variable, then a system of two first-order differential equations is given by

$$\begin{aligned} \frac{dx}{dt} &= f(t, x, y) \\ \frac{dy}{dt} &= g(t, x, y). \end{aligned} \tag{9}$$

A **solution** of a system such as (9) is a pair of differentiable functions $x = \phi_1(t)$, $y = \phi_2(t)$, defined on a common interval I, that satisfy each equation of the system on this interval.

REMARKS

(*i*) A few last words about implicit solutions of differential equations are in order. In Example 5 we were able to solve the relation $x^2 + y^2 = 25$ for y in terms of x to get two explicit solutions, $\phi_1(x) = \sqrt{25 - x^2}$ and $\phi_2(x) = -\sqrt{25 - x^2}$, of the differential equation (8). But don't read too much into this one example. Unless it is easy or important or you are instructed to, there is usually no need to try to solve an implicit solution $G(x, y) = 0$ for y explicitly in terms of x. Also do not misinterpret the second sentence following Definition 1.1.3. An implicit solution $G(x, y) = 0$ can define a perfectly good differentiable function ϕ that is a solution of a DE, yet we might not be able to solve $G(x, y) = 0$ using analytical methods such as algebra. The solution curve of ϕ may be a segment or piece of the graph of $G(x, y) = 0$. See Problems 45 and 46 in Exercises 1.1. Also, read the discussion following Example 4 in Section 2.2.

(*ii*) Although the concept of a solution has been emphasized in this section, you should also be aware that a DE does not necessarily have to possess a solution. See Problem 39 in Exercises 1.1. The question of whether a solution exists will be touched on in the next section.

(*iii*) It might not be apparent whether a first-order ODE written in differential form $M(x, y)dx + N(x, y)dy = 0$ is linear or nonlinear because there is nothing in this form that tells us which symbol denotes the dependent variable. See Problems 9 and 10 in Exercises 1.1.

(*iv*) It might not seem like a big deal to assume that $F(x, y, y', \ldots, y^{(n)}) = 0$ can be solved for $y^{(n)}$, but one should be a little bit careful here. There are exceptions, and there certainly are some problems connected with this assumption. See Problems 52 and 53 in Exercises 1.1.

(*v*) You may run across the term *closed form solutions* in DE texts or in lectures in courses in differential equations. Translated, this phrase usually refers to explicit solutions that are expressible in terms of *elementary* (or familiar) *functions:* finite combinations of integer powers of x, roots, exponential and logarithmic functions, and trigonometric and inverse trigonometric functions.

(*vi*) If *every* solution of an nth-order ODE $F(x, y, y', \ldots, y^{(n)}) = 0$ on an interval I can be obtained from an n-parameter family $G(x, y, c_1, c_2, \ldots, c_n) = 0$ by appropriate choices of the parameters c_i, $i = 1, 2, \ldots, n$, we then say that the family is the **general solution** of the DE. In solving linear ODEs, we shall impose relatively simple restrictions on the coefficients of the equation; with these restrictions one can be assured that not only does a solution exist on an interval but also that a family of solutions yields all possible solutions. Nonlinear ODEs, with the exception of some first-order equations, are usually difficult or impossible to solve in terms of elementary functions. Furthermore, if we happen to obtain a family of solutions for a nonlinear equation, it is not obvious whether this family contains all solutions. On a practical level, then, the designation "general solution" is applied only to linear ODEs. Don't be concerned about this concept at this point, but store the words "general solution" in the back of your mind—we will come back to this notion in Section 2.3 and again in Chapter 4.

EXERCISES 1.1

Answers to selected odd-numbered problems begin on page ANS-1.

In Problems 1–8 state the order of the given ordinary differential equation. Determine whether the equation is linear or nonlinear by matching it with (6).

1. $(1 - x)y'' - 4xy' + 5y = \cos x$
2. $x\dfrac{d^3y}{dx^3} - \left(\dfrac{dy}{dx}\right)^4 + y = 0$
3. $t^5y^{(4)} - t^3y'' + 6y = 0$
4. $\dfrac{d^2u}{dr^2} + \dfrac{du}{dr} + u = \cos(r + u)$
5. $\dfrac{d^2y}{dx^2} = \sqrt{1 + \left(\dfrac{dy}{dx}\right)^2}$
6. $\dfrac{d^2R}{dt^2} = -\dfrac{k}{R^2}$
7. $(\sin\theta)y''' - (\cos\theta)y' = 2$
8. $\ddot{x} - \left(1 - \dfrac{\dot{x}^2}{3}\right)\dot{x} + x = 0$

In Problems 9 and 10 determine whether the given first-order differential equation is linear in the indicated dependent variable by matching it with the first differential equation given in (7).

9. $(y^2 - 1)\,dx + x\,dy = 0$; in y; in x
10. $u\,dv + (v + uv - ue^u)\,du = 0$; in v; in u

In Problems 11–14 verify that the indicated function is an explicit solution of the given differential equation. Assume an appropriate interval I of definition for each solution.

11. $2y' + y = 0$; $y = e^{-x/2}$
12. $\dfrac{dy}{dt} + 20y = 24$; $y = \dfrac{6}{5} - \dfrac{6}{5}e^{-20t}$
13. $y'' - 6y' + 13y = 0$; $y = e^{3x}\cos 2x$
14. $y'' + y = \tan x$; $y = -(\cos x)\ln(\sec x + \tan x)$

In Problems 15–18 verify that the indicated function $y = \phi(x)$ is an explicit solution of the given first-order differential equation. Proceed as in Example 2, by considering ϕ simply as a *function,* give its domain. Then by considering ϕ as a *solution* of the differential equation, give at least one interval I of definition.

15. $(y - x)y' = y - x + 8$; $y = x + 4\sqrt{x + 2}$
16. $y' = 25 + y^2$; $y = 5\tan 5x$
17. $y' = 2xy^2$; $y = 1/(4 - x^2)$
18. $2y' = y^3\cos x$; $y = (1 - \sin x)^{-1/2}$

In Problems 19 and 20 verify that the indicated expression is an implicit solution of the given first-order differential equation. Find at least one explicit solution $y = \phi(x)$ in each case.

Use a graphing utility to obtain the graph of an explicit solution. Give an interval I of definition of each solution ϕ.

19. $\dfrac{dX}{dt} = (X - 1)(1 - 2X); \quad \ln\left(\dfrac{2X - 1}{X - 1}\right) = t$

20. $2xy\,dx + (x^2 - y)\,dy = 0; \quad -2x^2y + y^2 = 1$

In Problems 21–24 verify that the indicated family of functions is a solution of the given differential equation. Assume an appropriate interval I of definition for each solution.

21. $\dfrac{dP}{dt} = P(1 - P); \quad P = \dfrac{c_1e^t}{1 + c_1e^t}$

22. $\dfrac{dy}{dx} + 2xy = 1; \quad y = e^{-x^2}\displaystyle\int_0^x e^{t^2}\,dt + c_1e^{-x^2}$

23. $\dfrac{d^2y}{dx^2} - 4\dfrac{dy}{dx} + 4y = 0; \quad y = c_1e^{2x} + c_2xe^{2x}$

24. $x^3\dfrac{d^3y}{dx^3} + 2x^2\dfrac{d^2y}{dx^2} - x\dfrac{dy}{dx} + y = 12x^2;$

$y = c_1x^{-1} + c_2x + c_3x\ln x + 4x^2$

25. Verify that the piecewise-defined function

$$y = \begin{cases} -x^2, & x < 0 \\ x^2, & x \geq 0 \end{cases}$$

is a solution of the differential equation $xy' - 2y = 0$ on $(-\infty, \infty)$.

26. In Example 5 we saw that $y = \phi_1(x) = \sqrt{25 - x^2}$ and $y = \phi_2(x) = -\sqrt{25 - x^2}$ are solutions of $dy/dx = -x/y$ on the interval $(-5, 5)$. Explain why the piecewise-defined function

$$y = \begin{cases} \sqrt{25 - x^2}, & -5 < x < 0 \\ -\sqrt{25 - x^2}, & 0 \leq x < 5 \end{cases}$$

is *not* a solution of the differential equation on the interval $(-5, 5)$.

In Problems 27–30 find values of m so that the function $y = e^{mx}$ is a solution of the given differential equation.

27. $y' + 2y = 0$

28. $5y' = 2y$

29. $y'' - 5y' + 6y = 0$

30. $2y'' + 7y' - 4y = 0$

In Problems 31 and 32 find values of m so that the function $y = x^m$ is a solution of the given differential equation.

31. $xy'' + 2y' = 0$

32. $x^2y'' - 7xy' + 15y = 0$

In Problems 33–36 use the concept that $y = c$, $-\infty < x < \infty$, is a constant function if and only if $y' = 0$ to determine whether the given differential equation possesses constant solutions.

33. $3xy' + 5y = 10$

34. $y' = y^2 + 2y - 3$

35. $(y - 1)y' = 1$

36. $y'' + 4y' + 6y = 10$

In Problems 37 and 38 verify that the indicated pair of functions is a solution of the given system of differential equations on the interval $(-\infty, \infty)$.

37. $\dfrac{dx}{dt} = x + 3y$

$\dfrac{dy}{dt} = 5x + 3y;$

$x = e^{-2t} + 3e^{6t},$

$y = -e^{-2t} + 5e^{6t}$

38. $\dfrac{d^2x}{dt^2} = 4y + e^t$

$\dfrac{d^2y}{dt^2} = 4x - e^t;$

$x = \cos 2t + \sin 2t + \frac{1}{5}e^t,$

$y = -\cos 2t - \sin 2t - \frac{1}{5}e^t$

Discussion Problems

39. Make up a differential equation that does not possess any real solutions.

40. Make up a differential equation that you feel confident possesses only the trivial solution $y = 0$. Explain your reasoning.

41. What function do you know from calculus is such that its first derivative is itself? Its first derivative is a constant multiple k of itself? Write each answer in the form of a first-order differential equation with a solution.

42. What function (or functions) do you know from calculus is such that its second derivative is itself? Its second derivative is the negative of itself? Write each answer in the form of a second-order differential equation with a solution.

43. Given that $y = \sin x$ is an explicit solution of the first-order differential equation $\dfrac{dy}{dx} = \sqrt{1 - y^2}$. Find an interval I of definition. [*Hint*: I is *not* the interval $(-\infty, \infty)$.]

44. Discuss why it makes intuitive sense to presume that the linear differential equation $y'' + 2y' + 4y = 5\sin t$ has a solution of the form $y = A\sin t + B\cos t$, where A and B are constants. Then find specific constants A and B so that $y = A\sin t + B\cos t$ is a particular solution of the DE.

In Problems 45 and 46 the given figure represents the graph of an implicit solution $G(x, y) = 0$ of a differential equation $dy/dx = f(x, y)$. In each case the relation $G(x, y) = 0$ implicitly defines several solutions of the DE. Carefully reproduce each figure on a piece of paper. Use different colored pencils to mark off segments, or pieces, on each graph that correspond to graphs of solutions. Keep in mind that a solution ϕ must be a function and differentiable. Use the solution curve to estimate an interval I of definition of each solution ϕ.

45.

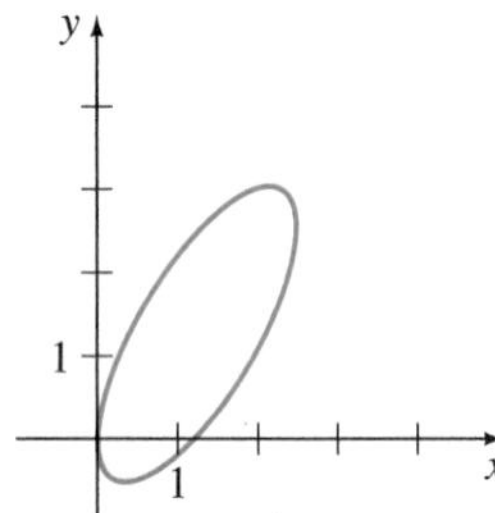

FIGURE 1.1.6 Graph for Problem 45

46.

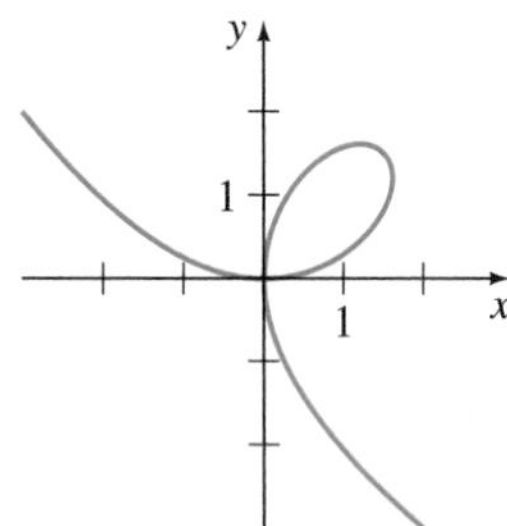

FIGURE 1.1.7 Graph for Problem 46

47. The graphs of members of the one-parameter family $x^3 + y^3 = 3cxy$ are called **folia of Descartes.** Verify that this family is an implicit solution of the first-order differential equation

$$\frac{dy}{dx} = \frac{y(y^3 - 2x^3)}{x(2y^3 - x^3)}.$$

48. The graph in Figure 1.1.7 is the member of the family of folia in Problem 47 corresponding to $c = 1$. Discuss: How can the DE in Problem 47 help in finding points on the graph of $x^3 + y^3 = 3xy$ where the tangent line is vertical? How does knowing where a tangent line is vertical help in determining an interval I of definition of a solution ϕ of the DE? Carry out your ideas, and compare with your estimates of the intervals in Problem 46.

49. In Example 5 the largest interval I over which the explicit solutions $y = \phi_1(x)$ and $y = \phi_2(x)$ are defined is the open interval $(-5, 5)$. Why can't the interval I of definition be the closed interval $[-5, 5]$?

50. In Problem 21 a one-parameter family of solutions of the DE $P' = P(1 - P)$ is given. Does any solution curve pass through the point $(0, 3)$? Through the point $(0, 1)$?

51. Discuss, and illustrate with examples, how to solve differential equations of the forms $dy/dx = f(x)$ and $d^2y/dx^2 = f(x)$.

52. The differential equation $x(y')^2 - 4y' - 12x^3 = 0$ has the form given in (4). Determine whether the equation can be put into the normal form $dy/dx = f(x, y)$.

53. The normal form (5) of an nth-order differential equation is equivalent to (4) whenever both forms have exactly the same solutions. Make up a first-order differential equation for which $F(x, y, y') = 0$ is not equivalent to the normal form $dy/dx = f(x, y)$.

54. Find a linear second-order differential equation $F(x, y, y', y'') = 0$ for which $y = c_1x + c_2x^2$ is a two-parameter family of solutions. Make sure that your equation is free of the arbitrary parameters c_1 and c_2.

Qualitative information about a solution $y = \phi(x)$ of a differential equation can often be obtained from the equation itself. Before working Problems 55–58, recall the geometric significance of the derivatives dy/dx and d^2y/dx^2.

55. Consider the differential equation $dy/dx = e^{-x^2}$.

(a) Explain why a solution of the DE must be an increasing function on any interval of the x-axis.

(b) What are $\lim_{x\to-\infty} dy/dx$ and $\lim_{x\to\infty} dy/dx$? What does this suggest about a solution curve as $x \to \pm\infty$?

(c) Determine an interval over which a solution curve is concave down and an interval over which the curve is concave up.

(d) Sketch the graph of a solution $y = \phi(x)$ of the differential equation whose shape is suggested by parts (a)–(c).

56. Consider the differential equation $dy/dx = 5 - y$.

(a) Either by inspection or by the method suggested in Problems 33–36, find a constant solution of the DE.

(b) Using only the differential equation, find intervals on the y-axis on which a nonconstant solution $y = \phi(x)$ is increasing. Find intervals on the y-axis on which $y = \phi(x)$ is decreasing.

57. Consider the differential equation $dy/dx = y(a - by)$, where a and b are positive constants.

(a) Either by inspection or by the method suggested in Problems 33–36, find two constant solutions of the DE.

(b) Using only the differential equation, find intervals on the y-axis on which a nonconstant solution $y = \phi(x)$ is increasing. Find intervals on which $y = \phi(x)$ is decreasing.

(c) Using only the differential equation, explain why $y = a/2b$ is the y-coordinate of a point of inflection of the graph of a nonconstant solution $y = \phi(x)$.

(d) On the same coordinate axes, sketch the graphs of the two constant solutions found in part (a). These constant solutions partition the xy-plane into three regions. In each region, sketch the graph of a nonconstant solution $y = \phi(x)$ whose shape is suggested by the results in parts (b) and (c).

58. Consider the differential equation $y' = y^2 + 4$.

(a) Explain why there exist no constant solutions of the DE.

(b) Describe the graph of a solution $y = \phi(x)$. For example, can a solution curve have any relative extrema?

(c) Explain why $y = 0$ is the y-coordinate of a point of inflection of a solution curve.

(d) Sketch the graph of a solution $y = \phi(x)$ of the differential equation whose shape is suggested by parts (a)–(c).

Computer Lab Assignments

In Problems 59 and 60 use a CAS to compute all derivatives and to carry out the simplifications needed to verify that the indicated function is a particular solution of the given differential equation.

59. $y^{(4)} - 20y''' + 158y'' - 580y' + 841y = 0;$
$y = xe^{5x}\cos 2x$

60. $x^3y''' + 2x^2y'' + 20xy' - 78y = 0;$
$$y = 20\frac{\cos(5\ln x)}{x} - 3\frac{\sin(5\ln x)}{x}$$

1.2 INITIAL-VALUE PROBLEMS

REVIEW MATERIAL

- Normal form of a DE
- Solution of a DE
- Family of solutions

INTRODUCTION We are often interested in problems in which we seek a solution $y(x)$ of a differential equation so that $y(x)$ also satisfies certain prescribed side conditions—that is, conditions that are imposed on the unknown function $y(x)$ and its derivatives at a point x_0. On some interval I containing x_0 the problem of solving an nth-order differential equation subject to n side conditions specified at x_0:

$$\textit{Solve:} \quad \frac{d^ny}{dx^n} = f\left(x, y, y', \ldots, y^{(n-1)}\right)$$
$$\textit{Subject to:} \quad y(x_0) = y_0,\ y'(x_0) = y_1,\ \ldots,\ y^{(n-1)}(x_0) = y_{n-1}, \tag{1}$$

where $y_0, y_1, \ldots, y_{n-1}$ are arbitrary real constants, is called an ***n*th-order initial-value problem (IVP).** The values of $y(x)$ and its first $n - 1$ derivatives at x_0, $y(x_0) = y_0$, $y'(x_0) = y_1, \ldots, y^{(n-1)}(x_0) = y_{n-1}$ are called **initial conditions (IC).**

Solving an nth-order initial-value problem such as (1) frequently entails first finding an n-parameter family of solutions of the given differential equation and then using the initial-conditions at x_0 to determine the n constants in this family. The resulting particular solution is defined on some interval I containing the initial point x_0.

Geometric Interpretation of IVPs

The cases $n = 1$ and $n = 2$ in (1),

$$\textit{Solve:} \quad \frac{dy}{dx} = f(x, y)$$
$$\textit{Subject to:} \quad y(x_0) = y_0 \tag{2}$$

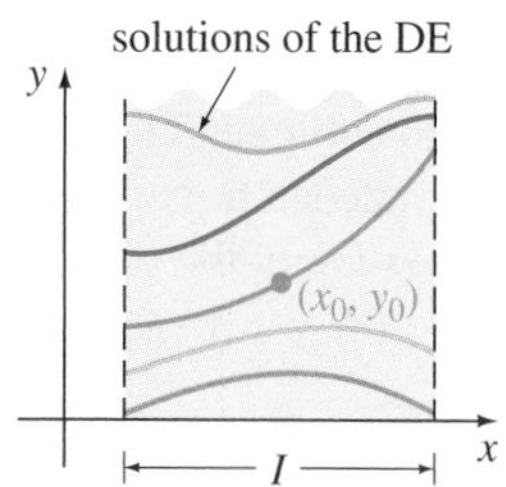

FIGURE 1.2.1 Solution curve of first-order IVP

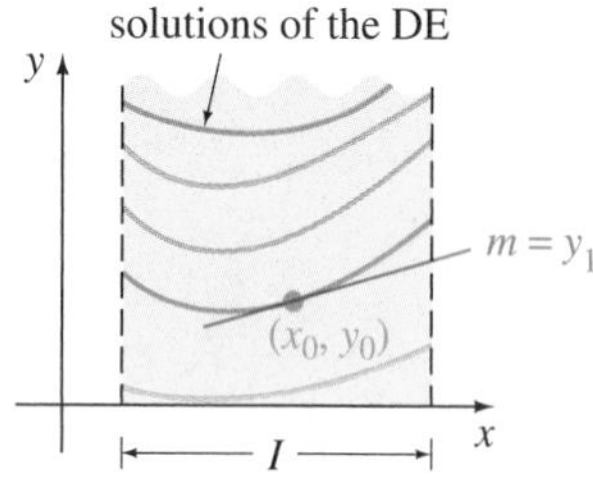

FIGURE 1.2.2 Solution curve of second-order IVP

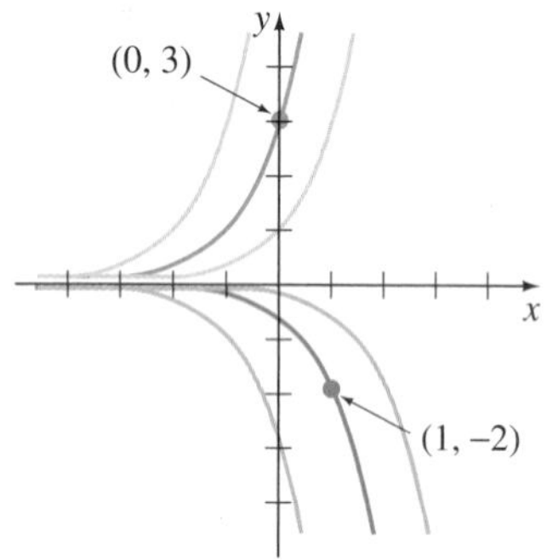

FIGURE 1.2.3 Solution curves of two IVPs in Example 1

and

$$\textit{Solve:} \quad \frac{d^2y}{dx^2} = f(x, y, y')$$
$$\textit{Subject to:} \quad y(x_0) = y_0, \; y'(x_0) = y_1 \tag{3}$$

are examples of **first-** and **second-order** initial-value problems, respectively. These two problems are easy to interpret in geometric terms. For (2) we are seeking a solution $y(x)$ of the differential equation $y' = f(x, y)$ on an interval I containing x_0 so that its graph passes through the specified point (x_0, y_0). A solution curve is shown in blue in Figure 1.2.1. For (3) we want to find a solution $y(x)$ of the differential equation $y'' = f(x, y, y')$ on an interval I containing x_0 so that its graph not only passes through (x_0, y_0) but the slope of the curve at this point is the number y_1. A solution curve is shown in blue in Figure 1.2.2. The words *initial conditions* derive from physical systems where the independent variable is time t and where $y(t_0) = y_0$ and $y'(t_0) = y_1$ represent the position and velocity, respectively, of an object at some beginning, or initial, time t_0.

EXAMPLE 1 Two First-Order IVPs

(a) In Problem 41 in Exercises 1.1 you were asked to deduce that $y = ce^x$ is a one-parameter family of solutions of the simple first-order equation $y' = y$. All the solutions in this family are defined on the interval $(-\infty, \infty)$. If we impose an initial condition, say, $y(0) = 3$, then substituting $x = 0$, $y = 3$ in the family determines the constant $3 = ce^0 = c$. Thus $y = 3e^x$ is a solution of the IVP

$$y' = y, \quad y(0) = 3.$$

(b) Now if we demand that a solution curve pass through the point $(1, -2)$ rather than $(0, 3)$, then $y(1) = -2$ will yield $-2 = ce$ or $c = -2e^{-1}$. In this case $y = -2e^{x-1}$ is a solution of the IVP

$$y' = y, \quad y(1) = -2.$$

The two solution curves are shown in dark blue and dark red in Figure 1.2.3. ≡

The next example illustrates another first-order initial-value problem. In this example notice how the interval I of definition of the solution $y(x)$ depends on the initial condition $y(x_0) = y_0$.

EXAMPLE 2 Interval *I* of Definition of a Solution

In Problem 6 of Exercises 2.2 you will be asked to show that a one-parameter family of solutions of the first-order differential equation $y' + 2xy^2 = 0$ is $y = 1/(x^2 + c)$. If we impose the initial condition $y(0) = -1$, then substituting $x = 0$ and $y = -1$ into the family of solutions gives $-1 = 1/c$ or $c = -1$. Thus $y = 1/(x^2 - 1)$. We now emphasize the following three distinctions:

- Considered as a *function*, the domain of $y = 1/(x^2 - 1)$ is the set of real numbers x for which $y(x)$ is defined; this is the set of all real numbers except $x = -1$ and $x = 1$. See Figure 1.2.4(a).
- Considered as a *solution of the differential equation* $y' + 2xy^2 = 0$, the interval I of definition of $y = 1/(x^2 - 1)$ could be taken to be any interval over which $y(x)$ is defined and differentiable. As can be seen in Figure 1.2.4(a), the largest intervals on which $y = 1/(x^2 - 1)$ is a solution are $(-\infty, -1)$, $(-1, 1)$, and $(1, \infty)$.

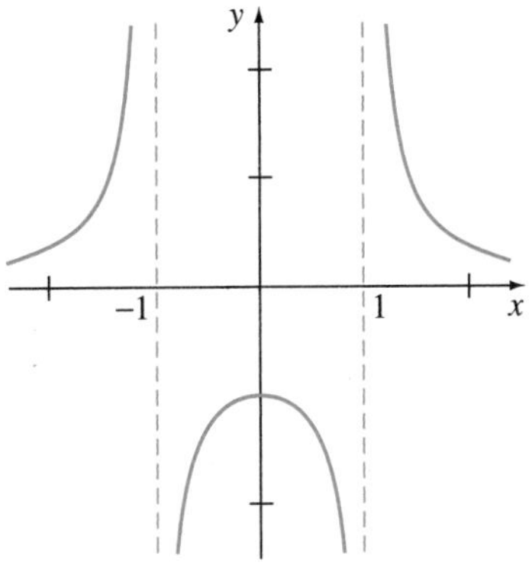

(a) function defined for all x except $x = \pm 1$

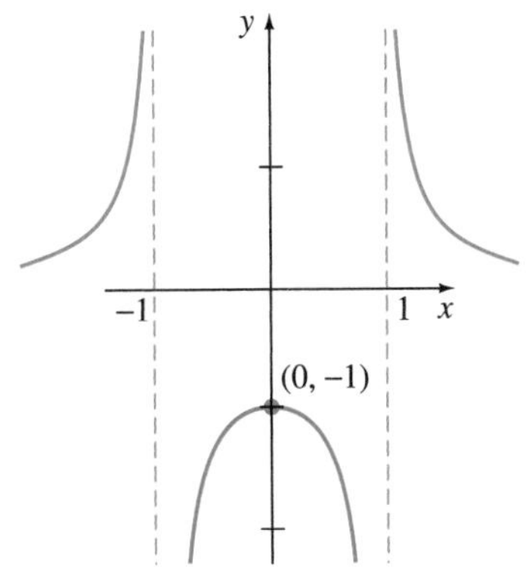

(b) solution defined on interval containing $x = 0$

FIGURE 1.2.4 Graphs of function and solution of IVP in Example 2

- Considered as a *solution of the initial-value problem* $y' + 2xy^2 = 0$, $y(0) = -1$, the interval I of definition of $y = 1/(x^2 - 1)$ could be taken to be any interval over which $y(x)$ is defined, differentiable, *and* contains the initial point $x = 0$; the largest interval for which this is true is $(-1, 1)$. See the red curve in Figure 1.2.4(b). ≡

See Problems 3–6 in Exercises 1.2 for a continuation of Example 2.

EXAMPLE 3 Second-Order IVP

In Example 7 of Section 1.1 we saw that $x = c_1 \cos 4t + c_2 \sin 4t$ is a two-parameter family of solutions of $x'' + 16x = 0$. Find a solution of the initial-value problem

$$x'' + 16x = 0, \quad x\left(\frac{\pi}{2}\right) = -2, \quad x'\left(\frac{\pi}{2}\right) = 1. \tag{4}$$

SOLUTION We first apply $x(\pi/2) = -2$ to the given family of solutions: $c_1 \cos 2\pi + c_2 \sin 2\pi = -2$. Since $\cos 2\pi = 1$ and $\sin 2\pi = 0$, we find that $c_1 = -2$. We next apply $x'(\pi/2) = 1$ to the one-parameter family $x(t) = -2 \cos 4t + c_2 \sin 4t$. Differentiating and then setting $t = \pi/2$ and $x' = 1$ gives $8 \sin 2\pi + 4c_2 \cos 2\pi = 1$, from which we see that $c_2 = \frac{1}{4}$. Hence $x = -2 \cos 4t + \frac{1}{4} \sin 4t$ is a solution of (4). ≡

≡ Existence and Uniqueness Two fundamental questions arise in considering an initial-value problem:

Does a solution of the problem exist?
If a solution exists, is it unique?

For the first-order initial-value problem (2) we ask:

Existence { *Does the differential equation $dy/dx = f(x, y)$ possess solutions?* *Do any of the solution curves pass through the point (x_0, y_0)?*

Uniqueness { *When can we be certain that there is precisely one solution curve passing through the point (x_0, y_0)?*

Note that in Examples 1 and 3 the phrase "*a* solution" is used rather than "*the* solution" of the problem. The indefinite article "a" is used deliberately to suggest the possibility that other solutions may exist. At this point it has not been demonstrated that there is a single solution of each problem. The next example illustrates an initial-value problem with two solutions.

EXAMPLE 4 An IVP Can Have Several Solutions

Each of the functions $y = 0$ and $y = \frac{1}{16}x^4$ satisfies the differential equation $dy/dx = xy^{1/2}$ and the initial condition $y(0) = 0$, so the initial-value problem

$$\frac{dy}{dx} = xy^{1/2}, \quad y(0) = 0$$

has at least two solutions. As illustrated in Figure 1.2.5, the graphs of both functions, shown in red and blue pass through the same point (0, 0). ≡

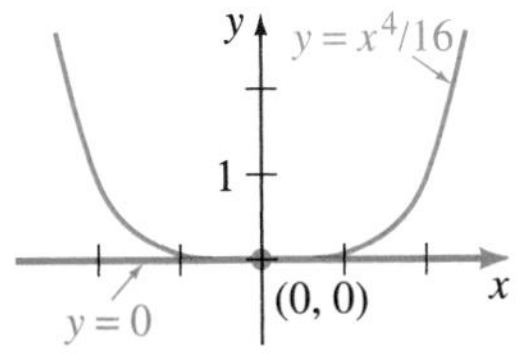

FIGURE 1.2.5 Two solutions curves of the same IVP in Example 4

Within the safe confines of a formal course in differential equations one can be fairly confident that *most* differential equations will have solutions and that solutions of initial-value problems will *probably* be unique. Real life, however, is not so idyllic. Therefore it is desirable to know in advance of trying to solve an initial-value problem whether a solution exists and, when it does, whether it is the only solution of the problem. Since we are going to consider first-order differential equations in the next two chapters, we state here without proof a straightforward theorem that gives conditions that are sufficient to guarantee the existence and uniqueness of a solution of a first-order initial-value problem of the form given in (2). We shall wait until Chapter 4 to address the question of existence and uniqueness of a second-order initial-value problem.

THEOREM 1.2.1 Existence of a Unique Solution

Let R be a rectangular region in the xy-plane defined by $a \le x \le b$, $c \le y \le d$ that contains the point (x_0, y_0) in its interior. If $f(x, y)$ and $\partial f/\partial y$ are continuous on R, then there exists some interval I_0: $(x_0 - h, x_0 + h)$, $h > 0$, contained in $[a, b]$, and a unique function $y(x)$, defined on I_0, that is a solution of the initial-value problem (2).

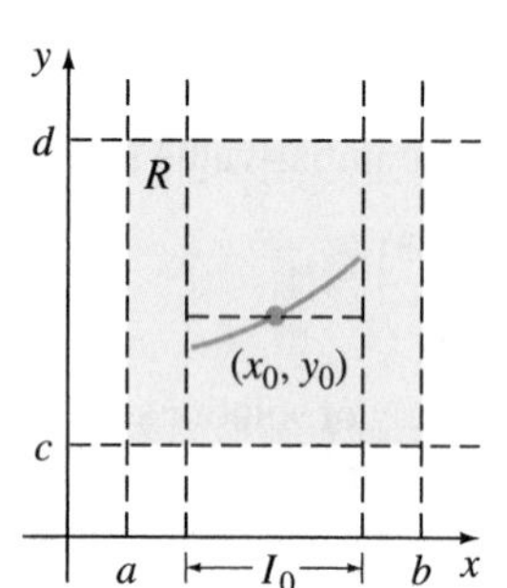

FIGURE 1.2.6 Rectangular region R

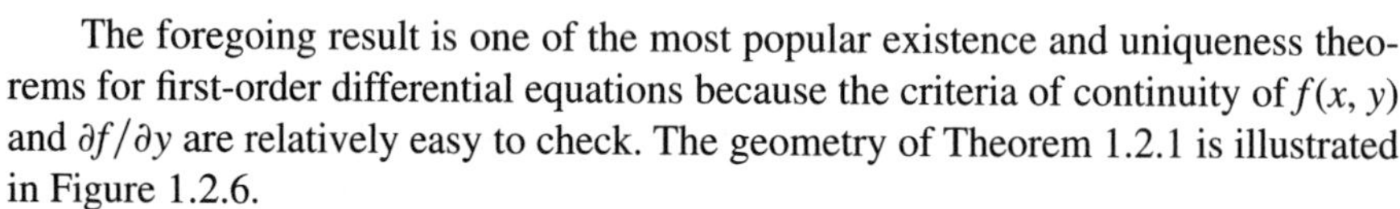

The foregoing result is one of the most popular existence and uniqueness theorems for first-order differential equations because the criteria of continuity of $f(x, y)$ and $\partial f/\partial y$ are relatively easy to check. The geometry of Theorem 1.2.1 is illustrated in Figure 1.2.6.

EXAMPLE 5 Example 4 Revisited

We saw in Example 4 that the differential equation $dy/dx = xy^{1/2}$ possesses at least two solutions whose graphs pass through (0, 0). Inspection of the functions

$$f(x, y) = xy^{1/2} \qquad \text{and} \qquad \frac{\partial f}{\partial y} = \frac{x}{2y^{1/2}}$$

shows that they are continuous in the upper half-plane defined by $y > 0$. Hence Theorem 1.2.1 enables us to conclude that through any point (x_0, y_0), $y_0 > 0$ in the upper half-plane there is some interval centered at x_0 on which the given differential equation has a unique solution. Thus, for example, even without solving it, we know that there exists some interval centered at 2 on which the initial-value problem $dy/dx = xy^{1/2}$, $y(2) = 1$ has a unique solution. ≡

In Example 1, Theorem 1.2.1 guarantees that there are no other solutions of the initial-value problems $y' = y$, $y(0) = 3$ and $y' = y$, $y(1) = -2$ other than $y = 3e^x$ and $y = -2e^{x-1}$, respectively. This follows from the fact that $f(x, y) = y$ and $\partial f/\partial y = 1$ are continuous throughout the entire xy-plane. It can be further shown that the interval I on which each solution is defined is $(-\infty, \infty)$.

≡ **Interval of Existence/Uniqueness** Suppose $y(x)$ represents a solution of the initial-value problem (2). The following three sets on the real x-axis may not be the same: the domain of the function $y(x)$, the interval I over which the solution $y(x)$ is defined or exists, and the interval I_0 of existence *and* uniqueness. Example 2 of Section 1.1 illustrated the difference between the domain of a function and the interval I of definition. Now suppose (x_0, y_0) is a point in the interior of the rectangular region R in Theorem 1.2.1. It turns out that the continuity of the function $f(x, y)$ on R by itself is sufficient to guarantee the existence of at least one solution of $dy/dx = f(x, y)$, $y(x_0) = y_0$, defined on some interval I. The interval I of definition for this initial-value problem is usually taken to be the largest interval containing x_0 over which the solution $y(x)$ is defined and differentiable. The interval I depends on both $f(x, y)$ and the initial condition $y(x_0) = y_0$. See Problems 31–34 in Exercises 1.2. The extra condition of continuity of the first partial derivative $\partial f/\partial y$

on R enables us to say that not only does a solution exist on some interval I_0 containing x_0, but it is the *only* solution satisfying $y(x_0) = y_0$. However, Theorem 1.2.1 does not give any indication of the sizes of intervals I and I_0; *the interval I of definition need not be as wide as the region R, and the interval I_0 of existence and uniqueness may not be as large as I.* The number $h > 0$ that defines the interval I_0: $(x_0 - h, x_0 + h)$ could be very small, so it is best to think that the solution $y(x)$ is *unique in a local sense*—that is, a solution defined near the point (x_0, y_0). See Problem 50 in Exercises 1.2.

REMARKS

(*i*) The conditions in Theorem 1.2.1 are sufficient but not necessary. This means that when $f(x, y)$ and $\partial f/\partial y$ are continuous on a rectangular region R, it must always follow that a solution of (2) exists and is unique whenever (x_0, y_0) is a point interior to R. However, if the conditions stated in the hypothesis of Theorem 1.2.1 do not hold, then anything could happen: Problem (2) *may* still have a solution and this solution *may* be unique, or (2) may have several solutions, or it may have no solution at all. A rereading of Example 5 reveals that the hypotheses of Theorem 1.2.1 do not hold on the line $y = 0$ for the differential equation $dy/dx = xy^{1/2}$, so it is not surprising, as we saw in Example 4 of this section, that there are two solutions defined on a common interval $-h < x < h$ satisfying $y(0) = 0$. On the other hand, the hypotheses of Theorem 1.2.1 do not hold on the line $y = 1$ for the differential equation $dy/dx = |y - 1|$. Nevertheless it can be proved that the solution of the initial-value problem $dy/dx = |y - 1|$, $y(0) = 1$, is unique. Can you guess this solution?

(*ii*) You are encouraged to read, think about, work, and then keep in mind Problem 49 in Exercises 1.2.

(*iii*) Initial conditions are prescribed at a *single* point x_0. But we are also interested in solving differential equations that are subject to conditions specified on $y(x)$ or its derivative at *two* different points x_0 and x_1. Conditions such as

$$y(1) = 0, \quad y(5) = 0 \qquad \text{or} \qquad y(\pi/2) = 0, \quad y'(\pi) = 1$$

and called **boundary conditions**. A differential equation together with boundary conditions is called a **boundary-value problem (BVP)**. For example,

$$y'' + \lambda y = 0, \quad y'(0) = 0, \quad y'(\pi) = 0$$

is a boundary-value problem. See Problems 39–44 in Exercises 1.2.

When we start to solve differential equations in Chapter 2 we will solve only first-order equations and first-order initial-value problems. The mathematical description of many problems in science and engineering involve second-order IVPs or two-point BVPs. We will examine some of these problems in Chapters 4 and 5.

EXERCISES 1.2

Answers to selected odd-numbered problems begin on page ANS-1.

In Problems 1 and 2, $y = 1/(1 + c_1e^{-x})$ is a one-parameter family of solutions of the first-order DE $y' = y - y^2$. Find a solution of the first-order IVP consisting of this differential equation and the given initial condition.

1. $y(0) = -\frac{1}{3}$

2. $y(-1) = 2$

In Problems 3–6, $y = 1/(x^2 + c)$ is a one-parameter family of solutions of the first-order DE $y' + 2xy^2 = 0$. Find a solution of the first-order IVP consisting of this differential equation and the given initial condition. Give the largest interval I over which the solution is defined.

3. $y(2) = \frac{1}{3}$

4. $y(-2) = \frac{1}{2}$

5. $y(0) = 1$

6. $y(\frac{1}{2}) = -4$

In Problems 7–10, $x = c_1 \cos t + c_2 \sin t$ is a two-parameter family of solutions of the second-order DE $x'' + x = 0$. Find a solution of the second-order IVP consisting of this differential equation and the given initial conditions.

7. $x(0) = -1, \quad x'(0) = 8$

8. $x(\pi/2) = 0, \quad x'(\pi/2) = 1$

9. $x(\pi/6) = \frac{1}{2}, \quad x'(\pi/6) = 0$

10. $x(\pi/4) = \sqrt{2}, \quad x'(\pi/4) = 2\sqrt{2}$

In Problems 11–14, $y = c_1e^x + c_2e^{-x}$ is a two-parameter family of solutions of the second-order DE $y'' - y = 0$. Find a solution of the second-order IVP consisting of this differential equation and the given initial conditions.

11. $y(0) = 1, \quad y'(0) = 2$

12. $y(1) = 0, \quad y'(1) = e$

13. $y(-1) = 5, \quad y'(-1) = -5$

14. $y(0) = 0, \quad y'(0) = 0$

In Problems 15 and 16 determine by inspection at least two solutions of the given first-order IVP.

15. $y' = 3y^{2/3}, \quad y(0) = 0$

16. $xy' = 2y, \quad y(0) = 0$

In Problems 17–24 determine a region of the xy-plane for which the given differential equation would have a unique solution whose graph passes through a point (x_0, y_0) in the region.

17. $\dfrac{dy}{dx} = y^{2/3}$

18. $\dfrac{dy}{dx} = \sqrt{xy}$

19. $x\dfrac{dy}{dx} = y$

20. $\dfrac{dy}{dx} - y = x$

21. $(4 - y^2)y' = x^2$

22. $(1 + y^3)y' = x^2$

23. $(x^2 + y^2)y' = y^2$

24. $(y - x)y' = y + x$

In Problems 25–28 determine whether Theorem 1.2.1 guarantees that the differential equation $y' = \sqrt{y^2 - 9}$ possesses a unique solution through the given point.

25. $(1, 4)$

26. $(5, 3)$

27. $(2, -3)$

28. $(-1, 1)$

29. **(a)** By inspection find a one-parameter family of solutions of the differential equation $xy' = y$. Verify that each member of the family is a solution of the initial-value problem $xy' = y$, $y(0) = 0$.

(b) Explain part (a) by determining a region R in the xy-plane for which the differential equation $xy' = y$ would have a unique solution through a point (x_0, y_0) in R.

(c) Verify that the piecewise-defined function

$$y = \begin{cases} 0, & x < 0 \\ x, & x \geq 0 \end{cases}$$

satisfies the condition $y(0) = 0$. Determine whether this function is also a solution of the initial-value problem in part (a).

30. **(a)** Verify that $y = \tan(x + c)$ is a one-parameter family of solutions of the differential equation $y' = 1 + y^2$.

(b) Since $f(x, y) = 1 + y^2$ and $\partial f/\partial y = 2y$ are continuous everywhere, the region R in Theorem 1.2.1 can be taken to be the entire xy-plane. Use the family of solutions in part (a) to find an explicit solution of the first-order initial-value problem $y' = 1 + y^2$, $y(0) = 0$. Even though $x_0 = 0$ is in the interval $(-2, 2)$, explain why the solution is not defined on this interval.

(c) Determine the largest interval I of definition for the solution of the initial-value problem in part (b).

31. **(a)** Verify that $y = -1/(x + c)$ is a one-parameter family of solutions of the differential equation $y' = y^2$.

(b) Since $f(x, y) = y^2$ and $\partial f/\partial y = 2y$ are continuous everywhere, the region R in Theorem 1.2.1 can be taken to be the entire xy-plane. Find a solution from the family in part (a) that satisfies $y(0) = 1$. Then find a solution from the family in part (a) that satisfies $y(0) = -1$. Determine the largest interval I of definition for the solution of each initial-value problem.

(c) Determine the largest interval I of definition for the solution of the first-order initial-value problem $y' = y^2$, $y(0) = 0$. [*Hint*: The solution is not a member of the family of solutions in part (a).]

32. **(a)** Show that a solution from the family in part (a) of Problem 31 that satisfies $y' = y^2$, $y(1) = 1$, is $y = 1/(2 - x)$.

(b) Then show that a solution from the family in part (a) of Problem 31 that satisfies $y' = y^2$, $y(3) = -1$, is $y = 1/(2 - x)$.

(c) Are the solutions in parts (a) and (b) the same?

33. **(a)** Verify that $3x^2 - y^2 = c$ is a one-parameter family of solutions of the differential equation $y\,dy/dx = 3x$.

(b) By hand, sketch the graph of the implicit solution $3x^2 - y^2 = 3$. Find all explicit solutions $y = \phi(x)$ of the DE in part (a) defined by this relation. Give the interval I of definition of each explicit solution.

(c) The point $(-2, 3)$ is on the graph of $3x^2 - y^2 = 3$, but which of the explicit solutions in part (b) satisfies $y(-2) = 3$?

34. **(a)** Use the family of solutions in part (a) of Problem 33 to find an implicit solution of the initial-value

problem $y\,dy/dx = 3x$, $y(2) = -4$. Then, by hand, sketch the graph of the explicit solution of this problem and give its interval I of definition.

(b) Are there any explicit solutions of $y\,dy/dx = 3x$ that pass through the origin?

In Problems 35–38 the graph of a member of a family of solutions of a second-order differential equation $d^2y/dx^2 = f(x, y, y')$ is given. Match the solution curve with at least one pair of the following initial conditions.

(a) $y(1) = 1, \quad y'(1) = -2$

(b) $y(-1) = 0, \quad y'(-1) = -4$

(c) $y(1) = 1, \quad y'(1) = 2$

(d) $y(0) = -1, \quad y'(0) = 2$

(e) $y(0) = -1, \quad y'(0) = 0$

(f) $y(0) = -4, \quad y'(0) = -2$

35.

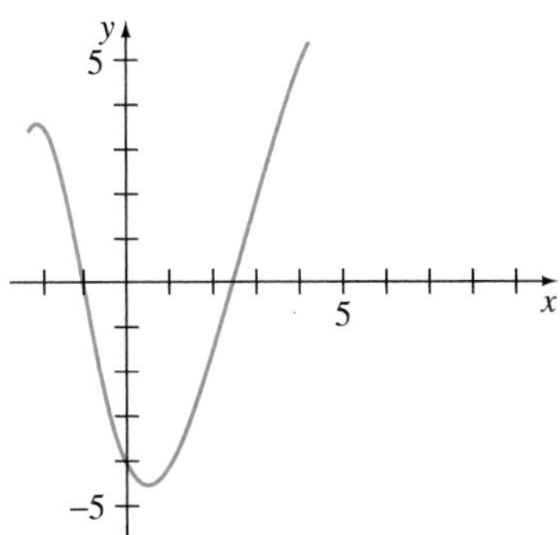

FIGURE 1.2.7 Graph for Problem 35

36.

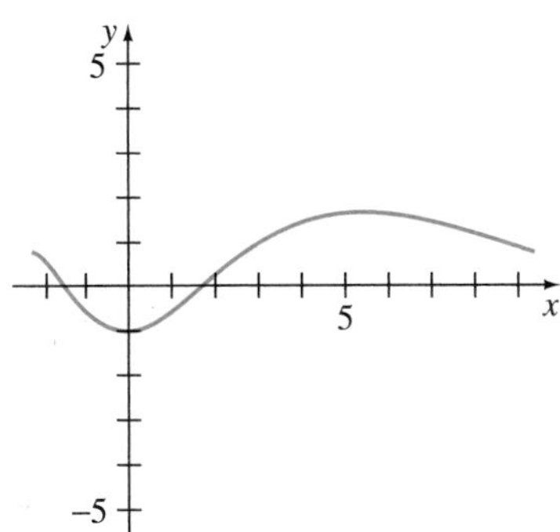

FIGURE 1.2.8 Graph for Problem 36

37.

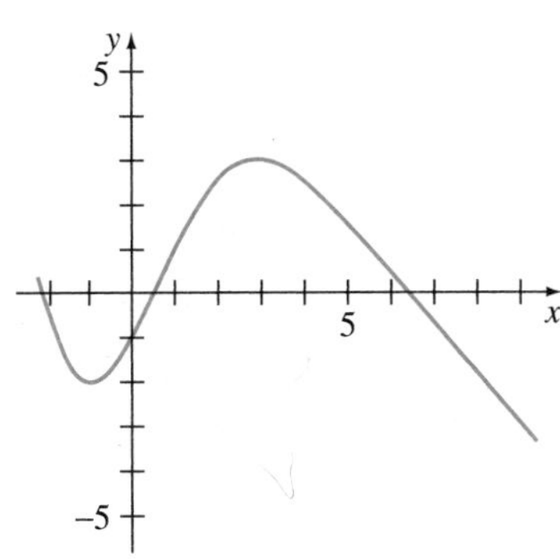

FIGURE 1.2.9 Graph for Problem 37

38.

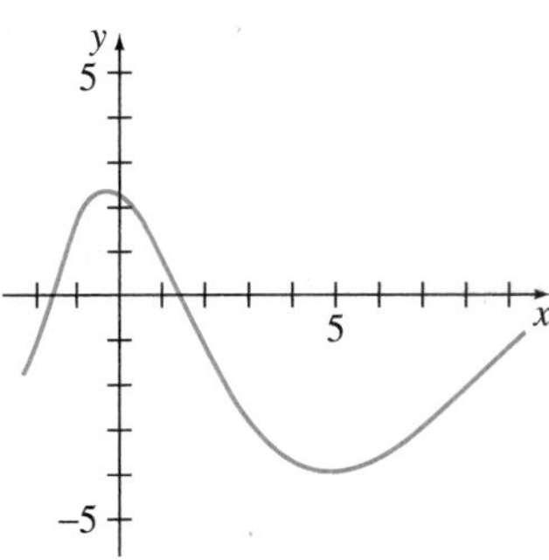

FIGURE 1.2.10 Graph for Problem 38

In Problems 39–44, $y = c_1 \cos 2x + c_2 \sin 2x$ is a two-parameter family of solutions of the second-order DE $y'' + 4y = 0$. If possible, find a solution of the differential equation that satisfies the given side conditions. The conditions specified at two different points are called boundary conditions.

39. $y(0) = 0, y(\pi/4) = 3$

40. $y(0) = 0, y(\pi) = 0$

41. $y'(0) = 0, y'(\pi/6) = 0$

42. $y(0) = 1, y'(\pi) = 5$

43. $y(0) = 0, y(\pi) = 2$

44. $y'(\pi/2) = 1, y'(\pi) = 0$

Discussion Problems

In Problems 45 and 46 use Problem 51 in Exercises 1.1 and (2) and (3) of this section.

45. Find a function $y = f(x)$ whose graph at each point (x, y) has the slope given by $8e^{2x} + 6x$ and has the y-intercept $(0, 9)$.

46. Find a function $y = f(x)$ whose second derivative is $y'' = 12x - 2$ at each point (x, y) on its graph and $y = -x + 5$ is tangent to the graph at the point corresponding to $x = 1$.

47. Consider the initial-value problem $y' = x - 2y$, $y(0) = \frac{1}{2}$. Determine which of the two curves shown in Figure 1.2.11 is the only plausible solution curve. Explain your reasoning.

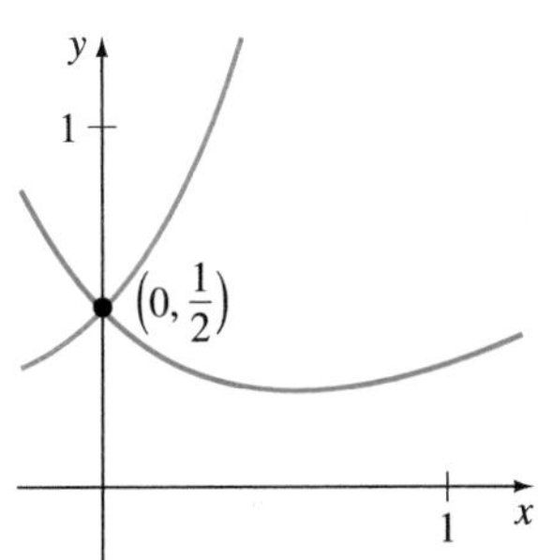

FIGURE 1.2.11 Graphs for Problem 47

48. Determine a plausible value of x_0 for which the graph of the solution of the initial-value problem $y' + 2y = 3x - 6$, $y(x_0) = 0$ is tangent to the x-axis at $(x_0, 0)$. Explain your reasoning.

49. Suppose that the first-order differential equation $dy/dx = f(x, y)$ possesses a one-parameter family of solutions and that $f(x, y)$ satisfies the hypotheses of Theorem 1.2.1 in some rectangular region R of the xy-plane. Explain why two different solution curves cannot intersect or be tangent to each other at a point (x_0, y_0) in R.

50. The functions $y(x) = \frac{1}{16}x^4$, $-\infty < x < \infty$ and

$$y(x) = \begin{cases} 0, & x < 0 \\ \frac{1}{16}x^4, & x \geq 0 \end{cases}$$

have the same domain but are clearly different. See Figures 1.2.12(a) and 1.2.12(b), respectively. Show that both functions are solutions of the initial-value problem $dy/dx = xy^{1/2}$, $y(2) = 1$ on the interval $(-\infty, \infty)$. Resolve the apparent contradiction between this fact and the last sentence in Example 5.

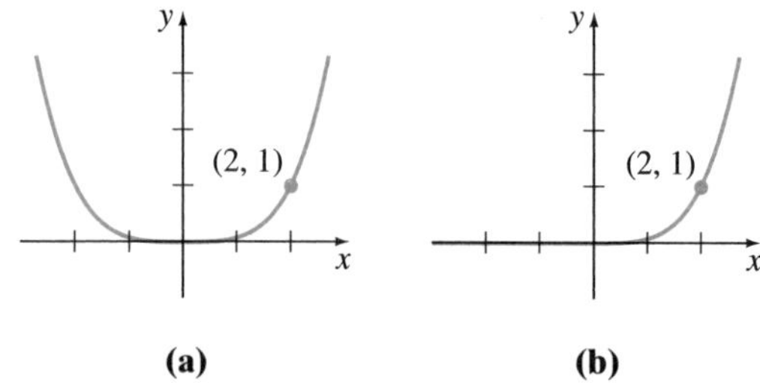

FIGURE 1.2.12 Two solutions of the IVP in Problem 50

Mathematical Model

51. Population Growth Beginning in the next section we will see that differential equations can be used to describe or *model* many different physical systems. In this problem suppose that a model of the growing population of a small community is given by the initial-value problem

$$\frac{dP}{dt} = 0.15P(t) + 20, \quad P(0) = 100,$$

where P is the number of individuals in the community and time t is measured in years. How fast—that is, at what *rate*—is the population increasing at $t = 0$? How fast is the population increasing when the population is 500?

1.3 DIFFERENTIAL EQUATIONS AS MATHEMATICAL MODELS

REVIEW MATERIAL

- Units of measurement for weight, mass, and density
- Newton's second law of motion
- Hooke's law
- Kirchhoff's laws
- Archimedes' principle

INTRODUCTION In this section we introduce the notion of a differential equation as a mathematical model and discuss some specific models in biology, chemistry, and physics. Once we have studied some methods for solving DEs in Chapters 2 and 4, we return to, and solve, some of these models in Chapters 3 and 5.

≡ Mathematical Models It is often desirable to describe the behavior of some real-life system or phenomenon, whether physical, sociological, or even economic, in mathematical terms. The mathematical description of a system of phenomenon is called a **mathematical model** and is constructed with certain goals in mind. For example, we may wish to understand the mechanisms of a certain ecosystem by studying the growth of animal populations in that system, or we may wish to date fossils by analyzing the decay of a radioactive substance, either in the fossil or in the stratum in which it was discovered.

Construction of a mathematical model of a system starts with

(*i*) identification of the variables that are responsible for changing the system. We may choose not to incorporate all these variables into the model at first. In this step we are specifying the **level of resolution** of the model.

Next

(*ii*) we make a set of reasonable assumptions, or hypotheses, about the system we are trying to describe. These assumptions will also include any empirical laws that may be applicable to the system.

For some purposes it may be perfectly within reason to be content with low-resolution models. For example, you may already be aware that in beginning physics courses, the retarding force of air friction is sometimes ignored in modeling the motion of a body falling near the surface of the Earth, but if you are a scientist whose job it is to accurately predict the flight path of a long-range projectile, you have to take into account air resistance and other factors such as the curvature of the Earth.

Since the assumptions made about a system frequently involve *a rate of change* of one or more of the variables, the mathematical depiction of all these assumptions may be one or more equations involving *derivatives*. In other words, the mathematical model may be a differential equation or a system of differential equations.

Once we have formulated a mathematical model that is either a differential equation or a system of differential equations, we are faced with the not insignificant problem of trying to solve it. *If* we can solve it, then we deem the model to be reasonable if its solution is consistent with either experimental data or known facts about the behavior of the system. But if the predictions produced by the solution are poor, we can either increase the level of resolution of the model or make alternative assumptions about the mechanisms for change in the system. The steps of the modeling process are then repeated, as shown in the diagram in Figure 1.3.1.

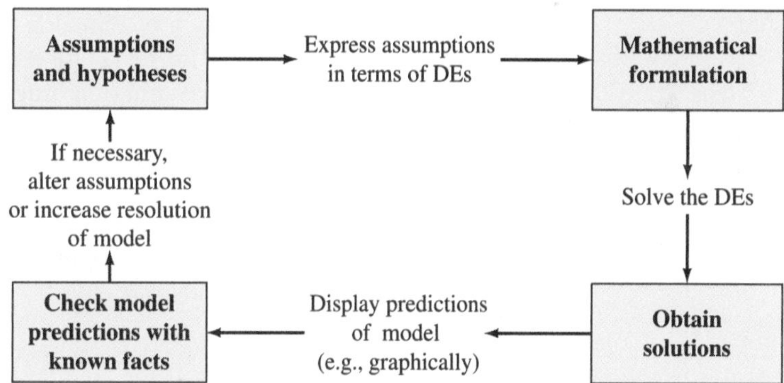

FIGURE 1.3.1 Steps in the modeling process with differential equations

Of course, by increasing the resolution, we add to the complexity of the mathematical model and increase the likelihood that we cannot obtain an explicit solution.

A mathematical model of a physical system will often involve the variable time t. A solution of the model then gives the **state of the system**; in other words, the values of the dependent variable (or variables) for appropriate values of t describe the system in the past, present, and future.

Population Dynamics One of the earliest attempts to model human **population growth** by means of mathematics was by the English clergyman and economist Thomas Malthus in 1798. Basically, the idea behind the Malthusian model is the assumption that the rate at which the population of a country grows at a certain time is proportional[*] to the total population of the country at that time. In other words, the more people there are at time t, the more there are going to be in the future. In mathematical terms, if $P(t)$ denotes the total population at time t, then this assumption can be expressed as

$$\frac{dP}{dt} \propto P \quad \text{or} \quad \frac{dP}{dt} = kP, \tag{1}$$

where k is a constant of proportionality. This simple model, which fails to take into account many factors that can influence human populations to either grow or decline (immigration and emigration, for example), nevertheless turned out to be fairly accurate in predicting the population of the United States during the years 1790–1860. Populations that grow at a rate described by (1) are rare; nevertheless, (1) is still used to model *growth of small populations over short intervals of time* (bacteria growing in a petri dish, for example).

Radioactive Decay The nucleus of an atom consists of combinations of protons and neutrons. Many of these combinations of protons and neutrons are unstable—that is, the atoms decay or transmute into atoms of another substance. Such nuclei are said to be radioactive. For example, over time the highly radioactive radium, Ra-226, transmutes into the radioactive gas radon, Rn-222. To model the phenomenon of **radioactive decay,** it is assumed that the rate dA/dt at which the nuclei of a substance decay is proportional to the amount (more precisely, the number of nuclei) $A(t)$ of the substance remaining at time t:

$$\frac{dA}{dt} \propto A \quad \text{or} \quad \frac{dA}{dt} = kA. \tag{2}$$

Of course, equations (1) and (2) are exactly the same; the difference is only in the interpretation of the symbols and the constants of proportionality. For growth, as we expect in (1), $k > 0$, and for decay, as in (2), $k < 0$.

The model (1) for growth can also be seen as the equation $dS/dt = rS$, which describes the growth of capital S when an annual rate of interest r is compounded continuously. The model (2) for decay also occurs in biological applications such as determining the half-life of a drug—the time that it takes for 50% of a drug to be eliminated from a body by excretion or metabolism. In chemistry the decay model (2) appears in the mathematical description of a first-order chemical reaction. The point is this:

> *A single differential equation can serve as a mathematical model for many different phenomena.*

Mathematical models are often accompanied by certain side conditions. For example, in (1) and (2) we would expect to know, in turn, the initial population P_0 and the initial amount of radioactive substance A_0 on hand. If the initial point in time is taken to be $t = 0$, then we know that $P(0) = P_0$ and $A(0) = A_0$. In other words, a mathematical model can consist of either an initial-value problem or, as we shall see later on in Section 5.2, a boundary-value problem.

Newton's Law of Cooling/Warming According to Newton's empirical law of cooling/warming, the rate at which the temperature of a body changes is proportional to the difference between the temperature of the body and the temperature of the surrounding medium, the so-called ambient temperature. If $T(t)$ represents the temperature of a body at time t, T_m the temperature of the surrounding

[*]If two quantities u and v are proportional, we write $u \propto v$. This means that one quantity is a constant multiple of the other: $u = kv$.

medium, and dT/dt the rate at which the temperature of the body changes, then Newton's law of cooling/warming translates into the mathematical statement

$$\frac{dT}{dt} \propto T - T_m \quad \text{or} \quad \frac{dT}{dt} = k(T - T_m), \tag{3}$$

where k is a constant of proportionality. In either case, cooling or warming, if T_m is a constant, it stands to reason that $k < 0$.

Spread of a Disease A contagious disease—for example, a flu virus—is spread throughout a community by people coming into contact with other people. Let $x(t)$ denote the number of people who have contracted the disease and $y(t)$ denote the number of people who have not yet been exposed. It seems reasonable to assume that the rate dx/dt at which the disease spreads is proportional to the number of encounters, or *interactions,* between these two groups of people. If we assume that the number of interactions is jointly proportional to $x(t)$ and $y(t)$—that is, proportional to the product xy—then

$$\frac{dx}{dt} = kxy, \tag{4}$$

where k is the usual constant of proportionality. Suppose a small community has a fixed population of n people. If one infected person is introduced into this community, then it could be argued that $x(t)$ and $y(t)$ are related by $x + y = n + 1$. Using this last equation to eliminate y in (4) gives us the model

$$\frac{dx}{dt} = kx(n + 1 - x). \tag{5}$$

An obvious initial condition accompanying equation (5) is $x(0) = 1$.

Chemical Reactions The disintegration of a radioactive substance, governed by the differential equation (1), is said to be a **first-order reaction.** In chemistry a few reactions follow this same empirical law: If the molecules of substance A decompose into smaller molecules, it is a natural assumption that the rate at which this decomposition takes place is proportional to the amount of the first substance that has not undergone conversion; that is, if $X(t)$ is the amount of substance A remaining at any time, then $dX/dt = kX$, where k is a negative constant since X is decreasing. An example of a first-order chemical reaction is the conversion of *t*-butyl chloride, $(CH_3)_3CCl$, into *t*-butyl alcohol, $(CH_3)_3COH$:

$$(CH_3)_3CCl + NaOH \rightarrow (CH_3)_3COH + NaCl.$$

Only the concentration of the *t*-butyl chloride controls the rate of reaction. But in the reaction

$$CH_3Cl + NaOH \rightarrow CH_3OH + NaCl$$

one molecule of sodium hydroxide, NaOH, is consumed for every molecule of methyl chloride, CH_3Cl, thus forming one molecule of methyl alcohol, CH_3OH, and one molecule of sodium chloride, NaCl. In this case the rate at which the reaction proceeds is proportional to the product of the remaining concentrations of CH_3Cl and NaOH. To describe this second reaction in general, let us suppose *one* molecule of a substance A combines with *one* molecule of a substance B to form *one* molecule of a substance C. If X denotes the amount of chemical C formed at time t and if α and β are, in turn, the amounts of the two chemicals A and B at $t = 0$ (the initial amounts), then the instantaneous amounts of A and B not converted to chemical C are $\alpha - X$ and $\beta - X$, respectively. Hence the rate of formation of C is given by

$$\frac{dX}{dt} = k(\alpha - X)(\beta - X), \tag{6}$$

where k is a constant of proportionality. A reaction whose model is equation (6) is said to be a **second-order reaction.**

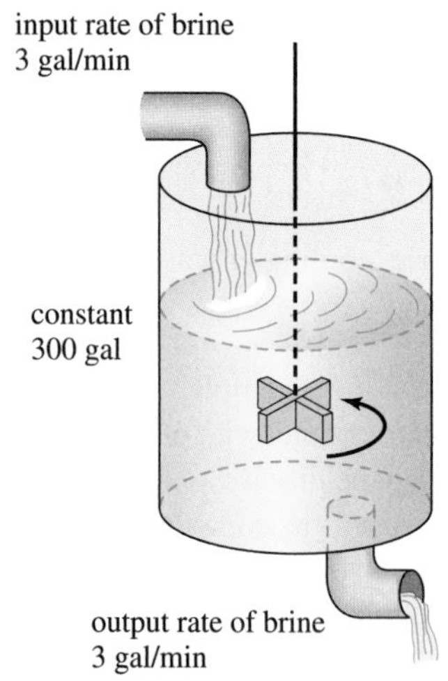

FIGURE 1.3.2 Mixing tank

Mixtures The mixing of two salt solutions of differing concentrations gives rise to a first-order differential equation for the amount of salt contained in the mixture. Let us suppose that a large mixing tank initially holds 300 gallons of brine (that is, water in which a certain number of pounds of salt has been dissolved). Another brine solution is pumped into the large tank at a rate of 3 gallons per minute; the concentration of the salt in this inflow is 2 pounds per gallon. When the solution in the tank is well stirred, it is pumped out at the same rate as the entering solution. See Figure 1.3.2. If $A(t)$ denotes the amount of salt (measured in pounds) in the tank at time t, then the rate at which $A(t)$ changes is a net rate:

$$\frac{dA}{dt} = \begin{pmatrix} \text{input rate} \\ \text{of salt} \end{pmatrix} - \begin{pmatrix} \text{output rate} \\ \text{of salt} \end{pmatrix} = R_{in} - R_{out}. \tag{7}$$

The input rate R_{in} at which salt enters the tank is the product of the inflow concentration of salt and the inflow rate of fluid. Note that R_{in} is measured in pounds per minute:

$$\underset{\substack{\text{concentration}\\ \text{of salt}\\ \text{in inflow}}}{\downarrow} \qquad \underset{\substack{\text{input rate}\\ \text{of brine}}}{\downarrow} \qquad \underset{\substack{\text{input rate}\\ \text{of salt}}}{\downarrow}$$

$$R_{in} = (2 \text{ lb/gal}) \cdot (3 \text{ gal/min}) = (6 \text{ lb/min}).$$

Now, since the solution is being pumped out of the tank at the same rate that it is pumped in, the number of gallons of brine in the tank at time t is a constant 300 gallons. Hence the concentration of the salt in the tank as well as in the outflow is $c(t) = A(t)/300$ lb/gal, so the output rate R_{out} of salt is

$$\underset{\substack{\text{concentration}\\ \text{of salt}\\ \text{in outflow}}}{\downarrow} \qquad \underset{\substack{\text{output rate}\\ \text{of brine}}}{\downarrow} \qquad \underset{\substack{\text{output rate}\\ \text{of salt}}}{\downarrow}$$

$$R_{out} = \left(\frac{A(t)}{300} \text{ lb/gal}\right) \cdot (3 \text{ gal/min}) = \frac{A(t)}{100} \text{ lb/min}.$$

The net rate (7) then becomes

$$\frac{dA}{dt} = 6 - \frac{A}{100} \qquad \text{or} \qquad \frac{dA}{dt} + \frac{1}{100}A = 6. \tag{8}$$

If r_{in} and r_{out} denote general input and output rates of the brine solutions,* then there are three possibilities: $r_{in} = r_{out}$, $r_{in} > r_{out}$, and $r_{in} < r_{out}$. In the analysis leading to (8) we have assumed that $r_{in} = r_{out}$. In the latter two cases the number of gallons of brine in the tank is either increasing ($r_{in} > r_{out}$) or decreasing ($r_{in} < r_{out}$) at the net rate $r_{in} - r_{out}$. See Problems 10–12 in Exercises 1.3.

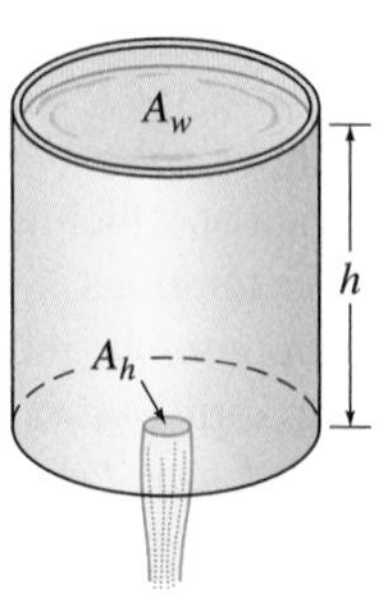

FIGURE 1.3.3 Draining tank

Draining a Tank In hydrodynamics, **Torricelli's law** states that the speed v of efflux of water though a sharp-edged hole at the bottom of a tank filled to a depth h is the same as the speed that a body (in this case a drop of water) would acquire in falling freely from a height h—that is, $v = \sqrt{2gh}$, where g is the acceleration due to gravity. This last expression comes from equating the kinetic energy $\frac{1}{2}mv^2$ with the potential energy mgh and solving for v. Suppose a tank filled with water is allowed to drain through a hole under the influence of gravity. We would like to find the depth h of water remaining in the tank at time t. Consider the tank shown in Figure 1.3.3. If the area of the hole is A_h (in ft^2) and the speed of the water leaving the tank is $v = \sqrt{2gh}$ (in ft/s), then the volume of water leaving the tank per second is $A_h\sqrt{2gh}$ (in ft^3/s). Thus if $V(t)$ denotes the volume of water in the tank at time t, then

$$\frac{dV}{dt} = -A_h\sqrt{2gh}, \tag{9}$$

*Don't confuse these symbols with R_{in} and R_{out}, which are input and output rates of *salt*.

where the minus sign indicates that V is decreasing. Note here that we are ignoring the possibility of friction at the hole that might cause a reduction of the rate of flow there. Now if the tank is such that the volume of water in it at time t can be written $V(t) = A_w h$, where A_w (in ft^2) is the *constant* area of the upper surface of the water (see Figure 1.3.3), then $dV/dt = A_w\, dh/dt$. Substituting this last expression into (9) gives us the desired differential equation for the height of the water at time t:

$$\frac{dh}{dt} = -\frac{A_h}{A_w}\sqrt{2gh}. \tag{10}$$

It is interesting to note that (10) remains valid even when A_w is not constant. In this case we must express the upper surface area of the water as a function of h—that is, $A_w = A(h)$. See Problem 14 in Exercises 1.3.

≡ Series Circuits

Consider the single-loop *LRC*-series circuit shown in Figure 1.3.4(a), containing an inductor, resistor, and capacitor. The current in a circuit after a switch is closed is denoted by $i(t)$; the charge on a capacitor at time t is denoted by $q(t)$. The letters L, R, and C are known as inductance, resistance, and capacitance, respectively, and are generally constants. Now according to **Kirchhoff's second law,** the impressed voltage $E(t)$ on a closed loop must equal the sum of the voltage drops in the loop. Figure 1.3.4(b) shows the symbols and the formulas for the respective voltage drops across an inductor, a capacitor, and a resistor. Since current $i(t)$ is related to charge $q(t)$ on the capacitor by $i = dq/dt$, adding the three voltages

$$\overset{\text{inductor}}{L\frac{di}{dt} = L\frac{d^2q}{dt^2}}, \qquad \overset{\text{resistor}}{iR = R\frac{dq}{dt}}, \qquad \text{and} \qquad \overset{\text{capacitor}}{\frac{1}{C}q}$$

and equating the sum to the impressed voltage yields a second-order differential equation

$$L\frac{d^2q}{dt^2} + R\frac{dq}{dt} + \frac{1}{C}q = E(t). \tag{11}$$

We will examine a differential equation analogous to (11) in great detail in Section 5.1.

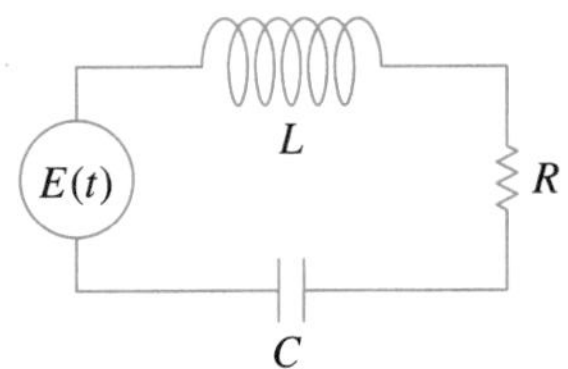

(a) *LRC*-series circuit

Inductor
inductance L: henries (h)
voltage drop across: $L\dfrac{di}{dt}$

$i \rightarrow$ L

Resistor
resistance R: ohms (Ω)
voltage drop across: iR

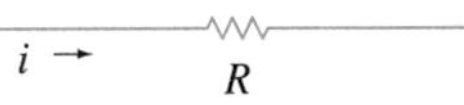

Capacitor
capacitance C: farads (f)
voltage drop across: $\dfrac{1}{C}q$

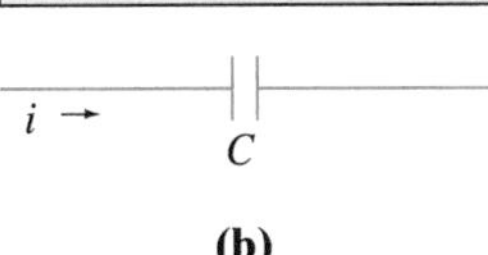

(b)

FIGURE 1.3.4 Symbols, units, and voltages. Current $i(t)$ and charge $q(t)$ are measured in amperes (A) and coulombs (C), respectively

≡ Falling Bodies

To construct a mathematical model of the motion of a body moving in a force field, one often starts with the laws of motion formulated by the English mathematician **Isaac Newton** (1643–1727). Recall from elementary physics that **Newton's first law of motion** states that a body either will remain at rest or will continue to move with a constant velocity unless acted on by an external force. In each case this is equivalent to saying that when the sum of the forces $F = \sum F_k$—that is, the *net* or resultant force—acting on the body is zero, then the acceleration a of the body is zero. **Newton's second law of motion** indicates that when the net force acting on a body is not zero, then the net force is proportional to its acceleration a or, more precisely, $F = ma$, where m is the mass of the body.

Now suppose a rock is tossed upward from the roof of a building as illustrated in Figure 1.3.5. What is the position $s(t)$ of the rock relative to the ground at time t? The acceleration of the rock is the second derivative d^2s/dt^2. If we assume that the upward direction is positive and that no force acts on the rock other than the force of gravity, then Newton's second law gives

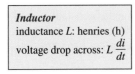

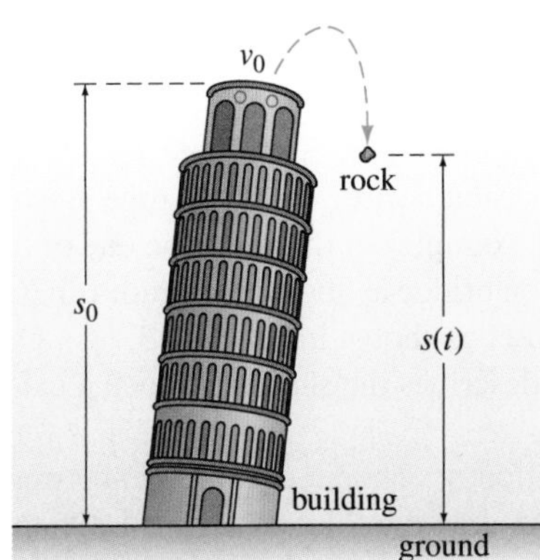

FIGURE 1.3.5 Position of rock measured from ground level

$$m\frac{d^2s}{dt^2} = -mg \qquad \text{or} \qquad \frac{d^2s}{dt^2} = -g. \tag{12}$$

In other words, the net force is simply the weight $F = F_1 = -W$ of the rock near the surface of the Earth. Recall that the magnitude of the weight is $W = mg$, where m is the mass of the body and g is the acceleration due to gravity. The minus sign in (12) is used because the weight of the rock is a force directed downward, which is opposite to the positive direction. If the height of the building is s_0 and the initial velocity of the rock is v_0, then s is determined from the second-order initial-value problem

$$\frac{d^2s}{dt^2} = -g, \quad s(0) = s_0, \quad s'(0) = v_0. \tag{13}$$

Although we have not been stressing solutions of the equations we have constructed, note that (13) can be solved by integrating the constant $-g$ twice with respect to t. The initial conditions determine the two constants of integration. From elementary physics you might recognize the solution of (13) as the formula $s(t) = -\frac{1}{2}gt^2 + v_0t + s_0$.

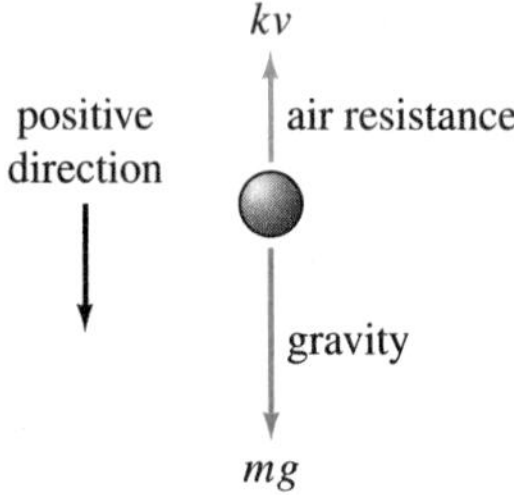

FIGURE 1.3.6 Falling body of mass m

Falling Bodies and Air Resistance

Before the famous experiment by the Italian mathematician and physicist **Galileo Galilei** (1564–1642) from the leaning tower of Pisa, it was generally believed that heavier objects in free fall, such as a cannonball, fell with a greater acceleration than lighter objects, such as a feather. Obviously, a cannonball and a feather when dropped simultaneously from the same height *do* fall at different rates, but it is not because a cannonball is heavier. The difference in rates is due to air resistance. The resistive force of air was ignored in the model given in (13). Under some circumstances a falling body of mass m, such as a feather with low density and irregular shape, encounters air resistance proportional to its instantaneous velocity v. If we take, in this circumstance, the positive direction to be oriented downward, then the net force acting on the mass is given by $F = F_1 + F_2 = mg - kv$, where the weight $F_1 = mg$ of the body is force acting in the positive direction and air resistance $F_2 = -kv$ is a force, called **viscous damping,** acting in the opposite or upward direction. See Figure 1.3.6. Now since v is related to acceleration a by $a = dv/dt$, Newton's second law becomes $F = ma = m\,dv/dt$. By equating the net force to this form of Newton's second law, we obtain a first-order differential equation for the velocity $v(t)$ of the body at time t,

$$m\frac{dv}{dt} = mg - kv. \tag{14}$$

Here k is a positive constant of proportionality. If $s(t)$ is the distance the body falls in time t from its initial point of release, then $v = ds/dt$ and $a = dv/dt = d^2s/dt^2$. In terms of s, (14) is a second-order differential equation

$$m\frac{d^2s}{dt^2} = mg - k\frac{ds}{dt} \quad \text{or} \quad m\frac{d^2s}{dt^2} + k\frac{ds}{dt} = mg. \tag{15}$$

(a) suspension bridge cable

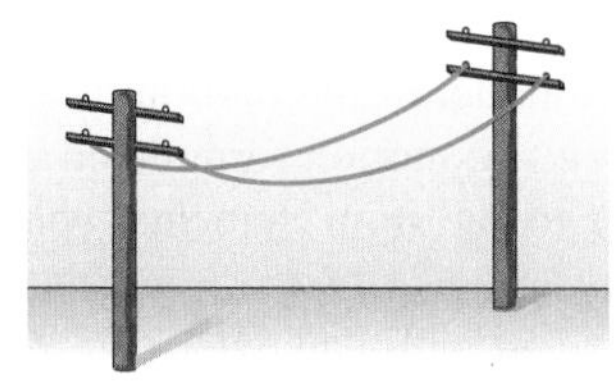

(b) telephone wires

FIGURE 1.3.7 Cables suspended between vertical supports

Suspended Cables

Suppose a flexible cable, wire, or heavy rope is suspended between two vertical supports. Physical examples of this could be one of the two cables supporting the roadbed of a suspension bridge as shown in Figure 1.3.7(a) or a long telephone wire strung between two posts as shown in Figure 1.3.7(b). Our goal is to construct a mathematical model that describes the shape that such a cable assumes.

To begin, let's agree to examine only a portion or element of the cable between its lowest point P_1 and any arbitrary point P_2. As drawn in blue in Figure 1.3.8, this element of the cable is the curve in a rectangular coordinate system with y-axis chosen to pass through the lowest point P_1 on the curve and the x-axis chosen a units below P_1. Three forces are acting on the cable: the tensions $\mathbf{T}_1$ and $\mathbf{T}_2$ in the cable that are tangent to the cable at P_1 and P_2, respectively, and the portion $\mathbf{W}$ of the total vertical load between the points P_1 and P_2. Let $T_1 = |\mathbf{T}_1|$, $T_2 = |\mathbf{T}_2|$, and $W = |\mathbf{W}|$ denote the magnitudes of these vectors. Now the tension $\mathbf{T}_2$ resolves into horizontal and vertical components (scalar quantities) $T_2 \cos\theta$ and $T_2 \sin\theta$.

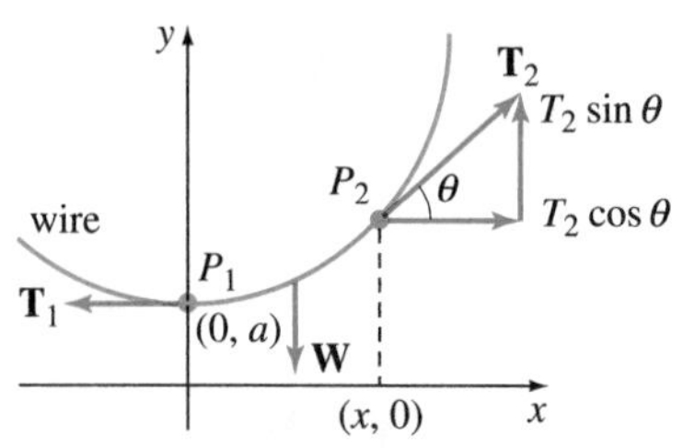

FIGURE 1.3.8 Element of cable

Because of static equilibrium we can write

$$T_1 = T_2 \cos \theta \qquad \text{and} \qquad W = T_2 \sin \theta.$$

By dividing the last equation by the first, we eliminate T_2 and get $\tan \theta = W/T_1$. But because $dy/dx = \tan \theta$, we arrive at

$$\frac{dy}{dx} = \frac{W}{T_1}. \tag{16}$$

This simple first-order differential equation serves as a model for both the shape of a flexible wire such as a telephone wire hanging under its own weight and the shape of the cables that support the roadbed of a suspension bridge. We will come back to equation (16) in Exercises 2.2 and Section 5.3.

What Lies Ahead Throughout this text you will see three different types of approaches to, or analyses of, differential equations. Over the centuries differential equations would often spring from the efforts of a scientist or engineer to describe some physical phenomenon or to translate an empirical or experimental law into mathematical terms. As a consequence, a scientist, engineer, or mathematician would often spend many years of his or her life trying to find the solutions of a DE. With a solution in hand, the study of its properties then followed. This quest for solutions is called by some the *analytical approach* to differential equations. Once they realized that explicit solutions are at best difficult to obtain and at worst impossible to obtain, mathematicians learned that a differential equation itself could be a font of valuable information. It is possible, in some instances, to glean directly from the differential equation answers to questions such as *Does the DE actually have solutions? If a solution of the DE exists and satisfies an initial condition, is it the only such solution? What are some of the properties of the unknown solutions? What can we say about the geometry of the solution curves?* Such an approach is *qualitative analysis.* Finally, if a differential equation cannot be solved by analytical methods, yet we can prove that a solution exists, the next logical query is *Can we somehow approximate the values of an unknown solution?* Here we enter the realm of *numerical analysis.* An affirmative answer to the last question stems from the fact that a differential equation can be used as a cornerstone for constructing very accurate approximation algorithms. In Chapter 2 we start with qualitative considerations of first-order ODEs, then examine analytical stratagems for solving some special first-order equations, and conclude with an introduction to an elementary numerical method. See Figure 1.3.9.

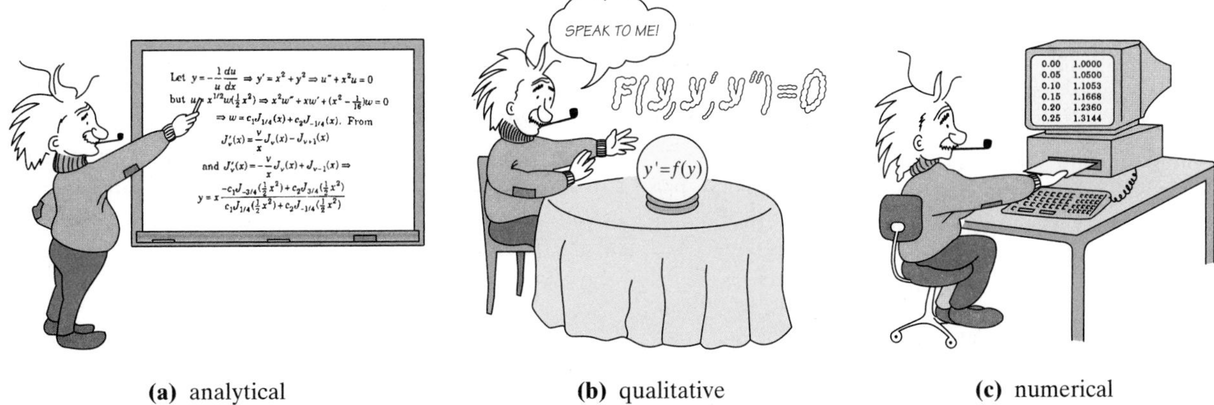

(a) analytical **(b)** qualitative **(c)** numerical

FIGURE 1.3.9 Different approaches to the study of differential equations

REMARKS

Each example in this section has described a dynamical system—a system that changes or evolves with the flow of time t. Since the study of dynamical systems is a branch of mathematics currently in vogue, we shall occasionally relate the terminology of that field to the discussion at hand.

In more precise terms, a **dynamical system** consists of a set of time-dependent variables, called **state variables,** together with a rule that enables us to determine (without ambiguity) the state of the system (this may be a past, present, or future state) in terms of a state prescribed at some time t_0. Dynamical systems are classified as either discrete-time systems or continuous-time systems. In this course we shall be concerned only with continuous-time systems—systems in which *all* variables are defined over a continuous range of time. The rule, or mathematical model, in a continuous-time dynamical system is a differential equation or a system of differential equations. The **state of the system** at a time t is the value of the state variables at that time; the specified state of the system at a time t_0 is simply the initial conditions that accompany the mathematical model. The solution of the initial-value problem is referred to as the **response of the system.** For example, in the case of radioactive decay, the rule is $dA/dt = kA$. Now if the quantity of a radioactive substance at some time t_0 is known, say $A(t_0) = A_0$, then by solving the rule we find that the response of the system for $t \geq t_0$ is $A(t) = A_0e^{(t-t_0)}$ (see Section 3.1). The response $A(t)$ is the single state variable for this system. In the case of the rock tossed from the roof of a building, the response of the system—the solution of the differential equation $d^2s/dt^2 = -g$, subject to the initial state $s(0) = s_0$, $s'(0) = v_0$, is the function $s(t) = -\frac{1}{2}gt^2 + v_0t + s_0$, $0 \leq t \leq T$, where T represents the time when the rock hits the ground. The state variables are $s(t)$ and $s'(t)$, which are the vertical position of the rock above ground and its velocity at time t, respectively. The acceleration $s''(t)$ is *not* a state variable, since we have to know only any initial position and initial velocity at a time t_0 to uniquely determine the rock's position $s(t)$ and velocity $s'(t) = v(t)$ for any time in the interval $t_0 \leq t \leq T$. The acceleration $s''(t) = a(t)$ is, of course, given by the differential equation $s''(t) = -g$, $0 < t < T$.

One last point: Not every system studied in this text is a dynamical system. We shall also examine some static systems in which the model is a differential equation.

EXERCISES 1.3

Answers to selected odd-numbered problems begin on page ANS-1.

Population Dynamics

1. Under the same assumptions that underlie the model in (1), determine a differential equation for the population $P(t)$ of a country when individuals are allowed to immigrate into the country at a constant rate $r > 0$. What is the differential equation for the population $P(t)$ of the country when individuals are allowed to emigrate from the country at a constant rate $r > 0$?

2. The population model given in (1) fails to take death into consideration; the growth rate equals the birth rate. In another model of a changing population of a community it is assumed that the rate at which the population changes is a *net* rate—that is, the difference between the rate of births and the rate of deaths in the community. Determine a model for the population $P(t)$ if both the birth rate and the death rate are proportional to the population present at time $t > 0$.

3. Using the concept of net rate introduced in Problem 2, determine a model for a population $P(t)$ if the birth rate is proportional to the population present at time t but the death rate is proportional to the square of the population present at time t.

4. Modify the model in Problem 3 for net rate at which the population $P(t)$ of a certain kind of fish changes by also assuming that the fish are harvested at a constant rate $h > 0$.

Newton's Law of Cooling/Warming

5. A cup of coffee cools according to Newton's law of cooling (3). Use data from the graph of the temperature $T(t)$ in Figure 1.3.10 to estimate the constants T_m, T_0, and k in a model of the form of a first-order initial-value problem: $dT/dt = k(T - T_m)$, $T(0) = T_0$.

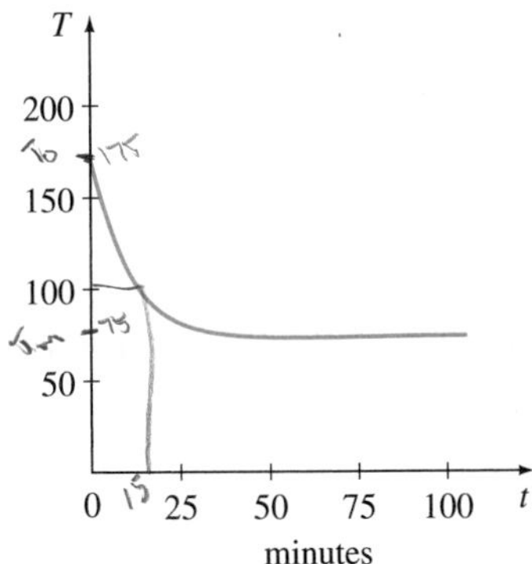

FIGURE 1.3.10 Cooling curve in Problem 5

6. The ambient temperature T_m in (3) could be a function of time t. Suppose that in an artificially controlled environment, $T_m(t)$ is periodic with a 24-hour period, as illustrated in Figure 1.3.11. Devise a mathematical model for the temperature $T(t)$ of a body within this environment.

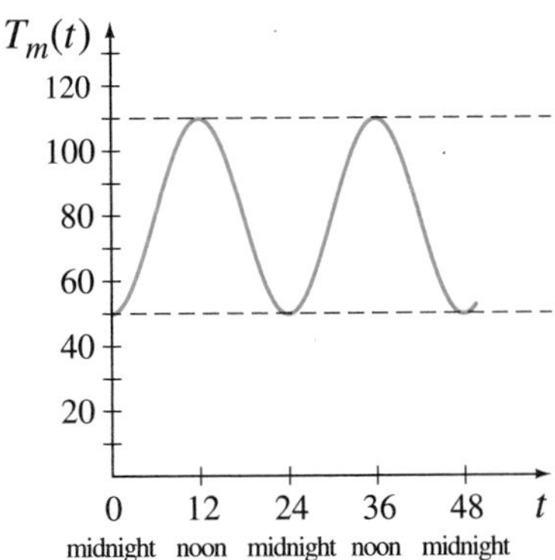

FIGURE 1.3.11 Ambient temperature in Problem 6

Spread of a Disease/Technology

7. Suppose a student carrying a flu virus returns to an isolated college campus of 1000 students. Determine a differential equation for the number of people $x(t)$ who have contracted the flu if the rate at which the disease spreads is proportional to the number of interactions between the number of students who have the flu and the number of students who have not yet been exposed to it.

8. At a time denoted as $t = 0$ a technological innovation is introduced into a community that has a fixed population of n people. Determine a differential equation for the number of people $x(t)$ who have adopted the innovation at time t if it is assumed that the rate at which the innovations spread through the community is jointly proportional to the number of people who have adopted it and the number of people who have not adopted it.

Mixtures

9. Suppose that a large mixing tank initially holds 300 gallons of water in which 50 pounds of salt have been dissolved. Pure water is pumped into the tank at a rate of 3 gal/min, and when the solution is well stirred, it is then pumped out at the same rate. Determine a differential equation for the amount of salt $A(t)$ in the tank at time $t > 0$. What is $A(0)$?

10. Suppose that a large mixing tank initially holds 300 gallons of water is which 50 pounds of salt have been dissolved. Another brine solution is pumped into the tank at a rate of 3 gal/min, and when the solution is well stirred, it is then pumped out at a *slower* rate of 2 gal/min. If the concentration of the solution entering is 2 lb/gal, determine a differential equation for the amount of salt $A(t)$ in the tank at time $t > 0$.

11. What is the differential equation in Problem 10, if the well-stirred solution is pumped out at a *faster* rate of 3.5 gal/min?

12. Generalize the model given in equation (8) on page 24 by assuming that the large tank initially contains N_0 number of gallons of brine, r_{in} and r_{out} are the input and output rates of the brine, respectively (measured in gallons per minute), c_{in} is the concentration of the salt in the inflow, $c(t)$ the concentration of the salt in the tank as well as in the outflow at time t (measured in pounds of salt per gallon), and $A(t)$ is the amount of salt in the tank at time $t > 0$.

Draining a Tank

13. Suppose water is leaking from a tank through a circular hole of area A_h at its bottom. When water leaks through a hole, friction and contraction of the stream near the hole reduce the volume of water leaving the tank per second to $cA_h\sqrt{2gh}$, where c $(0 < c < 1)$ is an empirical constant. Determine a differential equation for the height h of water at time t for the cubical tank shown in Figure 1.3.12. The radius of the hole is 2 in., and $g = 32$ ft/s^2.

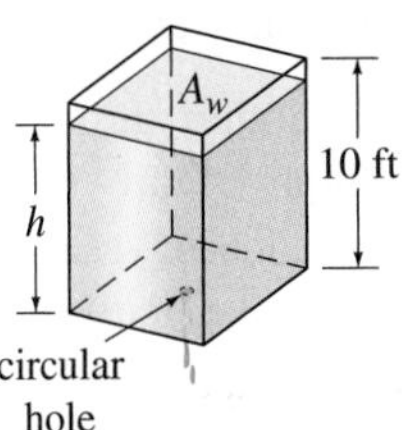

FIGURE 1.3.12 Cubical tank in Problem 13

14. The right-circular conical tank shown in Figure 1.3.13 loses water out of a circular hole at its bottom. Determine a differential equation for the height of the water h at time $t > 0$. The radius of the hole is 2 in., $g = 32$ ft/s^2, and the friction/contraction factor introduced in Problem 13 is $c = 0.6$.

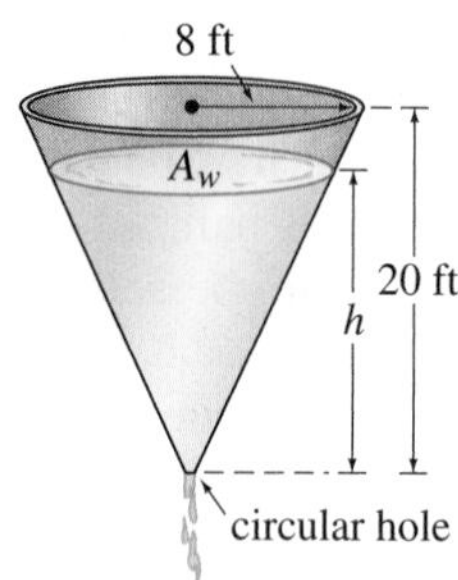

FIGURE 1.3.13 Conical tank in Problem 14

Series Circuits

15. A series circuit contains a resistor and an inductor as shown in Figure 1.3.14. Determine a differential equation for the current $i(t)$ if the resistance is R, the inductance is L, and the impressed voltage is $E(t)$.

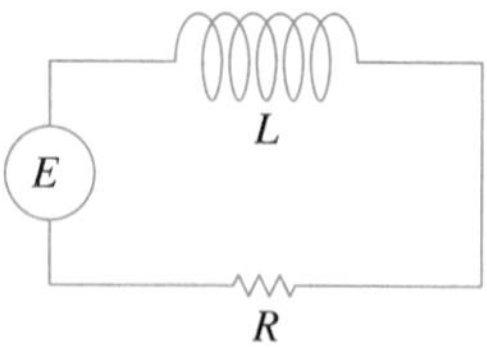

FIGURE 1.3.14 *LR*-series circuit in Problem 15

16. A series circuit contains a resistor and a capacitor as shown in Figure 1.3.15. Determine a differential equation for the charge $q(t)$ on the capacitor if the resistance is R, the capacitance is C, and the impressed voltage is $E(t)$.

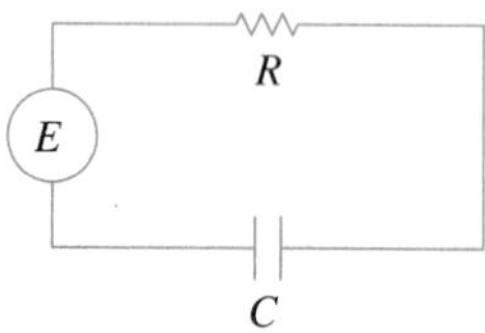

FIGURE 1.3.15 *RC*-series circuit in Problem 16

Falling Bodies and Air Resistance

17. For high-speed motion through the air—such as the skydiver shown in Figure 1.3.16, falling before the parachute is opened—air resistance is closer to a power of the instantaneous velocity $v(t)$. Determine a differential equation for the velocity $v(t)$ of a falling body of mass m if air resistance is proportional to the square of the instantaneous velocity.

FIGURE 1.3.16 Air resistance proportional to square of velocity in Problem 17

Newton's Second Law and Archimedes' Principle

18. A cylindrical barrel s feet in diameter of weight w lb is floating in water as shown in Figure 1.3.17(a). After an initial depression the barrel exhibits an up-and-down bobbing motion along a vertical line. Using Figure 1.3.17(b), determine a differential equation for the vertical displacement $y(t)$ if the origin is taken to be on the vertical axis at the surface of the water when the barrel is at rest. Use **Archimedes' principle:** Buoyancy, or upward force of the water on the barrel, is equal to the weight of the water displaced. Assume that the downward direction is positive, that the weight density of water is 62.4 lb/ft^3, and that there is no resistance between the barrel and the water.

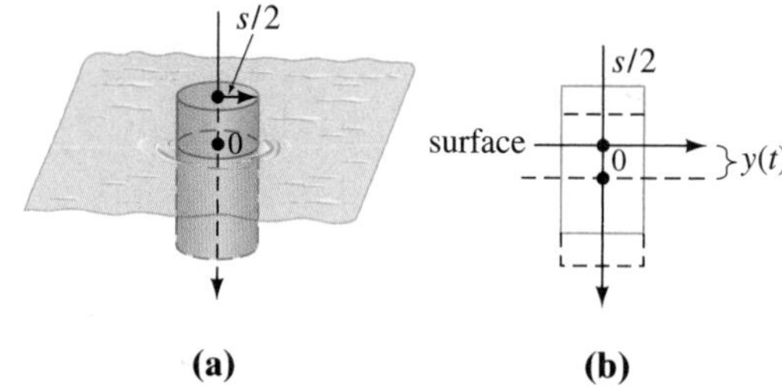

FIGURE 1.3.17 Bobbing motion of floating barrel in Problem 18

Newton's Second Law and Hooke's Law

19. After a mass m is attached to a spring, it stretches it s units and then hangs at rest in the equilibrium position as shown in Figure 1.3.18(b). After the spring/mass

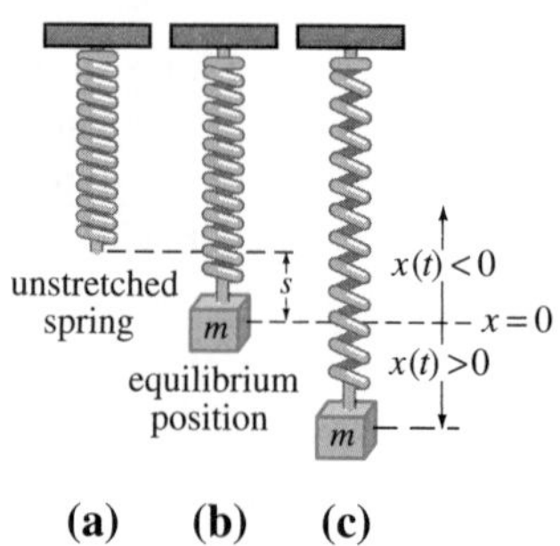

FIGURE 1.3.18 Spring/mass system in Problem 19

system has been set in motion, let $x(t)$ denote the directed distance of the mass beyond the equilibrium position. As indicated in Figure 1.3.17(c), assume that the downward direction is positive, that the motion takes place in a vertical straight line through the center of gravity of the mass, and that the only forces acting on the system are the weight of the mass and the restoring force of the stretched spring. Use **Hooke's law:** The restoring force of a spring is proportional to its total elongation. Determine a differential equation for the displacement $x(t)$ at time $t > 0$.

20. In Problem 19, what is a differential equation for the displacement $x(t)$ if the motion takes place in a medium that imparts a damping force on the spring/mass system that is proportional to the instantaneous velocity of the mass and acts in a direction opposite to that of motion?

Newton's Second Law and Rocket Motion

When the mass m of a body is changing with time, Newton's second law of motion becomes

$$F = \frac{d}{dt}(mv), \tag{17}$$

where F is the net force acting on the body and mv is its momentum. Use (17) in Problems 21 and 22.

21. A small single-stage rocket is launched vertically as shown in Figure 1.3.19. Once launched, the rocket consumes its fuel, and so its total mass $m(t)$ varies with time $t > 0$. If it is assumed that the positive direction is upward, air resistance is proportional to the instantaneous velocity v of the rocket, and R is the upward thrust or force generated by the propulsion system, then construct a mathematical model for the velocity $v(t)$ of the rocket. [*Hint*: See (14) in Section 1.3.]

FIGURE 1.3.19 Single-stage rocket in Problem 21

22. In Problem 21, the mass $m(t)$ is the sum of three different masses: $m(t) = m_p + m_v + m_f(t)$, where m_p is the constant mass of the payload, m_v is the constant mass of the vehicle, and $m_f(t)$ is the variable amount of fuel.

(a) Show that the rate at which the total mass $m(t)$ of the rocket changes is the same as the rate at which the mass $m_f(t)$ of the fuel changes.

(b) If the rocket consumes its fuel at a constant rate λ, find $m(t)$. Then rewrite the differential equation in Problem 21 in terms of λ and the initial total mass $m(0) = m_0$.

(c) Under the assumption in part (b), show that the burnout time $t_b > 0$ of the rocket, or the time at which all the fuel is consumed, is $t_b = m_f(0)/\lambda$, where $m_f(0)$ is the initial mass of the fuel.

Newton's Second Law and the Law of Universal Gravitation

23. By **Newton's universal law of gravitation** the free-fall acceleration a of a body, such as the satellite shown in Figure 1.3.20, falling a great distance to the surface is *not* the constant g. Rather, the acceleration a is inversely proportional to the square of the distance from the center of the Earth, $a = k/r^2$, where k is the constant of proportionality. Use the fact that at the surface of the Earth $r = R$ and $a = g$ to determine k. If the positive direction is upward, use Newton's second law and his universal law of gravitation to find a differential equation for the distance r.

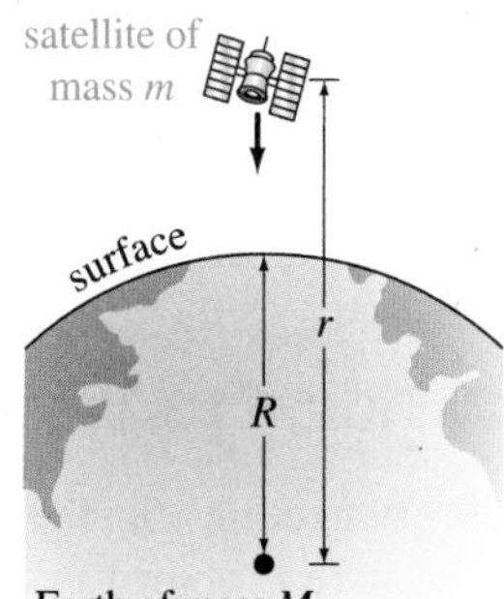

FIGURE 1.3.20 Satellite in Problem 23

24. Suppose a hole is drilled through the center of the Earth and a bowling ball of mass m is dropped into the hole, as shown in Figure 1.3.21. Construct a mathematical model that describes the motion of the ball. At time t let r denote the distance from the center of the Earth to the mass m, M denote the mass of the Earth, M_r denote the mass of that portion of the Earth within a sphere of radius r, and δ denote the constant density of the Earth.

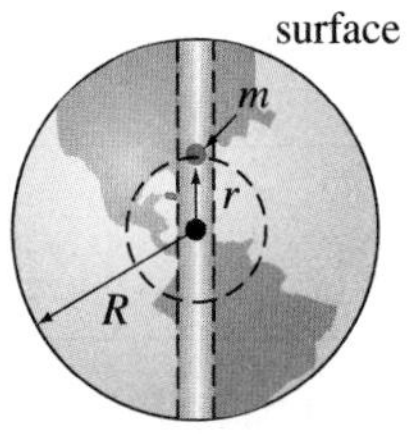

FIGURE 1.3.21 Hole through Earth in Problem 24

Additional Mathematical Models

25. **Learning Theory** In the theory of learning, the rate at which a subject is memorized is assumed to be proportional to the amount that is left to be memorized. Suppose M denotes the total amount of a subject to be memorized and $A(t)$ is the amount memorized in time $t > 0$. Determine a differential equation for the amount $A(t)$.

26. **Forgetfulness** In Problem 25 assume that the rate at which material is *forgotten* is proportional to the amount memorized in time $t > 0$. Determine a differential equation for the amount $A(t)$ when forgetfulness is taken into account.

27. **Infusion of a Drug** A drug is infused into a patient's bloodstream at a constant rate of r grams per second. Simultaneously, the drug is removed at a rate proportional to the amount $x(t)$ of the drug present at time t. Determine a differential equation for the amount $x(t)$.

28. **Tractrix** A person P, starting at the origin, moves in the direction of the positive x-axis, pulling a weight along the curve C, called a **tractrix,** as shown in Figure 1.3.22. The weight, initially located on the y-axis at $(0, s)$, is pulled by a rope of constant length s, which is kept taut throughout the motion. Determine a differential equation for the path C of motion. Assume that the rope is always tangent to C.

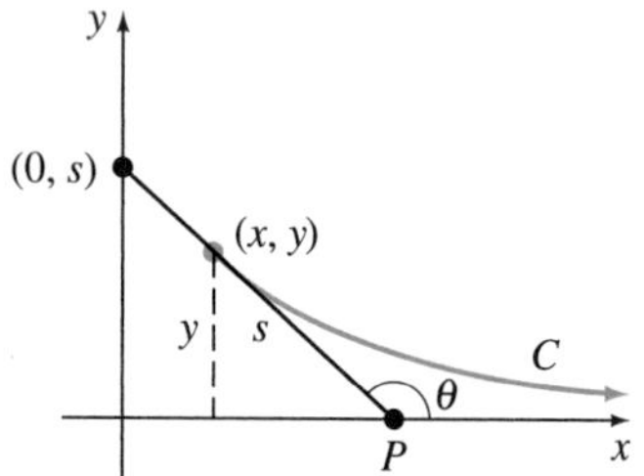

FIGURE 1.3.22 Tractrix curve in Problem 28

29. **Reflecting Surface** Assume that when the plane curve C shown in Figure 1.3.23 is revolved about the x-axis, it generates a surface of revolution with the property that all light rays L parallel to the x-axis striking the surface are reflected to a single point O (the origin). Use the fact that the angle of incidence is equal to the angle of reflection to determine a differential equation that describes the shape of the curve C. Such a curve C is important in applications ranging from construction of telescopes to satellite antennas, automobile headlights, and solar collectors. [*Hint*: Inspection of the figure shows that we can write $\phi = 2\theta$. Why? Now use an appropriate trigonometric identity.]

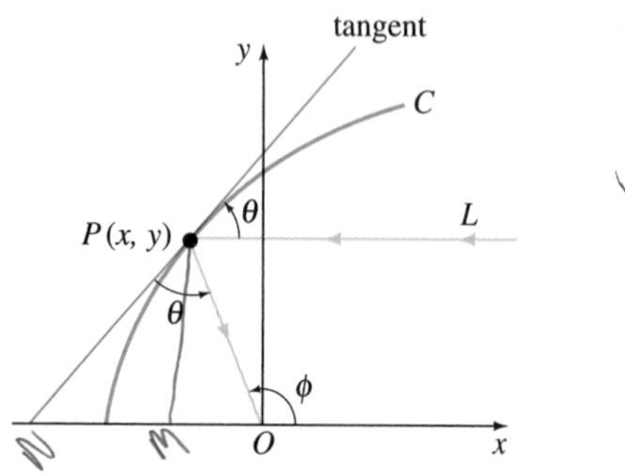

FIGURE 1.3.23 Reflecting surface in Problem 29

Discussion Problems

30. Reread Problem 41 in Exercises 1.1 and then give an explicit solution $P(t)$ for equation (1). Find a one-parameter family of solutions of (1).

31. Reread the sentence following equation (3) and assume that T_m is a positive constant. Discuss why we would expect $k < 0$ in (3) in both cases of cooling and warming. You might start by interpreting, say, $T(t) > T_m$ in a graphical manner.

32. Reread the discussion leading up to equation (8). If we assume that initially the tank holds, say, 50 lb of salt, it stands to reason that because salt is being added to the tank continuously for $t > 0$, $A(t)$ should be an increasing function. Discuss how you might determine from the DE, without actually solving it, the number of pounds of salt in the tank after a long period of time.

33. **Population Model** The differential equation $\dfrac{dP}{dt} = (k \cos t)P$, where k is a positive constant, is a model of human population $P(t)$ of a certain community. Discuss an interpretation for the solution of this equation. In other words, what kind of population do you think the differential equation describes?

34. **Rotating Fluid** As shown in Figure 1.3.24(a), a right-circular cylinder partially filled with fluid is rotated with a constant angular velocity ω about a vertical y-axis through its center. The rotating fluid forms a surface of revolution S. To identify S, we first establish a coordinate system consisting of a vertical plane determined by the y-axis and an x-axis drawn perpendicular to the y-axis such that the point of intersection of the axes (the origin) is located at the lowest point on the surface S. We then seek a function $y = f(x)$ that represents the curve C of intersection of the surface S and the vertical coordinate plane. Let the point $P(x, y)$ denote the position of a particle of the rotating fluid of mass m in the coordinate plane. See Figure 1.3.23(b).

    (a) At P there is a reaction force of magnitude F due to the other particles of the fluid which is normal to the surface S. By Newton's second law the magnitude of the net force acting on the particle is $m\omega^2 x$. What is this force? Use Figure 1.3.24(b) to discuss the nature and origin of the equations

    $$F \cos \theta = mg, \qquad F \sin \theta = m\omega^2 x.$$

    (b) Use part (a) to find a first-order differential equation that defines the function $y = f(x)$.

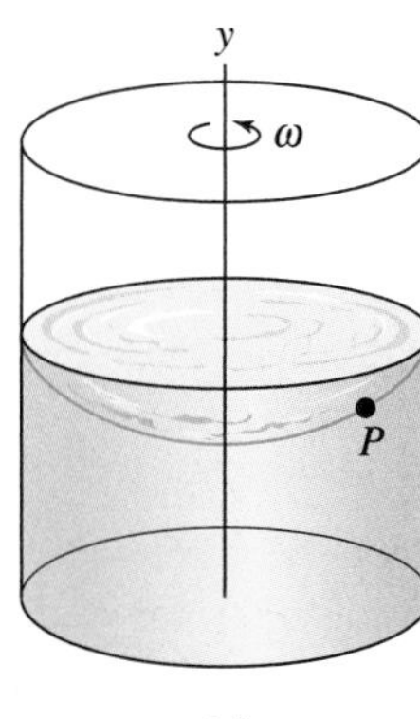

(a)

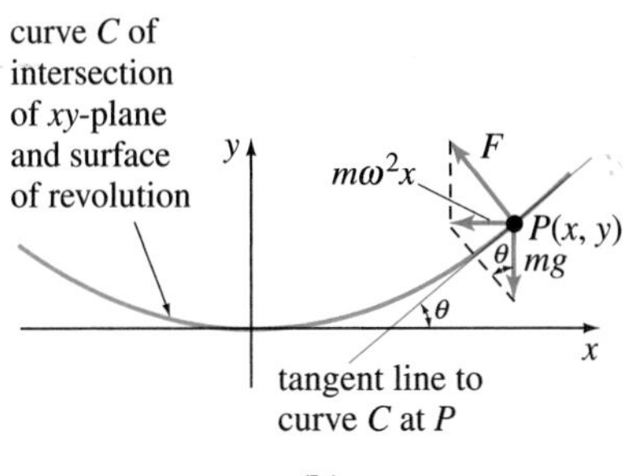

(b)

FIGURE 1.3.24 Rotating fluid in Problem 34

35. Falling Body In Problem 23, suppose $r = R + s$, where s is the distance from the surface of the Earth to the falling body. What does the differential equation obtained in Problem 23 become when s is very small in comparison to R? [*Hint*: Think binomial series for

$$(R + s)^{-2} = R^{-2}(1 + s/R)^{-2}.]$$

36. Raindrops Keep Falling In meteorology the term *virga* refers to falling raindrops or ice particles that evaporate before they reach the ground. Assume that a typical raindrop is spherical. Starting at some time, which we can designate as $t = 0$, the raindrop of radius r_0 falls from rest from a cloud and begins to evaporate.

(a) If it is assumed that a raindrop evaporates in such a manner that its shape remains spherical, then it also makes sense to assume that the rate at which the raindrop evaporates—that is, the rate at which it loses mass—is proportional to its surface area. Show that this latter assumption implies that the rate at which the radius r of the raindrop decreases is a constant. Find $r(t)$. [*Hint*: See Problem 51 in Exercises 1.1.]

(b) If the positive direction is downward, construct a mathematical model for the velocity v of the falling raindrop at time $t > 0$. Ignore air resistance. [*Hint*: Use the form of Newton's second law given in (17).]

37. Let It Snow The "snowplow problem" is a classic and appears in many differential equations texts, but it was probably made famous by Ralph Palmer Agnew:

> *One day it started snowing at a heavy and steady rate. A snowplow started out at noon, going 2 miles the first hour and 1 mile the second hour. What time did it start snowing?*

Find the textbook *Differential Equations,* Ralph Palmer Agnew, McGraw-Hill Book Co., and then discuss the construction and solution of the mathematical model.

38. Reread this section and classify each mathematical model as linear or nonlinear.

CHAPTER 1 IN REVIEW

Answers to selected odd-numbered problems begin on page ANS-1.

In Problems 1 and 2 fill in the blank and then write this result as a linear first-order differential equation that is free of the symbol c_1 and has the form $dy/dx = f(x, y)$. The symbol c_1 represents a constant.

1. $\dfrac{d}{dx} c_1 e^{10x} =$ ________

2. $\dfrac{d}{dx}(5 + c_1 e^{-2x}) =$ ________

In Problems 3 and 4 fill in the blank and then write this result as a linear second-order differential equation that is free of the symbols c_1 and c_2 and has the form $F(y, y'') = 0$. The symbols c_1, c_2, and k represent constants.

3. $\dfrac{d^2}{dx^2}(c_1 \cos kx + c_2 \sin kx) =$ ________

4. $\dfrac{d^2}{dx^2}(c_1 \cosh kx + c_2 \sinh kx) =$ ________

In Problems 5 and 6 compute y' and y'' and then combine these derivatives with y as a linear second-order differential equation that is free of the symbols c_1 and c_2 and has the form $F(y, y' y'') = 0$. The symbols c_1 and c_2 represent constants.

5. $y = c_1 e^x + c_2 x e^x$ **6.** $y = c_1 e^x \cos x + c_2 e^x \sin x$

In Problems 7–12 match each of the given differential equations with one or more of these solutions:

(a) $y = 0$, **(b)** $y = 2$, **(c)** $y = 2x$, **(d)** $y = 2x^2$.

7. $xy' = 2y$ **8.** $y' = 2$

9. $y' = 2y - 4$ **10.** $xy' = y$

11. $y'' + 9y = 18$ **12.** $xy'' - y' = 0$

In Problems 13 and 14 determine by inspection at least one solution of the given differential equation.

13. $y'' = y'$ **14.** $y' = y(y - 3)$

In Problems 15 and 16 interpret each statement as a differential equation.

15. On the graph of $y = \phi(x)$ the slope of the tangent line at a point $P(x, y)$ is the square of the distance from $P(x, y)$ to the origin.

16. On the graph of $y = \phi(x)$ the rate at which the slope changes with respect to x at a point $P(x, y)$ is the negative of the slope of the tangent line at $P(x, y)$.

17. **(a)** Give the domain of the function $y = x^{2/3}$.

(b) Give the largest interval I of definition over which $y = x^{2/3}$ is solution of the differential equation $3xy' - 2y = 0$.

18. **(a)** Verify that the one-parameter family $y^2 - 2y = x^2 - x + c$ is an implicit solution of the differential equation $(2y - 2)y' = 2x - 1$.

(b) Find a member of the one-parameter family in part (a) that satisfies the initial condition $y(0) = 1$.

(c) Use your result in part (b) to find an explicit *function* $y = \phi(x)$ that satisfies $y(0) = 1$. Give the domain of the function ϕ. Is $y = \phi(x)$ a *solution* of the initial-value problem? If so, give its interval I of definition; if not, explain.

19. Given that $y = x - 2/x$ is a solution of the DE $xy' + y = 2x$. Find x_0 and the largest interval I for which $y(x)$ is a solution of the first-order IVP $xy' + y = 2x$, $y(x_0) = 1$.

20. Suppose that $y(x)$ denotes a solution of the first-order IVP $y' = x^2 + y^2$, $y(1) = -1$ and that $y(x)$ possesses at least a second derivative at $x = 1$. In some neighborhood of $x = 1$ use the DE to determine whether $y(x)$ is increasing or decreasing and whether the graph $y(x)$ is concave up or concave down.

21. A differential equation may possess more than one family of solutions.

(a) Plot different members of the families $y = \phi_1(x) = x^2 + c_1$ and $y = \phi_2(x) = -x^2 + c_2$.

(b) Verify that $y = \phi_1(x)$ and $y = \phi_2(x)$ are two solutions of the nonlinear first-order differential equation $(y')^2 = 4x^2$.

(c) Construct a piecewise-defined function that is a solution of the nonlinear DE in part (b) but is not a member of either family of solutions in part (a).

22. What is the slope of the tangent line to the graph of a solution of $y' = 6\sqrt{y} + 5x^3$ that passes through $(-1, 4)$?

In Problems 23–26 verify that the indicated function is an explicit solution of the given differential equation. Give an interval of definition I for each solution.

23. $y'' + y = 2\cos x - 2\sin x;\quad y = x\sin x + x\cos x$

24. $y'' + y = \sec x;\quad y = x\sin x + (\cos x)\ln(\cos x)$

25. $x^2y'' + xy' + y = 0;\quad y = \sin(\ln x)$

26. $x^2y'' + xy' + y = \sec(\ln x);$
$y = \cos(\ln x)\ln(\cos(\ln x)) + (\ln x)\sin(\ln x)$

In Problems 27–30 verify that the indicated expression is an implicit solution of the given differential equation.

27. $x\dfrac{dy}{dx} + y = \dfrac{1}{y^2};\quad x^3y^3 = x^3 + 1$

28. $\left(\dfrac{dy}{dx}\right)^2 + 1 = \dfrac{1}{y^2};\quad (x - 5)^2 + y^2 = 1$

29. $y'' = 2y(y')^3;\quad y^3 + 3y = 1 - 3x$

30. $(1 - xy)y' = y^2;\quad y = e^{xy}$

In Problems 31–34, $y = c_1e^{3x} + c_2e^{-x} - 2x$ is a two-parameter family of the second-order DE $y'' - 2y' - 3y = 6x + 4$. Find a solution of the second-order IVP consisting of this differential equation and the given initial conditions.

31. $y(0) = 0, y'(0) = 0$

32. $y(0) = 1, y'(0) = -3$

33. $y(1) = 4, y'(1) = -2$

34. $y(-1) = 0, y'(-1) = 1$

35. The graph of a solution of a second-order initial-value problem $d^2y/dx^2 = f(x, y, y')$, $y(2) = y_0$, $y'(2) = y_1$, is given in Figure 1.R.1. Use the graph to estimate the values of y_0 and y_1.

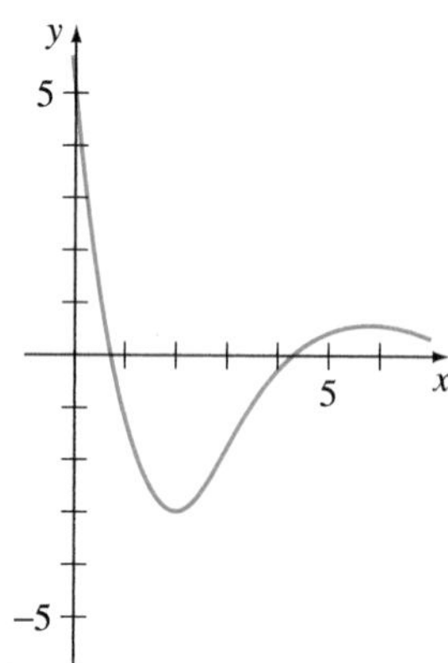

FIGURE 1.R.1 Graph for Problem 35

36. A tank in the form of a right-circular cylinder of radius 2 feet and height 10 feet is standing on end. If the tank is initially full of water and water leaks from a circular hole of radius $\frac{1}{2}$ inch at its bottom, determine a differential equation for the height h of the water at time $t > 0$. Ignore friction and contraction of water at the hole.

37. The number of field mice in a certain pasture is given by the function $200 - 10t$, where time t is measured in years. Determine a differential equation governing a population of owls that feed on the mice if the rate at which the owl population grows is proportional to the difference between the number of owls at time t and number of field mice at time $t > 0$.

38. Suppose that $dA/dt = -0.0004332\,A(t)$ represents a mathematical model for the radioactive decay of radium-226, where $A(t)$ is the amount of radium (measured in grams) remaining at time t (measured in years). How much of the radium sample remains at the time t when the sample is decaying at a rate of 0.002 gram per year?

2 First-Order Differential Equations

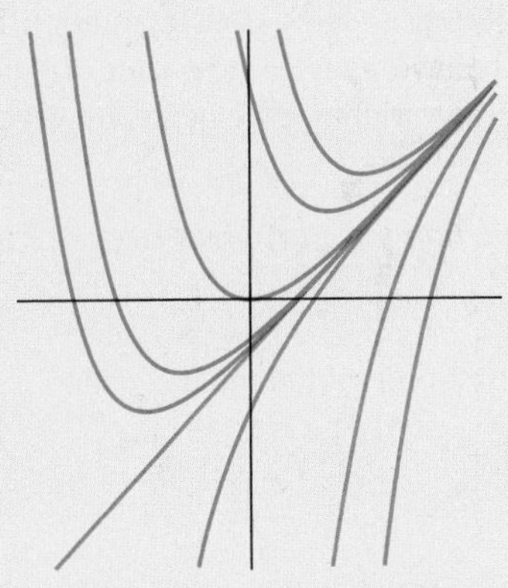

The history of mathematics is rife with stories of people who devoted much of their lives to solving equations—algebraic equations at first and then eventually differential equations. In Sections 2.2–2.5 we will study some of the more important analytical methods for solving first-order DEs. However, before we start solving anything, you should be aware of two facts: It is possible for a differential equation to have no solutions, and a differential equation can possess solutions, yet there might not exist any analytical method for solving it. In Sections 2.1 and 2.6 we do not solve any DEs but show how to glean information about solutions directly from the equation itself. In Section 2.1 we see how the DE yields qualitative information about graphs that enables us to sketch renditions of solution curves. In Section 2.6 we use the differential equation to construct a procedure, called a numerical method, for approximating solutions.

2.1 SOLUTION CURVES WITHOUT A SOLUTION

REVIEW MATERIAL

- The first derivative as slope of a tangent line
- The algebraic sign of the first derivative indicates increasing or decreasing

INTRODUCTION Let us imagine for the moment that we have in front of us a first-order differential equation $dy/dx = f(x, y)$, and let us further imagine that we can neither find nor invent a method for solving it analytically. This is not as bad a predicament as one might think, since the differential equation itself can sometimes "tell" us specifics about how its solutions "behave."

We begin our study of first-order differential equations with two ways of analyzing a DE qualitatively. Both these ways enable us to determine, in an approximate sense, what a solution curve must look like without actually solving the equation.

2.1.1 DIRECTION FIELDS

Some Fundamental Questions We saw in Section 1.2 that whenever $f(x, y)$ and $\partial f/\partial y$ satisfy certain continuity conditions, qualitative questions about existence and uniqueness of solutions can be answered. In this section we shall see that other qualitative questions about properties of solutions—How does a solution behave near a certain point? How does a solution behave as $x \to \infty$?—can often be answered when the function f depends solely on the variable y. We begin, however, with a simple concept from calculus:

A derivative dy/dx of a differentiable function $y = y(x)$ gives slopes of tangent lines at points on its graph.

Slope Because a solution $y = y(x)$ of a first-order differential equation

$$\frac{dy}{dx} = f(x, y) \tag{1}$$

is necessarily a differentiable function on its interval I of definition, it must also be continuous on I. Thus the corresponding solution curve on I must have no breaks and must possess a tangent line at each point $(x, y(x))$. The function f in the normal form (1) is called the **slope function** or **rate function.** The slope of the tangent line at $(x, y(x))$ on a solution curve is the value of the first derivative dy/dx at this point, and we know from (1) that this is the value of the slope function $f(x, y(x))$. Now suppose that (x, y) represents any point in a region of the xy-plane over which the function f is defined. The value $f(x, y)$ that the function f assigns to the point represents the slope of a line or, as we shall envision it, a line segment called a **lineal element.** For example, consider the equation $dy/dx = 0.2xy$, where $f(x, y) = 0.2xy$. At, say, the point (2, 3) the slope of a lineal element is $f(2, 3) = 0.2(2)(3) = 1.2$. Figure 2.1.1(a) shows a line segment with slope 1.2 passing though (2, 3). As shown in Figure 2.1.1(b), *if* a solution curve also passes through the point (2, 3), it does so tangent to this line segment; in other words, the lineal element is a miniature tangent line at that point.

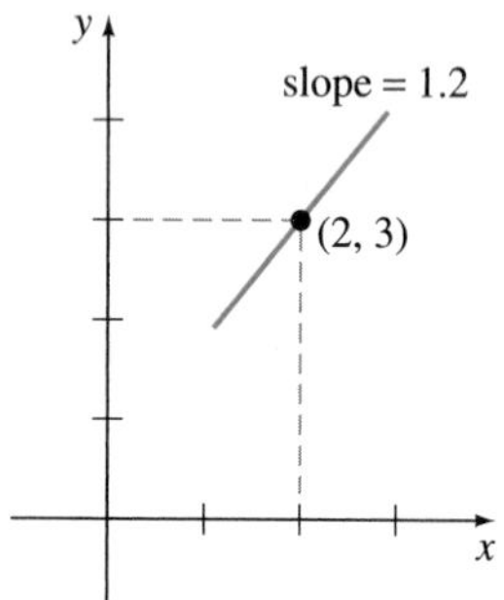

(a) lineal element at a point

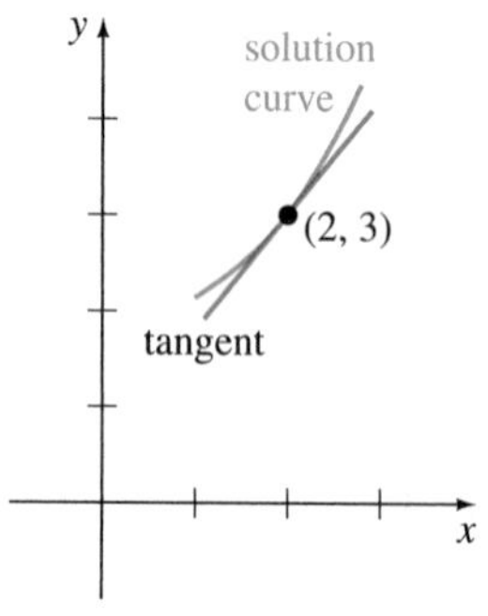

(b) lineal element is tangent to solution curve that passes through the point

FIGURE 2.1.1 A solution curve is tangent to lineal element at (2, 3)

Direction Field If we systematically evaluate f over a rectangular grid of points in the xy-plane and draw a line element at each point (x, y) of the grid with slope $f(x, y)$, then the collection of all these line elements is called a **direction field** or a **slope field** of the differential equation $dy/dx = f(x, y)$. Visually, the direction field suggests the appearance or shape of a family of solution curves of the differential equation, and consequently, it may be possible to see at a glance certain qualitative aspects of the solutions—regions in the plane, for example, in which a

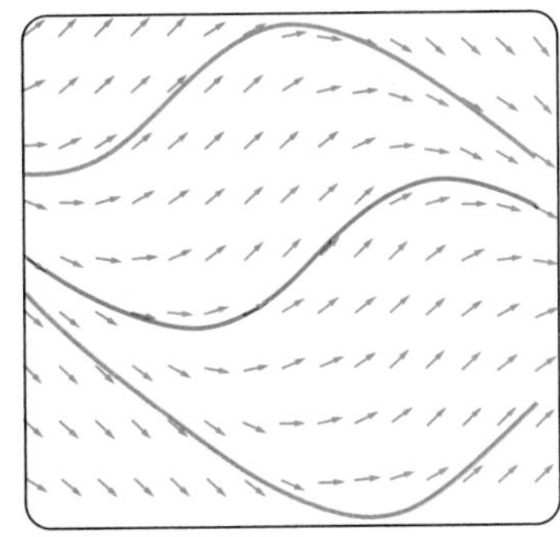

FIGURE 2.1.2 Solution curves following flow of a direction field

solution exhibits an unusual behavior. A single solution curve that passes through a direction field must follow the flow pattern of the field; it is tangent to a lineal element when it intersects a point in the grid. Figure 2.1.2 shows a computer-generated direction field of the differential equation $dy/dx = \sin(x + y)$ over a region of the xy-plane. Note how the three solution curves shown in color follow the flow of the field.

EXAMPLE 1 Direction Field

The direction field for the differential equation $dy/dx = 0.2xy$ shown in Figure 2.1.3(a) was obtained by using computer software in which a 5×5 grid of points (mh, nh), m and n integers, was defined by letting $-5 \le m \le 5$, $-5 \le n \le 5$, and $h = 1$. Notice in Figure 2.1.3(a) that at any point along the x-axis ($y = 0$) and the y-axis ($x = 0$), the slopes are $f(x, 0) = 0$ and $f(0, y) = 0$, respectively, so the lineal elements are horizontal. Moreover, observe in the first quadrant that for a fixed value of x the values of $f(x, y) = 0.2xy$ increase as y increases; similarly, for a fixed y the values of $f(x, y) = 0.2xy$ increase as x increases. This means that as both x and y increase, the lineal elements almost become vertical and have positive slope ($f(x, y) = 0.2xy > 0$ for $x > 0$, $y > 0$). In the second quadrant, $|f(x, y)|$ increases as $|x|$ and y increase, so the lineal elements again become almost vertical but this time have negative slope ($f(x, y) = 0.2xy < 0$ for $x < 0$, $y > 0$). Reading from left to right, imagine a solution curve that starts at a point in the second quadrant, moves steeply downward, becomes flat as it passes through the y-axis, and then, as it enters the first quadrant, moves steeply upward—in other words, its shape would be concave upward and similar to a horseshoe. From this it could be surmised that $y \to \infty$ as $x \to \pm\infty$. Now in the third and fourth quadrants, since $f(x, y) = 0.2xy > 0$ and $f(x, y) = 0.2xy < 0$, respectively, the situation is reversed: A solution curve increases and then decreases as we move from left to right. We saw in (1) of Section 1.1 that $y = e^{0.1x^2}$ is an explicit solution of the differential equation $dy/dx = 0.2xy$; you should verify that a one-parameter family of solutions of the same equation is given by $y = ce^{0.1x^2}$. For purposes of comparison with Figure 2.1.3(a) some representative graphs of members of this family are shown in Figure 2.1.3(b). ≡

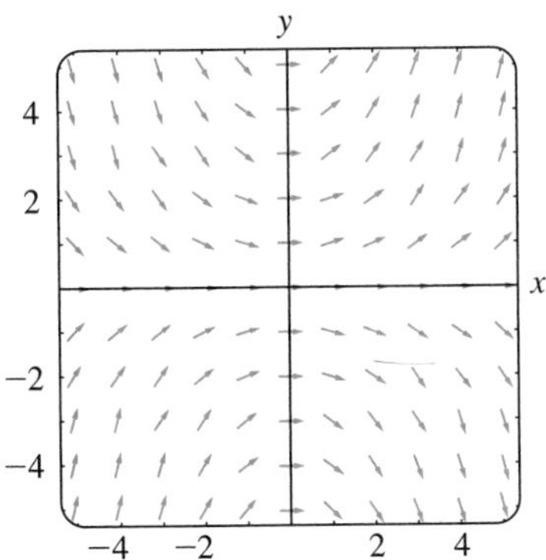

(a) direction field for $dy/dx = 0.2xy$

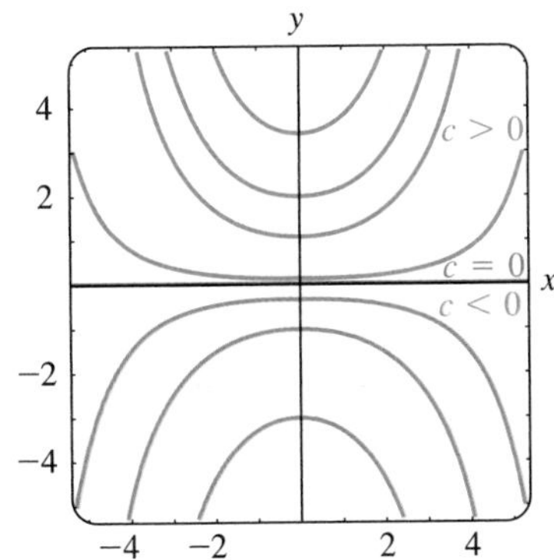

(b) some solution curves in the family $y = ce^{0.1x^2}$

FIGURE 2.1.3 Direction field and solution curves in Example 1

EXAMPLE 2 Direction Field

Use a direction field to sketch an approximate solution curve for the initial-value problem $dy/dx = \sin y$, $y(0) = -\frac{3}{2}$.

SOLUTION Before proceeding, recall that from the continuity of $f(x, y) = \sin y$ and $\partial f/\partial y = \cos y$, Theorem 1.2.1 guarantees the existence of a unique solution curve passing through any specified point (x_0, y_0) in the plane. Now we set our computer software again for a 5×5 rectangular region and specify (because of the initial condition) points in that region with vertical and horizontal separation of $\frac{1}{2}$ unit—that is, at points (mh, nh), $h = \frac{1}{2}$, m and n integers such that $-10 \le m \le 10$, $-10 \le n \le 10$. The result is shown in Figure 2.1.4. Because the right-hand side of $dy/dx = \sin y$ is 0 at $y = 0$, and at $y = -\pi$, the lineal elements are horizontal at all points whose second coordinates are $y = 0$ or $y = -\pi$. It makes sense then that a solution curve passing through the initial point $(0, -\frac{3}{2})$ has the shape shown in the figure. ≡

≡ **Increasing/Decreasing** Interpretation of the derivative dy/dx as a function that gives slope plays the key role in the construction of a direction field. Another telling property of the first derivative will be used next, namely, if $dy/dx > 0$ (or $dy/dx < 0$) for all x in an interval I, then a differentiable function $y = y(x)$ is increasing (or decreasing) on I.

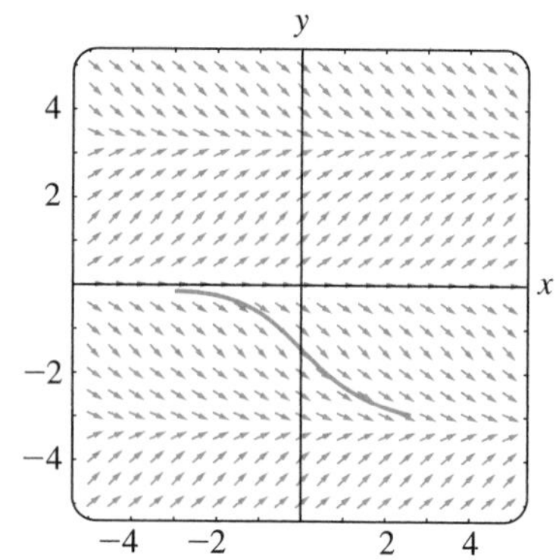

FIGURE 2.1.4 Direction field in Example 2 on page 37

REMARKS

Sketching a direction field by hand is straightforward but time consuming; it is probably one of those tasks about which an argument can be made for doing it once or twice in a lifetime, but it is overall most efficiently carried out by means of computer software. Before calculators, PCs, and software the **method of isoclines** was used to facilitate sketching a direction field by hand. For the DE $dy/dx = f(x, y)$, any member of the family of curves $f(x, y) = c$, c a constant, is called an **isocline.** Lineal elements drawn through points on a specific isocline, say, $f(x, y) = c_1$ all have the same slope c_1. In Problem 15 in Exercises 2.1 you have your two opportunities to sketch a direction field by hand.

2.1.2 AUTONOMOUS FIRST-ORDER DEs

Autonomous First-Order DEs In Section 1.1 we divided the class of ordinary differential equations into two types: linear and nonlinear. We now consider briefly another kind of classification of ordinary differential equations, a classification that is of particular importance in the qualitative investigation of differential equations. An ordinary differential equation in which the independent variable does not appear explicitly is said to be **autonomous.** If the symbol x denotes the independent variable, then an autonomous first-order differential equation can be written as $f(y, y') = 0$ or in normal form as

$$\frac{dy}{dx} = f(y). \tag{2}$$

We shall assume throughout that the function f in (2) and its derivative f' are continuous functions of y on some interval I. The first-order equations

$$\overset{f(y)}{\downarrow} \qquad\qquad \overset{f(x,\,y)}{\downarrow}$$

$$\frac{dy}{dx} = 1 + y^2 \quad \text{and} \quad \frac{dy}{dx} = 0.2xy$$

are autonomous and nonautonomous, respectively.

Many differential equations encountered in applications or equations that are models of physical laws that do not change over time are autonomous. As we have already seen in Section 1.3, in an applied context, symbols other than y and x are routinely used to represent the dependent and independent variables. For example, if t represents time then inspection of

$$\frac{dA}{dt} = kA, \qquad \frac{dx}{dt} = kx(n + 1 - x), \qquad \frac{dT}{dt} = k(T - T_m), \qquad \frac{dA}{dt} = 6 - \frac{1}{100}A,$$

where k, n, and T_m are constants, shows that each equation is time independent. Indeed, *all* of the first-order differential equations introduced in Section 1.3 are time independent and so are autonomous.

Critical Points The zeros of the function f in (2) are of special importance. We say that a real number c is a **critical point** of the autonomous differential equation (2) if it is a zero of f—that is, $f(c) = 0$. A critical point is also called an **equilibrium point** or **stationary point.** Now observe that if we substitute the constant function $y(x) = c$ into (2), then both sides of the equation are zero. This means:

If c is a critical point of (2), then $y(x) = c$ is a constant solution of the autonomous differential equation.

A constant solution $y(x) = c$ of (2) is called an **equilibrium solution;** equilibria are the *only* constant solutions of (2).

As was already mentioned, we can tell when a nonconstant solution $y = y(x)$ of (2) is increasing or decreasing by determining the algebraic sign of the derivative dy/dx; in the case of (2) we do this by identifying intervals on the y-axis over which the function $f(y)$ is positive or negative.

EXAMPLE 3 An Autonomous DE

The differential equation

$$\frac{dP}{dt} = P(a - bP),$$

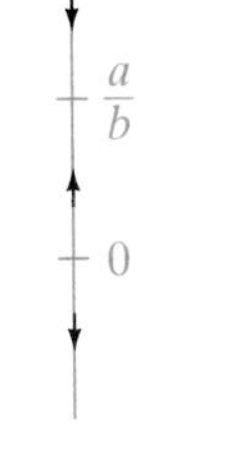

FIGURE 2.1.5 Phase portrait of DE in Example 3

where a and b are positive constants, has the normal form $dP/dt = f(P)$, which is (2) with t and P playing the parts of x and y, respectively, and hence is autonomous. From $f(P) = P(a - bP) = 0$ we see that 0 and a/b are critical points of the equation, so the equilibrium solutions are $P(t) = 0$ and $P(t) = a/b$. By putting the critical points on a vertical line, we divide the line into three intervals defined by $-\infty < P < 0$, $0 < P < a/b$, $a/b < P < \infty$. The arrows on the line shown in Figure 2.1.5 indicate the algebraic sign of $f(P) = P(a - bP)$ on these intervals and whether a nonconstant solution $P(t)$ is increasing or decreasing on an interval. The following table explains the figure.

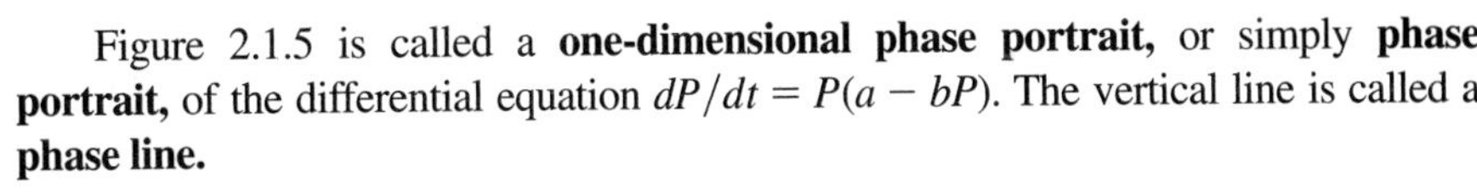

Interval	Sign of $f(P)$	$P(t)$	Arrow
$(-\infty, 0)$	minus	decreasing	points down
$(0, a/b)$	plus	increasing	points up
$(a/b, \infty)$	minus	decreasing	points down

Figure 2.1.5 is called a **one-dimensional phase portrait,** or simply **phase portrait,** of the differential equation $dP/dt = P(a - bP)$. The vertical line is called a **phase line.**

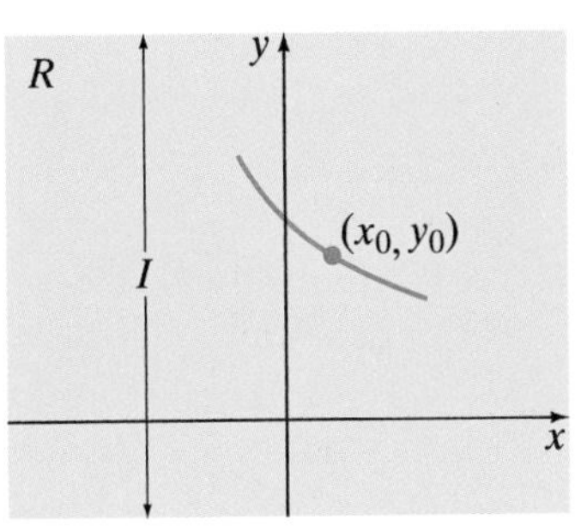

(a) region R

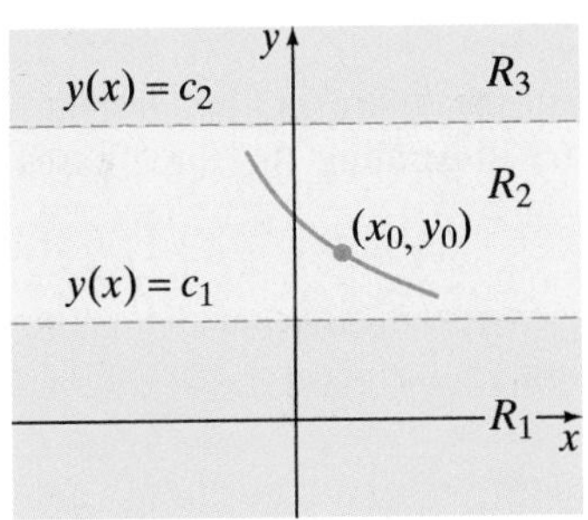

(b) subregions R_1, R_2, and R_3 of R

FIGURE 2.1.6 Lines $y(x) = c_1$ and $y(x) = c_2$ partition R into three horizontal subregions

Solution Curves Without solving an autonomous differential equation, we can usually say a great deal about its solution curves. Since the function f in (2) is independent of the variable x, we may consider f defined for $-\infty < x < \infty$ or for $0 \le x < \infty$. Also, since f and its derivative f' are continuous functions of y on some interval I of the y-axis, the fundamental results of Theorem 1.2.1 hold in some horizontal strip or region R in the xy-plane corresponding to I, and so through any point (x_0, y_0) in R there passes only one solution curve of (2). See Figure 2.1.6(a). For the sake of discussion, let us suppose that (2) possesses exactly two critical points c_1 and c_2 and that $c_1 < c_2$. The graphs of the equilibrium solutions $y(x) = c_1$ and $y(x) = c_2$ are horizontal lines, and these lines partition the region R into three subregions R_1, R_2, and R_3, as illustrated in Figure 2.1.6(b). Without proof here are some conclusions that we can draw about a nonconstant solution $y(x)$ of (2):

- If (x_0, y_0) is in a subregion R_i, $i = 1, 2, 3$, and $y(x)$ is a solution whose graph passes through this point, then $y(x)$ remains in the subregion R_i for all x. As illustrated in Figure 2.1.6(b), the solution $y(x)$ in R_2 is bounded below by c_1 and above by c_2, that is, $c_1 < y(x) < c_2$ for all x. The solution curve stays within R_2 for all x because the graph of a nonconstant solution of (2) cannot cross the graph of either equilibrium solution $y(x) = c_1$ or $y(x) = c_2$. See Problem 33 in Exercises 2.1.
- By continuity of f we must then have either $f(y) > 0$ or $f(y) < 0$ for all x in a subregion R_i, $i = 1, 2, 3$. In other words, $f(y)$ cannot change signs in a subregion. See Problem 33 in Exercises 2.1.

- Since $dy/dx = f(y(x))$ is either positive or negative in a subregion R_i, $i = 1$, 2, 3, a solution $y(x)$ is strictly monotonic—that is, $y(x)$ is either increasing or decreasing in the subregion R_i. Therefore $y(x)$ cannot be oscillatory, nor can it have a relative extremum (maximum or minimum). See Problem 33 in Exercises 2.1.
- If $y(x)$ is *bounded above* by a critical point c_1 (as in subregion R_1 where $y(x) < c_1$ for all x), then the graph of $y(x)$ must approach the graph of the equilibrium solution $y(x) = c_1$ either as $x \to \infty$ or as $x \to -\infty$. If $y(x)$ is *bounded*—that is, bounded above and below by two consecutive critical points (as in subregion R_2 where $c_1 < y(x) < c_2$ for all x)—then the graph of $y(x)$ must approach the graphs of the equilibrium solutions $y(x) = c_1$ and $y(x) = c_2$, one as $x \to \infty$ and the other as $x \to -\infty$. If $y(x)$ is *bounded below* by a critical point (as in subregion R_3 where $c_2 < y(x)$ for all x), then the graph of $y(x)$ must approach the graph of the equilibrium solution $y(x) = c_2$ either as $x \to \infty$ or as $x \to -\infty$. See Problem 34 in Exercises 2.1.

With the foregoing facts in mind, let us reexamine the differential equation in Example 3.

EXAMPLE 4 Example 3 Revisited

The three intervals determined on the P-axis or phase line by the critical points 0 and a/b now correspond in the tP-plane to three subregions defined by:

$$R_1\colon -\infty < P < 0, \qquad R_2\colon 0 < P < a/b, \qquad \text{and} \qquad R_3\colon a/b < P < \infty,$$

where $-\infty < t < \infty$. The phase portrait in Figure 2.1.7 tells us that $P(t)$ is decreasing in R_1, increasing in R_2, and decreasing in R_3. If $P(0) = P_0$ is an initial value, then in R_1, R_2, and R_3 we have, respectively, the following:

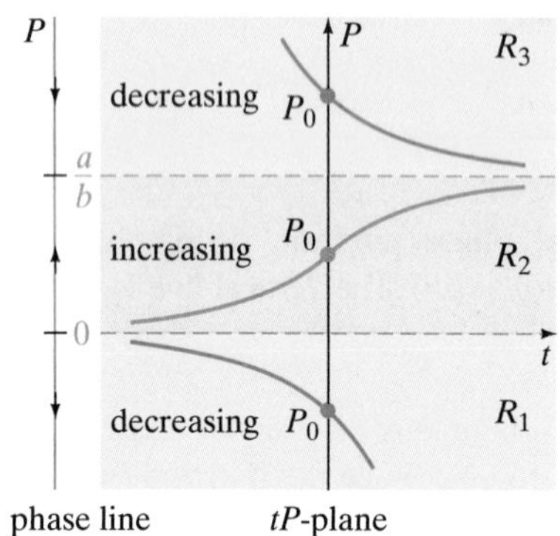

FIGURE 2.1.7 Phase portrait and solution curves in Example 4

(*i*) For $P_0 < 0$, $P(t)$ is bounded above. Since $P(t)$ is decreasing, $P(t)$ decreases without bound for increasing t, and so $P(t) \to 0$ as $t \to -\infty$. This means that the negative t-axis, the graph of the equilibrium solution $P(t) = 0$, is a horizontal asymptote for a solution curve.

(*ii*) For $0 < P_0 < a/b$, $P(t)$ is bounded. Since $P(t)$ is increasing, $P(t) \to a/b$ as $t \to \infty$ and $P(t) \to 0$ as $t \to -\infty$. The graphs of the two equilibrium solutions, $P(t) = 0$ and $P(t) = a/b$, are horizontal lines that are horizontal asymptotes for any solution curve starting in this subregion.

(*iii*) For $P_0 > a/b$, $P(t)$ is bounded below. Since $P(t)$ is decreasing, $P(t) \to a/b$ as $t \to \infty$. The graph of the equilibrium solution $P(t) = a/b$ is a horizontal asymptote for a solution curve.

In Figure 2.1.7 the phase line is the P-axis in the tP-plane. For clarity the original phase line from Figure 2.1.5 is reproduced to the left of the plane in which the subregions R_1, R_2, and R_3 are shaded. The graphs of the equilibrium solutions $P(t) = a/b$ and $P(t) = 0$ (the t-axis) are shown in the figure as blue dashed lines; the solid graphs represent typical graphs of $P(t)$ illustrating the three cases just discussed. ≡

In a subregion such as R_1 in Example 4, where $P(t)$ is decreasing and unbounded below, we must necessarily have $P(t) \to -\infty$. Do *not* interpret this last statement to mean $P(t) \to -\infty$ as $t \to \infty$; we could have $P(t) \to -\infty$ as $t \to T$, where $T > 0$ is a finite number that depends on the initial condition $P(t_0) = P_0$. Thinking in dynamic terms, $P(t)$ could "blow up" in finite time; thinking graphically, $P(t)$ could have a vertical asymptote at $t = T > 0$. A similar remark holds for the subregion R_3.

The differential equation $dy/dx = \sin y$ in Example 2 is autonomous and has an infinite number of critical points, since $\sin y = 0$ at $y = n\pi$, n an integer. Moreover, we now know that because the solution $y(x)$ that passes through $(0, -\frac{3}{2})$ is bounded above and below by two consecutive critical points ($-\pi < y(x) < 0$) and is decreasing ($\sin y < 0$ for $-\pi < y < 0$), the graph of $y(x)$ must approach the graphs of the equilibrium solutions as horizontal asymptotes: $y(x) \to -\pi$ as $x \to \infty$ and $y(x) \to 0$ as $x \to -\infty$.

EXAMPLE 5 Solution Curves of an Autonomous DE

The autonomous equation $dy/dx = (y - 1)^2$ possesses the single critical point 1. From the phase portrait in Figure 2.1.8(a) we conclude that a solution $y(x)$ is an increasing function in the subregions defined by $-\infty < y < 1$ and $1 < y < \infty$, where $-\infty < x < \infty$. For an initial condition $y(0) = y_0 < 1$, a solution $y(x)$ is increasing and bounded above by 1, and so $y(x) \to 1$ as $x \to \infty$; for $y(0) = y_0 > 1$ a solution $y(x)$ is increasing and unbounded.

Now $y(x) = 1 - 1/(x + c)$ is a one-parameter family of solutions of the differential equation. (See Problem 4 in Exercises 2.2.) A given initial condition determines a value for c. For the initial conditions, say, $y(0) = -1 < 1$ and $y(0) = 2 > 1$, we find, in turn, that $y(x) = 1 - 1/(x + \frac{1}{2})$, and $y(x) = 1 - 1/(x - 1)$. As shown in Figures 2.1.8(b) and 2.1.8(c), the graph of each of these rational functions possesses

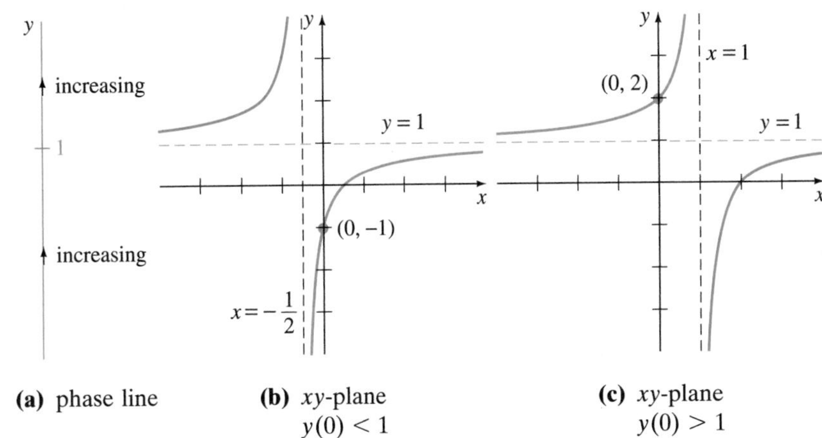

FIGURE 2.1.8 Behavior of solutions near $y = 1$ in Example 5

a vertical asymptote. But bear in mind that the solutions of the IVPs

$$\frac{dy}{dx} = (y - 1)^2, \quad y(0) = -1 \qquad \text{and} \qquad \frac{dy}{dx} = (y - 1)^2, \quad y(0) = 2$$

are defined on special intervals. They are, respectively,

$$y(x) = 1 - \frac{1}{x + \frac{1}{2}}, \quad -\tfrac{1}{2} < x < \infty \qquad \text{and} \qquad y(x) = 1 - \frac{1}{x - 1}, \quad -\infty < x < 1.$$

The solution curves are the portions of the graphs in Figures 2.1.8(b) and 2.1.8(c) shown in blue. As predicted by the phase portrait, for the solution curve in Figure 2.1.8(b), $y(x) \to 1$ as $x \to \infty$; for the solution curve in Figure 2.1.8(c), $y(x) \to \infty$ as $x \to 1$ from the left. ≡

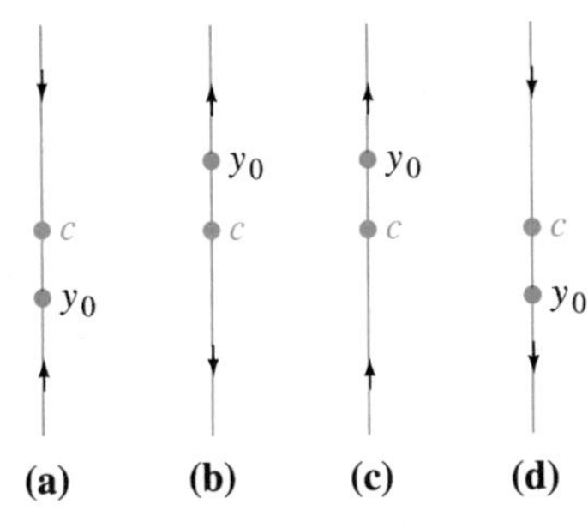

FIGURE 2.1.9 Critical point c is an attractor in (a), a repeller in (b), and semi-stable in (c) and (d).

≡ Attractors and Repellers Suppose that $y(x)$ is a nonconstant solution of the autonomous differential equation given in (1) and that c is a critical point of the DE. There are basically three types of behavior that $y(x)$ can exhibit near c. In Figure 2.1.9 we have placed c on four vertical phase lines. When both arrowheads on either side of the dot labeled c point *toward* c, as in Figure 2.1.9(a), all solutions $y(x)$ of (1) that start from an initial point (x_0, y_0) sufficiently near c exhibit the asymptotic behavior $\lim_{x\to\infty} y(x) = c$. For this reason the critical point c is said to be

asymptotically stable. Using a physical analogy, a solution that starts near c is like a charged particle that, over time, is drawn to a particle of opposite charge, and so c is also referred to as an **attractor.** When both arrowheads on either side of the dot labeled c point *away* from c, as in Figure 2.1.9(b), all solutions $y(x)$ of (1) that start from an initial point (x_0, y_0) move away from c as x increases. In this case the critical point c is said to be **unstable.** An unstable critical point is also called a **repeller,** for obvious reasons. The critical point c illustrated in Figures 2.1.9(c) and 2.1.9(d) is neither an attractor nor a repeller. But since c exhibits characteristics of both an attractor and a repeller—that is, a solution starting from an initial point (x_0, y_0) sufficiently near c is attracted to c from one side and repelled from the other side—we say that the critical point c is **semi-stable.** In Example 3 the critical point a/b is asymptotically stable (an attractor) and the critical point 0 is unstable (a repeller). The critical point 1 in Example 5 is semi-stable.

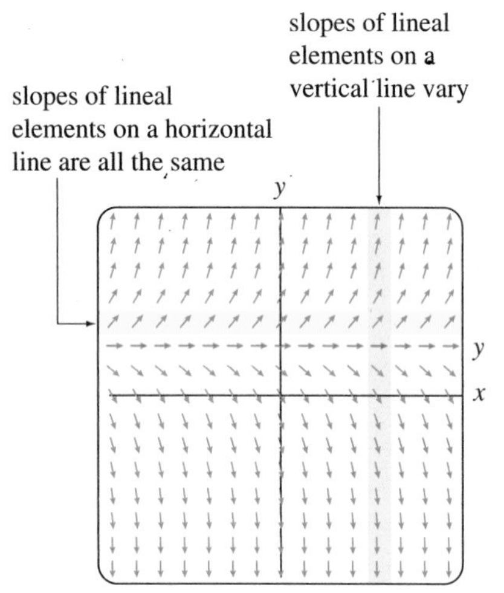

FIGURE 2.1.10 Direction field for an autonomous DE

Autonomous DEs and Direction Fields If a first-order differential equation is autonomous, then we see from the right-hand side of its normal form $dy/dx = f(y)$ that slopes of lineal elements through points in the rectangular grid used to construct a direction field for the DE depend solely on the y-coordinate of the points. Put another way, lineal elements passing through points on any *horizontal* line must all have the same slope and therefore are parallel; slopes of lineal elements along any *vertical* line will, of course, vary. These facts are apparent from inspection of the horizontal yellow strip and vertical blue strip in Figure 2.1.10. The figure exhibits a direction field for the autonomous equation $dy/dx = 2(y - 1)$. The red lineal elements in Figure 2.1.10 have zero slope because they lie along the graph of the equilibrium solution $y = 1$.

Translation Property You may recall from precalculus mathematics that the graph of a function $y = f(x - k)$, where k is a constant, is the graph of $y = f(x)$ rigidly translated or shifted horizontally along the x-axis by an amount $|k|$; the translation is to the right if $k > 0$ and to the left if $k < 0$. It turns out that under the conditions stipulated for (2), solution curves of an autonomous first-order DE are related by the concept of translation. To see this, let's consider the differential equation $dy/dx = y(3 - y)$, which is a special case of the autonomous equation considered in Examples 3 and 4. Because $y = 0$ and $y = 3$ are equilibrium solutions of the DE, their graphs divide the xy-plane into three subregions R_1, R_2, and R_3:

$$R_1: -\infty < y < 0, \quad R_2: 0 < y < 3, \quad \text{and} \quad R_3: 3 < y < \infty.$$

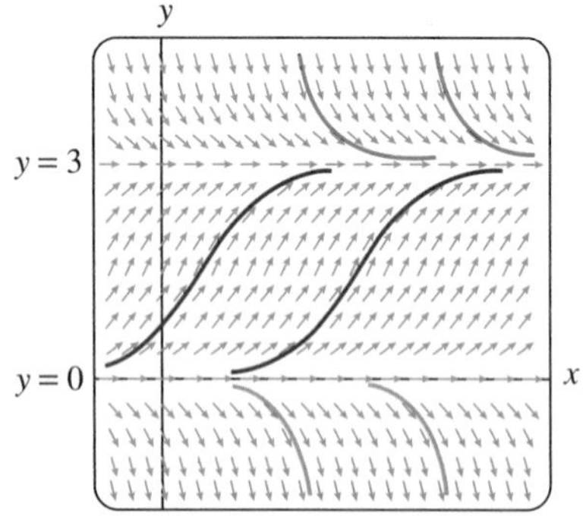

FIGURE 2.1.11 Translated solution curves of an autonomous DE

In Figure 2.1.11 we have superimposed on a direction field of the DE six solutions curves. The figure illustrates that all solution curves of the same color, that is, solution curves lying within a particular subregion R_i, all look alike. This is no coincidence but is a natural consequence of the fact that lineal elements passing through points on any horizontal line are parallel. That said, the following **translation property** of an automonous DE should make sense:

If $y(x)$ is a solution of an autonomous differential equation $dy/dx = f(y)$, then $y_1(x) = y(x - k)$, k a constant, is also a solution.

Thus, if $y(x)$ is a solution of the initial-value problem $dy/dx = f(y)$, $y(0) = y_0$, then $y_1(x) = y(x - x_0)$ is a solution of the IVP $dy/dx = f(y)$, $y(x_0) = y_0$. For example, it is easy to verify that $y(x) = e^x$, $-\infty < x < \infty$, is a solution of the IVP $dy/dx = y$, $y(0) = 1$ and so a solution $y_1(x)$ of, say, $dy/dx = y$, $y(5) = 1$ is $y(x) = e^x$ translated 5 units to the right:

$$y_1(x) = y(x - 5) = e^{x-5}, \ -\infty < x < \infty.$$

EXERCISES 2.1

Answers to selected odd-numbered problems begin on page ANS-1.

2.1.1 DIRECTION FIELDS

In Problems 1–4 reproduce the given computer-generated direction field. Then sketch, by hand, an approximate solution curve that passes through each of the indicated points. Use different colored pencils for each solution curve.

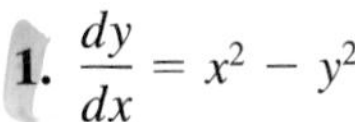

1. $\dfrac{dy}{dx} = x^2 - y^2$

(a) $y(-2) = 1$ **(b)** $y(3) = 0$

(c) $y(0) = 2$ **(d)** $y(0) = 0$

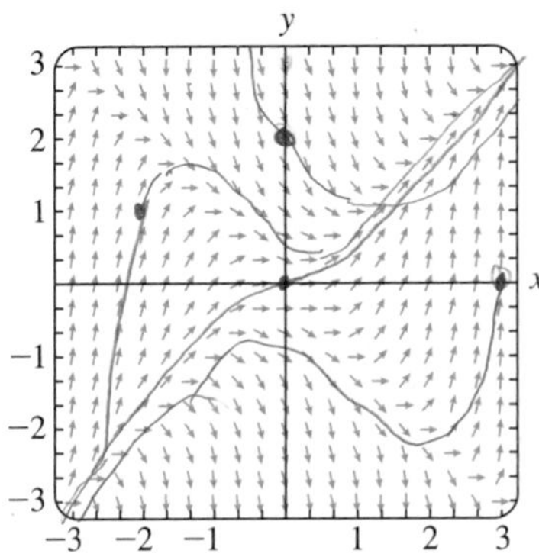

FIGURE 2.1.12 Direction field for Problem 1

2. $\dfrac{dy}{dx} = e^{-0.01xy^2}$

(a) $y(-6) = 0$ **(b)** $y(0) = 1$

(c) $y(0) = -4$ **(d)** $y(8) = -4$

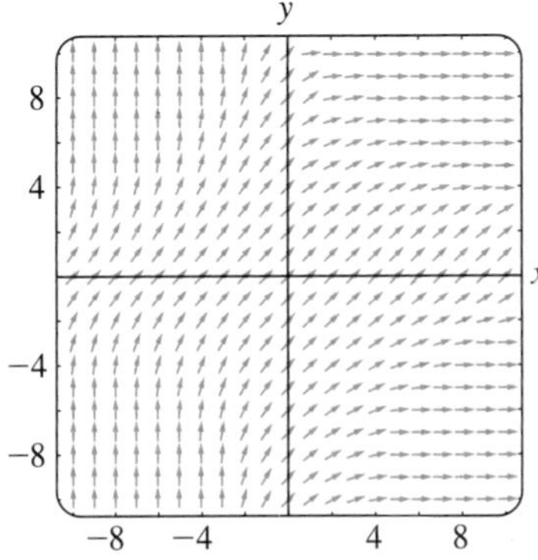

FIGURE 2.1.13 Direction field for Problem 2

3. $\dfrac{dy}{dx} = 1 - xy$

(a) $y(0) = 0$ **(b)** $y(-1) = 0$

(c) $y(2) = 2$ **(d)** $y(0) = -4$

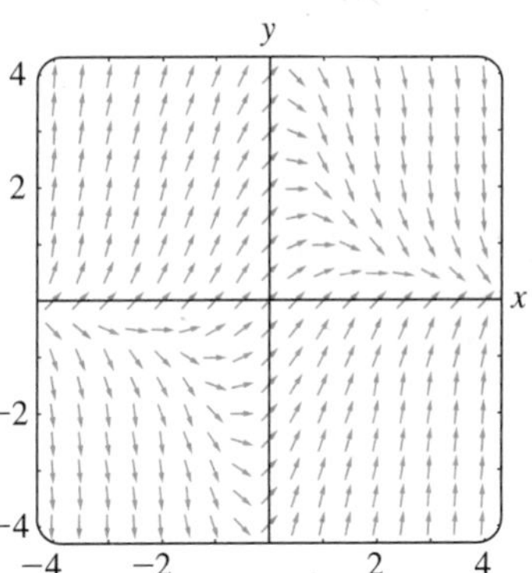

FIGURE 2.1.14 Direction field for Problem 3

4. $\dfrac{dy}{dx} = (\sin x)\cos y$

(a) $y(0) = 1$ **(b)** $y(1) = 0$

(c) $y(3) = 3$ **(d)** $y(0) = -\frac{5}{2}$

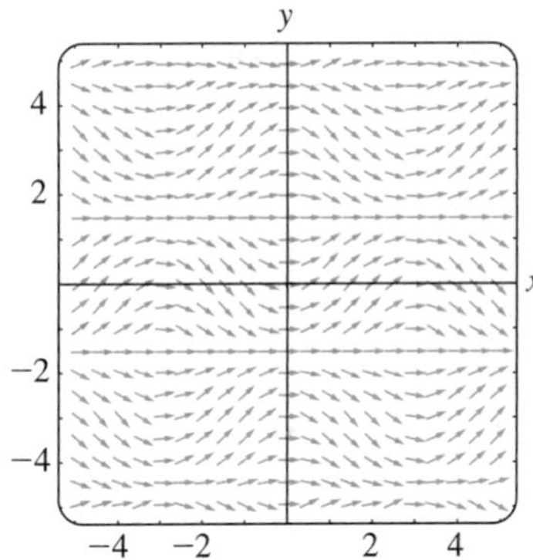

FIGURE 2.1.15 Direction field for Problem 4

In Problems 5–12 use computer software to obtain a direction field for the given differential equation. By hand, sketch an approximate solution curve passing through each of the given points.

5. $y' = x$

(a) $y(0) = 0$

(b) $y(0) = -3$

6. $y' = x + y$

(a) $y(-2) = 2$

(b) $y(1) = -3$

7. $y\dfrac{dy}{dx} = -x$

(a) $y(1) = 1$

(b) $y(0) = 4$

8. $\dfrac{dy}{dx} = \dfrac{1}{y}$

(a) $y(0) = 1$

(b) $y(-2) = -1$

9. $\dfrac{dy}{dx} = 0.2x^2 + y$

(a) $y(0) = \frac{1}{2}$

(b) $y(2) = -1$

10. $\dfrac{dy}{dx} = xe^y$

(a) $y(0) = -2$

(b) $y(1) = 2.5$

11. $y' = y - \cos\frac{\pi}{2}x$

(a) $y(2) = 2$

(b) $y(-1) = 0$

12. $\frac{dy}{dx} = 1 - \frac{y}{x}$

(a) $y\left(-\frac{1}{2}\right) = 2$

(b) $y\left(\frac{3}{2}\right) = 0$

In Problems 13 and 14 the given figure represents the graph of $f(y)$ and $f(x)$, respectively. By hand, sketch a direction field over an appropriate grid for $dy/dx = f(y)$ (Problem 13) and then for $dy/dx = f(x)$ (Problem 14).

13.

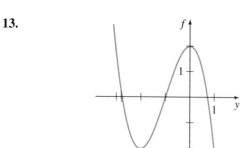

FIGURE 2.1.16 Graph for Problem 13

14.

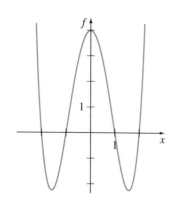

FIGURE 2.1.17 Graph for Problem 14

15. In parts (a) and (b) sketch **isoclines** $f(x, y) = c$ (see the *Remarks* on page 38) for the given differential equation using the indicated values of c. Construct a direction field over a grid by carefully drawing lineal elements with the appropriate slope at chosen points on each isocline. In each case, use this rough direction field to sketch an approximate solution curve for the IVP consisting of the DE and the initial condition $y(0) = 1$.

(a) $dy/dx = x + y$; c an integer satisfying $-5 \le c \le 5$

(b) $dy/dx = x^2 + y^2$; $c = \frac{1}{4}$, $c = 1$, $c = \frac{9}{4}$, $c = 4$

Discussion Problems

16. **(a)** Consider the direction field of the differential equation $dy/dx = x(y - 4)^2 - 2$, but do not use technology to obtain it. Describe the slopes of the lineal elements on the lines $x = 0$, $y = 3$, $y = 4$, and $y = 5$.

(b) Consider the IVP $dy/dx = x(y - 4)^2 - 2$, $y(0) = y_0$, where $y_0 < 4$. Can a solution $y(x) \to \infty$ as $x \to \infty$? Based on the information in part (a), discuss.

17. For a first-order DE $dy/dx = f(x, y)$ a curve in the plane defined by $f(x, y) = 0$ is called a **nullcline** of the equation, since a lineal element at a point on the curve has zero slope. Use computer software to obtain a direction field over a rectangular grid of points for $dy/dx = x^2 - 2y$, and then superimpose the graph of the nullcline $y = \frac{1}{2}x^2$ over the direction field. Discuss the behavior of solution curves in regions of the plane defined by $y < \frac{1}{2}x^2$ and by $y > \frac{1}{2}x^2$. Sketch some approximate solution curves. Try to generalize your observations.

18. **(a)** Identify the nullclines (see Problem 17) in Problems 1, 3, and 4. With a colored pencil, circle any lineal elements in Figures 2.1.12, 2.1.14, and 2.1.15 that you think may be a lineal element at a point on a nullcline.

(b) What are the nullclines of an autonomous first-order DE?

2.1.2 AUTONOMOUS FIRST-ORDER DEs

19. Consider the autonomous first-order differential equation $dy/dx = y - y^3$ and the initial condition $y(0) = y_0$. By hand, sketch the graph of a typical solution $y(x)$ when y_0 has the given values.

(a) $y_0 > 1$ **(b)** $0 < y_0 < 1$

(c) $-1 < y_0 < 0$ **(d)** $y_0 < -1$

20. Consider the autonomous first-order differential equation $dy/dx = y^2 - y^4$ and the initial condition $y(0) = y_0$. By hand, sketch the graph of a typical solution $y(x)$ when y_0 has the given values.

(a) $y_0 > 1$ **(b)** $0 < y_0 < 1$

(c) $-1 < y_0 < 0$ **(d)** $y_0 < -1$

In Problems 21–28 find the critical points and phase portrait of the given autonomous first-order differential equation. Classify each critical point as asymptotically stable, unstable, or semi-stable. By hand, sketch typical solution curves in the regions in the xy-plane determined by the graphs of the equilibrium solutions.

21. $\frac{dy}{dx} = y^2 - 3y$

22. $\frac{dy}{dx} = y^2 - y^3$

23. $\frac{dy}{dx} = (y - 2)^4$

24. $\frac{dy}{dx} = 10 + 3y - y^2$

25. $\frac{dy}{dx} = y^2(4 - y^2)$

26. $\frac{dy}{dx} = y(2 - y)(4 - y)$

27. $\frac{dy}{dx} = y\ln(y + 2)$

28. $\frac{dy}{dx} = \frac{ye^y - 9y}{e^y}$

In Problems 29 and 30 consider the autonomous differential equation $dy/dx = f(y)$, where the graph of f is given. Use the graph to locate the critical points of each differential equation. Sketch a phase portrait of each differential equation. By hand, sketch typical solution curves in the subregions in the xy-plane determined by the graphs of the equilibrium solutions.

29.

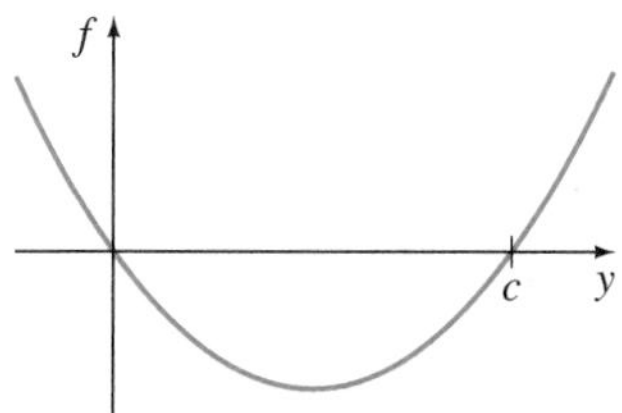

FIGURE 2.1.18 Graph for Problem 29

30.

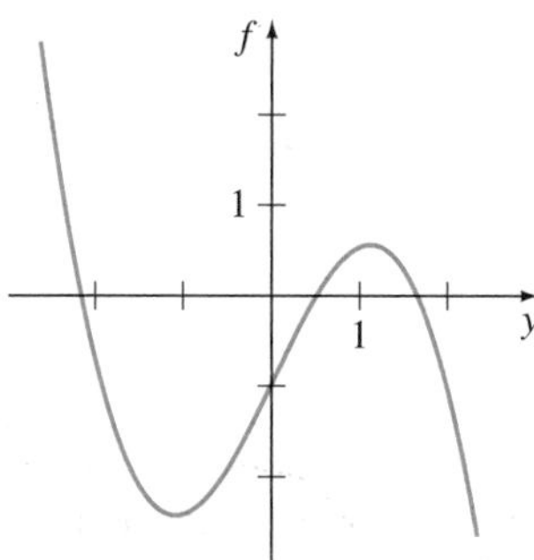

FIGURE 2.1.19 Graph for Problem 30

Discussion Problems

31. Consider the autonomous DE $dy/dx = (2/\pi)y - \sin y$. Determine the critical points of the equation. Discuss a way of obtaining a phase portrait of the equation. Classify the critical points as asymptotically stable, unstable, or semi-stable.

32. A critical point c of an autonomous first-order DE is said to be **isolated** if there exists some open interval that contains c but no other critical point. Can there exist an autonomous DE of the form given in (2) for which *every* critical point is nonisolated? Discuss; do not think profound thoughts.

33. Suppose that $y(x)$ is a nonconstant solution of the autonomous equation $dy/dx = f(y)$ and that c is a critical point of the DE. Discuss: Why can't the graph of $y(x)$ cross the graph of the equilibrium solution $y = c$? Why can't $f(y)$ change signs in one of the subregions discussed on page 39? Why can't $y(x)$ be oscillatory or have a relative extremum (maximum or minimum)?

34. Suppose that $y(x)$ is a solution of the autonomous equation $dy/dx = f(y)$ and is bounded above and below by two consecutive critical points $c_1 < c_2$, as in subregion R_2 of Figure 2.1.6(b). If $f(y) > 0$ in the region, then $\lim_{x\to\infty} y(x) = c_2$. Discuss why there cannot exist a number $L < c_2$ such that $\lim_{x\to\infty} y(x) = L$. As part of your discussion, consider what happens to $y'(x)$ as $x \to \infty$.

35. Using the autonomous equation (2), discuss how it is possible to obtain information about the location of points of inflection of a solution curve.

36. Consider the autonomous DE $dy/dx = y^2 - y - 6$. Use your ideas from Problem 35 to find intervals on the y-axis for which solution curves are concave up and intervals for which solution curves are concave down. Discuss why *each* solution curve of an initial-value problem of the form $dy/dx = y^2 - y - 6$, $y(0) = y_0$, where $-2 < y_0 < 3$, has a point of inflection with the same y-coordinate. What is that y-coordinate? Carefully sketch the solution curve for which $y(0) = -1$. Repeat for $y(2) = 2$.

37. Suppose the autonomous DE in (2) has no critical points. Discuss the behavior of the solutions.

Mathematical Models

38. Population Model The differential equation in Example 3 is a well-known population model. Suppose the DE is changed to

$$\frac{dP}{dt} = P(aP - b),$$

where a and b are positive constants. Discuss what happens to the population P as time t increases.

39. Population Model Another population model is given by

$$\frac{dP}{dt} = kP - h,$$

where h and k are positive constants. For what initial values $P(0) = P_0$ does this model predict that the population will go extinct?

40. Terminal Velocity In Section 1.3 we saw that the autonomous differential equation

$$m\frac{dv}{dt} = mg - kv,$$

where k is a positive constant and g is the acceleration due to gravity, is a model for the velocity v of a body of mass m that is falling under the influence of gravity. Because the term $-kv$ represents air resistance, the velocity of a body falling from a great height does not increase without bound as time t increases. Use a phase portrait of the differential equation to find the limiting, or terminal, velocity of the body. Explain your reasoning.

41. Suppose the model in Problem 40 is modified so that air resistance is proportional to v^2, that is,

$$m\frac{dv}{dt} = mg - kv^2.$$

See Problem 17 in Exercises 1.3. Use a phase portrait to find the terminal velocity of the body. Explain your reasoning.

42. Chemical Reactions When certain kinds of chemicals are combined, the rate at which the new compound is formed is modeled by the autonomous differential equation

$$\frac{dX}{dt} = k(\alpha - X)(\beta - X),$$

where $k > 0$ is a constant of proportionality and $\beta > \alpha > 0$. Here $X(t)$ denotes the number of grams of the new compound formed in time t.

(a) Use a phase portrait of the differential equation to predict the behavior of $X(t)$ as $t \to \infty$.

(b) Consider the case when $\alpha = \beta$. Use a phase portrait of the differential equation to predict the behavior of $X(t)$ as $t \to \infty$ when $X(0) < \alpha$. When $X(0) > \alpha$.

(c) Verify that an explicit solution of the DE in the case when $k = 1$ and $\alpha = \beta$ is $X(t) = \alpha - 1/(t + c)$. Find a solution that satisfies $X(0) = \alpha/2$. Then find a solution that satisfies $X(0) = 2\alpha$. Graph these two solutions. Does the behavior of the solutions as $t \to \infty$ agree with your answers to part (b)?

2.2 SEPARABLE EQUATIONS

REVIEW MATERIAL

- Basic integration formulas (See inside front cover)
- Techniques of integration: integration by parts and partial fraction decomposition
- See also the *Student Resource Manual.*

INTRODUCTION We begin our study of how to solve differential equations with the simplest of all differential equations: first-order equations with separable variables. Because the method in this section and many techniques for solving differential equations involve integration, you are urged to refresh your memory on important formulas (such as $\int du/u$) and techniques (such as integration by parts) by consulting a calculus text.

Solution by Integration Consider the first-order differential equation $dy/dx = f(x, y)$. When f does not depend on the variable y, that is, $f(x, y) = g(x)$, the differential equation

$$\frac{dy}{dx} = g(x) \tag{1}$$

can be solved by integration. If $g(x)$ is a continuous function, then integrating both sides of (1) gives $y = \int g(x)\,dx = G(x) + c$, where $G(x)$ is an antiderivative (indefinite integral) of $g(x)$. For example, if $dy/dx = 1 + e^{2x}$, then its solution is $y = \int(1 + e^{2x})\,dx$ or $y = x + \frac{1}{2}e^{2x} + c$.

A Definition Equation (1), as well as its method of solution, is just a special case when the function f in the normal form $dy/dx = f(x, y)$ can be factored into a function of x times a function of y.

DEFINITION 2.2.1 Separable Equation

A first-order differential equation of the form

$$\frac{dy}{dx} = g(x)h(y)$$

is said to be **separable** or to have **separable variables.**

For example, the equations

$$\frac{dy}{dx} = y^2xe^{3x+4y} \quad \text{and} \quad \frac{dy}{dx} = y + \sin x$$

are separable and nonseparable, respectively. In the first equation we can factor $f(x, y) = y^2xe^{3x+4y}$ as

$$f(x, y) = y^2xe^{3x+4y} = \overset{g(x)}{\overset{\downarrow}{(xe^{3x})}}\overset{h(y)}{\overset{\downarrow}{(y^2e^{4y})}},$$

but in the second equation there is no way of expressing $y + \sin x$ as a product of a function of x times a function of y.

Observe that by dividing by the function $h(y)$, we can write a separable equation $dy/dx = g(x)h(y)$ as

$$p(y)\frac{dy}{dx} = g(x), \tag{2}$$

where, for convenience, we have denoted $1/h(y)$ by $p(y)$. From this last form we can see immediately that (2) reduces to (1) when $h(y) = 1$.

Now if $y = \phi(x)$ represents a solution of (2), we must have $p(\phi(x))\phi'(x) = g(x)$, and therefore

$$\int p(\phi(x))\phi'(x)\,dx = \int g(x)\,dx. \tag{3}$$

But $dy = \phi'(x)\,dx$, and so (3) is the same as

$$\int p(y)\,dy = \int g(x)\,dx \qquad \text{or} \qquad H(y) = G(x) + c, \tag{4}$$

where $H(y)$ and $G(x)$ are antiderivatives of $p(y) = 1/h(y)$ and $g(x)$, respectively.

Method of Solution Equation (4) indicates the procedure for solving separable equations. A one-parameter family of solutions, usually given implicitly, is obtained by integrating both sides of $p(y)\,dy = g(x)\,dx$.

Note There is no need to use two constants in the integration of a separable equation, because if we write $H(y) + c_1 = G(x) + c_2$, then the difference $c_2 - c_1$ can be replaced by a single constant c, as in (4). In many instances throughout the chapters that follow, we will relabel constants in a manner convenient to a given equation. For example, multiples of constants or combinations of constants can sometimes be replaced by a single constant.

EXAMPLE 1 Solving a Separable DE

Solve $(1 + x)\,dy - y\,dx = 0$.

SOLUTION Dividing by $(1 + x)y$, we can write $dy/y = dx/(1 + x)$, from which it follows that

$$\int \frac{dy}{y} = \int \frac{dx}{1+x}$$

$$\ln|y| = \ln|1 + x| + c_1$$

$$y = e^{\ln|1+x|+c_1} = e^{\ln|1+x|} \cdot e^{c_1} \qquad \leftarrow \text{laws of exponents}$$

$$= |1 + x|\,e^{c_1} \qquad \leftarrow \begin{cases} |1+x| = 1 + x, & x \geq -1 \\ |1+x| = -(1+x), & x < -1 \end{cases}$$

$$= \pm e^{c_1}(1 + x).$$

Relabeling $\pm e^{c_1}$ as c then gives $y = c(1 + x)$.

ALTERNATIVE SOLUTION Because each integral results in a logarithm, a judicious choice for the constant of integration is $\ln|c|$ rather than c. Rewriting the second line of the solution as $\ln|y| = \ln|1 + x| + \ln|c|$ enables us to combine the terms on the right-hand side by the properties of logarithms. From $\ln|y| = \ln|c(1 + x)|$ we immediately get $y = c(1 + x)$. Even if the indefinite integrals are not *all* logarithms, it may still be advantageous to use $\ln|c|$. However, no firm rule can be given. ≡

In Section 1.1 we saw that a solution curve may be only a segment or an arc of the graph of an implicit solution $G(x, y) = 0$.

EXAMPLE 2 Solution Curve

Solve the initial-value problem $\dfrac{dy}{dx} = -\dfrac{x}{y}$, $y(4) = -3$.

SOLUTION Rewriting the equation as $y\,dy = -x\,dx$, we get

$$\int y\,dy = -\int x\,dx \quad \text{and} \quad \frac{y^2}{2} = -\frac{x^2}{2} + c_1.$$

We can write the result of the integration as $x^2 + y^2 = c^2$ by replacing the constant $2c_1$ by c^2. This solution of the differential equation represents a family of concentric circles centered at the origin.

Now when $x = 4$, $y = -3$, so $16 + 9 = 25 = c^2$. Thus the initial-value problem determines the circle $x^2 + y^2 = 25$ with radius 5. Because of its simplicity we can solve this implicit solution for an explicit solution that satisfies the initial condition. We saw this solution as $y = \phi_2(x)$ or $y = -\sqrt{25 - x^2}$, $-5 < x < 5$ in Example 3 of Section 1.1. A solution curve is the graph of a differentiable function. In this case the solution curve is the lower semicircle, shown in dark blue in Figure 2.2.1 containing the point $(4, -3)$. ≡

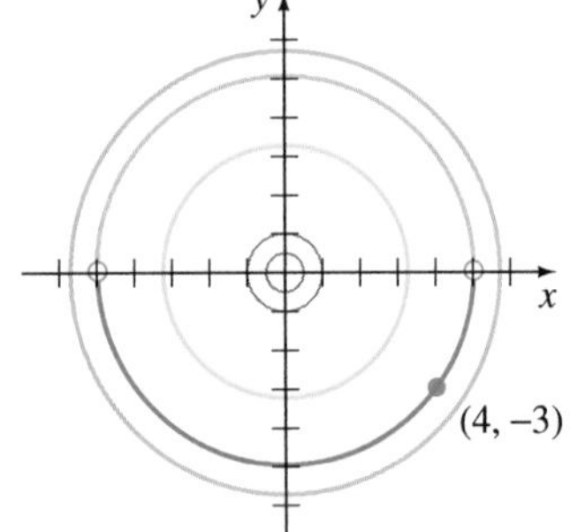

FIGURE 2.2.1 Solution curve for the IVP in Example 2

≡ **Losing a Solution** Some care should be exercised in separating variables, since the variable divisors could be zero at a point. Specifically, if r is a zero of the function $h(y)$, then substituting $y = r$ into $dy/dx = g(x)h(y)$ makes both sides zero; in other words, $y = r$ is a constant solution of the differential equation. But after variables are separated, the left-hand side of $\dfrac{dy}{h(y)} = g(x)\,dx$ is undefined at r. As a consequence, $y = r$ might not show up in the family of solutions that are obtained after integration and simplification. Recall that such a solution is called a singular solution.

EXAMPLE 3 Losing a Solution

Solve $\dfrac{dy}{dx} = y^2 - 4$.

SOLUTION We put the equation in the form

$$\frac{dy}{y^2 - 4} = dx \quad \text{or} \quad \left[\frac{\frac{1}{4}}{y - 2} - \frac{\frac{1}{4}}{y + 2}\right] dy = dx. \tag{5}$$

The second equation in (5) is the result of using partial fractions on the left-hand side of the first equation. Integrating and using the laws of logarithms gives

$$\frac{1}{4}\ln|y - 2| - \frac{1}{4}\ln|y + 2| = x + c_1$$

$$\ln\left|\frac{y - 2}{y + 2}\right| = 4x + c_2 \quad \text{or} \quad \frac{y - 2}{y + 2} = \pm e^{4x + c_2}.$$

Here we have replaced $4c_1$ by c_2. Finally, after replacing $\pm e^{c_2}$ by c and solving the last equation for y, we get the one-parameter family of solutions

$$y = 2\frac{1 + ce^{4x}}{1 - ce^{4x}}. \tag{6}$$

Now if we factor the right-hand side of the differential equation as $dy/dx = (y - 2)(y + 2)$, we know from the discussion of critical points in Section 2.1 that $y = 2$ and $y = -2$ are two constant (equilibrium) solutions. The solution $y = 2$ is a member of the family of solutions defined by (6) corresponding to the value $c = 0$. However, $y = -2$ is a singular solution; it cannot be obtained from (6) for any choice of the parameter c. This latter solution was lost early on in the solution process. Inspection of (5) clearly indicates that we must preclude $y = \pm 2$ in these steps. ≡

EXAMPLE 4 An Initial-Value Problem

Solve $(e^{2y} - y)\cos x \dfrac{dy}{dx} = e^y \sin 2x, \quad y(0) = 0.$

SOLUTION Dividing the equation by $e^y \cos x$ gives

$$\frac{e^{2y} - y}{e^y}\,dy = \frac{\sin 2x}{\cos x}\,dx.$$

Before integrating, we use termwise division on the left-hand side and the trigonometric identity $\sin 2x = 2 \sin x \cos x$ on the right-hand side. Then

integration by parts → $$\int (e^y - ye^{-y})\,dy = 2\int \sin x\,dx$$

yields $$e^y + ye^{-y} + e^{-y} = -2\cos x + c. \tag{7}$$

The initial condition $y = 0$ when $x = 0$ implies $c = 4$. Thus a solution of the initial-value problem is

$$e^y + ye^{-y} + e^{-y} = 4 - 2\cos x. \tag{8}$$ ≡

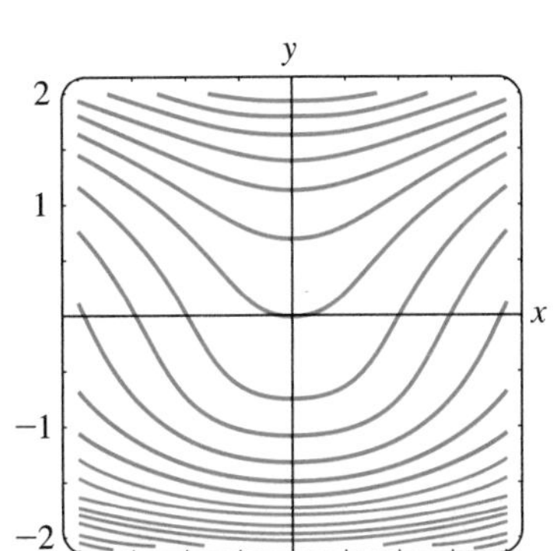

FIGURE 2.2.2 Level curves of $G(x, y) = e^y + ye^{-y} + e^{-y} + 2\cos x$

≡ **Use of Computers** The *Remarks* at the end of Section 1.1 mentioned that it may be difficult to use an implicit solution $G(x, y) = 0$ to find an explicit solution $y = \phi(x)$. Equation (8) shows that the task of solving for y in terms of x may present more problems than just the drudgery of symbol pushing—sometimes it simply cannot be done! Implicit solutions such as (8) are somewhat frustrating; neither the graph of the equation nor an interval over which a solution satisfying $y(0) = 0$ is defined is apparent. The problem of "seeing" what an implicit solution looks like can be overcome in some cases by means of technology. One way* of proceeding is to use the contour plot application of a computer algebra system (CAS). Recall from multivariate calculus that for a function of two variables $z = G(x, y)$ the *two-dimensional* curves defined by $G(x, y) = c$, where c is constant, are called the *level curves* of the function. With the aid of a CAS, some of the level curves of the function $G(x, y) = e^y + ye^{-y} + e^{-y} + 2\cos x$ have been reproduced in Figure 2.2.2. The family of solutions defined by (7) is the level curves $G(x, y) = c$. Figure 2.2.3 illustrates the level curve $G(x, y) = 4$, which is the particular solution (8), in blue color. The other curve in Figure 2.2.3 is the level curve $G(x, y) = 2$, which is the member of the family $G(x, y) = c$ that satisfies $y(\pi/2) = 0$.

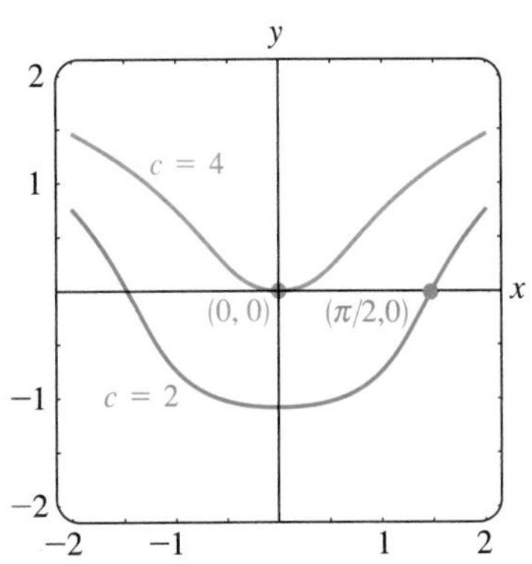

FIGURE 2.2.3 Level curves $c = 2$ and $c = 4$

If an initial condition leads to a particular solution by yielding a specific value of the parameter c in a family of solutions for a first-order differential equation, there is

*In Section 2.6 we will discuss several other ways of proceeding that are based on the concept of a numerical solver.

a natural inclination for most students (and instructors) to relax and be content. However, a solution of an initial-value problem might not be unique. We saw in Example 4 of Section 1.2 that the initial-value problem

$$\frac{dy}{dx} = xy^{1/2}, \quad y(0) = 0 \tag{9}$$

has at least two solutions, $y = 0$ and $y = \frac{1}{16}x^4$. We are now in a position to solve the equation. Separating variables and integrating $y^{-1/2}\,dy = x\,dx$ gives

$$2y^{1/2} = \frac{x^2}{2} + c_1 \quad \text{or} \quad y = \left(\frac{x^2}{4} + c\right)^2, \quad c \geq 0.$$

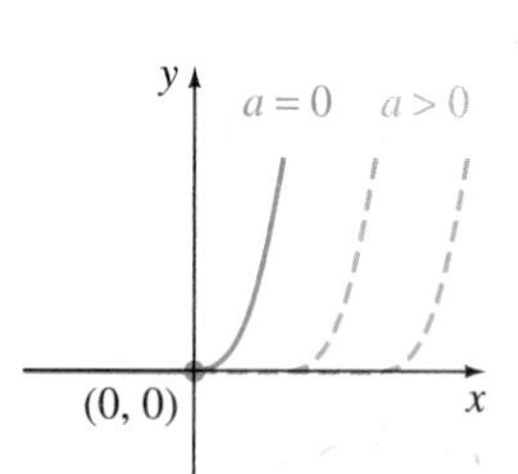

FIGURE 2.2.4 Piecewise-defined solutions of (9)

When $x = 0$, then $y = 0$, so necessarily, $c = 0$. Therefore $y = \frac{1}{16}x^4$. The trivial solution $y = 0$ was lost by dividing by $y^{1/2}$. In addition, the initial-value problem (9) possesses infinitely many more solutions, since for any choice of the parameter $a \geq 0$ the piecewise-defined function

$$y = \begin{cases} 0, & x < a \\ \frac{1}{16}(x^2 - a^2)^2, & x \geq a \end{cases}$$

satisfies both the differential equation and the initial condition. See Figure 2.2.4.

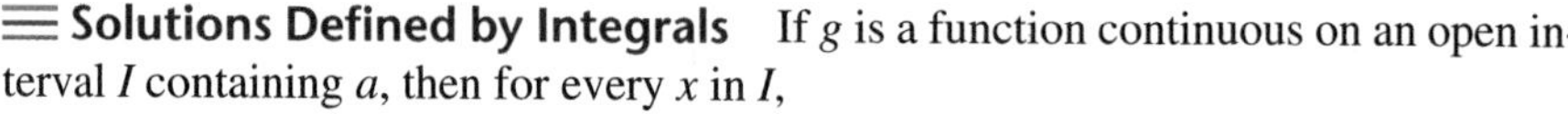

Solutions Defined by Integrals If g is a function continuous on an open interval I containing a, then for every x in I,

$$\frac{d}{dx}\int_a^x g(t)\,dt = g(x).$$

You might recall that the foregoing result is one of the two forms of the fundamental theorem of calculus. In other words, $\int_a^x g(t)\,dt$ is an antiderivative of the function g. There are times when this form is convenient in solving DEs. For example, if g is continuous on an interval I containing x_0 and x, then a solution of the simple initial-value problem $dy/dx = g(x)$, $y(x_0) = y_0$, that is defined on I is given by

$$y(x) = y_0 + \int_{x_0}^x g(t)\,dt$$

You should verify that $y(x)$ defined in this manner satisfies the initial condition. Since an antiderivative of a continuous function g cannot always be expressed in terms of elementary functions, this might be the best we can do in obtaining an explicit solution of an IVP. The next example illustrates this idea.

EXAMPLE 5 An Initial-Value Problem

Solve $\dfrac{dy}{dx} = e^{-x^2}$, $y(3) = 5$.

SOLUTION The function $g(x) = e^{-x^2}$ is continuous on $(-\infty, \infty)$, but its antiderivative is not an elementary function. Using t as dummy variable of integration, we can write

$$\int_3^x \frac{dy}{dt}\,dt = \int_3^x e^{-t^2}\,dt$$

$$y(t)\Big]_3^x = \int_3^x e^{-t^2}\,dt$$

$$y(x) - y(3) = \int_3^x e^{-t^2}\,dt$$

$$y(x) = y(3) + \int_3^x e^{-t^2}\,dt.$$

Using the initial condition $y(3) = 5$, we obtain the solution

$$y(x) = 5 + \int_3^x e^{-t^2}\,dt.$$

The procedure demonstrated in Example 5 works equally well on separable equations $dy/dx = g(x)f(y)$ where, say, $f(y)$ possesses an elementary antiderivative but $g(x)$ does not possess an elementary antiderivative. See Problems 29 and 30 in Exercises 2.2.

REMARKS

(*i*) As we have just seen in Example 5, some simple functions do not possess an antiderivative that is an elementary function. Integrals of these kinds of functions are called **nonelementary.** For example, $\int_3^x e^{-t^2}\,dt$ and $\int \sin x^2\,dx$ are nonelementary integrals. We will run into this concept again in Section 2.3.

(*ii*) In some of the preceding examples we saw that the constant in the one-parameter family of solutions for a first-order differential equation can be relabeled when convenient. Also, it can easily happen that two individuals solving the same equation correctly arrive at dissimilar expressions for their answers. For example, by separation of variables we can show that one-parameter families of solutions for the DE $(1 + y^2)\,dx + (1 + x^2)\,dy = 0$ are

$$\arctan x + \arctan y = c \qquad \text{or} \qquad \frac{x + y}{1 - xy} = c.$$

As you work your way through the next several sections, bear in mind that families of solutions may be equivalent in the sense that one family may be obtained from another by either relabeling the constant or applying algebra and trigonometry. See Problems 27 and 28 in Exercises 2.2.

EXERCISES 2.2

Answers to selected odd-numbered problems begin on page ANS-2.

In Problems 1–22 solve the given differential equation by separation of variables.

1. $\dfrac{dy}{dx} = \sin 5x$
2. $\dfrac{dy}{dx} = (x + 1)^2$
3. $dx + e^{3x}dy = 0$
4. $dy - (y - 1)^2dx = 0$
5. $x\dfrac{dy}{dx} = 4y$
6. $\dfrac{dy}{dx} + 2xy^2 = 0$
7. $\dfrac{dy}{dx} = e^{3x+2y}$
8. $e^xy\dfrac{dy}{dx} = e^{-y} + e^{-2x-y}$
9. $y \ln x \dfrac{dx}{dy} = \left(\dfrac{y + 1}{x}\right)^2$
10. $\dfrac{dy}{dx} = \left(\dfrac{2y + 3}{4x + 5}\right)^2$
11. $\csc y\,dx + \sec^2x\,dy = 0$
12. $\sin 3x\,dx + 2y\cos^3 3x\,dy = 0$
13. $(e^y + 1)^2e^{-y}\,dx + (e^x + 1)^3e^{-x}\,dy = 0$
14. $x(1 + y^2)^{1/2}\,dx = y(1 + x^2)^{1/2}\,dy$
15. $\dfrac{dS}{dr} = kS$
16. $\dfrac{dQ}{dt} = k(Q - 70)$
17. $\dfrac{dP}{dt} = P - P^2$
18. $\dfrac{dN}{dt} + N = Nte^{t+2}$
19. $\dfrac{dy}{dx} = \dfrac{xy + 3x - y - 3}{xy - 2x + 4y - 8}$
20. $\dfrac{dy}{dx} = \dfrac{xy + 2y - x - 2}{xy - 3y + x - 3}$
21. $\dfrac{dy}{dx} = x\sqrt{1 - y^2}$
22. $(e^x + e^{-x})\dfrac{dy}{dx} = y^2$

In Problems 23–28 find an explicit solution of the given initial-value problem.

23. $\dfrac{dx}{dt} = 4(x^2 + 1), \quad x(\pi/4) = 1$
24. $\dfrac{dy}{dx} = \dfrac{y^2 - 1}{x^2 - 1}, \quad y(2) = 2$
25. $x^2\dfrac{dy}{dx} = y - xy, \quad y(-1) = -1$

26. $\dfrac{dy}{dt} + 2y = 1, \quad y(0) = \frac{5}{2}$

27. $\sqrt{1 - y^2}\,dx - \sqrt{1 - x^2}\,dy = 0, \quad y(0) = \dfrac{\sqrt{3}}{2}$

28. $(1 + x^4)\,dy + x(1 + 4y^2)\,dx = 0, \quad y(1) = 0$

In Problems 29 and 30 proceed as in Example 5 and find an explicit solution of the given initial-value problem.

29. $\dfrac{dy}{dx} = ye^{-x^2}, \quad y(4) = 1$

30. $\dfrac{dy}{dx} = y^2 \sin x^2, \quad y(-2) = \frac{1}{3}$

In Problems 31–34 find an explicit solution of the given initial-value problem. Determine the exact interval I of definition by analytical methods. Use a graphing utility to plot the graph of the solution.

31. $\dfrac{dy}{dx} = \dfrac{2x + 1}{2y}, \quad y(-2) = -1$

32. $(2y - 2)\dfrac{dy}{dx} = 3x^2 + 4x + 2, \quad y(1) = -2$

33. $e^y\,dx - e^{-x}\,dy = 0, \quad y(0) = 0$

34. $\sin x\,dx + y\,dy = 0, \quad y(0) = 1$

35. (a) Find a solution of the initial-value problem consisting of the differential equation in Example 3 and each of the initial-conditions: $y(0) = 2$, $y(0) = -2$, and $y\left(\frac{1}{4}\right) = 1$.

(b) Find the solution of the differential equation in Example 4 when $\ln c_1$ is used as the constant of integration on the *left-hand* side in the solution and $4 \ln c_1$ is replaced by $\ln c$. Then solve the same initial-value problems in part (a).

36. Find a solution of $x\dfrac{dy}{dx} = y^2 - y$ that passes through the indicated points.

(a) $(0, 1)$ (b) $(0, 0)$ (c) $\left(\frac{1}{2}, \frac{1}{2}\right)$ (d) $\left(2, \frac{1}{4}\right)$

37. Find a singular solution of Problem 21. Of Problem 22.

38. Show that an implicit solution of

$$2x \sin^2 y\,dx - (x^2 + 10)\cos y\,dy = 0$$

is given by $\ln(x^2 + 10) + \csc y = c$. Find the constant solutions, if any, that were lost in the solution of the differential equation.

Often a radical change in the form of the solution of a differential equation corresponds to a very small change in either the initial condition or the equation itself. In Problems 39–42 find an explicit solution of the given initial-value problem. Use a graphing utility to plot the graph of each solution. Compare each solution curve in a neighborhood of $(0, 1)$.

39. $\dfrac{dy}{dx} = (y - 1)^2, \quad y(0) = 1$

40. $\dfrac{dy}{dx} = (y - 1)^2, \quad y(0) = 1.01$

41. $\dfrac{dy}{dx} = (y - 1)^2 + 0.01, \quad y(0) = 1$

42. $\dfrac{dy}{dx} = (y - 1)^2 - 0.01, \quad y(0) = 1$

43. Every autonomous first-order equation $dy/dx = f(y)$ is separable. Find explicit solutions $y_1(x)$, $y_2(x)$, $y_3(x)$, and $y_4(x)$ of the differential equation $dy/dx = y - y^3$ that satisfy, in turn, the initial conditions $y_1(0) = 2$, $y_2(0) = \frac{1}{2}$, $y_3(0) = -\frac{1}{2}$, and $y_4(0) = -2$. Use a graphing utility to plot the graphs of each solution. Compare these graphs with those predicted in Problem 19 of Exercises 2.1. Give the exact interval of definition for each solution.

44. (a) The autonomous first-order differential equation $dy/dx = 1/(y - 3)$ has no critical points. Nevertheless, place 3 on the phase line and obtain a phase portrait of the equation. Compute d^2y/dx^2 to determine where solution curves are concave up and where they are concave down (see Problems 35 and 36 in Exercises 2.1). Use the phase portrait and concavity to sketch, by hand, some typical solution curves.

(b) Find explicit solutions $y_1(x)$, $y_2(x)$, $y_3(x)$, and $y_4(x)$ of the differential equation in part (a) that satisfy, in turn, the initial conditions $y_1(0) = 4$, $y_2(0) = 2$, $y_3(1) = 2$, and $y_4(-1) = 4$. Graph each solution and compare with your sketches in part (a). Give the exact interval of definition for each solution.

In Problems 45–50 use a technique of integration or a substitution to find an explicit solution of the given differential equation or initial-value problem.

45. $\dfrac{dy}{dx} = \dfrac{1}{1 + \sin x}$

46. $\dfrac{dy}{dx} = \dfrac{\sin\sqrt{x}}{\sqrt{y}}$

47. $(\sqrt{x} + x)\dfrac{dy}{dx} = \sqrt{y} + y$

48. $\dfrac{dy}{dx} = y^{2/3} - y$

49. $\dfrac{dy}{dx} = \dfrac{e^{\sqrt{x}}}{y}, \quad y(1) = 4$

50. $\dfrac{dy}{dx} = \dfrac{x \tan^{-1} x}{y}, \quad y(0) = 3$

Discussion Problems

51. (a) Explain why the interval of definition of the explicit solution $y = \phi_2(x)$ of the initial-value problem in Example 2 is the *open* interval $(-5, 5)$.

(b) Can any solution of the differential equation cross the x-axis? Do you think that $x^2 + y^2 = 1$ is an implicit solution of the initial-value problem $dy/dx = -x/y$, $y(1) = 0$?

52. (a) If $a > 0$, discuss the differences, if any, between the solutions of the initial-value problems consisting of the differential equation $dy/dx = x/y$ and

each of the initial conditions $y(a) = a$, $y(a) = -a$, $y(-a) = a$, and $y(-a) = -a$.

(b) Does the initial-value problem $dy/dx = x/y$, $y(0) = 0$ have a solution?

(c) Solve $dy/dx = x/y$, $y(1) = 2$ and give the exact interval I of definition of its solution.

53. In Problems 43 and 44 we saw that every autonomous first-order differential equation $dy/dx = f(y)$ is separable. Does this fact help in the solution of the initial-value problem $\dfrac{dy}{dx} = \sqrt{1 + y^2}\sin^2 y,\quad y(0) = \tfrac{1}{2}$? Discuss. Sketch, by hand, a plausible solution curve of the problem.

54. (a) Solve the two initial-value problems:

$$\frac{dy}{dx} = y, \quad y(0) = 1$$

and

$$\frac{dy}{dx} = y + \frac{y}{x \ln x}, \quad y(e) = 1.$$

(b) Show that there are more than 1.65 million digits in the y-coordinate of the point of intersection of the two solution curves in part (a).

55. Find a function whose square plus the square of its derivative is 1.

56. (a) The differential equation in Problem 27 is equivalent to the normal form

$$\frac{dy}{dx} = \sqrt{\frac{1 - y^2}{1 - x^2}}$$

in the square region in the xy-plane defined by $|x| < 1$, $|y| < 1$. But the quantity under the radical is nonnegative also in the regions defined by $|x| > 1$, $|y| > 1$. Sketch all regions in the xy-plane for which this differential equation possesses real solutions.

(b) Solve the DE in part (a) in the regions defined by $|x| > 1$, $|y| > 1$. Then find an implicit and an explicit solution of the differential equation subject to $y(2) = 2$.

Mathematical Model

57. Suspension Bridge In (16) of Section 1.3 we saw that a mathematical model for the shape of a flexible cable strung between two vertical supports is

$$\frac{dy}{dx} = \frac{W}{T_1}, \qquad (10)$$

where W denotes the portion of the total vertical load between the points P_1 and P_2 shown in Figure 1.3.7. The DE (10) is separable under the following conditions that describe a suspension bridge.

Let us assume that the x- and y-axes are as shown in Figure 2.2.5—that is, the x-axis runs along the horizontal roadbed, and the y-axis passes through $(0, a)$, which is the lowest point on one cable over the span of the bridge, coinciding with the interval $[-L/2, L/2]$. In the case of a suspension bridge, the usual assumption is that the vertical load in (10) is only a uniform roadbed distributed along the horizontal axis. In other words, it is assumed that the weight of all cables is negligible in comparison to the weight of the roadbed and that the weight per unit length of the roadbed (say, pounds per horizontal foot) is a constant ρ. Use this information to set up and solve an appropriate initial-value problem from which the shape (a curve with equation $y = \phi(x)$) of each of the two cables in a suspension bridge is determined. Express your solution of the IVP in terms of the sag h and span L. See Figure 2.2.5.

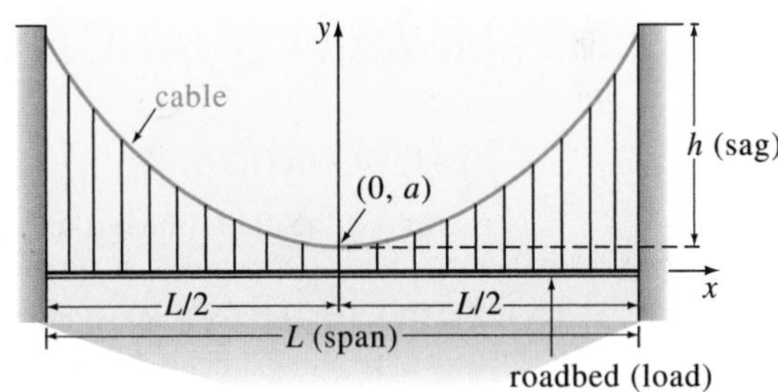

FIGURE 2.2.5 Shape of a cable in Problem 57

Computer Lab Assignments

58. (a) Use a CAS and the concept of level curves to plot representative graphs of members of the family of solutions of the differential equation $\dfrac{dy}{dx} = -\dfrac{8x + 5}{3y^2 + 1}$. Experiment with different numbers of level curves as well as various rectangular regions defined by $a \le x \le b$, $c \le y \le d$.

(b) On separate coordinate axes plot the graphs of the particular solutions corresponding to the initial conditions: $y(0) = -1$; $y(0) = 2$; $y(-1) = 4$; $y(-1) = -3$.

59. (a) Find an implicit solution of the IVP

$$(2y + 2)\,dy - (4x^3 + 6x)\,dx = 0, \quad y(0) = -3.$$

(b) Use part (a) to find an explicit solution $y = \phi(x)$ of the IVP.

(c) Consider your answer to part (b) as a *function* only. Use a graphing utility or a CAS to graph this function, and then use the graph to estimate its domain.

(d) With the aid of a root-finding application of a CAS, determine the approximate largest interval I of definition of the *solution* $y = \phi(x)$ in part (b). Use a graphing utility or a CAS to graph the solution curve for the IVP on this interval.

60. (a) Use a CAS and the concept of level curves to plot representative graphs of members of the family of solutions of the differential equation $\dfrac{dy}{dx} = \dfrac{x(1-x)}{y(-2+y)}$. Experiment with different numbers of level curves as well as various rectangular regions in the xy-plane until your result resembles Figure 2.2.6.

(b) On separate coordinate axes, plot the graph of the implicit solution corresponding to the initial condition $y(0) = \frac{3}{2}$. Use a colored pencil to mark off that segment of the graph that corresponds to the solution curve of a solution ϕ that satisfies the initial condition. With the aid of a root-finding application of a CAS, determine the approximate largest interval I of definition of the solution ϕ. [*Hint*: First find the points on the curve in part (a) where the tangent is vertical.]

(c) Repeat part (b) for the initial condition $y(0) = -2$.

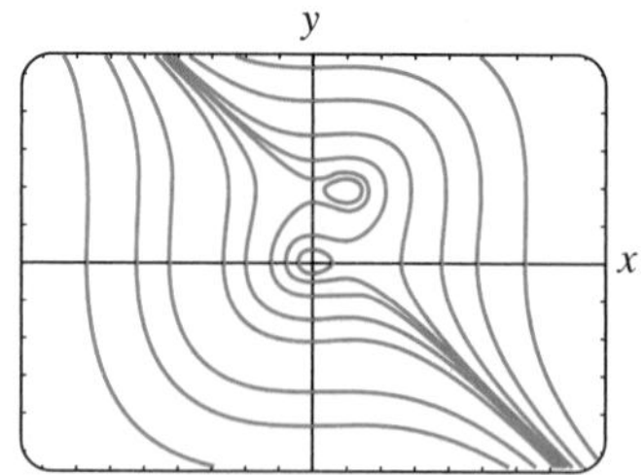

FIGURE 2.2.6 Level curves in Problem 60

2.3 LINEAR EQUATIONS

REVIEW MATERIAL

- Review the definitions of linear DEs in (6) and (7) of Section 1.1.

INTRODUCTION We continue our quest for solutions of first-order differential equations by next examining linear equations. Linear differential equations are an especially "friendly" family of differential equations, in that, given a linear equation, whether first order or a higher-order kin, there is always a good possibility that we can find some sort of solution of the equation that we can examine.

A Definition The form of a linear first-order DE was given in (7) of Section 1.1. This form, the case when $n = 1$ in (6) of that section, is reproduced here for convenience.

DEFINITION 2.3.1 Linear Equation

A first-order differential equation of the form

$$a_1(x)\frac{dy}{dx} + a_0(x)y = g(x), \tag{1}$$

is said to be a linear equation in the variable y.

Standard Form By dividing both sides of (1) by the lead coefficient $a_1(x)$, we obtain a more useful form, the **standard form**, of a linear equation:

$$\frac{dy}{dx} + P(x)y = f(x). \tag{2}$$

We seek a solution of (2) on an interval I for which both coefficient functions P and f are continuous.

Before we examine a general procedure for solving equations of form (2) we note that in some instances (2) can be solved by separation of variables. For example, you should verify that the equations

We match each equation with (2). In the first equation $P(x) = 2x$, $f(x) = 0$ and in the second $P(x) = -1$, $f(x) = 5$.

$$\frac{dy}{dx} + 2xy = 0 \quad \text{and} \quad \frac{dy}{dx} = y + 5$$

are both linear and separable, but that the linear equation

$$\frac{dy}{dx} + y = x$$

is not separable.

Method of Solution The method for solving (2) hinges on a remarkable fact that the *left-hand side* of the equation can be recast into the form of the exact derivative of a product by multiplying the both sides of (2) by a special function $\mu(x)$. It is relatively easy to find the function $\mu(x)$ because we want

$$\frac{d}{dx}\underbrace{[\mu(x)y]}_{\text{product}} = \underbrace{\mu\frac{dy}{dx} + \frac{d\mu}{dx}y}_{\text{product rule}} = \underbrace{\mu\frac{dy}{dx} + \mu P y}_{\substack{\text{left hand side of}\\ \text{(2) multipled by } \mu(x)}}.$$

these must be equal

The equality is true provided that

$$\frac{d\mu}{dx} = \mu P.$$

The last equation can be solved by separation of variables. Integrating

$$\frac{d\mu}{\mu} = Pdx \quad \text{and solving} \quad \ln|\mu(x)| = \int P(x)dx + c_1$$

See Problem 50 in Exercises 2.3

gives $\mu(x) = c_2 e^{\int P(x)dx}$. Even though there are an infinite choices of $\mu(x)$ (all constant multiples of $e^{\int P(x)dx}$), all produce the same desired result. Hence we can simplify life and choose $c_2 = 1$. The function

$$\mu(x) = e^{\int P(x)dx} \tag{3}$$

is called an **integrating factor** for equation (2).

Here is what we have so far: We multiplied both sides of (2) by (3) and, by construction, the left-hand side is the derivative of a product of the integrating factor and y:

$$e^{\int P(x)dx}\frac{dy}{dx} + P(x)e^{\int P(x)dx}y = e^{\int P(x)dx}f(x)$$

$$\frac{d}{dx}\left[e^{\int P(x)dx}y\right] = e^{\int P(x)dx}f(x).$$

Finally, we discover why (3) is called an *integrating* factor. We can integrate both sides of the last equation,

$$e^{\int P(x)dx}y = \int e^{\int P(x)dx}f(x) + c$$

and solve for y. The result is a one-parameter family of solutions of (2):

$$y = e^{-\int P(x)dx}\int e^{\int P(x)dx}f(x)dx + ce^{-\int P(x)dx}. \tag{4}$$

We emphasize that you should **not memorize** formula (4). The following procedure should be worked through each time.

SOLVING A LINEAR FIRST-ORDER EQUATION

(*i*) Remember to put a linear equation into the standard form (2).

(*ii*) From the standard form of the equation identify $P(x)$ and then find the integrating factor $e^{\int P(x)dx}$. No constant need be used in evaluating the indefinite integral $\int P(x)dx$.

(*iii*) Multiply the both sides of the standard form equation by the integrating factor. The left-hand side of the resulting equation is automatically the derivative of the product of the integrating factor $e^{\int P(x)dx}$ and y:

$$\frac{d}{dx}\left[e^{\int P(x)dx}y\right] = e^{\int P(x)dx}f(x).$$

(*iv*) Integrate both sides of the last equation and solve for y.

EXAMPLE 1 Solving a Linear Equation

Solve $\dfrac{dy}{dx} - 3y = 0$.

SOLUTION This linear equation can be solved by separation of variables. Alternatively, since the differential equation is already in standard form (2), we identify $P(x) = -3$, and so the integrating factor is $e^{\int(-3)dx} = e^{-3x}$. We then multiply the given equation by this factor and recognize that

$$e^{-3x}\frac{dy}{dx} - 3e^{-3x}y = e^{-3x}\cdot 0 \qquad \text{is the same as} \qquad \frac{d}{dx}[e^{-3x}y] = 0.$$

Integration of the last equation,

$$\int \frac{d}{dx}[e^{-3x}y]\,dx = \int 0\,dx$$

then yields $e^{-3x}y = c$ or $y = ce^{3x}$, $-\infty < x < \infty$. ≡

EXAMPLE 2 Solving a Linear Equation

Solve $\dfrac{dy}{dx} - 3y = 6$.

SOLUTION This linear equation, like the one in Example 1, is already in standard form with $P(x) = -3$. Thus the integrating factor is again e^{-3x}. This time multiplying the given equation by this factor gives

$$e^{-3x}\frac{dy}{dx} - 3e^{-3x}y = 6e^{-3x} \qquad \text{and so} \qquad \frac{d}{dx}[e^{-3x}y] = 6e^{-3x}.$$

Integrating the last equation,

$$\int \frac{d}{dx}[e^{-3x}y]\,dx = 6\int e^{-3x}\,dx \qquad \text{gives} \qquad e^{-3x}y = -6\left(\frac{e^{-3x}}{3}\right) + c,$$

or $y = -2 + ce^{3x}$, $-\infty < x < \infty$. ≡

When a_1, a_0, and g in (1) are constants, the differential equation is autonomous. In Example 2 you can verify from the normal form $dy/dx = 3(y + 2)$ that -2 is a critical point and that it is unstable (a repeller). Thus a solution curve with an

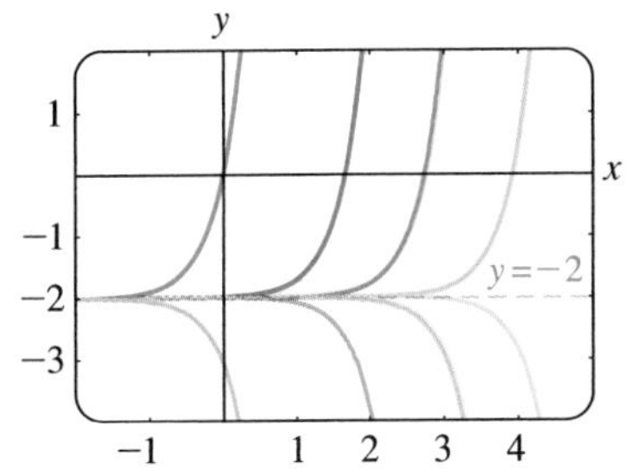

FIGURE 2.3.1 Solution curves of DE in Example 2

initial point either above or below the graph of the equilibrium solution $y = -2$ pushes away from this horizontal line as x increases. Figure 2.3.1, obtained with the aid of a graphing utility, shows the graph of $y = -2$ along with some additional solution curves.

General Solution Suppose again that the functions P and f in (2) are continuous on a common interval I. In the steps leading to (4) we showed that *if* (2) has a solution on I, then it must be of the form given in (4). Conversely, it is a straightforward exercise in differentiation to verify that any function of the form given in (4) is a solution of the differential equation (2) on I. In other words, (4) is a one-parameter family of solutions of equation (2) and *every solution of* (2) *defined on I is a member of this family.* Therefore we call (4) the **general solution** of the differential equation on the interval I. (See the *Remarks* at the end of Section 1.1.) Now by writing (2) in the normal form $y' = F(x, y)$, we can identify $F(x, y) = -P(x)y + f(x)$ and $\partial F/\partial y = -P(x)$. From the continuity of P and f on the interval I we see that F and $\partial F/\partial y$ are also continuous on I. With Theorem 1.2.1 as our justification, we conclude that there exists one and only one solution of the initial-value problem

$$\frac{dy}{dx} + P(x)y = f(x), \quad y(x_0) = y_0 \tag{5}$$

defined on *some* interval I_0 containing x_0. But when x_0 is in I, finding a solution of (5) is just a matter of finding an appropriate value of c in (4)—that is, to each x_0 in I there corresponds a distinct c. In other words, the interval I_0 of existence and uniqueness in Theorem 1.2.1 for the initial-value problem (5) is the entire interval I.

EXAMPLE 3 General Solution

Solve $x\dfrac{dy}{dx} - 4y = x^6e^x$.

SOLUTION Dividing by x, the standard form of the given DE is

$$\frac{dy}{dx} - \frac{4}{x}y = x^5e^x. \tag{6}$$

From this form we identify $P(x) = -4/x$ and $f(x) = x^5e^x$ and further observe that P and f are continuous on $(0, \infty)$. Hence the integrating factor is

we can use $\ln x$ instead of $\ln|x|$ since $x > 0$
↓
$$e^{-4\int dx/x} = e^{-4\ln x} = e^{\ln x^{-4}} = x^{-4}.$$

Here we have used the basic identity $b^{\log_b N} = N$, $N > 0$. Now we multiply (6) by x^{-4} and rewrite

$$x^{-4}\frac{dy}{dx} - 4x^{-5}y = xe^x \quad \text{as} \quad \frac{d}{dx}[x^{-4}y] = xe^x.$$

It follows from integration by parts that the general solution defined on the interval $(0, \infty)$ is $x^{-4}y = xe^x - e^x + c$ or $y = x^5e^x - x^4e^x + cx^4$. ≡

In case you are wondering why the interval $(0, \infty)$ is important in Example 3, read this paragraph and the paragraph following Example 4. ▶

Except in the case in which the lead coefficient is 1, the recasting of equation (1) into the standard form (2) requires division by $a_1(x)$. Values of x for which $a_1(x) = 0$ are called **singular points** of the equation. Singular points are potentially troublesome. Specifically, in (2), if $P(x)$ (formed by dividing $a_0(x)$ by $a_1(x)$) is discontinuous at a point, the discontinuity may carry over to solutions of the differential equation.

EXAMPLE 4 General Solution

Find the general solution of $(x^2 - 9)\dfrac{dy}{dx} + xy = 0$.

SOLUTION We write the differential equation in standard form

$$\frac{dy}{dx} + \frac{x}{x^2 - 9}y = 0 \tag{7}$$

and identify $P(x) = x/(x^2 - 9)$. Although P is continuous on $(-\infty, -3)$, $(-3, 3)$, and $(3, \infty)$, we shall solve the equation on the first and third intervals. On these intervals the integrating factor is

$$e^{\int x\,dx/(x^2-9)} = e^{\frac{1}{2}\int 2x\,dx/(x^2-9)} = e^{\frac{1}{2}\ln|x^2-9|} = \sqrt{x^2 - 9}.$$

After multiplying the standard form (7) by this factor, we get

$$\frac{d}{dx}\left[\sqrt{x^2 - 9}\,y\right] = 0.$$

Integrating both sides of the last equation gives $\sqrt{x^2 - 9}\,y = c$. Thus for either $x > 3$ or $x < -3$ the general solution of the equation is $y = \dfrac{c}{\sqrt{x^2 - 9}}$. ≡

Notice in Example 4 that $x = 3$ and $x = -3$ are singular points of the equation and that every function in the general solution $y = c/\sqrt{x^2 - 9}$ is discontinuous at these points. On the other hand, $x = 0$ is a singular point of the differential equation in Example 3, but the general solution $y = x^5e^x - x^4e^x + cx^4$ is noteworthy in that every function in this one-parameter family is continuous at $x = 0$ and is defined on the interval $(-\infty, \infty)$ and not just on $(0, \infty)$, as stated in the solution. However, the family $y = x^5e^x - x^4e^x + cx^4$ defined on $(-\infty, \infty)$ cannot be considered the general solution of the DE, since the singular point $x = 0$ still causes a problem. See Problems 45 and 46 in Exercises 2.3.

EXAMPLE 5 An Initial-Value Problem

Solve $\dfrac{dy}{dx} + y = x$, $y(0) = 4$.

SOLUTION The equation is in standard form, and $P(x) = 1$ and $f(x) = x$ are continuous on $(-\infty, \infty)$. The integrating factor is $e^{\int dx} = e^x$, so integrating

$$\frac{d}{dx}[e^x y] = xe^x$$

gives $e^x y = xe^x - e^x + c$. Solving this last equation for y yields the general solution $y = x - 1 + ce^{-x}$. But from the initial condition we know that $y = 4$ when $x = 0$. Substituting these values into the general solution implies that $c = 5$. Hence the solution of the problem is

$$y = x - 1 + 5e^{-x}, \quad -\infty < x < \infty. \tag{8}$$ ≡

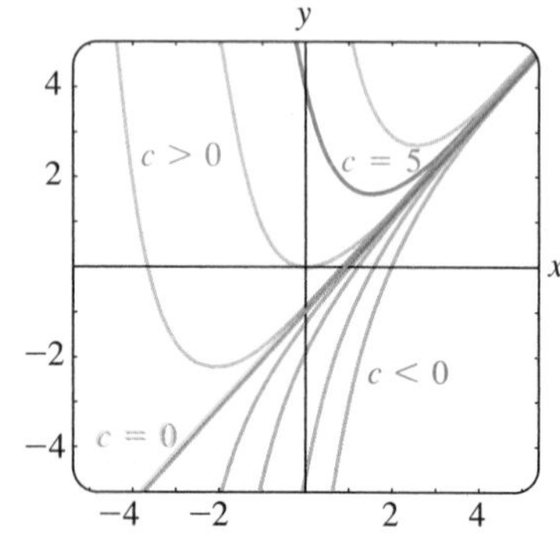

FIGURE 2.3.2 Solution curves of DE in Example 5

Figure 2.3.2, obtained with the aid of a graphing utility, shows the graph of the solution (8) in dark blue along with the graphs of other members of the one-parameter family of solutions $y = x - 1 + ce^{-x}$. It is interesting to observe that as x increases, the graphs of *all* members of this family are close to the graph of the solution $y = x - 1$. The last solution corresponds to $c = 0$ in the family and is shown in dark green in Figure 2.3.2. This asymptotic behavior of solutions is due to the fact that the contribution of ce^{-x}, $c \neq 0$, becomes negligible for increasing values of x. We say that ce^{-x} is a **transient term,** since $e^{-x} \rightarrow 0$ as $x \rightarrow \infty$. While this behavior is not characteristic of all general solutions of linear equations (see Example 2), the notion of a transient is often important in applied problems.

≡ Discontinuous Coefficients In applications, the coefficients $P(x)$ and $f(x)$ in (2) may be piecewise continuous. In the next example $f(x)$ is piecewise continuous on $[0, \infty)$ with a single discontinuity, namely, a (finite) jump discontinuity at $x = 1$. We solve the problem in two parts corresponding to the two intervals over which f is defined. It is then possible to piece together the two solutions at $x = 1$ so that $y(x)$ is continuous on $[0, \infty)$.

EXAMPLE 6 An Initial-Value Problem

Solve $\dfrac{dy}{dx} + y = f(x)$, $y(0) = 0$ where $f(x) = \begin{cases} 1, & 0 \leq x \leq 1, \\ 0, & x > 1. \end{cases}$

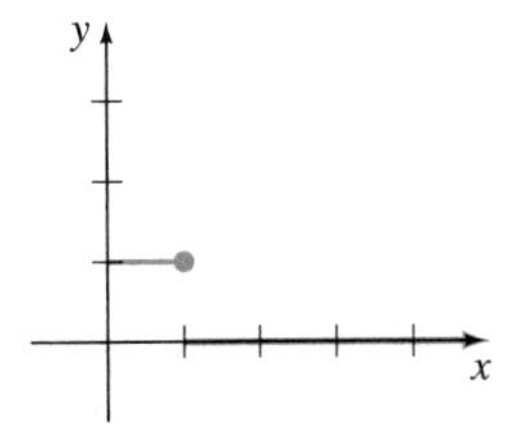

FIGURE 2.3.3 Discontinuous $f(x)$ in Example 6

SOLUTION The graph of the discontinuous function f is shown in Figure 2.3.3. We solve the DE for $y(x)$ first on the interval $[0, 1]$ and then on the interval $(1, \infty)$. For $0 \leq x \leq 1$ we have

$$\frac{dy}{dx} + y = 1 \qquad \text{or, equivalently,} \qquad \frac{d}{dx}[e^x y] = e^x.$$

Integrating this last equation and solving for y gives $y = 1 + c_1 e^{-x}$. Since $y(0) = 0$, we must have $c_1 = -1$, and therefore $y = 1 - e^{-x}$, $0 \leq x \leq 1$. Then for $x > 1$ the equation

$$\frac{dy}{dx} + y = 0$$

leads to $y = c_2 e^{-x}$. Hence we can write

$$y = \begin{cases} 1 - e^{-x}, & 0 \leq x \leq 1, \\ c_2 e^{-x}, & x > 1. \end{cases}$$

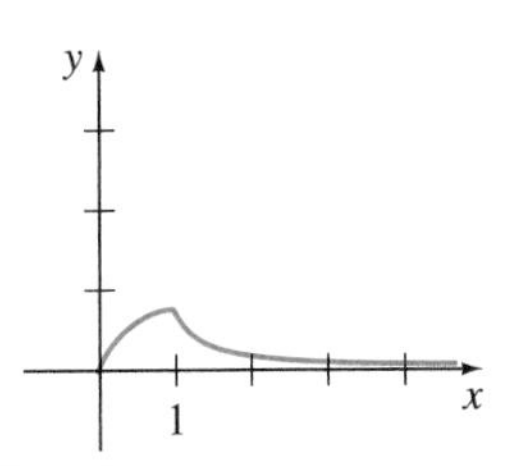

FIGURE 2.3.4 Graph of (9) in Example 6

By appealing to the definition of continuity at a point, it is possible to determine c_2 so that the foregoing function is continuous at $x = 1$. The requirement that $\lim_{x \rightarrow 1^+} y(x) = y(1)$ implies that $c_2 e^{-1} = 1 - e^{-1}$ or $c_2 = e - 1$. As seen in Figure 2.3.4, the function

$$y = \begin{cases} 1 - e^{-x}, & 0 \leq x \leq 1, \\ (e - 1)e^{-x}, & x > 1 \end{cases} \tag{9}$$

is continuous on $(0, \infty)$. ≡

It is worthwhile to think about (9) and Figure 2.3.4 a little bit; you are urged to read and answer Problem 48 in Exercises 2.3.

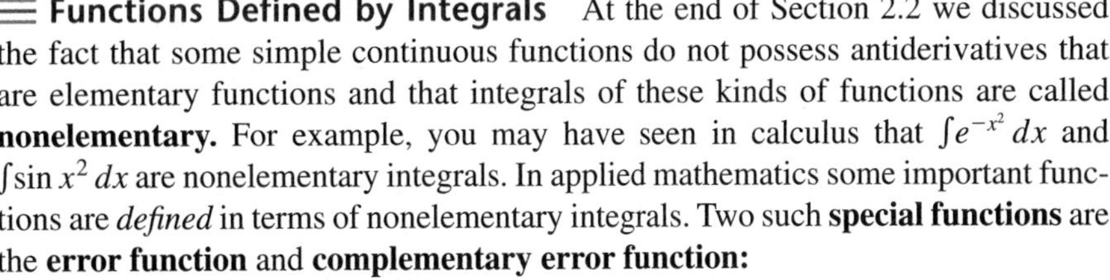

≡ Functions Defined by Integrals At the end of Section 2.2 we discussed the fact that some simple continuous functions do not possess antiderivatives that are elementary functions and that integrals of these kinds of functions are called **nonelementary.** For example, you may have seen in calculus that $\int e^{-x^2} dx$ and $\int \sin x^2 \, dx$ are nonelementary integrals. In applied mathematics some important functions are *defined* in terms of nonelementary integrals. Two such **special functions** are the **error function** and **complementary error function:**

$$\operatorname{erf}(x) = \frac{2}{\sqrt{\pi}} \int_0^x e^{-t^2} dt \qquad \text{and} \qquad \operatorname{erfc}(x) = \frac{2}{\sqrt{\pi}} \int_x^\infty e^{-t^2} dt. \tag{10}$$

From the known result $\int_0^\infty e^{-t^2}\,dt = \sqrt{\pi}/2$* we can write $(2/\sqrt{\pi})\int_0^\infty e^{-t^2}\,dt = 1$. Then from $\int_0^\infty = \int_0^x + \int_x^\infty$ it is seen from (10) that the complementary error function erfc(x) is related to erf(x) by $\text{erf}(x) + \text{erfc}(x) = 1$. Because of its importance in probability, statistics, and applied partial differential equations, the error function has been extensively tabulated. Note that erf(0) = 0 is one obvious function value. Values of erf(x) can also be found by using a CAS.

EXAMPLE 7 The Error Function

Solve the initial-value problem $\dfrac{dy}{dx} - 2xy = 2, \quad y(0) = 1.$

SOLUTION Since the equation is already in standard form, we see that the integrating factor is e^{-x^2}, so from

$$\frac{d}{dx}[e^{-x^2}y] = 2e^{-x^2} \qquad \text{we get} \qquad y = 2e^{x^2}\int_0^x e^{-t^2}\,dt + ce^{x^2}. \tag{11}$$

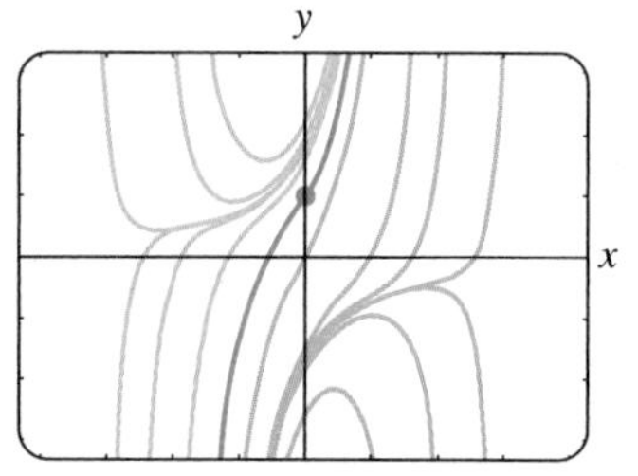

FIGURE 2.3.5 Solution curves of DE in Example 7

Applying $y(0) = 1$ to the last expression then gives $c = 1$. Hence the solution of the problem is

$$y = 2e^{x^2}\int_0^x e^{-t^2}\,dt + e^{x^2} \qquad \text{or} \qquad y = e^{x^2}[1 + \sqrt{\pi}\,\text{erf}(x)].$$

The graph of this solution on the interval $(-\infty, \infty)$, shown in dark blue in Figure 2.3.5 among other members of the family defined in (11), was obtained with the aid of a computer algebra system. ≡

≡ **Use of Computers** The computer algebra systems *Mathematica* and *Maple* are capable of producing implicit or explicit solutions for some kinds of differential equations using their *dsolve* commands.†

REMARKS

(*i*) A linear first-order differential equation

$$a_1(x)\frac{dy}{dx} + a_0(x)y = 0$$

is said to be **homogeneous,** whereas the equation

$$a_1(x)\frac{dy}{dx} + a_0(x)y = g(x)$$

with $g(x)$ not identically zero is said to be **nonhomogeneous.** For example, the linear equations $xy' + y = 0$ and $xy' + y = e^x$ are, in turn, homogeneous and nonhomogeneous. As can be seen in this example the trivial solution $y = 0$ is always a solution of a homogeneous linear DE. Store this terminology in the back of your mind because it becomes important when we study linear higher-order ordinary differential equations in Chapter 4.

---

*This result is usually proved in the third semester of calculus.

†Certain commands have the same spelling, but in *Mathematica* commands begin with a capital letter **(DSolve),** whereas in *Maple* the same command begins with a lower case letter **(dsolve).** When discussing such common syntax, we compromise and write, for example, *dsolve*. See the *Student Resource Manual* for the complete input commands used to solve a linear first-order DE.

(ii) Occasionally, a first-order differential equation is not linear in one variable but in linear in the other variable. For example, the differential equation

$$\frac{dy}{dx} = \frac{1}{x + y^2}$$

is not linear in the variable y. But its reciprocal

$$\frac{dx}{dy} = x + y^2 \quad \text{or} \quad \frac{dx}{dy} - x = y^2$$

is recognized as linear in the variable x. You should verify that the integrating factor $e^{\int(-1)dy} = e^{-y}$ and integration by parts yield the explicit solution $x = -y^2 - 2y - 2 + ce^y$ for the second equation. This expression is then an implicit solution of the first equation.

(iii) Mathematicians have adopted as their own certain words from engineering, which they found appropriately descriptive. The word *transient,* used earlier, is one of these terms. In future discussions the words *input* and *output* will occasionally pop up. The function f in (2) is called the **input** or **driving function;** a solution $y(x)$ of the differential equation for a given input is called the **output** or **response.**

(iv) The term **special functions** mentioned in conjunction with the error function also applies to the **sine integral function** and the **Fresnel sine integral** introduced in Problems 55 and 56 in Exercises 2.3. "Special Functions" is actually a well-defined branch of mathematics. More special functions are studied in Section 6.4.

EXERCISES 2.3

Answers to selected odd-numbered problems begin on page ANS-2.

In Problems 1–24 find the general solution of the given differential equation. Give the largest interval I over which the general solution is defined. Determine whether there are any transient terms in the general solution.

1. $\dfrac{dy}{dx} = 5y$
2. $\dfrac{dy}{dx} + 2y = 0$
3. $\dfrac{dy}{dx} + y = e^{3x}$
4. $3\dfrac{dy}{dx} + 12y = 4$
5. $y' + 3x^2y = x^2$
6. $y' + 2xy = x^3$
7. $x^2y' + xy = 1$
8. $y' = 2y + x^2 + 5$
9. $x\dfrac{dy}{dx} - y = x^2 \sin x$
10. $x\dfrac{dy}{dx} + 2y = 3$
11. $x\dfrac{dy}{dx} + 4y = x^3 - x$
12. $(1 + x)\dfrac{dy}{dx} - xy = x + x^2$
13. $x^2y' + x(x + 2)y = e^x$
14. $xy' + (1 + x)y = e^{-x} \sin 2x$
15. $y\,dx - 4(x + y^6)\,dy = 0$
16. $y\,dx = (ye^y - 2x)\,dy$
17. $\cos x\dfrac{dy}{dx} + (\sin x)y = 1$
18. $\cos^2x \sin x\dfrac{dy}{dx} + (\cos^3x)y = 1$
19. $(x + 1)\dfrac{dy}{dx} + (x + 2)y = 2xe^{-x}$
20. $(x + 2)^2\dfrac{dy}{dx} = 5 - 8y - 4xy$
21. $\dfrac{dr}{d\theta} + r \sec \theta = \cos \theta$
22. $\dfrac{dP}{dt} + 2tP = P + 4t - 2$
23. $x\dfrac{dy}{dx} + (3x + 1)y = e^{-3x}$
24. $(x^2 - 1)\dfrac{dy}{dx} + 2y = (x + 1)^2$

In Problems 25–36 solve the given initial-value problem. Give the largest interval I over which the solution is defined.

25. $\dfrac{dy}{dx} = x + 5y, \quad y(0) = 3$
26. $\dfrac{dy}{dx} = 2x - 3y, \quad y(0) = \frac{1}{3}$
27. $xy' + y = e^x, \quad y(1) = 2$

28. $y\dfrac{dx}{dy} - x = 2y^2, \quad y(1) = 5$

29. $L\dfrac{di}{dt} + Ri = E, \quad i(0) = i_0, \quad L, R, E, i_0$ constants

30. $\dfrac{dT}{dt} = k(T - T_m), \quad T(0) = T_0, \quad k, T_m, T_0$ constants

31. $x\dfrac{dy}{dx} + y = 4x + 1, \quad y(1) = 8$

32. $y' + 4xy = x^3e^{x^2}, \quad y(0) = -1$

33. $(x + 1)\dfrac{dy}{dx} + y = \ln x, \quad y(1) = 10$

34. $x(x + 1)\dfrac{dy}{dx} + xy = 1, \quad y(e) = 1$

35. $y' - (\sin x)y = 2\sin x, \quad y(\pi/2) = 1$

36. $y' + (\tan x)y = \cos^2 x, \quad y(0) = -1$

In Problems 37–40 proceed as in Example 6 to solve the given initial-value problem. Use a graphing utility to graph the continuous function $y(x)$.

37. $\dfrac{dy}{dx} + 2y = f(x), y(0) = 0$, where

$$f(x) = \begin{cases} 1, & 0 \le x \le 3 \\ 0, & x > 3 \end{cases}$$

38. $\dfrac{dy}{dx} + y = f(x), y(0) = 1$, where

$$f(x) = \begin{cases} 1, & 0 \le x \le 1 \\ -1, & x > 1 \end{cases}$$

39. $\dfrac{dy}{dx} + 2xy = f(x), y(0) = 2$, where

$$f(x) = \begin{cases} x, & 0 \le x < 1 \\ 0, & x \ge 1 \end{cases}$$

40. $(1 + x^2)\dfrac{dy}{dx} + 2xy = f(x), y(0) = 0$, where

$$f(x) = \begin{cases} x, & 0 \le x < 1 \\ -x, & x \ge 1 \end{cases}$$

41. Proceed in a manner analogous to Example 6 to solve the initial-value problem $y' + P(x)y = 4x$, $y(0) = 3$, where

$$P(x) = \begin{cases} 2, & 0 \le x \le 1, \\ -2/x, & x > 1. \end{cases}$$

Use a graphing utility to graph the continuous function $y(x)$.

42. Consider the initial-value problem $y' + e^x y = f(x)$, $y(0) = 1$. Express the solution of the IVP for $x > 0$ as a nonelementary integral when $f(x) = 1$. What is the solution when $f(x) = 0$? When $f(x) = e^x$?

43. Express the solution of the initial-value problem $y' - 2xy = 1, y(1) = 1$, in terms of $\operatorname{erf}(x)$.

Discussion Problems

44. Reread the discussion following Example 2. Construct a linear first-order differential equation for which all nonconstant solutions approach the horizontal asymptote $y = 4$ as $x \to \infty$.

45. Reread Example 3 and then discuss, with reference to Theorem 1.2.1, the existence and uniqueness of a solution of the initial-value problem consisting of $xy' - 4y = x^6e^x$ and the given initial condition.

   (a) $y(0) = 0$ (b) $y(0) = y_0, y_0 > 0$

   (c) $y(x_0) = y_0, x_0 > 0, y_0 > 0$

46. Reread Example 4 and then find the general solution of the differential equation on the interval $(-3, 3)$.

47. Reread the discussion following Example 5. Construct a linear first-order differential equation for which all solutions are asymptotic to the line $y = 3x - 5$ as $x \to \infty$.

48. Reread Example 6 and then discuss why it is technically incorrect to say that the function in (9) is a "solution" of the IVP on the interval $[0, \infty)$.

49. (a) Construct a linear first-order differential equation of the form $xy' + a_0(x)y = g(x)$ for which $y_c = c/x^3$ and $y_p = x^3$. Give an interval on which $y = x^3 + c/x^3$ is the general solution of the DE.

   (b) Give an initial condition $y(x_0) = y_0$ for the DE found in part (a) so that the solution of the IVP is $y = x^3 - 1/x^3$. Repeat if the solution is $y = x^3 + 2/x^3$. Give an interval I of definition of each of these solutions. Graph the solution curves. Is there an initial-value problem whose solution is defined on $(-\infty, \infty)$?

   (c) Is each IVP found in part (b) unique? That is, can there be more than one IVP for which, say, $y = x^3 - 1/x^3$, x in some interval I, is the solution?

50. In determining the integrating factor (3), we did not use a constant of integration in the evaluation of $\int P(x)\,dx$. Explain why using $\int P(x)\,dx + c_1$ has no effect on the solution of (2).

51. Suppose $P(x)$ is continuous on some interval I and a is a number in I. What can be said about the solution of the initial-value problem $y' + P(x)y = 0, y(a) = 0$?

Mathematical Models

52. **Radioactive Decay Series** The following system of differential equations is encountered in the study of the decay of a special type of radioactive series of elements:

$$\frac{dx}{dt} = -\lambda_1 x$$

$$\frac{dy}{dt} = \lambda_1 x - \lambda_2 y,$$

where λ_1 and λ_2 are constants. Discuss how to solve this system subject to $x(0) = x_0$, $y(0) = y_0$. Carry out your ideas.

53. Heart Pacemaker A heart pacemaker consists of a switch, a battery of constant voltage E_0, a capacitor with constant capacitance C, and the heart as a resistor with constant resistance R. When the switch is closed, the capacitor charges; when the switch is open, the capacitor discharges, sending an electrical stimulus to the heart. During the time the heart is being stimulated, the voltage E across the heart satisfies the linear differential equation

$$\frac{dE}{dt} = -\frac{1}{RC}E.$$

Solve the DE, subject to $E(4) = E_0$.

Computer Lab Assignments

54. **(a)** Express the solution of the initial-value problem $y' - 2xy = -1$, $y(0) = \sqrt{\pi}/2$, in terms of erfc(x).

(b) Use tables or a CAS to find the value of $y(2)$. Use a CAS to graph the solution curve for the IVP on $(-\infty, \infty)$.

55. **(a)** The **sine integral function** is defined by $\text{Si}(x) = \int_0^x (\sin t/t)\,dt$, where the integrand is defined to be 1 at $t = 0$. Express the solution $y(x)$ of the initial-value problem $x^3y' + 2x^2y = 10 \sin x$, $y(1) = 0$ in terms of Si(x).

(b) Use a CAS to graph the solution curve for the IVP for $x > 0$.

(c) Use a CAS to find the value of the absolute maximum of the solution $y(x)$ for $x > 0$.

56. **(a)** The **Fresnel sine integral** is defined by $S(x) = \int_0^x \sin(\pi t^2/2)\,dt$. Express the solution $y(x)$ of the initial-value problem $y' - (\sin x^2)y = 0$, $y(0) = 5$, in terms of $S(x)$.

(b) Use a CAS to graph the solution curve for the IVP on $(-\infty, \infty)$.

(c) It is known that $S(x) \to \frac{1}{2}$ as $x \to \infty$ and $S(x) \to -\frac{1}{2}$ as $x \to -\infty$. What does the solution $y(x)$ approach as $x \to \infty$? As $x \to -\infty$?

(d) Use a CAS to find the values of the absolute maximum and the absolute minimum of the solution $y(x)$.

2.4 EXACT EQUATIONS

REVIEW MATERIAL

- Multivariate calculus
- Partial differentiation and partial integration
- Differential of a function of two variables

INTRODUCTION Although the simple first-order equation

$$y\,dx + x\,dy = 0$$

is separable, we can solve the equation in an alternative manner by recognizing that the expression on the left-hand side of the equality is the differential of the function $f(x, y) = xy$; that is,

$$d(xy) = y\,dx + x\,dy.$$

In this section we examine first-order equations in differential form $M(x, y)\,dx + N(x, y)\,dy = 0$. By applying a simple test to M and N, we can determine whether $M(x, y)\,dx + N(x, y)\,dy$ is a differential of a function $f(x, y)$. If the answer is yes, we can construct f by partial integration.

Differential of a Function of Two Variables If $z = f(x, y)$ is a function of two variables with continuous first partial derivatives in a region R of the xy-plane, then its differential is

$$dz = \frac{\partial f}{\partial x}dx + \frac{\partial f}{\partial y}dy. \tag{1}$$

In the special case when $f(x, y) = c$, where c is a constant, then (1) implies

$$\frac{\partial f}{\partial x}dx + \frac{\partial f}{\partial y}dy = 0. \tag{2}$$

In other words, given a one-parameter family of functions $f(x, y) = c$, we can generate a first-order differential equation by computing the differential of both sides of the equality. For example, if $x^2 - 5xy + y^3 = c$, then (2) gives the first-order DE

$$(2x - 5y)\,dx + (-5x + 3y^2)\,dy = 0. \tag{3}$$

A Definition Of course, not every first-order DE written in differential form $M(x, y)\,dx + N(x, y)\,dy = 0$ corresponds to a differential of $f(x, y) = c$. So for our purposes it is more important to turn the foregoing example around; namely, if we are given a first-order DE such as (3), is there some way we can recognize that the differential expression $(2x - 5y)\,dx + (-5x + 3y^2)\,dy$ is the differential $d(x^2 - 5xy + y^3)$? If there is, then an implicit solution of (3) is $x^2 - 5xy + y^3 = c$. We answer this question after the next definition.

DEFINITION 2.4.1 Exact Equation

A differential expression $M(x, y)\,dx + N(x, y)\,dy$ is an **exact differential** in a region R of the xy-plane if it corresponds to the differential of some function $f(x, y)$ defined in R. A first-order differential equation of the form

$$M(x, y)\,dx + N(x, y)\,dy = 0$$

is said to be an **exact equation** if the expression on the left-hand side is an exact differential.

For example, $x^2y^3\,dx + x^3y^2\,dy = 0$ is an exact equation, because its left-hand side is an exact differential:

$$d\left(\tfrac{1}{3}x^3y^3\right) = x^2y^3\,dx + x^3y^2\,dy.$$

Notice that if we make the identifications $M(x, y) = x^2y^3$ and $N(x, y) = x^3y^2$, then $\partial M/\partial y = 3x^2y^2 = \partial N/\partial x$. Theorem 2.4.1, given next, shows that the equality of the partial derivatives $\partial M/\partial y$ and $\partial N/\partial x$ is no coincidence.

THEOREM 2.4.1 Criterion for an Exact Differential

Let $M(x, y)$ and $N(x, y)$ be continuous and have continuous first partial derivatives in a rectangular region R defined by $a < x < b$, $c < y < d$. Then a necessary and sufficient condition that $M(x, y)\,dx + N(x, y)\,dy$ be an exact differential is

$$\frac{\partial M}{\partial y} = \frac{\partial N}{\partial x}. \tag{4}$$

PROOF OF THE NECESSITY For simplicity let us assume that $M(x, y)$ and $N(x, y)$ have continuous first partial derivatives for all (x, y). Now if the expression $M(x, y)\,dx + N(x, y)\,dy$ is exact, there exists some function f such that for all x in R,

$$M(x, y)\,dx + N(x, y)\,dy = \frac{\partial f}{\partial x}dx + \frac{\partial f}{\partial y}dy.$$

Therefore
$$M(x, y) = \frac{\partial f}{\partial x}, \qquad N(x, y) = \frac{\partial f}{\partial y},$$

and
$$\frac{\partial M}{\partial y} = \frac{\partial}{\partial y}\left(\frac{\partial f}{\partial x}\right) = \frac{\partial^2 f}{\partial y\, \partial x} = \frac{\partial}{\partial x}\left(\frac{\partial f}{\partial y}\right) = \frac{\partial N}{\partial x}.$$

The equality of the mixed partials is a consequence of the continuity of the first partial derivatives of $M(x, y)$ and $N(x, y)$. ≡

The sufficiency part of Theorem 2.4.1 consists of showing that there exists a function f for which $\partial f/\partial x = M(x, y)$ and $\partial f/\partial y = N(x, y)$ whenever (4) holds. The construction of the function f actually reflects a basic procedure for solving exact equations.

≡ **Method of Solution** Given an equation in the differential form $M(x, y)\, dx + N(x, y)\, dy = 0$, determine whether the equality in (4) holds. If it does, then there exists a function f for which

$$\frac{\partial f}{\partial x} = M(x, y).$$

We can find f by integrating $M(x, y)$ with respect to x while holding y constant:

$$f(x, y) = \int M(x, y)\, dx + g(y), \tag{5}$$

where the arbitrary function $g(y)$ is the "constant" of integration. Now differentiate (5) with respect to y and assume that $\partial f/\partial y = N(x, y)$:

$$\frac{\partial f}{\partial y} = \frac{\partial}{\partial y}\int M(x, y)\, dx + g'(y) = N(x, y).$$

This gives
$$g'(y) = N(x, y) - \frac{\partial}{\partial y}\int M(x, y)\, dx. \tag{6}$$

Finally, integrate (6) with respect to y and substitute the result in (5). The implicit solution of the equation is $f(x, y) = c$.

Some observations are in order. First, it is important to realize that the expression $N(x, y) - (\partial/\partial y) \int M(x, y)\, dx$ in (6) is independent of x, because

$$\frac{\partial}{\partial x}\left[N(x, y) - \frac{\partial}{\partial y}\int M(x, y)\, dx\right] = \frac{\partial N}{\partial x} - \frac{\partial}{\partial y}\left(\frac{\partial}{\partial x}\int M(x, y)\, dx\right) = \frac{\partial N}{\partial x} - \frac{\partial M}{\partial y} = 0.$$

Second, we could just as well start the foregoing procedure with the assumption that $\partial f/\partial y = N(x, y)$. After integrating N with respect to y and then differentiating that result, we would find the analogues of (5) and (6) to be, respectively,

$$f(x, y) = \int N(x, y)\, dy + h(x) \quad \text{and} \quad h'(x) = M(x, y) - \frac{\partial}{\partial x}\int N(x, y)\, dy.$$

In either case *none of these formulas should be memorized.*

EXAMPLE 1 Solving an Exact DE

Solve $2xy\, dx + (x^2 - 1)\, dy = 0$.

SOLUTION With $M(x, y) = 2xy$ and $N(x, y) = x^2 - 1$ we have

$$\frac{\partial M}{\partial y} = 2x = \frac{\partial N}{\partial x}.$$

Thus the equation is exact, and so by Theorem 2.4.1 there exists a function $f(x, y)$ such that

$$\frac{\partial f}{\partial x} = 2xy \quad \text{and} \quad \frac{\partial f}{\partial y} = x^2 - 1.$$

From the first of these equations we obtain, after integrating,

$$f(x, y) = x^2y + g(y).$$

Taking the partial derivative of the last expression with respect to y and setting the result equal to $N(x, y)$ gives

$$\frac{\partial f}{\partial y} = x^2 + g'(y) = x^2 - 1. \quad \leftarrow N(x, y)$$

It follows that $g'(y) = -1$ and $g(y) = -y$. Hence $f(x, y) = x^2y - y$, so the solution of the equation in implicit form is $x^2y - y = c$. The explicit form of the solution is easily seen to be $y = c/(1 - x^2)$ and is defined on any interval not containing either $x = 1$ or $x = -1$. ≡

≡ **Note** The solution of the DE in Example 1 is *not* $f(x, y) = x^2y - y$. Rather, it is $f(x, y) = c$; if a constant is used in the integration of $g'(y)$, we can then write the solution as $f(x, y) = 0$. Note, too, that the equation could be solved by separation of variables.

EXAMPLE 2 Solving an Exact DE

Solve $(e^{2y} - y\cos xy)\,dx + (2xe^{2y} - x\cos xy + 2y)\,dy = 0$.

SOLUTION The equation is exact because

$$\frac{\partial M}{\partial y} = 2e^{2y} + xy\sin xy - \cos xy = \frac{\partial N}{\partial x}.$$

Hence a function $f(x, y)$ exists for which

$$M(x, y) = \frac{\partial f}{\partial x} \quad \text{and} \quad N(x, y) = \frac{\partial f}{\partial y}.$$

Now, for variety, we shall start with the assumption that $\partial f/\partial y = N(x, y)$; that is,

$$\frac{\partial f}{\partial y} = 2xe^{2y} - x\cos xy + 2y$$

$$f(x, y) = 2x\int e^{2y}\,dy - x\int \cos xy\,dy + 2\int y\,dy.$$

Remember, the reason x can come out in front of the symbol $\int$ is that in the integration with respect to y, x is treated as an ordinary constant. It follows that

$$f(x, y) = xe^{2y} - \sin xy + y^2 + h(x)$$

$$\frac{\partial f}{\partial x} = e^{2y} - y\cos xy + h'(x) = e^{2y} - y\cos xy, \quad \leftarrow M(x, y)$$

and so $h'(x) = 0$ or $h(x) = c$. Hence a family of solutions is

$$xe^{2y} - \sin xy + y^2 + c = 0.$$ ≡

EXAMPLE 3 An Initial-Value Problem

Solve $\dfrac{dy}{dx} = \dfrac{xy^2 - \cos x \sin x}{y(1 - x^2)}$, $y(0) = 2$.

SOLUTION By writing the differential equation in the form

$$(\cos x \sin x - xy^2)\, dx + y(1 - x^2)\, dy = 0,$$

we recognize that the equation is exact because

$$\frac{\partial M}{\partial y} = -2xy = \frac{\partial N}{\partial x}.$$

Now
$$\frac{\partial f}{\partial y} = y(1 - x^2)$$

$$f(x, y) = \frac{y^2}{2}(1 - x^2) + h(x)$$

$$\frac{\partial f}{\partial x} = -xy^2 + h'(x) = \cos x \sin x - xy^2.$$

The last equation implies that $h'(x) = \cos x \sin x$. Integrating gives

$$h(x) = -\int (\cos x)(-\sin x\, dx) = -\frac{1}{2}\cos^2 x.$$

Thus
$$\frac{y^2}{2}(1 - x^2) - \frac{1}{2}\cos^2 x = c_1 \quad \text{or} \quad y^2(1 - x^2) - \cos^2 x = c, \tag{7}$$

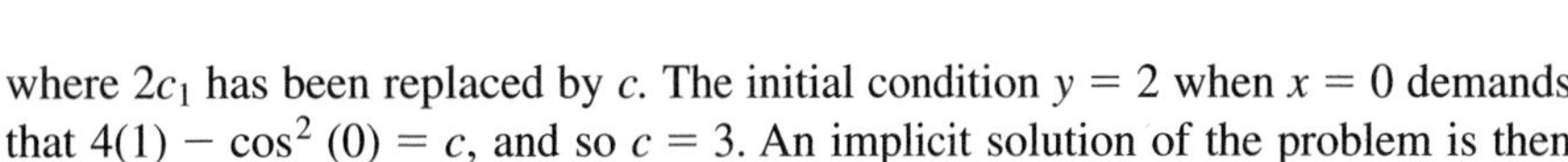

where $2c_1$ has been replaced by c. The initial condition $y = 2$ when $x = 0$ demands that $4(1) - \cos^2(0) = c$, and so $c = 3$. An implicit solution of the problem is then $y^2(1 - x^2) - \cos^2 x = 3$.

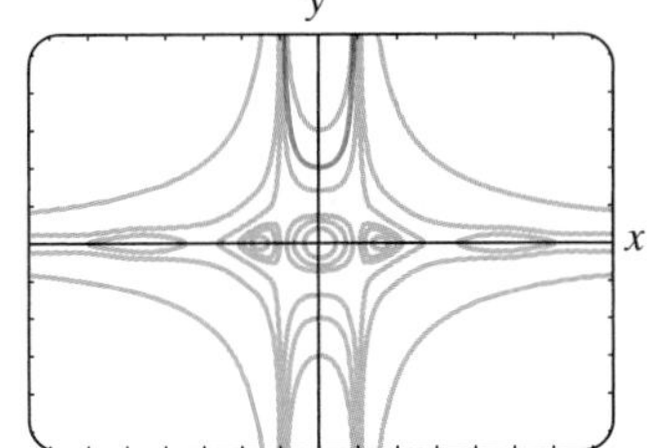

FIGURE 2.4.1 Solution curves of DE in Example 3

The solution curve of the IVP is the curve drawn in blue in Figure 2.4.1; it is part of an interesting family of curves. The graphs of the members of the one-parameter family of solutions given in (7) can be obtained in several ways, two of which are using software to graph level curves (as discussed in Section 2.2) and using a graphing utility to carefully graph the explicit functions obtained for various values of c by solving $y^2 = (c + \cos^2 x)/(1 - x^2)$ for y. ≡

≡ Integrating Factors

Recall from Section 2.3 that the left-hand side of the linear equation $y' + P(x)y = f(x)$ can be transformed into a derivative when we multiply the equation by an integrating factor. The same basic idea sometimes works for a nonexact differential equation $M(x, y)\, dx + N(x, y)\, dy = 0$. That is, it is sometimes possible to find an **integrating factor** $\mu(x, y)$ so that after multiplying, the left-hand side of

$$\mu(x, y)M(x, y)\, dx + \mu(x, y)N(x, y)\, dy = 0 \tag{8}$$

is an exact differential. In an attempt to find μ, we turn to the criterion (4) for exactness. Equation (8) is exact if and only if $(\mu M)_y = (\mu N)_x$, where the subscripts denote partial derivatives. By the Product Rule of differentiation the last equation is the same as $\mu M_y + \mu_y M = \mu N_x + \mu_x N$ or

$$\mu_x N - \mu_y M = (M_y - N_x)\mu. \tag{9}$$

Although M, N, M_y, and N_x are known functions of x and y, the difficulty here in determining the unknown $\mu(x, y)$ from (9) is that we must solve a partial differential equation. Since we are not prepared to do that, we make a simplifying assumption. Suppose μ is a function of one variable; for example, say that μ depends only on x. In this case, $\mu_x = d\mu/dx$ and $\mu_y = 0$, so (9) can be written as

$$\frac{d\mu}{dx} = \frac{M_y - N_x}{N}\mu. \tag{10}$$

We are still at an impasse if the quotient $(M_y - N_x)/N$ depends on both x and y. However, if after all obvious algebraic simplifications are made, the quotient $(M_y - N_x)/N$ turns out to depend solely on the variable x, then (10) is a first-order ordinary differential equation. We can finally determine μ because (10) is *separable* as well as *linear.* It follows from either Section 2.2 or Section 2.3 that $\mu(x) = e^{\int((M_y - N_x)/N)dx}$. In like manner, it follows from (9) that if μ depends only on the variable y, then

$$\frac{d\mu}{dy} = \frac{N_x - M_y}{M}\mu. \tag{11}$$

In this case, if $(N_x - M_y)/M$ is a function of y only, then we can solve (11) for μ.

We summarize the results for the differential equation

$$M(x, y)\,dx + N(x, y)\,dy = 0. \tag{12}$$

- If $(M_y - N_x)/N$ is a function of x alone, then an integrating factor for (12) is

$$\mu(x) = e^{\int \frac{M_y - N_x}{N}dx}. \tag{13}$$

- If $(N_x - M_y)/M$ is a function of y alone, then an integrating factor for (12) is

$$\mu(y) = e^{\int \frac{N_x - M_y}{M}dy}. \tag{14}$$

EXAMPLE 4 A Nonexact DE Made Exact

The nonlinear first-order differential equation

$$xy\,dx + (2x^2 + 3y^2 - 20)\,dy = 0$$

is not exact. With the identifications $M = xy$, $N = 2x^2 + 3y^2 - 20$, we find the partial derivatives $M_y = x$ and $N_x = 4x$. The first quotient from (13) gets us nowhere, since

$$\frac{M_y - N_x}{N} = \frac{x - 4x}{2x^2 + 3y^2 - 20} = \frac{-3x}{2x^2 + 3y^2 - 20}$$

depends on x and y. However, (14) yields a quotient that depends only on y:

$$\frac{N_x - M_y}{M} = \frac{4x - x}{xy} = \frac{3x}{xy} = \frac{3}{y}.$$

The integrating factor is then $e^{\int 3dy/y} = e^{3\ln y} = e^{\ln y^3} = y^3$. After we multiply the given DE by $\mu(y) = y^3$, the resulting equation is

$$xy^4\,dx + (2x^2y^3 + 3y^5 - 20y^3)\,dy = 0.$$

You should verify that the last equation is now exact as well as show, using the method of this section, that a family of solutions is $\frac{1}{2}x^2y^4 + \frac{1}{2}y^6 - 5y^4 = c$. ≡

REMARKS

(*i*) When testing an equation for exactness, make sure it is of the precise form $M(x, y)\ dx + N(x, y)\ dy = 0$. Sometimes a differential equation is written $G(x, y)\ dx = H(x, y)\ dy$. In this case, first rewrite it as $G(x, y)\ dx - H(x, y)\ dy = 0$ and then identify $M(x, y) = G(x, y)$ and $N(x, y) = -H(x, y)$ before using (4).

(*ii*) In some texts on differential equations the study of exact equations precedes that of linear DEs. Then the method for finding integrating factors just discussed can be used to derive an integrating factor for $y' + P(x)y = f(x)$. By rewriting the last equation in the differential form $(P(x)y - f(x))\,dx + dy = 0$, we see that

$$\frac{M_y - N_x}{N} = P(x).$$

From (13) we arrive at the already familiar integrating factor $e^{\int P(x)dx}$ used in Section 2.3.

EXERCISES 2.4

Answers to selected odd-numbered problems begin on page ANS-2.

In Problems 1–20 determine whether the given differential equation is exact. If it is exact, solve it.

1. $(2x - 1)\,dx + (3y + 7)\,dy = 0$

2. $(2x + y)\,dx - (x + 6y)\,dy = 0$

3. $(5x + 4y)\,dx + (4x - 8y^3)\,dy = 0$

4. $(\sin y - y \sin x)\,dx + (\cos x + x \cos y - y)\,dy = 0$

5. $(2xy^2 - 3)\,dx + (2x^2y + 4)\,dy = 0$

6. $\left(2y - \dfrac{1}{x} + \cos 3x\right)\dfrac{dy}{dx} + \dfrac{y}{x^2} - 4x^3 + 3y \sin 3x = 0$

7. $(x^2 - y^2)\,dx + (x^2 - 2xy)\,dy = 0$

8. $\left(1 + \ln x + \dfrac{y}{x}\right)dx = (1 - \ln x)\,dy$

9. $(x - y^3 + y^2 \sin x)\,dx = (3xy^2 + 2y \cos x)\,dy$

10. $(x^3 + y^3)\,dx + 3xy^2\,dy = 0$

11. $(y \ln y - e^{-xy})\,dx + \left(\dfrac{1}{y} + x \ln y\right)dy = 0$

12. $(3x^2y + e^y)\,dx + (x^3 + xe^y - 2y)\,dy = 0$

13. $x\dfrac{dy}{dx} = 2xe^x - y + 6x^2$

14. $\left(1 - \dfrac{3}{y} + x\right)\dfrac{dy}{dx} + y = \dfrac{3}{x} - 1$

15. $\left(x^2y^3 - \dfrac{1}{1 + 9x^2}\right)\dfrac{dx}{dy} + x^3y^2 = 0$

16. $(5y - 2x)y' - 2y = 0$

17. $(\tan x - \sin x \sin y)\,dx + \cos x \cos y\,dy = 0$

18. $(2y \sin x \cos x - y + 2y^2e^{xy^2})\,dx$
$\quad = (x - \sin^2 x - 4xye^{xy^2})\,dy$

19. $(4t^3y - 15t^2 - y)\,dt + (t^4 + 3y^2 - t)\,dy = 0$

20. $\left(\dfrac{1}{t} + \dfrac{1}{t^2} - \dfrac{y}{t^2 + y^2}\right)dt + \left(ye^y + \dfrac{t}{t^2 + y^2}\right)dy = 0$

In Problems 21–26 solve the given initial-value problem.

21. $(x + y)^2\,dx + (2xy + x^2 - 1)\,dy = 0,\quad y(1) = 1$

22. $(e^x + y)\,dx + (2 + x + ye^y)\,dy = 0,\quad y(0) = 1$

23. $(4y + 2t - 5)\,dt + (6y + 4t - 1)\,dy = 0,\quad y(-1) = 2$

24. $\left(\dfrac{3y^2 - t^2}{y^5}\right)\dfrac{dy}{dt} + \dfrac{t}{2y^4} = 0,\quad y(1) = 1$

25. $(y^2 \cos x - 3x^2y - 2x)\,dx$
$\quad + (2y \sin x - x^3 + \ln y)\,dy = 0,\quad y(0) = e$

26. $\left(\dfrac{1}{1 + y^2} + \cos x - 2xy\right)\dfrac{dy}{dx} = y(y + \sin x),\ y(0) = 1$

In Problems 27 and 28 find the value of k so that the given differential equation is exact.

27. $(y^3 + kxy^4 - 2x)\,dx + (3xy^2 + 20x^2y^3)\,dy = 0$

28. $(6xy^3 + \cos y)\,dx + (2kx^2y^2 - x \sin y)\,dy = 0$

In Problems 29 and 30 verify that the given differential equation is not exact. Multiply the given differential equation by the indicated integrating factor $\mu(x, y)$ and verify that the new equation is exact. Solve.

29. $(-xy \sin x + 2y \cos x)\,dx + 2x \cos x\,dy = 0;$
$\quad \mu(x, y) = xy$

30. $(x^2 + 2xy - y^2)\,dx + (y^2 + 2xy - x^2)\,dy = 0;$
$\quad \mu(x, y) = (x + y)^{-2}$

In Problems 31–36 solve the given differential equation by finding, as in Example 4, an appropriate integrating factor.

31. $(2y^2 + 3x)\,dx + 2xy\,dy = 0$

32. $y(x + y + 1)\,dx + (x + 2y)\,dy = 0$

33. $6xy\,dx + (4y + 9x^2)\,dy = 0$

34. $\cos x\,dx + \left(1 + \dfrac{2}{y}\right)\sin x\,dy = 0$

35. $(10 - 6y + e^{-3x})\,dx - 2\,dy = 0$

36. $(y^2 + xy^3)\,dx + (5y^2 - xy + y^3 \sin y)\,dy = 0$

In Problems 37 and 38 solve the given initial-value problem by finding, as in Example 4, an appropriate integrating factor.

37. $x\,dx + (x^2y + 4y)\,dy = 0, \quad y(4) = 0$

38. $(x^2 + y^2 - 5)\,dx = (y + xy)\,dy, \quad y(0) = 1$

39. **(a)** Show that a one-parameter family of solutions of the equation

$$(4xy + 3x^2)\,dx + (2y + 2x^2)\,dy = 0$$

is $x^3 + 2x^2y + y^2 = c$.

(b) Show that the initial conditions $y(0) = -2$ and $y(1) = 1$ determine the same implicit solution.

(c) Find explicit solutions $y_1(x)$ and $y_2(x)$ of the differential equation in part (a) such that $y_1(0) = -2$ and $y_2(1) = 1$. Use a graphing utility to graph $y_1(x)$ and $y_2(x)$.

Discussion Problems

40. Consider the concept of an integrating factor used in Problems 29–38. Are the two equations $M\,dx + N\,dy = 0$ and $\mu M\,dx + \mu N\,dy = 0$ necessarily equivalent in the sense that a solution of one is also a solution of the other? Discuss.

41. Reread Example 3 and then discuss why we can conclude that the interval of definition of the explicit solution of the IVP (the blue curve in Figure 2.4.1) is $(-1, 1)$.

42. Discuss how the functions $M(x, y)$ and $N(x, y)$ can be found so that each differential equation is exact. Carry out your ideas.

(a) $M(x, y)\,dx + \left(xe^{xy} + 2xy + \dfrac{1}{x}\right)dy = 0$

(b) $\left(x^{-1/2}y^{1/2} + \dfrac{x}{x^2 + y}\right)dx + N(x, y)\,dy = 0$

43. Differential equations are sometimes solved by having a clever idea. Here is a little exercise in cleverness: Although the differential equation $(x - \sqrt{x^2 + y^2})\,dx + y\,dy = 0$ is not exact, show how the rearrangement $(x\,dx + y\,dy)/\sqrt{x^2 + y^2} = dx$ and the observation $\frac{1}{2}d(x^2 + y^2) = x\,dx + y\,dy$ can lead to a solution.

44. True or False: Every separable first-order equation $dy/dx = g(x)h(y)$ is exact.

Mathematical Model

45. Falling Chain A portion of a uniform chain of length 8 ft is loosely coiled around a peg at the edge of a high horizontal platform, and the remaining portion of the chain hangs at rest over the edge of the platform. See Figure 2.4.2. Suppose that the length of the overhanging chain is 3 ft, that the chain weighs 2 lb/ft, and that the positive direction is downward. Starting at $t = 0$ seconds, the weight of the overhanging portion causes the chain on the table to uncoil smoothly and to fall to the floor. If $x(t)$ denotes the length of the chain overhanging the table at time $t > 0$, then $v = dx/dt$ is its velocity. When all resistive forces are ignored, it can be shown that a mathematical model relating v to x is given by

$$xv\frac{dv}{dx} + v^2 = 32x.$$

(a) Rewrite this model in differential form. Proceed as in Problems 31–36 and solve the DE for v in terms of x by finding an appropriate integrating factor. Find an explicit solution $v(x)$.

(b) Determine the velocity with which the chain leaves the platform.

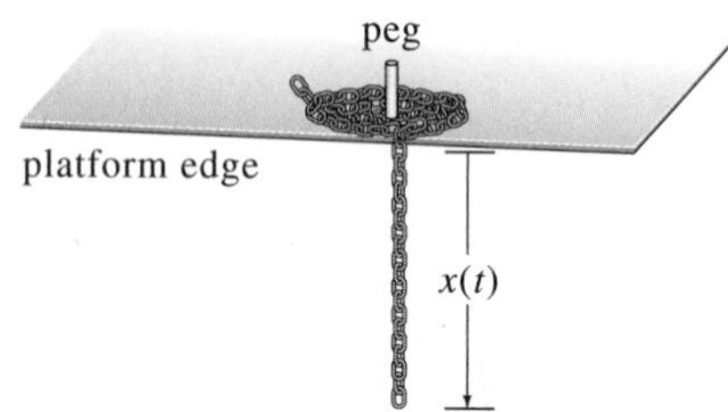

FIGURE 2.4.2 Uncoiling chain in Problem 45

Computer Lab Assignments

46. Streamlines

(a) The solution of the differential equation

$$\frac{2xy}{(x^2 + y^2)^2}\,dx + \left[1 + \frac{y^2 - x^2}{(x^2 + y^2)^2}\right]dy = 0$$

is a family of curves that can be interpreted as streamlines of a fluid flow around a circular object whose boundary is described by the equation $x^2 + y^2 = 1$. Solve this DE and note the solution $f(x, y) = c$ for $c = 0$.

(b) Use a CAS to plot the streamlines for $c = 0, \pm 0.2, \pm 0.4, \pm 0.6$, and ± 0.8 in three different ways. First, use the *contourplot* of a CAS. Second, solve for x in terms of the variable y. Plot the resulting two functions of y for the given values of c, and then combine the graphs. Third, use the CAS to solve a cubic equation for y in terms of x.

2.5 SOLUTIONS BY SUBSTITUTIONS

REVIEW MATERIAL

- Techniques of integration
- Separation of variables
- Solution of linear DEs

INTRODUCTION We usually solve a differential equation by recognizing it as a certain kind of equation (say, separable, linear, or exact) and then carrying out a procedure, consisting of *equation-specific mathematical steps,* that yields a solution of the equation. But it is not uncommon to be stumped by a differential equation because it does not fall into one of the classes of equations that we know how to solve. The procedures that are discussed in this section may be helpful in this situation.

Substitutions Often the first step in solving a differential equation consists of transforming it into another differential equation by means of a **substitution.** For example, suppose we wish to transform the first-order differential equation $dy/dx = f(x, y)$ by the substitution $y = g(x, u)$, where u is regarded as a function of the variable x. If g possesses first-partial derivatives, then the Chain Rule

$$\frac{dy}{dx} = \frac{\partial g}{\partial x}\frac{dx}{dx} + \frac{\partial g}{\partial u}\frac{du}{dx} \qquad \text{gives} \qquad \frac{dy}{dx} = g_x(x, u) + g_u(x, u)\frac{du}{dx}.$$

If we replace dy/dx by the foregoing derivative and replace y in $f(x, y)$ by $g(x, u)$, then the DE $dy/dx = f(x, y)$ becomes $g_x(x, u) + g_u(x, u)\dfrac{du}{dx} = f(x, g(x, u))$, which, solved for du/dx, has the form $\dfrac{du}{dx} = F(x, u)$. If we can determine a solution $u = \phi(x)$ of this last equation, then a solution of the original differential equation is $y = g(x, \phi(x))$.

In the discussion that follows we examine three different kinds of first-order differential equations that are solvable by means of a substitution.

Homogeneous Equations If a function f possesses the property $f(tx, ty) = t^{\alpha}f(x, y)$ for some real number α, then f is said to be a **homogeneous function** of degree α. For example, $f(x, y) = x^3 + y^3$ is a homogeneous function of degree 3, since

$$f(tx, ty) = (tx)^3 + (ty)^3 = t^3(x^3 + y^3) = t^3f(x, y),$$

whereas $f(x, y) = x^3 + y^3 + 1$ is not homogeneous. A first-order DE in differential form

$$M(x, y)\,dx + N(x, y)\,dy = 0 \tag{1}$$

is said to be **homogeneous*** if both coefficient functions M and N are homogeneous functions of the *same* degree. In other words, (1) is homogeneous if

$$M(tx, ty) = t^{\alpha}M(x, y) \qquad \text{and} \qquad N(tx, ty) = t^{\alpha}N(x, y).$$

In addition, if M and N are homogeneous functions of degree α, we can also write

$$M(x, y) = x^{\alpha}M(1, u) \qquad \text{and} \qquad N(x, y) = x^{\alpha}N(1, u), \quad \text{where } u = y/x, \tag{2}$$

*Here the word *homogeneous* does not mean the same as it did in the *Remarks* at the end of Section 2.3. Recall that a linear first-order equation $a_1(x)y' + a_0(x)y = g(x)$ is homogeneous when $g(x) = 0$.

and

$$M(x, y) = y^{\alpha}M(v, 1) \qquad \text{and} \qquad N(x, y) = y^{\alpha}N(v, 1), \quad \text{where } v = x/y. \tag{3}$$

See Problem 31 in Exercises 2.5. Properties (2) and (3) suggest the substitutions that can be used to solve a homogeneous differential equation. Specifically, *either* of the substitutions $y = ux$ or $x = vy$, where u and v are new dependent variables, will reduce a homogeneous equation to a *separable* first-order differential equation. To show this, observe that as a consequence of (2) a homogeneous equation $M(x, y)\,dx + N(x, y)\,dy = 0$ can be rewritten as

$$x^{\alpha}M(1, u)\,dx + x^{\alpha}N(1, u)\,dy = 0 \qquad \text{or} \qquad M(1, u)\,dx + N(1, u)\,dy = 0,$$

where $u = y/x$ or $y = ux$. By substituting the differential $dy = u\,dx + x\,du$ into the last equation and gathering terms, we obtain a separable DE in the variables u and x:

$$M(1, u)\,dx + N(1, u)[u\,dx + x\,du] = 0$$

$$[M(1, u) + uN(1, u)]\,dx + xN(1, u)\,du = 0$$

or

$$\frac{dx}{x} + \frac{N(1, u)\,du}{M(1, u) + uN(1, u)} = 0.$$

At this point we offer the same advice as in the preceding sections: Do not memorize anything here (especially the last formula); rather, *work through the procedure each time.* The proof that the substitutions $x = vy$ and $dx = v\,dy + y\,dv$ also lead to a separable equation follows in an analogous manner from (3).

EXAMPLE 1 Solving a Homogeneous DE

Solve $(x^2 + y^2)\,dx + (x^2 - xy)\,dy = 0$.

SOLUTION Inspection of $M(x, y) = x^2 + y^2$ and $N(x, y) = x^2 - xy$ shows that these coefficients are homogeneous functions of degree 2. If we let $y = ux$, then $dy = u\,dx + x\,du$, so after substituting, the given equation becomes

$$(x^2 + u^2x^2)\,dx + (x^2 - ux^2)[u\,dx + x\,du] = 0$$

$$x^2(1 + u)\,dx + x^3(1 - u)\,du = 0$$

$$\frac{1 - u}{1 + u}\,du + \frac{dx}{x} = 0$$

$$\left[-1 + \frac{2}{1 + u}\right]du + \frac{dx}{x} = 0. \quad \leftarrow \text{long division}$$

After integration the last line gives

$$-u + 2\ln|1 + u| + \ln|x| = \ln|c|$$

$$-\frac{y}{x} + 2\ln\left|1 + \frac{y}{x}\right| + \ln|x| = \ln|c|. \quad \leftarrow \text{resubstituting } u = y/x$$

Using the properties of logarithms, we can write the preceding solution as

$$\ln\left|\frac{(x + y)^2}{cx}\right| = \frac{y}{x} \qquad \text{or} \qquad (x + y)^2 = cxe^{y/x}.$$

Although either of the indicated substitutions can be used for every homogeneous differential equation, in practice we try $x = vy$ whenever the function $M(x, y)$ is simpler than $N(x, y)$. Also it could happen that after using one substitution, we may encounter integrals that are difficult or impossible to evaluate in closed form; switching substitutions may result in an easier problem.

Bernoulli's Equation The differential equation

$$\frac{dy}{dx} + P(x)y = f(x)y^n, \tag{4}$$

where n is any real number, is called **Bernoulli's equation.** Note that for $n = 0$ and $n = 1$, equation (4) is linear. For $n \neq 0$ and $n \neq 1$ the substitution $u = y^{1-n}$ reduces any equation of form (4) to a linear equation.

EXAMPLE 2 **Solving a Bernoulli DE**

Solve $x\dfrac{dy}{dx} + y = x^2y^2$.

SOLUTION We first rewrite the equation as

$$\frac{dy}{dx} + \frac{1}{x}y = xy^2$$

by dividing by x. With $n = 2$ we have $u = y^{-1}$ or $y = u^{-1}$. We then substitute

$$\frac{dy}{dx} = \frac{dy}{du}\frac{du}{dx} = -u^{-2}\frac{du}{dx} \qquad \leftarrow \text{Chain Rule}$$

into the given equation and simplify. The result is

$$\frac{du}{dx} - \frac{1}{x}u = -x.$$

The integrating factor for this linear equation on, say, $(0, \infty)$ is

$$e^{-\int dx/x} = e^{-\ln x} = e^{\ln x^{-1}} = x^{-1}.$$

Integrating $\qquad \dfrac{d}{dx}[x^{-1}u] = -1$

gives $x^{-1}u = -x + c$ or $u = -x^2 + cx$. Since $u = y^{-1}$, we have $y = 1/u$, so a solution of the given equation is $y = 1/(-x^2 + cx)$. ≡

Note that we have not obtained the general solution of the original nonlinear differential equation in Example 2, since $y = 0$ is a singular solution of the equation.

Reduction to Separation of Variables A differential equation of the form

$$\frac{dy}{dx} = f(Ax + By + C) \tag{5}$$

can always be reduced to an equation with separable variables by means of the substitution $u = Ax + By + C$, $B \neq 0$. Example 3 illustrates the technique.

EXAMPLE 3 **An Initial-Value Problem**

Solve $\dfrac{dy}{dx} = (-2x + y)^2 - 7, \quad y(0) = 0.$

SOLUTION If we let $u = -2x + y$, then $du/dx = -2 + dy/dx$, so the differential equation is transformed into

$$\frac{du}{dx} + 2 = u^2 - 7 \qquad \text{or} \qquad \frac{du}{dx} = u^2 - 9.$$

The last equation is separable. Using partial fractions

$$\frac{du}{(u-3)(u+3)} = dx \quad \text{or} \quad \frac{1}{6}\left[\frac{1}{u-3} - \frac{1}{u+3}\right] du = dx$$

and then integrating yields

$$\frac{1}{6}\ln\left|\frac{u-3}{u+3}\right| = x + c_1 \quad \text{or} \quad \frac{u-3}{u+3} = e^{6x+6c_1} = ce^{6x}. \quad \leftarrow \text{replace } e^{6c_1} \text{ by } c$$

Solving the last equation for u and then resubstituting gives the solution

$$u = \frac{3(1+ce^{6x})}{1-ce^{6x}} \quad \text{or} \quad y = 2x + \frac{3(1+ce^{6x})}{1-ce^{6x}}. \tag{6}$$

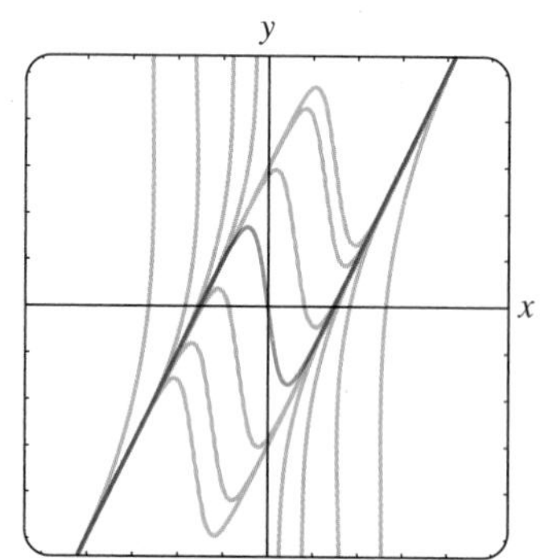

FIGURE 2.5.1 Solutions of DE in Example 3

Finally, applying the initial condition $y(0) = 0$ to the last equation in (6) gives $c = -1$. Figure 2.5.1, obtained with the aid of a graphing utility, shows the graph of the particular solution $y = 2x + \dfrac{3(1-e^{6x})}{1+e^{6x}}$ in dark blue, along with the graphs of some other members of the family of solutions (6). ≡

EXERCISES 2.5

Answers to selected odd-numbered problems begin on page ANS-2.

Each DE in Problems 1–14 is homogeneous.

In Problems 1–10 solve the given differential equation by using an appropriate substitution.

1. $(x - y)\,dx + x\,dy = 0$

2. $(x + y)\,dx + x\,dy = 0$

3. $x\,dx + (y - 2x)\,dy = 0$

4. $y\,dx = 2(x + y)\,dy$

5. $(y^2 + yx)\,dx - x^2\,dy = 0$

6. $(y^2 + yx)\,dx + x^2\,dy = 0$

7. $\dfrac{dy}{dx} = \dfrac{y - x}{y + x}$

8. $\dfrac{dy}{dx} = \dfrac{x + 3y}{3x + y}$

9. $-y\,dx + \left(x + \sqrt{xy}\right)dy = 0$

10. $x\dfrac{dy}{dx} = y + \sqrt{x^2 - y^2}, \quad x > 0$

In Problems 11–14 solve the given initial-value problem.

11. $xy^2\dfrac{dy}{dx} = y^3 - x^3, \quad y(1) = 2$

12. $(x^2 + 2y^2)\dfrac{dx}{dy} = xy, \quad y(-1) = 1$

13. $(x + ye^{y/x})\,dx - xe^{y/x}\,dy = 0, \quad y(1) = 0$

14. $y\,dx + x(\ln x - \ln y - 1)\,dy = 0, \quad y(1) = e$

Each DE in Problems 15–22 is a Bernoulli equation.

In Problems 15–20 solve the given differential equation by using an appropriate substitution.

15. $x\dfrac{dy}{dx} + y = \dfrac{1}{y^2}$

16. $\dfrac{dy}{dx} - y = e^x y^2$

17. $\dfrac{dy}{dx} = y(xy^3 - 1)$

18. $x\dfrac{dy}{dx} - (1 + x)y = xy^2$

19. $t^2\dfrac{dy}{dt} + y^2 = ty$

20. $3(1 + t^2)\dfrac{dy}{dt} = 2ty(y^3 - 1)$

In Problems 21 and 22 solve the given initial-value problem.

21. $x^2\dfrac{dy}{dx} - 2xy = 3y^4, \quad y(1) = \tfrac{1}{2}$

22. $y^{1/2}\dfrac{dy}{dx} + y^{3/2} = 1, \quad y(0) = 4$

Each DE in Problems 23–30 is of the form given in (5).

In Problems 23–28 solve the given differential equation by using an appropriate substitution.

23. $\dfrac{dy}{dx} = (x + y + 1)^2$

24. $\dfrac{dy}{dx} = \dfrac{1 - x - y}{x + y}$

25. $\dfrac{dy}{dx} = \tan^2(x + y)$

26. $\dfrac{dy}{dx} = \sin(x + y)$

27. $\dfrac{dy}{dx} = 2 + \sqrt{y - 2x + 3}$

28. $\dfrac{dy}{dx} = 1 + e^{y-x+5}$

In Problems 29 and 30 solve the given initial-value problem.

29. $\dfrac{dy}{dx} = \cos(x + y), \quad y(0) = \pi/4$

30. $\dfrac{dy}{dx} = \dfrac{3x + 2y}{3x + 2y + 2}, \quad y(-1) = -1$

Discussion Problems

31. Explain why it is always possible to express any homogeneous differential equation $M(x, y)\,dx + N(x, y)\,dy = 0$ in the form

$$\frac{dy}{dx} = F\left(\frac{y}{x}\right).$$

You might start by proving that

$$M(x, y) = x^{\alpha}M(1, y/x) \quad \text{and} \quad N(x, y) = x^{\alpha}N(1, y/x).$$

32. Put the homogeneous differential equation

$$(5x^2 - 2y^2)\,dx - xy\,dy = 0$$

into the form given in Problem 31.

33. **(a)** Determine two singular solutions of the DE in Problem 10.

(b) If the initial condition $y(5) = 0$ is as prescribed in Problem 10, then what is the largest interval I over which the solution is defined? Use a graphing utility to graph the solution curve for the IVP.

34. In Example 3 the solution $y(x)$ becomes unbounded as $x \to \pm\infty$. Nevertheless, $y(x)$ is asymptotic to a curve as $x \to -\infty$ and to a different curve as $x \to \infty$. What are the equations of these curves?

35. The differential equation $dy/dx = P(x) + Q(x)y + R(x)y^2$ is known as **Riccati's equation.**

(a) A Riccati equation can be solved by a succession of two substitutions *provided* that we know a particular solution y_1 of the equation. Show that the substitution $y = y_1 + u$ reduces Riccati's equation to a Bernoulli equation (4) with $n = 2$. The Bernoulli equation can then be reduced to a linear equation by the substitution $w = u^{-1}$.

(b) Find a one-parameter family of solutions for the differential equation

$$\frac{dy}{dx} = -\frac{4}{x^2} - \frac{1}{x}y + y^2$$

where $y_1 = 2/x$ is a known solution of the equation.

36. Determine an appropriate substitution to solve

$$xy' = y\ln(xy).$$

Mathematical Models

37. Falling Chain In Problem 45 in Exercises 2.4 we saw that a mathematical model for the velocity v of a chain slipping off the edge of a high horizontal platform is

$$xv\frac{dv}{dx} + v^2 = 32x.$$

In that problem you were asked to solve the DE by converting it into an exact equation using an integrating factor. This time solve the DE using the fact that it is a Bernoulli equation.

38. Population Growth In the study of population dynamics one of the most famous models for a growing but bounded population is the **logistic equation**

$$\frac{dP}{dt} = P(a - bP),$$

where a and b are positive constants. Although we will come back to this equation and solve it by an alternative method in Section 3.2, solve the DE this first time using the fact that it is a Bernoulli equation.

2.6 A NUMERICAL METHOD

INTRODUCTION A first-order differential equation $dy/dx = f(x, y)$ is a source of information. We started this chapter by observing that we could garner *qualitative* information from a first-order DE about its solutions even before we attempted to solve the equation. Then in Sections 2.2–2.5 we examined first-order DEs *analytically*—that is, we developed some procedures for obtaining explicit and implicit solutions. But a differential equation can a possess a solution, yet we may not be able to obtain it analytically. So to round out the picture of the different types of analyses of differential equations, we conclude this chapter with a method by which we can "solve" the differential equation *numerically*—this means that the DE is used as the cornerstone of an algorithm for approximating the unknown solution.

In this section we are going to develop only the simplest of numerical methods—a method that utilizes the idea that a tangent line can be used to approximate the values of a function in a small neighborhood of the point of tangency. A more extensive treatment of numerical methods for ordinary differential equations is given in Chapter 9.

Using the Tangent Line Let us assume that the first-order initial-value problem

$$y' = f(x, y), \quad y(x_0) = y_0 \tag{1}$$

possesses a solution. One way of approximating this solution is to use tangent lines. For example, let $y(x)$ denote the unknown solution of the first-order initial-value problem $y' = 0.1\sqrt{y} + 0.4x^2$, $y(2) = 4$. The nonlinear differential equation in this IVP cannot be solved directly by any of the methods considered in Sections 2.2, 2.4, and 2.5; nevertheless, we can still find approximate numerical values of the unknown $y(x)$. Specifically, suppose we wish to know the value of $y(2.5)$. The IVP has a solution, and as the flow of the direction field of the DE in Figure 2.6.1(a) suggests, a solution curve must have a shape similar to the curve shown in blue.

The direction field in Figure 2.6.1(a) was generated with lineal elements passing through points in a grid with integer coordinates. As the solution curve passes through the initial point (2, 4), the lineal element at this point is a tangent line with slope given by $f(2, 4) = 0.1\sqrt{4} + 0.4(2)^2 = 1.8$. As is apparent in Figure 2.6.1(a) and the "zoom in" in Figure 2.6.1(b), when x is close to 2, the points on the solution curve are close to the points on the tangent line (the lineal element). Using the point (2, 4), the slope $f(2, 4) = 1.8$, and the point-slope form of a line, we find that an equation of the tangent line is $y = L(x)$, where $L(x) = 1.8x + 0.4$. This last equation, called a **linearization** of $y(x)$ at $x = 2$, can be used to approximate values of $y(x)$ within a small neighborhood of $x = 2$. If $y_1 = L(x_1)$ denotes the y-coordinate on the tangent line and $y(x_1)$ is the y-coordinate on the solution curve corresponding to an x-coordinate x_1 that is close to $x = 2$, then $y(x_1) \approx y_1$. If we choose, say, $x_1 = 2.1$, then $y_1 = L(2.1) = 1.8(2.1) + 0.4 = 4.18$, so $y(2.1) \approx 4.18$.

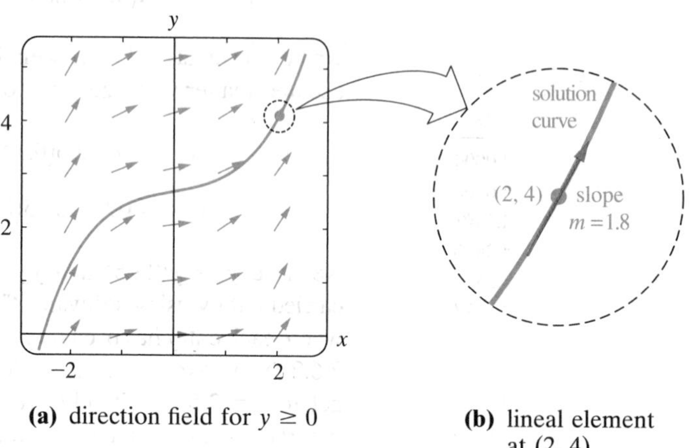

(a) direction field for $y \geq 0$ **(b)** lineal element at (2, 4)

FIGURE 2.6.1 Magnification of a neighborhood about the point (2, 4)

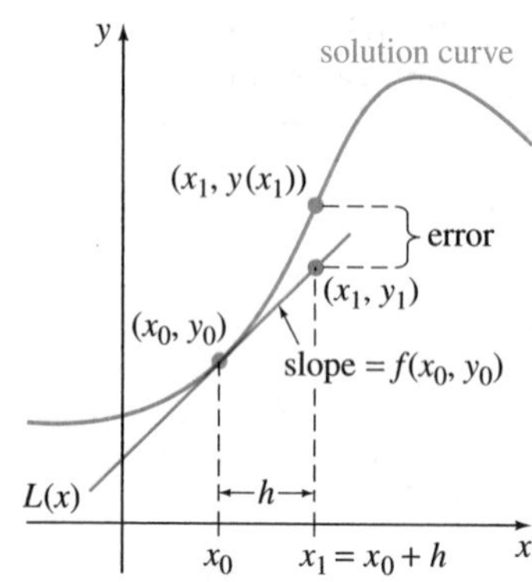

FIGURE 2.6.2 Approximating $y(x_1)$ using a tangent line

Euler's Method To generalize the procedure just illustrated, we use the linearization of the unknown solution $y(x)$ of (1) at $x = x_0$:

$$L(x) = y_0 + f(x_0, y_0)(x - x_0). \tag{2}$$

The graph of this linearization is a straight line tangent to the graph of $y = y(x)$ at the point (x_0, y_0). We now let h be a positive increment of the x-axis, as shown in Figure 2.6.2. Then by replacing x by $x_1 = x_0 + h$ in (2), we get

$$L(x_1) = y_0 + f(x_0, y_0)(x_0 + h - x_0) \quad \text{or} \quad y_1 = y_0 + hf(x_1, y_1),$$

where $y_1 = L(x_1)$. The point (x_1, y_1) on the tangent line is an approximation to the point $(x_1, y(x_1))$ on the solution curve. Of course, the accuracy of the approximation $L(x_1) \approx y(x_1)$ or $y_1 \approx y(x_1)$ depends heavily on the size of the increment h. Usually, we must choose this **step size** to be "reasonably small." We now repeat the process using a second "tangent line" at (x_1, y_1).* By identifying the new starting

*This is not an actual tangent line, since (x_1, y_1) lies on the first tangent and not on the solution curve.

15. The number 0 is a critical point of the autonomous differential equation $dx/dt = x^n$, where n is a positive integer. For what values of n is 0 asymptotically stable? Semi-stable? Unstable? Repeat for the differential equation $dx/dt = -x^n$.

16. Consider the differential equation $dP/dt = f(P)$, where

$$f(P) = -0.5P^3 - 1.7P + 3.4.$$

The function $f(P)$ has one real zero, as shown in Figure 2.R.3. Without attempting to solve the differential equation, estimate the value of $\lim_{t\to\infty} P(t)$.

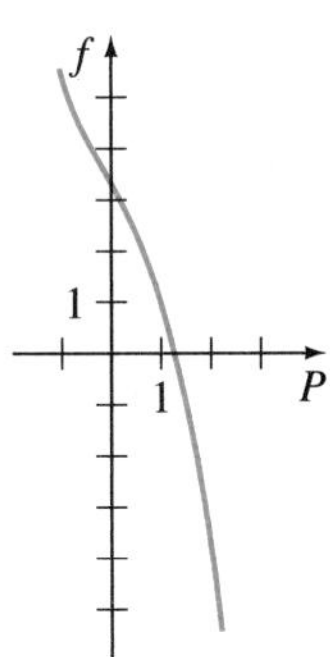

FIGURE 2.R.3 Graph for Problem 16

17. Figure 2.R.4 is a portion of a direction field of a differential equation $dy/dx = f(x, y)$. By hand, sketch two different solution curves—one that is tangent to the lineal element shown in black and one that is tangent to the lineal element shown in red.

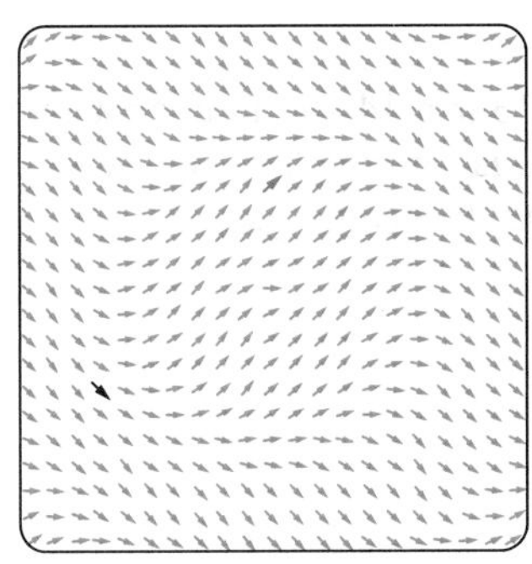

FIGURE 2.R.4 Portion of a direction field for Problem 17

18. Classify each differential equation as separable, exact, linear, homogeneous, or Bernoulli. Some equations may be more than one kind. Do not solve.

(a) $\dfrac{dy}{dx} = \dfrac{x - y}{x}$ **(b)** $\dfrac{dy}{dx} = \dfrac{1}{y - x}$

(c) $(x + 1)\dfrac{dy}{dx} = -y + 10$ **(d)** $\dfrac{dy}{dx} = \dfrac{1}{x(x - y)}$

(e) $\dfrac{dy}{dx} = \dfrac{y^2 + y}{x^2 + x}$ **(f)** $\dfrac{dy}{dx} = 5y + y^2$

(g) $y\,dx = (y - xy^2)\,dy$ **(h)** $x\dfrac{dy}{dx} = ye^{x/y} - x$

(i) $xy\,y' + y^2 = 2x$ **(j)** $2xy\,y' + y^2 = 2x^2$

(k) $y\,dx + x\,dy = 0$

(l) $\left(x^2 + \dfrac{2y}{x}\right)dx = (3 - \ln x^2)\,dy$

(m) $\dfrac{dy}{dx} = \dfrac{x}{y} + \dfrac{y}{x} + 1$ **(n)** $\dfrac{y}{x^2}\dfrac{dy}{dx} + e^{2x^3+y^2} = 0$

In Problems 19–26 solve the given differential equation.

19. $(y^2 + 1)\,dx = y\sec^2 x\,dy$

20. $y(\ln x - \ln y)\,dx = (x\ln x - x\ln y - y)\,dy$

21. $(6x + 1)y^2\dfrac{dy}{dx} + 3x^2 + 2y^3 = 0$

22. $\dfrac{dx}{dy} = -\dfrac{4y^2 + 6xy}{3y^2 + 2x}$

23. $t\dfrac{dQ}{dt} + Q = t^4\ln t$

24. $(2x + y + 1)y' = 1$

25. $(x^2 + 4)\,dy = (2x - 8xy)\,dx$

26. $(2r^2\cos\theta\sin\theta + r\cos\theta)\,d\theta + (4r + \sin\theta - 2r\cos^2\theta)\,dr = 0$

In Problems 27 and 28 solve the given initial-value problem and give the largest interval I on which the solution is defined.

27. $\sin x\dfrac{dy}{dx} + (\cos x)y = 0, \quad y(7\pi/6) = -2$

28. $\dfrac{dy}{dt} + 2(t + 1)y^2 = 0, \quad y(0) = -\frac{1}{8}$

29. **(a)** Without solving, explain why the initial-value problem

$$\frac{dy}{dx} = \sqrt{y}, \quad y(x_0) = y_0$$

has no solution for $y_0 < 0$.

(b) Solve the initial-value problem in part (a) for $y_0 > 0$ and find the largest interval I on which the solution is defined.

30. **(a)** Find an implicit solution of the initial-value problem

$$\frac{dy}{dx} = \frac{y^2 - x^2}{xy}, \quad y(1) = -\sqrt{2}.$$

(b) Find an explicit solution of the problem in part (a) and give the largest interval I over which the solution is defined. A graphing utility may be helpful here.

31. Graphs of some members of a family of solutions for a first-order differential equation $dy/dx = f(x, y)$ are shown in Figure 2.R.5. The graphs of two implicit solutions, one that passes through the point $(1, -1)$ and one that passes through $(-1, 3)$, are shown in blue. Reproduce the figure on a piece of paper. With colored pencils trace out the solution curves for the solutions $y = y_1(x)$ and $y = y_2(x)$ defined by the implicit solutions such that $y_1(1) = -1$ and $y_2(-1) = 3$, respectively. Estimate the intervals on which the solutions $y = y_1(x)$ and $y = y_2(x)$ are defined.

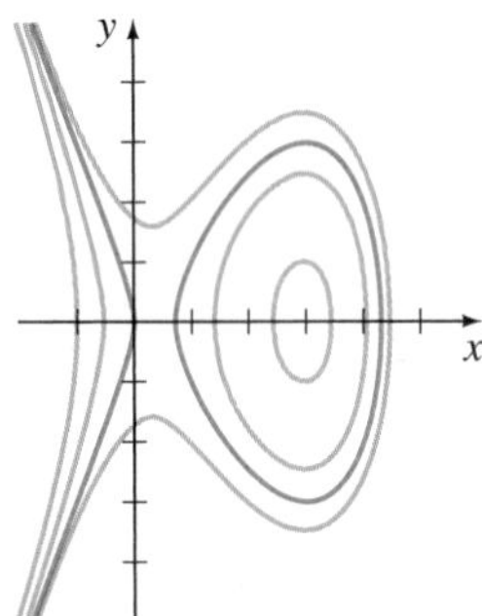

FIGURE 2.R.5 Graph for Problem 31

32. Use Euler's method with step size $h = 0.1$ to approximate $y(1.2)$, where $y(x)$ is a solution of the initial-value problem $y' = 1 + x\sqrt{y}$, $y(1) = 9$.

In Problems 33 and 34 each figure represents a portion of a direction field of an autonomous first-order differential equation $dy/dx = f(y)$. Reproduce the figure on a separate piece of paper and then complete the direction field over the grid. The points of the grid are (mh, nh), where $h = \frac{1}{2}$, m and n integers, $-7 \leq m \leq 7$, $-7 \leq n \leq 7$. In each direction field, sketch by hand an approximate solution curve that passes through each of the solid points shown in red. Discuss: Does it appear that the DE possesses critical points in the interval $-3.5 \leq y \leq 3.5$? If so, classify the critical points as asymptotically stable, unstable, or semi-stable.

33.

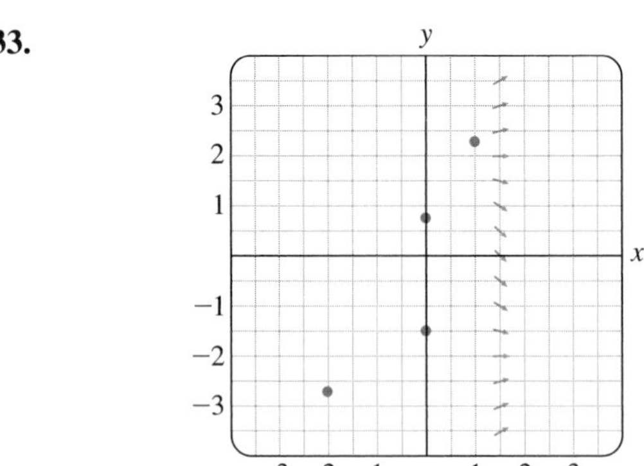

FIGURE 2.R.6 Portion of a direction field for Problem 33

34.

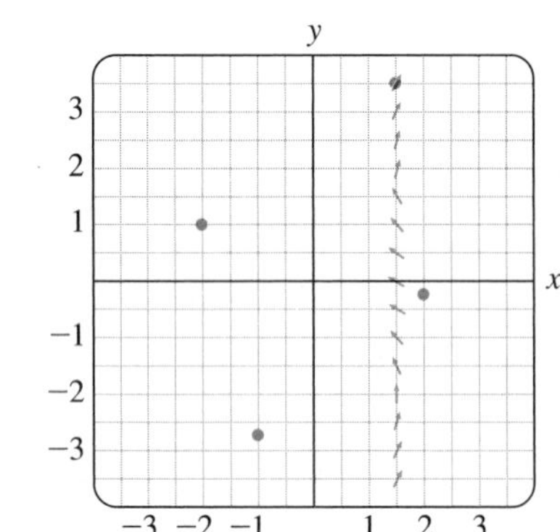

FIGURE 2.R.7 Portion of a direction field for Problem 34

point as (x_1, y_1) with (x_0, y_0) in the above discussion, we obtain an approximation $y_2 \approx y(x_2)$ corresponding to two steps of length h from x_0, that is, $x_2 = x_1 + h = x_0 + 2h$, and

$$y(x_2) = y(x_0 + 2h) = y(x_1 + h) \approx y_2 = y_1 + hf(x_1, y_1).$$

Continuing in this manner, we see that $y_1, y_2, y_3, \ldots$, can be defined recursively by the general formula

$$y_{n+1} = y_n + hf(x_n, y_n), \tag{3}$$

where $x_n = x_0 + nh$, $n = 0, 1, 2, \ldots$. This procedure of using successive "tangent lines" is called **Euler's method.**

EXAMPLE 1 Euler's Method

Consider the initial-value problem $y' = 0.1\sqrt{y} + 0.4x^2$, $y(2) = 4$. Use Euler's method to obtain an approximation of $y(2.5)$ using first $h = 0.1$ and then $h = 0.05$.

SOLUTION With the identification $f(x, y) = 0.1\sqrt{y} + 0.4x^2$, (3) becomes

$$y_{n+1} = y_n + h\left(0.1\sqrt{y_n} + 0.4x_n^2\right).$$

Then for $h = 0.1$, $x_0 = 2$, $y_0 = 4$, and $n = 0$ we find

$$y_1 = y_0 + h\left(0.1\sqrt{y_0} + 0.4x_0^2\right) = 4 + 0.1\left(0.1\sqrt{4} + 0.4(2)^2\right) = 4.18,$$

which, as we have already seen, is an estimate to the value of $y(2.1)$. However, if we use the smaller step size $h = 0.05$, it takes two steps to reach $x = 2.1$. From

$$y_1 = 4 + 0.05\left(0.1\sqrt{4} + 0.4(2)^2\right) = 4.09$$

$$y_2 = 4.09 + 0.05\left(0.1\sqrt{4.09} + 0.4(2.05)^2\right) = 4.18416187$$

we have $y_1 \approx y(2.05)$ and $y_2 \approx y(2.1)$. The remainder of the calculations were carried out by using software. The results are summarized in Tables 2.6.1 and 2.6.2, where each entry has been rounded to four decimal places. We see in Tables 2.6.1 and 2.6.2 that it takes five steps with $h = 0.1$ and 10 steps with $h = 0.05$, respectively, to get to $x = 2.5$. Intuitively, we would expect that $y_{10} = 5.0997$ corresponding to $h = 0.05$ is the better approximation of $y(2.5)$ than the value $y_5 = 5.0768$ corresponding to $h = 0.1$. ≡

TABLE 2.6.1 $h = 0.1$

x_n	y_n
2.00	4.0000
2.10	4.1800
2.20	4.3768
2.30	4.5914
2.40	4.8244
2.50	5.0768

TABLE 2.6.2 $h = 0.05$

x_n	y_n
2.00	4.0000
2.05	4.0900
2.10	4.1842
2.15	4.2826
2.20	4.3854
2.25	4.4927
2.30	4.6045
2.35	4.7210
2.40	4.8423
2.45	4.9686
2.50	5.0997

In Example 2 we apply Euler's method to a differential equation for which we have already found a solution. We do this to compare the values of the approximations y_n at each step with the true or actual values of the solution $y(x_n)$ of the initial-value problem.

EXAMPLE 2 Comparison of Approximate and Actual Values

Consider the initial-value problem $y' = 0.2xy$, $y(1) = 1$. Use Euler's method to obtain an approximation of $y(1.5)$ using first $h = 0.1$ and then $h = 0.05$.

SOLUTION With the identification $f(x, y) = 0.2xy$, (3) becomes

$$y_{n+1} = y_n + h(0.2x_n y_n)$$

where $x_0 = 1$ and $y_0 = 1$. Again with the aid of computer software we obtain the values in Tables 2.6.3 and 2.6.4 on page 78.

TABLE 2.6.3 $h = 0.1$

x_n	y_n	Actual value	Abs. error	% Rel. error
1.00	1.0000	1.0000	0.0000	0.00
1.10	1.0200	1.0212	0.0012	0.12
1.20	1.0424	1.0450	0.0025	0.24
1.30	1.0675	1.0714	0.0040	0.37
1.40	1.0952	1.1008	0.0055	0.50
1.50	1.1259	1.1331	0.0073	0.64

TABLE 2.6.4 $h = 0.05$

x_n	y_n	Actual value	Abs. error	% Rel. error
1.00	1.0000	1.0000	0.0000	0.00
1.05	1.0100	1.0103	0.0003	0.03
1.10	1.0206	1.0212	0.0006	0.06
1.15	1.0318	1.0328	0.0009	0.09
1.20	1.0437	1.0450	0.0013	0.12
1.25	1.0562	1.0579	0.0016	0.16
1.30	1.0694	1.0714	0.0020	0.19
1.35	1.0833	1.0857	0.0024	0.22
1.40	1.0980	1.1008	0.0028	0.25
1.45	1.1133	1.1166	0.0032	0.29
1.50	1.1295	1.1331	0.0037	0.32

In Example 1 the true or actual values were calculated from the known solution $y = e^{0.1(x^2-1)}$. (Verify.) The **absolute error** is defined to be

$$|\textit{actual value} - \textit{approximation}|.$$

The **relative error** and **percentage relative error** are, in turn,

$$\frac{\textit{absolute error}}{|\textit{actual value}|} \quad \text{and} \quad \frac{\textit{absolute error}}{|\textit{actual value}|} \times 100.$$

It is apparent from Tables 2.6.3 and 2.6.4 that the accuracy of the approximations improves as the step size h decreases. Also, we see that even though the percentage relative error is growing with each step, it does not appear to be that bad. But you should not be deceived by one example. If we simply change the coefficient of the right side of the DE in Example 2 from 0.2 to 2, then at $x_n = 1.5$ the percentage relative errors increase dramatically. See Problem 4 in Exercises 2.6.

A Caveat Euler's method is just one of many different ways in which a solution of a differential equation can be approximated. Although attractive for its simplicity, *Euler's method is seldom used in serious calculations.* It was introduced here simply to give you a first taste of numerical methods. We will go into greater detail in discussing numerical methods that give significantly greater accuracy, notably the **fourth order Runge-Kutta method,** referred to as the **RK4 method,** in Chapter 9.

Numerical Solvers Regardless of whether we can actually find an explicit or implicit solution, if a solution of a differential equation exists, it represents a smooth curve in the Cartesian plane. The basic idea behind *any* numerical method for first-order ordinary differential equations is to somehow approximate the y-values of a solution for preselected values of x. We start at a specified initial point (x_0, y_0) on a solution curve and proceed to calculate in a step-by-step fashion a sequence of points $(x_1, y_1), (x_2, y_2), \ldots, (x_n, y_n)$ whose y-coordinates y_i approximate the y-coordinates $y(x_i)$ of points $(x_1, y(x_1)), (x_2, y(x_2)), \ldots, (x_n, y(x_n))$ that lie on the graph of the usually unknown solution $y(x)$. By taking the x-coordinates close together (that is, for small values of h) and by joining the points (x_1, y_1), $(x_2, y_2), \ldots, (x_n, y_n)$ with short line segments, we obtain a polygonal curve whose qualitative characteristics we hope are close to those of an actual solution curve. Drawing curves is something that is well suited to a computer. A computer program written to either implement a numerical method or render a visual representation of an approximate solution curve fitting the numerical data produced by this method is referred to as a **numerical solver.** Many different numerical solvers are commercially available, either embedded in a larger software package, such as a computer

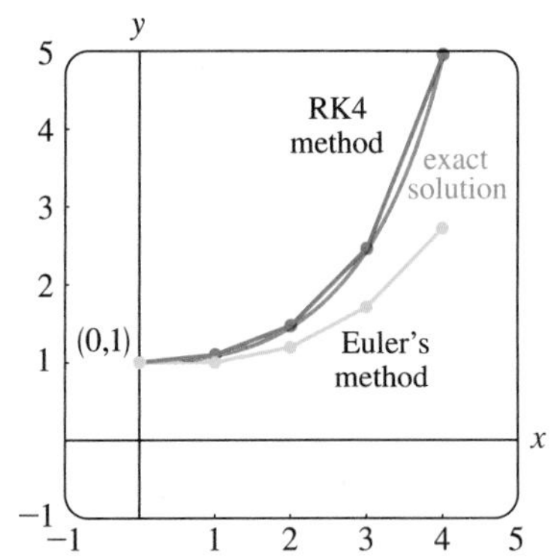

FIGURE 2.6.3 Comparison of the Runge-Kutta (RK4) and Euler methods

algebra system, or provided as a stand-alone package. Some software packages simply plot the generated numerical approximations, whereas others generate hard numerical data as well as the corresponding approximate or **numerical solution curves.** By way of illustration of the connect-the-dots nature of the graphs produced by a numerical solver, the two colored polygonal graphs in Figure 2.6.3 are the numerical solution curves for the initial-value problem $y' = 0.2xy$, $y(0) = 1$ on the interval $[0, 4]$ obtained from Euler's method and the RK4 method using the step size $h = 1$. The blue smooth curve is the graph of the exact solution $y = e^{0.1x^2}$ of the IVP. Notice in Figure 2.6.3 that, even with the ridiculously large step size of $h = 1$, the RK4 method produces the more believable "solution curve." The numerical solution curve obtained from the RK4 method is indistinguishable from the actual solution curve on the interval $[0, 4]$ when a more typical step size of $h = 0.1$ is used.

Using a Numerical Solver Knowledge of the various numerical methods is not necessary in order to use a numerical solver. A solver usually requires that the differential equation be expressed in normal form $dy/dx = f(x, y)$. Numerical solvers that generate only curves usually require that you supply $f(x, y)$ and the initial data x_0 and y_0 and specify the desired numerical method. If the idea is to approximate the numerical value of $y(a)$, then a solver may additionally require that you state a value for h or, equivalently, give the number of steps that you want to take to get from $x = x_0$ to $x = a$. For example, if we wanted to approximate $y(4)$ for the IVP illustrated in Figure 2.6.3, then, starting at $x = 0$ it would take four steps to reach $x = 4$ with a step size of $h = 1$; 40 steps is equivalent to a step size of $h = 0.1$. Although we will not delve here into the many problems that one can encounter when attempting to approximate mathematical quantities, you should at least be aware of the fact that a numerical solver may break down near certain points or give an incomplete or misleading picture when applied to some first-order differential equations in the normal form. Figure 2.6.4 illustrates the graph obtained by applying Euler's method to a certain first-order initial-value problem $dy/dx = f(x, y)$, $y(0) = 1$. Equivalent results were obtained using three different commercial numerical solvers, yet the graph is hardly a plausible solution curve. (Why?) There are several avenues of recourse when a numerical solver has difficulties; three of the more obvious are decrease the step size, use another numerical method, and try a different numerical solver.

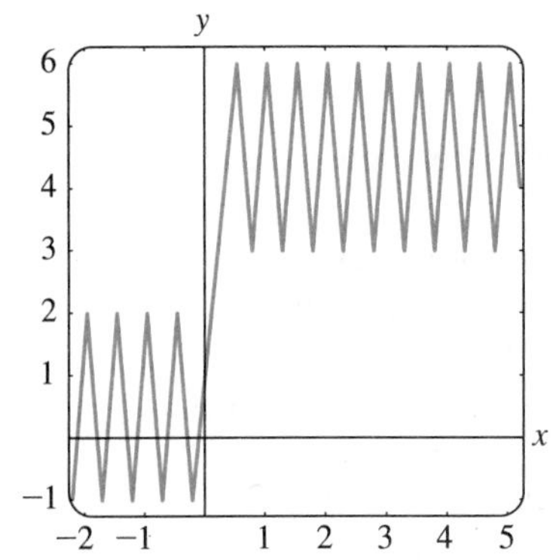

FIGURE 2.6.4 A not-very helpful numerical solution curve

EXERCISES 2.6

Answers to selected odd-numbered problems begin on page ANS-3.

In Problems 1 and 2 use Euler's method to obtain a four-decimal approximation of the indicated value. Carry out the recursion of (3) by hand, first using $h = 0.1$ and then using $h = 0.05$.

1. $y' = 2x - 3y + 1$, $y(1) = 5$; $y(1.2)$

2. $y' = x + y^2$, $y(0) = 0$; $y(0.2)$

In Problems 3 and 4 use Euler's method to obtain a four-decimal approximation of the indicated value. First use $h = 0.1$ and then use $h = 0.05$. Find an explicit solution for each initial-value problem and then construct tables similar to Tables 2.6.3 and 2.6.4.

3. $y' = y$, $y(0) = 1$; $y(1.0)$

4. $y' = 2xy$, $y(1) = 1$; $y(1.5)$

In Problems 5–10 use a numerical solver and Euler's method to obtain a four-decimal approximation of the indicated value. First use $h = 0.1$ and then use $h = 0.05$.

5. $y' = e^{-y}$, $y(0) = 0$; $y(0.5)$

6. $y' = x^2 + y^2$, $y(0) = 1$; $y(0.5)$

7. $y' = (x - y)^2$, $y(0) = 0.5$; $y(0.5)$

8. $y' = xy + \sqrt{y}$, $y(0) = 1$; $y(0.5)$

9. $y' = xy^2 - \dfrac{y}{x}$, $y(1) = 1$; $y(1.5)$

10. $y' = y - y^2$, $y(0) = 0.5$; $y(0.5)$

In Problems 11 and 12 use a numerical solver to obtain a numerical solution curve for the given initial-value problem. First use Euler's method and then the RK4 method. Use

$h = 0.25$ in each case. Superimpose both solution curves on the same coordinate axes. If possible, use a different color for each curve. Repeat, using $h = 0.1$ and $h = 0.05$.

11. $y' = 2(\cos x)y, \quad y(0) = 1$

12. $y' = y(10 - 2y), \quad y(0) = 1$

Discussion Problems

13. Use a numerical solver and Euler's method to approximate $y(1.0)$, where $y(x)$ is the solution to $y' = 2xy^2$, $y(0) = 1$. First use $h = 0.1$ and then use $h = 0.05$. Repeat, using the RK4 method. Discuss what might cause the approximations to $y(1.0)$ to differ so greatly.

Computer Lab Assignments

14. (a) Use a numerical solver and the RK4 method to graph the solution of the initial-value problem $y' = -2xy + 1, y(0) = 0$.

(b) Solve the initial-value problem by one of the analytic procedures developed earlier in this chapter.

(c) Use the analytic solution $y(x)$ found in part (b) and a CAS to find the coordinates of all relative extrema.

CHAPTER 2 IN REVIEW

Answers to selected odd-numbered problems begin on page ANS-3.

Answer Problems 1–12 without referring back to the text. Fill in the blanks or answer true or false.

1. The linear DE, $y' - ky = A$, where k and A are constants, is autonomous. The critical point ________ of the equation is a(n) ________ (attractor or repeller) for $k > 0$ and a(n) ________ (attractor or repeller) for $k < 0$.

2. The initial-value problem $x\dfrac{dy}{dx} - 4y = 0, y(0) = k$, has an infinite number of solutions for $k =$ ________ and no solution for $k =$ ________.

3. The linear DE, $y' + k_1y = k_2$, where k_1 and k_2 are nonzero constants, always possesses a constant solution. ________

4. The linear DE, $a_1(x)y' + a_0(x)y = 0$ is also separable. ________

5. An example of a nonlinear third-order differential equation in normal form is ________.

6. The first-order DE $\dfrac{dr}{d\theta} = r\theta + r + \theta + 1$ is not separable. ________

7. Every autonomous DE $dy/dx = f(y)$ is separable. ________

8. By inspection, two solutions of the differential equation $y' + |y| = 2$ are ________.

9. If $y' = e^xy$, then $y =$ ________.

10. If a differentiable function $y(x)$ satisfies $y' = |x|$, $y(-1) = 2$, then $y(x) =$ ________.

11. $y = e^{\cos x}\displaystyle\int_0^x te^{-\cos t}\,dt$ is a solution of the linear first-order differential equation ________.

12. An example of an autonomous linear first-order DE with a single critical point -3 is ________, whereas an autonomous nonlinear first-order DE with a single critical point -3 is ________.

In Problems 13 and 14 construct an autonomous first-order differential equation $dy/dx = f(y)$ whose phase portrait is consistent with the given figure.

13.

FIGURE 2.R.1 Graph for Problem 13

14.

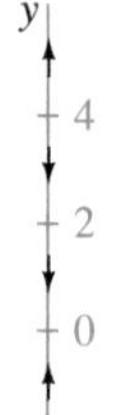

FIGURE 2.R.2 Graph for Problem 14

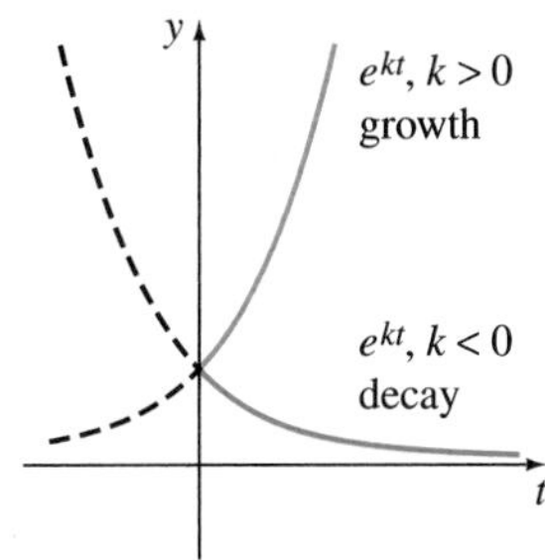

FIGURE 3.1.2 Growth ($k > 0$) and decay ($k < 0$)

Notice in Example 1 that the actual number P_0 of bacteria present at time $t = 0$ played no part in determining the time required for the number in the culture to triple. The time necessary for an initial population of, say, 100 or 1,000,000 bacteria to triple is still approximately 2.71 hours.

As shown in Figure 3.1.2, the exponential function e^{kt} increases as t increases for $k > 0$ and decreases as t increases for $k < 0$. Thus problems describing growth (whether of populations, bacteria, or even capital) are characterized by a positive value of k, whereas problems involving decay (as in radioactive disintegration) yield a negative k value. Accordingly, we say that k is either a **growth constant** ($k > 0$) or a **decay constant** ($k < 0$).

Half-Life In physics the **half-life** is a measure of the stability of a radioactive substance. The half-life is simply the time it takes for one-half of the atoms in an initial amount A_0 to disintegrate, or transmute, into the atoms of another element. The longer the half-life of a substance, the more stable it is. For example, the half-life of highly radioactive radium, Ra-226, is about 1700 years. In 1700 years one-half of a given quantity of Ra-226 is transmuted into radon, Rn-222. The most commonly occurring uranium isotope, U-238, has a half-life of approximately 4,500,000,000 years. In about 4.5 billion years, one-half of a quantity of U-238 is transmuted into lead, Pb-206.

EXAMPLE 2 Half-Life of Plutonium

A breeder reactor converts relatively stable uranium-238 into the isotope plutonium-239. After 15 years it is determined that 0.043% of the initial amount A_0 of plutonium has disintegrated. Find the half-life of this isotope if the rate of disintegration is proportional to the amount remaining.

SOLUTION Let $A(t)$ denote the amount of plutonium remaining at time t. As in Example 1 the solution of the initial-value problem

$$\frac{dA}{dt} = kA, \quad A(0) = A_0$$

is $A(t) = A_0e^{kt}$. If 0.043% of the atoms of A_0 have disintegrated, then 99.957% of the substance remains. To find the decay constant k, we use $0.99957A_0 = A(15)$—that is, $0.99957A_0 = A_0e^{15k}$. Solving for k then gives $k = \frac{1}{15}\ln 0.99957 = -0.00002867$. Hence $A(t) = A_0e^{-0.00002867t}$. Now the half-life is the corresponding value of time at which $A(t) = \frac{1}{2}A_0$. Solving for t gives $\frac{1}{2}A_0 = A_0e^{-0.00002867t}$, or $\frac{1}{2} = e^{-0.00002867t}$. The last equation yields

$$t = \frac{\ln 2}{0.00002867} \approx 24{,}180 \text{ yr.}$$

Carbon Dating About 1950, a team of scientists at the University of Chicago led by the chemist Willard Libby devised a method using a radioactive isotope of carbon as a means of determining the approximate ages of carbonaceous fossilized matter. The theory of **carbon dating** is based on the fact that the radioisotope carbon-14 is produced in the atmosphere by the action of cosmic radiation on nitrogen-14. The ratio of the amount of C-14 to the stable C-12 in the atmosphere appears to be a constant, and as a consequence the proportionate amount of the isotope present in all living organisms is the same as that in the atmosphere. When a living organism dies, the absorption of C-14, by breathing, eating, or photosynthesis, ceases. By comparing the proportionate amount of C-14, say, in a fossil with the constant amount ratio found in the atmosphere, it is possible to obtain a reasonable estimation of its age. The method is based on the knowledge of the half-life of C-14. Libby's calculated

FIGURE 3.1.3 A page of the Gnostic Gospel of Judas

value of the half-life of C-14 was approximately 5600 years, but today the commonly accepted value of the half-life is approximately 5730 years. For his work, Libby was awarded the Nobel Prize for chemistry in 1960. Libby's method has been used to date wooden furniture found in Egyptian tombs, the woven flax wrappings of the Dead Sea Scrolls, a recently discovered copy of the Gnostic Gospel of Judas written on papyrus, and the cloth of the enigmatic Shroud of Turin. See Figure 3.1.3 and Problem 12 in Exercises 3.1.

EXAMPLE 3 Age of a Fossil

A fossilized bone is found to contain 0.1% of its original amount of C-14. Determine the age of the fossil.

SOLUTION The starting point is again $A(t) = A_0e^{kt}$. To determine the value of the decay constant k we use the fact that $\frac{1}{2}A_0 = A(5730)$ or $\frac{1}{2}A_0 = A_0e^{5730k}$. The last equation implies $5730k = \ln\frac{1}{2} = -\ln 2$ and so we get $k = -(\ln 2)/5730 = -0.00012097$. Therefore $A(t) = A_0e^{-0.00012097t}$. With $A(t) = 0.001A_0$ we have $0.001A_0 = A_0e^{-0.00012097t}$ and $-0.00012097t = \ln(0.001) = -\ln 1000$. Thus

$$t = \frac{\ln 1000}{0.00012097} \approx 57{,}100 \text{ years.}$$

The date found in Example 3 is really at the border of accuracy for this method. The usual carbon-14 technique is limited to about 10 half-lives of the isotope, or roughly 60,000 years. One reason for this limitation is that the chemical analysis needed to obtain an accurate measurement of the remaining C-14 becomes somewhat formidable around the point $0.001A_0$. Also, this analysis demands the destruction of a rather large sample of the specimen. If this measurement is accomplished indirectly, based on the actual radioactivity of the specimen, then it is very difficult to distinguish between the radiation from the specimen and the normal background radiation.* But recently the use of a particle accelerator has enabled scientists to separate the C-14 from the stable C-12 directly. When the precise value of the ratio of C-14 to C-12 is computed, the accuracy can be extended to 70,000 to 100,000 years. Other isotopic techniques, such as using potassium-40 and argon-40, can give dates of several million years. Nonisotopic methods based on the use of amino acids are also sometimes possible.

Newton's Law of Cooling/Warming In equation (3) of Section 1.3 we saw that the mathematical formulation of Newton's empirical law of cooling/warming of an object is given by the linear first-order differential equation

$$\frac{dT}{dt} = k(T - T_m), \tag{2}$$

where k is a constant of proportionality, $T(t)$ is the temperature of the object for $t > 0$, and T_m is the ambient temperature—that is, the temperature of the medium around the object. In Example 4 we assume that T_m is constant.

EXAMPLE 4 Cooling of a Cake

When a cake is removed from an oven, its temperature is measured at 300° F. Three minutes later its temperature is 200° F. How long will it take for the cake to cool off to a room temperature of 70° F?

*The number of disintegrations per minute per gram of carbon is recorded by using a Geiger counter. The lower level of detectability is about 0.1 disintegrations per minute per gram.

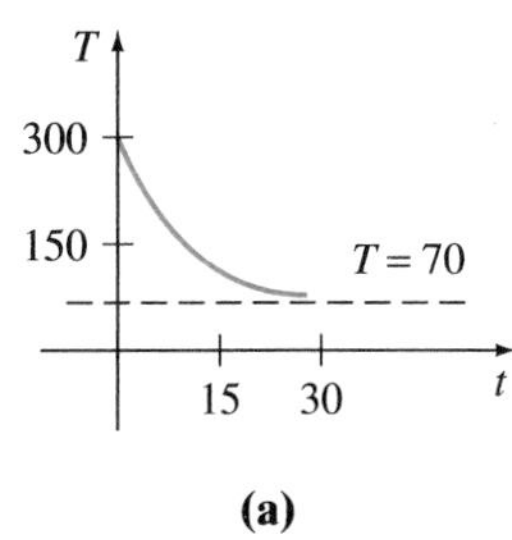

(a)

$T(t)$	t (min)
75°	20.1
74°	21.3
73°	22.8
72°	24.9
71°	28.6
70.5°	32.3

(b)

FIGURE 3.1.4 Temperature of cooling cake in Example 4

SOLUTION In (2) we make the identification $T_m = 70$. We must then solve the initial-value problem

$$\frac{dT}{dt} = k(T - 70), \quad T(0) = 300 \tag{3}$$

and determine the value of k so that $T(3) = 200$.

Equation (3) is both linear and separable. If we separate variables,

$$\frac{dT}{T - 70} = k\,dt,$$

yields $\ln|T - 70| = kt + c_1$, and so $T = 70 + c_2e^{kt}$. When $t = 0$, $T = 300$, so $300 = 70 + c_2$ gives $c_2 = 230$; therefore $T = 70 + 230e^{kt}$. Finally, the measurement $T(3) = 200$ leads to $e^{3k} = \frac{13}{23}$, or $k = \frac{1}{3}\ln\frac{13}{23} = -0.19018$. Thus

$$T(t) = 70 + 230e^{-0.19018t}. \tag{4}$$

We note that (4) furnishes no finite solution to $T(t) = 70$, since $\lim_{t\to\infty} T(t) = 70$. Yet we intuitively expect the cake to reach room temperature after a reasonably long period of time. How long is "long"? Of course, we should not be disturbed by the fact that the model (3) does not quite live up to our physical intuition. Parts (a) and (b) of Figure 3.1.4 clearly show that the cake will be approximately at room temperature in about one-half hour. ≡

The ambient temperature in (2) need not be a constant but could be a function $T_m(t)$ of time t. See Problem 18 in Exercises 3.1.

≡ **Mixtures** The mixing of two fluids sometimes gives rise to a linear first-order differential equation. When we discussed the mixing of two brine solutions in Section 1.3, we assumed that the rate $A'(t)$ at which the amount of salt in the mixing tank changes was a net rate:

$$\frac{dA}{dt} = (\text{input rate of salt}) - (\text{output rate of salt}) = R_{in} - R_{out}. \tag{5}$$

In Example 5 we solve equation (8) of Section 1.3.

EXAMPLE 5 Mixture of Two Salt Solutions

Recall that the large tank considered in Section 1.3 held 300 gallons of a brine solution. Salt was entering and leaving the tank; a brine solution was being pumped into the tank at the rate of 3 gal/min; it mixed with the solution there, and then the mixture was pumped out at the rate of 3 gal/min. The concentration of the salt in the inflow, or solution entering, was 2 lb/gal, so salt was entering the tank at the rate $R_{in} = (2\text{ lb/gal})\cdot(3\text{ gal/min}) = 6\text{ lb/min}$ and leaving the tank at the rate $R_{out} = (A/300\text{ lb/gal})\cdot(3\text{ gal/min}) = A/100\text{ lb/min}$. From this data and (5) we get equation (8) of Section 1.3. Let us pose the question: If 50 pounds of salt were dissolved initially in the 300 gallons, how much salt is in the tank after a long time?

SOLUTION To find the amount of salt $A(t)$ in the tank at time t, we solve the initial-value problem

$$\frac{dA}{dt} + \frac{1}{100}A = 6, \quad A(0) = 50.$$

Note here that the side condition is the initial amount of salt $A(0) = 50$ in the tank and *not* the initial amount of liquid in the tank. Now since the integrating factor of the linear differential equation is $e^{t/100}$, we can write the equation as

$$\frac{d}{dt}[e^{t/100}A] = 6e^{t/100}.$$

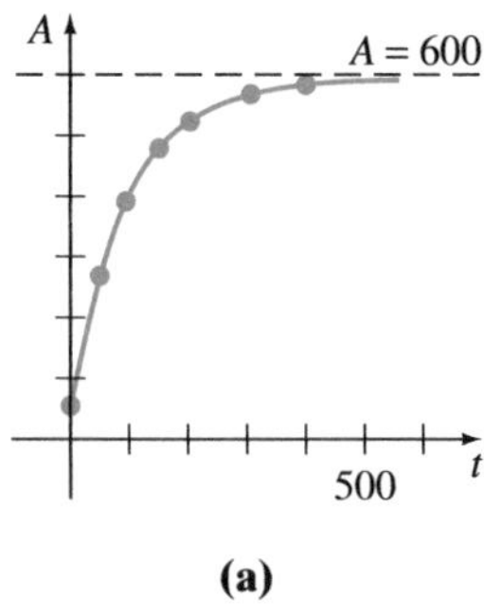

(a)

t (min)	A (lb)
50	266.41
100	397.67
150	477.27
200	525.57
300	572.62
400	589.93

(b)

FIGURE 3.1.5 Pounds of salt in the tank in Example 5

Integrating the last equation and solving for A gives the general solution $A(t) = 600 + ce^{-t/100}$. When $t = 0$, $A = 50$, so we find that $c = -550$. Thus the amount of salt in the tank at time t is given by

$$A(t) = 600 - 550e^{-t/100}. \tag{6}$$

The solution (6) was used to construct the table in Figure 3.1.5(b). Also, it can be seen from (6) and Figure 3.1.5(a) that $A(t) \to 600$ as $t \to \infty$. Of course, this is what we would intuitively expect; over a long time the number of pounds of salt in the solution must be (300 gal)(2 lb/gal) = 600 lb. ≡

In Example 5 we assumed that the rate at which the solution was pumped in was the same as the rate at which the solution was pumped out. However, this need not be the case; the mixed brine solution could be pumped out at a rate r_{out} that is faster or slower than the rate r_{in} at which the other brine solution is pumped in. The next example illustrates the case when the mixture is pumped out at rate that is *slower* than the rate at which the brine solution is being pumped into the tank.

EXAMPLE 6 **Example 5 Revisited**

If the well-stirred solution in Example 5 is pumped out at a slower rate of, say, $r_{out} = 2$ gal/min, then liquid will accumulate in the tank at the rate of $r_{in} - r_{out} = (3 - 2)$ gal/min $= 1$ gal/min. After t minutes,

$$(1 \text{ gal/min}) \cdot (t \text{ min}) = t \text{ gal}$$

will accumulate, so the tank will contain $300 + t$ gallons of brine. The concentration of the outflow is then $c(t) = A/(300 + t)$ lb/gal, and the output rate of salt is $R_{out} = c(t) \cdot r_{out}$, or

$$R_{out} = \left(\frac{A}{300 + t} \text{ lb/gal}\right) \cdot (2 \text{ gal/min}) = \frac{2A}{300 + t} \text{ lb/min}.$$

Hence equation (5) becomes

$$\frac{dA}{dt} = 6 - \frac{2A}{300 + t} \quad \text{or} \quad \frac{dA}{dt} + \frac{2}{300 + t}A = 6.$$

The integrating factor for the last equation is

$$e^{\int 2dt/(300+t)} = e^{2\ln(300+t)} = e^{\ln(300+t)^2} = (300 + t)^2$$

and so after multiplying by the factor the equation is cast into the form

$$\frac{d}{dt}\left[(300 + t)^2 A\right] = 6(300 + t)^2.$$

Integrating the last equation gives $(300 + t)^2 A = 2(300 + t)^3 + c$. By applying the initial condition $A(0) = 50$ and solving for A yields the solution $A(t) = 600 + 2t - (4.95 \times 10^7)(300 + t)^{-2}$. As Figure 3.1.6 shows, not unexpectedly, salt builds up in the tank over time, that is, $A \to \infty$ as $t \to \infty$. ≡

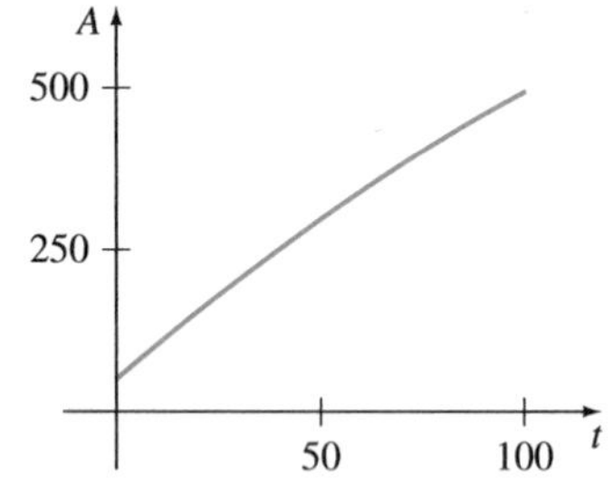

FIGURE 3.1.6 Graph of $A(t)$ in Example 6

≡ **Series Circuits** For a series circuit containing only a resistor and an inductor, Kirchhoff's second law states that the sum of the voltage drop across the inductor $(L(di/dt))$ and the voltage drop across the resistor (iR) is the same as the impressed voltage $(E(t))$ on the circuit. See Figure 3.1.7.

Thus we obtain the linear differential equation for the current $i(t)$,

$$L\frac{di}{dt} + Ri = E(t), \tag{7}$$

where L and R are constants known as the inductance and the resistance, respectively. The current $i(t)$ is also called the **response** of the system.

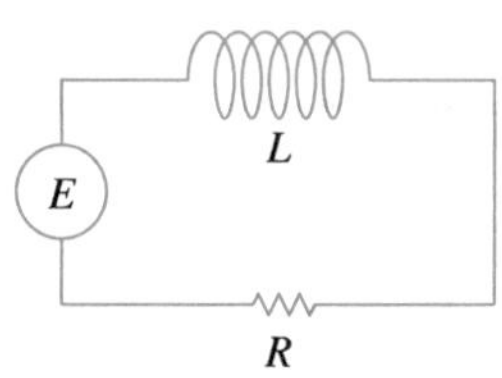

FIGURE 3.1.7 *LR*-series circuit

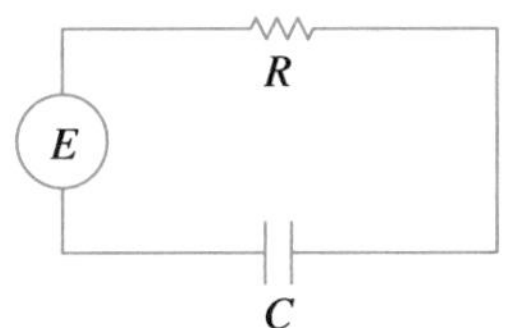

FIGURE 3.1.8 *RC*-series circuit

The voltage drop across a capacitor with capacitance C is given by $q(t)/C$, where q is the charge on the capacitor. Hence, for the series circuit shown in Figure 3.1.8, Kirchhoff's second law gives

$$Ri + \frac{1}{C}q = E(t). \tag{8}$$

But current i and charge q are related by $i = dq/dt$, so (8) becomes the linear differential equation

$$R\frac{dq}{dt} + \frac{1}{C}q = E(t). \tag{9}$$

EXAMPLE 7 Series Circuit

A 12-volt battery is connected to a series circuit in which the inductance is $\frac{1}{2}$ henry and the resistance is 10 ohms. Determine the current i if the initial current is zero.

SOLUTION From (7) we see that we must solve

$$\frac{1}{2}\frac{di}{dt} + 10i = 12,$$

subject to $i(0) = 0$. First, we multiply the differential equation by 2 and read off the integrating factor e^{20t}. We then obtain

$$\frac{d}{dt}[e^{20t}i] = 24e^{20t}.$$

Integrating each side of the last equation and solving for i gives $i(t) = \frac{6}{5} + ce^{-20t}$. Now $i(0) = 0$ implies that $0 = \frac{6}{5} + c$ or $c = -\frac{6}{5}$. Therefore the response is $i(t) = \frac{6}{5} - \frac{6}{5}e^{-20t}$. ≡

From (4) of Section 2.3 we can write a general solution of (7):

$$i(t) = \frac{e^{-(R/L)t}}{L}\int e^{(R/L)t}E(t)\,dt + ce^{-(R/L)t}. \tag{10}$$

In particular, when $E(t) = E_0$ is a constant, (10) becomes

$$i(t) = \frac{E_0}{R} + ce^{-(R/L)t}. \tag{11}$$

Note that as $t \to \infty$, the second term in equation (11) approaches zero. Such a term is usually called a **transient term;** any remaining terms are called the **steady-state** part of the solution. In this case E_0/R is also called the **steady-state current;** for large values of time it appears that the current in the circuit is simply governed by Ohm's law ($E = iR$).

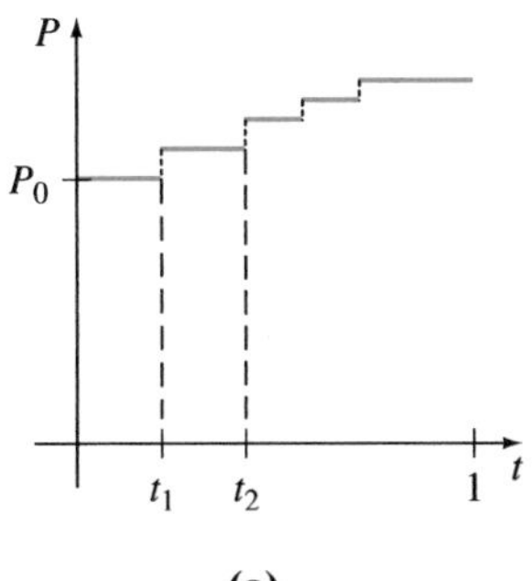

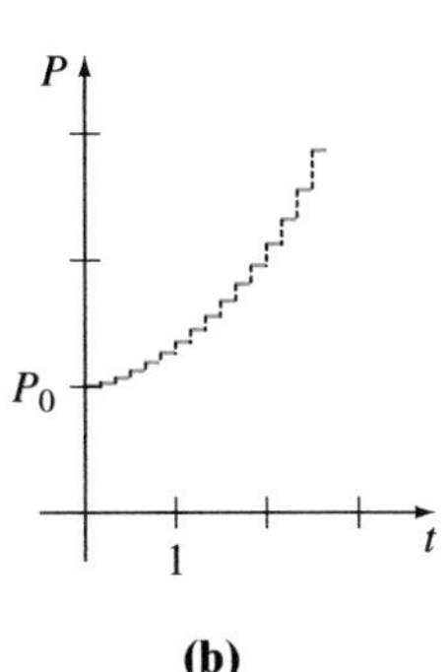

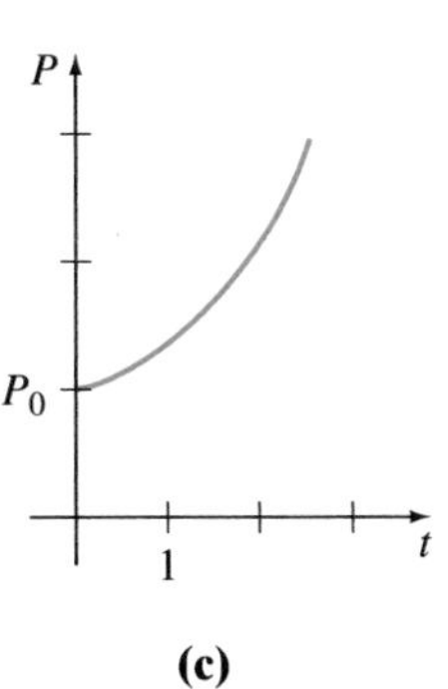

FIGURE 3.1.9 Population growth is a discrete process

REMARKS

The solution $P(t) = P_0e^{0.4055t}$ of the initial-value problem in Example 1 described the population of a colony of bacteria at any time $t > 0$. Of course, $P(t)$ is a continuous function that takes on *all* real numbers in the interval $P_0 \le P < \infty$. But since we are talking about a population, common sense dictates that P can take on only positive integer values. Moreover, we would not expect the population to grow continuously—that is, every second, every microsecond, and so on—as predicted by our solution; there may be intervals of time $[t_1, t_2]$ over which there is no growth at all. Perhaps, then, the graph shown in Figure 3.1.9(a) is a more realistic description of P than is the graph of an exponential function. Using a continuous function to describe a discrete phenomenon is often more a matter of convenience than of accuracy. However, for some purposes we may be satisfied if our model describes the system fairly closely when viewed macroscopically in time, as in Figures 3.1.9(b) and 3.1.9(c), rather than microscopically, as in Figure 3.1.9(a).

EXERCISES 3.1

Answers to selected odd-numbered problems begin on page ANS-3.

Growth and Decay

1. The population of a community is known to increase at a rate proportional to the number of people present at time t. If an initial population P_0 has doubled in 5 years, how long will it take to triple? To quadruple?

2. Suppose it is known that the population of the community in Problem 1 is 10,000 after 3 years. What was the initial population P_0? What will be the population in 10 years? How fast is the population growing at $t = 10$?

3. The population of a town grows at a rate proportional to the population present at time t. The initial population of 500 increases by 15% in 10 years. What will be the population in 30 years? How fast is the population growing at $t = 30$?

4. The population of bacteria in a culture grows at a rate proportional to the number of bacteria present at time t. After 3 hours it is observed that 400 bacteria are present. After 10 hours 2000 bacteria are present. What was the initial number of bacteria?

5. The radioactive isotope of lead, Pb-209, decays at a rate proportional to the amount present at time t and has a half-life of 3.3 hours. If 1 gram of this isotope is present initially, how long will it take for 90% of the lead to decay?

6. Initially 100 milligrams of a radioactive substance was present. After 6 hours the mass had decreased by 3%. If the rate of decay is proportional to the amount of the substance present at time t, find the amount remaining after 24 hours.

7. Determine the half-life of the radioactive substance described in Problem 6.

8. **(a)** Consider the initial-value problem $dA/dt = kA$, $A(0) = A_0$ as the model for the decay of a radioactive substance. Show that, in general, the half-life T of the substance is $T = -(\ln 2)/k$.

(b) Show that the solution of the initial-value problem in part (a) can be written $A(t) = A_0 2^{-t/T}$.

(c) If a radioactive substance has the half-life T given in part (a), how long will it take an initial amount A_0 of the substance to decay to $\frac{1}{8}A_0$?

9. When a vertical beam of light passes through a transparent medium, the rate at which its intensity I decreases is proportional to $I(t)$, where t represents the thickness of the medium (in feet). In clear seawater, the intensity 3 feet below the surface is 25% of the initial intensity I_0 of the incident beam. What is the intensity of the beam 15 feet below the surface?

10. When interest is compounded continuously, the amount of money increases at a rate proportional to the amount S present at time t, that is, $dS/dt = rS$, where r is the annual rate of interest.

(a) Find the amount of money accrued at the end of 5 years when \$5000 is deposited in a savings account drawing $5\frac{3}{4}\%$ annual interest compounded continuously.

(b) In how many years will the initial sum deposited have doubled?

(c) Use a calculator to compare the amount obtained in part (a) with the amount $S = 5000(1 + \frac{1}{4}(0.0575))^{5(4)}$ that is accrued when interest is compounded quarterly.

Carbon Dating

11. Archaeologists used pieces of burned wood, or charcoal, found at the site to date prehistoric paintings and drawings on walls and ceilings of a cave in Lascaux, France. See Figure 3.1.10. Use the information on page 86 to determine the approximate age of a piece of burned wood, if it was found that 85.5% of the C-14 found in living trees of the same type had decayed.

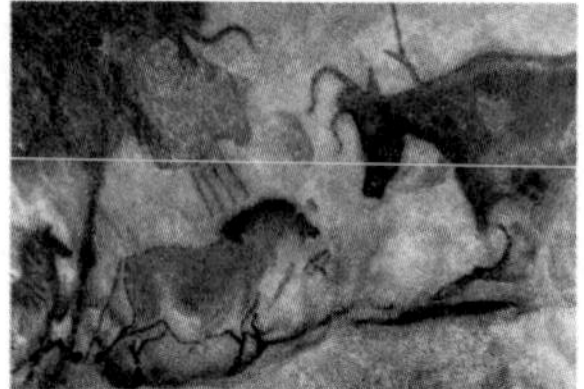

FIGURE 3.1.10 Cave wall painting in Problem 11

12. The Shroud of Turin, which shows the negative image of the body of a man who appears to have been crucified, is believed by many to be the burial shroud of Jesus of Nazareth. See Figure 3.1.11. In 1988 the Vatican granted permission to have the shroud carbon-dated. Three independent scientific laboratories analyzed the cloth and concluded that the shroud was approximately 660 years old,* an age consistent with its historical appearance.

FIGURE 3.1.11 Shroud image in Problem 12

*Some scholars have disagreed with this finding. For more information on this fascinating mystery see the Shroud of Turin home page at http://www.shroud.com/.

Using this age, determine what percentage of the original amount of C-14 remained in the cloth as of 1988.

Newton's Law of Cooling/Warming

13. A thermometer is removed from a room where the temperature is 70° F and is taken outside, where the air temperature is 10° F. After one-half minute the thermometer reads 50° F. What is the reading of the thermometer at $t = 1$ min? How long will it take for the thermometer to reach 15° F?

14. A thermometer is taken from an inside room to the outside, where the air temperature is 5° F. After 1 minute the thermometer reads 55° F, and after 5 minutes it reads 30° F. What is the initial temperature of the inside room?

15. A small metal bar, whose initial temperature was 20° C, is dropped into a large container of boiling water. How long will it take the bar to reach 90° C if it is known that its temperature increases 2° in 1 second? How long will it take the bar to reach 98° C?

16. Two large containers A and B of the same size are filled with different fluids. The fluids in containers A and B are maintained at 0° C and 100° C, respectively. A small metal bar, whose initial temperature is 100° C, is lowered into container A. After 1 minute the temperature of the bar is 90° C. After 2 minutes the bar is removed and instantly transferred to the other container. After 1 minute in container B the temperature of the bar rises 10°. How long, measured from the start of the entire process, will it take the bar to reach 99.9° C?

17. A thermometer reading 70° F is placed in an oven preheated to a constant temperature. Through a glass window in the oven door, an observer records that the thermometer reads 110° F after $\frac{1}{2}$ minute and 145° F after 1 minute. How hot is the oven?

18. At $t = 0$ a sealed test tube containing a chemical is immersed in a liquid bath. The initial temperature of the chemical in the test tube is 80° F. The liquid bath has a controlled temperature (measured in degrees Fahrenheit) given by $T_m(t) = 100 - 40e^{-0.1t}$, $t \geq 0$, where t is measured in minutes.

(a) Assume that $k = -0.1$ in (2). Before solving the IVP, describe in words what you expect the temperature $T(t)$ of the chemical to be like in the short term. In the long term.

(b) Solve the initial-value problem. Use a graphing utility to plot the graph of $T(t)$ on time intervals of various lengths. Do the graphs agree with your predictions in part (a)?

19. A dead body was found within a closed room of a house where the temperature was a constant 70° F. At the time of discovery the core temperature of the body was determined to be 85° F. One hour later a second measurement showed that the core temperature of the body was 80° F. Assume that the time of death corresponds to $t = 0$ and that the core temperature at that time was 98.6° F. Determine how many hours elapsed before the body was found. [*Hint*: Let $t_1 > 0$ denote the time that the body was discovered.]

20. The rate at which a body cools also depends on its exposed surface area S. If S is a constant, then a modification of (2) is

$$\frac{dT}{dt} = kS(T - T_m),$$

where $k < 0$ and T_m is a constant. Suppose that two cups A and B are filled with coffee at the same time. Initially, the temperature of the coffee is 150° F. The exposed surface area of the coffee in cup B is twice the surface area of the coffee in cup A. After 30 min the temperature of the coffee in cup A is 100° F. If $T_m = 70°$ F, then what is the temperature of the coffee in cup B after 30 min?

Mixtures

21. A tank contains 200 liters of fluid in which 30 grams of salt is dissolved. Brine containing 1 gram of salt per liter is then pumped into the tank at a rate of 4 L/min; the well-mixed solution is pumped out at the same rate. Find the number $A(t)$ of grams of salt in the tank at time t.

22. Solve Problem 21 assuming that pure water is pumped into the tank.

23. A large tank is filled to capacity with 500 gallons of pure water. Brine containing 2 pounds of salt per gallon is pumped into the tank at a rate of 5 gal/min. The well-mixed solution is pumped out at the same rate. Find the number $A(t)$ of pounds of salt in the tank at time t.

24. In Problem 23, what is the concentration $c(t)$ of the salt in the tank at time t? At $t = 5$ min? What is the concentration of the salt in the tank after a long time, that is, as $t \to \infty$? At what time is the concentration of the salt in the tank equal to one-half this limiting value?

25. Solve Problem 23 under the assumption that the solution is pumped out at a faster rate of 10 gal/min. When is the tank empty?

26. Determine the amount of salt in the tank at time t in Example 5 if the concentration of salt in the inflow is variable and given by $c_{in}(t) = 2 + \sin(t/4)$ lb/gal. Without actually graphing, conjecture what the solution curve of the IVP should look like. Then use a graphing utility to plot the graph of the solution on the interval [0, 300]. Repeat for the interval [0, 600] and compare your graph with that in Figure 3.1.5(a).

27. A large tank is partially filled with 100 gallons of fluid in which 10 pounds of salt is dissolved. Brine containing

$\frac{1}{2}$ pound of salt per gallon is pumped into the tank at a rate of 6 gal/min. The well-mixed solution is then pumped out at a slower rate of 4 gal/min. Find the number of pounds of salt in the tank after 30 minutes.

28. In Example 5 the size of the tank containing the salt mixture was not given. Suppose, as in the discussion following Example 5, that the rate at which brine is pumped into the tank is 3 gal/min but that the well-stirred solution is pumped out at a rate of 2 gal/min. It stands to reason that since brine is accumulating in the tank at the rate of 1 gal/min, any finite tank must eventually overflow. Now suppose that the tank has an open top and has a total capacity of 400 gallons.

(a) When will the tank overflow?

(b) What will be the number of pounds of salt in the tank at the instant it overflows?

(c) Assume that although the tank is overflowing, brine solution continues to be pumped in at a rate of 3 gal/min and the well-stirred solution continues to be pumped out at a rate of 2 gal/min. Devise a method for determining the number of pounds of salt in the tank at $t = 150$ minutes.

(d) Determine the number of pounds of salt in the tank as $t \to \infty$. Does your answer agree with your intuition?

(e) Use a graphing utility to plot the graph of $A(t)$ on the interval $[0, 500)$.

Series Circuits

29. A 30-volt electromotive force is applied to an *LR*-series circuit in which the inductance is 0.1 henry and the resistance is 50 ohms. Find the current $i(t)$ if $i(0) = 0$. Determine the current as $t \to \infty$.

30. Solve equation (7) under the assumption that $E(t) = E_0 \sin \omega t$ and $i(0) = i_0$.

31. A 100-volt electromotive force is applied to an *RC*-series circuit in which the resistance is 200 ohms and the capacitance is 10^{-4} farad. Find the charge $q(t)$ on the capacitor if $q(0) = 0$. Find the current $i(t)$.

32. A 200-volt electromotive force is applied to an *RC*-series circuit in which the resistance is 1000 ohms and the capacitance is 5×10^{-6} farad. Find the charge $q(t)$ on the capacitor if $i(0) = 0.4$. Determine the charge and current at $t = 0.005$ s. Determine the charge as $t \to \infty$.

33. An electromotive force

$$E(t) = \begin{cases} 120, & 0 \le t \le 20 \\ 0, & t > 20 \end{cases}$$

is applied to an *LR*-series circuit in which the inductance is 20 henries and the resistance is 2 ohms. Find the current $i(t)$ if $i(0) = 0$.

34. Suppose an *RC*-series circuit has a variable resistor. If the resistance at time t is given by $R = k_1 + k_2 t$, where k_1 and k_2 are known positive constants, then (9) becomes

$$(k_1 + k_2 t)\frac{dq}{dt} + \frac{1}{C}q = E(t).$$

If $E(t) = E_0$ and $q(0) = q_0$, where E_0 and q_0 are constants, show that

$$q(t) = E_0 C + (q_0 - E_0 C)\left(\frac{k_1}{k_1 + k_2 t}\right)^{1/Ck_2}.$$

Additional Linear Models

35. Air Resistance In (14) of Section 1.3 we saw that a differential equation describing the velocity v of a falling mass subject to air resistance proportional to the instantaneous velocity is

$$m\frac{dv}{dt} = mg - kv,$$

where $k > 0$ is a constant of proportionality. The positive direction is downward.

(a) Solve the equation subject to the initial condition $v(0) = v_0$.

(b) Use the solution in part (a) to determine the limiting, or terminal, velocity of the mass. We saw how to determine the terminal velocity without solving the DE in Problem 40 in Exercises 2.1.

(c) If the distance s, measured from the point where the mass was released above ground, is related to velocity v by $ds/dt = v(t)$, find an explicit expression for $s(t)$ if $s(0) = 0$.

36. How High?—No Air Resistance Suppose a small cannonball weighing 16 pounds is shot vertically upward, as shown in Figure 3.1.12, with an initial velocity $v_0 = 300$ ft/s. The answer to the question "How high does the cannonball go?" depends on whether we take air resistance into account.

(a) Suppose air resistance is ignored. If the positive direction is upward, then a model for the state of the cannonball is given by $d^2s/dt^2 = -g$ (equation (12) of Section 1.3). Since $ds/dt = v(t)$ the last

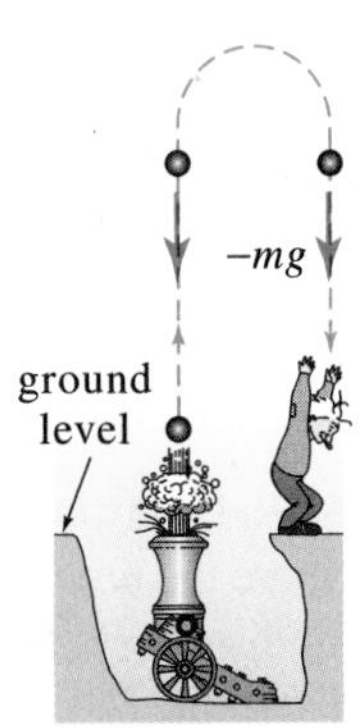

FIGURE 3.1.12 Find the maximum height of the cannonball in Problem 36

differential equation is the same as $dv/dt = -g$, where we take $g = 32$ ft/s^2. Find the velocity $v(t)$ of the cannonball at time t.

(b) Use the result obtained in part (a) to determine the height $s(t)$ of the cannonball measured from ground level. Find the maximum height attained by the cannonball.

37. How High?—Linear Air Resistance Repeat Problem 36, but this time assume that air resistance is proportional to instantaneous velocity. It stands to reason that the maximum height attained by the cannonball must be *less* than that in part (b) of Problem 36. Show this by supposing that the constant of proportionality is $k = 0.0025$. [*Hint*: Slightly modify the DE in Problem 35.]

38. Skydiving A skydiver weighs 125 pounds, and her parachute and equipment combined weigh another 35 pounds. After exiting from a plane at an altitude of 15,000 feet, she waits 15 seconds and opens her parachute. Assume that the constant of proportionality in the model in Problem 35 has the value $k = 0.5$ during free fall and $k = 10$ after the parachute is opened. Assume that her initial velocity on leaving the plane is zero. What is her velocity and how far has she traveled 20 seconds after leaving the plane? See Figure 3.1.13. How does her velocity at 20 seconds compare with her terminal velocity? How long does it take her to reach the ground? [*Hint*: Think in terms of two distinct IVPs.]

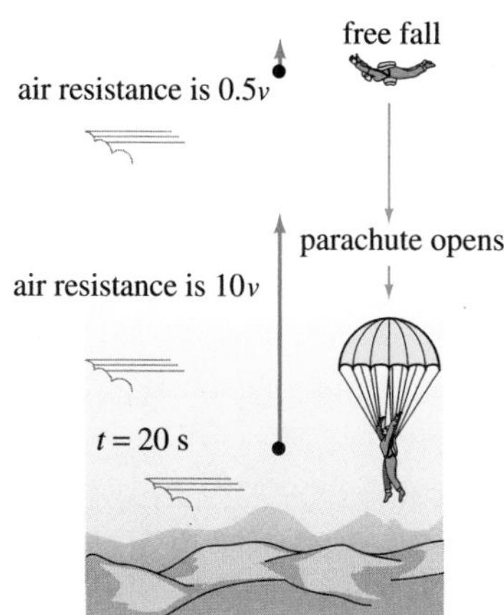

FIGURE 3.1.13 Find the time to reach the ground in Problem 38

39. Evaporating Raindrop As a raindrop falls, it evaporates while retaining its spherical shape. If we make the further assumptions that the rate at which the raindrop evaporates is proportional to its surface area and that air resistance is negligible, then a model for the velocity $v(t)$ of the raindrop is

$$\frac{dv}{dt} + \frac{3(k/\rho)}{(k/\rho)t + r_0}v = g.$$

Here ρ is the density of water, r_0 is the radius of the raindrop at $t = 0$, $k < 0$ is the constant of proportionality, and the downward direction is taken to be the positive direction.

(a) Solve for $v(t)$ if the raindrop falls from rest.

(b) Reread Problem 36 of Exercises 1.3 and then show that the radius of the raindrop at time t is $r(t) = (k/\rho)t + r_0$.

(c) If $r_0 = 0.01$ ft and $r = 0.007$ ft 10 seconds after the raindrop falls from a cloud, determine the time at which the raindrop has evaporated completely.

40. Fluctuating Population The differential equation $dP/dt = (k \cos t)P$, where k is a positive constant, is a mathematical model for a population $P(t)$ that undergoes yearly seasonal fluctuations. Solve the equation subject to $P(0) = P_0$. Use a graphing utility to graph the solution for different choices of P_0.

41. Population Model In one model of the changing population $P(t)$ of a community, it is assumed that

$$\frac{dP}{dt} = \frac{dB}{dt} - \frac{dD}{dt},$$

where dB/dt and dD/dt are the birth and death rates, respectively.

(a) Solve for $P(t)$ if $dB/dt = k_1P$ and $dD/dt = k_2P$.

(b) Analyze the cases $k_1 > k_2$, $k_1 = k_2$, and $k_1 < k_2$.

42. Constant-Harvest Model A model that describes the population of a fishery in which harvesting takes place at a constant rate is given by

$$\frac{dP}{dt} = kP - h,$$

where k and h are positive constants.

(a) Solve the DE subject to $P(0) = P_0$.

(b) Describe the behavior of the population $P(t)$ for increasing time in the three cases $P_0 > h/k$, $P_0 = h/k$, and $0 < P_0 < h/k$.

(c) Use the results from part (b) to determine whether the fish population will ever go extinct in finite time, that is, whether there exists a time $T > 0$ such that $P(T) = 0$. If the population goes extinct, then find T.

43. Drug Dissemination A mathematical model for the rate at which a drug disseminates into the bloodstream is given by

$$\frac{dx}{dt} = r - kx,$$

where r and k are positive constants. The function $x(t)$ describes the concentration of the drug in the bloodstream at time t.

(a) Since the DE is autonomous, use the phase portrait concept of Section 2.1 to find the limiting value of $x(t)$ as $t \to \infty$.

(b) Solve the DE subject to $x(0) = 0$. Sketch the graph of $x(t)$ and verify your prediction in part (a). At what time is the concentration one-half this limiting value?

44. Memorization When forgetfulness is taken into account, the rate of memorization of a subject is given by

$$\frac{dA}{dt} = k_1(M - A) - k_2A,$$

where $k_1 > 0$, $k_2 > 0$, $A(t)$ is the amount memorized in time t, M is the total amount to be memorized, and $M - A$ is the amount remaining to be memorized.

(a) Since the DE is autonomous, use the phase portrait concept of Section 2.1 to find the limiting value of $A(t)$ as $t \to \infty$. Interpret the result.

(b) Solve the DE subject to $A(0) = 0$. Sketch the graph of $A(t)$ and verify your prediction in part (a).

45. Heart Pacemaker A heart pacemaker, shown in Figure 3.1.14, consists of a switch, a battery, a capacitor, and the heart as a resistor. When the switch S is at P, the capacitor charges; when S is at Q, the capacitor discharges, sending an electrical stimulus to the heart. In Problem 53 in Exercises 2.3 we saw that during this time the electrical stimulus is being applied to the heart, the voltage E across the heart satisfies the linear DE

$$\frac{dE}{dt} = -\frac{1}{RC}E.$$

(a) Let us assume that over the time interval of length t_1, $0 < t < t_1$, the switch S is at position P shown in Figure 3.1.14 and the capacitor is being charged. When the switch is moved to position Q at time t_1 the capacitor discharges, sending an impulse to the heart over the time interval of length t_2: $t_1 \le t < t_1 + t_2$. Thus over the initial charging/discharging interval $0 < t < t_1 + t_2$ the voltage to the heart is actually modeled by the piecewise-defined differential equation

$$\frac{dE}{dt} = \begin{cases} 0, & 0 \le t < t_1 \\ -\dfrac{1}{RC}E, & t_1 \le t < t_1 + t_2. \end{cases}$$

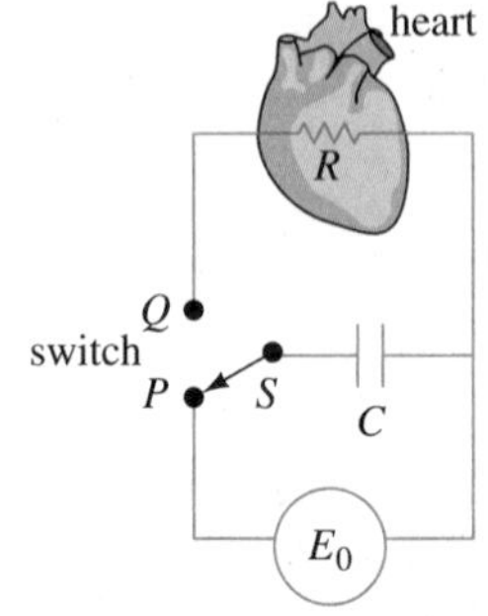

FIGURE 3.1.14 Model of a pacemaker in Problem 45

By moving S between P and Q, the charging and discharging over time intervals of lengths t_1 and t_2 is repeated indefinitely. Suppose $t_1 = 4$ s, $t_2 = 2$ s, $E_0 = 12$ V, and $E(0) = 0$, $E(4) = 12$, $E(6) = 0$, $E(10) = 12$, $E(12) = 0$, and so on. Solve for $E(t)$ for $0 \le t \le 24$.

(b) Suppose for the sake of illustration that $R = C = 1$. Use a graphing utility to graph the solution for the IVP in part (a) for $0 \le t \le 24$.

46. Sliding Box (a) A box of mass m slides down an inclined plane that makes an angle θ with the horizontal as shown in Figure 3.1.15. Find a differential equation for the velocity $v(t)$ of the box at time t in each of the following three cases:

(*i*) No sliding friction and no air resistance
(*ii*) With sliding friction and no air resistance
(*iii*) With sliding friction and air resistance

In cases (*ii*) and (*iii*), use the fact that the force of friction opposing the motion of the box is μN, where μ is the coefficient of sliding friction and N is the normal component of the weight of the box. In case (*iii*) assume that air resistance is proportional to the instantaneous velocity.

(b) In part (a), suppose that the box weighs 96 pounds, that the angle of inclination of the plane is $\theta = 30°$, that the coefficient of sliding friction is $\mu = \sqrt{3}/4$, and that the additional retarding force due to air resistance is numerically equal to $\frac{1}{4}v$. Solve the differential equation in each of the three cases, assuming that the box starts from rest from the highest point 50 ft above ground.

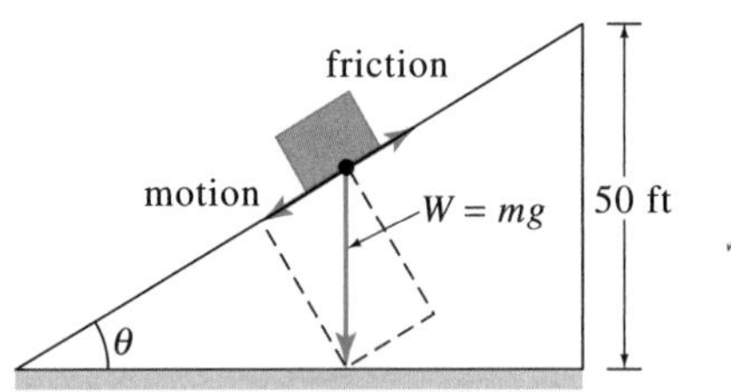

FIGURE 3.1.15 Box sliding down inclined plane in Problem 46

47. Sliding Box—Continued (a) In Problem 46 let $s(t)$ be the distance measured down the inclined plane from the highest point. Use $ds/dt = v(t)$ and the solution for each of the three cases in part (b) of Problem 46 to find the time that it takes the box to slide completely down the inclined plane. A root-finding application of a CAS may be useful here.

(b) In the case in which there is friction ($\mu \ne 0$) but no air resistance, explain why the box will not slide down the plane starting from *rest* from the highest point above ground when the inclination angle θ satisfies $\tan\theta \leq \mu$.

(c) The box *will* slide downward on the plane when $\tan\theta \leq \mu$ if it is given an initial velocity $v(0) = v_0 > 0$. Suppose that $\mu = \sqrt{3}/4$ and $\theta = 23°$. Verify that $\tan\theta \leq \mu$. How far will the box slide down the plane if $v_0 = 1$ ft/s?

(d) Using the values $\mu = \sqrt{3}/4$ and $\theta = 23°$, approximate the smallest initial velocity v_0 that can be given to the box so that, starting at the highest point 50 ft above ground, it will slide completely down the inclined plane. Then find the corresponding time it takes to slide down the plane.

48. What Goes Up . . . (a) It is well known that the model in which air resistance is ignored, part (a) of Problem 36, predicts that the time t_a it takes the cannonball to attain its maximum height is the same as the time t_d it takes the cannonball to fall from the maximum height to the ground. Moreover, the magnitude of the impact velocity v_i will be the same as the initial velocity v_0 of the cannonball. Verify both of these results.

(b) Then, using the model in Problem 37 that takes air resistance into account, compare the value of t_a with t_d and the value of the magnitude of v_i with v_0. A root-finding application of a CAS (or graphic calculator) may be useful here.

3.2 NONLINEAR MODELS

REVIEW MATERIAL

- Equations (5), (6), and (10) of Section 1.3 and Problems 7, 8, 13, 14, and 17 of Exercises 1.3
- Separation of variables in Section 2.2

INTRODUCTION We finish our study of single first-order differential equations with an examination of some nonlinear models.

Population Dynamics If $P(t)$ denotes the size of a population at time t, the model for exponential growth begins with the assumption that $dP/dt = kP$ for some $k > 0$. In this model, the **relative,** or **specific, growth rate** defined by

$$\frac{dP/dt}{P} \tag{1}$$

is a constant k. True cases of exponential growth over long periods of time are hard to find because the limited resources of the environment will at some time exert restrictions on the growth of a population. Thus for other models, (1) can be expected to decrease as the population P increases in size.

The assumption that the rate at which a population grows (or decreases) is dependent only on the number P present and not on any time-dependent mechanisms such as seasonal phenomena (see Problem 33 in Exercises 1.3) can be stated as

$$\frac{dP/dt}{P} = f(P) \quad \text{or} \quad \frac{dP}{dt} = Pf(P). \tag{2}$$

The differential equation in (2), which is widely assumed in models of animal populations, is called the **density-dependent hypothesis.**

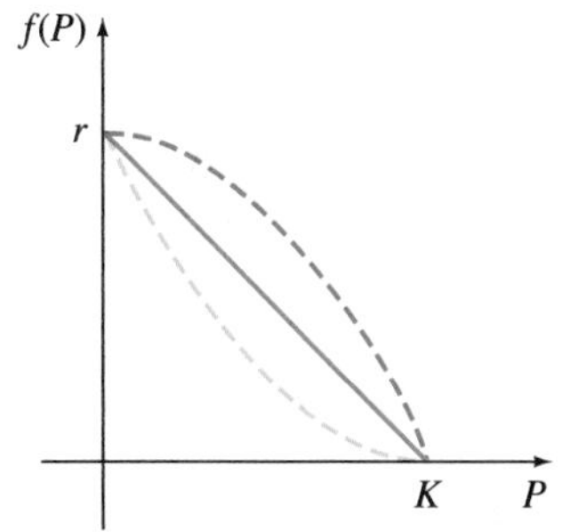

FIGURE 3.2.1 Simplest assumption for $f(P)$ is a straight line (blue color)

Logistic Equation Suppose an environment is capable of sustaining no more than a fixed number K of individuals in its population. The quantity K is called the **carrying capacity** of the environment. Hence for the function f in (2) we have $f(K) = 0$, and we simply let $f(0) = r$. Figure 3.2.1 shows three functions f that satisfy these two conditions. The simplest assumption that we can make is that $f(P)$ is linear—that is, $f(P) = c_1P + c_2$. If we use the conditions $f(0) = r$ and $f(K) = 0$, we find, in turn, $c_2 = r$ and $c_1 = -r/K$, and so f takes on the form $f(P) = r - (r/K)P$. Equation (2) becomes

$$\frac{dP}{dt} = P\left(r - \frac{r}{K}P\right). \tag{3}$$

With constants relabeled, the nonlinear equation (3) is the same as

$$\frac{dP}{dt} = P(a - bP). \tag{4}$$

Around 1840 the Belgian mathematician-biologist **P. F. Verhulst** (1804–1849) was concerned with mathematical models for predicting the human populations of various countries. One of the equations he studied was (4), where $a > 0$ and $b > 0$. Equation (4) came to be known as the **logistic equation,** and its solution is called the **logistic function.** The graph of a logistic function is called a **logistic curve.**

The linear differential equation $dP/dt = kP$ does not provide a very accurate model for population when the population itself is very large. Overcrowded conditions, with the resulting detrimental effects on the environment such as pollution and excessive and competitive demands for food and fuel, can have an inhibiting effect on population growth. As we shall now see, the solution of (4) is bounded as $t \to \infty$. If we rewrite (4) as $dP/dt = aP - bP^2$, the nonlinear term $-bP^2$, $b > 0$, can be interpreted as an "inhibition" or "competition" term. Also, in most applications the positive constant a is much larger than the constant b.

Logistic curves have proved to be quite accurate in predicting the growth patterns, in a limited space, of certain types of bacteria, protozoa, water fleas (*Daphnia*), and fruit flies (*Drosophila*).

Solution of the Logistic Equation One method of solving (4) is separation of variables. Decomposing the left side of $dP/P(a - bP) = dt$ into partial fractions and integrating gives

$$\left(\frac{1/a}{P} + \frac{b/a}{a - bP}\right)dP = dt$$

$$\frac{1}{a}\ln|P| - \frac{1}{a}\ln|a - bP| = t + c$$

$$\ln\left|\frac{P}{a - bP}\right| = at + ac$$

$$\frac{P}{a - bP} = c_1e^{at}.$$

It follows from the last equation that

$$P(t) = \frac{ac_1e^{at}}{1 + bc_1e^{at}} = \frac{ac_1}{bc_1 + e^{-at}}.$$

If $P(0) = P_0$, $P_0 \neq a/b$, we find $c_1 = P_0/(a - bP_0)$, and so after substituting and simplifying, the solution becomes

$$P(t) = \frac{aP_0}{bP_0 + (a - bP_0)e^{-at}}. \tag{5}$$

Graphs of $P(t)$ The basic shape of the graph of the logistic function $P(t)$ can be obtained without too much effort. Although the variable t usually represents time and we are seldom concerned with applications in which $t < 0$, it is nonetheless of some interest to include this interval in displaying the various graphs of P. From (5) we see that

$$P(t) \to \frac{aP_0}{bP_0} = \frac{a}{b} \quad \text{as} \quad t \to \infty \qquad \text{and} \qquad P(t) \to 0 \quad \text{as} \quad t \to -\infty.$$

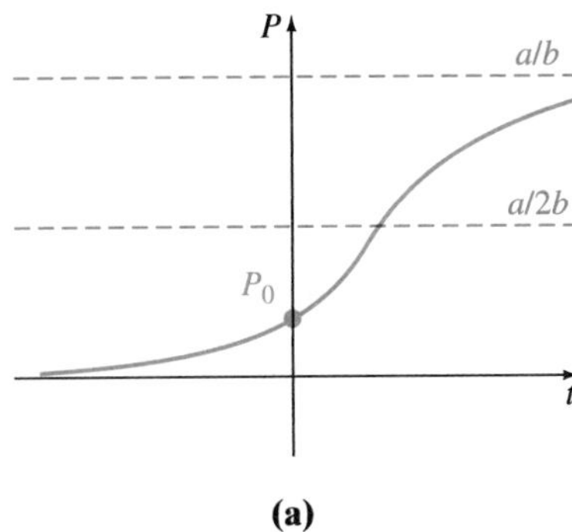

(a)

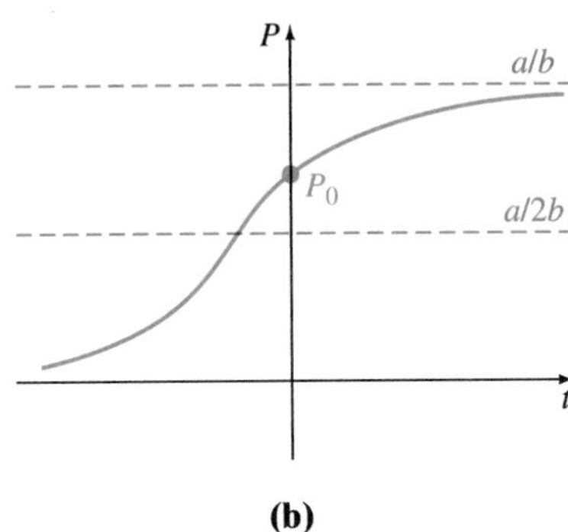

(b)

FIGURE 3.2.2 Logistic curves for different initial conditions

The dashed line $P = a/2b$ shown in Figure 3.2.2 corresponds to the ordinate of a point of inflection of the logistic curve. To show this, we differentiate (4) by the Product Rule:

$$\begin{aligned}\frac{d^2P}{dt^2} &= P\left(-b\frac{dP}{dt}\right) + (a - bP)\frac{dP}{dt} = \frac{dP}{dt}(a - 2bP)\\ &= P(a - bP)(a - 2bP)\\ &= 2b^2P\left(P - \frac{a}{b}\right)\left(P - \frac{a}{2b}\right).\end{aligned}$$

From calculus recall that the points where $d^2P/dt^2 = 0$ are possible points of inflection, but $P = 0$ and $P = a/b$ can obviously be ruled out. Hence $P = a/2b$ is the only possible ordinate value at which the concavity of the graph can change. For $0 < P < a/2b$ it follows that $P'' > 0$, and $a/2b < P < a/b$ implies that $P'' < 0$. Thus, as we read from left to right, the graph changes from concave up to concave down at the point corresponding to $P = a/2b$. When the initial value satisfies $0 < P_0 < a/2b$, the graph of $P(t)$ assumes the shape of an S, as we see in Figure 3.2.2(a). For $a/2b < P_0 < a/b$ the graph is still S-shaped, but the point of inflection occurs at a negative value of t, as shown in Figure 3.2.2(b).

We have already seen equation (4) in (5) of Section 1.3 in the form $dx/dt = kx(n + 1 - x)$, $k > 0$. This differential equation provides a reasonable model for describing the spread of an epidemic brought about initially by introducing an infected individual into a static population. The solution $x(t)$ represents the number of individuals infected with the disease at time t.

EXAMPLE 1 Logistic Growth

Suppose a student carrying a flu virus returns to an isolated college campus of 1000 students. If it is assumed that the rate at which the virus spreads is proportional not only to the number x of infected students but also to the number of students not infected, determine the number of infected students after 6 days if it is further observed that after 4 days $x(4) = 50$.

SOLUTION Assuming that no one leaves the campus throughout the duration of the disease, we must solve the initial-value problem

$$\frac{dx}{dt} = kx(1000 - x), \quad x(0) = 1.$$

By making the identification $a = 1000k$ and $b = k$, we have immediately from (5) that

$$x(t) = \frac{1000k}{k + 999ke^{-1000kt}} = \frac{1000}{1 + 999e^{-1000kt}}.$$

Now, using the information $x(4) = 50$, we determine k from

$$50 = \frac{1000}{1 + 999e^{-4000k}}.$$

We find $-1000k = \frac{1}{4}\ln\frac{19}{999} = -0.9906$. Thus

$$x(t) = \frac{1000}{1 + 999e^{-0.9906t}}.$$

Finally, $$x(6) = \frac{1000}{1 + 999e^{-5.9436}} = 276 \text{ students}.$$

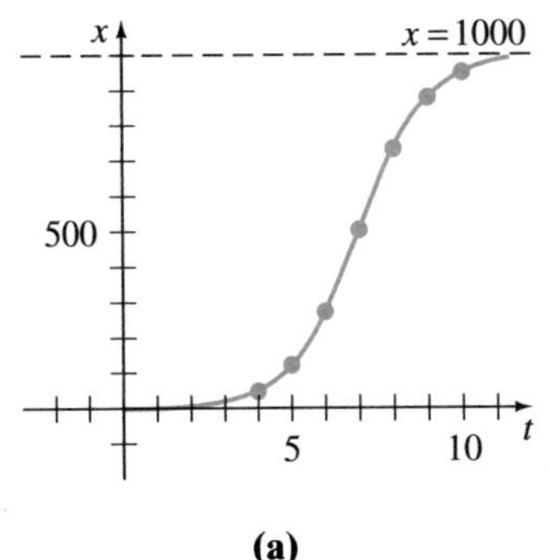

(a)

t (days)	x (number infected)
4	50 (observed)
5	124
6	276
7	507
8	735
9	882
10	953

(b)

FIGURE 3.2.3 Number of infected students in Example 1

Additional calculated values of $x(t)$ are given in the table in Figure 3.2.3(b). Note that the number of infected students $x(t)$ approaches 1000 as t increases. ≡

Modifications of the Logistic Equation There are many variations of the logistic equation. For example, the differential equations

$$\frac{dP}{dt} = P(a - bP) - h \quad \text{and} \quad \frac{dP}{dt} = P(a - bP) + h \tag{6}$$

could serve, in turn, as models for the population in a fishery where fish are **harvested** or are **restocked** at rate h. When $h > 0$ is a constant, the DEs in (6) can be readily analyzed qualitatively or solved analytically by separation of variables. The equations in (6) could also serve as models of the human population decreased by **emigration** or increased by immigration, respectively. The rate h in (6) could be a function of time t or could be population dependent; for example, harvesting might be done periodically over time or might be done at a rate proportional to the population P at time t. In the latter instance, the model would look like $P' = P(a - bP) - cP$, $c > 0$. The human population of a community might change because of **immigration** in such a manner that the contribution due to immigration was large when the population P of the community was itself small but small when P was large; a reasonable model for the population of the community would then be $P' = P(a - bP) + ce^{-kP}$, $c > 0, k > 0$. See Problem 24 in Exercises 3.2. Another equation of the form given in (2),

$$\frac{dP}{dt} = P(a - b \ln P), \tag{7}$$

is a modification of the logistic equation known as the **Gompertz differential equation** named after the English mathematician **Benjamin Gompertz** (1779–1865). This DE is sometimes used as a model in the study of the growth or decline of populations, the growth of solid tumors, and certain kinds of actuarial predictions. See Problem 8 in Exercises 3.2.

Chemical Reactions Suppose that a grams of chemical A are combined with b grams of chemical B. If there are M parts of A and N parts of B formed in the compound and $X(t)$ is the number of grams of chemical C formed, then the number of grams of chemical A and the number of grams of chemical B remaining at time t are, respectively,

$$a - \frac{M}{M + N}X \quad \text{and} \quad b - \frac{N}{M + N}X.$$

The law of mass action states that when no temperature change is involved, the rate at which the two substances react is proportional to the product of the amounts of A and B that are untransformed (remaining) at time t:

$$\frac{dX}{dt} \propto \left(a - \frac{M}{M + N}X\right)\left(b - \frac{N}{M + N}X\right). \tag{8}$$

If we factor out $M/(M + N)$ from the first factor and $N/(M + N)$ from the second and introduce a constant of proportionality $k > 0$, (8) has the form

$$\frac{dX}{dt} = k(\alpha - X)(\beta - X), \tag{9}$$

where $\alpha = a(M + N)/M$ and $\beta = b(M + N)/N$. Recall from (6) of Section 1.3 that a chemical reaction governed by the nonlinear differential equation (9) is said to be a **second-order reaction.**

EXAMPLE 2 **Second-Order Chemical Reaction**

A compound C is formed when two chemicals A and B are combined. The resulting reaction between the two chemicals is such that for each gram of A, 4 grams of B is used. It is observed that 30 grams of the compound C is formed in 10 minutes.

Determine the amount of C at time t if the rate of the reaction is proportional to the amounts of A and B remaining and if initially there are 50 grams of A and 32 grams of B. How much of the compound C is present at 15 minutes? Interpret the solution as $t \to \infty$.

SOLUTION Let $X(t)$ denote the number of grams of the compound C present at time t. Clearly, $X(0) = 0$ g and $X(10) = 30$ g.

If, for example, 2 grams of compound C is present, we must have used, say, a grams of A and b grams of B, so $a + b = 2$ and $b = 4a$. Thus we must use $a = \frac{2}{5} = 2\left(\frac{1}{5}\right)$ g of chemical A and $b = \frac{8}{5} = 2\left(\frac{4}{5}\right)$ g of B. In general, for X grams of C we must use

$$\frac{1}{5}X \text{ grams of } A \qquad \text{and} \qquad \frac{4}{5}X \text{ grams of } B.$$

The amounts of A and B remaining at time t are then

$$50 - \frac{1}{5}X \quad \text{and} \quad 32 - \frac{4}{5}X,$$

respectively.

Now we know that the rate at which compound C is formed satisfies

$$\frac{dX}{dt} \propto \left(50 - \frac{1}{5}X\right)\left(32 - \frac{4}{5}X\right).$$

To simplify the subsequent algebra, we factor $\frac{1}{5}$ from the first term and $\frac{4}{5}$ from the second and then introduce the constant of proportionality:

$$\frac{dX}{dt} = k(250 - X)(40 - X).$$

By separation of variables and partial fractions we can write

$$-\frac{\frac{1}{210}}{250 - X}\,dX + \frac{\frac{1}{210}}{40 - X}\,dX = k\,dt.$$

Integrating gives

$$\ln\frac{250 - X}{40 - X} = 210kt + c_1 \qquad \text{or} \qquad \frac{250 - X}{40 - X} = c_2 e^{210kt}. \tag{10}$$

When $t = 0$, $X = 0$, so it follows at this point that $c_2 = \frac{25}{4}$. Using $X = 30$ g at $t = 10$, we find $210k = \frac{1}{10}\ln\frac{88}{25} = 0.1258$. With this information we solve the last equation in (10) for X:

$$X(t) = 1000\frac{1 - e^{-0.1258t}}{25 - 4e^{-0.1258t}}. \tag{11}$$

From (11) we find $X(15) = 34.78$ grams. The behavior of X as a function of time is displayed in Figure 3.2.4. It is clear from the accompanying table and (11) that $X \to 40$ as $t \to \infty$. This means that 40 grams of compound C is formed, leaving

$$50 - \frac{1}{5}(40) = 42 \text{ g of } A \qquad \text{and} \qquad 32 - \frac{4}{5}(40) = 0 \text{ g of } B. \qquad \equiv$$

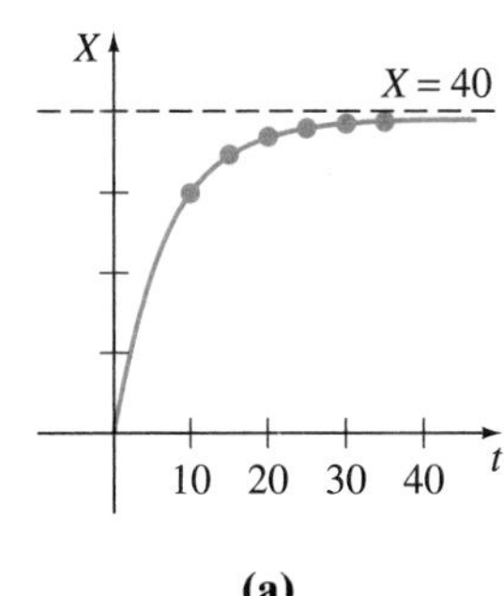

(a)

t (min)	X (g)
10	30 (measured)
15	34.78
20	37.25
25	38.54
30	39.22
35	39.59

(b)

FIGURE 3.2.4 Number of grams of compound C in Example 2

REMARKS

The indefinite integral $\int du/(a^2 - u^2)$ can be evaluated in terms of logarithms, the inverse hyperbolic tangent, or the inverse hyperbolic cotangent. For example, of the two results

$$\int \frac{du}{a^2 - u^2} = \frac{1}{a}\tanh^{-1}\frac{u}{a} + c, \quad |u| < a \tag{12}$$

$$\int \frac{du}{a^2 - u^2} = \frac{1}{2a}\ln\left|\frac{a + u}{a - u}\right| + c, \quad |u| \neq a, \tag{13}$$

(12) may be convenient in Problems 15 and 26 in Exercises 3.2, whereas (13) may be preferable in Problem 27.

EXERCISES 3.2

Answers to selected odd-numbered problems begin on page ANS-3.

Logistic Equation

1. The number $N(t)$ of supermarkets throughout the country that are using a computerized checkout system is described by the initial-value problem

$$\frac{dN}{dt} = N(1 - 0.0005N), \quad N(0) = 1.$$

(a) Use the phase portrait concept of Section 2.1 to predict how many supermarkets are expected to adopt the new procedure over a long period of time. By hand, sketch a solution curve of the given initial-value problem.

(b) Solve the initial-value problem and then use a graphing utility to verify the solution curve in part (a). How many companies are expected to adopt the new technology when $t = 10$?

2. The number $N(t)$ of people in a community who are exposed to a particular advertisement is governed by the logistic equation. Initially, $N(0) = 500$, and it is observed that $N(1) = 1000$. Solve for $N(t)$ if it is predicted that the limiting number of people in the community who will see the advertisement is 50,000.

3. A model for the population $P(t)$ in a suburb of a large city is given by the initial-value problem

$$\frac{dP}{dt} = P(10^{-1} - 10^{-7}P), \quad P(0) = 5000,$$

where t is measured in months. What is the limiting value of the population? At what time will the population be equal to one-half of this limiting value?

4. (a) Census data for the United States between 1790 and 1950 are given in Table 3.2.1. Construct a logistic population model using the data from 1790, 1850, and 1910.

(b) Construct a table comparing actual census population with the population predicted by the model in part (a). Compute the error and the percentage error for each entry pair.

TABLE 3.2.1

Year	Population (in millions)
1790	3.929
1800	5.308
1810	7.240
1820	9.638
1830	12.866
1840	17.069
1850	23.192
1860	31.433
1870	38.558
1880	50.156
1890	62.948
1900	75.996
1910	91.972
1920	105.711
1930	122.775
1940	131.669
1950	150.697

Modifications of the Logistic Model

5. (a) If a constant number h of fish are harvested from a fishery per unit time, then a model for the population $P(t)$ of the fishery at time t is given by

$$\frac{dP}{dt} = P(a - bP) - h, \quad P(0) = P_0,$$

where a, b, h, and P_0 are positive constants. Suppose $a = 5$, $b = 1$, and $h = 4$. Since the DE is autonomous, use the phase portrait concept of Section 2.1 to sketch representative solution curves corresponding to the cases $P_0 > 4$, $1 < P_0 < 4$, and $0 < P_0 < 1$. Determine the long-term behavior of the population in each case.

(b) Solve the IVP in part (a). Verify the results of your phase portrait in part (a) by using a graphing utility to plot the graph of $P(t)$ with an initial condition taken from each of the three intervals given.

(c) Use the information in parts (a) and (b) to determine whether the fishery population becomes extinct in finite time. If so, find that time.

6. Investigate the harvesting model in Problem 5 both qualitatively and analytically in the case $a = 5$, $b = 1$, $h = \frac{25}{4}$. Determine whether the population becomes extinct in finite time. If so, find that time.

7. Repeat Problem 6 in the case $a = 5$, $b = 1$, $h = 7$.

8. **(a)** Suppose $a = b = 1$ in the Gompertz differential equation (7). Since the DE is autonomous, use the phase portrait concept of Section 2.1 to sketch representative solution curves corresponding to the cases $P_0 > e$ and $0 < P_0 < e$.

(b) Suppose $a = 1$, $b = -1$ in (7). Use a new phase portrait to sketch representative solution curves corresponding to the cases $P_0 > e^{-1}$ and $0 < P_0 < e^{-1}$.

(c) Find an explicit solution of (7) subject to $P(0) = P_0$.

Chemical Reactions

9. Two chemicals A and B are combined to form a chemical C. The rate, or velocity, of the reaction is proportional to the product of the instantaneous amounts of A and B not converted to chemical C. Initially, there are 40 grams of A and 50 grams of B, and for each gram of B, 2 grams of A is used. It is observed that 10 grams of C is formed in 5 minutes. How much is formed in 20 minutes? What is the limiting amount of C after a long time? How much of chemicals A and B remains after a long time?

10. Solve Problem 9 if 100 grams of chemical A is present initially. At what time is chemical C half-formed?

Additional Nonlinear Models

11. Leaking Cylindrical Tank A tank in the form of a right-circular cylinder standing on end is leaking water through a circular hole in its bottom. As we saw in (10) of Section 1.3, when friction and contraction of water at the hole are ignored, the height h of water in the tank is described by

$$\frac{dh}{dt} = -\frac{A_h}{A_w}\sqrt{2gh},$$

where A_w and A_h are the cross-sectional areas of the water and the hole, respectively.

(a) Solve the DE if the initial height of the water is H. By hand, sketch the graph of $h(t)$ and give its interval I of definition in terms of the symbols A_w, A_h, and H. Use $g = 32$ ft/s^2.

(b) Suppose the tank is 10 feet high and has radius 2 feet and the circular hole has radius $\frac{1}{2}$ inch. If the tank is initially full, how long will it take to empty?

12. Leaking Cylindrical Tank—Continued When friction and contraction of the water at the hole are taken into account, the model in Problem 11 becomes

$$\frac{dh}{dt} = -c\frac{A_h}{A_w}\sqrt{2gh},$$

where $0 < c < 1$. How long will it take the tank in Problem 11(b) to empty if $c = 0.6$? See Problem 13 in Exercises 1.3.

13. Leaking Conical Tank A tank in the form of a right-circular cone standing on end, vertex down, is leaking water through a circular hole in its bottom.

(a) Suppose the tank is 20 feet high and has radius 8 feet and the circular hole has radius 2 inches. In Problem 14 in Exercises 1.3 you were asked to show that the differential equation governing the height h of water leaking from a tank is

$$\frac{dh}{dt} = -\frac{5}{6h^{3/2}}.$$

In this model, friction and contraction of the water at the hole were taken into account with $c = 0.6$, and g was taken to be 32 ft/s^2. See Figure 1.3.12. If the tank is initially full, how long will it take the tank to empty?

(b) Suppose the tank has a vertex angle of 60° and the circular hole has radius 2 inches. Determine the differential equation governing the height h of water. Use $c = 0.6$ and $g = 32$ ft/s^2. If the height of the water is initially 9 feet, how long will it take the tank to empty?

14. Inverted Conical Tank Suppose that the conical tank in Problem 13(a) is inverted, as shown in Figure 3.2.5, and that water leaks out a circular hole of radius 2 inches in the center of its circular base. Is the time it takes to empty a full tank the same as for the tank with vertex down in Problem 13? Take the friction/contraction coefficient to be $c = 0.6$ and $g = 32$ ft/s^2.

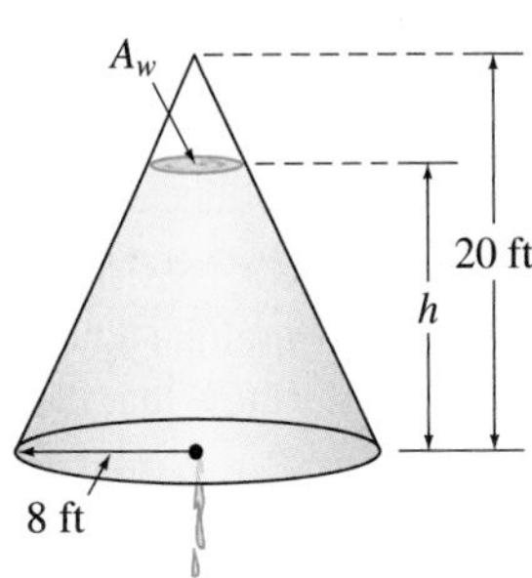

FIGURE 3.2.5 Inverted conical tank in Problem 14

15. Air Resistance A differential equation for the velocity v of a falling mass m subjected to air resistance proportional to the square of the instantaneous velocity is

$$m\frac{dv}{dt} = mg - kv^2,$$

where $k > 0$ is a constant of proportionality. The positive direction is downward.

(a) Solve the equation subject to the initial condition $v(0) = v_0$.

(b) Use the solution in part (a) to determine the limiting, or terminal, velocity of the mass. We saw how to determine the terminal velocity without solving the DE in Problem 41 in Exercises 2.1.

(c) If the distance s, measured from the point where the mass was released above ground, is related to velocity v by $ds/dt = v(t)$, find an explicit expression for $s(t)$ if $s(0) = 0$.

16. How High?—Nonlinear Air Resistance Consider the 16-pound cannonball shot vertically upward in Problems 36 and 37 in Exercises 3.1 with an initial velocity $v_0 = 300$ ft/s. Determine the maximum height attained by the cannonball if air resistance is assumed to be proportional to the square of the instantaneous velocity. Assume that the positive direction is upward and take $k = 0.0003$. [*Hint*: Slightly modify the DE in Problem 15.]

17. That Sinking Feeling **(a)** Determine a differential equation for the velocity $v(t)$ of a mass m sinking in water that imparts a resistance proportional to the square of the instantaneous velocity and also exerts an upward buoyant force whose magnitude is given by Archimedes' principle. See Problem 18 in Exercises 1.3. Assume that the positive direction is downward.

(b) Solve the differential equation in part (a).

(c) Determine the limiting, or terminal, velocity of the sinking mass.

18. Solar Collector The differential equation

$$\frac{dy}{dx} = \frac{-x + \sqrt{x^2 + y^2}}{y}$$

describes the shape of a plane curve C that will reflect all incoming light beams to the same point and could be a model for the mirror of a reflecting telescope, a satellite antenna, or a solar collector. See Problem 29 in Exercises 1.3. There are several ways of solving this DE.

(a) Verify that the differential equation is homogeneous (see Section 2.5). Show that the substitution $y = ux$ yields

$$\frac{u\,du}{\sqrt{1 + u^2}\left(1 - \sqrt{1 + u^2}\right)} = \frac{dx}{x}.$$

Use a CAS (or another judicious substitution) to integrate the left-hand side of the equation. Show that the curve C must be a parabola with focus at the origin and is symmetric with respect to the x-axis.

(b) Show that the first differential equation can also be solved by means of the substitution $u = x^2 + y^2$.

19. Tsunami **(a)** A simple model for the shape of a tsunami is given by

$$\frac{dW}{dx} = W\sqrt{4 - 2W},$$

where $W(x) > 0$ is the height of the wave expressed as a function of its position relative to a point offshore. By inspection, find all constant solutions of the DE.

(b) Solve the differential equation in part (a). A CAS may be useful for integration.

(c) Use a graphing utility to obtain the graphs of all solutions that satisfy the initial condition $W(0) = 2$.

20. Evaporation An outdoor decorative pond in the shape of a hemispherical tank is to be filled with water pumped into the tank through an inlet in its bottom. Suppose that the radius of the tank is $R = 10$ ft, that water is pumped in at a rate of π ft^3/min, and that the tank is initially empty. See Figure 3.2.6. As the tank fills, it loses water through evaporation. Assume that the rate of evaporation is proportional to the area A of the surface of the water and that the constant of proportionality is $k = 0.01$.

(a) The rate of change dV/dt of the volume of the water at time t is a net rate. Use this net rate to determine a differential equation for the height h of the water at time t. The volume of the water shown in the figure is $V = \pi Rh^2 - \frac{1}{3}\pi h^3$, where $R = 10$. Express the area of the surface of the water $A = \pi r^2$ in terms of h.

(b) Solve the differential equation in part (a). Graph the solution.

(c) If there were no evaporation, how long would it take the tank to fill?

(d) With evaporation, what is the depth of the water at the time found in part (c)? Will the tank ever be filled? Prove your assertion.

Output: water evaporates at rate proportional to area A of surface

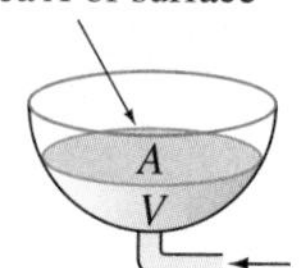

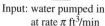

Input: water pumped in at rate π ft^3/min

(a) hemispherical tank

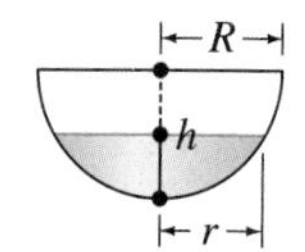

(b) cross-section of tank

FIGURE 3.2.6 Decorative pond in Problem 20

21. Doomsday Equation Consider the differential equation

$$\frac{dP}{dt} = kP^{1+c},$$

where $k > 0$ and $c \geq 0$. In Section 3.1 we saw that in the case $c = 0$ the linear differential equation $dP/dt = kP$ is a mathematical model of a population $P(t)$ that exhibits unbounded growth over the infinite time interval $[0, \infty)$, that is, $P(t) \to \infty$ as $t \to \infty$. See Example 1 on page 84.

(a) Suppose for $c = 0.01$ that the nonlinear differential equation

$$\frac{dP}{dt} = kP^{1.01}, \ k > 0,$$

is a mathematical model for a population of small animals, where time t is measured in months. Solve the differential equation subject to the initial condition $P(0) = 10$ and the fact that the animal population has doubled in 5 months.

(b) The differential equation in part (a) is called a **doomsday equation** because the population $P(t)$ exhibits unbounded growth over a finite time interval $(0, T)$, that is, there is some time T such $P(t) \to \infty$ as $t \to T^-$. Find T.

(c) From part (a), what is $P(50)$? $P(100)$?

22. Doomsday or Extinction Suppose the population model (4) is modified to be

$$\frac{dP}{dt} = P(bP - a).$$

(a) If $a > 0$, $b > 0$ show by means of a phase portrait (see page 39) that, depending on the initial condition $P(0) = P_0$, the mathematical model could include a doomsday scenario ($P(t) \to \infty$) or an extinction scenario ($P(t) \to 0$).

(b) Solve the initial-value problem

$$\frac{dP}{dt} = P(0.0005P - 0.1), \ P(0) = 300.$$

Show that this model predicts a doomsday for the population in a finite time T.

(c) Solve the differential equation in part (b) subject to the initial condition $P(0) = 100$. Show that this model predicts extinction for the population as $t \to \infty$.

Project Problems

23. Regression Line Read the documentation for your CAS on *scatter plots* (or *scatter diagrams*) and *least-squares linear fit*. The straight line that best fits a set of data points is called a **regression line** or a **least squares line.** Your task is to construct a logistic model for the population of the United States, defining $f(P)$ in (2) as an equation of a regression line based on the population data in the table in Problem 4. One way of doing this is to approximate the left-hand side $\dfrac{1}{P}\dfrac{dP}{dt}$ of the first equation in (2), using the forward difference quotient in place of dP/dt:

$$Q(t) = \frac{1}{P(t)} \frac{P(t + h) - P(t)}{h}.$$

(a) Make a table of the values t, $P(t)$, and $Q(t)$ using $t = 0, 10, 20, \ldots, 160$ and $h = 10$. For example, the first line of the table should contain $t = 0$, $P(0)$, and $Q(0)$. With $P(0) = 3.929$ and $P(10) = 5.308$,

$$Q(0) = \frac{1}{P(0)} \frac{P(10) - P(0)}{10} = 0.035.$$

Note that $Q(160)$ depends on the 1960 census population $P(170)$. Look up this value.

(b) Use a CAS to obtain a scatter plot of the data $(P(t), Q(t))$ computed in part (a). Also use a CAS to find an equation of the regression line and to superimpose its graph on the scatter plot.

(c) Construct a logistic model $dP/dt = Pf(P)$, where $f(P)$ is the equation of the regression line found in part (b).

(d) Solve the model in part (c) using the initial condition $P(0) = 3.929$.

(e) Use a CAS to obtain another scatter plot, this time of the ordered pairs $(t, P(t))$ from your table in part (a). Use your CAS to superimpose the graph of the solution in part (d) on the scatter plot.

(f) Look up the U.S. census data for 1970, 1980, and 1990. What population does the logistic model in part (c) predict for these years? What does the model predict for the U.S. population $P(t)$ as $t \to \infty$?

24. Immigration Model **(a)** In Examples 3 and 4 of Section 2.1 we saw that any solution $P(t)$ of (4) possesses the asymptotic behavior $P(t) \to a/b$ as $t \to \infty$ for $P_0 > a/b$ and for $0 < P_0 < a/b$; as a consequence the equilibrium solution $P = a/b$ is called an attractor. Use a root-finding application of a CAS (or a graphic calculator) to approximate the equilibrium solution of the immigration model

$$\frac{dP}{dt} = P(1 - P) + 0.3e^{-P}.$$

(b) Use a graphing utility to graph the function $F(P) = P(1 - P) + 0.3e^{-P}$. Explain how this graph can be used to determine whether the number found in part (a) is an attractor.

(c) Use a numerical solver to compare the solution curves for the IVPs

$$\frac{dP}{dt} = P(1 - P), \quad P(0) = P_0$$

for $P_0 = 0.2$ and $P_0 = 1.2$ with the solution curves for the IVPs

$$\frac{dP}{dt} = P(1 - P) + 0.3e^{-P}, \quad P(0) = P_0$$

for $P_0 = 0.2$ and $P_0 = 1.2$. Superimpose all curves on the same coordinate axes but, if possible, use a different color for the curves of the second initial-value problem. Over a long period of time, what percentage increase does the immigration model predict in the population compared to the logistic model?

25. What Goes Up . . . In Problem 16 let t_a be the time it takes the cannonball to attain its maximum height and let t_d be the time it takes the cannonball to fall from the maximum height to the ground. Compare the value of t_a with the value of t_d and compare the magnitude of the impact velocity v_i with the initial velocity v_0. See Problem 48 in Exercises 3.1. A root-finding application of a CAS might be useful here. [*Hint*: Use the model in Problem 15 when the cannonball is falling.]

26. Skydiving A skydiver is equipped with a stopwatch and an altimeter. As shown in Figure 3.2.7, he opens his parachute 25 seconds after exiting a plane flying at an altitude of 20,000 feet and observes that his altitude is 14,800 feet. Assume that air resistance is proportional to the square of the instantaneous velocity, his initial velocity on leaving the plane is zero, and $g = 32$ ft/s^2.

(a) Find the distance $s(t)$, measured from the plane, the skydiver has traveled during freefall in time t. [*Hint*: The constant of proportionality k in the model given in Problem 15 is not specified. Use the expression for terminal velocity v_t obtained in part (b) of Problem 15 to eliminate k from the IVP. Then eventually solve for v_t.]

(b) How far does the skydiver fall and what is his velocity at $t = 15$ s?

FIGURE 3.2.7 Skydiver in Problem 26

27. Hitting Bottom A helicopter hovers 500 feet above a large open tank full of liquid (not water). A dense compact object weighing 160 pounds is dropped (released from rest) from the helicopter into the liquid. Assume that air resistance is proportional to instantaneous velocity v while the object is in the air and that viscous damping is proportional to v^2 after the object has entered the liquid. For air take $k = \frac{1}{4}$, and for the liquid take $k = 0.1$. Assume that the positive direction is downward. If the tank is 75 feet high, determine the time and the impact velocity when the object hits the bottom of the tank. [*Hint*: Think in terms of two distinct IVPs. If you use (13), be careful in removing the absolute value sign. You might compare the velocity when the object hits the liquid—the initial velocity for the second problem—with the terminal velocity v_t of the object falling through the liquid.]

28. Old Man River . . . In Figure 3.2.8(a) suppose that the y-axis and the dashed vertical line $x = 1$ represent, respectively, the straight west and east beaches of a river that is 1 mile wide. The river flows northward with a velocity $\mathbf{v}_r$, where $|\mathbf{v}_r| = v_r$ mi/h is a constant. A man enters the current at the point $(1, 0)$ on the east shore and swims in a direction and rate relative to the river given by the vector $\mathbf{v}_s$, where the speed $|\mathbf{v}_s| = v_s$ mi/h is a constant. The man wants to reach the west beach exactly at $(0, 0)$ and so swims in such a manner that keeps his velocity vector $\mathbf{v}_s$ always directed toward the point $(0, 0)$. Use Figure 3.2.8(b) as an aid in showing that a mathematical model for the path of the swimmer in the river is

$$\frac{dy}{dx} = \frac{v_s y - v_r\sqrt{x^2 + y^2}}{v_s x}.$$

[*Hint*: The velocity $\mathbf{v}$ of the swimmer along the path or curve shown in Figure 3.2.8 is the resultant $\mathbf{v} = \mathbf{v}_s + \mathbf{v}_r$. Resolve $\mathbf{v}_s$ and $\mathbf{v}_r$ into components in the x- and

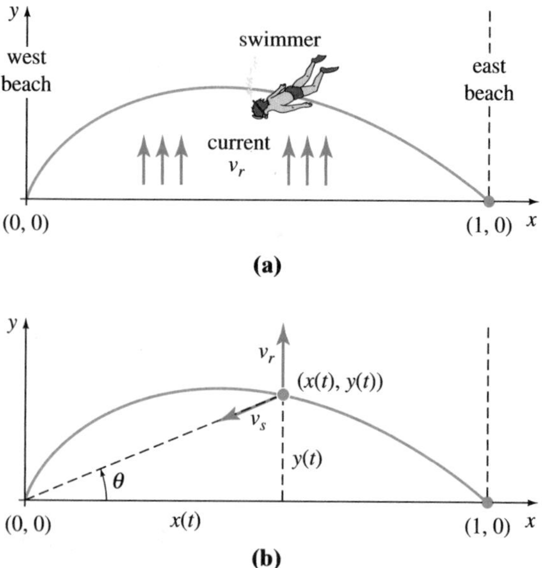

FIGURE 3.2.8 Path of swimmer in Problem 28

y-directions. If $x = x(t)$, $y = y(t)$ are parametric equations of the swimmer's path, then $\mathbf{v} = (dx/dt, dy/dt)$.]

29. **(a)** Solve the DE in Problem 28 subject to $y(1) = 0$. For convenience let $k = v_r/v_s$.

(b) Determine the values of v_s for which the swimmer will reach the point (0, 0) by examining $\lim_{x\to 0^+} y(x)$ in the cases $k = 1$, $k > 1$, and $0 < k < 1$.

30. Old Man River Keeps Moving . . . Suppose the man in Problem 28 again enters the current at (1, 0) but this time decides to swim so that his velocity vector $\mathbf{v}_s$ is always directed toward the west beach. Assume that the speed $|\mathbf{v}_s| = v_s$ mi/h is a constant. Show that a mathematical model for the path of the swimmer in the river is now

$$\frac{dy}{dx} = -\frac{v_r}{v_s}.$$

31. The current speed v_r of a straight river such as that in Problem 28 is usually not a constant. Rather, an approximation to the current speed (measured in miles per hour) could be a function such as $v_r(x) = 30x(1 - x)$, $0 \le x \le 1$, whose values are small at the shores (in this case, $v_r(0) = 0$ and $v_r(1) = 0$) and largest in the middle of the river. Solve the DE in Problem 30 subject to $y(1) = 0$, where $v_s = 2$ mi/h and $v_r(x)$ is as given. When the swimmer makes it across the river, how far will he have to walk along the beach to reach the point (0, 0)?

32. Raindrops Keep Falling . . . When a bottle of liquid refreshment was opened recently, the following factoid was found inside the bottle cap:

The average velocity of a falling raindrop is 7 miles/hour.

A quick search of the Internet found that meteorologist Jeff Haby offers the additional information that an "average" spherical raindrop has a radius of 0.04 in. and an approximate volume of 0.000000155 ft^3. Use this data and, if need be, dig up other data and make other reasonable assumptions to determine whether *"average velocity of . . . 7 mi/h"* is consistent with the models in Problems 35 and 36 in Exercises 3.1 and Problem 15 in this exercise set. Also see Problem 36 in Exercises 1.3.

33. Time Drips By The **clepsydra,** or water clock, was a device that the ancient Egyptians, Greeks, Romans, and Chinese used to measure the passage of time by observing the change in the height of water that was permitted to flow out of a small hole in the bottom of a container or tank.

(a) Suppose a tank is made of glass and has the shape of a right-circular cylinder of radius 1 ft. Assume that $h(0) = 2$ ft corresponds to water filled to the top of the tank, a hole in the bottom is circular with radius $\frac{1}{32}$ in., $g = 32$ ft/s^2, and $c = 0.6$. Use the differential equation in Problem 12 to find the height $h(t)$ of the water.

(b) For the tank in part (a), how far up from its bottom should a mark be made on its side, as shown in Figure 3.2.9, that corresponds to the passage of one hour? Next determine where to place the marks corresponding to the passage of 2 hr, 3 hr, . . . , 12 hr. Explain why these marks are not evenly spaced.

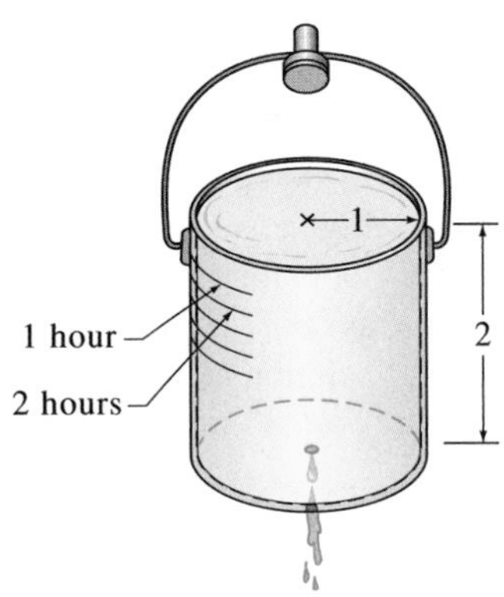

FIGURE 3.2.9 Clepsydra in Problem 33

34. **(a)** Suppose that a glass tank has the shape of a cone with circular cross section as shown in Figure 3.2.10. As in part (a) of Problem 33, assume that $h(0) = 2$ ft corresponds to water filled to the top of the tank, a hole in the bottom is circular with radius $\frac{1}{32}$ in., $g = 32$ ft/s^2, and $c = 0.6$. Use the differential equation in Problem 12 to find the height $h(t)$ of the water.

(b) Can this water clock measure 12 time intervals of length equal to 1 hour? Explain using sound mathematics.

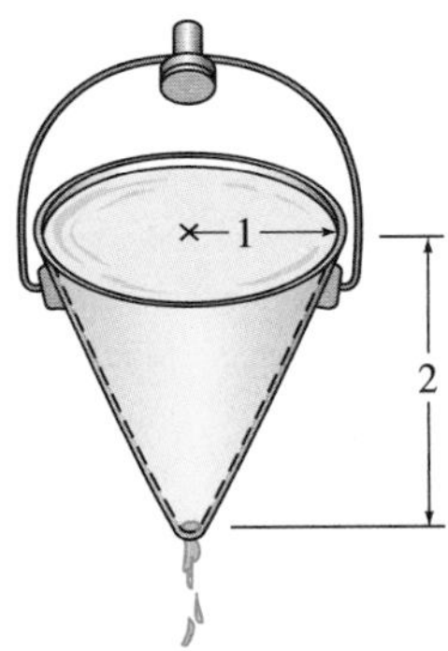

FIGURE 3.2.10 Clepsydra in Problem 34

35. Suppose that $r = f(h)$ defines the shape of a water clock for which the time marks are equally spaced. Use the differential equation in Problem 12 to find $f(h)$ and sketch a typical graph of h as a function of r. Assume that the cross-sectional area A_h of the hole is constant. [*Hint*: In this situation $dh/dt = -a$, where $a > 0$ is a constant.]

3.3 MODELING WITH SYSTEMS OF FIRST-ORDER DEs

REVIEW MATERIAL

- Section 1.3

INTRODUCTION This section is similar to Section 1.3 in that we are just going to discuss certain mathematical models, but instead of a single differential equation the models will be systems of first-order differential equations. Although some of the models will be based on topics that we explored in the preceding two sections, we are not going to develop any general methods for solving these systems. There are reasons for this: First, we do not possess the necessary mathematical tools for solving systems at this point. Second, some of the systems that we discuss—notably the systems of *nonlinear* first-order DEs—simply cannot be solved analytically. We shall examine solution methods for systems of *linear* DEs in Chapters 4, 7, and 8.

Linear/Nonlinear Systems We have seen that a single differential equation can serve as a mathematical model for a single population in an environment. But if there are, say, two interacting and perhaps competing species living in the same environment (for example, rabbits and foxes), then a model for their populations $x(t)$ and $y(t)$ might be a system of two first-order differential equations such as

$$\begin{aligned}\frac{dx}{dt} &= g_1(t, x, y)\\ \frac{dy}{dt} &= g_2(t, x, y).\end{aligned} \tag{1}$$

When g_1 and g_2 are linear in the variables x and y—that is, g_1 and g_2 have the forms

$$g_1(t, x, y) = c_1x + c_2y + f_1(t) \quad \text{and} \quad g_2(t, x, y) = c_3x + c_4y + f_2(t),$$

where the coefficients c_i could depend on t—then (1) is said to be a **linear system.** A system of differential equations that is not linear is said to be **nonlinear.**

Radioactive Series In the discussion of radioactive decay in Sections 1.3 and 3.1 we assumed that the rate of decay was proportional to the number $A(t)$ of nuclei of the substance present at time t. When a substance decays by radioactivity, it usually doesn't just transmute in one step into a stable substance; rather, the first substance decays into another radioactive substance, which in turn decays into a third substance, and so on. This process, called a **radioactive decay series,** continues until a stable element is reached. For example, the uranium decay series is U-238 → Th-234 → · · · → Pb-206, where Pb-206 is a stable isotope of lead. The half-lives of the various elements in a radioactive series can range from billions of years (4.5×10^9 years for U-238) to a fraction of a second. Suppose a radioactive series is described schematically by $X \xrightarrow{-\lambda_1} Y \xrightarrow{-\lambda_2} Z$, where $k_1 = -\lambda_1 < 0$ and $k_2 = -\lambda_2 < 0$ are the decay constants for substances X and Y, respectively, and Z is a stable element. Suppose, too, that $x(t)$, $y(t)$, and $z(t)$ denote amounts of substances X, Y, and Z, respectively, remaining at time t. The decay of element X is described by

$$\frac{dx}{dt} = -\lambda_1 x,$$

whereas the rate at which the second element Y decays is the net rate

$$\frac{dy}{dt} = \lambda_1 x - \lambda_2 y,$$

since Y is *gaining* atoms from the decay of X and at the same time *losing* atoms because of its own decay. Since Z is a stable element, it is simply gaining atoms from the decay of element Y:

$$\frac{dz}{dt} = \lambda_2 y.$$

In other words, a model of the radioactive decay series for three elements is the linear system of three first-order differential equations

$$\begin{aligned}\frac{dx}{dt} &= -\lambda_1 x \\ \frac{dy}{dt} &= \lambda_1 x - \lambda_2 y \\ \frac{dz}{dt} &= \lambda_2 y.\end{aligned} \tag{2}$$

Mixtures Consider the two tanks shown in Figure 3.3.1. Let us suppose for the sake of discussion that tank A contains 50 gallons of water in which 25 pounds of salt is dissolved. Suppose tank B contains 50 gallons of pure water. Liquid is pumped into and out of the tanks as indicated in the figure; the mixture exchanged between the two tanks and the liquid pumped out of tank B are assumed to be well stirred. We wish to construct a mathematical model that describes the number of pounds $x_1(t)$ and $x_2(t)$ of salt in tanks A and B, respectively, at time t.

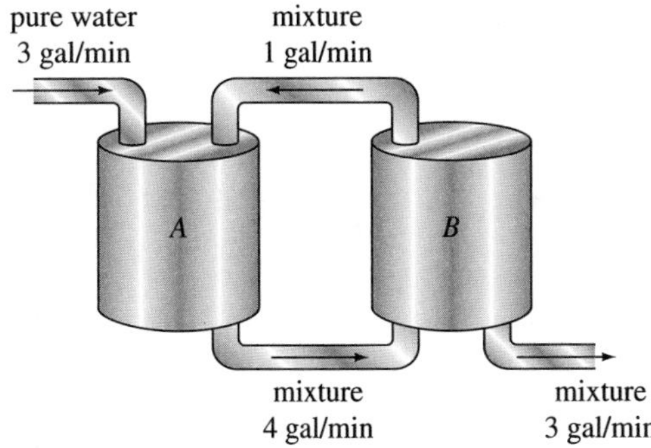

FIGURE 3.3.1 Connected mixing tanks

By an analysis similar to that on page 24 in Section 1.3 and Example 5 of Section 3.1 we see that the net rate of change of $x_1(t)$ for tank A is

$$\begin{aligned}\frac{dx_1}{dt} &= \overbrace{(3\text{ gal/min})\cdot(0\text{ lb/gal}) + (1\text{ gal/min})\cdot\left(\frac{x_2}{50}\text{ lb/gal}\right)}^{\text{input rate of salt}} - \overbrace{(4\text{ gal/min})\cdot\left(\frac{x_1}{50}\text{ lb/gal}\right)}^{\text{output rate of salt}} \\ &= -\frac{2}{25}x_1 + \frac{1}{50}x_2.\end{aligned}$$

Similarly, for tank B the net rate of change of $x_2(t)$ is

$$\begin{aligned}\frac{dx_2}{dt} &= 4\cdot\frac{x_1}{50} - 3\cdot\frac{x_2}{50} - 1\cdot\frac{x_2}{50} \\ &= \frac{2}{25}x_1 - \frac{2}{25}x_2.\end{aligned}$$

Thus we obtain the linear system

$$\begin{aligned}\frac{dx_1}{dt} &= -\frac{2}{25}x_1 + \frac{1}{50}x_2\\ \frac{dx_2}{dt} &= \frac{2}{25}x_1 - \frac{2}{25}x_2.\end{aligned} \tag{3}$$

Observe that the foregoing system is accompanied by the initial conditions $x_1(0) = 25$, $x_2(0) = 0$.

A Predator-Prey Model Suppose that two different species of animals interact within the same environment or ecosystem, and suppose further that the first species eats only vegetation and the second eats only the first species. In other words, one species is a predator, and the other is a prey. For example, wolves hunt grass-eating caribou, sharks devour little fish, and the snowy owl pursues an arctic rodent called the lemming. For the sake of discussion, let us imagine that the predators are foxes and the prey are rabbits.

Let $x(t)$ and $y(t)$ denote the fox and rabbit populations, respectively, at time t. If there were no rabbits, then one might expect that the foxes, lacking an adequate food supply, would decline in number according to

$$\frac{dx}{dt} = -ax, \qquad a > 0. \tag{4}$$

When rabbits are present in the environment, however, it seems reasonable that the number of encounters or interactions between these two species per unit time is jointly proportional to their populations x and y—that is, proportional to the product xy. Thus when rabbits are present, there is a supply of food, so foxes are added to the system at a rate bxy, $b > 0$. Adding this last rate to (4) gives a model for the fox population:

$$\frac{dx}{dt} = -ax + bxy. \tag{5}$$

On the other hand, if there were no foxes, then the rabbits would, with an added assumption of unlimited food supply, grow at a rate that is proportional to the number of rabbits present at time t:

$$\frac{dy}{dt} = dy, \qquad d > 0. \tag{6}$$

But when foxes are present, a model for the rabbit population is (6) decreased by cxy, $c > 0$—that is, decreased by the rate at which the rabbits are eaten during their encounters with the foxes:

$$\frac{dy}{dt} = dy - cxy. \tag{7}$$

Equations (5) and (7) constitute a system of nonlinear differential equations

$$\begin{aligned}\frac{dx}{dt} &= -ax + bxy = x(-a + by)\\ \frac{dy}{dt} &= dy - cxy = y(d - cx),\end{aligned} \tag{8}$$

where a, b, c, and d are positive constants. This famous system of equations is known as the **Lotka-Volterra predator-prey model.**

Except for two constant solutions, $x(t) = 0$, $y(t) = 0$ and $x(t) = d/c$, $y(t) = a/b$, the nonlinear system (8) cannot be solved in terms of elementary functions. However, we can analyze such systems quantitatively and qualitatively. See Chapter 9, "Numerical Solutions of Ordinary Differential Equations," and Chapter 10, "Plane Autonomous Systems."*

*Chapters 10–15 are in the expanded version of this text, *Differential Equations with Boundary-Value Problems.*

EXAMPLE 1 Predator-Prey Model

Suppose

$$\frac{dx}{dt} = -0.16x + 0.08xy$$

$$\frac{dy}{dt} = 4.5y - 0.9xy$$

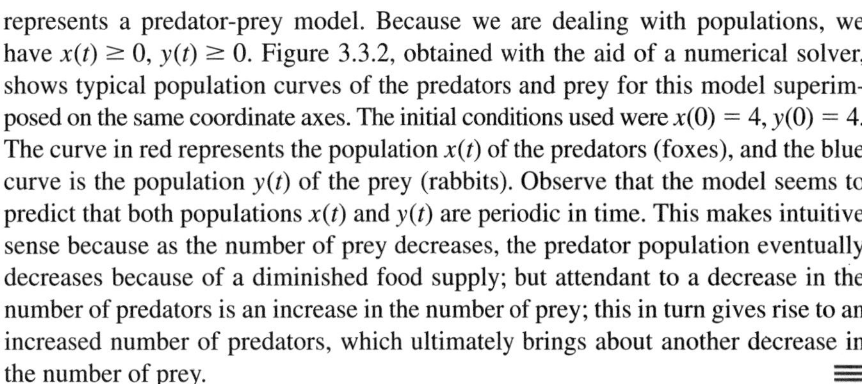

represents a predator-prey model. Because we are dealing with populations, we have $x(t) \geq 0$, $y(t) \geq 0$. Figure 3.3.2, obtained with the aid of a numerical solver, shows typical population curves of the predators and prey for this model superimposed on the same coordinate axes. The initial conditions used were $x(0) = 4$, $y(0) = 4$. The curve in red represents the population $x(t)$ of the predators (foxes), and the blue curve is the population $y(t)$ of the prey (rabbits). Observe that the model seems to predict that both populations $x(t)$ and $y(t)$ are periodic in time. This makes intuitive sense because as the number of prey decreases, the predator population eventually decreases because of a diminished food supply; but attendant to a decrease in the number of predators is an increase in the number of prey; this in turn gives rise to an increased number of predators, which ultimately brings about another decrease in the number of prey. ≡

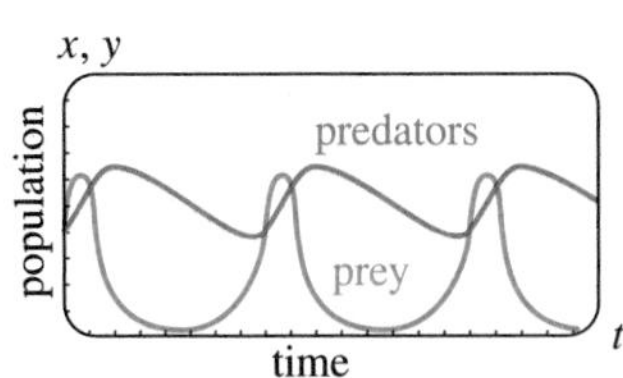

FIGURE 3.3.2 Populations of predators (red) and prey (blue) in Example 1

Competition Models Now suppose two different species of animals occupy the same ecosystem, not as predator and prey but rather as competitors for the same resources (such as food and living space) in the system. In the absence of the other, let us assume that the rate at which each population grows is given by

$$\frac{dx}{dt} = ax \quad \text{and} \quad \frac{dy}{dt} = cy, \tag{9}$$

respectively.

Since the two species compete, another assumption might be that each of these rates is diminished simply by the influence, or existence, of the other population. Thus a model for the two populations is given by the linear system

$$\begin{aligned} \frac{dx}{dt} &= ax - by \\ \frac{dy}{dt} &= cy - dx, \end{aligned} \tag{10}$$

where a, b, c, and d are positive constants.

On the other hand, we might assume, as we did in (5), that each growth rate in (9) should be reduced by a rate proportional to the number of interactions between the two species:

$$\begin{aligned} \frac{dx}{dt} &= ax - bxy \\ \frac{dy}{dt} &= cy - dxy. \end{aligned} \tag{11}$$

Inspection shows that this nonlinear system is similar to the Lotka-Volterra predator-prey model. Finally, it might be more realistic to replace the rates in (9), which indicate that the population of each species in isolation grows exponentially, with rates indicating that each population grows logistically (that is, over a long time the population is bounded):

$$\frac{dx}{dt} = a_1x - b_1x^2 \quad \text{and} \quad \frac{dy}{dt} = a_2y - b_2y^2. \tag{12}$$

When these new rates are decreased by rates proportional to the number of interactions, we obtain another nonlinear model:

$$\begin{aligned}\frac{dx}{dt} &= a_1x - b_1x^2 - c_1xy = x(a_1 - b_1x - c_1y)\\ \frac{dy}{dt} &= a_2y - b_2y^2 - c_2xy = y(a_2 - b_2y - c_2x),\end{aligned} \tag{13}$$

where all coefficients are positive. The linear system (10) and the nonlinear systems (11) and (13) are, of course, called **competition models.**

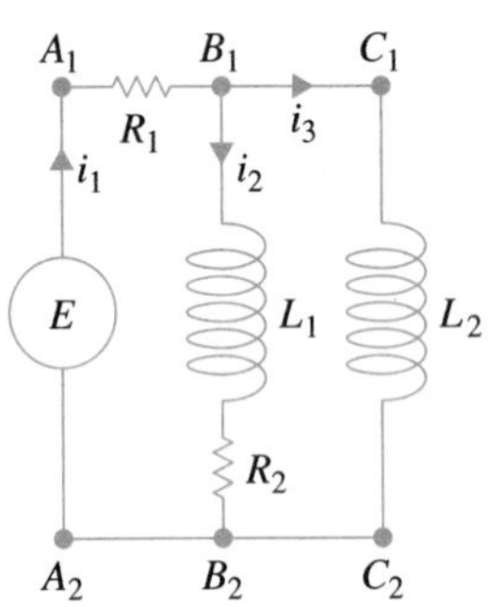

FIGURE 3.3.3 Network whose model is given in (17)

Networks An electrical network having more than one loop also gives rise to simultaneous differential equations. As shown in Figure 3.3.3, the current $i_1(t)$ splits in the directions shown at point B_1, called a *branch point* of the network. By **Kirchhoff's first law** we can write

$$i_1(t) = i_2(t) + i_3(t). \tag{14}$$

We can also apply **Kirchhoff's second law** to each loop. For loop $A_1B_1B_2A_2A_1$, summing the voltage drops across each part of the loop gives

$$E(t) = i_1R_1 + L_1\frac{di_2}{dt} + i_2R_2. \tag{15}$$

Similarly, for loop $A_1B_1C_1C_2B_2A_2A_1$ we find

$$E(t) = i_1R_1 + L_2\frac{di_3}{dt}. \tag{16}$$

Using (14) to eliminate i_1 in (15) and (16) yields two linear first-order equations for the currents $i_2(t)$ and $i_3(t)$:

$$\begin{aligned}L_1\frac{di_2}{dt} + (R_1 + R_2)i_2 + R_1i_3 &= E(t)\\ L_2\frac{di_3}{dt} + R_1i_2 + R_1i_3 &= E(t).\end{aligned} \tag{17}$$

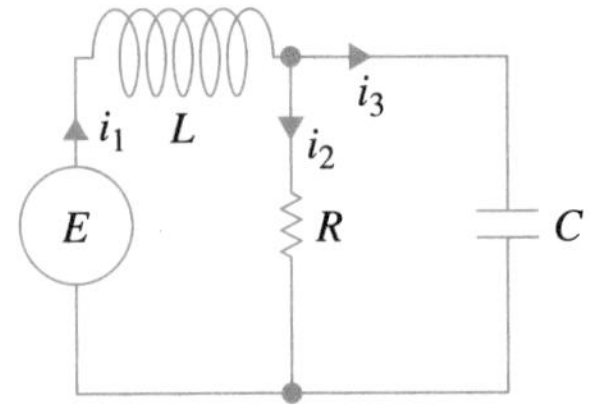

FIGURE 3.3.4 Network whose model is given in (18)

We leave it as an exercise (see Problem 14 in Exercises 3.3) to show that the system of differential equations describing the currents $i_1(t)$ and $i_2(t)$ in the network containing a resistor, an inductor, and a capacitor shown in Figure 3.3.4 is

$$\begin{aligned}L\frac{di_1}{dt} + Ri_2 &= E(t)\\ RC\frac{di_2}{dt} + i_2 - i_1 &= 0.\end{aligned} \tag{18}$$

EXERCISES 3.3

Answers to selected odd-numbered problems begin on page ANS-4.

Radioactive Series

1. We have not discussed methods by which systems of first-order differential equations can be solved. Nevertheless, systems such as (2) can be solved with no knowledge other than how to solve a single linear first-order equation. Find a solution of (2) subject to the initial conditions $x(0) = x_0$, $y(0) = 0$, $z(0) = 0$.

2. In Problem 1 suppose that time is measured in days, that the decay constants are $k_1 = -0.138629$ and $k_2 = -0.004951$, and that $x_0 = 20$. Use a graphing utility to obtain the graphs of the solutions $x(t)$, $y(t)$, and $z(t)$ on the same set of coordinate axes. Use the graphs to approximate the half-lives of substances X and Y.

3. Use the graphs in Problem 2 to approximate the times when the amounts $x(t)$ and $y(t)$ are the same, the times when the amounts $x(t)$ and $z(t)$ are the same, and the times when the amounts $y(t)$ and $z(t)$ are the same. Why does the time that is determined when the amounts $y(t)$ and $z(t)$ are the same make intuitive sense?

4. Construct a mathematical model for a radioactive series of four elements W, X, Y, and Z, where Z is a stable element.

Mixtures

5. Consider two tanks A and B, with liquid being pumped in and out at the same rates, as described by the system of equations (3). What is the system of differential equations if, instead of pure water, a brine solution containing 2 pounds of salt per gallon is pumped into tank A?

6. Use the information given in Figure 3.3.5 to construct a mathematical model for the number of pounds of salt $x_1(t)$, $x_2(t)$, and $x_3(t)$ at time t in tanks A, B, and C, respectively.

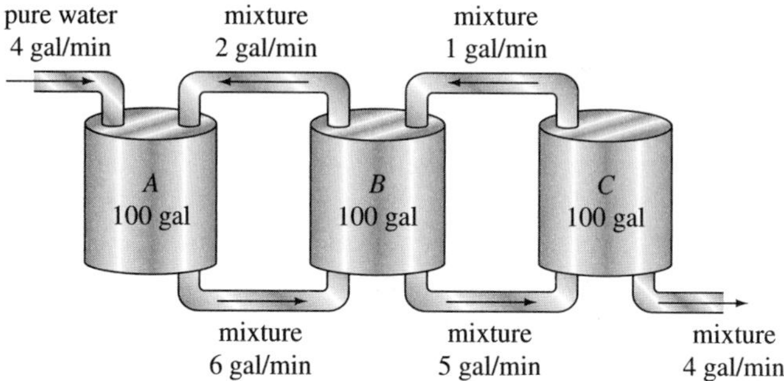

FIGURE 3.3.5 Mixing tanks in Problem 6

7. Two very large tanks A and B are each partially filled with 100 gallons of brine. Initially, 100 pounds of salt is dissolved in the solution in tank A and 50 pounds of salt is dissolved in the solution in tank B. The system is closed in that the well-stirred liquid is pumped only between the tanks, as shown in Figure 3.3.6.

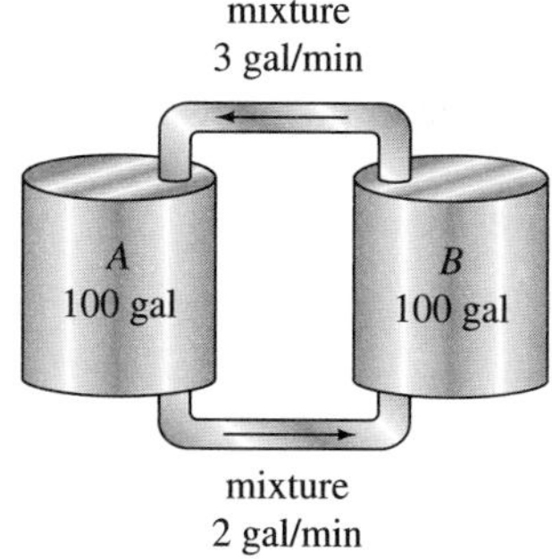

FIGURE 3.3.6 Mixing tanks in Problem 7

(a) Use the information given in the figure to construct a mathematical model for the number of pounds of salt $x_1(t)$ and $x_2(t)$ at time t in tanks A and B, respectively.

(b) Find a relationship between the variables $x_1(t)$ and $x_2(t)$ that holds at time t. Explain why this relationship makes intuitive sense. Use this relationship to help find the amount of salt in tank B at $t = 30$ min.

8. Three large tanks contain brine, as shown in Figure 3.3.7. Use the information in the figure to construct a mathematical model for the number of pounds of salt $x_1(t)$, $x_2(t)$, and $x_3(t)$ at time t in tanks A, B, and C, respectively. Without solving the system, predict limiting values of $x_1(t)$, $x_2(t)$, and $x_3(t)$ as $t \to \infty$.

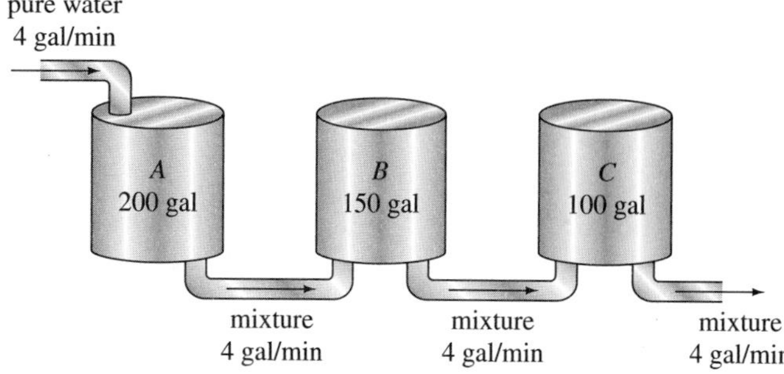

FIGURE 3.3.7 Mixing tanks in Problem 8

Predator-Prey Models

9. Consider the Lotka-Volterra predator-prey model defined by

$$\frac{dx}{dt} = -0.1x + 0.02xy$$

$$\frac{dy}{dt} = 0.2y - 0.025xy,$$

where the populations $x(t)$ (predators) and $y(t)$ (prey) are measured in thousands. Suppose $x(0) = 6$ and $y(0) = 6$. Use a numerical solver to graph $x(t)$ and $y(t)$. Use the graphs to approximate the time $t > 0$ when the two populations are first equal. Use the graphs to approximate the period of each population.

Competition Models

10. Consider the competition model defined by

$$\frac{dx}{dt} = x(2 - 0.4x - 0.3y)$$

$$\frac{dy}{dt} = y(1 - 0.1y - 0.3x),$$

where the populations $x(t)$ and $y(t)$ are measured in thousands and t in years. Use a numerical solver to analyze the populations over a long period of time for each of the following cases:
(a) $x(0) = 1.5, \quad y(0) = 3.5$
(b) $x(0) = 1, \quad y(0) = 1$
(c) $x(0) = 2, \quad y(0) = 7$
(d) $x(0) = 4.5, \quad y(0) = 0.5$

11. Consider the competition model defined by

$$\frac{dx}{dt} = x(1 - 0.1x - 0.05y)$$

$$\frac{dy}{dt} = y(1.7 - 0.1y - 0.15x),$$

where the populations $x(t)$ and $y(t)$ are measured in thousands and t in years. Use a numerical solver to analyze the populations over a long period of time for each of the following cases:

(a) $x(0) = 1, \quad y(0) = 1$
(b) $x(0) = 4, \quad y(0) = 10$
(c) $x(0) = 9, \quad y(0) = 4$
(d) $x(0) = 5.5, \quad y(0) = 3.5$

Networks

12. Show that a system of differential equations that describes the currents $i_2(t)$ and $i_3(t)$ in the electrical network shown in Figure 3.3.8 is

$$L\frac{di_2}{dt} + L\frac{di_3}{dt} + R_1 i_2 = E(t)$$

$$-R_1\frac{di_2}{dt} + R_2\frac{di_3}{dt} + \frac{1}{C}i_3 = 0.$$

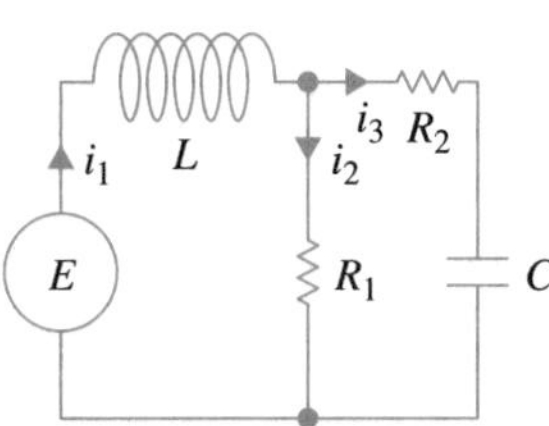

FIGURE 3.3.8 Network in Problem 12

13. Determine a system of first-order differential equations that describes the currents $i_2(t)$ and $i_3(t)$ in the electrical network shown in Figure 3.3.9.

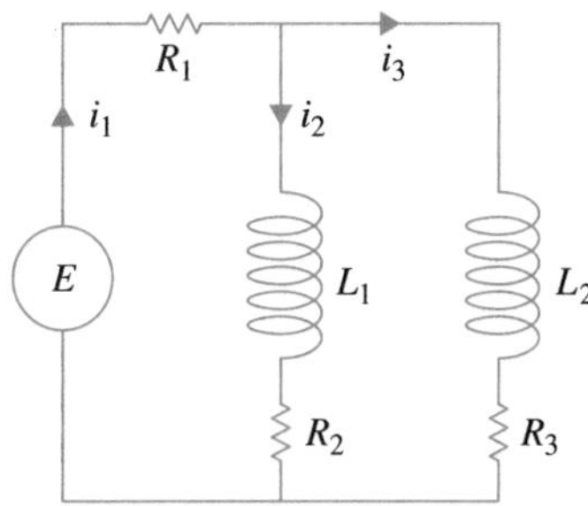

FIGURE 3.3.9 Network in Problem 13

14. Show that the linear system given in (18) describes the currents $i_1(t)$ and $i_2(t)$ in the network shown in Figure 3.3.4. [*Hint*: $dq/dt = i_3$.]

Additional Nonlinear Models

15. SIR Model A communicable disease is spread throughout a small community, with a fixed population of n people, by contact between infected individuals and people who are susceptible to the disease. Suppose that everyone is initially susceptible to the disease and that no one leaves the community while the epidemic is spreading. At time t, let $s(t)$, $i(t)$, and $r(t)$ denote, in turn, the number of people in the community (measured in hundreds) who are *susceptible* to the disease but not yet infected with it, the number of people who are *infected* with the disease, and the number of people who have ***recovered*** from the disease. Explain why the system of differential equations

$$\frac{ds}{dt} = -k_1 si$$

$$\frac{di}{dt} = -k_2 i + k_1 si$$

$$\frac{dr}{dt} = k_2 i,$$

where k_1 (called the *infection rate*) and k_2 (called the *removal rate*) are positive constants, is a reasonable mathematical model, commonly called a **SIR model,** for the spread of the epidemic throughout the community. Give plausible initial conditions associated with this system of equations.

16. (a) In Problem 15, explain why it is sufficient to analyze only

$$\frac{ds}{dt} = -k_1 si$$

$$\frac{di}{dt} = -k_2 i + k_1 si.$$

(b) Suppose $k_1 = 0.2$, $k_2 = 0.7$, and $n = 10$. Choose various values of $i(0) = i_0$, $0 < i_0 < 10$. Use a numerical solver to determine what the model predicts about the epidemic in the two cases $s_0 > k_2/k_1$ and $s_0 \leq k_2/k_1$. In the case of an epidemic, estimate the number of people who are eventually infected.

Project Problems

17. Concentration of a Nutrient Suppose compartments A and B shown in Figure 3.3.10 are filled with fluids and are separated by a permeable membrane. The figure is a compartmental representation of the exterior and interior of a cell. Suppose, too, that a nutrient necessary for cell growth passes through the membrane. A model

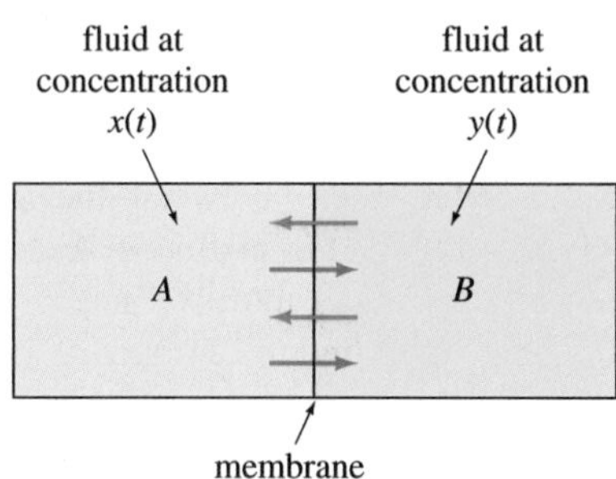

FIGURE 3.3.10 Nutrient flow through a membrane in Problem 17

for the concentrations $x(t)$ and $y(t)$ of the nutrient in compartments A and B, respectively, at time t is given by the linear system of differential equations

$$\frac{dx}{dt} = \frac{\kappa}{V_A}(y - x)$$

$$\frac{dy}{dt} = \frac{\kappa}{V_B}(x - y),$$

where V_A and V_B are the volumes of the compartments, and $\kappa > 0$ is a permeability factor. Let $x(0) = x_0$ and $y(0) = y_0$ denote the initial concentrations of the nutrient. Solely on the basis of the equations in the system and the assumption $x_0 > y_0 > 0$, sketch, on the same set of coordinate axes, possible solution curves of the system. Explain your reasoning. Discuss the behavior of the solutions over a long period of time.

18. The system in Problem 17, like the system in (2), can be solved with no advanced knowledge. Solve for $x(t)$ and $y(t)$ and compare their graphs with your sketches in Problem 17. Determine the limiting values of $x(t)$ and $y(t)$ as $t \to \infty$. Explain why the answer to the last question makes intuitive sense.

19. **Mixtures** Solely on the basis of the physical description of the mixture problem on page 107 and in Figure 3.3.1, discuss the nature of the functions $x_1(t)$ and $x_2(t)$. What is the behavior of each function over a long period of time? Sketch possible graphs of $x_1(t)$ and $x_2(t)$. Check your conjectures by using a numerical solver to obtain numerical solution curves of (3) subject to the initial conditions $x_1(0) = 25, x_2(0) = 0$.

20. **Newton's Law of Cooling/Warming** As shown in Figure 3.3.11, a small metal bar is placed inside container A, and container A then is placed within a much larger container B. As the metal bar cools, the ambient temperature $T_A(t)$ of the medium within container A changes according to Newton's law of cooling. As container A cools, the temperature of the medium inside container B does not change significantly and can be considered to be a constant T_B. Construct a mathematical model for the temperatures $T(t)$ and $T_A(t)$, where $T(t)$ is the temperature of the metal bar inside container A. As in Problems 1 and 18, this model can be solved by using prior knowledge. Find a solution of the system subject to the initial conditions $T(0) = T_0$, $T_A(0) = T_1$.

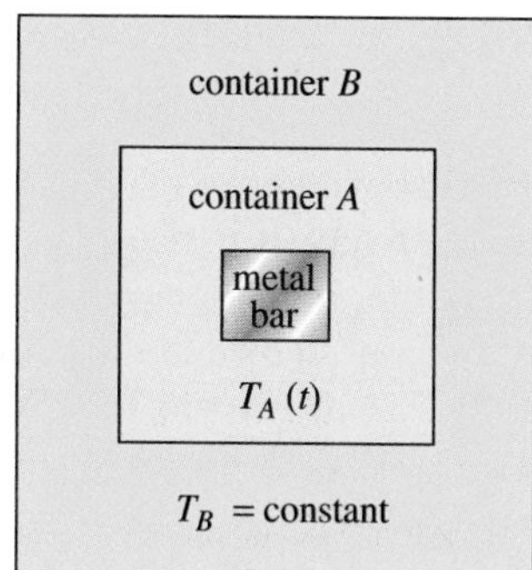

FIGURE 3.3.11 Container within a container in Problem 20

CHAPTER 3 IN REVIEW

Answers to selected odd-numbered problems begin on page ANS-4.

Answer Problems 1 and 2 without referring back to the text. Fill in the blank or answer true or false.

1. If $P(t) = P_0e^{0.15t}$ gives the population in an environment at time t, then a differential equation satisfied by $P(t)$ is ________.

2. If the rate of decay of a radioactive substance is proportional to the amount $A(t)$ remaining at time t, then the half-life of the substance is necessarily $T = -(\ln 2)/k$. The rate of decay of the substance at time $t = T$ is one-half the rate of decay at $t = 0$. ________

3. In March 1976 the world population reached 4 billion. At that time, a popular news magazine predicted that with an average yearly growth rate of 1.8%, the world population would be 8 billion in 45 years. How does this value compare with the value predicted by the model that assumes that the rate of increase in population is proportional to the population present at time t?

4. Air containing 0.06% carbon dioxide is pumped into a room whose volume is 8000 ft^3. The air is pumped in at a rate of 2000 ft^3/min, and the circulated air is then pumped out at the same rate. If there is an initial concentration of 0.2% carbon dioxide in the room, determine the subsequent amount in the room at time t. What is the concentration of carbon dioxide at 10 minutes? What is the steady-state, or equilibrium, concentration of carbon dioxide?

5. Solve the differential equation

$$\frac{dy}{dx} = -\frac{y}{\sqrt{s^2 - y^2}}$$

of the tractrix. See Problem 28 in Exercises 1.3. Assume that the initial point on the y-axis in (0, 10) and that the length of the rope is $x = 10$ ft.

6. Suppose a cell is suspended in a solution containing a solute of constant concentration C_s. Suppose further that the cell has constant volume V and that the area of its permeable membrane is the constant A. By **Fick's law** the rate of change of its mass m is directly proportional to the area A and the difference $C_s - C(t)$, where $C(t)$ is the concentration of the solute inside the cell at time t. Find $C(t)$ if $m = V \cdot C(t)$ and $C(0) = C_0$. See Figure 3.R.1.

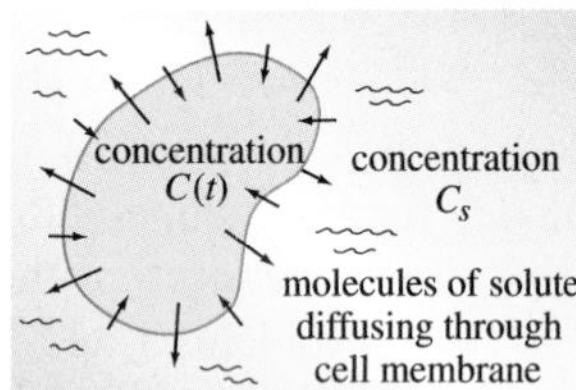

FIGURE 3.R.1 Cell in Problem 6

7. Suppose that as a body cools, the temperature of the surrounding medium increases because it completely absorbs the heat being lost by the body. Let $T(t)$ and $T_m(t)$ be the temperatures of the body and the medium at time t, respectively. If the initial temperature of the body is T_1 and the initial temperature of the medium is T_2, then it can be shown in this case that Newton's law of cooling is $dT/dt = k(T - T_m)$, $k < 0$, where $T_m = T_2 + B(T_1 - T)$, $B > 0$ is a constant.
   (a) The foregoing DE is autonomous. Use the phase portrait concept of Section 2.1 to determine the limiting value of the temperature $T(t)$ as $t \to \infty$. What is the limiting value of $T_m(t)$ as $t \to \infty$?
   (b) Verify your answers in part (a) by actually solving the differential equation.
   (c) Discuss a physical interpretation of your answers in part (a).

8. According to **Stefan's law of radiation** the absolute temperature T of a body cooling in a medium at constant absolute temperature T_m is given by

$$\frac{dT}{dt} = k(T^4 - T_m^4),$$

where k is a constant. Stefan's law can be used over a greater temperature range than Newton's law of cooling.
   (a) Solve the differential equation.
   (b) Show that when $T - T_m$ is small in comparison to T_m then Newton's law of cooling approximates Stefan's law. [*Hint*: Think binomial series of the right-hand side of the DE.]

9. An LR-series circuit has a variable inductor with the inductance defined by

$$L(t) = \begin{cases} 1 - \dfrac{1}{10}t, & 0 \le t < 10 \\ 0, & t \ge 10. \end{cases}$$

Find the current $i(t)$ if the resistance is 0.2 ohm, the impressed voltage is $E(t) = 4$, and $i(0) = 0$. Graph $i(t)$.

10. A classical problem in the calculus of variations is to find the shape of a curve $\mathscr{C}$ such that a bead, under the influence of gravity, will slide from point $A(0, 0)$ to point $B(x_1, y_1)$ in the least time. See Figure 3.R.2. It can be shown that a nonlinear differential for the shape $y(x)$ of the path is $y[1 + (y')^2] = k$, where k is a constant. First solve for dx in terms of y and dy, and then use the substitution $y = k \sin^2\theta$ to obtain a parametric form of the solution. The curve $\mathscr{C}$ turns out to be a **cycloid**.

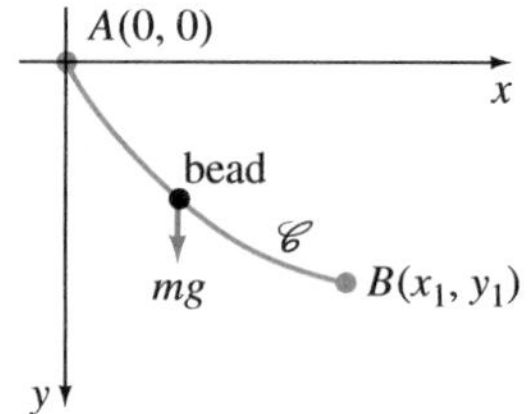

FIGURE 3.R.2 Sliding bead in Problem 10

11. A model for the populations of two interacting species of animals is

$$\frac{dx}{dt} = k_1 x(\alpha - x)$$
$$\frac{dy}{dt} = k_2 xy.$$

Solve for x and y in terms of t.

12. Initially, two large tanks A and B each hold 100 gallons of brine. The well-stirred liquid is pumped between the tanks as shown in Figure 3.R.3. Use the information given in the figure to construct a mathematical model for the number of pounds of salt $x_1(t)$ and $x_2(t)$ at time t in tanks A and B, respectively.

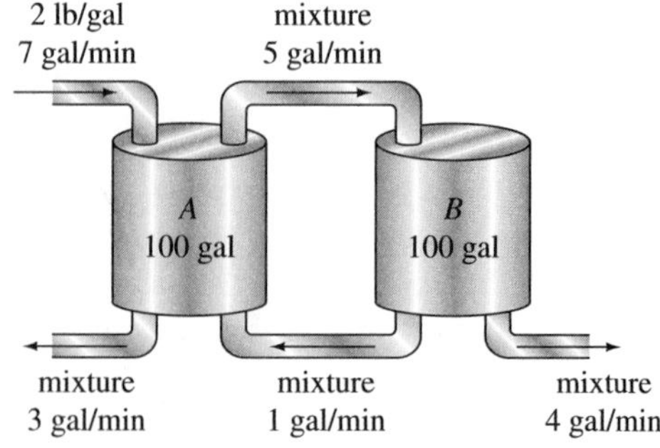

FIGURE 3.R.3 Mixing tanks in Problem 12

When all the curves in a family $G(x, y, c_1) = 0$ intersect orthogonally all the curves in another family $H(x, y, c_2) = 0$, the families are said to be **orthogonal trajectories** of each other. See Figure 3.R.4. If $dy/dx = f(x, y)$ is the differential equation of one family, then the differential equation for the orthogonal trajectories of this family is $dy/dx = -1/f(x, y)$. In Problems 13 and 14 find the differential equation of the given family. Find the orthogonal trajectories of this family. Use a graphing utility to graph both families on the same set of coordinate axes.

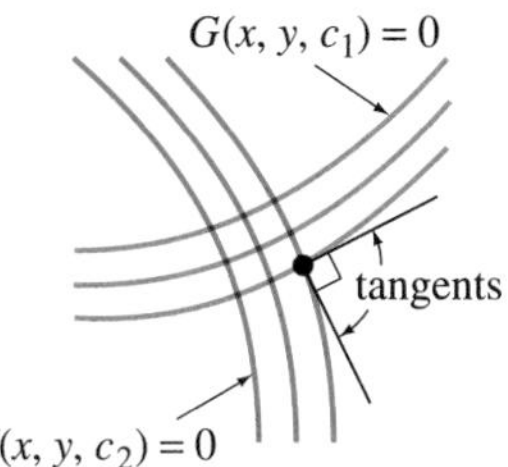

FIGURE 3.R.4 Orthogonal trajectories

13. $y = -x - 1 + c_1e^x$ **14.** $y = \dfrac{1}{x + c_1}$

15. Potassium-40 Decay One of the most abundant metals found throughout the Earth's crust and oceans is potassium. Although potassium occurs naturally in the form of three isotopes, only the isotope potassium-40 (K-40) is radioactive. This isotope is a bit unusual in that it decays by two different nuclear reactions. Over time, by emitting a beta particle, a great percentage of an initial amount of K-40 decays into the stable isotope calcium-40 (Ca-40), whereas by electron capture a smaller percentage of K-40 decays into the stable isotope argon-40 (Ar-40).* Because the rates at which the amounts $C(t)$ of Ca-40 and $A(t)$ of Ar-40 increase are proportional to the amount $K(t)$ of potassium present, and the rate at which potassium decreases is also proportional to $K(t)$ we obtain the system of linear first-order equations

$$\frac{dC}{dt} = \lambda_1 K$$
$$\frac{dA}{dt} = \lambda_2 K$$
$$\frac{dK}{dt} = -(\lambda_1 + \lambda_2)K,$$

where λ_1 and λ_2 are positive constants of proportionality.

(a) From the foregoing system of differential equations find $K(t)$ if $K(0) = K_0$. Then find $C(t)$ and $A(t)$ if $C(0) = 0$ and $A(0) = 0$.

(b) It is known that $\lambda_1 = 4.7526 \times 10^{-10}$ and $\lambda_2 = 0.5874 \times 10^{-10}$. Find the half-life of K-40.

(c) Use your solutions for $C(t)$ and $A(t)$ to determine the percentage of an initial amount K_0 of K-40 that decays into Ca-40 and the percentage that decays into Ar-40 over a very long period of time.

---

* The knowledge of how K-40 decays is the basis for the **potassium-argon dating method**. This method can be used to find the age of very old igneous rocks. Fossils can sometimes be dated indirectly by dating the igneous rocks in the substrata in which the fossils are found.

4 Higher-Order Differential Equations

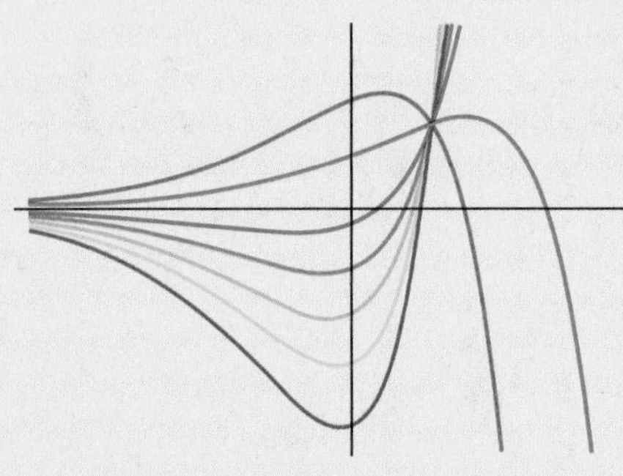

We turn now to the solution of ordinary differential equation of order two or higher. In the first seven sections of this chapter we examine the underlying theory and solution methods for certain kinds of linear equations. In the new, but optional, Section 4.8 we build on the material of Section 4.6 to construct Green's functions for solving linear initial-value and boundary-value problems. The elimination method of solving systems of linear equations is introduced in Section 4.9 because this method simply uncouples a system into individual linear equations in each dependent variable. The chapter concludes with a brief examination of nonlinear higher-order equations in Section 4.10.

4.1 PRELIMINARY THEORY—LINEAR EQUATIONS

REVIEW MATERIAL

- Reread the *Remarks* at the end of Section 1.1
- Section 2.3 (especially page 57)

INTRODUCTION In Chapter 2 we saw that we could solve a few first-order differential equations by recognizing them as separable, linear, exact, homogeneous, or perhaps Bernoulli equations. Even though the solutions of these equations were in the form of a one-parameter family, this family, with one exception, did not represent the general solution of the differential equation. Only in the case of *linear* first-order differential equations were we able to obtain general solutions, by paying attention to certain continuity conditions imposed on the coefficients. Recall that a **general solution** is a family of solutions defined on some interval I that contains *all* solutions of the DE that are defined on I. Because our primary goal in this chapter is to find general solutions of linear higher-order DEs, we first need to examine some of the theory of linear equations.

4.1.1 INITIAL-VALUE AND BOUNDARY-VALUE PROBLEMS

Initial-Value Problem In Section 1.2 we defined an initial-value problem for a general nth-order differential equation. For a linear differential equation an **nth-order initial-value problem** is

$$\text{Solve:} \quad a_n(x)\frac{d^n y}{dx^n} + a_{n-1}(x)\frac{d^{n-1} y}{dx^{n-1}} + \cdots + a_1(x)\frac{dy}{dx} + a_0(x)y = g(x) \tag{1}$$

$$\text{Subject to:} \quad y(x_0) = y_0, \quad y'(x_0) = y_1, \ldots, \quad y^{(n-1)}(x_0) = y_{n-1}.$$

Recall that for a problem such as this one we seek a function defined on some interval I, containing x_0, that satisfies the differential equation and the n initial conditions specified at x_0: $y(x_0) = y_0, y'(x_0) = y_1, \ldots, y^{(n-1)}(x_0) = y_{n-1}$. We have already seen that in the case of a second-order initial-value problem a solution curve must pass through the point (x_0, y_0) and have slope y_1 at this point.

Existence and Uniqueness In Section 1.2 we stated a theorem that gave conditions under which the existence and uniqueness of a solution of a first-order initial-value problem were guaranteed. The theorem that follows gives sufficient conditions for the existence of a unique solution of the problem in (1).

THEOREM 4.1.1 Existence of a Unique Solution

Let $a_n(x), a_{n-1}(x), \ldots, a_1(x), a_0(x)$ and $g(x)$ be continuous on an interval I and let $a_n(x) \neq 0$ for every x in this interval. If $x = x_0$ is any point in this interval, then a solution $y(x)$ of the initial-value problem (1) exists on the interval and is unique.

EXAMPLE 1 Unique Solution of an IVP

The initial-value problem

$$3y''' + 5y'' - y' + 7y = 0, \quad y(1) = 0, \quad y'(1) = 0, \quad y''(1) = 0$$

possesses the trivial solution $y = 0$. Because the third-order equation is linear with constant coefficients, it follows that all the conditions of Theorem 4.1.1 are fulfilled. Hence $y = 0$ is the *only* solution on any interval containing $x = 1$. ≡

EXAMPLE 2 Unique Solution of an IVP

You should verify that the function $y = 3e^{2x} + e^{-2x} - 3x$ is a solution of the initial-value problem

$$y'' - 4y = 12x, \quad y(0) = 4, \quad y'(0) = 1.$$

Now the differential equation is linear, the coefficients as well as $g(x) = 12x$ are continuous, and $a_2(x) = 1 \neq 0$ on any interval I containing $x = 0$. We conclude from Theorem 4.1.1 that the given function is the unique solution on I. ≡

The requirements in Theorem 4.1.1 that $a_i(x)$, $i = 0, 1, 2, \ldots, n$ be continuous and $a_n(x) \neq 0$ for every x in I are both important. Specifically, if $a_n(x) = 0$ for some x in the interval, then the solution of a linear initial-value problem may not be unique or even exist. For example, you should verify that the function $y = cx^2 + x + 3$ is a solution of the initial-value problem

$$x^2y'' - 2xy' + 2y = 6, \quad y(0) = 3, \quad y'(0) = 1$$

on the interval $(-\infty, \infty)$ for any choice of the parameter c. In other words, there is no unique solution of the problem. Although most of the conditions of Theorem 4.1.1 are satisfied, the obvious difficulties are that $a_2(x) = x^2$ is zero at $x = 0$ and that the initial conditions are also imposed at $x = 0$.

≡ Boundary-Value Problem

Another type of problem consists of solving a linear differential equation of order two or greater in which the dependent variable y or its derivatives are specified at *different points*. A problem such as

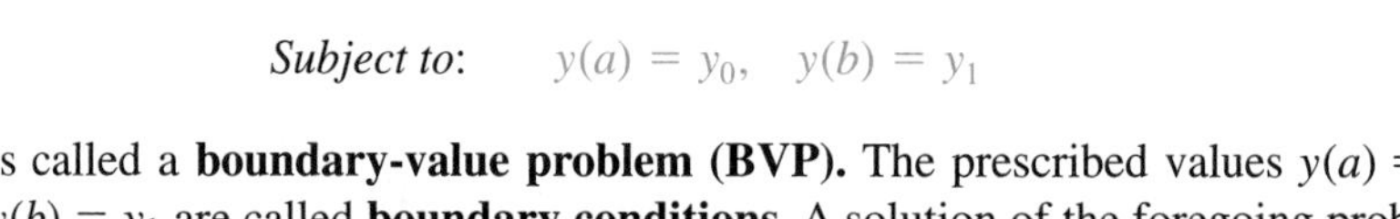

Solve: $$a_2(x)\frac{d^2y}{dx^2} + a_1(x)\frac{dy}{dx} + a_0(x)y = g(x)$$

Subject to: $$y(a) = y_0, \quad y(b) = y_1$$

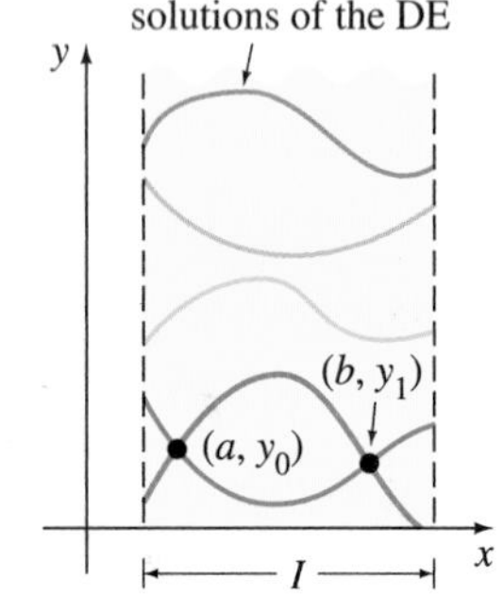

FIGURE 4.1.1 Solution curves of a BVP that pass through two points

is called a **boundary-value problem (BVP).** The prescribed values $y(a) = y_0$ and $y(b) = y_1$ are called **boundary conditions.** A solution of the foregoing problem is a function satisfying the differential equation on some interval I, containing a and b, whose graph passes through the two points (a, y_0) and (b, y_1). See Figure 4.1.1.

For a second-order differential equation other pairs of boundary conditions could be

$$\begin{aligned} y'(a) &= y_0, & y(b) &= y_1 \\ y(a) &= y_0, & y'(b) &= y_1 \\ y'(a) &= y_0, & y'(b) &= y_1, \end{aligned}$$

where y_0 and y_1 denote arbitrary constants. These three pairs of conditions are just special cases of the general boundary conditions

$$\begin{aligned} \alpha_1 y(a) + \beta_1 y'(a) &= \gamma_1 \\ \alpha_2 y(b) + \beta_2 y'(b) &= \gamma_2. \end{aligned}$$

The next example shows that even when the conditions of Theorem 4.1.1 are fulfilled, a boundary-value problem may have several solutions (as suggested in Figure 4.1.1), a unique solution, or no solution at all.

EXAMPLE 3 A BVP Can Have Many, One, or No Solutions

In Example 7 of Section 1.1 we saw that the two-parameter family of solutions of the differential equation $x'' + 16x = 0$ is

$$x = c_1 \cos 4t + c_2 \sin 4t. \tag{2}$$

(a) Suppose we now wish to determine the solution of the equation that further satisfies the boundary conditions $x(0) = 0$, $x(\pi/2) = 0$. Observe that the first condition $0 = c_1 \cos 0 + c_2 \sin 0$ implies that $c_1 = 0$, so $x = c_2 \sin 4t$. But when $t = \pi/2$, $0 = c_2 \sin 2\pi$ is satisfied for any choice of c_2, since $\sin 2\pi = 0$. Hence the boundary-value problem

$$x'' + 16x = 0, \quad x(0) = 0, \quad x\left(\frac{\pi}{2}\right) = 0 \tag{3}$$

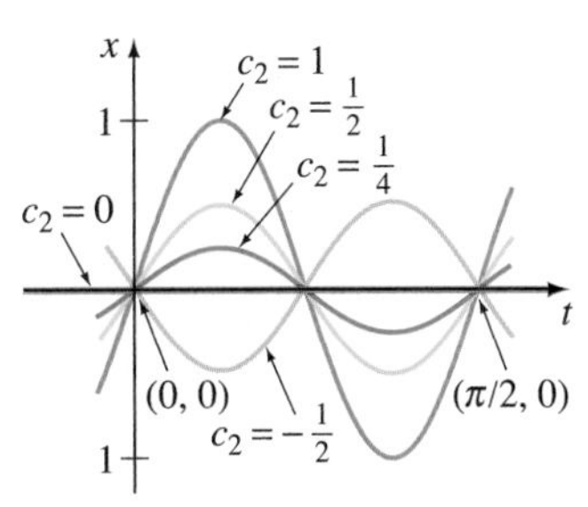

FIGURE 4.1.2 Solution curves for BVP in part (a) of Example 3

has infinitely many solutions. Figure 4.1.2 shows the graphs of some of the members of the one-parameter family $x = c_2 \sin 4t$ that pass through the two points (0, 0) and $(\pi/2, 0)$.

(b) If the boundary-value problem in (3) is changed to

$$x'' + 16x = 0, \quad x(0) = 0, \quad x\left(\frac{\pi}{8}\right) = 0, \tag{4}$$

then $x(0) = 0$ still requires $c_1 = 0$ in the solution (2). But applying $x(\pi/8) = 0$ to $x = c_2 \sin 4t$ demands that $0 = c_2 \sin(\pi/2) = c_2 \cdot 1$. Hence $x = 0$ is a solution of this new boundary-value problem. Indeed, it can be proved that $x = 0$ is the *only* solution of (4).

(c) Finally, if we change the problem to

$$x'' + 16x = 0, \quad x(0) = 0, \quad x\left(\frac{\pi}{2}\right) = 1, \tag{5}$$

we find again from $x(0) = 0$ that $c_1 = 0$, but applying $x(\pi/2) = 1$ to $x = c_2 \sin 4t$ leads to the contradiction $1 = c_2 \sin 2\pi = c_2 \cdot 0 = 0$. Hence the boundary-value problem (5) has no solution. ≡

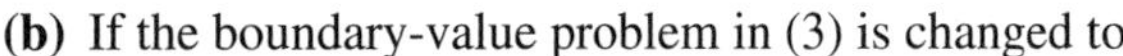

4.1.2 HOMOGENEOUS EQUATIONS

A linear nth-order differential equation of the form

$$a_n(x)\frac{d^ny}{dx^n} + a_{n-1}(x)\frac{d^{n-1}y}{dx^{n-1}} + \cdots + a_1(x)\frac{dy}{dx} + a_0(x)y = 0 \tag{6}$$

is said to be **homogeneous,** whereas an equation

$$a_n(x)\frac{d^ny}{dx^n} + a_{n-1}(x)\frac{d^{n-1}y}{dx^{n-1}} + \cdots + a_1(x)\frac{dy}{dx} + a_0(x)y = g(x), \tag{7}$$

with $g(x)$ not identically zero, is said to be **nonhomogeneous.** For example, $2y'' + 3y' - 5y = 0$ is a homogeneous linear second-order differential equation, whereas $x^3y''' + 6y' + 10y = e^x$ is a nonhomogeneous linear third-order differential equation. The word *homogeneous* in this context does not refer to coefficients that are homogeneous functions, as in Section 2.5.

We shall see that to solve a nonhomogeneous linear equation (7), we must first be able to solve the **associated homogeneous equation** (6).

To avoid needless repetition throughout the remainder of this text, we shall, as a matter of course, make the following important assumptions when stating definitions and theorems about linear equations (1). On some common interval I,

Please remember these two assumptions.

- the coefficient functions $a_i(x)$, $i = 0, 1, 2, \ldots, n$ and $g(x)$ are continuous;
- $a_n(x) \neq 0$ for every x in the interval.

Differential Operators In calculus differentiation is often denoted by the capital letter D—that is, $dy/dx = Dy$. The symbol D is called a **differential operator** because it transforms a differentiable function into another function. For example, $D(\cos 4x) = -4 \sin 4x$ and $D(5x^3 - 6x^2) = 15x^2 - 12x$. Higher-order derivatives can be expressed in terms of D in a natural manner:

$$\frac{d}{dx}\left(\frac{dy}{dx}\right) = \frac{d^2y}{dx^2} = D(Dy) = D^2y \qquad \text{and, in general,} \qquad \frac{d^ny}{dx^n} = D^ny,$$

where y represents a sufficiently differentiable function. Polynomial expressions involving D, such as $D + 3$, $D^2 + 3D - 4$, and $5x^3D^3 - 6x^2D^2 + 4xD + 9$, are also differential operators. In general, we define an ***n*th-order differential operator** or **polynomial operator** to be

$$L = a_n(x)D^n + a_{n-1}(x)D^{n-1} + \cdots + a_1(x)D + a_0(x). \tag{8}$$

As a consequence of two basic properties of differentiation, $D(cf(x)) = cDf(x)$, c is a constant, and $D\{f(x) + g(x)\} = Df(x) + Dg(x)$, the differential operator L possesses a linearity property; that is, L operating on a linear combination of two differentiable functions is the same as the linear combination of L operating on the individual functions. In symbols this means that

$$L\{\alpha f(x) + \beta g(x)\} = \alpha L(f(x)) + \beta L(g(x)), \tag{9}$$

where α and β are constants. Because of (9) we say that the nth-order differential operator L is a **linear operator.**

Differential Equations Any linear differential equation can be expressed in terms of the D notation. For example, the differential equation $y'' + 5y' + 6y = 5x - 3$ can be written as $D^2y + 5Dy + 6y = 5x - 3$ or $(D^2 + 5D + 6)y = 5x - 3$. Using (8), we can write the linear nth-order differential equations (6) and (7) compactly as

$$L(y) = 0 \qquad \text{and} \qquad L(y) = g(x),$$

respectively.

Superposition Principle In the next theorem we see that the sum, or **superposition,** of two or more solutions of a homogeneous linear differential equation is also a solution.

THEOREM 4.1.2 Superposition Principle—Homogeneous Equations

Let $y_1, y_2, \ldots, y_k$ be solutions of the homogeneous nth-order differential equation (6) on an interval I. Then the linear combination

$$y = c_1y_1(x) + c_2y_2(x) + \cdots + c_ky_k(x),$$

where the c_i, $i = 1, 2, \ldots, k$ are arbitrary constants, is also a solution on the interval.

PROOF We prove the case $k = 2$. Let L be the differential operator defined in (8), and let $y_1(x)$ and $y_2(x)$ be solutions of the homogeneous equation $L(y) = 0$. If we define $y = c_1y_1(x) + c_2y_2(x)$, then by linearity of L we have

$$L(y) = L\{c_1y_1(x) + c_2y_2(x)\} = c_1\,L(y_1) + c_2\,L(y_2) = c_1 \cdot 0 + c_2 \cdot 0 = 0.$$

COROLLARIES TO THEOREM 4.1.2

(A) A constant multiple $y = c_1y_1(x)$ of a solution $y_1(x)$ of a homogeneous linear differential equation is also a solution.
(B) A homogeneous linear differential equation always possesses the trivial solution $y = 0$.

EXAMPLE 4 Superposition—Homogeneous DE

The functions $y_1 = x^2$ and $y_2 = x^2 \ln x$ are both solutions of the homogeneous linear equation $x^3y''' - 2xy' + 4y = 0$ on the interval $(0, \infty)$. By the superposition principle the linear combination

$$y = c_1x^2 + c_2x^2 \ln x$$

is also a solution of the equation on the interval. ≡

The function $y = e^{7x}$ is a solution of $y'' - 9y' + 14y = 0$. Because the differential equation is linear and homogeneous, the constant multiple $y = ce^{7x}$ is also a solution. For various values of c we see that $y = 9e^{7x}$, $y = 0$, $y = -\sqrt{5}e^{7x}, \ldots$ are all solutions of the equation.

≡ Linear Dependence and Linear Independence

The next two concepts are basic to the study of linear differential equations.

DEFINITION 4.1.1 Linear Dependence/Independence

A set of functions $f_1(x), f_2(x), \ldots, f_n(x)$ is said to be **linearly dependent** on an interval I if there exist constants $c_1, c_2, \ldots, c_n$, not all zero, such that

$$c_1f_1(x) + c_2f_2(x) + \cdots + c_nf_n(x) = 0$$

for every x in the interval. If the set of functions is not linearly dependent on the interval, it is said to be **linearly independent.**

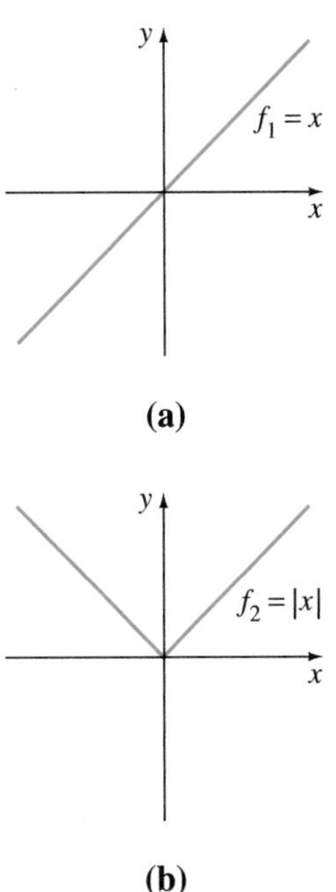

FIGURE 4.1.3 Set consisting of f_1 and f_2 is linearly independent on $(-\infty, \infty)$

In other words, a set of functions is linearly independent on an interval I if the only constants for which

$$c_1f_1(x) + c_2f_2(x) + \cdots + c_nf_n(x) = 0$$

for every x in the interval are $c_1 = c_2 = \cdots = c_n = 0$.

It is easy to understand these definitions for a set consisting of two functions $f_1(x)$ and $f_2(x)$. If the set of functions is linearly dependent on an interval, then there exist constants c_1 and c_2 that are not both zero such that for every x in the interval, $c_1f_1(x) + c_2f_2(x) = 0$. Therefore if we assume that $c_1 \neq 0$, it follows that $f_1(x) = (-c_2/c_1)f_2(x)$; that is, *if a set of two functions is linearly dependent, then one function is simply a constant multiple of the other.* Conversely, if $f_1(x) = c_2f_2(x)$ for some constant c_2, then $(-1) \cdot f_1(x) + c_2f_2(x) = 0$ for every x in the interval. Hence the set of functions is linearly dependent because at least one of the constants (namely, $c_1 = -1$) is not zero. We conclude that *a set of two functions $f_1(x)$ and $f_2(x)$ is linearly independent when neither function is a constant multiple of the other* on the interval. For example, the set of functions $f_1(x) = \sin 2x$, $f_2(x) = \sin x \cos x$ is linearly dependent on $(-\infty, \infty)$ because $f_1(x)$ is a constant multiple of $f_2(x)$. Recall from the double-angle formula for the sine that $\sin 2x = 2 \sin x \cos x$. On the other hand, the set of functions $f_1(x) = x$, $f_2(x) = |x|$ is linearly independent on $(-\infty, \infty)$. Inspection of Figure 4.1.3 should convince you that neither function is a constant multiple of the other on the interval.

It follows from the preceding discussion that the quotient $f_2(x)/f_1(x)$ is not a constant on an interval on which the set $f_1(x), f_2(x)$ is linearly independent. This little fact will be used in the next section.

EXAMPLE 5 Linearly Dependent Set of Functions

The set of functions $f_1(x) = \cos^2 x$, $f_2(x) = \sin^2 x$, $f_3(x) = \sec^2 x$, $f_4(x) = \tan^2 x$ is linearly dependent on the interval $(-\pi/2, \pi/2)$ because

$$c_1 \cos^2 x + c_2 \sin^2 x + c_3 \sec^2 x + c_4 \tan^2 x = 0$$

when $c_1 = c_2 = 1$, $c_3 = -1$, $c_4 = 1$. We used here $\cos^2 x + \sin^2 x = 1$ and $1 + \tan^2 x = \sec^2 x$. ≡

A set of functions $f_1(x), f_2(x), \ldots, f_n(x)$ is linearly dependent on an interval if at least one function can be expressed as a linear combination of the remaining functions.

EXAMPLE 6 Linearly Dependent Set of Functions

The set of functions $f_1(x) = \sqrt{x} + 5$, $f_2(x) = \sqrt{x} + 5x$, $f_3(x) = x - 1$, $f_4(x) = x^2$ is linearly dependent on the interval $(0, \infty)$ because f_2 can be written as a linear combination of f_1, f_3, and f_4. Observe that

$$f_2(x) = 1 \cdot f_1(x) + 5 \cdot f_3(x) + 0 \cdot f_4(x)$$

for every x in the interval $(0, \infty)$. ≡

≡ Solutions of Differential Equations We are primarily interested in linearly independent functions or, more to the point, linearly independent *solutions* of a linear differential equation. Although we could always appeal directly to Definition 4.1.1, it turns out that the question of whether the set of n solutions $y_1, y_2, \ldots, y_n$ of a homogeneous linear nth-order differential equation (6) is linearly independent can be settled somewhat mechanically by using a determinant.

DEFINITION 4.1.2 Wronskian

Suppose each of the functions $f_1(x), f_2(x), \ldots, f_n(x)$ possesses at least $n - 1$ derivatives. The determinant

$$W(f_1, f_2, \ldots, f_n) = \begin{vmatrix} f_1 & f_2 & \cdots & f_n \\ f_1' & f_2' & \cdots & f_n' \\ \vdots & \vdots & & \vdots \\ f_1^{(n-1)} & f_2^{(n-1)} & \cdots & f_n^{(n-1)} \end{vmatrix},$$

where the primes denote derivatives, is called the **Wronskian** of the functions.

THEOREM 4.1.3 Criterion for Linearly Independent Solutions

Let $y_1, y_2, \ldots, y_n$ be n solutions of the homogeneous linear nth-order differential equation (6) on an interval I. Then the set of solutions is **linearly independent** on I if and only if $W(y_1, y_2, \ldots, y_n) \neq 0$ for every x in the interval.

It follows from Theorem 4.1.3 that when $y_1, y_2, \ldots, y_n$ are n solutions of (6) on an interval I, the Wronskian $W(y_1, y_2, \ldots, y_n)$ is either identically zero or never zero on the interval.

A set of n linearly independent solutions of a homogeneous linear nth-order differential equation is given a special name.

DEFINITION 4.1.3 Fundamental Set of Solutions

Any set $y_1, y_2, \ldots, y_n$ of n linearly independent solutions of the homogeneous linear nth-order differential equation (6) on an interval I is said to be a **fundamental set of solutions** on the interval.

The basic question of whether a fundamental set of solutions exists for a linear equation is answered in the next theorem.

THEOREM 4.1.4 Existence of a Fundamental Set

There exists a fundamental set of solutions for the homogeneous linear nth-order differential equation (6) on an interval I.

Analogous to the fact that any vector in three dimensions can be expressed as a linear combination of the *linearly independent* vectors **i, j, k,** any solution of an nth-order homogeneous linear differential equation on an interval I can be expressed as a linear combination of n linearly independent solutions on I. In other words, n linearly independent solutions $y_1, y_2, \ldots, y_n$ are the basic building blocks for the general solution of the equation.

THEOREM 4.1.5 General Solution—Homogeneous Equations

Let $y_1, y_2, \ldots, y_n$ be a fundamental set of solutions of the homogeneous linear nth-order differential equation (6) on an interval I. Then the **general solution** of the equation on the interval is

$$y = c_1y_1(x) + c_2y_2(x) + \cdots + c_ny_n(x),$$

where $c_i, i = 1, 2, \ldots, n$ are arbitrary constants.

Theorem 4.1.5 states that if $Y(x)$ is any solution of (6) on the interval, then constants $C_1, C_2, \ldots, C_n$ can always be found so that

$$Y(x) = C_1y_1(x) + C_2y_2(x) + \cdots + C_ny_n(x).$$

We will prove the case when $n = 2$.

PROOF Let Y be a solution and let y_1 and y_2 be linearly independent solutions of $a_2y'' + a_1y' + a_0y = 0$ on an interval I. Suppose that $x = t$ is a point in I for which $W(y_1(t), y_2(t)) \neq 0$. Suppose also that $Y(t) = k_1$ and $Y'(t) = k_2$. If we now examine the equations

$$\begin{aligned} C_1y_1(t) + C_2y_2(t) &= k_1 \\ C_1y_1'(t) + C_2y_2'(t) &= k_2, \end{aligned}$$

it follows that we can determine C_1 and C_2 uniquely, provided that the determinant of the coefficients satisfies

$$\begin{vmatrix} y_1(t) & y_2(t) \\ y_1'(t) & y_2'(t) \end{vmatrix} \neq 0.$$

But this determinant is simply the Wronskian evaluated at $x = t$, and by assumption, $W \neq 0$. If we define $G(x) = C_1y_1(x) + C_2y_2(x)$, we observe that $G(x)$ satisfies the differential equation since it is a superposition of two known solutions; $G(x)$ satisfies the initial conditions

$$G(t) = C_1y_1(t) + C_2y_2(t) = k_1 \quad \text{and} \quad G'(t) = C_1y_1'(t) + C_2y_2'(t) = k_2;$$

and $Y(x)$ satisfies the *same* linear equation and the *same* initial conditions. Because the solution of this linear initial-value problem is unique (Theorem 4.1.1), we have $Y(x) = G(x)$ or $Y(x) = C_1y_1(x) + C_2y_2(x)$. ≡

EXAMPLE 7 General Solution of a Homogeneous DE

The functions $y_1 = e^{3x}$ and $y_2 = e^{-3x}$ are both solutions of the homogeneous linear equation $y'' - 9y = 0$ on the interval $(-\infty, \infty)$. By inspection the solutions are linearly independent on the x-axis. This fact can be corroborated by observing that the Wronskian

$$W(e^{3x}, e^{-3x}) = \begin{vmatrix} e^{3x} & e^{-3x} \\ 3e^{3x} & -3e^{-3x} \end{vmatrix} = -6 \neq 0$$

for every x. We conclude that y_1 and y_2 form a fundamental set of solutions, and consequently, $y = c_1e^{3x} + c_2e^{-3x}$ is the general solution of the equation on the interval. ≡

EXAMPLE 8 A Solution Obtained from a General Solution

The function $y = 4\sinh 3x - 5e^{-3x}$ is a solution of the differential equation in Example 7. (Verify this.) In view of Theorem 4.1.5 we must be able to obtain this solution from the general solution $y = c_1e^{3x} + c_2e^{-3x}$. Observe that if we choose $c_1 = 2$ and $c_2 = -7$, then $y = 2e^{3x} - 7e^{-3x}$ can be rewritten as

$$y = 2e^{3x} - 2e^{-3x} - 5e^{-3x} = 4\left(\frac{e^{3x} - e^{-3x}}{2}\right) - 5e^{-3x}.$$

The last expression is recognized as $y = 4\sinh 3x - 5e^{-3x}$. ≡

EXAMPLE 9 General Solution of a Homogeneous DE

The functions $y_1 = e^x$, $y_2 = e^{2x}$, and $y_3 = e^{3x}$ satisfy the third-order equation $y''' - 6y'' + 11y' - 6y = 0$. Since

$$W(e^x, e^{2x}, e^{3x}) = \begin{vmatrix} e^x & e^{2x} & e^{3x} \\ e^x & 2e^{2x} & 3e^{3x} \\ e^x & 4e^{2x} & 9e^{3x} \end{vmatrix} = 2e^{6x} \neq 0$$

for every real value of x, the functions y_1, y_2, and y_3 form a fundamental set of solutions on $(-\infty, \infty)$. We conclude that $y = c_1e^x + c_2e^{2x} + c_3e^{3x}$ is the general solution of the differential equation on the interval. ≡

4.1.3 NONHOMOGENEOUS EQUATIONS

Any function y_p, free of arbitrary parameters, that satisfies (7) is said to be a **particular solution** or **particular integral** of the equation. For example, it is a straightforward task to show that the constant function $y_p = 3$ is a particular solution of the nonhomogeneous equation $y'' + 9y = 27$.

Now if $y_1, y_2, \ldots, y_k$ are solutions of (6) on an interval I and y_p is any particular solution of (7) on I, then the linear combination

$$y = c_1y_1(x) + c_2y_2(x) + \cdots + c_ky_k(x) + y_p \tag{10}$$

is also a solution of the nonhomogeneous equation (7). If you think about it, this makes sense, because the linear combination $c_1y_1(x) + c_2y_2(x) + \cdots + c_ky_k(x)$ is transformed into 0 by the operator $L = a_nD^n + a_{n-1}D^{n-1} + \cdots + a_1D + a_0$, whereas y_p is transformed into $g(x)$. If we use $k = n$ linearly independent solutions of the nth-order equation (6), then the expression in (10) becomes the general solution of (7).

THEOREM 4.1.6 General Solution—Nonhomogeneous Equations

Let y_p be any particular solution of the nonhomogeneous linear nth-order differential equation (7) on an interval I, and let $y_1, y_2, \ldots, y_n$ be a fundamental set of solutions of the associated homogeneous differential equation (6) on I. Then the **general solution** of the equation on the interval is

$$y = c_1y_1(x) + c_2y_2(x) + \cdots + c_ny_n(x) + y_p,$$

where the c_i, $i = 1, 2, \ldots, n$ are arbitrary constants.

PROOF Let L be the differential operator defined in (8) and let $Y(x)$ and $y_p(x)$ be particular solutions of the nonhomogeneous equation $L(y) = g(x)$. If we define $u(x) = Y(x) - y_p(x)$, then by linearity of L we have

$$L(u) = L\{Y(x) - y_p(x)\} = L(Y(x)) - L(y_p(x)) = g(x) - g(x) = 0.$$

This shows that $u(x)$ is a solution of the homogeneous equation $L(y) = 0$. Hence by Theorem 4.1.5, $u(x) = c_1y_1(x) + c_2y_2(x) + \cdots + c_ny_n(x)$, and so

$$Y(x) - y_p(x) = c_1y_1(x) + c_2y_2(x) + \cdots + c_ny_n(x)$$

or

$$Y(x) = c_1y_1(x) + c_2y_2(x) + \cdots + c_ny_n(x) + y_p(x). \qquad \equiv$$

≡ Complementary Function

We see in Theorem 4.1.6 that the general solution of a nonhomogeneous linear equation consists of the sum of two functions:

$$y = c_1y_1(x) + c_2y_2(x) + \cdots + c_ny_n(x) + y_p(x) = y_c(x) + y_p(x).$$

The linear combination $y_c(x) = c_1y_1(x) + c_2y_2(x) + \cdots + c_ny_n(x)$, which is the general solution of (6), is called the **complementary function** for equation (7). In other words, to solve a nonhomogeneous linear differential equation, we first solve the associated homogeneous equation and then find any particular solution of the nonhomogeneous equation. The general solution of the nonhomogeneous equation is then

$$\begin{aligned} y &= \textit{complementary function} + \textit{any particular solution} \\ &= y_c + y_p. \end{aligned}$$

EXAMPLE 10 General Solution of a Nonhomogeneous DE

By substitution the function $y_p = -\frac{11}{12} - \frac{1}{2}x$ is readily shown to be a particular solution of the nonhomogeneous equation

$$y''' - 6y'' + 11y' - 6y = 3x. \tag{11}$$

To write the general solution of (11), we must also be able to solve the associated homogeneous equation

$$y''' - 6y'' + 11y' - 6y = 0.$$

But in Example 9 we saw that the general solution of this latter equation on the interval $(-\infty, \infty)$ was $y_c = c_1e^x + c_2e^{2x} + c_3e^{3x}$. Hence the general solution of (11) on the interval is

$$y = y_c + y_p = c_1e^x + c_2e^{2x} + c_3e^{3x} - \frac{11}{12} - \frac{1}{2}x.$$ ≡

≡ **Another Superposition Principle** The last theorem of this discussion will be useful in Section 4.4 when we consider a method for finding particular solutions of nonhomogeneous equations.

THEOREM 4.1.7 Superposition Principle—Nonhomogeneous Equations

Let $y_{p_1}, y_{p_2}, \ldots, y_{p_k}$ be k particular solutions of the nonhomogeneous linear nth-order differential equation (7) on an interval I corresponding, in turn, to k distinct functions $g_1, g_2, \ldots, g_k$. That is, suppose y_{p_i} denotes a particular solution of the corresponding differential equation

$$a_n(x)y^{(n)} + a_{n-1}(x)y^{(n-1)} + \cdots + a_1(x)y' + a_0(x)y = g_i(x), \quad (12)$$

where $i = 1, 2, \ldots, k$. Then

$$y_p = y_{p_1}(x) + y_{p_2}(x) + \cdots + y_{p_k}(x) \quad (13)$$

is a particular solution of

$$a_n(x)y^{(n)} + a_{n-1}(x)y^{(n-1)} + \cdots + a_1(x)y' + a_0(x)y$$
$$= g_1(x) + g_2(x) + \cdots + g_k(x). \quad (14)$$

PROOF We prove the case $k = 2$. Let L be the differential operator defined in (8) and let $y_{p_1}(x)$ and $y_{p_2}(x)$ be particular solutions of the nonhomogeneous equations $L(y) = g_1(x)$ and $L(y) = g_2(x)$, respectively. If we define $y_p = y_{p_1}(x) + y_{p_2}(x)$, we want to show that y_p is a particular solution of $L(y) = g_1(x) + g_2(x)$. The result follows again by the linearity of the operator L:

$$L(y_p) = L\{y_{p_1}(x) + y_{p_2}(x)\} = L(y_{p_1}(x)) + L(y_{p_2}(x)) = g_1(x) + g_2(x).$$ ≡

EXAMPLE 11 **Superposition—Nonhomogeneous DE**

You should verify that

$y_{p_1} = -4x^2$ is a particular solution of $y'' - 3y' + 4y = -16x^2 + 24x - 8$,

$y_{p_2} = e^{2x}$ is a particular solution of $y'' - 3y' + 4y = 2e^{2x}$,

$y_{p_3} = xe^x$ is a particular solution of $y'' - 3y' + 4y = 2xe^x - e^x$.

It follows from (13) of Theorem 4.1.7 that the superposition of y_{p_1}, y_{p_2}, and y_{p_3},

$$y = y_{p_1} + y_{p_2} + y_{p_3} = -4x^2 + e^{2x} + xe^x,$$

is a solution of

$$y'' - 3y' + 4y = \underbrace{-16x^2 + 24x - 8}_{g_1(x)} + \underbrace{2e^{2x}}_{g_2(x)} + \underbrace{2xe^x - e^x}_{g_3(x)}.$$ ≡

Note If the y_{p_i} are particular solutions of (12) for $i = 1, 2, \ldots, k$, then the linear combination

$$y_p = c_1 y_{p_1} + c_2 y_{p_2} + \cdots + c_k y_{p_k},$$

where the c_i are constants, is also a particular solution of (14) when the right-hand member of the equation is the linear combination

$$c_1 g_1(x) + c_2 g_2(x) + \cdots + c_k g_k(x).$$

Before we actually start solving homogeneous and nonhomogeneous linear differential equations, we need one additional bit of theory, which is presented in the next section.

REMARKS

This remark is a continuation of the brief discussion of dynamical systems given at the end of Section 1.3.

A dynamical system whose rule or mathematical model is a linear nth-order differential equation

$$a_n(t)y^{(n)} + a_{n-1}(t)y^{(n-1)} + \cdots + a_1(t)y' + a_0(t)y = g(t)$$

is said to be an nth-order **linear system.** The n time-dependent functions $y(t)$, $y'(t), \ldots, y^{(n-1)}(t)$ are the **state variables** of the system. Recall that their values at some time t give the **state of the system.** The function g is variously called the **input function, forcing function,** or **excitation function.** A solution $y(t)$ of the differential equation is said to be the **output** or **response of the system.** Under the conditions stated in Theorem 4.1.1, the output or response $y(t)$ is uniquely determined by the input and the state of the system prescribed at a time t_0—that is, by the initial conditions $y(t_0), y'(t_0), \ldots, y^{(n-1)}(t_0)$.

For a dynamical system to be a linear system, it is necessary that the superposition principle (Theorem 4.1.7) holds in the system; that is, the response of the system to a superposition of inputs is a superposition of outputs. We have already examined some simple linear systems in Section 3.1 (linear first-order equations); in Section 5.1 we examine linear systems in which the mathematical models are second-order differential equations.

EXERCISES 4.1

Answers to selected odd-numbered problems begin on page ANS-4.

4.1.1 INITIAL-VALUE AND BOUNDARY-VALUE PROBLEMS

In Problems 1–4 the given family of functions is the general solution of the differential equation on the indicated interval. Find a member of the family that is a solution of the initial-value problem.

1. $y = c_1e^x + c_2e^{-x}$, $(-\infty, \infty)$;
$y'' - y = 0, \quad y(0) = 0, \quad y'(0) = 1$

2. $y = c_1e^{4x} + c_2e^{-x}$, $(-\infty, \infty)$;
$y'' - 3y' - 4y = 0, \quad y(0) = 1, \quad y'(0) = 2$

3. $y = c_1x + c_2x \ln x$, $(0, \infty)$;
$x^2y'' - xy' + y = 0, \quad y(1) = 3, \quad y'(1) = -1$

4. $y = c_1 + c_2 \cos x + c_3 \sin x$, $(-\infty, \infty)$;
$y''' + y' = 0, \quad y(\pi) = 0, \quad y'(\pi) = 2, \quad y''(\pi) = -1$

5. Given that $y = c_1 + c_2x^2$ is a two-parameter family of solutions of $xy'' - y' = 0$ on the interval $(-\infty, \infty)$, show that constants c_1 and c_2 cannot be found so that a member of the family satisfies the initial conditions $y(0) = 0$, $y'(0) = 1$. Explain why this does not violate Theorem 4.1.1.

6. Find two members of the family of solutions in Problem 5 that satisfy the initial conditions $y(0) = 0$, $y'(0) = 0$.

7. Given that $x(t) = c_1 \cos \omega t + c_2 \sin \omega t$ is the general solution of $x'' + \omega^2x = 0$ on the interval $(-\infty, \infty)$, show that a solution satisfying the initial conditions $x(0) = x_0$, $x'(0) = x_1$ is given by

$$x(t) = x_0 \cos \omega t + \frac{x_1}{\omega} \sin \omega t.$$

8. Use the general solution of $x'' + \omega^2 x = 0$ given in Problem 7 to show that a solution satisfying the initial conditions $x(t_0) = x_0, x'(t_0) = x_1$ is the solution given in Problem 7 shifted by an amount t_0:

$$x(t) = x_0 \cos \omega(t - t_0) + \frac{x_1}{\omega} \sin \omega(t - t_0).$$

In Problems 9 and 10 find an interval centered about $x = 0$ for which the given initial-value problem has a unique solution.

9. $(x - 2)y'' + 3y = x, \quad y(0) = 0, \quad y'(0) = 1$

10. $y'' + (\tan x)y = e^x, \quad y(0) = 1, \quad y'(0) = 0$

11. **(a)** Use the family in Problem 1 to find a solution of $y'' - y = 0$ that satisfies the boundary conditions $y(0) = 0, y(1) = 1$.

(b) The DE in part (a) has the alternative general solution $y = c_3 \cosh x + c_4 \sinh x$ on $(-\infty, \infty)$. Use this family to find a solution that satisfies the boundary conditions in part (a).

(c) Show that the solutions in parts (a) and (b) are equivalent

12. Use the family in Problem 5 to find a solution of $xy'' - y' = 0$ that satisfies the boundary conditions $y(0) = 1, y'(1) = 6$.

In Problems 13 and 14 the given two-parameter family is a solution of the indicated differential equation on the interval $(-\infty, \infty)$. Determine whether a member of the family can be found that satisfies the boundary conditions.

13. $y = c_1 e^x \cos x + c_2 e^x \sin x; \quad y'' - 2y' + 2y = 0$

(a) $y(0) = 1, \quad y'(\pi) = 0$

(b) $y(0) = 1, \quad y(\pi) = -1$

(c) $y(0) = 1, \quad y(\pi/2) = 1$

(d) $y(0) = 0, \quad y(\pi) = 0.$

14. $y = c_1 x^2 + c_2 x^4 + 3; \quad x^2 y'' - 5xy' + 8y = 24$

(a) $y(-1) = 0, \quad y(1) = 4$

(b) $y(0) = 1, \quad y(1) = 2$

(c) $y(0) = 3, \quad y(1) = 0$

(d) $y(1) = 3, \quad y(2) = 15$

4.1.2 HOMOGENEOUS EQUATIONS

In Problems 15–22 determine whether the given set of functions is linearly independent on the interval $(-\infty, \infty)$.

15. $f_1(x) = x, \quad f_2(x) = x^2, \quad f_3(x) = 4x - 3x^2$

16. $f_1(x) = 0, \quad f_2(x) = x, \quad f_3(x) = e^x$

17. $f_1(x) = 5, \quad f_2(x) = \cos^2 x, \quad f_3(x) = \sin^2 x$

18. $f_1(x) = \cos 2x, \quad f_2(x) = 1, \quad f_3(x) = \cos^2 x$

19. $f_1(x) = x, \quad f_2(x) = x - 1, \quad f_3(x) = x + 3$

20. $f_1(x) = 2 + x, \quad f_2(x) = 2 + |x|$

21. $f_1(x) = 1 + x, \quad f_2(x) = x, \quad f_3(x) = x^2$

22. $f_1(x) = e^x, \quad f_2(x) = e^{-x}, \quad f_3(x) = \sinh x$

In Problems 23–30 verify that the given functions form a fundamental set of solutions of the differential equation on the indicated interval. Form the general solution.

23. $y'' - y' - 12y = 0; \quad e^{-3x}, e^{4x}, (-\infty, \infty)$

24. $y'' - 4y = 0; \quad \cosh 2x, \sinh 2x, (-\infty, \infty)$

25. $y'' - 2y' + 5y = 0; \quad e^x \cos 2x, e^x \sin 2x, (-\infty, \infty)$

26. $4y'' - 4y' + y = 0; \quad e^{x/2}, xe^{x/2}, (-\infty, \infty)$

27. $x^2 y'' - 6xy' + 12y = 0; \quad x^3, x^4, (0, \infty)$

28. $x^2 y'' + xy' + y = 0; \quad \cos(\ln x), \sin(\ln x), (0, \infty)$

29. $x^3 y''' + 6x^2 y'' + 4xy' - 4y = 0; \quad x, x^{-2}, x^{-2} \ln x, (0, \infty)$

30. $y^{(4)} + y'' = 0; \quad 1, x, \cos x, \sin x, (-\infty, \infty)$

4.1.3 NONHOMOGENEOUS EQUATIONS

In Problems 31–34 verify that the given two-parameter family of functions is the general solution of the nonhomogeneous differential equation on the indicated interval.

31. $y'' - 7y' + 10y = 24e^x;$
$y = c_1 e^{2x} + c_2 e^{5x} + 6e^x, (-\infty, \infty)$

32. $y'' + y = \sec x;$
$y = c_1 \cos x + c_2 \sin x + x \sin x + (\cos x) \ln(\cos x),$
$(-\pi/2, \pi/2)$

33. $y'' - 4y' + 4y = 2e^{2x} + 4x - 12;$
$y = c_1 e^{2x} + c_2 x e^{2x} + x^2 e^{2x} + x - 2, (-\infty, \infty)$

34. $2x^2 y'' + 5xy' + y = x^2 - x;$
$y = c_1 x^{-1/2} + c_2 x^{-1} + \frac{1}{15}x^2 - \frac{1}{6}x, (0, \infty)$

35. **(a)** Verify that $y_{p_1} = 3e^{2x}$ and $y_{p_2} = x^2 + 3x$ are, respectively, particular solutions of

$$y'' - 6y' + 5y = -9e^{2x}$$

and $$y'' - 6y' + 5y = 5x^2 + 3x - 16.$$

(b) Use part (a) to find particular solutions of

$$y'' - 6y' + 5y = 5x^2 + 3x - 16 - 9e^{2x}$$

and $$y'' - 6y' + 5y = -10x^2 - 6x + 32 + e^{2x}.$$

36. **(a)** By inspection find a particular solution of

$$y'' + 2y = 10.$$

(b) By inspection find a particular solution of

$$y'' + 2y = -4x.$$

(c) Find a particular solution of $y'' + 2y = -4x + 10$.

(d) Find a particular solution of $y'' + 2y = 8x + 5$.

Discussion Problems

37. Let $n = 1, 2, 3, \ldots$. Discuss how the observations $D^n x^{n-1} = 0$ and $D^n x^n = n!$ can be used to find the general solutions of the given differential equations.

(a) $y'' = 0$ **(b)** $y''' = 0$ **(c)** $y^{(4)} = 0$

(d) $y'' = 2$ **(e)** $y''' = 6$ **(f)** $y^{(4)} = 24$

38. Suppose that $y_1 = e^x$ and $y_2 = e^{-x}$ are two solutions of a homogeneous linear differential equation. Explain why $y_3 = \cosh x$ and $y_4 = \sinh x$ are also solutions of the equation.

39. **(a)** Verify that $y_1 = x^3$ and $y_2 = |x|^3$ are linearly independent solutions of the differential equation $x^2y'' - 4xy' + 6y = 0$ on the interval $(-\infty, \infty)$.

(b) Show that $W(y_1, y_2) = 0$ for every real number x. Does this result violate Theorem 4.1.3? Explain.

(c) Verify that $Y_1 = x^3$ and $Y_2 = x^2$ are also linearly independent solutions of the differential equation in part (a) on the interval $(-\infty, \infty)$.

(d) Find a solution of the differential equation satisfying $y(0) = 0$, $y'(0) = 0$.

(e) By the superposition principle, Theorem 4.1.2, both linear combinations $y = c_1y_1 + c_2y_2$ and $Y = c_1Y_1 + c_2Y_2$ are solutions of the differential equation. Discuss whether one, both, or neither of the linear combinations is a general solution of the differential equation on the interval $(-\infty, \infty)$.

40. Is the set of functions $f_1(x) = e^{x+2}$, $f_2(x) = e^{x-3}$ linearly dependent or linearly independent on $(-\infty, \infty)$? Discuss.

41. Suppose $y_1, y_2, \ldots, y_k$ are k linearly independent solutions on $(-\infty, \infty)$ of a homogeneous linear nth-order differential equation with constant coefficients. By Theorem 4.1.2 it follows that $y_{k+1} = 0$ is also a solution of the differential equation. Is the set of solutions $y_1, y_2, \ldots, y_k, y_{k+1}$ linearly dependent or linearly independent on $(-\infty, \infty)$? Discuss.

42. Suppose that $y_1, y_2, \ldots, y_k$ are k nontrivial solutions of a homogeneous linear nth-order differential equation with constant coefficients and that $k = n + 1$. Is the set of solutions $y_1, y_2, \ldots, y_k$ linearly dependent or linearly independent on $(-\infty, \infty)$? Discuss.

4.2 REDUCTION OF ORDER

REVIEW MATERIAL

- Section 2.5 (using a substitution)
- Section 4.1

INTRODUCTION In the preceding section we saw that the general solution of a homogeneous linear second-order differential equation

$$a_2(x)y'' + a_1(x)y' + a_0(x)y = 0 \tag{1}$$

is a linear combination $y = c_1y_1 + c_2y_2$, where y_1 and y_2 are solutions that constitute a linearly independent set on some interval I. Beginning in the next section, we examine a method for determining these solutions when the coefficients of the differential equation in (1) are constants. This method, which is a straightforward exercise in algebra, breaks down in a few cases and yields only a single solution y_1 of the DE. It turns out that we can construct a second solution y_2 of a homogeneous equation (1) (even when the coefficients in (1) are variable) provided that we know a nontrivial solution y_1 of the DE. The basic idea described in this section is that *equation* (1) *can be reduced to a linear first-order DE by means of a substitution* involving the known solution y_1. A second solution y_2 of (1) is apparent after this first-order differential equation is solved.

Reduction of Order Suppose that y_1 denotes a nontrivial solution of (1) and that y_1 is defined on an interval I. We seek a second solution y_2 so that the set consisting of y_1 and y_2 is linearly independent on I. Recall from Section 4.1 that if y_1 and y_2 are linearly independent, then their quotient y_2/y_1 is nonconstant on I—that is, $y_2(x)/y_1(x) = u(x)$ or $y_2(x) = u(x)y_1(x)$. The function $u(x)$ can be found by substituting $y_2(x) = u(x)y_1(x)$ into the given differential equation. This method is called **reduction of order** because we must solve a linear first-order differential equation to find u.

EXAMPLE 1 **A Second Solution by Reduction of Order**

Given that $y_1 = e^x$ is a solution of $y'' - y = 0$ on the interval $(-\infty, \infty)$, use reduction of order to find a second solution y_2.

SOLUTION If $y = u(x)y_1(x) = u(x)e^x$, then the Product Rule gives

$$y' = ue^x + e^x u', \quad y'' = ue^x + 2e^x u' + e^x u'',$$

and so

$$y'' - y = e^x(u'' + 2u') = 0.$$

Since $e^x \neq 0$, the last equation requires $u'' + 2u' = 0$. If we make the substitution $w = u'$, this linear second-order equation in u becomes $w' + 2w = 0$, which is a linear first-order equation in w. Using the integrating factor e^{2x}, we can write $\dfrac{d}{dx}[e^{2x}w] = 0$. After integrating, we get $w = c_1e^{-2x}$ or $u' = c_1e^{-2x}$. Integrating again then yields $u = -\frac{1}{2}c_1e^{-2x} + c_2$. Thus

$$y = u(x)e^x = -\frac{c_1}{2}e^{-x} + c_2e^x. \tag{2}$$

By picking $c_2 = 0$ and $c_1 = -2$, we obtain the desired second solution, $y_2 = e^{-x}$. Because $W(e^x, e^{-x}) \neq 0$ for every x, the solutions are linearly independent on $(-\infty, \infty)$. ≡

Since we have shown that $y_1 = e^x$ and $y_2 = e^{-x}$ are linearly independent solutions of a linear second-order equation, the expression in (2) is actually the general solution of $y'' - y = 0$ on $(-\infty, \infty)$.

≡ General Case Suppose we divide by $a_2(x)$ to put equation (1) in the **standard form**

$$y'' + P(x)y' + Q(x)y = 0, \tag{3}$$

where $P(x)$ and $Q(x)$ are continuous on some interval I. Let us suppose further that $y_1(x)$ is a known solution of (3) on I and that $y_1(x) \neq 0$ for every x in the interval. If we define $y = u(x)y_1(x)$, it follows that

$$y' = uy_1' + y_1u', \quad y'' = uy_1'' + 2y_1'u' + y_1u''$$

$$y'' + Py' + Qy = u\underbrace{[y_1'' + Py_1' + Qy_1]}_{\text{zero}} + y_1u'' + (2y_1' + Py_1)u' = 0.$$

This implies that we must have

$$y_1u'' + (2y_1' + Py_1)u' = 0 \qquad \text{or} \qquad y_1w' + (2y_1' + Py_1)w = 0, \tag{4}$$

where we have let $w = u'$. Observe that the last equation in (4) is both linear and separable. Separating variables and integrating, we obtain

$$\frac{dw}{w} + 2\frac{y_1'}{y_1}dx + P\,dx = 0$$

$$\ln|wy_1^2| = -\int P\,dx + c \qquad \text{or} \qquad wy_1^2 = c_1e^{-\int P\,dx}.$$

We solve the last equation for w, use $w = u'$, and integrate again:

$$u = c_1\int \frac{e^{-\int P\,dx}}{y_1^2}\,dx + c_2.$$

By choosing $c_1 = 1$ and $c_2 = 0$, we find from $y = u(x)y_1(x)$ that a second solution of equation (3) is

$$y_2 = y_1(x) \int \frac{e^{-\int P(x)\,dx}}{y_1^2(x)}\,dx. \tag{5}$$

It makes a good review of differentiation to verify that the function $y_2(x)$ defined in (5) satisfies equation (3) and that y_1 and y_2 are linearly independent on any interval on which $y_1(x)$ is not zero.

EXAMPLE 2 A Second Solution by Formula (5)

The function $y_1 = x^2$ is a solution of $x^2y'' - 3xy' + 4y = 0$. Find the general solution of the differential equation on the interval $(0, \infty)$.

SOLUTION From the standard form of the equation,

$$y'' - \frac{3}{x}y' + \frac{4}{x^2}y = 0,$$

we find from (5)

$$y_2 = x^2 \int \frac{e^{3\int dx/x}}{x^4}\,dx \qquad \leftarrow e^{3\int dx/x} = e^{\ln x^3} = x^3$$

$$= x^2 \int \frac{dx}{x} = x^2 \ln x.$$

The general solution on the interval $(0, \infty)$ is given by $y = c_1y_1 + c_2y_2$; that is, $y = c_1x^2 + c_2x^2 \ln x$. ≡

REMARKS

(i) The derivation and use of formula (5) have been illustrated here because this formula appears again in the next section and in Sections 4.7 and 6.3. We use (5) simply to save time in obtaining a desired result. Your instructor will tell you whether you should memorize (5) or whether you should know the first principles of reduction of order.

(ii) Reduction of order can be used to find the general solution of a nonhomogeneous equation $a_2(x)y'' + a_1(x)y' + a_0(x)y = g(x)$ whenever a solution y_1 of the associated homogeneous equation is known. See Problems 17–20 in Exercises 4.2.

EXERCISES 4.2

Answers to selected odd-numbered problems begin on page ANS-5.

In Problems 1–16 the indicated function $y_1(x)$ is a solution of the given differential equation. Use reduction of order or formula (5), as instructed, to find a second solution $y_2(x)$.

1. $y'' - 4y' + 4y = 0; \quad y_1 = e^{2x}$
2. $y'' + 2y' + y = 0; \quad y_1 = xe^{-x}$
3. $y'' + 16y = 0; \quad y_1 = \cos 4x$
4. $y'' + 9y = 0; \quad y_1 = \sin 3x$
5. $y'' - y = 0; \quad y_1 = \cosh x$
6. $y'' - 25y = 0; \quad y_1 = e^{5x}$
7. $9y'' - 12y' + 4y = 0; \quad y_1 = e^{2x/3}$
8. $6y'' + y' - y = 0; \quad y_1 = e^{x/3}$
9. $x^2y'' - 7xy' + 16y = 0; \quad y_1 = x^4$
10. $x^2y'' + 2xy' - 6y = 0; \quad y_1 = x^2$
11. $xy'' + y' = 0; \quad y_1 = \ln x$
12. $4x^2y'' + y = 0; \quad y_1 = x^{1/2} \ln x$
13. $x^2y'' - xy' + 2y = 0; \quad y_1 = x \sin(\ln x)$
14. $x^2y'' - 3xy' + 5y = 0; \quad y_1 = x^2 \cos(\ln x)$

15. $(1 - 2x - x^2)y'' + 2(1 + x)y' - 2y = 0;\quad y_1 = x + 1$

16. $(1 - x^2)y'' + 2xy' = 0;\quad y_1 = 1$

In Problems 17–20 the indicated function $y_1(x)$ is a solution of the associated homogeneous equation. Use the method of reduction of order to find a second solution $y_2(x)$ of the homogeneous equation and a particular solution of the given nonhomogeneous equation.

17. $y'' - 4y = 2;\quad y_1 = e^{-2x}$

18. $y'' + y' = 1;\quad y_1 = 1$

19. $y'' - 3y' + 2y = 5e^{3x};\quad y_1 = e^x$

20. $y'' - 4y' + 3y = x;\quad y_1 = e^x$

Discussion Problems

21. (a) Give a convincing demonstration that the second-order equation $ay'' + by' + cy = 0$, a, b, and c constants, always possesses at least one solution of the form $y_1 = e^{m_1x}$, m_1 a constant.

(b) Explain why the differential equation in part (a) must then have a second solution either of the form $y_2 = e^{m_2x}$ or of the form $y_2 = xe^{m_1x}$, m_1 and m_2 constants.

(c) Reexamine Problems 1–8. Can you explain why the statements in parts (a) and (b) above are not contradicted by the answers to Problems 3–5?

22. Verify that $y_1(x) = x$ is a solution of $xy'' - xy' + y = 0$. Use reduction of order to find a second solution $y_2(x)$ in the form of an infinite series. Conjecture an interval of definition for $y_2(x)$.

Computer Lab Assignments

23. (a) Verify that $y_1(x) = e^x$ is a solution of

$$xy'' - (x + 10)y' + 10y = 0.$$

(b) Use (5) to find a second solution $y_2(x)$. Use a CAS to carry out the required integration.

(c) Explain, using Corollary (A) of Theorem 4.1.2, why the second solution can be written compactly as

$$y_2(x) = \sum_{n=0}^{10} \frac{1}{n!}x^n.$$

4.3 HOMOGENEOUS LINEAR EQUATIONS WITH CONSTANT COEFFICIENTS

REVIEW MATERIAL

- Review Problems 27–30 in Exercises 1.1 and Theorem 4.1.5
- Review the algebra of solving polynomial equations (see the *Student Resource Manual*)

INTRODUCTION As a means of motivating the discussion in this section, let us return to first-order differential equations—more specifically, to *homogeneous* linear equations $ay' + by = 0$, where the coefficients $a \neq 0$ and b are constants. This type of equation can be solved either by separation of variables or with the aid of an integrating factor, but there is another solution method, one that uses only algebra. Before illustrating this alternative method, we make one observation: Solving $ay' + by = 0$ for y' yields $y' = ky$, where k is a constant. This observation reveals the nature of the unknown solution y; the only nontrivial elementary function whose derivative is a constant multiple of itself is an exponential function e^{mx}. Now the new solution method: If we substitute $y = e^{mx}$ and $y' = me^{mx}$ into $ay' + by = 0$, we get

$$ame^{mx} + be^{mx} = 0 \quad \text{or} \quad e^{mx}(am + b) = 0.$$

Since e^{mx} is never zero for real values of x, the last equation is satisfied only when m is a solution or root of the first-degree polynomial equation $am + b = 0$. For this single value of m, $y = e^{mx}$ is a solution of the DE. To illustrate, consider the constant-coefficient equation $2y' + 5y = 0$. It is not necessary to go through the differentiation and substitution of $y = e^{mx}$ into the DE; we merely have to form the equation $2m + 5 = 0$ and solve it for m. From $m = -\frac{5}{2}$ we conclude that $y = e^{-5x/2}$ is a solution of $2y' + 5y = 0$, and its general solution on the interval $(-\infty, \infty)$ is $y = c_1e^{-5x/2}$.

In this section we will see that the foregoing procedure can produce exponential solutions for homogeneous linear higher-order DEs,

$$a_ny^{(n)} + a_{n-1}y^{(n-1)} + \cdots + a_2y'' + a_1y' + a_0y = 0, \tag{1}$$

where the coefficients a_i, $i = 0, 1, \ldots, n$ are real constants and $a_n \neq 0$.

Auxiliary Equation We begin by considering the special case of the second-order equation

$$ay'' + by' + cy = 0, \tag{2}$$

where a, b, and c are constants. If we try to find a solution of the form $y = e^{mx}$, then after substitution of $y' = me^{mx}$ and $y'' = m^2e^{mx}$, equation (2) becomes

$$am^2e^{mx} + bme^{mx} + ce^{mx} = 0 \quad \text{or} \quad e^{mx}(am^2 + bm + c) = 0.$$

As in the introduction we argue that because $e^{mx} \neq 0$ for all x, it is apparent that the only way $y = e^{mx}$ can satisfy the differential equation (2) is when m is chosen as a root of the quadratic equation

$$am^2 + bm + c = 0. \tag{3}$$

This last equation is called the **auxiliary equation** of the differential equation (2). Since the two roots of (3) are $m_1 = (-b + \sqrt{b^2 - 4ac})/2a$ and $m_2 = (-b - \sqrt{b^2 - 4ac})/2a$, there will be three forms of the general solution of (2) corresponding to the three cases:

- m_1 and m_2 real and distinct $(b^2 - 4ac > 0)$,
- m_1 and m_2 real and equal $(b^2 - 4ac = 0)$, and
- m_1 and m_2 conjugate complex numbers $(b^2 - 4ac < 0)$.

We discuss each of these cases in turn.

Case I: Distinct Real Roots Under the assumption that the auxiliary equation (3) has two unequal real roots m_1 and m_2, we find two solutions, $y_1 = e^{m_1x}$ and $y_2 = e^{m_2x}$. We see that these functions are linearly independent on $(-\infty, \infty)$ and hence form a fundamental set. It follows that the general solution of (2) on this interval is

$$y = c_1e^{m_1x} + c_2e^{m_2x}. \tag{4}$$

Case II: Repeated Real Roots When $m_1 = m_2$, we necessarily obtain only one exponential solution, $y_1 = e^{m_1x}$. From the quadratic formula we find that $m_1 = -b/2a$ since the only way to have $m_1 = m_2$ is to have $b^2 - 4ac = 0$. It follows from (5) in Section 4.2 that a second solution of the equation is

$$y_2 = e^{m_1x}\int \frac{e^{2m_1x}}{e^{2m_1x}}\,dx = e^{m_1x}\int dx = xe^{m_1x}. \tag{5}$$

In (5) we have used the fact that $-b/a = 2m_1$. The general solution is then

$$y = c_1e^{m_1x} + c_2xe^{m_1x}. \tag{6}$$

Case III: Conjugate Complex Roots If m_1 and m_2 are complex, then we can write $m_1 = \alpha + i\beta$ and $m_2 = \alpha - i\beta$, where α and $\beta > 0$ are real and $i^2 = -1$. Formally, there is no difference between this case and Case I, and hence

$$y = C_1e^{(\alpha+i\beta)x} + C_2e^{(\alpha-i\beta)x}.$$

However, in practice we prefer to work with real functions instead of complex exponentials. To this end we use **Euler's formula:**

$$e^{i\theta} = \cos\theta + i\sin\theta,$$

where θ is any real number.* It follows from this formula that

$$e^{i\beta x} = \cos\beta x + i\sin\beta x \quad \text{and} \quad e^{-i\beta x} = \cos\beta x - i\sin\beta x, \tag{7}$$

*A formal derivation of Euler's formula can be obtained from the Maclaurin series $e^x = \sum_{n=0}^{\infty}\frac{x^n}{n!}$ by substituting $x = i\theta$, using $i^2 = -1$, $i^3 = -i$, . . . , and then separating the series into real and imaginary parts. The plausibility thus established, we can adopt $\cos\theta + i\sin\theta$ as the *definition* of $e^{i\theta}$.

where we have used $\cos(-\beta x) = \cos \beta x$ and $\sin(-\beta x) = -\sin \beta x$. Note that by first adding and then subtracting the two equations in (7), we obtain, respectively,

$$e^{i\beta x} + e^{-i\beta x} = 2\cos \beta x \qquad \text{and} \qquad e^{i\beta x} - e^{-i\beta x} = 2i \sin \beta x.$$

Since $y = C_1 e^{(\alpha+i\beta)x} + C_2 e^{(\alpha-i\beta)x}$ is a solution of (2) for any choice of the constants C_1 and C_2, the choices $C_1 = C_2 = 1$ and $C_1 = 1$, $C_2 = -1$ give, in turn, two solutions:

$$y_1 = e^{(\alpha+i\beta)x} + e^{(\alpha-i\beta)x} \qquad \text{and} \qquad y_2 = e^{(\alpha+i\beta)x} - e^{(\alpha-i\beta)x}.$$

But
$$y_1 = e^{\alpha x}(e^{i\beta x} + e^{-i\beta x}) = 2e^{\alpha x}\cos \beta x$$

and
$$y_2 = e^{\alpha x}(e^{i\beta x} - e^{-i\beta x}) = 2ie^{\alpha x}\sin \beta x.$$

Hence from Corollary (A) of Theorem 4.1.2 the last two results show that $e^{\alpha x}\cos \beta x$ and $e^{\alpha x}\sin \beta x$ are *real* solutions of (2). Moreover, these solutions form a fundamental set on $(-\infty, \infty)$. Consequently, the general solution is

$$y = c_1 e^{\alpha x}\cos \beta x + c_2 e^{\alpha x}\sin \beta x = e^{\alpha x}(c_1 \cos \beta x + c_2 \sin \beta x). \tag{8}$$

EXAMPLE 1 Second-Order DEs

Solve the following differential equations.

(a) $2y'' - 5y' - 3y = 0$ **(b)** $y'' - 10y' + 25y = 0$ **(c)** $y'' + 4y' + 7y = 0$

SOLUTION We give the auxiliary equations, the roots, and the corresponding general solutions.

(a) $2m^2 - 5m - 3 = (2m + 1)(m - 3) = 0,\quad m_1 = -\frac{1}{2}, m_2 = 3$

From (4), $y = c_1 e^{-x/2} + c_2 e^{3x}$.

(b) $m^2 - 10m + 25 = (m - 5)^2 = 0,\quad m_1 = m_2 = 5$

From (6), $y = c_1 e^{5x} + c_2 x e^{5x}$.

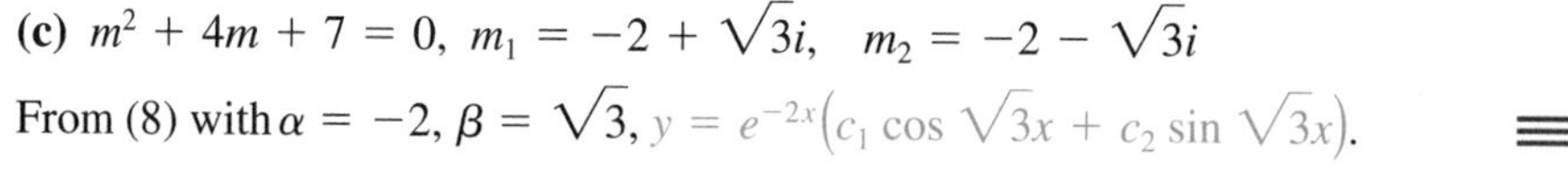

(c) $m^2 + 4m + 7 = 0,\ m_1 = -2 + \sqrt{3}i,\quad m_2 = -2 - \sqrt{3}i$

From (8) with $\alpha = -2$, $\beta = \sqrt{3}$, $y = e^{-2x}(c_1 \cos \sqrt{3}x + c_2 \sin \sqrt{3}x)$. ≡

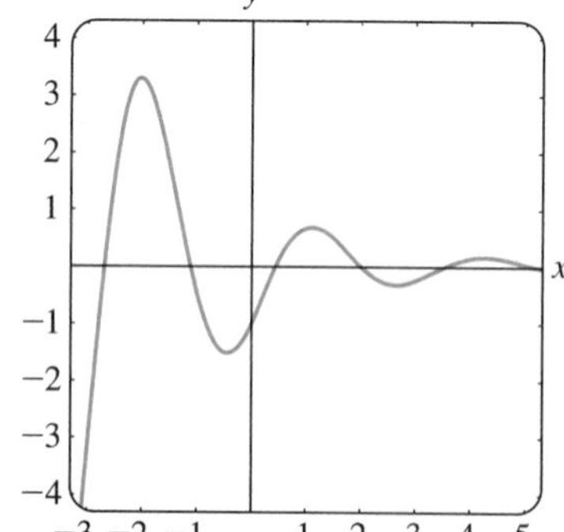

FIGURE 4.3.1 Solution curve of IVP in Example 2

EXAMPLE 2 An Initial-Value Problem

Solve $4y'' + 4y' + 17y = 0$, $y(0) = -1$, $y'(0) = 2$.

SOLUTION By the quadratic formula we find that the roots of the auxiliary equation $4m^2 + 4m + 17 = 0$ are $m_1 = -\frac{1}{2} + 2i$ and $m_2 = -\frac{1}{2} - 2i$. Thus from (8) we have $y = e^{-x/2}(c_1 \cos 2x + c_2 \sin 2x)$. Applying the condition $y(0) = -1$, we see from $e^0(c_1 \cos 0 + c_2 \sin 0) = -1$ that $c_1 = -1$. Differentiating $y = e^{-x/2}(-\cos 2x + c_2 \sin 2x)$ and then using $y'(0) = 2$ gives $2c_2 + \frac{1}{2} = 2$ or $c_2 = \frac{3}{4}$. Hence the solution of the IVP is $y = e^{-x/2}(-\cos 2x + \frac{3}{4}\sin 2x)$. In Figure 4.3.1 we see that the solution is oscillatory, but $y \to 0$ as $x \to \infty$. ≡

≡ Two Equations Worth Knowing

The two differential equations

$$y'' + k^2 y = 0 \qquad \text{and} \qquad y'' - k^2 y = 0,$$

where k is real, are important in applied mathematics. For $y'' + k^2y = 0$ the auxiliary equation $m^2 + k^2 = 0$ has imaginary roots $m_1 = ki$ and $m_2 = -ki$. With $\alpha = 0$ and $\beta = k$ in (8) the general solution of the DE is seen to be

$$y = c_1 \cos kx + c_2 \sin kx. \tag{9}$$

On the other hand, the auxiliary equation $m^2 - k^2 = 0$ for $y'' - k^2y = 0$ has distinct real roots $m_1 = k$ and $m_2 = -k$, and so by (4) the general solution of the DE is

$$y = c_1e^{kx} + c_2e^{-kx}. \tag{10}$$

Notice that if we choose $c_1 = c_2 = \frac{1}{2}$ and $c_1 = \frac{1}{2}, c_2 = -\frac{1}{2}$ in (10), we get the particular solutions $y = \frac{1}{2}(e^{kx} + e^{-kx}) = \cosh kx$ and $y = \frac{1}{2}(e^{kx} - e^{-kx}) = \sinh kx$. Since $\cosh kx$ and $\sinh kx$ are linearly independent on any interval of the x-axis, an alternative form for the general solution of $y'' - k^2y = 0$ is

$$y = c_1 \cosh kx + c_2 \sinh kx. \tag{11}$$

See Problems 41 and 42 in Exercises 4.3.

Higher-Order Equations In general, to solve an nth-order differential equation (1), where the a_i, $i = 0, 1, \ldots, n$ are real constants, we must solve an nth-degree polynomial equation

$$a_nm^n + a_{n-1}m^{n-1} + \cdots + a_2m^2 + a_1m + a_0 = 0. \tag{12}$$

If all the roots of (12) are real and distinct, then the general solution of (1) is

$$y = c_1e^{m_1x} + c_2e^{m_2x} + \cdots + c_ne^{m_nx}.$$

It is somewhat harder to summarize the analogues of Cases II and III because the roots of an auxiliary equation of degree greater than two can occur in many combinations. For example, a fifth-degree equation could have five distinct real roots, or three distinct real and two complex roots, or one real and four complex roots, or five real but equal roots, or five real roots but two of them equal, and so on. When m_1 is a root of multiplicity k of an nth-degree auxiliary equation (that is, k roots are equal to m_1), it can be shown that the linearly independent solutions are

$$e^{m_1x}, \quad xe^{m_1x}, \quad x^2e^{m_1x}, \ldots, \quad x^{k-1}e^{m_1x}$$

and the general solution must contain the linear combination

$$c_1e^{m_1x} + c_2xe^{m_1x} + c_3x^2e^{m_1x} + \cdots + c_kx^{k-1}e^{m_1x}.$$

Finally, it should be remembered that when the coefficients are real, complex roots of an auxiliary equation always appear in conjugate pairs. Thus, for example, a cubic polynomial equation can have at most two complex roots.

EXAMPLE 3 Third-Order DE

Solve $y''' + 3y'' - 4y = 0$.

SOLUTION It should be apparent from inspection of $m^3 + 3m^2 - 4 = 0$ that one root is $m_1 = 1$, so $m - 1$ is a factor of $m^3 + 3m^2 - 4$. By division we find

$$m^3 + 3m^2 - 4 = (m - 1)(m^2 + 4m + 4) = (m - 1)(m + 2)^2,$$

so the other roots are $m_2 = m_3 = -2$. Thus the general solution of the DE is $y = c_1e^x + c_2e^{-2x} + c_3xe^{-2x}$. ≡

EXAMPLE 4 Fourth-Order DE

Solve $\dfrac{d^4y}{dx^4} + 2\dfrac{d^2y}{dx^2} + y = 0.$

SOLUTION The auxiliary equation $m^4 + 2m^2 + 1 = (m^2 + 1)^2 = 0$ has roots $m_1 = m_3 = i$ and $m_2 = m_4 = -i$. Thus from Case II the solution is

$$y = C_1e^{ix} + C_2e^{-ix} + C_3xe^{ix} + C_4xe^{-ix}.$$

By Euler's formula the grouping $C_1e^{ix} + C_2e^{-ix}$ can be rewritten as

$$c_1 \cos x + c_2 \sin x$$

after a relabeling of constants. Similarly, $x(C_3e^{ix} + C_4e^{-ix})$ can be expressed as $x(c_3 \cos x + c_4 \sin x)$. Hence the general solution is

$$y = c_1 \cos x + c_2 \sin x + c_3x \cos x + c_4x \sin x.$$

Example 4 illustrates a special case when the auxiliary equation has repeated complex roots. In general, if $m_1 = \alpha + i\beta$, $\beta > 0$ is a complex root of multiplicity k of an auxiliary equation with real coefficients, then its conjugate $m_2 = \alpha - i\beta$ is also a root of multiplicity k. From the $2k$ complex-valued solutions

$$e^{(\alpha+i\beta)x}, \quad xe^{(\alpha+i\beta)x}, \quad x^2e^{(\alpha+i\beta)x}, \quad \ldots, \quad x^{k-1}e^{(\alpha+i\beta)x},$$

$$e^{(\alpha-i\beta)x}, \quad xe^{(\alpha-i\beta)x}, \quad x^2e^{(\alpha-i\beta)x}, \quad \ldots, \quad x^{k-1}e^{(\alpha-i\beta)x},$$

we conclude, with the aid of Euler's formula, that the general solution of the corresponding differential equation must then contain a linear combination of the $2k$ real linearly independent solutions

$$e^{\alpha x}\cos \beta x, \quad xe^{\alpha x}\cos \beta x, \quad x^2e^{\alpha x}\cos \beta x, \quad \ldots, \quad x^{k-1}e^{\alpha x}\cos \beta x,$$

$$e^{\alpha x}\sin \beta x, \quad xe^{\alpha x}\sin \beta x, \quad x^2e^{\alpha x}\sin \beta x, \quad \ldots, \quad x^{k-1}e^{\alpha x}\sin \beta x.$$

In Example 4 we identify $k = 2$, $\alpha = 0$, and $\beta = 1$.

Of course the most difficult aspect of solving constant-coefficient differential equations is finding roots of auxiliary equations of degree greater than two. For example, to solve $3y''' + 5y'' + 10y' - 4y = 0$, we must solve $3m^3 + 5m^2 + 10m - 4 = 0$. Something we can try is to test the auxiliary equation for rational roots. Recall that if $m_1 = p/q$ is a rational root (expressed in lowest terms) of an auxiliary equation $a_nm^n + \cdots + a_1m + a_0 = 0$ with integer coefficients, then p is a factor of a_0 and q is a factor of a_n. For our specific cubic auxiliary equation, all the factors of $a_0 = -4$ and $a_n = 3$ are p: $\pm1, \pm2, \pm4$ and q: $\pm1, \pm3$, so the possible rational roots are p/q: $\pm1, \pm2, \pm4, \pm\frac{1}{3}, \pm\frac{2}{3}, \pm\frac{4}{3}$. Each of these numbers can then be tested—say, by synthetic division. In this way we discover both the root $m_1 = \frac{1}{3}$ and the factorization

$$3m^3 + 5m^2 + 10m - 4 = \left(m - \tfrac{1}{3}\right)(3m^2 + 6m + 12).$$

The quadratic formula then yields the remaining roots $m_2 = -1 + \sqrt{3}i$ and $m_3 = -1 - \sqrt{3}i$. Therefore the general solution of $3y''' + 5y'' + 10y' - 4y = 0$ is $y = c_1e^{x/3} + e^{-x}(c_2 \cos \sqrt{3}x + c_3 \sin \sqrt{3}x)$.

Use of Computers Finding roots or approximation of roots of auxiliary equations is a routine problem with an appropriate calculator or computer software. Polynomial equations (in one variable) of degree less than five can be solved by means of algebraic formulas using the *solve* commands in *Mathematica* and *Maple*. For auxiliary equations of degree five or greater it might be necessary to resort to numerical commands such as **NSolve** and **FindRoot** in *Mathematica*. Because of their capability of solving polynomial equations, it is not surprising that these computer

There is more on this in the *SRM*. ▶

algebra systems are also able, by means of their *dsolve* commands, to provide explicit solutions of homogeneous linear constant-coefficient differential equations.

In the classic text *Differential Equations* by Ralph Palmer Agnew* (used by the author as a student) the following statement is made:

> *It is not reasonable to expect students in this course to have computing skill and equipment necessary for efficient solving of equations such as*
>
> $$4.317\frac{d^4y}{dx^4} + 2.179\frac{d^3y}{dx^3} + 1.416\frac{d^2y}{dx^2} + 1.295\frac{dy}{dx} + 3.169y = 0. \qquad (13)$$

Although it is debatable whether computing skills have improved in the intervening years, it is a certainty that technology has. If one has access to a computer algebra system, equation (13) could now be considered reasonable. After simplification and some relabeling of output, *Mathematica* yields the (approximate) general solution

$$y = c_1e^{-0.728852x}\cos(0.618605x) + c_2e^{-0.728852x}\sin(0.618605x)$$
$$+ c_3e^{0.476478x}\cos(0.759081x) + c_4e^{0.476478x}\sin(0.759081x).$$

Finally, if we are faced with an initial-value problem consisting of, say, a fourth-order equation, then to fit the general solution of the DE to the four initial conditions, we must solve four linear equations in four unknowns (the c_1, c_2, c_3, c_4 in the general solution). Using a CAS to solve the system can save lots of time. See Problems 69 and 70 in Exercises 4.3 and Problem 41 in Chapter 4 in Review.

*McGraw-Hill, New York, 1960.

EXERCISES 4.3

Answers to selected odd-numbered problems begin on page ANS-5.

In Problems 1–14 find the general solution of the given second-order differential equation.

1. $4y'' + y' = 0$
2. $y'' - 36y = 0$
3. $y'' - y' - 6y = 0$
4. $y'' - 3y' + 2y = 0$
5. $y'' + 8y' + 16y = 0$
6. $y'' - 10y' + 25y = 0$
7. $12y'' - 5y' - 2y = 0$
8. $y'' + 4y' - y = 0$
9. $y'' + 9y = 0$
10. $3y'' + y = 0$
11. $y'' - 4y' + 5y = 0$
12. $2y'' + 2y' + y = 0$
13. $3y'' + 2y' + y = 0$
14. $2y'' - 3y' + 4y = 0$

In Problems 15–28 find the general solution of the given higher-order differential equation.

15. $y''' - 4y'' - 5y' = 0$
16. $y''' - y = 0$
17. $y''' - 5y'' + 3y' + 9y = 0$
18. $y''' + 3y'' - 4y' - 12y = 0$
19. $\dfrac{d^3u}{dt^3} + \dfrac{d^2u}{dt^2} - 2u = 0$
20. $\dfrac{d^3x}{dt^3} - \dfrac{d^2x}{dt^2} - 4x = 0$
21. $y''' + 3y'' + 3y' + y = 0$
22. $y''' - 6y'' + 12y' - 8y = 0$
23. $y^{(4)} + y''' + y'' = 0$
24. $y^{(4)} - 2y'' + y = 0$
25. $16\dfrac{d^4y}{dx^4} + 24\dfrac{d^2y}{dx^2} + 9y = 0$
26. $\dfrac{d^4y}{dx^4} - 7\dfrac{d^2y}{dx^2} - 18y = 0$
27. $\dfrac{d^5u}{dr^5} + 5\dfrac{d^4u}{dr^4} - 2\dfrac{d^3u}{dr^3} - 10\dfrac{d^2u}{dr^2} + \dfrac{du}{dr} + 5u = 0$
28. $2\dfrac{d^5x}{ds^5} - 7\dfrac{d^4x}{ds^4} + 12\dfrac{d^3x}{ds^3} + 8\dfrac{d^2x}{ds^2} = 0$

In Problems 29–36 solve the given initial-value problem.

29. $y'' + 16y = 0, \quad y(0) = 2, y'(0) = -2$
30. $\dfrac{d^2y}{d\theta^2} + y = 0, \quad y(\pi/3) = 0, y'(\pi/3) = 2$

31. $\dfrac{d^2y}{dt^2} - 4\dfrac{dy}{dt} - 5y = 0, \quad y(1) = 0, y'(1) = 2$

32. $4y'' - 4y' - 3y = 0, \quad y(0) = 1, y'(0) = 5$

33. $y'' + y' + 2y = 0, \quad y(0) = y'(0) = 0$

34. $y'' - 2y' + y = 0, \quad y(0) = 5, y'(0) = 10$

35. $y''' + 12y'' + 36y' = 0, \quad y(0) = 0, y'(0) = 1, y''(0) = -7$

36. $y''' + 2y'' - 5y' - 6y = 0, \quad y(0) = y'(0) = 0, \ y''(0) = 1$

In Problems 37–40 solve the given boundary-value problem.

37. $y'' - 10y' + 25y = 0, \quad y(0) = 1, y(1) = 0$

38. $y'' + 4y = 0, \quad y(0) = 0, y(\pi) = 0$

39. $y'' + y = 0, \quad y'(0) = 0, y'(\pi/2) = 0$

40. $y'' - 2y' + 2y = 0, \quad y(0) = 1, y(\pi) = 1$

In Problems 41 and 42 solve the given problem first using the form of the general solution given in (10). Solve again, this time using the form given in (11).

41. $y'' - 3y = 0, \quad y(0) = 1, y'(0) = 5$

42. $y'' - y = 0, \quad y(0) = 1, y'(1) = 0$

In Problems 43–48 each figure represents the graph of a particular solution of one of the following differential equations:

(a) $y'' - 3y' - 4y = 0$ (b) $y'' + 4y = 0$
(c) $y'' + 2y' + y = 0$ (d) $y'' + y = 0$
(e) $y'' + 2y' + 2y = 0$ (f) $y'' - 3y' + 2y = 0$

Match a solution curve with one of the differential equations. Explain your reasoning.

43.

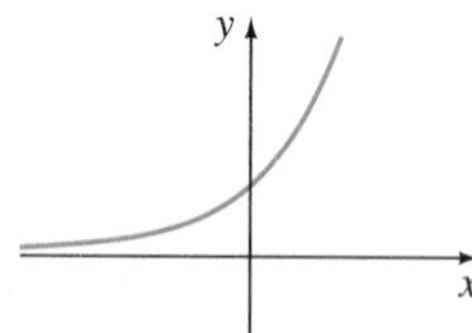

FIGURE 4.3.2 Graph for Problem 43

44.

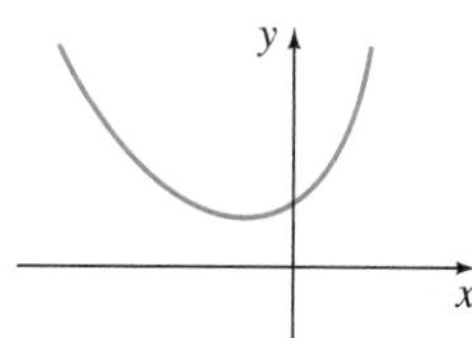

FIGURE 4.3.3 Graph for Problem 44

45.

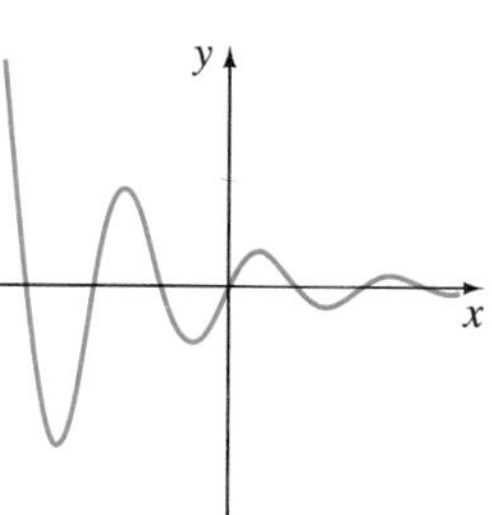

FIGURE 4.3.4 Graph for Problem 45

46.

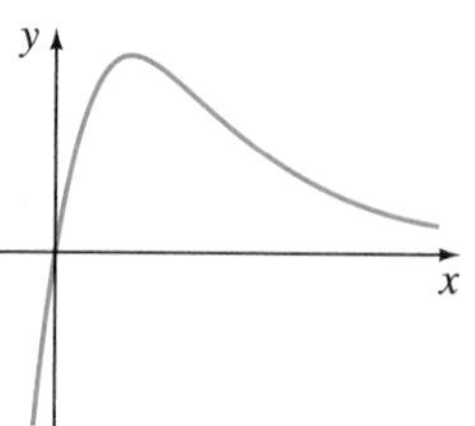

FIGURE 4.3.5 Graph for Problem 46

47.

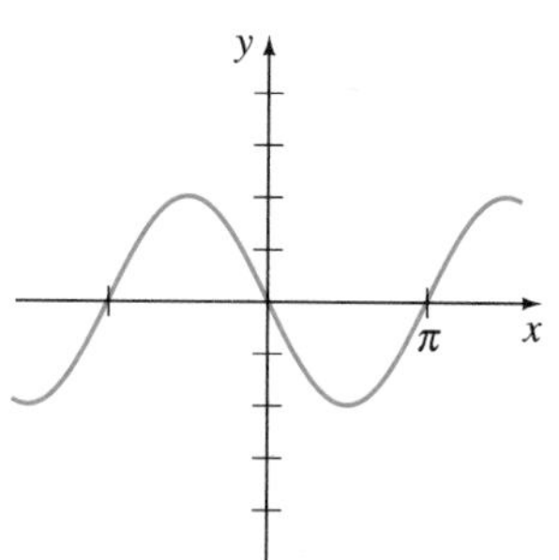

FIGURE 4.3.6 Graph for Problem 47

48.

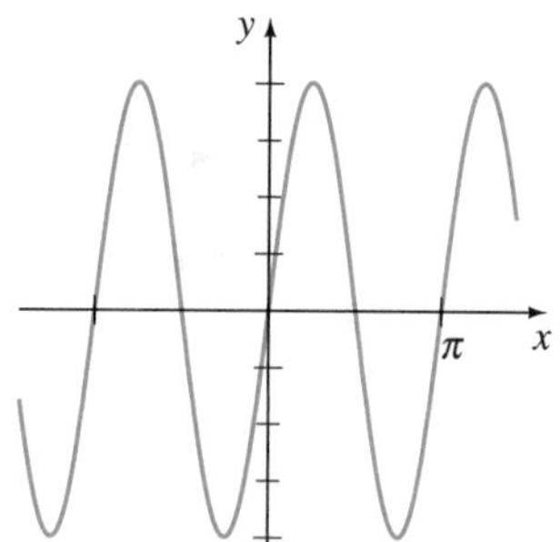

FIGURE 4.3.7 Graph for Problem 48

In Problems 49–58 find a homogeneous linear differential equation with constant coefficients whose general solution is given.

49. $y = c_1e^x + c_2e^{5x}$

50. $y = c_1e^{-4x} + c_2e^{-3x}$

51. $y = c_1 + c_2e^{2x}$

52. $y = c_1e^{10x} + c_2xe^{10x}$

53. $y = c_1 \cos 3x + c_2 \sin 3x$

54. $y = c_1 \cosh 7x + c_2 \sinh 7x$

55. $y = c_1e^{-x}\cos x + c_2e^{-x}\sin x$

56. $y = c_1 + c_2e^{2x}\cos 5x + c_3e^{2x}\sin 5x$

57. $y = c_1 + c_2x + c_3e^{8x}$

58. $y = c_1 \cos x + c_2 \sin x + c_3 \cos 2x + c_4 \sin 2x$

Discussion Problems

59. Two roots of a cubic auxiliary equation with real coefficients are $m_1 = -\frac{1}{2}$ and $m_2 = 3 + i$. What is the corresponding homogeneous linear differential equation? Discuss: Is your answer unique?

60. Find the general solution of $2y''' + 7y'' + 4y' - 4y = 0$ if $m_1 = \frac{1}{2}$ is one root of its auxiliary equation.

61. Find the general solution of $y''' + 6y'' + y' - 34y = 0$ if it is known that $y_1 = e^{-4x}\cos x$ is one solution.

62. To solve $y^{(4)} + y = 0$, we must find the roots of $m^4 + 1 = 0$. This is a trivial problem using a CAS but can also be done by hand working with complex numbers. Observe that $m^4 + 1 = (m^2 + 1)^2 - 2m^2$. How does this help? Solve the differential equation.

63. Verify that $y = \sinh x - 2\cos(x + \pi/6)$ is a particular solution of $y^{(4)} - y = 0$. Reconcile this particular solution with the general solution of the DE.

64. Consider the boundary-value problem $y'' + \lambda y = 0$, $y(0) = 0$, $y(\pi/2) = 0$. Discuss: Is it possible to determine values of λ so that the problem possesses **(a)** trivial solutions? **(b)** nontrivial solutions?

Computer Lab Assignments

In Problems 65–68 use a computer either as an aid in solving the auxiliary equation or as a means of directly obtaining the general solution of the given differential equation. If you use a CAS to obtain the general solution, simplify the output and, if necessary, write the solution in terms of real functions.

65. $y''' - 6y'' + 2y' + y = 0$

66. $6.11y''' + 8.59y'' + 7.93y' + 0.778y = 0$

67. $3.15y^{(4)} - 5.34y'' + 6.33y' - 2.03y = 0$

68. $y^{(4)} + 2y'' - y' + 2y = 0$

In Problems 69 and 70 use a CAS as an aid in solving the auxiliary equation. Form the general solution of the differential equation. Then use a CAS as an aid in solving the system of equations for the coefficients c_i, $i = 1, 2, 3, 4$ that results when the initial conditions are applied to the general solution.

69. $2y^{(4)} + 3y''' - 16y'' + 15y' - 4y = 0$,
$y(0) = -2,\ y'(0) = 6,\ y''(0) = 3,\ y'''(0) = \frac{1}{2}$

70. $y^{(4)} - 3y''' + 3y'' - y' = 0$,
$y(0) = y'(0) = 0,\ y''(0) = y'''(0) = 1$

4.4 UNDETERMINED COEFFICIENTS—SUPERPOSITION APPROACH*

REVIEW MATERIAL

- Review Theorems 4.1.6 and 4.1.7 (Section 4.1)

INTRODUCTION To solve a nonhomogeneous linear differential equation

$$a_n y^{(n)} + a_{n-1} y^{(n-1)} + \cdots + a_1 y' + a_0 y = g(x), \tag{1}$$

we must do two things:

- find the complementary function y_c and
- find *any* particular solution y_p of the nonhomogeneous equation (1).

Then, as was discussed in Section 4.1, the general solution of (1) is $y = y_c + y_p$. The complementary function y_c is the general solution of the associated homogeneous DE of (1), that is,

$$a_n y^{(n)} + a_{n-1} y^{(n-1)} + \cdots + a_1 y' + a_0 y = 0.$$

In Section 4.3 we saw how to solve these kinds of equations when the coefficients were constants. Our goal in the present section is to develop a method for obtaining particular solutions.

---

*__Note to the Instructor:__ In this section the method of undetermined coefficients is developed from the viewpoint of the superposition principle for nonhomogeneous equations (Theorem 4.7.1). In Section 4.5 an entirely different approach will be presented, one utilizing the concept of differential annihilator operators. Take your pick.

Method of Undetermined Coefficients The first of two ways we shall consider for obtaining a particular solution y_p for a nonhomogeneous linear DE is called the **method of undetermined coefficients.** The underlying idea behind this method is a conjecture about the form of y_p, an educated guess really, that is motivated by the kinds of functions that make up the input function $g(x)$. The general method is limited to linear DEs such as (1) where

- the coefficients a_i, $i = 0, 1, \ldots, n$ are constants and
- $g(x)$ is a constant k, a polynomial function, an exponential function $e^{\alpha x}$, a sine or cosine function $\sin \beta x$ or $\cos \beta x$, or finite sums and products of these functions.

Note Strictly speaking, $g(x) = k$ (constant) is a polynomial function. Since a constant function is probably not the first thing that comes to mind when you think of polynomial functions, for emphasis we shall continue to use the redundancy "constant functions, polynomials, ..."

The following functions are some examples of the types of inputs $g(x)$ that are appropriate for this discussion:

$$g(x) = 10, \quad g(x) = x^2 - 5x, \quad g(x) = 15x - 6 + 8e^{-x},$$

$$g(x) = \sin 3x - 5x \cos 2x, \quad g(x) = xe^x \sin x + (3x^2 - 1)e^{-4x}.$$

That is, $g(x)$ is a linear combination of functions of the type

$$P(x) = a_n x^n + a_{n-1} x^{n-1} + \cdots + a_1 x + a_0, \quad P(x)\, e^{\alpha x}, \quad P(x)\, e^{\alpha x} \sin \beta x, \quad \text{and} \quad P(x)\, e^{\alpha x} \cos \beta x,$$

where n is a nonnegative integer and α and β are real numbers. The method of undetermined coefficients is not applicable to equations of form (1) when

$$g(x) = \ln x, \quad g(x) = \frac{1}{x}, \quad g(x) = \tan x, \quad g(x) = \sin^{-1} x,$$

and so on. Differential equations in which the input $g(x)$ is a function of this last kind will be considered in Section 4.6.

The set of functions that consists of constants, polynomials, exponentials $e^{\alpha x}$, sines, and cosines has the remarkable property that derivatives of their sums and products are again sums and products of constants, polynomials, exponentials $e^{\alpha x}$, sines, and cosines. Because the linear combination of derivatives $a_n y_p^{(n)} + a_{n-1} y_p^{(n-1)} + \cdots + a_1 y_p' + a_0 y_p$ must be identical to $g(x)$, it seems reasonable to assume that y_p *has the same form as* $g(x)$.

The next two examples illustrate the basic method.

EXAMPLE 1 General Solution Using Undetermined Coefficients

Solve $y'' + 4y' - 2y = 2x^2 - 3x + 6.$ (2)

SOLUTION **Step 1.** We first solve the associated homogeneous equation $y'' + 4y' - 2y = 0$. From the quadratic formula we find that the roots of the auxiliary equation $m^2 + 4m - 2 = 0$ are $m_1 = -2 - \sqrt{6}$ and $m_2 = -2 + \sqrt{6}$. Hence the complementary function is

$$y_c = c_1 e^{-(2+\sqrt{6})x} + c_2 e^{(-2+\sqrt{6})x}.$$

Step 2. Now, because the function $g(x)$ is a quadratic polynomial, let us assume a particular solution that is also in the form of a quadratic polynomial:

$$y_p = Ax^2 + Bx + C.$$

We seek to determine *specific* coefficients A, B, and C for which y_p is a solution of (2). Substituting y_p and the derivatives

$$y'_p = 2Ax + B \quad \text{and} \quad y''_p = 2A$$

into the given differential equation (2), we get

$$y''_p + 4y'_p - 2y_p = 2A + 8Ax + 4B - 2Ax^2 - 2Bx - 2C = 2x^2 - 3x + 6.$$

Because the last equation is supposed to be an identity, the coefficients of like powers of x must be equal:

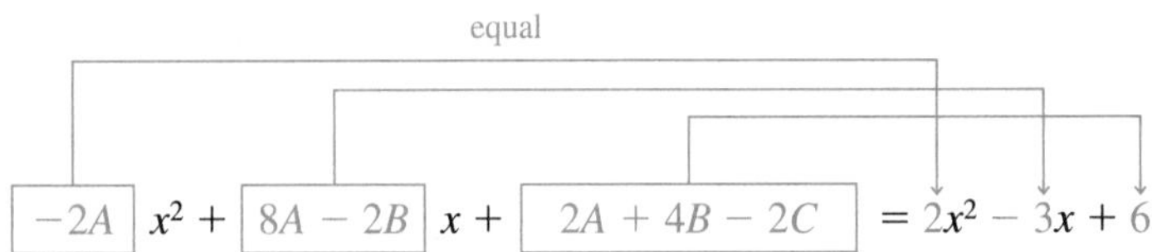

That is, $\quad -2A = 2, \quad 8A - 2B = -3, \quad 2A + 4B - 2C = 6.$

Solving this system of equations leads to the values $A = -1$, $B = -\frac{5}{2}$, and $C = -9$. Thus a particular solution is

$$y_p = -x^2 - \frac{5}{2}x - 9.$$

Step 3. The general solution of the given equation is

$$y = y_c + y_p = c_1e^{-(2+\sqrt{6})x} + c_2e^{(-2+\sqrt{6})x} - x^2 - \frac{5}{2}x - 9. \qquad \equiv$$

EXAMPLE 2 Particular Solution Using Undetermined Coefficients

Find a particular solution of $y'' - y' + y = 2 \sin 3x$.

SOLUTION A natural first guess for a particular solution would be $A \sin 3x$. But because successive differentiations of $\sin 3x$ produce $\sin 3x$ *and* $\cos 3x$, we are prompted instead to assume a particular solution that includes both of these terms:

$$y_p = A \cos 3x + B \sin 3x.$$

Differentiating y_p and substituting the results into the differential equation gives, after regrouping,

$$y''_p - y'_p + y_p = (-8A - 3B) \cos 3x + (3A - 8B) \sin 3x = 2 \sin 3x$$

or

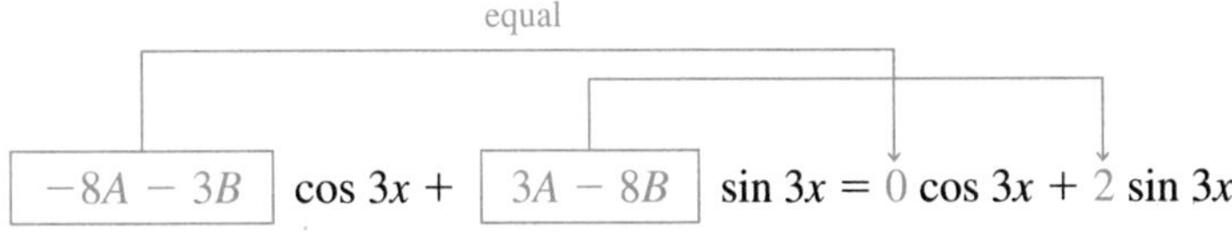

From the resulting system of equations,

$$-8A - 3B = 0, \qquad 3A - 8B = 2,$$

we get $A = \frac{6}{73}$ and $B = -\frac{16}{73}$. A particular solution of the equation is

$$y_p = \frac{6}{73} \cos 3x - \frac{16}{73} \sin 3x. \qquad \equiv$$

As we mentioned, the form that we assume for the particular solution y_p is an educated guess; it is not a blind guess. This educated guess must take into consideration not only the types of functions that make up $g(x)$ but also, as we shall see in Example 4, the functions that make up the complementary function y_c.

EXAMPLE 3 Forming y_p by Superposition

Solve $y'' - 2y' - 3y = 4x - 5 + 6xe^{2x}$. (3)

SOLUTION **Step 1.** First, the solution of the associated homogeneous equation $y'' - 2y' - 3y = 0$ is found to be $y_c = c_1e^{-x} + c_2e^{3x}$.

Step 2. Next, the presence of $4x - 5$ in $g(x)$ suggests that the particular solution includes a linear polynomial. Furthermore, because the derivative of the product xe^{2x} produces $2xe^{2x}$ and e^{2x}, we also assume that the particular solution includes both xe^{2x} and e^{2x}. In other words, g is the sum of two basic kinds of functions:

$$g(x) = g_1(x) + g_2(x) = \textit{polynomial} + \textit{exponentials}.$$

Correspondingly, the superposition principle for nonhomogeneous equations (Theorem 4.1.7) suggests that we seek a particular solution

$$y_p = y_{p_1} + y_{p_2},$$

where $y_{p_1} = Ax + B$ and $y_{p_2} = Cxe^{2x} + Ee^{2x}$. Substituting

$$y_p = Ax + B + Cxe^{2x} + Ee^{2x}$$

into the given equation (3) and grouping like terms gives

$$y_p'' - 2y_p' - 3y_p = -3Ax - 2A - 3B - 3Cxe^{2x} + (2C - 3E)e^{2x} = 4x - 5 + 6xe^{2x}. \quad (4)$$

From this identity we obtain the four equations

$$-3A = 4, \qquad -2A - 3B = -5, \qquad -3C = 6, \qquad 2C - 3E = 0.$$

The last equation in this system results from the interpretation that the coefficient of e^{2x} in the right member of (4) is zero. Solving, we find $A = -\frac{4}{3}$, $B = \frac{23}{9}$, $C = -2$, and $E = -\frac{4}{3}$. Consequently,

$$y_p = -\frac{4}{3}x + \frac{23}{9} - 2xe^{2x} - \frac{4}{3}e^{2x}.$$

Step 3. The general solution of the equation is

$$y = c_1e^{-x} + c_2e^{3x} - \frac{4}{3}x + \frac{23}{9} - \left(2x + \frac{4}{3}\right)e^{2x}. \qquad \equiv$$

In light of the superposition principle (Theorem 4.1.7) we can also approach Example 3 from the viewpoint of solving two simpler problems. You should verify that substituting

$$y_{p_1} = Ax + B \quad \text{into} \quad y'' - 2y' - 3y = 4x - 5$$

and

$$y_{p_2} = Cxe^{2x} + Ee^{2x} \quad \text{into} \quad y'' - 2y' - 3y = 6xe^{2x}$$

yields, in turn, $y_{p_1} = -\frac{4}{3}x + \frac{23}{9}$ and $y_{p_2} = -(2x + \frac{4}{3})e^{2x}$. A particular solution of (3) is then $y_p = y_{p_1} + y_{p_2}$.

The next example illustrates that sometimes the "obvious" assumption for the form of y_p is not a correct assumption.

EXAMPLE 4 A Glitch in the Method

Find a particular solution of $y'' - 5y' + 4y = 8e^x$.

SOLUTION Differentiation of e^x produces no new functions. Therefore proceeding as we did in the earlier examples, we can reasonably assume a particular solution of the form $y_p = Ae^x$. But substitution of this expression into the differential equation yields the contradictory statement $0 = 8e^x$, so we have clearly made the wrong guess for y_p.

The difficulty here is apparent on examining the complementary function $y_c = c_1e^x + c_2e^{4x}$. Observe that our assumption Ae^x is already present in y_c. This means that e^x is a solution of the associated homogeneous differential equation, and a constant multiple Ae^x when substituted into the differential equation necessarily produces zero.

What then should be the form of y_p? Inspired by Case II of Section 4.3, let's see whether we can find a particular solution of the form

$$y_p = Axe^x.$$

Substituting $y_p' = Axe^x + Ae^x$ and $y_p'' = Axe^x + 2Ae^x$ into the differential equation and simplifying gives

$$y_p'' - 5y_p' + 4y_p = -3Ae^x = 8e^x.$$

From the last equality we see that the value of A is now determined as $A = -\frac{8}{3}$. Therefore a particular solution of the given equation is $y_p = -\frac{8}{3}xe^x$. ≡

The difference in the procedures used in Examples 1–3 and in Example 4 suggests that we consider two cases. The first case reflects the situation in Examples 1–3.

≡ **Case I** No function in the assumed particular solution is a solution of the associated homogeneous differential equation.

In Table 4.4.1 we illustrate some specific examples of $g(x)$ in (1) along with the corresponding form of the particular solution. We are, of course, taking for granted that no function in the assumed particular solution y_p is duplicated by a function in the complementary function y_c.

TABLE 4.4.1 Trial Particular Solutions

$g(x)$	Form of y_p
1. 1 (any constant)	A
2. $5x + 7$	$Ax + B$
3. $3x^2 - 2$	$Ax^2 + Bx + C$
4. $x^3 - x + 1$	$Ax^3 + Bx^2 + Cx + E$
5. $\sin 4x$	$A\cos 4x + B\sin 4x$
6. $\cos 4x$	$A\cos 4x + B\sin 4x$
7. e^{5x}	Ae^{5x}
8. $(9x - 2)e^{5x}$	$(Ax + B)e^{5x}$
9. x^2e^{5x}	$(Ax^2 + Bx + C)e^{5x}$
10. $e^{3x}\sin 4x$	$Ae^{3x}\cos 4x + Be^{3x}\sin 4x$
11. $5x^2\sin 4x$	$(Ax^2 + Bx + C)\cos 4x + (Ex^2 + Fx + G)\sin 4x$
12. $xe^{3x}\cos 4x$	$(Ax + B)e^{3x}\cos 4x + (Cx + E)e^{3x}\sin 4x$

EXAMPLE 5 Forms of Particular Solutions—Case I

Determine the form of a particular solution of

(a) $y'' - 8y' + 25y = 5x^3e^{-x} - 7e^{-x}$ **(b)** $y'' + 4y = x\cos x$

SOLUTION **(a)** We can write $g(x) = (5x^3 - 7)e^{-x}$. Using entry 9 in Table 4.4.1 as a model, we assume a particular solution of the form

$$y_p = (Ax^3 + Bx^2 + Cx + E)e^{-x}.$$

Note that there is no duplication between the terms in y_p and the terms in the complementary function $y_c = e^{4x}(c_1\cos 3x + c_2\sin 3x)$.

(b) The function $g(x) = x \cos x$ is similar to entry 11 in Table 4.4.1 except, of course, that we use a linear rather than a quadratic polynomial and $\cos x$ and $\sin x$ instead of $\cos 4x$ and $\sin 4x$ in the form of y_p:

$$y_p = (Ax + B) \cos x + (Cx + E) \sin x.$$

Again observe that there is no duplication of terms between y_p and $y_c = c_1 \cos 2x + c_2 \sin 2x$. ≡

If $g(x)$ consists of a sum of, say, m terms of the kind listed in the table, then (as in Example 3) the assumption for a particular solution y_p consists of the sum of the trial forms $y_{p_1}, y_{p_2}, \ldots, y_{p_m}$ corresponding to these terms:

$$y_p = y_{p_1} + y_{p_2} + \cdots + y_{p_m}.$$

The foregoing sentence can be put another way.

Form Rule for Case I *The form of y_p is a linear combination of all linearly independent functions that are generated by repeated differentiations of $g(x)$.*

EXAMPLE 6 Forming y_p by Superposition—Case I

Determine the form of a particular solution of

$$y'' - 9y' + 14y = 3x^2 - 5 \sin 2x + 7xe^{6x}.$$

SOLUTION

Corresponding to $3x^2$ we assume $\qquad y_{p_1} = Ax^2 + Bx + C.$

Corresponding to $-5 \sin 2x$ we assume $\qquad y_{p_2} = E \cos 2x + F \sin 2x.$

Corresponding to $7xe^{6x}$ we assume $\qquad y_{p_3} = (Gx + H)e^{6x}.$

The assumption for the particular solution is then

$$y_p = y_{p_1} + y_{p_2} + y_{p_3} = Ax^2 + Bx + C + E \cos 2x + F \sin 2x + (Gx + H)e^{6x}.$$

No term in this assumption duplicates a term in $y_c = c_1 e^{2x} + c_2 e^{7x}$. ≡

≡ **Case II** A function in the assumed particular solution is also a solution of the associated homogeneous differential equation.

The next example is similar to Example 4.

EXAMPLE 7 Particular Solution—Case II

Find a particular solution of $y'' - 2y' + y = e^x$.

SOLUTION The complementary function is $y_c = c_1 e^x + c_2 xe^x$. As in Example 4, the assumption $y_p = Ae^x$ will fail, since it is apparent from y_c that e^x is a solution of the associated homogeneous equation $y'' - 2y' + y = 0$. Moreover, we will not be able to find a particular solution of the form $y_p = Axe^x$, since the term xe^x is also duplicated in y_c. We next try

$$y_p = Ax^2 e^x.$$

Substituting into the given differential equation yields $2Ae^x = e^x$, so $A = \frac{1}{2}$. Thus a particular solution is $y_p = \frac{1}{2}x^2 e^x$. ≡

Suppose again that $g(x)$ consists of m terms of the kind given in Table 4.4.1, and suppose further that the usual assumption for a particular solution is

$$y_p = y_{p_1} + y_{p_2} + \cdots + y_{p_m},$$

where the y_{p_i}, $i = 1, 2, \ldots, m$ are the trial particular solution forms corresponding to these terms. Under the circumstances described in Case II, we can make up the following general rule.

Multiplication Rule for Case II *If any y_{p_i} contains terms that duplicate terms in y_c, then that y_{p_i} must be multiplied by x^n, where n is the smallest positive integer that eliminates that duplication.*

EXAMPLE 8 An Initial-Value Problem

Solve $y'' + y = 4x + 10 \sin x$, $y(\pi) = 0$, $y'(\pi) = 2$.

SOLUTION The solution of the associated homogeneous equation $y'' + y = 0$ is $y_c = c_1 \cos x + c_2 \sin x$. Because $g(x) = 4x + 10 \sin x$ is the sum of a linear polynomial and a sine function, our normal assumption for y_p, from entries 2 and 5 of Table 4.4.1, would be the sum of $y_{p_1} = Ax + B$ and $y_{p_2} = C \cos x + E \sin x$:

$$y_p = Ax + B + C \cos x + E \sin x. \tag{5}$$

But there is an obvious duplication of the terms $\cos x$ and $\sin x$ in this assumed form and two terms in the complementary function. This duplication can be eliminated by simply multiplying y_{p_2} by x. Instead of (5) we now use

$$y_p = Ax + B + Cx \cos x + Ex \sin x. \tag{6}$$

Differentiating this expression and substituting the results into the differential equation gives

$$y_p'' + y_p = Ax + B - 2C \sin x + 2E \cos x = 4x + 10 \sin x,$$

and so $A = 4$, $B = 0$, $-2C = 10$, and $2E = 0$. The solutions of the system are immediate: $A = 4$, $B = 0$, $C = -5$, and $E = 0$. Therefore from (6) we obtain $y_p = 4x - 5x \cos x$. The general solution of the given equation is

$$y = y_c + y_p = c_1 \cos x + c_2 \sin x + 4x - 5x \cos x.$$

We now apply the prescribed initial conditions to the general solution of the equation. First, $y(\pi) = c_1 \cos \pi + c_2 \sin \pi + 4\pi - 5\pi \cos \pi = 0$ yields $c_1 = 9\pi$, since $\cos \pi = -1$ and $\sin \pi = 0$. Next, from the derivative

$$y' = -9\pi \sin x + c_2 \cos x + 4 + 5x \sin x - 5 \cos x$$

and $$y'(\pi) = -9\pi \sin \pi + c_2 \cos \pi + 4 + 5\pi \sin \pi - 5 \cos \pi = 2$$

we find $c_2 = 7$. The solution of the initial-value is then

$$y = 9\pi \cos x + 7 \sin x + 4x - 5x \cos x. \qquad \equiv$$

EXAMPLE 9 Using the Multiplication Rule

Solve $y'' - 6y' + 9y = 6x^2 + 2 - 12e^{3x}$.

SOLUTION The complementary function is $y_c = c_1 e^{3x} + c_2 x e^{3x}$. And so, based on entries 3 and 7 of Table 4.4.1, the usual assumption for a particular solution would be

$$y_p = \underbrace{Ax^2 + Bx + C}_{y_{p_1}} + \underbrace{Ee^{3x}}_{y_{p_2}}.$$

Inspection of these functions shows that the one term in y_{p_2} is duplicated in y_c. If we multiply y_{p_2} by x, we note that the term xe^{3x} is still part of y_c. But multiplying y_{p_2} by x^2 eliminates all duplications. Thus the operative form of a particular solution is

$$y_p = Ax^2 + Bx + C + Ex^2e^{3x}.$$

Differentiating this last form, substituting into the differential equation, and collecting like terms gives

$$y_p'' - 6y_p' + 9y_p = 9Ax^2 + (-12A + 9B)x + 2A - 6B + 9C + 2Ee^{3x} = 6x^2 + 2 - 12e^{3x}.$$

It follows from this identity that $A = \frac{2}{3}$, $B = \frac{8}{9}$, $C = \frac{2}{3}$, and $E = -6$. Hence the general solution $y = y_c + y_p$ is $y = c_1e^{3x} + c_2xe^{3x} + \frac{2}{3}x^2 + \frac{8}{9}x + \frac{2}{3} - 6x^2e^{3x}$. ≡

EXAMPLE 10 **Third-Order DE—Case I**

Solve $y''' + y'' = e^x \cos x$.

SOLUTION From the characteristic equation $m^3 + m^2 = 0$ we find $m_1 = m_2 = 0$ and $m_3 = -1$. Hence the complementary function of the equation is $y_c = c_1 + c_2x + c_3e^{-x}$. With $g(x) = e^x \cos x$, we see from entry 10 of Table 4.4.1 that we should assume that

$$y_p = Ae^x \cos x + Be^x \sin x.$$

Because there are no functions in y_p that duplicate functions in the complementary solution, we proceed in the usual manner. From

$$y_p''' + y_p'' = (-2A + 4B)e^x \cos x + (-4A - 2B)e^x \sin x = e^x \cos x$$

we get $-2A + 4B = 1$ and $-4A - 2B = 0$. This system gives $A = -\frac{1}{10}$ and $B = \frac{1}{5}$, so a particular solution is $y_p = -\frac{1}{10}e^x \cos x + \frac{1}{5}e^x \sin x$. The general solution of the equation is

$$y = y_c + y_p = c_1 + c_2x + c_3e^{-x} - \frac{1}{10}e^x \cos x + \frac{1}{5}e^x \sin x. \quad ≡$$

EXAMPLE 11 **Fourth-Order DE—Case II**

Determine the form of a particular solution of $y^{(4)} + y''' = 1 - x^2e^{-x}$.

SOLUTION Comparing $y_c = c_1 + c_2x + c_3x^2 + c_4e^{-x}$ with our normal assumption for a particular solution

$$y_p = \underbrace{A}_{y_{p_1}} + \underbrace{Bx^2e^{-x} + Cxe^{-x} + Ee^{-x}}_{y_{p_2}},$$

we see that the duplications between y_c and y_p are eliminated when y_{p_1} is multiplied by x^3 and y_{p_2} is multiplied by x. Thus the correct assumption for a particular solution is $y_p = Ax^3 + Bx^3e^{-x} + Cx^2e^{-x} + Exe^{-x}$. ≡

REMARKS

(*i*) In Problems 27–36 in Exercises 4.4 you are asked to solve initial-value problems, and in Problems 37–40 you are asked to solve boundary-value problems. As illustrated in Example 8, be sure to apply the initial conditions or the boundary conditions to the general solution $y = y_c + y_p$. Students often make the mistake of applying these conditions only to the complementary function y_c because it is that part of the solution that contains the constants $c_1, c_2, \ldots, c_n$.

(*ii*) From the "Form Rule for Case I" on page 144 of this section you see why the method of undetermined coefficients is not well suited to nonhomogeneous linear DEs when the input function $g(x)$ is something other than one of the four basic types highlighted in color on page 140. For example, if $P(x)$ is a polynomial, then continued differentiation of $P(x)e^{\alpha x}\sin\beta x$ will generate an independent set containing only a *finite* number of functions—all of the same type, namely, a polynomial times $e^{\alpha x}\sin\beta x$ or a polynomial times $e^{\alpha x}\cos\beta x$. On the other hand, repeated differentiation of input functions such as $g(x) = \ln x$ or $g(x) = \tan^{-1}x$ generates an independent set containing an *infinite* number of functions:

$$\textit{derivatives of } \ln x: \quad \frac{1}{x}, \frac{-1}{x^2}, \frac{2}{x^3}, \ldots,$$

$$\textit{derivatives of } \tan^{-1}x: \quad \frac{1}{1+x^2}, \frac{-2x}{(1+x^2)^2}, \frac{-2+6x^2}{(1+x^2)^3}, \ldots.$$

EXERCISES 4.4

Answers to selected odd-numbered problems begin on page ANS-5.

In Problems 1–26 solve the given differential equation by undetermined coefficients.

1. $y'' + 3y' + 2y = 6$
2. $4y'' + 9y = 15$
3. $y'' - 10y' + 25y = 30x + 3$
4. $y'' + y' - 6y = 2x$
5. $\frac{1}{4}y'' + y' + y = x^2 - 2x$
6. $y'' - 8y' + 20y = 100x^2 - 26xe^x$
7. $y'' + 3y = -48x^2e^{3x}$
8. $4y'' - 4y' - 3y = \cos 2x$
9. $y'' - y' = -3$
10. $y'' + 2y' = 2x + 5 - e^{-2x}$
11. $y'' - y' + \frac{1}{4}y = 3 + e^{x/2}$
12. $y'' - 16y = 2e^{4x}$
13. $y'' + 4y = 3\sin 2x$
14. $y'' - 4y = (x^2 - 3)\sin 2x$
15. $y'' + y = 2x\sin x$
16. $y'' - 5y' = 2x^3 - 4x^2 - x + 6$
17. $y'' - 2y' + 5y = e^x\cos 2x$
18. $y'' - 2y' + 2y = e^{2x}(\cos x - 3\sin x)$
19. $y'' + 2y' + y = \sin x + 3\cos 2x$
20. $y'' + 2y' - 24y = 16 - (x + 2)e^{4x}$
21. $y''' - 6y'' = 3 - \cos x$
22. $y''' - 2y'' - 4y' + 8y = 6xe^{2x}$
23. $y''' - 3y'' + 3y' - y = x - 4e^x$
24. $y''' - y'' - 4y' + 4y = 5 - e^x + e^{2x}$
25. $y^{(4)} + 2y'' + y = (x - 1)^2$
26. $y^{(4)} - y'' = 4x + 2xe^{-x}$

In Problems 27–36 solve the given initial-value problem.

27. $y'' + 4y = -2, \quad y(\pi/8) = \frac{1}{2}, y'(\pi/8) = 2$
28. $2y'' + 3y' - 2y = 14x^2 - 4x - 11, \quad y(0) = 0, y'(0) = 0$
29. $5y'' + y' = -6x, \quad y(0) = 0, y'(0) = -10$
30. $y'' + 4y' + 4y = (3 + x)e^{-2x}, \quad y(0) = 2, y'(0) = 5$
31. $y'' + 4y' + 5y = 35e^{-4x}, \quad y(0) = -3, y'(0) = 1$

32. $y'' - y = \cosh x, \quad y(0) = 2, y'(0) = 12$

33. $\dfrac{d^2x}{dt^2} + \omega^2 x = F_0 \sin \omega t, \quad x(0) = 0, x'(0) = 0$

34. $\dfrac{d^2x}{dt^2} + \omega^2 x = F_0 \cos \gamma t, \quad x(0) = 0, x'(0) = 0$

35. $y''' - 2y'' + y' = 2 - 24e^x + 40e^{5x}, \quad y(0) = \frac{1}{2}, y'(0) = \frac{5}{2}, y''(0) = -\frac{9}{2}$

36. $y''' + 8y = 2x - 5 + 8e^{-2x}, \quad y(0) = -5, y'(0) = 3, y''(0) = -4$

In Problems 37–40 solve the given boundary-value problem.

37. $y'' + y = x^2 + 1, \quad y(0) = 5, y(1) = 0$

38. $y'' - 2y' + 2y = 2x - 2, \quad y(0) = 0, y(\pi) = \pi$

39. $y'' + 3y = 6x, \quad y(0) = 0, y(1) + y'(1) = 0$

40. $y'' + 3y = 6x, \quad y(0) + y'(0) = 0, y(1) = 0$

In Problems 41 and 42 solve the given initial-value problem in which the input function $g(x)$ is discontinuous. [*Hint*: Solve each problem on two intervals, and then find a solution so that y and y' are continuous at $x = \pi/2$ (Problem 41) and at $x = \pi$ (Problem 42).]

41. $y'' + 4y = g(x), \quad y(0) = 1, y'(0) = 2, \quad$ where

$$g(x) = \begin{cases} \sin x, & 0 \le x \le \pi/2 \\ 0, & x > \pi/2 \end{cases}$$

42. $y'' - 2y' + 10y = g(x), \quad y(0) = 0, y'(0) = 0, \quad$ where

$$g(x) = \begin{cases} 20, & 0 \le x \le \pi \\ 0, & x > \pi \end{cases}$$

Discussion Problems

43. Consider the differential equation $ay'' + by' + cy = e^{kx}$, where a, b, c, and k are constants. The auxiliary equation of the associated homogeneous equation is $am^2 + bm + c = 0$.

(a) If k is not a root of the auxiliary equation, show that we can find a particular solution of the form $y_p = Ae^{kx}$, where $A = 1/(ak^2 + bk + c)$.

(b) If k is a root of the auxiliary equation of multiplicity one, show that we can find a particular solution of the form $y_p = Axe^{kx}$, where $A = 1/(2ak + b)$. Explain how we know that $k \neq -b/(2a)$.

(c) If k is a root of the auxiliary equation of multiplicity two, show that we can find a particular solution of the form $y = Ax^2e^{kx}$, where $A = 1/(2a)$.

44. Discuss how the method of this section can be used to find a particular solution of $y'' + y = \sin x \cos 2x$. Carry out your idea.

45. Without solving, match a solution curve of $y'' + y = f(x)$ shown in the figure with one of the following functions:

(*i*) $f(x) = 1$, (*ii*) $f(x) = e^{-x}$,
(*iii*) $f(x) = e^x$, (*iv*) $f(x) = \sin 2x$,
(*v*) $f(x) = e^x \sin x$, (*vi*) $f(x) = \sin x$.

Briefly discuss your reasoning.

(a)

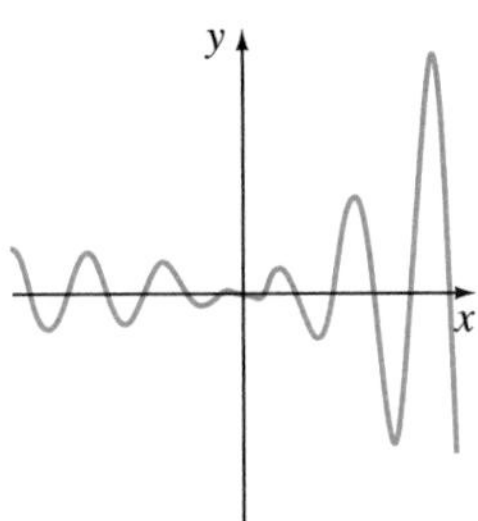

FIGURE 4.4.1 Solution curve

(b)

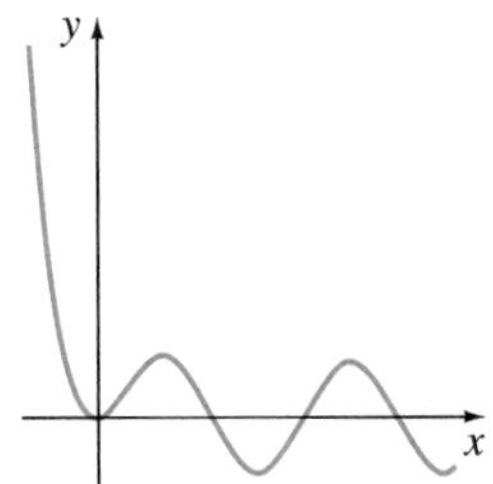

FIGURE 4.4.2 Solution curve

(c)

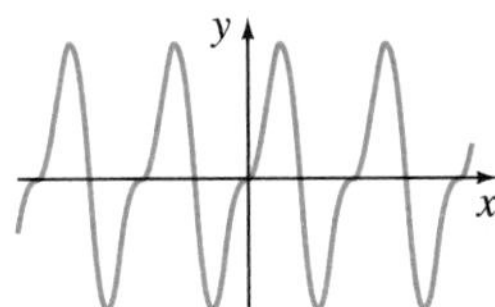

FIGURE 4.4.3 Solution curve

(d)

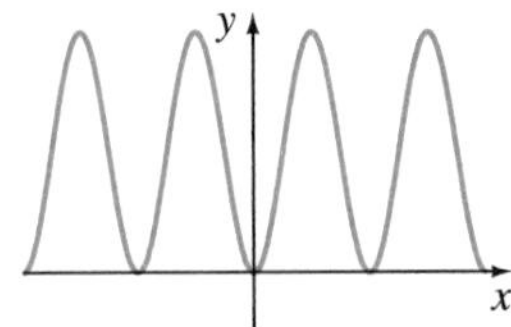

FIGURE 4.4.4 Solution curve

Computer Lab Assignments

In Problems 46 and 47 find a particular solution of the given differential equation. Use a CAS as an aid in carrying out differentiations, simplifications, and algebra.

46. $y'' - 4y' + 8y = (2x^2 - 3x)e^{2x} \cos 2x + (10x^2 - x - 1)e^{2x} \sin 2x$

47. $y^{(4)} + 2y'' + y = 2 \cos x - 3x \sin x$

4.5 UNDETERMINED COEFFICIENTS—ANNIHILATOR APPROACH

REVIEW MATERIAL

- Review Theorems 4.1.6 and 4.1.7 (Section 4.1)

INTRODUCTION We saw in Section 4.1 that an nth-order differential equation can be written

$$a_nD^ny + a_{n-1}D^{n-1}y + \cdots + a_1Dy + a_0y = g(x), \tag{1}$$

where $D^ky = d^ky/dx^k$, $k = 0, 1, \ldots, n$. When it suits our purpose, (1) is also written as $L(y) = g(x)$, where L denotes the linear nth-order differential, or polynomial, operator

$$a_nD^n + a_{n-1}D^{n-1} + \cdots + a_1D + a_0. \tag{2}$$

Not only is the operator notation a helpful shorthand, but also on a very practical level the application of differential operators enables us to justify the somewhat mind-numbing rules for determining the form of particular solution y_p that were presented in the preceding section. In this section there are no special rules; the form of y_p follows almost automatically once we have found an appropriate linear differential operator that *annihilates* $g(x)$ in (1). Before investigating how this is done, we need to examine two concepts.

Factoring Operators When the coefficients a_i, $i = 0, 1, \ldots, n$ are real constants, a linear differential operator (1) can be factored whenever the characteristic polynomial $a_nm^n + a_{n-1}m^{n-1} + \cdots + a_1m + a_0$ factors. In other words, if r_1 is a root of the auxiliary equation

$$a_nm^n + a_{n-1}m^{n-1} + \cdots + a_1m + a_0 = 0,$$

then $L = (D - r_1)\,P(D)$, where the polynomial expression $P(D)$ is a linear differential operator of order $n - 1$. For example, if we treat D as an algebraic quantity, then the operator $D^2 + 5D + 6$ can be factored as $(D + 2)(D + 3)$ or as $(D + 3)(D + 2)$. Thus if a function $y = f(x)$ possesses a second derivative, then

$$(D^2 + 5D + 6)y = (D + 2)(D + 3)y = (D + 3)(D + 2)y.$$

This illustrates a general property:

Factors of a linear differential operator with constant coefficients commute.

A differential equation such as $y'' + 4y' + 4y = 0$ can be written as

$$(D^2 + 4D + 4)y = 0 \quad \text{or} \quad (D + 2)(D + 2)y = 0 \quad \text{or} \quad (D + 2)^2y = 0.$$

Annihilator Operator If L is a linear differential operator with constant coefficients and f is a sufficiently differentiable function such that

$$L(f(x)) = 0,$$

then L is said to be an **annihilator** of the function. For example, a constant function $y = k$ is annihilated by D, since $Dk = 0$. The function $y = x$ is annihilated by the differential operator D^2 since the first and second derivatives of x are 1 and 0, respectively. Similarly, $D^3x^2 = 0$, and so on.

The differential operator D^n annihilates each of the functions

$$1, \quad x, \quad x^2, \quad \ldots, \quad x^{n-1}. \tag{3}$$

As an immediate consequence of (3) and the fact that differentiation can be done term by term, a polynomial

$$c_0 + c_1x + c_2x^2 + \cdots + c_{n-1}x^{n-1} \tag{4}$$

can be annihilated by finding an operator that annihilates the highest power of x.

The functions that are annihilated by a linear nth-order differential operator L are simply those functions that can be obtained from the general solution of the homogeneous differential equation $L(y) = 0$.

The differential operator $(D - \alpha)^n$ annihilates each of the functions

$$e^{\alpha x},\quad xe^{\alpha x},\quad x^2e^{\alpha x},\quad \ldots,\quad x^{n-1}e^{\alpha x}. \tag{5}$$

To see this, note that the auxiliary equation of the homogeneous equation $(D - \alpha)^n y = 0$ is $(m - \alpha)^n = 0$. Since α is a root of multiplicity n, the general solution is

$$y = c_1e^{\alpha x} + c_2xe^{\alpha x} + \cdots + c_nx^{n-1}e^{\alpha x}. \tag{6}$$

EXAMPLE 1 Annihilator Operators

Find a differential operator that annihilates the given function.

(a) $1 - 5x^2 + 8x^3$ **(b)** e^{-3x} **(c)** $4e^{2x} - 10xe^{2x}$

SOLUTION **(a)** From (3) we know that $D^4x^3 = 0$, so it follows from (4) that

$$D^4(1 - 5x^2 + 8x^3) = 0.$$

(b) From (5), with $\alpha = -3$ and $n = 1$, we see that

$$(D + 3)e^{-3x} = 0.$$

(c) From (5) and (6), with $\alpha = 2$ and $n = 2$, we have

$$(D - 2)^2(4e^{2x} - 10xe^{2x}) = 0.$$

≡

When α and β, $\beta > 0$ are real numbers, the quadratic formula reveals that $[m^2 - 2\alpha m + (\alpha^2 + \beta^2)]^n = 0$ has complex roots $\alpha + i\beta$, $\alpha - i\beta$, both of multiplicity n. From the discussion at the end of Section 4.3 we have the next result.

The differential operator $[D^2 - 2\alpha D + (\alpha^2 + \beta^2)]^n$ annihilates each of the functions

$$\begin{array}{lllll} e^{\alpha x}\cos\beta x, & xe^{\alpha x}\cos\beta x, & x^2e^{\alpha x}\cos\beta x, & \ldots, & x^{n-1}e^{\alpha x}\cos\beta x, \\ e^{\alpha x}\sin\beta x, & xe^{\alpha x}\sin\beta x, & x^2e^{\alpha x}\sin\beta x, & \ldots, & x^{n-1}e^{\alpha x}\sin\beta x. \end{array} \tag{7}$$

EXAMPLE 2 Annihilator Operator

Find a differential operator that annihilates $5e^{-x}\cos 2x - 9e^{-x}\sin 2x$.

SOLUTION Inspection of the functions $e^{-x}\cos 2x$ and $e^{-x}\sin 2x$ shows that $\alpha = -1$ and $\beta = 2$. Hence from (7) we conclude that $D^2 + 2D + 5$ will annihilate each function. Since $D^2 + 2D + 5$ is a linear operator, it will annihilate *any* linear combination of these functions such as $5e^{-x}\cos 2x - 9e^{-x}\sin 2x$.

≡

When $\alpha = 0$ and $n = 1$, a special case of (7) is

$$(D^2 + \beta^2)\begin{cases}\cos \beta x \\ \sin \beta x\end{cases} = 0. \tag{8}$$

For example, $D^2 + 16$ will annihilate any linear combination of $\sin 4x$ and $\cos 4x$.

We are often interested in annihilating the sum of two or more functions. As we have just seen in Examples 1 and 2, if L is a linear differential operator such that $L(y_1) = 0$ and $L(y_2) = 0$, then L will annihilate the linear combination $c_1y_1(x) + c_2y_2(x)$. This is a direct consequence of Theorem 4.1.2. Let us now suppose that L_1 and L_2 are linear differential operators with constant coefficients such that L_1 annihilates $y_1(x)$ and L_2 annihilates $y_2(x)$, but $L_1(y_2) \neq 0$ and $L_2(y_1) \neq 0$. Then the *product* of differential operators L_1L_2 annihilates the sum $c_1y_1(x) + c_2y_2(x)$. We can easily demonstrate this, using linearity and the fact that $L_1L_2 = L_2L_1$:

$$\begin{aligned} L_1L_2(y_1 + y_2) &= L_1L_2(y_1) + L_1L_2(y_2) \\ &= L_2L_1(y_1) + L_1L_2(y_2) \\ &= L_2[\underbrace{L_1(y_1)}_{\text{zero}}] + L_1[\underbrace{L_2(y_2)}_{\text{zero}}] = 0. \end{aligned}$$

For example, we know from (3) that D^2 annihilates $7 - x$ and from (8) that $D^2 + 16$ annihilates $\sin 4x$. Therefore the product of operators $D^2(D^2 + 16)$ will annihilate the linear combination $7 - x + 6 \sin 4x$.

Note The differential operator that annihilates a function is not unique. We saw in part (b) of Example 1 that $D + 3$ will annihilate e^{-3x}, but so will differential operators of higher order as long as $D + 3$ is one of the factors of the operator. For example, $(D + 3)(D + 1)$, $(D + 3)^2$, and $D^3(D + 3)$ all annihilate e^{-3x}. (Verify this.) As a matter of course, when we seek a differential annihilator for a function $y = f(x)$, we want the operator of *lowest possible order* that does the job.

Undetermined Coefficients This brings us to the point of the preceding discussion. Suppose that $L(y) = g(x)$ is a linear differential equation with constant coefficients and that the input $g(x)$ consists of finite sums and products of the functions listed in (3), (5), and (7)—that is, $g(x)$ is a linear combination of functions of the form

$$k \text{ (constant)}, \quad x^m, \quad x^m e^{\alpha x}, \quad x^m e^{\alpha x} \cos \beta x, \quad \text{and} \quad x^m e^{\alpha x} \sin \beta x,$$

where m is a nonnegative integer and α and β are real numbers. We now know that such a function $g(x)$ can be annihilated by a differential operator L_1 of lowest order, consisting of a product of the operators D^n, $(D - \alpha)^n$, and $(D^2 - 2\alpha D + \alpha^2 + \beta^2)^n$. Applying L_1 to both sides of the equation $L(y) = g(x)$ yields $L_1L(y) = L_1(g(x)) = 0$. By solving the *homogeneous higher-order* equation $L_1L(y) = 0$, we can discover the *form* of a particular solution y_p for the original *nonhomogeneous* equation $L(y) = g(x)$. We then substitute this assumed form into $L(y) = g(x)$ to find an explicit particular solution. This procedure for determining y_p, called the **method of undetermined coefficients,** is illustrated in the next several examples.

Before proceeding, recall that the general solution of a nonhomogeneous linear differential equation $L(y) = g(x)$ is $y = y_c + y_p$, where y_c is the complementary function—that is, the general solution of the associated homogeneous equation $L(y) = 0$. The general solution of each equation $L(y) = g(x)$ is defined on the interval $(-\infty, \infty)$.

EXAMPLE 3 General Solution Using Undetermined Coefficients

Solve $y'' + 3y' + 2y = 4x^2$. (9)

SOLUTION **Step 1.** First, we solve the homogeneous equation $y'' + 3y' + 2y = 0$. Then, from the auxiliary equation $m^2 + 3m + 2 = (m + 1)(m + 2) = 0$ we find $m_1 = -1$ and $m_2 = -2$, and so the complementary function is

$$y_c = c_1e^{-x} + c_2e^{-2x}.$$

Step 2. Now, since $4x^2$ is annihilated by the differential operator D^3, we see that $D^3(D^2 + 3D + 2)y = 4D^3x^2$ is the same as

$$D^3(D^2 + 3D + 2)y = 0. \tag{10}$$

The auxiliary equation of the fifth-order equation in (10),

$$m^3(m^2 + 3m + 2) = 0 \quad \text{or} \quad m^3(m + 1)(m + 2) = 0,$$

has roots $m_1 = m_2 = m_3 = 0$, $m_4 = -1$, and $m_5 = -2$. Thus its general solution must be

$$y = c_1 + c_2x + c_3x^2 + \boxed{c_4e^{-x} + c_5e^{-2x}}. \tag{11}$$

The terms in the shaded box in (11) constitute the complementary function of the original equation (9). We can then argue that a particular solution y_p of (9) should also satisfy equation (10). This means that the terms remaining in (11) must be the basic form of y_p:

$$y_p = A + Bx + Cx^2, \tag{12}$$

where, for convenience, we have replaced c_1, c_2, and c_3 by A, B, and C, respectively. For (12) to be a particular solution of (9), it is necessary to find *specific* coefficients A, B, and C. Differentiating (12), we have

$$y_p' = B + 2Cx, \qquad y_p'' = 2C,$$

and substitution into (9) then gives

$$y_p'' + 3y_p' + 2y_p = 2C + 3B + 6Cx + 2A + 2Bx + 2Cx^2 = 4x^2.$$

Because the last equation is supposed to be an identity, the coefficients of like powers of x must be equal:

equal

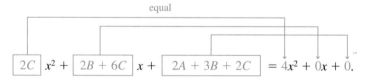

$$\boxed{2C}\,x^2 + \boxed{2B + 6C}\,x + \boxed{2A + 3B + 2C} = 4x^2 + 0x + 0.$$

That is $\quad 2C = 4, \quad 2B + 6C = 0, \quad 2A + 3B + 2C = 0.$ (13)

Solving the equations in (13) gives $A = 7$, $B = -6$, and $C = 2$. Thus $y_p = 7 - 6x + 2x^2$.

Step 3. The general solution of the equation in (9) is $y = y_c + y_p$ or

$$y = c_1e^{-x} + c_2e^{-2x} + 7 - 6x + 2x^2.$$

≡

EXAMPLE 4 General Solution Using Undetermined Coefficients

Solve $y'' - 3y' = 8e^{3x} + 4\sin x$. (14)

SOLUTION Step 1. The auxiliary equation for the associated homogeneous equation $y'' - 3y' = 0$ is $m^2 - 3m = m(m - 3) = 0$, so $y_c = c_1 + c_2e^{3x}$.

Step 2. Now, since $(D - 3)e^{3x} = 0$ and $(D^2 + 1)\sin x = 0$, we apply the differential operator $(D - 3)(D^2 + 1)$ to both sides of (14):

$$(D - 3)(D^2 + 1)(D^2 - 3D)y = 0. \tag{15}$$

The auxiliary equation of (15) is

$$(m - 3)(m^2 + 1)(m^2 - 3m) = 0 \qquad \text{or} \qquad m(m - 3)^2(m^2 + 1) = 0.$$

Thus $$y = \boxed{c_1 + c_2e^{3x}} + c_3xe^{3x} + c_4\cos x + c_5\sin x.$$

After excluding the linear combination of terms in the box that corresponds to y_c, we arrive at the form of y_p:

$$y_p = Axe^{3x} + B\cos x + C\sin x.$$

Substituting y_p in (14) and simplifying yield

$$y_p'' - 3y_p' = 3Ae^{3x} + (-B - 3C)\cos x + (3B - C)\sin x = 8e^{3x} + 4\sin x.$$

Equating coefficients gives $3A = 8$, $-B - 3C = 0$, and $3B - C = 4$. We find $A = \frac{8}{3}$, $B = \frac{6}{5}$, and $C = -\frac{2}{5}$, and consequently,

$$y_p = \frac{8}{3}xe^{3x} + \frac{6}{5}\cos x - \frac{2}{5}\sin x.$$

Step 3. The general solution of (14) is then

$$y = c_1 + c_2e^{3x} + \frac{8}{3}xe^{3x} + \frac{6}{5}\cos x - \frac{2}{5}\sin x. \qquad \equiv$$

EXAMPLE 5 General Solution Using Undetermined Coefficients

Solve $y'' + y = x\cos x - \cos x$. (16)

SOLUTION The complementary function is $y_c = c_1\cos x + c_2\sin x$. Now by comparing $\cos x$ and $x\cos x$ with the functions in the first row of (7), we see that $\alpha = 0$ and $n = 1$, and so $(D^2 + 1)^2$ is an annihilator for the right-hand member of the equation in (16). Applying this operator to the differential equation gives

$$(D^2 + 1)^2(D^2 + 1)y = 0 \qquad \text{or} \qquad (D^2 + 1)^3y = 0.$$

Since i and $-i$ are both complex roots of multiplicity 3 of the auxiliary equation of the last differential equation, we conclude that

$$y = \boxed{c_1\cos x + c_2\sin x} + c_3x\cos x + c_4x\sin x + c_5x^2\cos x + c_6x^2\sin x.$$

We substitute

$$y_p = Ax\cos x + Bx\sin x + Cx^2\cos x + Ex^2\sin x$$

into (16) and simplify:

$$\begin{aligned} y_p'' + y_p &= 4\,Ex\cos x - 4\,Cx\sin x + (2B + 2C)\cos x + (-2A + 2E)\sin x \\ &= x\cos x - \cos x. \end{aligned}$$

Equating coefficients gives the equations $4E = 1$, $-4C = 0$, $2B + 2C = -1$, and $-2A + 2E = 0$, from which we find $A = \frac{1}{4}$, $B = -\frac{1}{2}$, $C = 0$, and $E = \frac{1}{4}$. Hence the general solution of (16) is

$$y = c_1 \cos x + c_2 \sin x + \frac{1}{4} x \cos x - \frac{1}{2} x \sin x + \frac{1}{4} x^2 \sin x. \equiv$$

EXAMPLE 6 Form of a Particular Solution

Determine the form of a particular solution for

$$y'' - 2y' + y = 10e^{-2x} \cos x. \tag{17}$$

SOLUTION The complementary function for the given equation is $y_c = c_1 e^x + c_2 x e^x$.

Now from (7), with $\alpha = -2$, $\beta = 1$, and $n = 1$, we know that

$$(D^2 + 4D + 5)e^{-2x} \cos x = 0.$$

Applying the operator $D^2 + 4D + 5$ to (17) gives

$$(D^2 + 4D + 5)(D^2 - 2D + 1)y = 0. \tag{18}$$

Since the roots of the auxiliary equation of (18) are $-2 - i$, $-2 + i$, 1, and 1, we see from

$$y = c_1 e^x + c_2 x e^x + c_3 e^{-2x} \cos x + c_4 e^{-2x} \sin x$$

that a particular solution of (17) can be found with the form

$$y_p = Ae^{-2x} \cos x + Be^{-2x} \sin x. \equiv$$

EXAMPLE 7 Form of a Particular Solution

Determine the form of a particular solution for

$$y''' - 4y'' + 4y' = 5x^2 - 6x + 4x^2 e^{2x} + 3e^{5x}. \tag{19}$$

SOLUTION Observe that

$$D^3(5x^2 - 6x) = 0, \quad (D - 2)^3 x^2 e^{2x} = 0, \quad \text{and} \quad (D - 5)e^{5x} = 0.$$

Therefore $D^3(D - 2)^3(D - 5)$ applied to (19) gives

$$D^3(D - 2)^3(D - 5)(D^3 - 4D^2 + 4D)y = 0$$

or

$$D^4(D - 2)^5(D - 5)y = 0.$$

The roots of the auxiliary equation for the last differential equation are easily seen to be 0, 0, 0, 0, 2, 2, 2, 2, 2, and 5. Hence

$$y = c_1 + c_2 x + c_3 x^2 + c_4 x^3 + c_5 e^{2x} + c_6 x e^{2x} + c_7 x^2 e^{2x} + c_8 x^3 e^{2x} + c_9 x^4 e^{2x} + c_{10} e^{5x}. \tag{20}$$

Because the linear combination $c_1 + c_5 e^{2x} + c_6 x e^{2x}$ corresponds to the complementary function of (19), the remaining terms in (20) give the form of a particular solution of the differential equation:

$$y_p = Ax + Bx^2 + Cx^3 + Ex^2 e^{2x} + Fx^3 e^{2x} + Gx^4 e^{2x} + He^{5x}. \equiv$$

≡ **Summary of the Method** For your convenience the method of undetermined coefficients is summarized as follows.

UNDETERMINED COEFFICIENTS—ANNIHILATOR APPROACH

The differential equation $L(y) = g(x)$ has constant coefficients, and the function $g(x)$ consists of finite sums and products of constants, polynomials, exponential functions $e^{\alpha x}$, sines, and cosines.

(i) Find the complementary solution y_c for the homogeneous equation $L(y) = 0$.

(ii) Operate on both sides of the nonhomogeneous equation $L(y) = g(x)$ with a differential operator L_1 that annihilates the function $g(x)$.

(iii) Find the general solution of the higher-order homogeneous differential equation $L_1L(y) = 0$.

(iv) Delete from the solution in step *(iii)* all those terms that are duplicated in the complementary solution y_c found in step *(i)*. Form a linear combination y_p of the terms that remain. This is the form of a particular solution of $L(y) = g(x)$.

(v) Substitute y_p found in step *(iv)* into $L(y) = g(x)$. Match coefficients of the various functions on each side of the equality, and solve the resulting system of equations for the unknown coefficients in y_p.

(vi) With the particular solution found in step *(v)*, form the general solution $y = y_c + y_p$ of the given differential equation.

REMARKS

The method of undetermined coefficients is not applicable to linear differential equations with variable coefficients nor is it applicable to linear equations with constant coefficients when $g(x)$ is a function such as

$$g(x) = \ln x, \qquad g(x) = \frac{1}{x}, \qquad g(x) = \tan x, \qquad g(x) = \sin^{-1} x,$$

and so on. Differential equations in which the input $g(x)$ is a function of this last kind will be considered in the next section.

EXERCISES 4.5

Answers to selected odd-numbered problems begin on page ANS-5.

In Problems 1–10 write the given differential equation in the form $L(y) = g(x)$, where L is a linear differential operator with constant coefficients. If possible, factor L.

1. $9y'' - 4y = \sin x$

2. $y'' - 5y = x^2 - 2x$

3. $y'' - 4y' - 12y = x - 6$

4. $2y'' - 3y' - 2y = 1$

5. $y''' + 10y'' + 25y' = e^x$

6. $y''' + 4y' = e^x \cos 2x$

7. $y''' + 2y'' - 13y' + 10y = xe^{-x}$

8. $y''' + 4y'' + 3y' = x^2 \cos x - 3x$

9. $y^{(4)} + 8y' = 4$

10. $y^{(4)} - 8y'' + 16y = (x^3 - 2x)e^{4x}$

In Problems 11–14 verify that the given differential operator annihilates the indicated functions.

11. D^4; $y = 10x^3 - 2x$

12. $2D - 1$; $y = 4e^{x/2}$

13. $(D - 2)(D + 5)$; $y = e^{2x} + 3e^{-5x}$

14. $D^2 + 64$; $y = 2 \cos 8x - 5 \sin 8x$

In Problems 15–26 find a linear differential operator that annihilates the given function.

15. $1 + 6x - 2x^3$

16. $x^3(1 - 5x)$

17. $1 + 7e^{2x}$

18. $x + 3xe^{6x}$

19. $\cos 2x$

20. $1 + \sin x$

21. $13x + 9x^2 - \sin 4x$

22. $8x - \sin x + 10 \cos 5x$

23. $e^{-x} + 2xe^x - x^2e^x$

24. $(2 - e^x)^2$

25. $3 + e^x \cos 2x$

26. $e^{-x} \sin x - e^{2x} \cos x$

In Problems 27–34 find linearly independent functions that are annihilated by the given differential operator.

27. D^5

28. $D^2 + 4D$

29. $(D - 6)(2D + 3)$

30. $D^2 - 9D - 36$

31. $D^2 + 5$

32. $D^2 - 6D + 10$

33. $D^3 - 10D^2 + 25D$

34. $D^2(D - 5)(D - 7)$

In Problems 35–64 solve the given differential equation by undetermined coefficients.

35. $y'' - 9y = 54$

36. $2y'' - 7y' + 5y = -29$

37. $y'' + y' = 3$

38. $y''' + 2y'' + y' = 10$

39. $y'' + 4y' + 4y = 2x + 6$

40. $y'' + 3y' = 4x - 5$

41. $y''' + y'' = 8x^2$

42. $y'' - 2y' + y = x^3 + 4x$

43. $y'' - y' - 12y = e^{4x}$

44. $y'' + 2y' + 2y = 5e^{6x}$

45. $y'' - 2y' - 3y = 4e^x - 9$

46. $y'' + 6y' + 8y = 3e^{-2x} + 2x$

47. $y'' + 25y = 6 \sin x$

48. $y'' + 4y = 4 \cos x + 3 \sin x - 8$

49. $y'' + 6y' + 9y = -xe^{4x}$

50. $y'' + 3y' - 10y = x(e^x + 1)$

51. $y'' - y = x^2e^x + 5$

52. $y'' + 2y' + y = x^2e^{-x}$

53. $y'' - 2y' + 5y = e^x \sin x$

54. $y'' + y' + \frac{1}{4}y = e^x(\sin 3x - \cos 3x)$

55. $y'' + 25y = 20 \sin 5x$

56. $y'' + y = 4 \cos x - \sin x$

57. $y'' + y' + y = x \sin x$

58. $y'' + 4y = \cos^2 x$

59. $y''' + 8y'' = -6x^2 + 9x + 2$

60. $y''' - y'' + y' - y = xe^x - e^{-x} + 7$

61. $y''' - 3y'' + 3y' - y = e^x - x + 16$

62. $2y''' - 3y'' - 3y' + 2y = (e^x + e^{-x})^2$

63. $y^{(4)} - 2y''' + y'' = e^x + 1$

64. $y^{(4)} - 4y'' = 5x^2 - e^{2x}$

In Problems 65–72 solve the given initial-value problem.

65. $y'' - 64y = 16, \quad y(0) = 1, y'(0) = 0$

66. $y'' + y' = x, \quad y(0) = 1, y'(0) = 0$

67. $y'' - 5y' = x - 2, \quad y(0) = 0, y'(0) = 2$

68. $y'' + 5y' - 6y = 10e^{2x}, \quad y(0) = 1, y'(0) = 1$

69. $y'' + y = 8 \cos 2x - 4 \sin x, \quad y(\pi/2) = -1, y'(\pi/2) = 0$

70. $y''' - 2y'' + y' = xe^x + 5, \quad y(0) = 2, y'(0) = 2, y''(0) = -1$

71. $y'' - 4y' + 8y = x^3, \quad y(0) = 2, y'(0) = 4$

72. $y^{(4)} - y''' = x + e^x, \quad y(0) = 0, y'(0) = 0, y''(0) = 0, y'''(0) = 0$

Discussion Problems

73. Suppose L is a linear differential operator that factors but has variable coefficients. Do the factors of L commute? Defend your answer.

4.6 VARIATION OF PARAMETERS

REVIEW MATERIAL

- Basic integration formulas and techniques from calculus
- Review Section 2.3

INTRODUCTION We pointed out in the discussions in Sections 4.4 and 4.5 that the method of undetermined coefficients has two inherent weaknesses that limit its wider application to linear equations: The DE must have constant coefficients and the input function $g(x)$ must be of the type listed in Table 4.4.1. In this section we examine a method for determining a particular solution y_p of a nonhomogeneous linear DE that has, in theory, no such restrictions on it. This method, due to the eminent astronomer and mathematician **Joseph Louis Lagrange** (1736–1813), is known as **variation of parameters.**

Before examining this powerful method for higher-order equations we revisit the solution of linear first-order differential equations that have been put into standard form. The discussion under the first heading in this section is optional and is intended to motivate the main discussion of this section that starts under the second heading. If pressed for time this motivational material could be assigned for reading.

Linear First-Order DEs Revisited In Section 2.3 we saw that the general solution of a linear first-order differential equation $a_1(x)y' + a_0(x)y = g(x)$ can be found by first rewriting it in the standard form

$$\frac{dy}{dx} + P(x)y = f(x) \tag{1}$$

and assuming that $P(x)$ and $f(x)$are continuous on an common interval I. Using the integrating factor method, the general solution of (1) on the interval I, was found to be

See (4) of Section 2.3. ▶

$$y = c_1 e^{-\int P(x)dx} + e^{-\int P(x)dx} \int e^{\int P(x)dx} f(x)\,dx.$$

The foregoing solution has the same form as that given in Theorem 4.1.6, namely, $y = y_c + y_p$. In this case $y_c = c_1 e^{-\int P(x)dx}$ is a solution of the associated homogeneous equation

$$\frac{dy}{dx} + P(x)y = 0 \tag{2}$$

and

$$y_p = e^{-\int P(x)dx} \int e^{\int P(x)dx} f(x)\,dx \tag{3}$$

is a particular solution of the nonhomogeneous equation (1). As a means of motivating a method for solving nonhomogeneous linear equations of higher-order we propose to rederive the particular solution (3) by a method known as **variation of parameters.**

The basic procedure is that used in Section 4.2. ▶

Suppose that y_1 is a known solution of the homogeneous equation (2), that is,

$$\frac{dy_1}{dx} + P(x)y_1 = 0. \tag{4}$$

It is easily shown that $y_1 = e^{-\int P(x)dx}$ is a solution of (4) and because the equation is linear, $c_1 y_1(x)$ is its general solution. Variation of parameters consists of finding a particular solution of (1) of the form $y_p = u_1(x)y_1(x)$. In other words, we have replaced the *parameter* c_1 by a *function* u_1.

Substituting $y_p = u_1 y_1$ into (1) and using the Product Rule gives

$$\frac{d}{dx}\left[u_1 y_1\right] + P(x)u_1 y_1 = f(x)$$

$$u_1 \frac{dy_1}{dx} + y_1 \frac{du_1}{dx} + P(x)u_1 y_1 = f(x)$$

$$u_1 \overbrace{\left[\frac{dy_1}{dx} + P(x)y_1\right]}^{0 \text{ because of } (4)} + y_1 \frac{du_1}{dx} = f(x)$$

so

$$y_1 \frac{du_1}{dx} = f(x).$$

By separating variables and integrating, we find u_1:

$$du_1 = \frac{f(x)}{y_1(x)}dx \quad \text{yields} \quad u_1 = \int \frac{f(x)}{y_1(x)}dx.$$

Hence the sought-after particular solution is

$$y_p = u_1 y_1 = y_1 \int \frac{f(x)}{y_1(x)}dx.$$

From the fact that $y_1 = e^{-\int P(x)dx}$ we see the last result is identical to (3).

Linear Second-Order DEs Next we consider the case of a linear second-order equation

$$a_2(x)y'' + a_1(x)y' + a_0(x)y = g(x), \tag{5}$$

although, as we shall see, variation of parameters extends to higher-order equations. The method again begins by putting (5) into the standard form

$$y'' + P(x)y' + Q(x)y = f(x) \tag{6}$$

by dividing by the leading coefficient $a_2(x)$. In (6) we suppose that coefficient functions $P(x)$, $Q(x)$, and $f(x)$ are continuous on some common interval I. As we have already seen in Section 4.3, there is no difficulty in obtaining the complementary solution $y_c = c_1y_1(x) + c_2y_2(x)$, the general solution of the associated homogeneous equation of (6), when the coefficients are constants. Analogous to the preceding discussion, we now ask: Can the parameters c_1 and c_2 in y_c can be replaced with functions u_1 and u_2, or "variable parameters," so that

$$y = u_1(x)y_1(x) + u_2(x)y_2(x) \tag{7}$$

is a particular solution of (6)? To answer this question we substitute (7) into (6). Using the Product Rule to differentiate y_p twice, we get

$$y_p' = u_1y_1' + y_1u_1' + u_2y_2' + y_2u_2'$$

$$y_p'' = u_1y_1'' + y_1'u_1' + y_1u_1'' + u_1'y_1' + u_2y_2'' + y_2'u_2' + y_2u_2'' + u_2'y_2'.$$

Substituting (7) and the foregoing derivatives into (6) and grouping terms yields

$$\begin{aligned} y_p'' + P(x)y_p' + Q(x)y_p &= u_1\overbrace{[y_1'' + Py_1' + Qy_1]}^{\text{zero}} + u_2\overbrace{[y_2'' + Py_2' + Qy_2]}^{\text{zero}} + y_1u_1'' + u_1'y_1' \\ &\quad + y_2u_2'' + u_2'y_2' + P[y_1u_1' + y_2u_2'] + y_1'u_1' + y_2'u_2' \\ &= \frac{d}{dx}[y_1u_1'] + \frac{d}{dx}[y_2u_2'] + P[y_1u_1' + y_2u_2'] + y_1'u_1' + y_2'u_2' \\ &= \frac{d}{dx}[y_1u_1' + y_2u_2'] + P[y_1u_1' + y_2u_2'] + y_1'u_1' + y_2'u_2' = f(x). \end{aligned} \tag{8}$$

Because we seek to determine two unknown functions u_1 and u_2, reason dictates that we need two equations. We can obtain these equations by making the further assumption that the functions u_1 and u_2 satisfy $y_1u_1' + y_2u_2' = 0$. This assumption does not come out of the blue but is prompted by the first two terms in (8), since if we demand that $y_1u_1' + y_2u_2' = 0$, then (8) reduces to $y_1'u_1' + y_2'u_2' = f(x)$. We now have our desired two equations, albeit two equations for determining the derivatives u_1' and u_2'. By Cramer's Rule, the solution of the system

$$y_1u_1' + y_2u_2' = 0$$

$$y_1'u_1' + y_2'u_2' = f(x)$$

can be expressed in terms of determinants:

$$u_1' = \frac{W_1}{W} = -\frac{y_2f(x)}{W} \quad \text{and} \quad u_2' = \frac{W_2}{W} = \frac{y_1f(x)}{W}, \tag{9}$$

where
$$W = \begin{vmatrix} y_1 & y_2 \\ y_1' & y_2' \end{vmatrix}, \quad W_1 = \begin{vmatrix} 0 & y_2 \\ f(x) & y_2' \end{vmatrix}, \quad W_2 = \begin{vmatrix} y_1 & 0 \\ y_1' & f(x) \end{vmatrix}. \tag{10}$$

The functions u_1 and u_2 are found by integrating the results in (9). The determinant W is recognized as the Wronskian of y_1 and y_2. By linear independence of y_1 and y_2 on I, we know that $W(y_1(x), y_2(x)) \neq 0$ for every x in the interval.

Summary of the Method Usually, it is not a good idea to memorize formulas in lieu of understanding a procedure. However, the foregoing procedure is too long and complicated to use each time we wish to solve a differential equation. In this case it is more efficient to simply use the formulas in (9). Thus to solve $a_2y'' + a_1y' + a_0y = g(x)$, first find the complementary function $y_c = c_1y_1 + c_2y_2$ and then compute the Wronskian $W(y_1(x), y_2(x))$. By dividing by a_2, we put the equation into the standard form $y'' + Py' + Qy = f(x)$ to determine $f(x)$. We find u_1 and u_2 by integrating $u_1' = W_1/W$ and $u_2' = W_2/W$, where W_1 and W_2 are defined as in (10). A particular solution is $y_p = u_1y_1 + u_2y_2$. The general solution of the equation is then $y = y_c + y_p$.

EXAMPLE 1 General Solution Using Variation of Parameters

Solve $y'' - 4y' + 4y = (x + 1)e^{2x}$.

SOLUTION From the auxiliary equation $m^2 - 4m + 4 = (m - 2)^2 = 0$ we have $y_c = c_1e^{2x} + c_2xe^{2x}$. With the identifications $y_1 = e^{2x}$ and $y_2 = xe^{2x}$, we next compute the Wronskian:

$$W(e^{2x}, xe^{2x}) = \begin{vmatrix} e^{2x} & xe^{2x} \\ 2e^{2x} & 2xe^{2x} + e^{2x} \end{vmatrix} = e^{4x}.$$

Since the given differential equation is already in form (6) (that is, the coefficient of y'' is 1), we identify $f(x) = (x + 1)e^{2x}$. From (10) we obtain

$$W_1 = \begin{vmatrix} 0 & xe^{2x} \\ (x+1)e^{2x} & 2xe^{2x} + e^{2x} \end{vmatrix} = -(x + 1)xe^{4x}, \qquad W_2 = \begin{vmatrix} e^{2x} & 0 \\ 2e^{2x} & (x+1)e^{2x} \end{vmatrix} = (x + 1)e^{4x},$$

and so from (9)

$$u_1' = -\frac{(x + 1)xe^{4x}}{e^{4x}} = -x^2 - x, \qquad u_2' = \frac{(x + 1)e^{4x}}{e^{4x}} = x + 1.$$

It follows that $u_1 = -\frac{1}{3}x^3 - \frac{1}{2}x^2$ and $u_2 = \frac{1}{2}x^2 + x$. Hence

$$y_p = \left(-\frac{1}{3}x^3 - \frac{1}{2}x^2\right)e^{2x} + \left(\frac{1}{2}x^2 + x\right)xe^{2x} = \frac{1}{6}x^3e^{2x} + \frac{1}{2}x^2e^{2x}$$

and

$$y = y_c + y_p = c_1e^{2x} + c_2xe^{2x} + \frac{1}{6}x^3e^{2x} + \frac{1}{2}x^2e^{2x}. \qquad \equiv$$

EXAMPLE 2 General Solution Using Variation of Parameters

Solve $4y'' + 36y = \csc 3x$.

SOLUTION We first put the equation in the standard form (6) by dividing by 4:

$$y'' + 9y = \frac{1}{4}\csc 3x.$$

Because the roots of the auxiliary equation $m^2 + 9 = 0$ are $m_1 = 3i$ and $m_2 = -3i$, the complementary function is $y_c = c_1 \cos 3x + c_2 \sin 3x$. Using $y_1 = \cos 3x$, $y_2 = \sin 3x$, and $f(x) = \frac{1}{4}\csc 3x$, we obtain

$$W(\cos 3x, \sin 3x) = \begin{vmatrix} \cos 3x & \sin 3x \\ -3\sin 3x & 3\cos 3x \end{vmatrix} = 3,$$

$$W_1 = \begin{vmatrix} 0 & \sin 3x \\ \frac{1}{4}\csc 3x & 3\cos 3x \end{vmatrix} = -\frac{1}{4}, \qquad W_2 = \begin{vmatrix} \cos 3x & 0 \\ -3\sin 3x & \frac{1}{4}\csc 3x \end{vmatrix} = \frac{1}{4}\frac{\cos 3x}{\sin 3x}.$$

Integrating $\quad u_1' = \dfrac{W_1}{W} = -\dfrac{1}{12} \quad$ and $\quad u_2' = \dfrac{W_2}{W} = \dfrac{1}{12}\dfrac{\cos 3x}{\sin 3x}$

gives $u_1 = -\frac{1}{12}x$ and $u_2 = \frac{1}{36}\ln|\sin 3x|$. Thus a particular solution is

$$y_p = -\frac{1}{12}x\cos 3x + \frac{1}{36}(\sin 3x)\ln|\sin 3x|.$$

The general solution of the equation is

$$y = y_c + y_p = c_1\cos 3x + c_2\sin 3x - \frac{1}{12}x\cos 3x + \frac{1}{36}(\sin 3x)\ln|\sin 3x|. \qquad (11)$$

Equation (11) represents the general solution of the differential equation on, say, the interval $(0, \pi/6)$.

Constants of Integration When computing the indefinite integrals of u_1' and u_2', we need not introduce any constants. This is because

$$\begin{aligned} y = y_c + y_p &= c_1y_1 + c_2y_2 + (u_1 + a_1)y_1 + (u_2 + b_1)y_2 \\ &= (c_1 + a_1)y_1 + (c_2 + b_1)y_2 + u_1y_1 + u_2y_2 \\ &= C_1y_1 + C_2y_2 + u_1y_1 + u_2y_2. \end{aligned}$$

EXAMPLE 3 General Solution Using Variation of Parameters

Solve $y'' - y = \dfrac{1}{x}$.

SOLUTION The auxiliary equation $m^2 - 1 = 0$ yields $m_1 = -1$ and $m_2 = 1$. Therefore $y_c = c_1e^x + c_2e^{-x}$. Now $W(e^x, e^{-x}) = -2$, and

$$u_1' = -\frac{e^{-x}(1/x)}{-2}, \qquad u_1 = \frac{1}{2}\int_{x_0}^{x}\frac{e^{-t}}{t}\,dt,$$

$$u_2' = \frac{e^{x}(1/x)}{-2}, \qquad u_2 = -\frac{1}{2}\int_{x_0}^{x}\frac{e^{t}}{t}\,dt.$$

Since the foregoing integrals are nonelementary, we are forced to write

$$y_p = \frac{1}{2}e^x\int_{x_0}^{x}\frac{e^{-t}}{t}\,dt - \frac{1}{2}e^{-x}\int_{x_0}^{x}\frac{e^{t}}{t}\,dt,$$

and so $\quad y = y_c + y_p = c_1e^x + c_2e^{-x} + \dfrac{1}{2}e^x\displaystyle\int_{x_0}^{x}\frac{e^{-t}}{t}\,dt - \dfrac{1}{2}e^{-x}\int_{x_0}^{x}\frac{e^{t}}{t}\,dt. \qquad (12)$

In Example 3 we can integrate on any interval $[x_0, x]$ that does not contain the origin. We will solve the equation in Example 3 by an alternative method in Section 4.8.

Higher-Order Equations The method that we have just examined for non homogeneous second-order differential equations can be generalized to linear nth-order equations that have been put into the standard form

$$y^{(n)} + P_{n-1}(x)y^{(n-1)} + \cdots + P_1(x)y' + P_0(x)y = f(x). \qquad (13)$$

If $y_c = c_1y_1 + c_2y_2 + \cdots + c_ny_n$ is the complementary function for (13), then a particular solution is

$$y_p = u_1(x)y_1(x) + u_2(x)y_2(x) + \cdots + u_n(x)y_n(x),$$

where the u'_k, $k = 1, 2, \ldots, n$ are determined by the n equations

$$\begin{aligned} y_1u'_1 + \quad y_2u'_2 + \cdots + \quad y_nu'_n &= 0 \\ y'_1u'_1 + \quad y'_2u'_2 + \cdots + \quad y'_nu'_n &= 0 \\ \vdots \qquad\qquad\qquad\qquad &\vdots \\ y_1^{(n-1)}u'_1 + y_2^{(n-1)}u'_2 + \cdots + y_n^{(n-1)}u'_n &= f(x). \end{aligned} \tag{14}$$

The first $n - 1$ equations in this system, like $y_1u'_1 + y_2u'_2 = 0$ in (8), are assumptions that are made to simplify the resulting equation after $y_p = u_1(x)y_1(x) + \cdots + u_n(x)y_n(x)$ is substituted in (13). In this case Cramer's Rule gives

$$u'_k = \frac{W_k}{W}, \quad k = 1, 2, \ldots, n,$$

where W is the Wronskian of $y_1, y_2, \ldots, y_n$ and W_k is the determinant obtained by replacing the kth column of the Wronskian by the column consisting of the right-hand side of (14)—that is, the column consisting of $(0, 0, \ldots, f(x))$. When $n = 2$, we get (9). When $n = 3$, the particular solution is $y_p = u_1y_1 + u_2y_2 + u_3y_3$, where y_1, y_2, and y_3 constitute a linearly independent set of solutions of the associated homogeneous DE and u_1, u_2, u_3 are determined from

$$u'_1 = \frac{W_1}{W}, \qquad u'_2 = \frac{W_2}{W}, \qquad u'_3 = \frac{W_3}{W}, \tag{15}$$

$$W_1 = \begin{vmatrix} 0 & y_2 & y_3 \\ 0 & y'_2 & y'_3 \\ f(x) & y''_2 & y''_3 \end{vmatrix}, \quad W_2 = \begin{vmatrix} y_1 & 0 & y_3 \\ y'_1 & 0 & y'_3 \\ y''_1 & f(x) & y''_3 \end{vmatrix}, \quad W_3 = \begin{vmatrix} y_1 & y_2 & 0 \\ y'_1 & y'_2 & 0 \\ y''_1 & y''_2 & f(x) \end{vmatrix}, \quad \text{and} \quad W = \begin{vmatrix} y_1 & y_2 & y_3 \\ y'_1 & y'_2 & y'_3 \\ y''_1 & y''_2 & y''_3 \end{vmatrix}.$$

See Problems 25–28 in Exercises 4.6.

REMARKS

(*i*) Variation of parameters has a distinct advantage over the method of undetermined coefficients in that it will *always* yield a particular solution y_p provided that the associated homogeneous equation can be solved. The present method is not limited to a function $f(x)$ that is a combination of the four types listed on page 140. As we shall see in the next section, variation of parameters, unlike undetermined coefficients, is applicable to linear DEs with variable coefficients.

(*ii*) In the problems that follow, do not hesitate to simplify the form of y_p. Depending on how the antiderivatives of u'_1 and u'_2 are found, you might not obtain the same y_p as given in the answer section. For example, in Problem 3 in Exercises 4.6 both $y_p = \frac{1}{2}\sin x - \frac{1}{2}x\cos x$ and $y_p = \frac{1}{4}\sin x - \frac{1}{2}x\cos x$ are valid answers. In either case the general solution $y = y_c + y_p$ simplifies to $y = c_1\cos x + c_2\sin x - \frac{1}{2}x\cos x$. Why?

EXERCISES 4.6

Answers to selected odd-numbered problems begin on page ANS-6.

In Problems 1–18 solve each differential equation by variation of parameters.

1. $y'' + y = \sec x$

2. $y'' + y = \tan x$

3. $y'' + y = \sin x$

4. $y'' + y = \sec\theta\tan\theta$

5. $y'' + y = \cos^2 x$

6. $y'' + y = \sec^2 x$

7. $y'' - y = \cosh x$

8. $y'' - y = \sinh 2x$

9. $y'' - 4y = \dfrac{e^{2x}}{x}$

10. $y'' - 9y = \dfrac{9x}{e^{3x}}$

11. $y'' + 3y' + 2y = \dfrac{1}{1 + e^x}$

12. $y'' - 2y' + y = \dfrac{e^x}{1 + x^2}$

13. $y'' + 3y' + 2y = \sin e^x$

14. $y'' - 2y' + y = e^t \arctan t$

15. $y'' + 2y' + y = e^{-t} \ln t$

16. $2y'' + 2y' + y = 4\sqrt{x}$

17. $3y'' - 6y' + 6y = e^x \sec x$

18. $4y'' - 4y' + y = e^{x/2}\sqrt{1 - x^2}$

In Problems 19–22 solve each differential equation by variation of parameters, subject to the initial conditions $y(0) = 1$, $y'(0) = 0$.

19. $4y'' - y = xe^{x/2}$

20. $2y'' + y' - y = x + 1$

21. $y'' + 2y' - 8y = 2e^{-2x} - e^{-x}$

22. $y'' - 4y' + 4y = (12x^2 - 6x)e^{2x}$

In Problems 23 and 24 the indicated functions are known linearly independent solutions of the associated homogeneous differential equation on $(0, \infty)$. Find the general solution of the given nonhomogeneous equation.

23. $x^2y'' + xy' + \left(x^2 - \frac{1}{4}\right)y = x^{3/2}$;
$y_1 = x^{-1/2} \cos x$, $y_2 = x^{-1/2} \sin x$

24. $x^2y'' + xy' + y = \sec(\ln x)$;
$y_1 = \cos(\ln x)$, $y_2 = \sin(\ln x)$

In Problems 25–28 solve the given third-order differential equation by variation of parameters.

25. $y''' + y' = \tan x$

26. $y''' + 4y' = \sec 2x$

27. $y''' - 2y'' - y' + 2y = e^{4x}$

28. $y''' - 3y'' + 2y' = \dfrac{e^{2x}}{1 + e^x}$

Discussion Problems

In Problems 29 and 30 discuss how the methods of undetermined coefficients and variation of parameters can be combined to solve the given differential equation. Carry out your ideas.

29. $3y'' - 6y' + 30y = 15 \sin x + e^x \tan 3x$

30. $y'' - 2y' + y = 4x^2 - 3 + x^{-1}e^x$

31. What are the intervals of definition of the general solutions in Problems 1, 7, 9, and 18? Discuss why the interval of definition of the general solution in Problem 24 is *not* $(0, \infty)$.

32. Find the general solution of $x^4y'' + x^3y' - 4x^2y = 1$ given that $y_1 = x^2$ is a solution of the associated homogeneous equation.

4.7 CAUCHY-EULER EQUATION

REVIEW MATERIAL

- Review the concept of the auxiliary equation in Section 4.3.

INTRODUCTION The same relative ease with which we were able to find explicit solutions of higher-order linear differential equations with constant coefficients in the preceding sections does not, in general, carry over to linear equations with variable coefficients. We shall see in Chapter 6 that when a linear DE has variable coefficients, the best that we can *usually* expect is to find a solution in the form of an infinite series. However, the type of differential equation that we consider in this section is an exception to this rule; it is a linear equation with variable coefficients whose general solution can always be expressed in terms of powers of x, sines, cosines, and logarithmic functions. Moreover, its method of solution is quite similar to that for constant-coefficient equations in that an auxiliary equation must be solved.

Cauchy-Euler Equation A linear differential equation of the form

$$a_n x^n \frac{d^n y}{dx^n} + a_{n-1} x^{n-1} \frac{d^{n-1} y}{dx^{n-1}} + \cdots + a_1 x \frac{dy}{dx} + a_0 y = g(x),$$

where the coefficients $a_n, a_{n-1}, \ldots, a_0$ are constants, is known as a **Cauchy-Euler equation.** The differential equation is named in honor of two of the most prolific mathematicians of all time. **Augustin-Louis Cauchy** (French, 1789–1857) and **Leonhard Euler** (Swiss, 1707–1783). The observable characteristic of this type of equation is that the degree $k = n, n - 1, \ldots, 1, 0$ of the monomial coefficients x^k matches the order k of differentiation $d^k y/dx^k$:

$$\underbrace{a_n x^n \frac{d^n y}{dx^n}}_{\text{same}} + \underbrace{a_{n-1} x^{n-1} \frac{d^{n-1} y}{dx^{n-1}}}_{\text{same}} + \cdots.$$

As in Section 4.3, we start the discussion with a detailed examination of the forms of the general solutions of the homogeneous second-order equation

$$ax^2 \frac{d^2y}{dx^2} + bx\frac{dy}{dx} + cy = 0. \tag{1}$$

The solution of higher-order equations follows analogously. Also, we can solve the nonhomogeneous equation $ax^2y'' + bxy' + cy = g(x)$ by variation of parameters, once we have determined the complementary function y_c.

Note The coefficient ax^2 of y'' is zero at $x = 0$. Hence to guarantee that the fundamental results of Theorem 4.1.1 are applicable to the Cauchy-Euler equation, we focus our attention on finding the general solutions defined on the interval $(0, \infty)$.

Method of Solution We try a solution of the form $y = x^m$, where m is to be determined. Analogous to what happened when we substituted e^{mx} into a linear equation with constant coefficients, when we substitute x^m, each term of a Cauchy-Euler equation becomes a polynomial in m times x^m, since

$$a_k x^k \frac{d^k y}{dx^k} = a_k x^k m(m-1)(m-2)\cdots(m-k+1)x^{m-k} = a_k m(m-1)(m-2)\cdots(m-k+1)x^m.$$

For example, when we substitute $y = x^m$, the second-order equation becomes

$$ax^2\frac{d^2y}{dx^2} + bx\frac{dy}{dx} + cy = am(m-1)x^m + bmx^m + cx^m = (am(m-1) + bm + c)x^m.$$

Thus $y = x^m$ is a solution of the differential equation whenever m is a solution of the **auxiliary equation**

$$am(m-1) + bm + c = 0 \quad \text{or} \quad am^2 + (b-a)m + c = 0. \tag{2}$$

There are three different cases to be considered, depending on whether the roots of this quadratic equation are real and distinct, real and equal, or complex. In the last case the roots appear as a conjugate pair.

Case I: Distinct Real Roots Let m_1 and m_2 denote the real roots of (1) such that $m_1 \neq m_2$. Then $y_1 = x^{m_1}$ and $y_2 = x^{m_2}$ form a fundamental set of solutions. Hence the general solution is

$$y = c_1x^{m_1} + c_2x^{m_2}. \tag{3}$$

EXAMPLE 1 Distinct Roots

Solve $x^2\dfrac{d^2y}{dx^2} - 2x\dfrac{dy}{dx} - 4y = 0.$

SOLUTION Rather than just memorizing equation (2), it is preferable to assume $y = x^m$ as the solution a few times to understand the origin and the difference between this new form of the auxiliary equation and that obtained in Section 4.3. Differentiate twice,

$$\frac{dy}{dx} = mx^{m-1}, \qquad \frac{d^2y}{dx^2} = m(m-1)x^{m-2},$$

and substitute back into the differential equation:

$$x^2\frac{d^2y}{dx^2} - 2x\frac{dy}{dx} - 4y = x^2 \cdot m(m-1)x^{m-2} - 2x \cdot mx^{m-1} - 4x^m$$

$$= x^m(m(m-1) - 2m - 4) = x^m(m^2 - 3m - 4) = 0$$

if $m^2 - 3m - 4 = 0$. Now $(m+1)(m-4) = 0$ implies $m_1 = -1$, $m_2 = 4$, so $y = c_1x^{-1} + c_2x^4$. ≡

≡ **Case II: Repeated Real Roots** If the roots of (2) are repeated (that is, $m_1 = m_2$), then we obtain only one solution—namely, $y = x^{m_1}$. When the roots of the quadratic equation $am^2 + (b-a)m + c = 0$ are equal, the discriminant of the coefficients is necessarily zero. It follows from the quadratic formula that the root must be $m_1 = -(b-a)/2a$.

Now we can construct a second solution y_2, using (5) of Section 4.2. We first write the Cauchy-Euler equation in the standard form

$$\frac{d^2y}{dx^2} + \frac{b}{ax}\frac{dy}{dx} + \frac{c}{ax^2}y = 0$$

and make the identifications $P(x) = b/ax$ and $\int(b/ax)\,dx = (b/a)\ln x$. Thus

$$y_2 = x^{m_1}\int \frac{e^{-(b/a)\ln x}}{x^{2m_1}}\,dx$$

$$= x^{m_1}\int x^{-b/a} \cdot x^{-2m_1}dx \qquad \leftarrow e^{-(b/a)\ln x} = e^{\ln x^{-b/a}} = x^{-b/a}$$

$$= x^{m_1}\int x^{-b/a} \cdot x^{(b-a)/a}dx \qquad \leftarrow -2m_1 = (b-a)/a$$

$$= x^{m_1}\int \frac{dx}{x} = x^{m_1}\ln x.$$

The general solution is then

$$y = c_1x^{m_1} + c_2x^{m_1}\ln x. \tag{4}$$

EXAMPLE 2 **Repeated Roots**

Solve $4x^2\dfrac{d^2y}{dx^2} + 8x\dfrac{dy}{dx} + y = 0$.

SOLUTION The substitution $y = x^m$ yields

$$4x^2\frac{d^2y}{dx^2} + 8x\frac{dy}{dx} + y = x^m(4m(m-1) + 8m + 1) = x^m(4m^2 + 4m + 1) = 0$$

when $4m^2 + 4m + 1 = 0$ or $(2m+1)^2 = 0$. Since $m_1 = -\frac{1}{2}$, it follows from (4) that the general solution is $y = c_1x^{-1/2} + c_2x^{-1/2}\ln x$. ≡

For higher-order equations, if m_1 is a root of multiplicity k, then it can be shown that

$$x^{m_1}, \quad x^{m_1}\ln x, \quad x^{m_1}(\ln x)^2, \ldots, \quad x^{m_1}(\ln x)^{k-1}$$

are k linearly independent solutions. Correspondingly, the general solution of the differential equation must then contain a linear combination of these k solutions.

Case III: Conjugate Complex Roots If the roots of (2) are the conjugate pair $m_1 = \alpha + i\beta$, $m_2 = \alpha - i\beta$, where α and $\beta > 0$ are real, then a solution is

$$y = C_1 x^{\alpha + i\beta} + C_2 x^{\alpha - i\beta}.$$

But when the roots of the auxiliary equation are complex, as in the case of equations with constant coefficients, we wish to write the solution in terms of real functions only. We note the identity

$$x^{i\beta} = (e^{\ln x})^{i\beta} = e^{i\beta \ln x},$$

which, by Euler's formula, is the same as

$$x^{i\beta} = \cos(\beta \ln x) + i \sin(\beta \ln x).$$

Similarly, $$x^{-i\beta} = \cos(\beta \ln x) - i \sin(\beta \ln x).$$

Adding and subtracting the last two results yields

$$x^{i\beta} + x^{-i\beta} = 2 \cos(\beta \ln x) \quad \text{and} \quad x^{i\beta} - x^{-i\beta} = 2i \sin(\beta \ln x),$$

respectively. From the fact that $y = C_1 x^{\alpha + i\beta} + C_2 x^{\alpha - i\beta}$ is a solution for any values of the constants, we see, in turn, for $C_1 = C_2 = 1$ and $C_1 = 1$, $C_2 = -1$ that

$$y_1 = x^{\alpha}(x^{i\beta} + x^{-i\beta}) \quad \text{and} \quad y_2 = x^{\alpha}(x^{i\beta} - x^{-i\beta})$$

or $$y_1 = 2x^{\alpha} \cos(\beta \ln x) \quad \text{and} \quad y_2 = 2ix^{\alpha} \sin(\beta \ln x)$$

are also solutions. Since $W(x^{\alpha} \cos(\beta \ln x), x^{\alpha} \sin(\beta \ln x)) = \beta x^{2\alpha - 1} \neq 0$, $\beta > 0$ on the interval $(0, \infty)$, we conclude that

$$y_1 = x^{\alpha} \cos(\beta \ln x) \quad \text{and} \quad y_2 = x^{\alpha} \sin(\beta \ln x)$$

constitute a fundamental set of real solutions of the differential equation. Hence the general solution is

$$y = x^{\alpha}[c_1 \cos(\beta \ln x) + c_2 \sin(\beta \ln x)]. \tag{5}$$

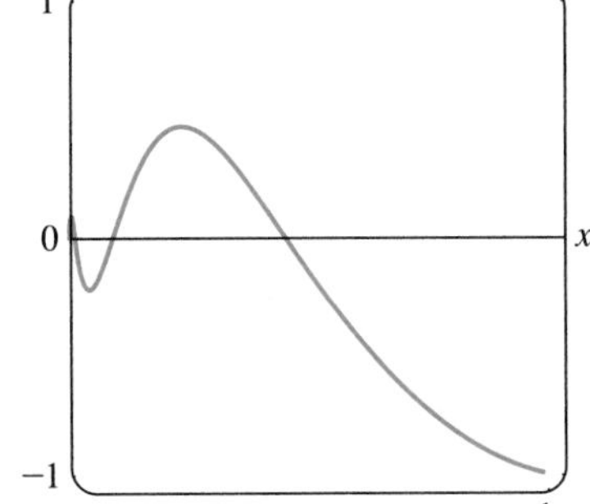

(a) solution for $0 < x \leq 1$

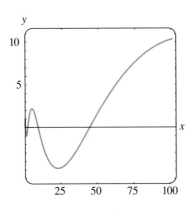

(b) solution for $0 < x \leq 100$

FIGURE 4.7.1 Solution curve of IVP in Example 3

EXAMPLE 3 **An Initial-Value Problem**

Solve $4x^2 y'' + 17y = 0$, $y(1) = -1$, $y'(1) = -\frac{1}{2}$.

SOLUTION The y' term is missing in the given Cauchy-Euler equation; nevertheless, the substitution $y = x^m$ yields

$$4x^2 y'' + 17y = x^m(4m(m - 1) + 17) = x^m(4m^2 - 4m + 17) = 0$$

when $4m^2 - 4m + 17 = 0$. From the quadratic formula we find that the roots are $m_1 = \frac{1}{2} + 2i$ and $m_2 = \frac{1}{2} - 2i$. With the identifications $\alpha = \frac{1}{2}$ and $\beta = 2$ we see from (5) that the general solution of the differential equation is

$$y = x^{1/2}[c_1 \cos(2 \ln x) + c_2 \sin(2 \ln x)].$$

By applying the initial conditions $y(1) = -1$, $y'(1) = -\frac{1}{2}$ to the foregoing solution and using $\ln 1 = 0$, we then find, in turn, that $c_1 = -1$ and $c_2 = 0$. Hence the solution of the initial-value problem is $y = -x^{1/2} \cos(2 \ln x)$. The graph of this function, obtained with the aid of computer software, is given in Figure 4.7.1. The particular solution is seen to be oscillatory and unbounded as $x \to \infty$. ≡

The next example illustrates the solution of a third-order Cauchy-Euler equation.

EXAMPLE 4 Third-Order Equation

Solve $x^3 \dfrac{d^3y}{dx^3} + 5x^2 \dfrac{d^2y}{dx^2} + 7x \dfrac{dy}{dx} + 8y = 0.$

SOLUTION The first three derivatives of $y = x^m$ are

$$\frac{dy}{dx} = mx^{m-1}, \qquad \frac{d^2y}{dx^2} = m(m-1)x^{m-2}, \qquad \frac{d^3y}{dx^3} = m(m-1)(m-2)x^{m-3},$$

so the given differential equation becomes

$$\begin{aligned} x^3 \frac{d^3y}{dx^3} + 5x^2 \frac{d^2y}{dx^2} + 7x \frac{dy}{dx} + 8y &= x^3 m(m-1)(m-2)x^{m-3} + 5x^2 m(m-1)x^{m-2} + 7xmx^{m-1} + 8x^m \\ &= x^m(m(m-1)(m-2) + 5m(m-1) + 7m + 8) \\ &= x^m(m^3 + 2m^2 + 4m + 8) = x^m(m+2)(m^2+4) = 0. \end{aligned}$$

In this case we see that $y = x^m$ will be a solution of the differential equation for $m_1 = -2$, $m_2 = 2i$, and $m_3 = -2i$. Hence the general solution is $y = c_1x^{-2} + c_2 \cos(2 \ln x) + c_3 \sin(2 \ln x)$. ≡

≡ Nonhomogeneous Equations The method of undetermined coefficients described in Sections 4.5 and 4.6 does not carry over, *in general*, to nonhomogeneous linear differential equations with variable coefficients. Consequently, in our next example the method of variation of parameters is employed.

EXAMPLE 5 Variation of Parameters

Solve $x^2y'' - 3xy' + 3y = 2x^4e^x$.

SOLUTION Since the equation is nonhomogeneous, we first solve the associated homogeneous equation. From the auxiliary equation $(m-1)(m-3) = 0$ we find $y_c = c_1x + c_2x^3$. Now before using variation of parameters to find a particular solution $y_p = u_1y_1 + u_2y_2$, recall that the formulas $u_1' = W_1/W$ and $u_2' = W_2/W$, where W_1, W_2, and W are the determinants defined on page 158, were derived under the assumption that the differential equation has been put into the standard form $y'' + P(x)y' + Q(x)y = f(x)$. Therefore we divide the given equation by x^2, and from

$$y'' - \frac{3}{x}y' + \frac{3}{x^2}y = 2x^2e^x$$

we make the identification $f(x) = 2x^2e^x$. Now with $y_1 = x$, $y_2 = x^3$, and

$$W = \begin{vmatrix} x & x^3 \\ 1 & 3x^2 \end{vmatrix} = 2x^3, \quad W_1 = \begin{vmatrix} 0 & x^3 \\ 2x^2e^x & 3x^2 \end{vmatrix} = -2x^5e^x, \quad W_2 = \begin{vmatrix} x & 0 \\ 1 & 2x^2e^x \end{vmatrix} = 2x^3e^x,$$

we find $\qquad u_1' = -\dfrac{2x^5e^x}{2x^3} = -x^2e^x \quad$ and $\quad u_2' = \dfrac{2x^3e^x}{2x^3} = e^x.$

The integral of the last function is immediate, but in the case of u_1' we integrate by parts twice. The results are $u_1 = -x^2e^x + 2xe^x - 2e^x$ and $u_2 = e^x$. Hence $y_p = u_1y_1 + u_2y_2$ is

$$y_p = (-x^2e^x + 2xe^x - 2e^x)x + e^xx^3 = 2x^2e^x - 2xe^x.$$

Finally, $y = y_c + y_p = c_1x + c_2x^3 + 2x^2e^x - 2xe^x$. ≡

Reduction to Constant Coefficients The similarities between the forms of solutions of Cauchy-Euler equations and solutions of linear equations with constant coefficients are not just a coincidence. For example, when the roots of the auxiliary equations for $ay'' + by' + cy = 0$ and $ax^2y'' + bxy' + cy = 0$ are distinct and real, the respective general solutions are

$$y = c_1e^{m_1x} + c_2e^{m_2x} \quad \text{and} \quad y = c_1x^{m_1} + c_2x^{m_2}, \quad x > 0. \tag{5}$$

In view of the identity $e^{\ln x} = x$, $x > 0$, the second solution given in (5) can be expressed in the same form as the first solution:

$$y = c_1e^{m_1 \ln x} + c_2e^{m_2 \ln x} = c_1e^{m_1t} + c_2e^{m_2t},$$

where $t = \ln x$. This last result illustrates the fact that any Cauchy-Euler equation can *always* be rewritten as a linear differential equation with constant coefficients by means of the substitution $x = e^t$. The idea is to solve the new differential equation in terms of the variable t, using the methods of the previous sections, and, once the general solution is obtained, resubstitute $t = \ln x$. This method, illustrated in the last example, requires the use of the Chain Rule of differentiation.

EXAMPLE 6 Changing to Constant Coefficients

Solve $x^2y'' - xy' + y = \ln x$.

SOLUTION With the substitution $x = e^t$ or $t = \ln x$, it follows that

$$\frac{dy}{dx} = \frac{dy}{dt}\frac{dt}{dx} = \frac{1}{x}\frac{dy}{dt} \qquad \leftarrow \text{Chain Rule}$$

$$\frac{d^2y}{dx^2} = \frac{1}{x}\frac{d}{dx}\left(\frac{dy}{dt}\right) + \frac{dy}{dt}\left(-\frac{1}{x^2}\right) \qquad \leftarrow \text{Product Rule and Chain Rule}$$

$$= \frac{1}{x}\left(\frac{d^2y}{dt^2}\frac{1}{x}\right) + \frac{dy}{dt}\left(-\frac{1}{x^2}\right) = \frac{1}{x^2}\left(\frac{d^2y}{dt^2} - \frac{dy}{dt}\right).$$

Substituting in the given differential equation and simplifying yields

$$\frac{d^2y}{dt^2} - 2\frac{dy}{dt} + y = t.$$

Since this last equation has constant coefficients, its auxiliary equation is $m^2 - 2m + 1 = 0$, or $(m - 1)^2 = 0$. Thus we obtain $y_c = c_1e^t + c_2te^t$.

By undetermined coefficients we try a particular solution of the form $y_p = A + Bt$. This assumption leads to $-2B + A + Bt = t$, so $A = 2$ and $B = 1$. Using $y = y_c + y_p$, we get

$$y = c_1e^t + c_2te^t + 2 + t.$$

By resubstituting $e^t = x$ and $t = \ln x$ we see that the general solution of the original differential equation on the interval $(0, \infty)$ is $y = c_1x + c_2x \ln x + 2 + \ln x$. ≡

Solutions For $x < 0$ In the preceding discussion we have solved Cauchy-Euler equations for $x > 0$. One way of solving a Cauchy-Euler equation for $x < 0$ is to change the independent variable by means of the substitution $t = -x$ (which implies $t > 0$) and using the Chain Rule:

$$\frac{dy}{dx} = \frac{dy}{dt}\frac{dt}{dx} = -\frac{dy}{dt} \quad \text{and} \quad \frac{d^2y}{dx^2} = \frac{d}{dt}\left(-\frac{dy}{dt}\right)\frac{dt}{dx} = \frac{d^2y}{dt^2}.$$

See Problems 37 and 38 in Exercises 4.7.

A Different Form A second-order equation of the form

$$a(x - x_0)^2 \frac{d^2y}{dx^2} + b(x - x_0)\frac{dy}{dx} + cy = 0 \tag{6}$$

is also a Cauchy-Euler equation. Observe that (6) reduces to (1) when $x_0 = 0$.

We can solve (6) as we did (1), namely, seeking solutions of $y = (x - x_0)^m$ and using

$$\frac{dy}{dx} = m(x - x_0)^{m-1} \quad \text{and} \quad \frac{d^2y}{dx^2} = m(m - 1)(x - x_0)^{m-2}.$$

Alternatively, we can reduce (6) to the familiar form (1) by means of the change of independent variable $t = x - x_0$, solving the reduced equation, and resubstituting. See Problems 39–42 in Exercises 4.7.

EXERCISES 4.7

Answers to selected odd-numbered problems begin on page ANS-6.

In Problems 1–18 solve the given differential equation.

1. $x^2y'' - 2y = 0$

2. $4x^2y'' + y = 0$

3. $xy'' + y' = 0$

4. $xy'' - 3y' = 0$

5. $x^2y'' + xy' + 4y = 0$

6. $x^2y'' + 5xy' + 3y = 0$

7. $x^2y'' - 3xy' - 2y = 0$

8. $x^2y'' + 3xy' - 4y = 0$

9. $25x^2y'' + 25xy' + y = 0$

10. $4x^2y'' + 4xy' - y = 0$

11. $x^2y'' + 5xy' + 4y = 0$

12. $x^2y'' + 8xy' + 6y = 0$

13. $3x^2y'' + 6xy' + y = 0$

14. $x^2y'' - 7xy' + 41y = 0$

15. $x^3y''' - 6y = 0$

16. $x^3y''' + xy' - y = 0$

17. $xy^{(4)} + 6y''' = 0$

18. $x^4y^{(4)} + 6x^3y''' + 9x^2y'' + 3xy' + y = 0$

In Problems 19–24 solve the given differential equation by variation of parameters.

19. $xy'' - 4y' = x^4$

20. $2x^2y'' + 5xy' + y = x^2 - x$

21. $x^2y'' - xy' + y = 2x$

22. $x^2y'' - 2xy' + 2y = x^4e^x$

23. $x^2y'' + xy' - y = \ln x$

24. $x^2y'' + xy' - y = \dfrac{1}{x + 1}$

In Problems 25–30 solve the given initial-value problem. Use a graphing utility to graph the solution curve.

25. $x^2y'' + 3xy' = 0, \quad y(1) = 0, y'(1) = 4$

26. $x^2y'' - 5xy' + 8y = 0, \quad y(2) = 32, y'(2) = 0$

27. $x^2y'' + xy' + y = 0, \quad y(1) = 1, y'(1) = 2$

28. $x^2y'' - 3xy' + 4y = 0, \quad y(1) = 5, y'(1) = 3$

29. $xy'' + y' = x, \quad y(1) = 1, y'(1) = -\frac{1}{2}$

30. $x^2y'' - 5xy' + 8y = 8x^6, \quad y(\frac{1}{2}) = 0, y'(\frac{1}{2}) = 0$

In Problems 31–36 use the substitution $x = e^t$ to transform the given Cauchy-Euler equation to a differential equation with constant coefficients. Solve the original equation by solving the new equation using the procedures in Sections 4.3–4.5.

31. $x^2y'' + 9xy' - 20y = 0$

32. $x^2y'' - 9xy' + 25y = 0$

33. $x^2y'' + 10xy' + 8y = x^2$

34. $x^2y'' - 4xy' + 6y = \ln x^2$

35. $x^2y'' - 3xy' + 13y = 4 + 3x$

36. $x^3y''' - 3x^2y'' + 6xy' - 6y = 3 + \ln x^3$

In Problems 37 and 38 use the substitution $t = -x$ to solve the given initial-value problem on the interval $(-\infty, 0)$.

37. $4x^2y'' + y = 0, \quad y(-1) = 2, y'(-1) = 4$

38. $x^2y'' - 4xy' + 6y = 0, \quad y(-2) = 8, y'(-2) = 0$

In Problems 39 and 40 use $y = (x - x_0)^m$ to solve the given differential equation.

39. $(x + 3)^2y'' - 8(x - 1)y' + 14y = 0$

40. $(x - 1)^2y'' - (x - 1)y' + 5y = 0$

In Problems 41 and 42 use the substitution $t = x - x_0$ to solve the given differential equation.

41. $(x + 2)^2y'' + (x + 2)y' + y = 0$

42. $(x - 4)^2y'' - 5(x - 4)y' + 9y = 0$

Discussion Problems

43. Give the largest interval over which the general solution of Problem 42 is defined.

44. Can a Cauchy-Euler differential equation of lowest order with real coefficients be found if it is known that 2 and $1 - i$ are roots of its auxiliary equation? Carry out your ideas.

45. The initial-conditions $y(0) = y_0$, $y'(0) = y_1$ apply to each of the following differential equations:

$$x^2y'' = 0,$$

$$x^2y'' - 2xy' + 2y = 0,$$

$$x^2y'' - 4xy' + 6y = 0.$$

For what values of y_0 and y_1 does each initial-value problem have a solution?

46. What are the x-intercepts of the solution curve shown in Figure 4.7.1? How many x-intercepts are there for $0 < x < \frac{1}{2}$?

Computer Lab Assignments

In Problems 47–50 solve the given differential equation by using a CAS to find the (approximate) roots of the auxiliary equation.

47. $2x^3y''' - 10.98x^2y'' + 8.5xy' + 1.3y = 0$

48. $x^3y''' + 4x^2y'' + 5xy' - 9y = 0$

49. $x^4y^{(4)} + 6x^3y''' + 3x^2y'' - 3xy' + 4y = 0$

50. $x^4y^{(4)} - 6x^3y''' + 33x^2y'' - 105xy' + 169y = 0$

51. Solve $x^3y''' - x^2y'' - 2xy' + 6y = x^2$ by variation of parameters. Use a CAS as an aid in computing roots of the auxiliary equation and the determinants given in (15) of Section 4.6.

4.8 GREEN'S FUNCTIONS

REVIEW MATERIAL

- See the *Remarks* at the end of Section 4.1 for the definitions of *response*, *input*, and *output*.
- Differential operators in Section 4.1 and Section 4.5
- The method of variation of parameters in Section 4.6

INTRODUCTION We will see in Chapter 5 that the linear second-order differential equation

$$a_2(x)\frac{d^2y}{dx^2} + a_1(x)\frac{dy}{dx} + a_0(x)y = g(x) \tag{1}$$

plays an important role in many applications. In the mathematical analysis of physical systems it is often desirable to express the **response** or **output** $y(x)$ of (1) subject to either initial conditions or boundary conditions directly in terms of the **forcing function** or **input** $g(x)$. In this manner the response of the system can quickly be analyzed for different forcing functions.

To see how this is done, we start by examining solutions of initial-value problems in which the DE (1) has been put into the standard form

$$y'' + P(x)y' + Q(x)y = f(x) \tag{2}$$

by dividing the equation by the lead coefficient $a_2(x)$. We also assume throughout this section that the coefficient functions $P(x)$, $Q(x)$, and $f(x)$ are continuous on some common interval I.

4.8.1 INITIAL-VALUE PROBLEMS

Three Initial-Value Problems We will see as the discussion unfolds that the solution $y(x)$ of the second order initial-value problem

$$y'' + P(x)y' + Q(x)y = f(x), \quad y(x_0) = y_0, \; y'(x_0) = y_1 \tag{3}$$

can be expressed as the superposition of two solutions:

$$y(x) = y_h(x) + y_p(x), \tag{4}$$

where $y_h(x)$ is the solution of the associated homogeneous DE with nonhomogeneous initial conditions

Here at least one of the numbers y_0 or y_1 is assumed to be nonzero. If both y_0 and y_1 are 0, then the solution of the IVP is $y = 0$.

$$y'' + P(x)y' + Q(x)y = 0, \quad y(x_0) = y_0, \; y'(x_0) = y_1 \tag{5}$$

and $y_p(x)$ is the solution of the nonhomogeneous DE with homogeneous (that is, zero) initial conditions

$$y'' + P(x)y' + Q(x)y = f(x), \quad y(x_0) = 0, \; y'(x_0) = 0. \tag{6}$$

In the case where the coefficients P and Q are constants the solution of the IVP (5) presents no difficulties: We use the method of Section 4.3 to find the general solution of the homogeneous DE and then use the given initial conditions to determine the two constants in that solution. So we will focus on the solution of the IVP (6). Because of the zero initial conditions, the solution of (6) could describe a physical system that is initially at rest and so is sometimes called a **rest solution**.

Green's Function If $y_1(x)$ and $y_2(x)$ form a fundamental set of solutions on the interval I of the associated homogeneous form of (2), then a particular solution of the nonhomogeneous equation (2) on the interval I can be found by variation of parameters. Recall from (3) of Section 4.6, the form of this solution is

$$y_p(x) = u_1(x)y_1(x) + u_2(x)y_2(x). \tag{7}$$

The variable coefficients $u_1(x)$ and $u_2(x)$ in (7) are defined by (9) of Section 4.6:

$$u_1'(x) = -\frac{y_2(x)f(x)}{W}, \quad u_2'(x) = \frac{y_1(x)f(x)}{W}. \tag{8}$$

The linear independence of $y_1(x)$ and $y_2(x)$ on the interval I guarantees that the Wronskian $W = W(y_1(x), y_2(x)) \neq 0$ for all x in I. If x and x_0 are numbers in I, then integrating the derivatives $u_1'(x)$ and $u_2'(x)$ in (8) on the interval $[x_0, x]$ and substituting the results into (7) give

Because $y_1(x)$ and $y_2(x)$ are constant with respect to the integration on t, we can move these functions inside the definite intergrals.

$$\begin{aligned} y_p(x) &= y_1(x)\int_{x_0}^{x} \frac{-y_2(t)f(t)}{W(t)}\,dt + y_2(x)\int_{x_0}^{x} \frac{y_1(t)f(t)}{W(t)}\,dt \\ &= \int_{x_0}^{x} \frac{-y_1(x)y_2(t)}{W(t)} f(t)\,dt + \int_{x_0}^{x} \frac{y_1(t)y_2(x)}{W(t)} f(t)\,dt, \end{aligned} \tag{9}$$

where $$W(t) = W(y_1(t), y_2(t)) = \begin{vmatrix} y_1(t) & y_2(t) \\ y_1'(t) & y_2'(t) \end{vmatrix}$$

From the properties of the definite integral, the two integrals in the second line of (9) can be rewritten as a single integral

$$y_p(x) = \int_{x_0}^{x} G(x, t)\, f(t)\, dt. \tag{10}$$

The function $G(x, t)$ in (10),

$$G(x, t) = \frac{y_1(t)y_2(x) - y_1(x)y_2(t)}{W(t)} \tag{11}$$

is called the **Green's function** for the differential equation (2).

Important. Read this paragraph a second time.

Observe that a Green's function (11) depends only on the fundamental solutions $y_1(x)$ and $y_2(x)$ of the associated homogeneous differential equation for (2) and *not* on the forcing function $f(x)$. Therefore all linear second-order differential equations (2) with the same left-hand side but with different forcing functions have the same the Green's function. So an alternative title for (11) is the **Green's function for the second-order differential operator** $L = D^2 + P(x)D + Q(x)$.

EXAMPLE 1 **Particular Solution**

Use (10) and (11) to find a particular solution of $y'' - y = f(x)$.

SOLUTION The solutions of the associated homogeneous equation $y'' - y = 0$ are $y_1 = e^x$, $y_2 = e^{-x}$, and $W(y_1(t), y_2(t)) = -2$. It follows from (11) that the Green's function is

$$G(x, t) = \frac{e^t e^{-x} - e^x e^{-t}}{-2} = \frac{e^{x-t} - e^{-(x-t)}}{2} = \sinh(x - t). \tag{12}$$

Thus from (10), a particular solution of the DE is

$$y_p(x) = \int_{x_0}^{x} \sinh(x - t) f(t)\, dt. \tag{13}$$

EXAMPLE 2 **General Solutions**

Find the general solution of following nonhomogeneous differential equations.

(a) $y'' - y = 1/x$ **(b)** $y'' - y = e^{2x}$

SOLUTION From Example 1, both DEs possess the same complementary function $y_c = c_1e^{-x} + c_2e^x$. Moreover, as pointed out in the paragraph preceding Example 1, the Green's function for both differential equations is (12).

(a) With the identifications $f(x) = 1/x$ and $f(t) = 1/t$ we see from (13) that a particular solution of $y'' - y = 1/x$ is $y_p(x) = \int_{x_0}^{x} \frac{\sinh(x - t)}{t}\, dt$. Thus the general solution $y = y_c + y_p$ of the given DE on any interval $[x_0, x]$ not containing the origin is

$$y = c_1e^x + c_2e^{-x} + \int_{x_0}^{x} \frac{\sinh(x - t)}{t}\, dt. \tag{14}$$

You should compare this solution with that found in Example 3 of Section 4.6.

(b) With $f(x) = e^{2x}$ in (13), a particular solution of $y'' - y = e^{2x}$ is $y_p(x) = \int_{x_0}^{x} \sinh(x - t)\, e^{2t}\, dt$. The general solution $y = y_c + y_p$ is then

$$y = c_1e^x + c_2e^{-x} + \int_{x_0}^{x} \sinh(x - t)\, e^{2t}\, dt. \tag{15}$$

Now consider the special initial-value problem (6) with homogeneous initial conditions. One way of solving the problem when $f(x) \neq 0$ has already been illustrated in Sections 4.4 and 4.6, that is, apply the initial conditions $y(x_0) = 0$, $y'(x_0) = 0$ to the general solution of the nonhomogeneous DE. But there is no actual need to do this because we already have a solution of the IVP at hand; it is the function defined in (10).

THEOREM 4.8.1 **Solution of the IVP (6)**

The function $y_p(x)$ defined in (10) is the solution of the initial-value problem (6).

PROOF By construction we know that $y_p(x)$ in (10) satisfies the nonhomogeneous DE. Next, because a definite integral has the property $\int_a^a = 0$ we have

$$y_p(x_0) = \int_{x_0}^{x_0} G(x_0, t) f(t)\, dt = 0.$$

Finally, to show that $y_p'(x_0) = 0$ we utilize the Leibniz formula* for the derivative of an integral:

$$y_p'(x) = \overbrace{G(x, x)}^{0 \text{ from } (11)} f(x) + \int_{x_0}^{x} \frac{y_1(t)y_2'(x) - y_1'(x)y_2(t)}{W(t)} f(t)\, dt.$$

Hence,
$$y_p'(x_0) = \int_{x_0}^{x_0} \frac{y_1(t)y_2'(x_0) - y_1'(x_0)y_2(t)}{W(t)} f(t)\, dt = 0.$$

≡

EXAMPLE 3 **Example 2 Revisited**

Solve the initial-value problems

(a) $y'' - y = 1/x,\ \ y(1) = 0, y'(1) = 0$ **(b)** $y'' - y = e^{2x},\ \ y(0) = 0, y'(0) = 0$

SOLUTION **(a)** With $x_0 = 1$ and $f(t) = 1/t$, it follows from (14) of Example 2 and Theorem 4.8.1 that the solution of the initial-value problem is

$$y_p(x) = \int_1^x \frac{\sinh(x - t)}{t} dt,$$

where $[1, x], x > 0$.

(b) Identifying $x_0 = 0$ and $f(t) = e^{2t}$, we see from (15) that the solution of the IVP is

$$y_p(x) = \int_0^x \sinh(x - t)\, e^{2t}\, dt. \tag{16}$$

≡

In part (b) of Example 3, we can carry out the integration in (16), but bear in mind that x is held constant throughout the integration with respect to t:

$$\begin{aligned} y_p(x) &= \int_0^x \sinh(x - t)\, e^{2t}\, dt = \int_0^x \frac{e^{x-t} - e^{-(x-t)}}{2} e^{2t}\, dt \\ &= \tfrac{1}{2}e^x \int_0^x e^t\, dt - \tfrac{1}{2}e^{-x} \int_0^x e^{3t}\, dt \\ &= \tfrac{1}{3}e^{2x} - \tfrac{1}{2}e^x + \tfrac{1}{6}e^{-x}. \end{aligned}$$

EXAMPLE 4 **Using (10) and (11)**

Solve the initial-value problem

$$y'' + 4y = x,\ \ y(0) = 0, y'(0) = 0.$$

SOLUTION We begin by constructing the Green's function for the given differential equation.

* This formula, usually discussed in advanced calculus, is given by

$$\frac{d}{dx}\int_{u(x)}^{v(x)} F(x, t) dt = F(x, v(x))v'(x) - F(x, u(x))u'(x) + \int_{u(x)}^{v(x)} \frac{\partial}{\partial x} F(x, t)\, dt.$$

The two linearly independent solutions of $y'' + 4y = 0$ are $y_1(x) = \cos 2x$ and $y_2(x) = \sin 2x$. From (11), with $W(\cos 2t, \sin 2t) = 2$, we find

Here we have used the trigonometric identity
$\sin(2x - 2t) = \sin 2x \cos 2t - \cos 2x \sin 2t$

$$G(x, t) = \frac{\cos 2t \sin 2x - \cos 2x \sin 2t}{2} = \tfrac{1}{2}\sin 2(x - t).$$

With the further identifications $x_0 = 0$ and $f(t) = t$ in (10) we see that a solution of the given initial-value problem is

$$y_p(x) = \tfrac{1}{2}\int_0^x t \sin 2(x - t)\, dt.$$

If we wish to evaluate the integral, we first write

$$y_p(x) = \tfrac{1}{2}\sin 2x \int_0^x t \cos 2t\, dt - \tfrac{1}{2}\cos 2x \int_0^x t \sin 2t\, dt$$

and then use integration by parts:

$$y_p(x) = \tfrac{1}{2}\sin 2x \left[\tfrac{1}{2}t \sin 2t + \tfrac{1}{4}\cos 2t\right]_0^x - \tfrac{1}{2}\cos 2x \left[-\tfrac{1}{2}t \cos 2t + \tfrac{1}{4}\sin 2t\right]_0^x$$

or

$$y_p(x) = \tfrac{1}{4}x - \tfrac{1}{8}\sin 2x.$$ ≡

≡ Initial-Value Problems—Continued Finally, we are now in a position to make use of Theorem 4.8.1 to find the solution of the initial-value problem posed in (3). It is simply the function already given in (4).

THEOREM 4.8.2 Solution of the IVP (3)

If $y_h(x)$ is the solution of the initial-value problem (5) and $y_p(x)$ is the solution (10) of the initial-value problem (6) on the interval I, then

$$y(x) = y_h(x) + y_p(x) \tag{17}$$

is the solution of the initial-value problem (3).

PROOF Because $y_h(x)$ is a linear combination of the fundamental solutions, it follows from (10) of Section 4.1 that $y = y_h + y_p$ is a solution of the nonhomogeneous DE. Moreover, since y_h satisfies the initial-conditions in (5) and y_p satisfies the initial conditions in (6), we have,

$$y(x_0) = y_h(x_0) + y_p(x_0) = y_0 + 0 = y_0$$
$$y'(x_0) = y_h'(x_0) + y_p'(x_0) = y_1 + 0 = y_1.$$ ≡

Keeping in mind the absence of a forcing function in (5) and the presence of such a term in (6), we see from (17) that the response $y(x)$ of a physical system described by the initial-value problem (3) can be separated into two different responses:

$$y(x) = \underbrace{y_h(x)}_{\substack{\text{response of system}\\ \text{due to initial conditions}\\ y(x_0) = y_0,\ y(x_0) = y_1}} + \underbrace{y_p(x)}_{\substack{\text{response of system}\\ \text{due to the forcing}\\ \text{function } f}} \tag{18}$$

If you wish to peek ahead, the following initial-value problem represents a pure resonance situation for a driven spring/mass system. See pages 200–202.

EXAMPLE 5 **Using Theorem 4.8.2**

Solve the initial-value problem

$$y'' + 4y = \sin 2x, \quad y(0) = 1, y'(0) = -2.$$

SOLUTION We solve two initial-value problems.

First, we solve $y'' + 4y = 0$, $y(0) = 1$, $y'(0) = -2$. By applying the initial conditions to the general solution $y(x) = c_1 \cos 2x + c_2 \sin 2x$ of the homogeneous DE, we find that $c_1 = 1$ and $c_2 = -1$. Therefore, $y_h(x) = \cos 2x - \sin 2x$.

Next we solve $y'' + 4y = \sin 2x$, $y(0) = 0$, $y'(0) = 0$. Since the left-hand side of the differential equation is the same as the DE in Example 4, the Green's function is the same, namely, $G(x, t) = \frac{1}{2}\sin 2(x - t)$. With $f(t) = \sin 2t$ we see from (10) that the solution of this second problem is $y_p(x) = \frac{1}{2}\int_0^x \sin 2(x - t)\sin 2t\, dt$.

Finally, in view of (17) in Theorem 4.8.2, the solution of the original IVP is

$$y(x) = y_h(x) + y_p(x) = \cos 2x - \sin 2x + \frac{1}{2}\int_0^x \sin 2(x - t)\sin 2t\, dt. \qquad (19)$$

If desired, we can integrate the definite integral in (19) by using the trigonometric identity

$$\sin A \sin B = \tfrac{1}{2}[\cos(A - B) - \cos(A + B)]$$

with $A = 2(x - t)$ and $B = 2t$:

$$\begin{aligned} y_p(x) &= \frac{1}{2}\int_0^x \sin 2(x - t)\sin 2t\, dt \\ &= \frac{1}{4}\int_0^x [\cos(2x - 4t) - \cos 2x]\, dt \qquad (20) \\ &= \frac{1}{4}\left[-\frac{1}{4}\sin(2x - 4t) - t\cos 2x\right]_0^x \\ &= \frac{1}{8}\sin 2x - \frac{1}{4}x\cos 2x. \end{aligned}$$

Hence, the solution (19) can be rewritten as

$$y(x) = y_h(x) + y_p(x) = \cos 2x - \sin 2x + \left(\tfrac{1}{8}\sin 2x - \tfrac{1}{4}x\cos 2x\right),$$

or

$$y(x) = \cos 2x - \tfrac{7}{8}\sin 2x - \tfrac{1}{4}x\cos 2x. \qquad (21)$$

Note that the physical significance indicated in (18) is lost in (21) after combining like terms in the two parts of the solution $y(x) = y_h(x) + y_p(x)$.

The beauty of the solution given in (19) is that we can immediately write down the response of a system if the initial conditions remain the same, but the forcing function is changed. For example, if the problem in Example 5 is changed to

$$y'' + 4y = x, \quad y(0) = 1, y'(0) = -2,$$

we simply replace $\sin 2t$ in the integral in (19) by t and the solution is then

$$\begin{aligned} y(x) &= y_h(x) + y_p(x) \\ &= \cos 2x - \sin 2x + \frac{1}{2}\int_0^x t\sin 2(x - t)\, dt \leftarrow \text{see Example 4} \\ &= \tfrac{1}{4}x + \cos 2x - \tfrac{9}{8}\sin 2x. \end{aligned}$$

Because the forcing function f is isolated in the particular solution $y_p(x) = \int_{x_0}^x G(x, t) f(t)\, dt$, the solution in (17) is useful when f is piecewise defined. The next example illustrates this idea.

EXAMPLE 6 **An Initial-Value Problem**

Solve the initial-value problem

$$y'' + 4y = f(x), \quad y(0) = 1, y'(0) = -2,$$

where the forcing function f is piecewise defined:

$$f(x) = \begin{cases} 0, & x < 0 \\ \sin 2x, & 0 \le x \le 2\pi \\ 0, & x > 2\pi. \end{cases}$$

SOLUTION From (19), with $\sin 2t$ replaced by $f(t)$, we can write

$$y(x) = \cos 2x - \sin 2x + \tfrac{1}{2}\int_0^x \sin 2(x-t)\, f(t)\, dt.$$

Because f is defined in three pieces, we consider three cases in the evaluation of the definite integral. For $x < 0$,

$$y_p(x) = \tfrac{1}{2}\int_0^x \sin 2(x-t)\, 0\, dt = 0,$$

for $0 \le x \le 2$,

$$y_p(x) = \tfrac{1}{2}\int_0^x \sin 2(x-t) \sin 2t\, dt \leftarrow \text{using the integration in (20)}$$
$$= \tfrac{1}{8}\sin 2x - \tfrac{1}{4}x\cos 2x,$$

and finally for $x > 2\pi$, we can use the integration following Example 5:

$$y_p(x) = \tfrac{1}{2}\int_0^{2\pi} \sin 2(x-t) \sin 2t\, dt + \tfrac{1}{2}\int_{2\pi}^x \sin 2(x-t)\, 0\, dt$$
$$= \tfrac{1}{2}\pi\int_0^{2\pi} \sin 2(x-t) \sin 2t\, dt$$
$$= \tfrac{1}{4}\Big[-\tfrac{1}{4}\sin(2x-4t) - t\cos 2x\Big]_0^{2\pi} \leftarrow \text{using the integration in (20)}$$
$$= -\tfrac{1}{16}\sin(2x-8\pi) - \tfrac{1}{2}\pi\cos 2x + \tfrac{1}{16}\sin 2x \leftarrow \sin(2x-8\pi) = \sin 2x$$
$$= -\tfrac{1}{2}\pi\cos 2x.$$

Hence $y_p(x)$ is

$$y_p(x) = \begin{cases} 0, & x < 0 \\ \tfrac{1}{8}\sin 2x - \tfrac{1}{4}x\cos 2x, & 0 \le x \le 2\pi \\ -\tfrac{1}{2}\pi\cos 2x, & x > 2\pi. \end{cases}$$

and so

$$y(x) = y_h(x) + y_p(x) = \cos 2x - \sin 2x + y_p(x).$$

Putting all the pieces together we get

$$y(x) = \begin{cases} \cos 2x - \sin 2x, & x < 0 \\ (1 - \tfrac{1}{4}x)\cos 2x - \tfrac{7}{8}\sin 2x, & 0 \le x \le 2\pi \\ (1 - \tfrac{1}{2}\pi)\cos 2x - \sin 2x, & x > 2\pi. \end{cases}$$

The three parts of $y(x)$ are shown in different colors in Figure 4.8.1. ≡

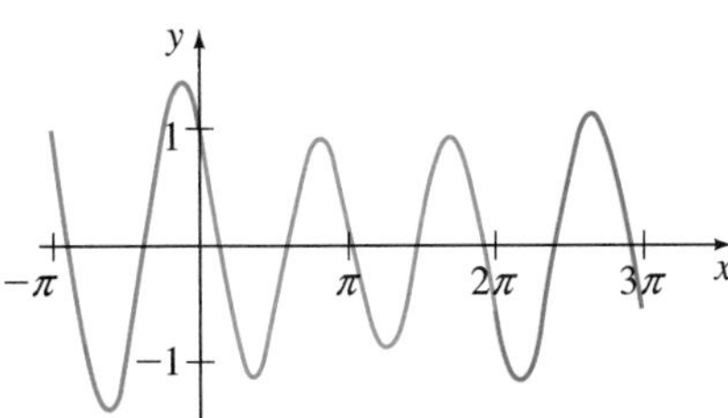

FIGURE 4.8.1 Graph of $y(x)$ in Example 6

We next examine how a boundary value problem (BVP) can be solved using a different kind of Green's function.

4.8.2 BOUNDARY-VALUE PROBLEMS

In contrast to a second-order IVP, in which $y(x)$ and $y'(x)$ are specified at the same point, a BVP for a second-order DE involves conditions on $y(x)$ and $y'(x)$ that are specified at two different points $x = a$ and $x = b$. Conditions such as

$$y(a) = 0, \quad y(b) = 0; \qquad y(a) = 0, \quad y'(b) = 0; \qquad y'(a) = 0, \quad y'(b) = 0.$$

are just special cases of the more general homogeneous boundary conditions:

$$A_1y(a) + B_1y'(a) = 0 \tag{22}$$

$$A_2y(b) + B_2y'(b) = 0, \tag{23}$$

where A_1, A_2, B_1, and B_2 are constants. Specifically, our goal is to find a integral solution $y_p(x)$ that is analogous to (10) for nonhomogeneous boundary-value problems of the form

$$\begin{aligned} y'' + P(x)y' + Q(x)y &= f(x), \\ A_1y(a) + B_1y'(a) &= 0 \\ A_2y(b) + B_2y'(b) &= 0. \end{aligned} \tag{24}$$

In addition to the usual assumptions that $P(x)$, $Q(x)$, and $f(x)$ are continuous on $[a, b]$, we assume that the homogeneous problem

$$\begin{aligned} y'' + P(x)y' + Q(x)y &= 0, \\ A_1y(a) + B_1y'(a) &= 0 \\ A_2y(b) + B_2y'(b) &= 0, \end{aligned}$$

possesses only the trivial solution $y = 0$. This latter assumption is sufficient to guarantee that a unique solution of (24) exists and is given by an integral $y_p(x) = \int_a^b G(x, t)\,f(t)dt$, where $G(x, t)$ is a Green's function.

The starting point in the construction of $G(x, t)$ is again the variation of parameters formulas (7) and (8).

Another Green's Function Suppose $y_1(x)$ and $y_2(x)$ are linearly independent solutions on $[a, b]$ of the associated homogeneous form of the DE in (24) and that x is a number in the interval $[a, b]$. Unlike the construction of (9) where we started by integrating the derivatives in (8) over the same interval, we now integrate the first equation in (8) on $[b, x]$ and the second equation in (8) on $[a, x]$:

$$u_1(x) = -\int_b^x \frac{y_2(t)f(t)}{W(t)}dt \quad \text{and} \quad u_2(x) = \int_a^x \frac{y_1(t)f(t)}{W(t)}dt. \tag{25}$$

The reason for integrating $u_1'(x)$ and $u_2'(x)$ over different intervals will become clear shortly. From (25), a particular solution $y_p(x) = u_1(x)y_1(x) + u_2(x)y_2(x)$ of the DE is

$$y_p(x) = y_1(x)\overbrace{\int_x^b \frac{y_2(t)f(t)}{W(t)}dt}^{\text{here we used the minus sign in (25) to reverse the limits of integration}} + y_2(x)\int_a^x \frac{y_1(t)f(t)}{W(t)}dt$$

or

$$y_p(x) = \int_a^x \frac{y_2(x)y_1(t)}{W(t)}f(t)\,dt + \int_x^b \frac{y_1(x)y_2(t)}{W(t)}f(t)dt. \tag{26}$$

The right-hand side of (26) can be written compactly as a single integral

$$y_p(x) = \int_a^b G(x, t)\,f(t)dt, \tag{27}$$

where the function $G(x, t)$ is

$$G(x, t) = \begin{cases} \dfrac{y_1(t)y_2(x)}{W(t)}, & a \le t \le x \\[2ex] \dfrac{y_1(x)y_2(t)}{W(t)}, & x \le t \le b. \end{cases} \tag{28}$$

The piecewise-defined function (28) is called a **Green's function** for the boundary-value problem (24). It can be proved that $G(x, t)$ is a continuous function of x on the interval $[a, b]$.

Now if the solutions $y_1(x)$ and $y_2(x)$ used in the construction of $G(x, t)$ in (28) are chosen in such a manner that at $x = a$, $y_1(x)$ satisfies $A_1y_1(a) + B_1y_1'(a) = 0$, and at $x = b$, $y_2(x)$ satisfies $A_2y_2(b) + B_2y_2'(b) = 0$, then, wondrously, $y_p(x)$ defined in (27) satisfies both homogeneous boundary conditions in (24).

To see this we will need

The second line in (30) results from the fact that

$y_1(x)u_1'(x) + y_2(x)u_2'(x) = 0.$

See the discussion in Section 4.6 following (4).

$$y_p(x) = u_1(x)y_1(x) + u_2(x)y_2(x) \tag{29}$$

and

$$\begin{aligned} y_p'(x) &= u_1(x)y_1'(x) + y_1(x)u_1'(x) + u_2(x)y_2'(x) + y_2(x)u_2'(x) \\ &= u_1(x)y_1'(x) + u_2(x)y_2'(x). \end{aligned} \tag{30}$$

Before proceeding, observe in (25) that $u_1(b) = 0$ and $u_2(a) = 0$. In view of the second of these two properties we can show that $y_p(x)$ satisfies (22) whenever $y_1(x)$ satisfies the same boundary condition. From (29) and (30) we have

$$\begin{aligned} A_1y_p(a) + B_1y_p'(a) &= A_1[u_1(a)y_1(a) + \overbrace{u_2(a)}^{0}y_2(a)] + B_1[u_1(a)y_1'(a) + \overbrace{u_2(a)}^{0}y_2'(a)] \\ &= u_1(a)\underbrace{[A_1y_1(a) + B_1y_1'(a)]}_{0 \text{ from } (22)} = 0. \end{aligned}$$

Likewise, $u_1(b) = 0$ implies that whenever $y_2(x)$ satisfies (23) so does $y_p(x)$:

$$\begin{aligned} A_2y_p(b) + B_2y_p'(b) &= A_2[\overbrace{u_1(b)}^{0}y_1(b) + u_2(b)y_2(b)] + B_2[\overbrace{u_1(b)}^{0}y_1'(b) + u_2(b)y_2'(b)] \\ &= u_2(b)\underbrace{[A_2y_2(b) + B_2y'_2(b)]}_{0 \text{ from } (22)} = 0. \end{aligned}$$

The next theorem summarizes these results.

THEOREM 4.8.3 Solution of the BVP (24)

Let $y_1(x)$ and $y_2(x)$ be linearly independent solutions of

$$y'' + P(x)y' + Q(x)y = 0$$

on $[a, b]$, and suppose $y_1(x)$ and $y_2(x)$ satisfy (22) and (23), respectively. Then the function $y_p(x)$ defined in (27) is a solution of the boundary-value problem (24).

EXAMPLE 7 Using Theorem 4.8.3

Solve the boundary-value problem

The boundary condition $y'(0) = 0$ is a special case of (22) with $a = 0$, $A_1 = 0$, and $B_1 = 1$. The boundary condition $y(\pi/2) = 0$ is a special case of (23) with $b = \pi/2$, $A_2 = 1$, $B_2 = 0$.

$$y'' + 4y = 3, \quad y'(0) = 0, \quad y(\pi/2) = 0.$$

SOLUTION The solutions of the associated homogeneous equation $y'' + 4y = 0$ are $y_1(x) = \cos 2x$ and $y_2(x) = \sin 2x$ and $y_1(x)$ satisfies $y'(0) = 0$ whereas $y_2(x)$ satisfies $y(\pi/2) = 0$. The Wronskian is $W(y_1, y_2) = 2$, and so from (28) we see that the Green's function for the boundary-value problem is

$$G(x, t) = \begin{cases} \frac{1}{2}\cos 2t \sin 2x, & 0 \le t \le x \\ \frac{1}{2}\cos 2x \sin 2t, & x \le t \le \pi/2. \end{cases}$$

It follows from Theorem 4.8.3 that a solution of the BVP is (27) with the identifications $a = 0$, $b = \pi/2$, and $f(t) = 3$:

$$y_p(x) = 3\int_0^{\pi/2} G(x, t)\, dt$$
$$= 3 \cdot \tfrac{1}{2}\sin 2x \int_0^x \cos 2t\, dt + 3 \cdot \tfrac{1}{2}\cos 2x \int_x^{\pi/2} \sin 2t\, dt,$$

or, after evaluating the definite integrals, $y_p(x) = \frac{3}{4} + \frac{3}{4}\cos 2x$. ≡

Don't infer from the preceding example that the demand that $y_1(x)$ satisfy (22) and $y_2(x)$ satisfy (23) uniquely determines these functions. As we see in the last example, there is a certain arbitrariness in the selection of these functions.

EXAMPLE 8 A Boundary-Value Problem

Solve the boundary-value problem

$$x^2y'' - 3xy' + 3y = 24x^5, \quad y(1) = 0, \quad y(2) = 0.$$

SOLUTION The differential equation is recognized as a Cauchy-Euler DE. From the auxiliary equation $m(m - 1) - 3m + 3 = (m - 1)(m - 3) = 0$ the general solution of the associated homogeneous equation is $y = c_1x + c_2x^3$. Applying $y(1) = 0$ to this solution implies $c_1 + c_2 = 0$ or $c_1 = -c_2$. By choosing $c_2 = -1$ we get $c_1 = 1$ and $y_1 = x - x^3$. On the other hand, $y(2) = 0$ applied to the general solution shows $2c_1 + 8c_2 = 0$ or $c_1 = -4c_2$. The choice $c_2 = -1$ now gives $c_1 = 4$ and so $y_2(x) = 4x - x^3$. The Wronskian of these two functions is

$$W(y_1(x), y_2(x)) = \begin{vmatrix} x - x^3 & 4x - x^3 \\ 1 - 3x^2 & 4 - 3x^2 \end{vmatrix} = 6x^3.$$

Hence the Green's function for the boundary-value problem is

$$G(x, t) = \begin{cases} \dfrac{(t - t^3)(4x - x^3)}{6t^3}, & 1 \le t \le x \\[2ex] \dfrac{(x - x^3)(4t - t^3)}{6t^3}, & x \le t \le 2. \end{cases}$$

In order to identify the correct forcing function f we must write the DE in standard form:

$$y'' - \frac{3}{x}y' + \frac{3}{x^2}y = 24x^3.$$

From this equation we see that $f(t) = 24t^3$ and so $y_p(x)$ in (27) becomes

$$y_p(x) = 24\int_1^2 G(x, t)\, t^3 dt$$
$$= 4(4x - x^3)\int_1^x (t - t^3)\, dt + 4(x - x^3)\int_x^2 (4t - t^3) dt.$$

Verify $y_p(x)$ that satisfies the differential equation and the two boundary conditions.

Straightforward definite integration and algebraic simplification yield the solution $y_p(x) = 3x^5 - 15x^3 + 12x$. ≡

REMARKS

We have barely scratched the surface of the elegant, albeit complicated, theory of Green's functions. Green's functions can also be constructed for linear second-order partial differential equations, but we leave coverage of the latter topic to an advanced course.

EXERCISES 4.8

Answers to selected odd-numbered problems begin on page ANS-6.

4.8.1 INITIAL-VALUE PROBLEMS

In Problems 1–6 proceed as in Example 1 to find a particular solution $y_p(x)$ of the given differential equation in the integral form (10).

1. $y'' - 16y = f(x)$
2. $y'' + 3y' - 10y = f(x)$
3. $y'' + 2y' + y = f(x)$
4. $4y'' - 4y' + y = f(x)$
5. $y'' + 9y = f(x)$
6. $y'' - 2y' + 2y = f(x)$

In Problems 7–12 proceed as in Example 2 to find the general solution of the given differential equation. Use the results obtained in Problems 1–6. Do not evaluate the integral that defines $y_p(x)$.

7. $y'' - 16y = xe^{-2x}$
8. $y'' + 3y' - 10y = x^2$
9. $y'' + 2y' + y = e^{-x}$
10. $4y'' - 4y' + y = \arctan x$
11. $y'' + 9y = x + \sin x$
12. $y'' - 2y' + 2y = \cos^2 x$

In Problems 13–18 proceed as in Example 3 to find a solution of the given initial-value problem. Evaluate the integral that defines $y_p(x)$.

13. $y'' - 4y = e^{2x}, y(0) = 0, y'(0) = 0$
14. $y'' - y' = 1, y(0) = 0, y'(0) = 0$
15. $y'' - 10y' + 25y = e^{5x}, y(0) = 0, y'(0) = 0$
16. $y'' + 6y' + 9y = x, y(0) = 0, y'(0) = 0$
17. $y'' + y = \csc x \cot x, y(\pi/2) = 0, y'(\pi/2) = 0$
18. $y'' + y = \sec^2 x, y(\pi) = 0, y'(\pi) = 0$

In Problems 19–30 proceed as in Example 5 to find a solution of the given initial-value problem.

19. $y'' - 4y = e^{2x}, y(0) = 1, y'(0) = -4$
20. $y'' - y' = 1, y(0) = 10, y'(0) = 1$
21. $y'' - 10y' + 25y = e^{5x}, y(0) = -1, y'(0) = 1$
22. $y'' + 6y' + 9y = x, y(0) = 1, y'(0) = -3$
23. $y'' + y = \csc x \cot x, y(\pi/2) = -\pi/2, y'(\pi/2) = -1$
24. $y'' + y = \sec^2 x, y(\pi) = \frac{1}{2}, y'(\pi) = -1$
25. $y'' + 3y' + 2y = \sin e^x, y(0) = -1, y'(0) = 0$
26. $y'' + 3y' + 2y = \dfrac{1}{1 + e^x}, y(0) = 0, y'(0) = 1$
27. $x^2y'' - 2xy' + 2y = x, y(1) = 2, y'(1) = -1$
28. $x^2y'' - 2xy' + 2y = x \ln x, y(1) = 1, y'(1) = 0$
29. $x^2y'' - 6y = \ln x, y(1) = 1, y'(1) = 3$
30. $x^2y'' - xy' + y = x^2, y(1) = 4, y'(1) = 3$

In Problems 31–34 proceed as in Example 6 to find a solution of the initial-value problem with the given piecewise-defined forcing function.

31. $y'' - y = f(x), y(0) = 8, y'(0) = 2,$

 where $f(x) = \begin{cases} -1, & x < 0 \\ 1, & x \ge 0 \end{cases}$

32. $y'' - y = f(x), y(0) = 3, y'(0) = 2,$

 where $f(x) = \begin{cases} 0, & x < 0 \\ x, & x \ge 0 \end{cases}$

33. $y'' + y = f(x), y(0) = 1, y'(0) = -1,$

 where $f(x) = \begin{cases} 0, & x < 0 \\ 10, & 0 \le x \le 3\pi \\ 0, & x > 3\pi \end{cases}$

34. $y'' + y = f(x), y(0) = 0, y'(0) = 1,$

 where $f(x) = \begin{cases} 0, & x < 0 \\ \cos x, & 0 \le x \le 4\pi \\ 0, & x > 4\pi \end{cases}$

4.8.2 BOUNDARY-VALUE PROBLEMS

In Problems 35 and 36, **(a)** use (27) and (28) to find a solution of the boundary-value problem. **(b)** Verify that the function $y_p(x)$ satisfies the differential equations and both boundary-conditions.

35. $y'' = f(x), y(0) = 0, y(1) = 0$
36. $y'' = f(x), y(0) = 0, y(1) + y'(1) = 0$
37. In Problem 35 find a solution of the BVP when $f(x) = 1$.
38. In Problem 36 find a solution of the BVP when $f(x) = x$.

In Problems 39–44 proceed as in Examples 7 and 8 to find a solution of the given boundary-value problem.

39. $y'' + y = 1, y(0) = 0, y(1) = 0$
40. $y'' + 9y = 1, y(0) = 0, y'(\pi) = 0$
41. $y'' - 2y' + 2y = e^x, y(0) = 0, y(\pi/2) = 0$
42. $y'' - y' = e^{2x}, y(0) = 0, y(1) = 0$
43. $x^2y'' + xy' = 1, y(e^{-1}) = 0, y(1) = 0$
44. $x^2y'' - 4xy' + 6y = x^4, y(1) - y'(1) = 0, y(3) = 0$

Discussion Problems

45. Suppose the solution of the boundary-value problem

$$y'' + Py' + Qy = f(x),\ y(a) = 0, y(b) = 0,$$

$a < b$, is given by $y_p(x) = \int_a^b G(x, t)f(t)\,dt$ where $y_1(x)$ and $y_2(x)$ are solutions of the associated homogeneous differential equation chosen in the construction of $G(x, t)$ so that $y_1(a) = 0$ and $y_2(b) = 0$. Prove that the solution of the boundary-value problem with nonhomogeneous DE and boundary conditions,

$$y'' + Py' + Qy = f(x),\ \ y(a) = A,\ y(b) = B$$

is given by

$$y(x) = y_p(x) + \frac{B}{y_1(b)}y_1(x) + \frac{A}{y_2(a)}y_2(x).$$

[*Hint*: In your proof, you will have to show that $y_1(b) \neq 0$ and $y_2(a) \neq 0$. Reread the assumptions following (24).]

46. Use the result in Problem 45 to solve

$$y'' + y = 1,\ y(0) = 5,\ y(1) = -10.$$

4.9 SOLVING SYSTEMS OF LINEAR DEs BY ELIMINATION

REVIEW MATERIAL

- Because the method of systematic elimination uncouples a system into distinct linear ODEs in each dependent variable, this section gives you an opportunity to practice what you learned in Sections 4.3, 4.4 (or 4.5), and 4.6.

INTRODUCTION Simultaneous ordinary differential equations involve two or more equations that contain derivatives of two or more dependent variables—the unknown functions—with respect to a single independent variable. The method of **systematic elimination** for solving systems of differential equations with constant coefficients is based on the algebraic principle of elimination of variables. We shall see that the analogue of *multiplying* an algebraic equation by a constant is *operating* on an ODE with some combination of derivatives.

Systematic Elimination The elimination of an unknown in a system of linear differential equations is expedited by rewriting each equation in the system in differential operator notation. Recall from Section 4.1 that a single linear equation

$$a_n y^{(n)} + a_{n-1}y^{(n-1)} + \cdots + a_1 y' + a_0 y = g(t),$$

where the a_i, $i = 0, 1, \ldots, n$ are constants, can be written as

$$(a_n D^n + a_{n-1}D^{(n-1)} + \cdots + a_1 D + a_0)y = g(t).$$

If the nth-order differential operator $a_n D^n + a_{n-1}D^{(n-1)} + \cdots + a_1 D + a_0$ factors into differential operators of lower order, then the factors commute. Now, for example, to rewrite the system

$$\begin{aligned} x'' + 2x' + y'' &= x + 3y + \sin t \\ x' + y' &= -4x + 2y + e^{-t} \end{aligned}$$

in terms of the operator D, we first bring all terms involving the dependent variables to one side and group the same variables:

$$\begin{aligned} x'' + 2x' - x + y'' - 3y &= \sin t \\ x' - 4x + y' - 2y &= e^{-t} \end{aligned} \qquad \text{is the same as} \qquad \begin{aligned} (D^2 + 2D - 1)x + (D^2 - 3)y &= \sin t \\ (D - 4)x + (D - 2)y &= e^{-t}. \end{aligned}$$

Solution of a System A **solution** of a system of differential equations is a set of sufficiently differentiable functions $x = \phi_1(t)$, $y = \phi_2(t)$, $z = \phi_3(t)$, and so on that satisfies each equation in the system on some common interval I.

Method of Solution Consider the simple system of linear first-order equations

$$\frac{dx}{dt} = 3y \qquad \qquad Dx - 3y = 0$$
$$\text{or, equivalently,} \qquad \qquad \qquad (1)$$
$$\frac{dy}{dt} = 2x \qquad \qquad 2x - Dy = 0.$$

Operating on the first equation in (1) by D while multiplying the second by -3 and then adding eliminates y from the system and gives $D^2x - 6x = 0$. Since the roots of the auxiliary equation of the last DE are $m_1 = \sqrt{6}$ and $m_2 = -\sqrt{6}$, we obtain

$$x(t) = c_1e^{-\sqrt{6}t} + c_2e^{\sqrt{6}t}. \qquad (2)$$

Multiplying the first equation in (1) by 2 while operating on the second by D and then subtracting gives the differential equation for y, $D^2y - 6y = 0$. It follows immediately that

$$y(t) = c_3e^{-\sqrt{6}t} + c_4e^{\sqrt{6}t}. \qquad (3)$$

Now (2) and (3) do not satisfy the system (1) for every choice of c_1, c_2, c_3, and c_4 because the system itself puts a constraint on the number of parameters in a solution that can be chosen arbitrarily. To see this, observe that substituting $x(t)$ and $y(t)$ into the first equation of the original system (1) gives, after simplification,

$$\left(-\sqrt{6}c_1 - 3c_3\right)e^{-\sqrt{6}t} + \left(\sqrt{6}c_2 - 3c_4\right)e^{\sqrt{6}t} = 0.$$

Since the latter expression is to be zero for all values of t, we must have $-\sqrt{6}c_1 - 3c_3 = 0$ and $\sqrt{6}c_2 - 3c_4 = 0$. These two equations enable us to write c_3 as a multiple of c_1 and c_4 as a multiple of c_2:

$$c_3 = -\frac{\sqrt{6}}{3}c_1 \quad \text{and} \quad c_4 = \frac{\sqrt{6}}{3}c_2. \qquad (4)$$

Hence we conclude that a solution of the system must be

$$x(t) = c_1e^{-\sqrt{6}t} + c_2e^{\sqrt{6}t}, \qquad y(t) = -\frac{\sqrt{6}}{3}c_1e^{-\sqrt{6}t} + \frac{\sqrt{6}}{3}c_2e^{\sqrt{6}t}.$$

You are urged to substitute (2) and (3) into the second equation of (1) and verify that the same relationship (4) holds between the constants.

EXAMPLE 1 **Solution by Elimination**

Solve
$$Dx + (D + 2)y = 0$$
$$(D - 3)x - 2y = 0. \qquad (5)$$

SOLUTION Operating on the first equation by $D - 3$ and on the second by D and then subtracting eliminates x from the system. It follows that the differential equation for y is

$$[(D - 3)(D + 2) + 2D]y = 0 \quad \text{or} \quad (D^2 + D - 6)y = 0.$$

Since the characteristic equation of this last differential equation is $m^2 + m - 6 = (m - 2)(m + 3) = 0$, we obtain the solution

$$y(t) = c_1e^{2t} + c_2e^{-3t}. \qquad (6)$$

Eliminating y in a similar manner yields $(D^2 + D - 6)x = 0$, from which we find

$$x(t) = c_3e^{2t} + c_4e^{-3t}. \qquad (7)$$

As we noted in the foregoing discussion, a solution of (5) does not contain four independent constants. Substituting (6) and (7) into the first equation of (5) gives

$$(4c_1 + 2c_3)e^{2t} + (-c_2 - 3c_4)e^{-3t} = 0.$$

From $4c_1 + 2c_3 = 0$ and $-c_2 - 3c_4 = 0$ we get $c_3 = -2c_1$ and $c_4 = -\frac{1}{3}c_2$. Accordingly, a solution of the system is

$$x(t) = -2c_1e^{2t} - \frac{1}{3}c_2e^{-3t}, \qquad y(t) = c_1e^{2t} + c_2e^{-3t}. \qquad \equiv$$

Because we could just as easily solve for c_3 and c_4 in terms of c_1 and c_2, the solution in Example 1 can be written in the alternative form

$$x(t) = c_3e^{2t} + c_4e^{-3t}, \qquad y(t) = -\frac{1}{2}c_3e^{2t} - 3c_4e^{-3t}.$$

This might save you some time. ▶

It sometimes pays to keep one's eyes open when solving systems. Had we solved for x first in Example 1, then y could be found, along with the relationship between the constants, using the last equation in the system (5). You should verify that substituting $x(t)$ into $y = \frac{1}{2}(Dx - 3x)$ yields $y = -\frac{1}{2}c_3e^{2t} - 3c_4e^{-3t}$. Also note in the initial discussion that the relationship given in (4) and the solution $y(t)$ of (1) could also have been obtained by using $x(t)$ in (2) and the first equation of (1) in the form

$$y = \tfrac{1}{3}Dx = -\tfrac{1}{3}\sqrt{6}c_1e^{-\sqrt{6}t} + \tfrac{1}{3}\sqrt{6}c_2e^{\sqrt{6}t}.$$

EXAMPLE 2 Solution by Elimination

Solve

$$\begin{aligned} x' - 4x + y'' &= t^2 \\ x' + x + y' &= 0. \end{aligned} \tag{8}$$

SOLUTION First we write the system in differential operator notation:

$$\begin{aligned} (D - 4)x + D^2y &= t^2 \\ (D + 1)x + Dy &= 0. \end{aligned} \tag{9}$$

Then, by eliminating x, we obtain

$$[(D + 1)D^2 - (D - 4)D]y = (D + 1)t^2 - (D - 4)0$$

or

$$(D^3 + 4D)y = t^2 + 2t.$$

Since the roots of the auxiliary equation $m(m^2 + 4) = 0$ are $m_1 = 0$, $m_2 = 2i$, and $m_3 = -2i$, the complementary function is $y_c = c_1 + c_2 \cos 2t + c_3 \sin 2t$. To determine the particular solution y_p, we use undetermined coefficients by assuming that $y_p = At^3 + Bt^2 + Ct$. Therefore $y_p' = 3At^2 + 2Bt + C$, $y_p'' = 6At + 2B$, $y_p''' = 6A$,

$$y_p''' + 4y_p' = 12At^2 + 8Bt + 6A + 4C = t^2 + 2t.$$

The last equality implies that $12A = 1$, $8B = 2$, and $6A + 4C = 0$; hence $A = \frac{1}{12}$, $B = \frac{1}{4}$, and $C = -\frac{1}{8}$. Thus

$$y = y_c + y_p = c_1 + c_2 \cos 2t + c_3 \sin 2t + \frac{1}{12}t^3 + \frac{1}{4}t^2 - \frac{1}{8}t. \tag{10}$$

Eliminating y from the system (9) leads to

$$[(D - 4) - D(D + 1)]x = t^2 \qquad \text{or} \qquad (D^2 + 4)x = -t^2.$$

It should be obvious that $x_c = c_4 \cos 2t + c_5 \sin 2t$ and that undetermined coefficients can be applied to obtain a particular solution of the form $x_p = At^2 + Bt + C$. In this case the usual differentiations and algebra yield $x_p = -\frac{1}{4}t^2 + \frac{1}{8}$, and so

$$x = x_c + x_p = c_4 \cos 2t + c_5 \sin 2t - \frac{1}{4}t^2 + \frac{1}{8}. \tag{11}$$

Now c_4 and c_5 can be expressed in terms of c_2 and c_3 by substituting (10) and (11) into either equation of (8). By using the second equation, we find, after combining terms,

$$(c_5 - 2c_4 - 2c_2)\sin 2t + (2c_5 + c_4 + 2c_3)\cos 2t = 0,$$

so $c_5 - 2c_4 - 2c_2 = 0$ and $2c_5 + c_4 + 2c_3 = 0$. Solving for c_4 and c_5 in terms of c_2 and c_3 gives $c_4 = -\frac{1}{5}(4c_2 + 2c_3)$ and $c_5 = \frac{1}{5}(2c_2 - 4c_3)$. Finally, a solution of (8) is found to be

$$x(t) = -\frac{1}{5}(4c_2 + 2c_3)\cos 2t + \frac{1}{5}(2c_2 - 4c_3)\sin 2t - \frac{1}{4}t^2 + \frac{1}{8},$$

$$y(t) = c_1 + c_2\cos 2t + c_3\sin 2t + \frac{1}{12}t^3 + \frac{1}{4}t^2 - \frac{1}{8}t.$$ ≡

EXAMPLE 3 A Mixture Problem Revisited

In (3) of Section 3.3 we saw that the system of linear first-order differential equations

$$\frac{dx_1}{dt} = -\frac{2}{25}x_1 + \frac{1}{50}x_2$$
$$\frac{dx_2}{dt} = \frac{2}{25}x_1 - \frac{2}{25}x_2$$

is a model for the number of pounds of salt $x_1(t)$ and $x_2(t)$ in brine mixtures in tanks A and B, respectively, shown in Figure 3.3.1. At that time we were not able to solve the system. But now, in terms of differential operators, the foregoing system can be written as

$$\left(D + \frac{2}{25}\right)x_1 - \frac{1}{50}x_2 = 0$$
$$-\frac{2}{25}x_1 + \left(D + \frac{2}{25}\right)x_2 = 0.$$

Operating on the first equation by $D + \frac{2}{25}$, multiplying the second equation by $\frac{1}{50}$, adding, and then simplifying gives $(625D^2 + 100D + 3)x_1 = 0$. From the auxiliary equation

$$625m^2 + 100m + 3 = (25m + 1)(25m + 3) = 0$$

we see immediately that $x_1(t) = c_1e^{-t/25} + c_2e^{-3t/25}$. We can now obtain $x_2(t)$ by using the first DE of the system in the form $x_2 = 50(D + \frac{2}{25})x_1$. In this manner we find the solution of the system to be

$$x_1(t) = c_1e^{-t/25} + c_2e^{-3t/25}, \qquad x_2(t) = 2c_1e^{-t/25} - 2c_2e^{-3t/25}.$$

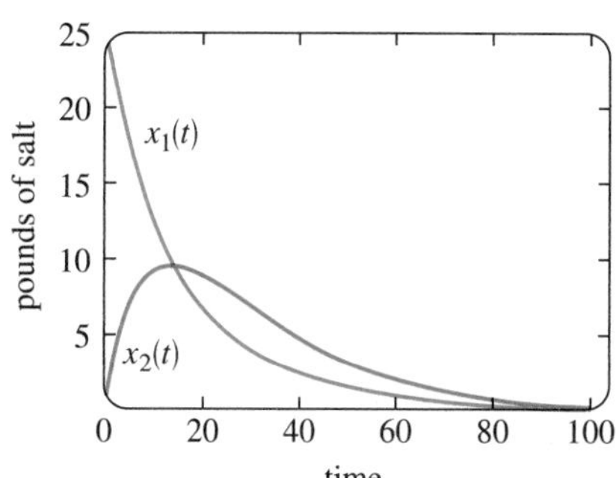

FIGURE 4.9.1 Pounds of salt in tanks A and B in Example 3

In the original discussion on page 108 we assumed that the initial conditions were $x_1(0) = 25$ and $x_2(0) = 0$. Applying these conditions to the solution yields $c_1 + c_2 = 25$ and $2c_1 - 2c_2 = 0$. Solving these equations simultaneously gives $c_1 = c_2 = \frac{25}{2}$. Finally, a solution of the initial-value problem is

$$x_1(t) = \frac{25}{2}e^{-t/25} + \frac{25}{2}e^{-3t/25}, \qquad x_2(t) = 25e^{-t/25} - 25e^{-3t/25}.$$

The graphs of both of these equations are given in Figure 4.9.1. Consistent with the fact that pure water is being pumped into tank A we see in the figure that $x_1(t) \to 0$ and $x_2(t) \to 0$ as $t \to \infty$. ≡

EXERCISES 4.9

Answers to selected odd-numbered problems begin on page ANS-6.

In Problems 1–20 solve the given system of differential equations by systematic elimination.

1. $\dfrac{dx}{dt} = 2x - y$, $\dfrac{dy}{dt} = x$

2. $\dfrac{dx}{dt} = 4x + 7y$, $\dfrac{dy}{dt} = x - 2y$

3. $\dfrac{dx}{dt} = -y + t$, $\dfrac{dy}{dt} = x - t$

4. $\dfrac{dx}{dt} - 4y = 1$, $\dfrac{dy}{dt} + x = 2$

5. $(D^2 + 5)x - 2y = 0$
$-2x + (D^2 + 2)y = 0$

6. $(D + 1)x + (D - 1)y = 2$
$3x + (D + 2)y = -1$

7. $\dfrac{d^2x}{dt^2} = 4y + e^t$, $\dfrac{d^2y}{dt^2} = 4x - e^t$

8. $\dfrac{d^2x}{dt^2} + \dfrac{dy}{dt} = -5x$, $\dfrac{dx}{dt} + \dfrac{dy}{dt} = -x + 4y$

9. $Dx + D^2y = e^{3t}$
$(D + 1)x + (D - 1)y = 4e^{3t}$

10. $D^2x - Dy = t$
$(D + 3)x + (D + 3)y = 2$

11. $(D^2 - 1)x - y = 0$
$(D - 1)x + Dy = 0$

12. $(2D^2 - D - 1)x - (2D + 1)y = 1$
$(D - 1)x + Dy = -1$

13. $2\dfrac{dx}{dt} - 5x + \dfrac{dy}{dt} = e^t$
$\dfrac{dx}{dt} - x + \dfrac{dy}{dt} = 5e^t$

14. $\dfrac{dx}{dt} + \dfrac{dy}{dt} = e^t$
$-\dfrac{d^2x}{dt^2} + \dfrac{dx}{dt} + x + y = 0$

15. $(D - 1)x + (D^2 + 1)y = 1$
$(D^2 - 1)x + (D + 1)y = 2$

16. $D^2x - 2(D^2 + D)y = \sin t$
$x + Dy = 0$

17. $Dx = y$
$Dy = z$
$Dz = x$

18. $Dx + z = e^t$
$(D - 1)x + Dy + Dz = 0$
$x + 2y + Dz = e^t$

19. $\dfrac{dx}{dt} = 6y$, $\dfrac{dy}{dt} = x + z$, $\dfrac{dz}{dt} = x + y$

20. $\dfrac{dx}{dt} = -x + z$, $\dfrac{dy}{dt} = -y + z$, $\dfrac{dz}{dt} = -x + y$

In Problems 21 and 22 solve the given initial-value problem.

21. $\dfrac{dx}{dt} = -5x - y$, $\dfrac{dy}{dt} = 4x - y$, $x(1) = 0,\ y(1) = 1$

22. $\dfrac{dx}{dt} = y - 1$, $\dfrac{dy}{dt} = -3x + 2y$, $x(0) = 0,\ y(0) = 0$

Mathematical Models

23. **Projectile Motion** A projectile shot from a gun has weight $w = mg$ and velocity $\mathbf{v}$ tangent to its path of motion. Ignoring air resistance and all other forces acting on the projectile except its weight, determine a system of differential equations that describes its path of motion. See Figure 4.9.2. Solve the system. [*Hint*: Use Newton's second law of motion in the x and y directions.]

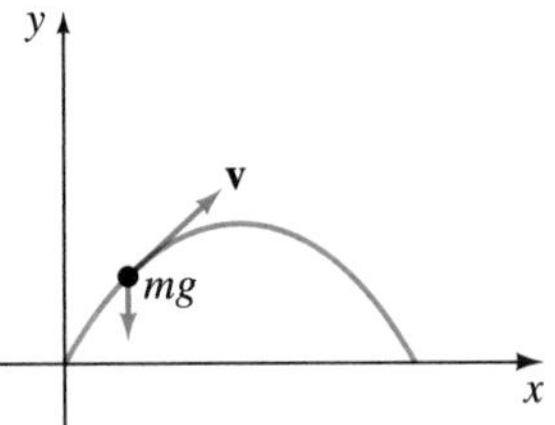

FIGURE 4.9.2 Path of projectile in Problem 23

24. **Projectile Motion with Air Resistance** Determine a system of differential equations that describes the path of motion in Problem 23 if air resistance is a retarding force $\mathbf{k}$ (of magnitude k) acting tangent to the path of the projectile but opposite to its motion. See Figure 4.9.3. Solve the system. [*Hint*: $\mathbf{k}$ is a multiple of velocity, say, $\beta\mathbf{v}$.]

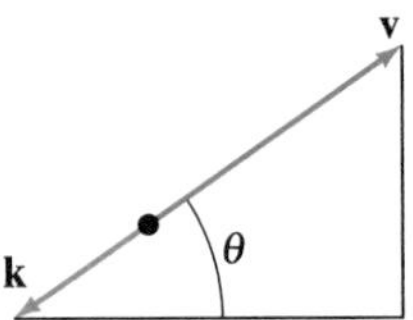

FIGURE 4.9.3 Forces in Problem 24

Discussion Problems

25. Examine and discuss the following system:

$$Dx - 2Dy = t^2$$
$$(D + 1)x - 2(D + 1)y = 1.$$

Computer Lab Assignments

26. Reexamine Figure 4.9.1 in Example 3. Then use a root-finding application to determine when tank B contains more salt than tank A.

27. **(a)** Reread Problem 8 of Exercises 3.3. In that problem you were asked to show that the system of differential equations

$$\frac{dx_1}{dt} = -\frac{1}{50}x_1$$
$$\frac{dx_2}{dt} = \frac{1}{50}x_1 - \frac{2}{75}x_2$$
$$\frac{dx_3}{dt} = \frac{2}{75}x_2 - \frac{1}{25}x_3$$

is a model for the amounts of salt in the connected mixing tanks A, B, and C shown in Figure 3.3.7. Solve the system subject to $x_1(0) = 15$, $x_2(t) = 10$, $x_3(t) = 5$.

(b) Use a CAS to graph $x_1(t)$, $x_2(t)$, and $x_3(t)$ in the same coordinate plane (as in Figure 4.9.1) on the interval [0, 200].

(c) Because only pure water is pumped into Tank A, it stands to reason that the salt will eventually be flushed out of all three tanks. Use a root-finding application of a CAS to determine the time when the amount of salt in each tank is less than or equal to 0.5 pound. When will the amounts of salt $x_1(t)$, $x_2(t)$, and $x_3(t)$ be simultaneously less than or equal to 0.5 pound?

4.10 NONLINEAR DIFFERENTIAL EQUATIONS

REVIEW MATERIAL

- Sections 2.2 and 2.5
- Section 4.2
- A review of Taylor series from calculus is also recommended.

INTRODUCTION The difficulties that surround higher-order *nonlinear* differential equations and the few methods that yield analytic solutions are examined next. Two of the solution methods considered in this section employ a change of variable to reduce a nonlinear second-order DE to a first-order DE. In that sense these methods are analogous to the material in Section 4.2.

Some Differences There are several significant differences between linear and nonlinear differential equations. We saw in Section 4.1 that homogeneous linear equations of order two or higher have the property that a linear combination of solutions is also a solution (Theorem 4.1.2). Nonlinear equations do not possess this property of superposability. See Problems 1 and 18 in Exercises 4.10. We can find general solutions of linear first-order DEs and higher-order equations with constant coefficients. Even when we can solve a nonlinear first-order differential equation in the form of a one-parameter family, this family does not, as a rule, represent a general solution. Stated another way, nonlinear first-order DEs can possess singular solutions, whereas linear equations cannot. But the major difference between linear and nonlinear equations of order two or higher lies in the realm of solvability. Given a linear equation, there is a chance that we can find some form of a solution that we can look at—an explicit solution or perhaps a solution in the form of an infinite series (see Chapter 6). On the other hand, nonlinear higher-order differential equations virtually defy solution by analytical methods. Although this might sound disheartening, there are still things that can be done. As was pointed out at the end of Section 1.3, we can always analyze a nonlinear DE qualitatively and numerically.

Let us make it clear at the outset that nonlinear higher-order differential equations are important—dare we say even more important than linear equations?—because as we fine-tune the mathematical model of, say, a physical system, we also increase the likelihood that this higher-resolution model will be nonlinear.

We begin by illustrating an analytical method that *occasionally* enables us to find explicit/implicit solutions of special kinds of nonlinear second-order differential equations.

Reduction of Order Nonlinear second-order differential equations $F(x, y', y'') = 0$, where the dependent variable y is missing, and $F(y, y', y'') = 0$, where the independent variable x is missing, can sometimes be solved by using first-order methods. Each equation can be reduced to a first-order equation by means of the substitution $u = y'$.

Dependent Variable Missing The next example illustrates the substitution technique for an equation of the form $F(x, y', y'') = 0$. If $u = y'$, then the differential equation becomes $F(x, u, u') = 0$. If we can solve this last equation for u, we can find y by integration. Note that since we are solving a second-order equation, its solution will contain two arbitrary constants.

EXAMPLE 1 Dependent Variable *y* Is Missing

Solve $y'' = 2x(y')^2$.

SOLUTION If we let $u = y'$, then $du/dx = y''$. After substituting, the second-order equation reduces to a first-order equation with separable variables; the independent variable is x and the dependent variable is u:

$$\frac{du}{dx} = 2xu^2 \quad \text{or} \quad \frac{du}{u^2} = 2x\,dx$$

$$\int u^{-2}\,du = \int 2x\,dx$$

$$-u^{-1} = x^2 + c_1^2.$$

The constant of integration is written as c_1^2 for convenience. The reason should be obvious in the next few steps. Because $u^{-1} = 1/y'$, it follows that

$$\frac{dy}{dx} = -\frac{1}{x^2 + c_1^2},$$

and so $$y = -\int \frac{dx}{x^2 + c_1^2} \quad \text{or} \quad y = -\frac{1}{c_1}\tan^{-1}\frac{x}{c_1} + c_2.$$ ≡

Independent Variable Missing Next we show how to solve an equation that has the form $F(y, y', y'') = 0$. Once more we let $u = y'$, but because the independent variable x is missing, we use this substitution to transform the differential equation into one in which the independent variable is y and the dependent variable is u. To this end we use the Chain Rule to compute the second derivative of y:

$$y'' = \frac{du}{dx} = \frac{du}{dy}\frac{dy}{dx} = u\frac{du}{dy}.$$

In this case the first-order equation that we must now solve is

$$F\left(y, u, u\frac{du}{dy}\right) = 0.$$

EXAMPLE 2 Independent Variable *x* Is Missing

Solve $yy'' = (y')^2$.

SOLUTION With the aid of $u = y'$, the Chain Rule shown above, and separation of variables, the given differential equation becomes

$$y\left(u\frac{du}{dy}\right) = u^2 \quad \text{or} \quad \frac{du}{u} = \frac{dy}{y}.$$

Integrating the last equation then yields $\ln|u| = \ln|y| + c_1$, which, in turn, gives $u = c_2 y$, where the constant $\pm e^{c_1}$ has been relabeled as c_2. We now resubstitute $u = dy/dx$, separate variables once again, integrate, and relabel constants a second time:

$$\int \frac{dy}{y} = c_2 \int dx \quad \text{or} \quad \ln|y| = c_2 x + c_3 \quad \text{or} \quad y = c_4 e^{c_2 x}.$$

Use of Taylor Series In some instances a solution of a nonlinear initial-value problem, in which the initial conditions are specified at x_0, can be approximated by a Taylor series centered at x_0.

EXAMPLE 3 Taylor Series Solution of an IVP

Let us assume that a solution of the initial-value problem

$$y'' = x + y - y^2, \quad y(0) = -1, \quad y'(0) = 1 \tag{1}$$

exists. If we further assume that the solution $y(x)$ of the problem is analytic at 0, then $y(x)$ possesses a Taylor series expansion centered at 0:

$$y(x) = y(0) + \frac{y'(0)}{1!}x + \frac{y''(0)}{2!}x^2 + \frac{y'''(0)}{3!}x^3 + \frac{y^{(4)}(0)}{4!}x^4 + \frac{y^{(5)}(0)}{5!}x^5 + \cdots. \tag{2}$$

Note that the values of the first and second terms in the series (2) are known since those values are the specified initial conditions $y(0) = -1$, $y'(0) = 1$. Moreover, the differential equation itself defines the value of the second derivative at 0: $y''(0) = 0 + y(0) - y(0)^2 = 0 + (-1) - (-1)^2 = -2$. We can then find expressions for the higher derivatives y''', $y^{(4)}, \ldots$ by calculating the successive derivatives of the differential equation:

$$y'''(x) = \frac{d}{dx}(x + y - y^2) = 1 + y' - 2yy' \tag{3}$$

$$y^{(4)}(x) = \frac{d}{dx}(1 + y' - 2yy') = y'' - 2yy'' - 2(y')^2 \tag{4}$$

$$y^{(5)}(x) = \frac{d}{dx}(y'' - 2yy'' - 2(y')^2) = y''' - 2yy''' - 6y'y'' \tag{5}$$

and so on. Now using $y(0) = -1$ and $y'(0) = 1$, we find from (3) that $y'''(0) = 4$. From the values $y(0) = -1$, $y'(0) = 1$, and $y''(0) = -2$ we find $y^{(4)}(0) = -8$ from (4). With the additional information that $y'''(0) = 4$, we then see from (5) that $y^{(5)}(0) = 24$. Hence from (2) the first six terms of a series solution of the initial-value problem (1) are

$$y(x) = -1 + x - x^2 + \frac{2}{3}x^3 - \frac{1}{3}x^4 + \frac{1}{5}x^5 + \cdots.$$

Use of a Numerical Solver Numerical methods, such as Euler's method or the Runge-Kutta method, are developed solely for first-order differential equations and then are extended to systems of first-order equations. To analyze an nth-order initial-value problem numerically, we express the nth-order ODE as a system of n first-order equations. In brief, here is how it is done for a second-order initial-value problem: First, solve for y''—that is, put the DE into normal form $y'' = f(x, y, y')$—and then let $y' = u$. For example, if we substitute $y' = u$ in

$$\frac{d^2y}{dx^2} = f(x, y, y'), \quad y(x_0) = y_0, \quad y'(x_0) = u_0, \tag{6}$$

then $y'' = u'$ and $y'(x_0) = u(x_0)$, so the initial-value problem (6) becomes

$$\textit{Solve:} \quad \begin{cases} y' = u \\ u' = f(x, y, u) \end{cases}$$
$$\textit{Subject to:} \quad y(x_0) = y_0,\ u(x_0) = u_0.$$

However, it should be noted that a commercial numerical solver *might not* require* that you supply the system.

EXAMPLE 4 Graphical Analysis of Example 3

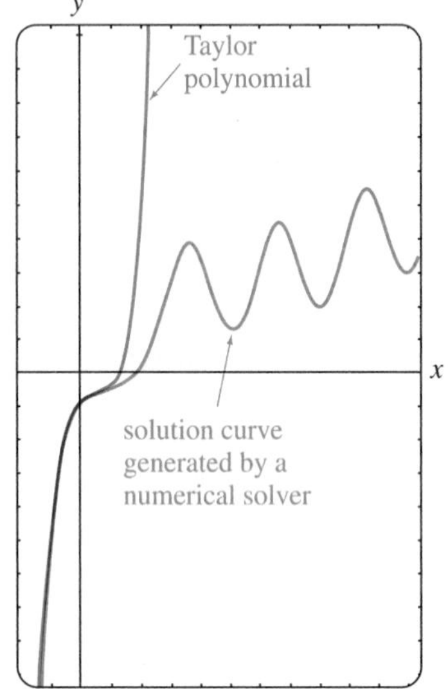

FIGURE 4.10.1 Comparison of two approximate solutions in Example 1

Following the foregoing procedure, we find that the second-order initial-value problem in Example 3 is equivalent to

$$\frac{dy}{dx} = u$$
$$\frac{du}{dx} = x + y - y^2$$

with initial conditions $y(0) = -1$, $u(0) = 1$. With the aid of a numerical solver we get the solution curve shown in blue in Figure 4.10.1. For comparison the graph of the fifth-degree Taylor polynomial $T_5(x) = -1 + x - x^2 + \frac{2}{3}x^3 - \frac{1}{3}x^4 + \frac{1}{5}x^5$ is shown in red. Although we do not know the interval of convergence of the Taylor series obtained in Example 3, the closeness of the two curves in a neighborhood of the origin suggests that the power series may converge on the interval $(-1, 1)$. ≡

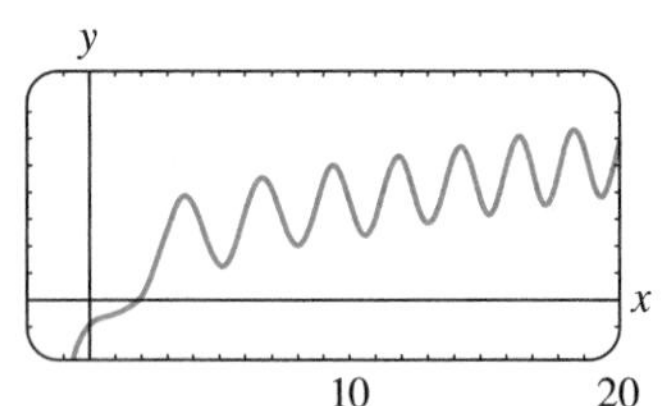

FIGURE 4.10.2 Numerical solution curve for the IVP in (1)

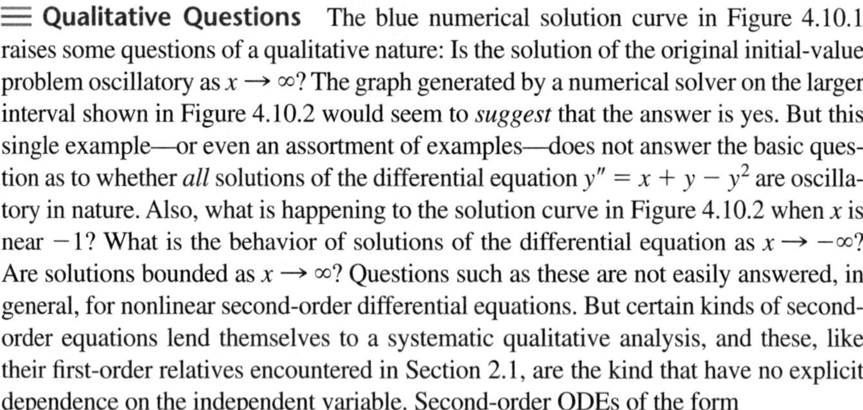

≡ **Qualitative Questions** The blue numerical solution curve in Figure 4.10.1 raises some questions of a qualitative nature: Is the solution of the original initial-value problem oscillatory as $x \to \infty$? The graph generated by a numerical solver on the larger interval shown in Figure 4.10.2 would seem to *suggest* that the answer is yes. But this single example—or even an assortment of examples—does not answer the basic question as to whether *all* solutions of the differential equation $y'' = x + y - y^2$ are oscillatory in nature. Also, what is happening to the solution curve in Figure 4.10.2 when x is near -1? What is the behavior of solutions of the differential equation as $x \to -\infty$? Are solutions bounded as $x \to \infty$? Questions such as these are not easily answered, in general, for nonlinear second-order differential equations. But certain kinds of second-order equations lend themselves to a systematic qualitative analysis, and these, like their first-order relatives encountered in Section 2.1, are the kind that have no explicit dependence on the independent variable. Second-order ODEs of the form

$$F(y, y', y'') = 0 \qquad \text{or} \qquad \frac{d^2y}{dx^2} = f(y, y'),$$

equations free of the independent variable x, are called **autonomous.** The differential equation in Example 2 is autonomous, and because of the presence of the x term on its right-hand side, the equation in Example 3 is nonautonomous. For an in-depth treatment of the topic of stability of autonomous second-order differential equations and autonomous systems of differential equations, refer to Chapter 10 in *Differential Equations with Boundary-Value Problems.*

*Some numerical solvers require only that a second-order differential equation be expressed in normal form $y'' = f(x, y, y')$. The translation of the single equation into a system of two equations is then built into the computer program, since the first equation of the system is always $y' = u$ and the second equation is $u' = f(x, y, u)$.

EXERCISES 4.10

Answers to selected odd-numbered problems begin on page ANS-7.

In Problems 1 and 2 verify that y_1 and y_2 are solutions of the given differential equation but that $y = c_1y_1 + c_2y_2$ is, in general, not a solution.

1. $(y'')^2 = y^2; \quad y_1 = e^x, y_2 = \cos x$

2. $yy'' = \dfrac{1}{2}(y')^2; \quad y_1 = 1, y_2 = x^2$

In Problems 3–8 solve the given differential equation by using the substitution $u = y'$.

3. $y'' + (y')^2 + 1 = 0$

4. $y'' = 1 + (y')^2$

5. $x^2y'' + (y')^2 = 0$

6. $(y + 1)y'' = (y')^2$

7. $y'' + 2y(y')^3 = 0$

8. $y^2y'' = y'$

In Problems 9 and 10 solve the given initial-value problem.

9. $2y'y'' = 1, y(0) = 2, y'(0) = 1$

10. $y'' + x(y')^2 = 0, y(1) = 4, y'(1) = 2$

11. Consider the initial-value problem

$$y'' + yy' = 0, \quad y(0) = 1, y'(0) = -1.$$

(a) Use the DE and a numerical solver to graph the solution curve.

(b) Find an explicit solution of the IVP. Use a graphing utility to graph this solution.

(c) Find an interval of definition for the solution in part (b).

12. Find two solutions of the initial-value problem

$$(y'')^2 + (y')^2 = 1, \quad y\left(\frac{\pi}{2}\right) = \frac{1}{2}, \quad y'\left(\frac{\pi}{2}\right) = \frac{\sqrt{3}}{2}.$$

Use a numerical solver to graph the solution curves.

In Problems 13 and 14 show that the substitution $u = y'$ leads to a Bernoulli equation. Solve this equation (see Section 2.5).

13. $xy'' = y' + (y')^3$

14. $xy'' = y' + x(y')^2$

In Problems 15–18 proceed as in Example 3 and obtain the first six nonzero terms of a Taylor series solution, centered at 0, of the given initial-value problem. Use a numerical solver and a graphing utility to compare the solution curve with the graph of the Taylor polynomial.

15. $y'' = x + y^2, \quad y(0) = 1, y'(0) = 1$

16. $y'' + y^2 = 1, \quad y(0) = 2, y'(0) = 3$

17. $y'' = x^2 + y^2 - 2y', \quad y(0) = 1, y'(0) = 1$

18. $y'' = e^y, \quad y(0) = 0, y'(0) = -1$

19. In calculus the curvature of a curve that is defined by a function $y = f(x)$ is defined as

$$\kappa = \frac{y''}{[1 + (y')^2]^{3/2}}.$$

Find $y = f(x)$ for which $\kappa = 1$. [*Hint*: For simplicity, ignore constants of integration.]

Discussion Problems

20. In Problem 1 we saw that $\cos x$ and e^x were solutions of the nonlinear equation $(y'')^2 - y^2 = 0$. Verify that $\sin x$ and e^{-x} are also solutions. Without attempting to solve the differential equation, discuss how these explicit solutions can be found by using knowledge about linear equations. Without attempting to verify, discuss why the linear combinations $y = c_1e^x + c_2e^{-x} + c_3 \cos x + c_4 \sin x$ and $y = c_2e^{-x} + c_4 \sin x$ are not, in general, solutions, but the two special linear combinations $y = c_1e^x + c_2e^{-x}$ and $y = c_3 \cos x + c_4 \sin x$ *must* satisfy the differential equation.

21. Discuss how the method of reduction of order considered in this section can be applied to the third-order differential equation $y''' = \sqrt{1 + (y'')^2}$. Carry out your ideas and solve the equation.

22. Discuss how to find an alternative two-parameter family of solutions for the nonlinear differential equation $y'' = 2x(y')^2$ in Example 1. [*Hint*: Suppose that $-c_1^2$ is used as the constant of integration instead of $+c_1^2$.]

Mathematical Models

23. Motion in a Force Field A mathematical model for the position $x(t)$ of a body moving rectilinearly on the x-axis in an inverse-square force field is given by

$$\frac{d^2x}{dt^2} = -\frac{k^2}{x^2}.$$

Suppose that at $t = 0$ the body starts from rest from the position $x = x_0$, $x_0 > 0$. Show that the velocity of the body at time t is given by $v^2 = 2k^2(1/x - 1/x_0)$. Use the last expression and a CAS to carry out the integration to express time t in terms of x.

24. A mathematical model for the position $x(t)$ of a moving object is

$$\frac{d^2x}{dt^2} + \sin x = 0.$$

Use a numerical solver to graphically investigate the solutions of the equation subject to $x(0) = 0$, $x'(0) = x_1$,

$x_1 \geq 0$. Discuss the motion of the object for $t \geq 0$ and for various choices of x_1. Investigate the equation

$$\frac{d^2x}{dt^2} + \frac{dx}{dt} + \sin x = 0$$

in the same manner. Give a possible physical interpretation of the dx/dt term.

CHAPTER 4 IN REVIEW

Answers to selected odd-numbered problems begin on page ANS-7.

Answer Problems 1–10 without referring back to the text. Fill in the blank or answer true or false.

1. The only solution of the initial-value problem $y'' + x^2y = 0$, $y(0) = 0$, $y'(0) = 0$ is __________.

2. For the method of undetermined coefficients, the assumed form of the particular solution y_p for $y'' - y = 1 + e^x$ is __________.

3. A constant multiple of a solution of a linear differential equation is also a solution. __________

4. If the set consisting of two functions f_1 and f_2 is linearly independent on an interval I, then the Wronskian $W(f_1, f_2) \neq 0$ for all x in I. __________

5. If $y = \sin 5x$ is a solution of a homogeneous linear second-order differential with constant coefficients, then the general solution of the DE is _______.

6. If $y = 1 - x + 6x^2 + 3e^x$ is a solution of a homogeneous fourth-order linear differential equation with constant coefficients, then the roots of the auxiliary equation are _______.

7. If $y = c_1x^2 + c_2x^2 \ln x$, $x > 0$, is the general solution of a homogeneous second-order Cauchy-Euler equation, then the DE is _______.

8. $y_p = Ax^2$ is particular solution of $y''' + y'' = 1$ for $A =$ _______.

9. If $y_{p_1} = x$ is a particular solution of $y'' + y = x$ and $y_{p_2} = x^2 - 2$ is a particular solution of $y'' + y = x^2$, then a particular solution of $y'' + y = x^2 + x$ is _______.

10. If $y_1 = e^x$ and $y_2 = e^{-x}$ are solutions of homogeneous linear differential equation, then necessarily $y = -5e^{-x} + 10e^x$ is also a solution of the DE. _______

11. Give an interval over which the set of two functions $f_1(x) = x^2$ and $f_2(x) = x\,|x|$ is linearly independent. Then give an interval over which the set consisting of f_1 and f_2 is linearly dependent.

12. Without the aid of the Wronskian, determine whether the given set of functions is linearly independent or linearly dependent on the indicated interval.

(a) $f_1(x) = \ln x, f_2(x) = \ln x^2, (0, \infty)$

(b) $f_1(x) = x^n, f_2(x) = x^{n+1}, n = 1, 2, \ldots, (-\infty, \infty)$

(c) $f_1(x) = x, f_2(x) = x + 1, (-\infty, \infty)$

(d) $f_1(x) = \cos\left(x + \dfrac{\pi}{2}\right), f_2(x) = \sin x, (-\infty, \infty)$

(e) $f_1(x) = 0, f_2(x) = x, (-5, 5)$

(f) $f_1(x) = 2, f_2(x) = 2x, (-\infty, \infty)$

(g) $f_1(x) = x^2, f_2(x) = 1 - x^2, f_3(x) = 2 + x^2, (-\infty, \infty)$

(h) $f_1(x) = xe^{x+1}, f_2(x) = (4x - 5)e^x,$ $f_3(x) = xe^x, (-\infty, \infty)$

13. Suppose $m_1 = 3$, $m_2 = -5$, and $m_3 = 1$ are roots of multiplicity one, two, and three, respectively, of an auxiliary equation. Write down the general solution of the corresponding homogeneous linear DE if it is

(a) an equation with constant coefficients,

(b) a Cauchy-Euler equation.

14. Consider the differential equation $ay'' + by' + cy = g(x)$, where a, b, and c are constants. Choose the input functions $g(x)$ for which the method of undetermined coefficients is applicable and the input functions for which the method of variation of parameters is applicable.

(a) $g(x) = e^x \ln x$ **(b)** $g(x) = x^3 \cos x$

(c) $g(x) = \dfrac{\sin x}{e^x}$ **(d)** $g(x) = 2x^{-2}e^x$

(e) $g(x) = \sin^2 x$ **(f)** $g(x) = \dfrac{e^x}{\sin x}$

In Problems 15–30 use the procedures developed in this chapter to find the general solution of each differential equation.

15. $y'' - 2y' - 2y = 0$

16. $2y'' + 2y' + 3y = 0$

17. $y''' + 10y'' + 25y' = 0$

18. $2y''' + 9y'' + 12y' + 5y = 0$

19. $3y''' + 10y'' + 15y' + 4y = 0$

20. $2y^{(4)} + 3y''' + 2y'' + 6y' - 4y = 0$

21. $y'' - 3y' + 5y = 4x^3 - 2x$

22. $y'' - 2y' + y = x^2e^x$

23. $y''' - 5y'' + 6y' = 8 + 2\sin x$

24. $y''' - y'' = 6$

25. $y'' - 2y' + 2y = e^x \tan x$

26. $y'' - y = \dfrac{2e^x}{e^x + e^{-x}}$

27. $6x^2y'' + 5xy' - y = 0$

28. $2x^3y''' + 19x^2y'' + 39xy' + 9y = 0$

29. $x^2y'' - 4xy' + 6y = 2x^4 + x^2$

30. $x^2y'' - xy' + y = x^3$

31. Write down the form of the general solution $y = y_c + y_p$ of the given differential equation in the two cases $\omega \neq \alpha$ and $\omega = \alpha$. Do not determine the coefficients in y_p.

(a) $y'' + \omega^2 y = \sin \alpha x$ **(b)** $y'' - \omega^2 y = e^{\alpha x}$

32. **(a)** Given that $y = \sin x$ is a solution of

$$y^{(4)} + 2y''' + 11y'' + 2y' + 10y = 0,$$

find the general solution of the DE *without the aid of a calculator or a computer.*

(b) Find a linear second-order differential equation with constant coefficients for which $y_1 = 1$ and $y_2 = e^{-x}$ are solutions of the associated homogeneous equation and $y_p = \frac{1}{2}x^2 - x$ is a particular solution of the nonhomogeneous equation.

33. **(a)** Write the general solution of the fourth-order DE $y^{(4)} - 2y'' + y = 0$ entirely in terms of hyperbolic functions.

(b) Write down the form of a particular solution of $y^{(4)} - 2y'' + y = \sinh x$.

34. Consider the differential equation

$$x^2y'' - (x^2 + 2x)y' + (x + 2)y = x^3.$$

Verify that $y_1 = x$ is one solution of the associated homogeneous equation. Then show that the method of reduction of order discussed in Section 4.2 leads to a second solution y_2 of the homogeneous equation as well as a particular solution y_p of the nonhomogeneous equation. Form the general solution of the DE on the interval $(0, \infty)$.

In Problems 35–40 solve the given differential equation subject to the indicated conditions.

35. $y'' - 2y' + 2y = 0, \quad y(\pi/2) = 0, y(\pi) = -1$

36. $y'' + 2y' + y = 0, \quad y(-1) = 0, y'(0) = 0$

37. $y'' - y = x + \sin x, \quad y(0) = 2, y'(0) = 3$

38. $y'' + y = \sec^3 x, \quad y(0) = 1, y'(0) = \frac{1}{2}$

39. $y'y'' = 4x, \quad y(1) = 5, y'(1) = 2$

40. $2y'' = 3y^2, \quad y(0) = 1, y'(0) = 1$

41. **(a)** Use a CAS as an aid in finding the roots of the auxiliary equation for

$$12y^{(4)} + 64y''' + 59y'' - 23y' - 12y = 0.$$

Give the general solution of the equation.

(b) Solve the DE in part (a) subject to the initial conditions $y(0) = -1$, $y'(0) = 2$, $y''(0) = 5$, $y'''(0) = 0$. Use a CAS as an aid in solving the resulting systems of four equations in four unknowns.

42. Find a member of the family of solutions of $xy'' + y' + \sqrt{x} = 0$ whose graph is tangent to the x-axis at $x = 1$. Use a graphing utility to graph the solution curve.

In Problems 43–46 use systematic elimination to solve the given system.

43. $\dfrac{dx}{dt} + \dfrac{dy}{dt} = 2x + 2y + 1$

$\dfrac{dx}{dt} + 2\dfrac{dy}{dt} = y + 3$

44. $\dfrac{dx}{dt} = 2x + y + t - 2$

$\dfrac{dy}{dt} = 3x + 4y - 4t$

45. $(D - 2)x - y = -e^t$

$-3x + (D - 4)y = -7e^t$

46. $(D + 2)x + (D + 1)y = \sin 2t$

$5x + (D + 3)y = \cos 2t$

5 Modeling with Higher-Order Differential Equations

We have seen that a single differential equation can serve as a mathematical model for diverse physical systems. For this reason we examine just one application, the motion of a mass attached to a spring, in great detail in Section 5.1. Except for terminology and physical interpretations of the four terms in the linear differential equation

$$a\frac{d^2y}{dt^2} + b\frac{dy}{dt} + cy = g(t),$$

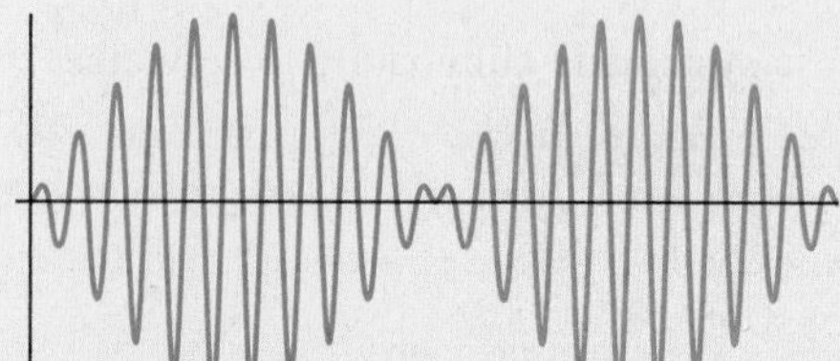

the mathematics of, say, an electrical series circuit is identical to that of a vibrating spring/mass system. Forms of this linear second-order equation appear in the analysis of problems in many different areas of science and engineering. In Section 5.1 we deal exclusively with initial-value problems, whereas in Section 5.2 we examine applications described by boundary-value problems. In Section 5.2 we also see how some boundary-value problems lead to the important concepts of *eigenvalues* and *eigenfunctions*. Section 5.3 begins with a discussion on the differences between linear and nonlinear springs; we then show how the simple pendulum and a suspended wire lead to nonlinear models.

5.1 LINEAR MODELS: INITIAL-VALUE PROBLEMS

REVIEW MATERIAL

- Sections 4.1, 4.3, and 4.4
- Problems 29–36 in Exercises 4.3
- Problems 27–36 in Exercises 4.4

INTRODUCTION In this section we are going to consider several linear dynamical systems in which each mathematical model is a second-order differential equation with constant coefficients along with initial conditions specified at a time that we shall take to be $t = 0$:

$$a\frac{d^2y}{dt^2} + b\frac{dy}{dt} + cy = g(t), \quad y(0) = y_0, \quad y'(0) = y_1.$$

Recall that the function g is the **input, driving function,** or **forcing function** of the system. A solution $y(t)$ of the differential equation on an interval I containing $t = 0$ that satisfies the initial conditions is called the **output** or **response** of the system.

5.1.1 SPRING/MASS SYSTEMS: FREE UNDAMPED MOTION

Hooke's Law Suppose that a flexible spring is suspended vertically from a rigid support and then a mass m is attached to its free end. The amount of stretch, or elongation, of the spring will of course depend on the mass; masses with different weights stretch the spring by differing amounts. By Hooke's law the spring itself exerts a restoring force F opposite to the direction of elongation and proportional to the amount of elongation s. Simply stated, $F = ks$, where k is a constant of proportionality called the **spring constant.** The spring is essentially characterized by the number k. For example, if a mass weighing 10 pounds stretches a spring $\frac{1}{2}$ foot, then $10 = k\left(\frac{1}{2}\right)$ implies $k = 20$ lb/ft. Necessarily then, a mass weighing, say, 8 pounds stretches the same spring only $\frac{2}{5}$ foot.

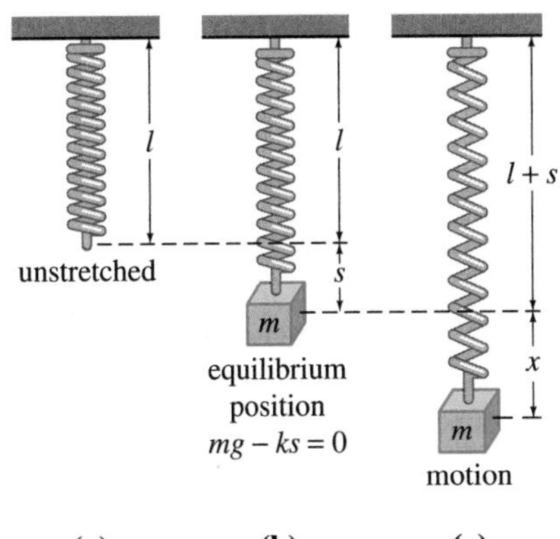

FIGURE 5.1.1 Spring/mass system

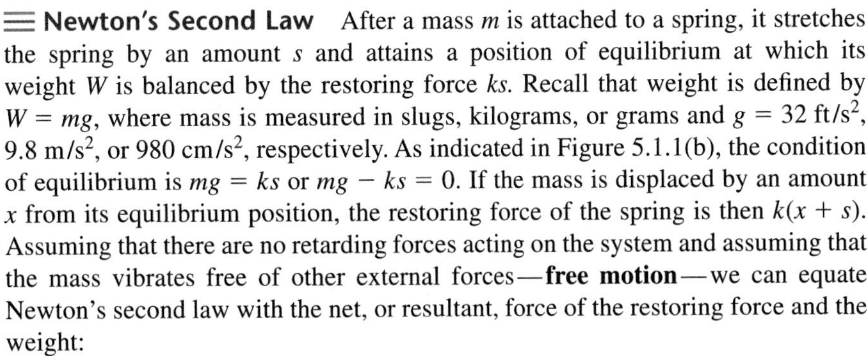

Newton's Second Law After a mass m is attached to a spring, it stretches the spring by an amount s and attains a position of equilibrium at which its weight W is balanced by the restoring force ks. Recall that weight is defined by $W = mg$, where mass is measured in slugs, kilograms, or grams and $g = 32$ ft/s^2, 9.8 m/s^2, or 980 cm/s^2, respectively. As indicated in Figure 5.1.1(b), the condition of equilibrium is $mg = ks$ or $mg - ks = 0$. If the mass is displaced by an amount x from its equilibrium position, the restoring force of the spring is then $k(x + s)$. Assuming that there are no retarding forces acting on the system and assuming that the mass vibrates free of other external forces—**free motion**—we can equate Newton's second law with the net, or resultant, force of the restoring force and the weight:

$$m\frac{d^2x}{dt^2} = -k(s + x) + mg = -kx + \underbrace{mg - ks}_{\text{zero}} = -kx. \tag{1}$$

The negative sign in (1) indicates that the restoring force of the spring acts opposite to the direction of motion. Furthermore, we adopt the convention that displacements measured *below* the equilibrium position $x = 0$ are positive. See Figure 5.1.2.

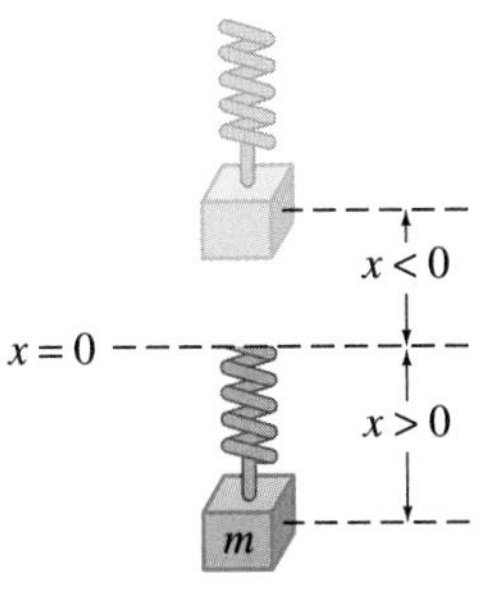

FIGURE 5.1.2 Direction below the equilibrium position is positive.

DE of Free Undamped Motion By dividing (1) by the mass m, we obtain the second-order differential equation $d^2x/dt^2 + (k/m)x = 0$, or

$$\frac{d^2x}{dt^2} + \omega^2 x = 0, \tag{2}$$

where $\omega^2 = k/m$. Equation (2) is said to describe **simple harmonic motion** or **free undamped motion.** Two obvious initial conditions associated with (2) are $x(0) = x_0$ and $x'(0) = x_1$, the initial displacement and initial velocity of the mass, respectively. For example, if $x_0 > 0$, $x_1 < 0$, the mass starts from a point *below* the equilibrium position with an imparted *upward* velocity. When $x'(0) = 0$, the mass is said to be released from rest. For example, if $x_0 < 0$, $x_1 = 0$, the mass is released from *rest* from a point $|x_0|$ units *above* the equilibrium position.

Equation of Motion To solve equation (2), we note that the solutions of its auxiliary equation $m^2 + \omega^2 = 0$ are the complex numbers $m_1 = \omega i$, $m_2 = -\omega i$. Thus from (8) of Section 4.3 we find the general solution of (2) to be

$$x(t) = c_1 \cos \omega t + c_2 \sin \omega t. \tag{3}$$

The **period** of motion described by (3) is $T = 2\pi/\omega$. The number T represents the time (measured in seconds) it takes the mass to execute one cycle of motion. A cycle is one complete oscillation of the mass, that is, the mass m moving from, say, the lowest point below the equilibrium position to the point highest above the equilibrium position and then back to the lowest point. From a graphical viewpoint $T = 2\pi/\omega$ seconds is the length of the time interval between two successive maxima (or minima) of $x(t)$. Keep in mind that a maximum of $x(t)$ is a positive displacement corresponding to the mass attaining its greatest distance below the equilibrium position, whereas a minimum of $x(t)$ is negative displacement corresponding to the mass attaining its greatest height above the equilibrium position. We refer to either case as an **extreme displacement** of the mass. The **frequency** of motion is $f = 1/T = \omega/2\pi$ and is the number of cycles completed each second. For example, if $x(t) = 2\cos 3\pi t - 4\sin 3\pi t$, then the period is $T = 2\pi/3\pi = 2/3$ s, and the frequency is $f = 3/2$ cycles/s. From a graphical viewpoint the graph of $x(t)$ repeats every $\frac{2}{3}$ second, that is, $x\left(t + \frac{2}{3}\right) = x(t)$, and $\frac{3}{2}$ cycles of the graph are completed each second (or, equivalently, three cycles of the graph are completed every 2 seconds). The number $\omega = \sqrt{k/m}$ (measured in radians per second) is called the **circular frequency** of the system. Depending on which text you read, both $f = \omega/2\pi$ and ω are also referred to as the **natural frequency** of the system. Finally, when the initial conditions are used to determine the constants c_1 and c_2 in (3), we say that the resulting particular solution or response is the **equation of motion.**

EXAMPLE 1 Free Undamped Motion

A mass weighing 2 pounds stretches a spring 6 inches. At $t = 0$ the mass is released from a point 8 inches below the equilibrium position with an upward velocity of $\frac{4}{3}$ ft/s. Determine the equation of motion.

SOLUTION Because we are using the engineering system of units, the measurements given in terms of inches must be converted into feet: 6 in. $= \frac{1}{2}$ ft; 8 in. $= \frac{2}{3}$ ft. In addition, we must convert the units of weight given in pounds into units of mass. From $m = W/g$ we have $m = \frac{2}{32} = \frac{1}{16}$ slug. Also, from Hooke's law, $2 = k\left(\frac{1}{2}\right)$ implies that the spring constant is $k = 4$ lb/ft. Hence (1) gives

$$\frac{1}{16}\frac{d^2x}{dt^2} = -4x \quad \text{or} \quad \frac{d^2x}{dt^2} + 64x = 0.$$

The initial displacement and initial velocity are $x(0) = \frac{2}{3}$, $x'(0) = -\frac{4}{3}$, where the negative sign in the last condition is a consequence of the fact that the mass is given an initial velocity in the negative, or upward, direction.

Now $\omega^2 = 64$ or $\omega = 8$, so the general solution of the differential equation is

$$x(t) = c_1 \cos 8t + c_2 \sin 8t. \tag{4}$$

Applying the initial conditions to $x(t)$ and $x'(t)$ gives $c_1 = \frac{2}{3}$ and $c_2 = -\frac{1}{6}$. Thus the equation of motion is

$$x(t) = \frac{2}{3}\cos 8t - \frac{1}{6}\sin 8t. \tag{5}$$

≡

≡ Alternative Forms of $x(t)$ When $c_1 \neq 0$ and $c_2 \neq 0$, the actual **amplitude** A of free vibrations is not obvious from inspection of equation (3). For example, although the mass in Example 1 is initially displaced $\frac{2}{3}$ foot beyond the equilibrium position, the amplitude of vibrations is a number larger than $\frac{2}{3}$. Hence it is often convenient to convert a solution of form (3) to the simpler form

$$x(t) = A\sin(\omega t + \phi), \tag{6}$$

where $A = \sqrt{c_1^2 + c_2^2}$ and ϕ is a **phase angle** defined by

$$\left.\begin{aligned} \sin\phi &= \frac{c_1}{A} \\ \cos\phi &= \frac{c_2}{A} \end{aligned}\right\} \tan\phi = \frac{c_1}{c_2}. \tag{7}$$

To verify this, we expand (6) by the addition formula for the sine function:

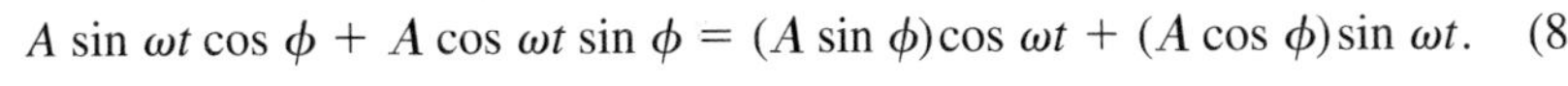

$$A\sin\omega t\cos\phi + A\cos\omega t\sin\phi = (A\sin\phi)\cos\omega t + (A\cos\phi)\sin\omega t. \tag{8}$$

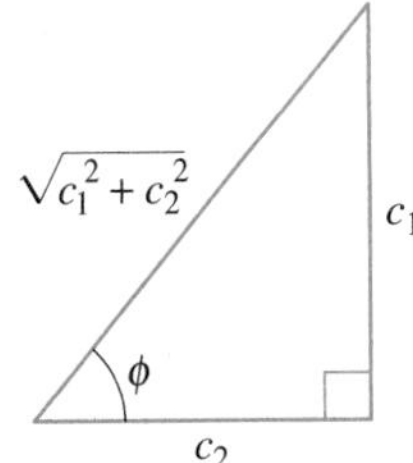

FIGURE 5.1.3 A relationship between $c_1 > 0$, $c_2 > 0$ and phase angle ϕ

It follows from Figure 5.1.3 that if ϕ is defined by

$$\sin\phi = \frac{c_1}{\sqrt{c_1^2 + c_2^2}} = \frac{c_1}{A}, \qquad \cos\phi = \frac{c_2}{\sqrt{c_1^2 + c_2^2}} = \frac{c_2}{A},$$

then (8) becomes

$$A\frac{c_1}{A}\cos\omega t + A\frac{c_2}{A}\sin\omega t = c_1\cos\omega t + c_2\sin\omega t = x(t).$$

EXAMPLE 2 Alternative Form of Solution (5)

In view of the foregoing discussion we can write solution (5) in the alternative form $x(t) = A\sin(8t + \phi)$. Computation of the amplitude is straightforward, $A = \sqrt{(\frac{2}{3})^2 + (-\frac{1}{6})^2} = \sqrt{\frac{17}{36}} \approx 0.69$ ft, but some care should be exercised in computing the phase angle ϕ defined by (7). With $c_1 = \frac{2}{3}$ and $c_2 = -\frac{1}{6}$ we find $\tan\phi = -4$, and a calculator then gives $\tan^{-1}(-4) = -1.326$ rad. This is *not* the phase angle, since $\tan^{-1}(-4)$ is located in the *fourth quadrant* and therefore contradicts the fact that $\sin\phi > 0$ and $\cos\phi < 0$ because $c_1 > 0$ and $c_2 < 0$. Hence we must take ϕ to be the *second-quadrant* angle $\phi = \pi + (-1.326) = 1.816$ rad. Thus (5) is the same as

$$x(t) = \frac{\sqrt{17}}{6}\sin(8t + 1.816). \tag{9}$$

The period of this function is $T = 2\pi/8 = \pi/4$ s. ≡

You should be aware that some instructors in science and engineering prefer that (3) be expressed as a shifted cosine function

$$x(t) = A\cos(\omega t - \phi), \tag{6'}$$

where $A = \sqrt{c_1^2 + c_2^2}$. In this case the radian measured angle ϕ is defined in slightly different manner than in (7):

$$\left.\begin{aligned} \sin\phi &= \frac{c_2}{A} \\ \cos\phi &= \frac{c_1}{A} \end{aligned}\right\} \tan\phi = \frac{c_2}{c_1}. \tag{7'}$$

For example, in Example 2 with $c_1 = \frac{2}{3}$ and $c_2 = -\frac{1}{6}$, (7′) indicates that $\tan\phi = -\frac{1}{4}$. Because $\sin\phi < 0$ and $\cos\phi > 0$ the angle ϕ lies in the fourth quadrant and so rounded to three decimal places $\phi = \tan^{-1}(-\frac{1}{4}) = -0.245$ rad. From (6′) we obtain a second alternative form of solution (5):

$$x(t) = \frac{\sqrt{17}}{6}\cos(8t - (-0.245)) \quad \text{or} \quad x(t) = \frac{\sqrt{17}}{6}\cos(8t + 0.245).$$

Graphical Interpretation Figure 5.1.4(a) illustrates the mass in Example 2 going through approximately two complete cycles of motion. Reading left to right, the first five positions (marked with black dots) correspond to the initial position of the mass below the equilibrium position $\left(x = \frac{2}{3}\right)$, the mass passing through the equilibrium position for the first time heading upward ($x = 0$), the mass at its extreme displacement above the equilibrium position ($x = -\sqrt{17}/6$), the mass at the equilibrium position for the second time heading downward ($x = 0$), and the mass at its extreme displacement below the equilibrium position ($x = \sqrt{17}/6$). The black dots on the graph of (9), given in Figure 5.1.4(b), also agree with the five positions just given. Note, however, that in Figure 5.1.4(b) the positive direction in the tx-plane is the usual upward

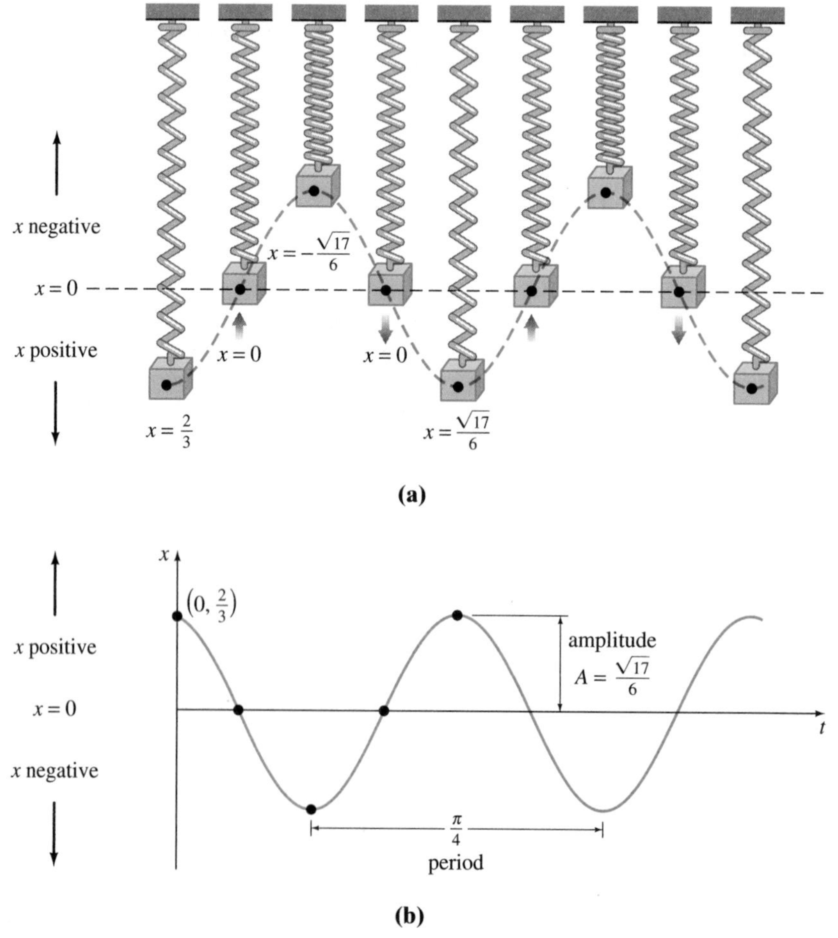

FIGURE 5.1.4 Simple harmonic motion

direction and so is opposite to the positive direction indicated in Figure 5.1.4(a). Hence the solid blue graph representing the motion of the mass in Figure 5.1.4(b) is the reflection through the t-axis of the blue dashed curve in Figure 5.1.4(a).

Form (6) is very useful because it is easy to find values of time for which the graph of $x(t)$ crosses the positive t-axis (the line $x = 0$). We observe that $\sin(\omega t + \phi) = 0$ when $\omega t + \phi = n\pi$, where n is a nonnegative integer.

Systems with Variable Spring Constants In the model discussed above we assumed an ideal world—a world in which the physical characteristics of the spring do not change over time. In the nonideal world, however, it seems reasonable to expect that when a spring/mass system is in motion for a long period, the spring will weaken; in other words, the "spring constant" will vary—or, more specifically, decay—with time. In one model for the **aging spring** the spring constant k in (1) is replaced by the decreasing function $K(t) = ke^{-\alpha t}$, $k > 0$, $\alpha > 0$. The linear differential equation $mx'' + ke^{-\alpha t}x = 0$ cannot be solved by the methods that were considered in Chapter 4. Nevertheless, we can obtain two linearly independent solutions using the methods in Chapter 6. See Problem 15 in Exercises 5.1, Example 4 in Section 6.4, and Problems 33 and 39 in Exercises 6.4.

When a spring/mass system is subjected to an environment in which the temperature is rapidly decreasing, it might make sense to replace the constant k with $K(t) = kt$, $k > 0$, a function that increases with time. The resulting model, $mx'' + ktx = 0$, is a form of **Airy's differential equation.** Like the equation for an aging spring, Airy's equation can be solved by the methods of Chapter 6. See Problem 16 in Exercises 5.1, Example 5 in Section 6.2, and Problems 34, 35, and 40 in Exercise 6.4.

5.1.2 SPRING/MASS SYSTEMS: FREE DAMPED MOTION

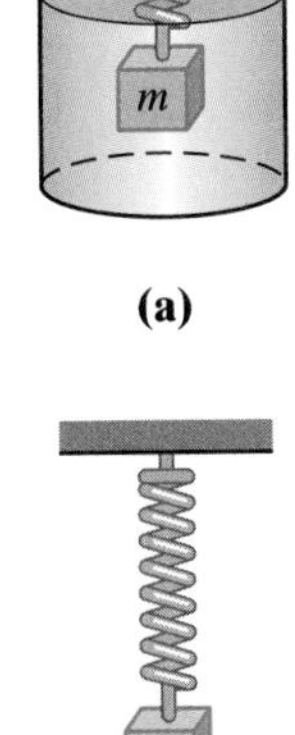

FIGURE 5.1.5 Damping devices

The concept of free harmonic motion is somewhat unrealistic, since the motion described by equation (1) assumes that there are no retarding forces acting on the moving mass. Unless the mass is suspended in a perfect vacuum, there will be at least a resisting force due to the surrounding medium. As Figure 5.1.5 shows, the mass could be suspended in a viscous medium or connected to a dashpot damping device.

DE of Free Damped Motion In the study of mechanics, damping forces acting on a body are considered to be proportional to a power of the instantaneous velocity. In particular, we shall assume throughout the subsequent discussion that this force is given by a constant multiple of dx/dt. When no other external forces are impressed on the system, it follows from Newton's second law that

$$m\frac{d^2x}{dt^2} = -kx - \beta\frac{dx}{dt}, \tag{10}$$

where β is a positive *damping constant* and the negative sign is a consequence of the fact that the damping force acts in a direction opposite to the motion.

Dividing (10) by the mass m, we find that the differential equation of **free damped motion** is

$$\frac{d^2x}{dt^2} + \frac{\beta}{m}\frac{dx}{dt} + \frac{k}{m}x = 0$$

or

$$\frac{d^2x}{dt^2} + 2\lambda\frac{dx}{dt} + \omega^2 x = 0, \tag{11}$$

where

$$2\lambda = \frac{\beta}{m}, \qquad \omega^2 = \frac{k}{m}. \tag{12}$$

The symbol 2λ is used only for algebraic convenience because the auxiliary equation is $m^2 + 2\lambda m + \omega^2 = 0$, and the corresponding roots are then

$$m_1 = -\lambda + \sqrt{\lambda^2 - \omega^2}, \qquad m_2 = -\lambda - \sqrt{\lambda^2 - \omega^2}.$$

We can now distinguish three possible cases depending on the algebraic sign of $\lambda^2 - \omega^2$. Since each solution contains the *damping factor* $e^{-\lambda t}$, $\lambda > 0$, the displacements of the mass become negligible as time t increases.

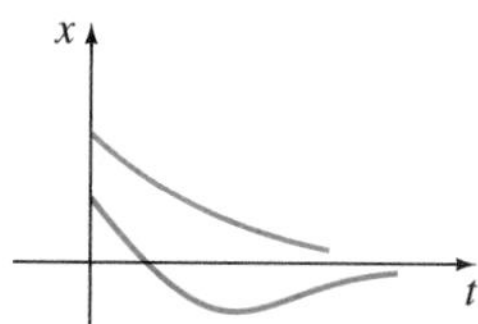

FIGURE 5.1.6 Motion of an overdamped system

Case I: $\lambda^2 - \omega^2 > 0$ In this situation the system is said to be **overdamped** because the damping coefficient β is large when compared to the spring constant k. The corresponding solution of (11) is $x(t) = c_1e^{m_1t} + c_2e^{m_2t}$ or

$$x(t) = e^{-\lambda t}\left(c_1e^{\sqrt{\lambda^2-\omega^2}\,t} + c_2e^{-\sqrt{\lambda^2-\omega^2}\,t}\right). \tag{13}$$

This equation represents a smooth and nonoscillatory motion. Figure 5.1.6 shows two possible graphs of $x(t)$.

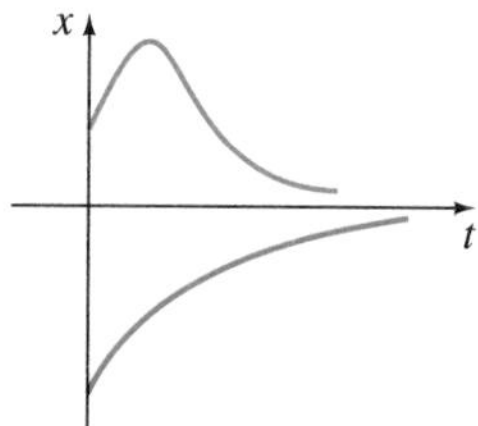

FIGURE 5.1.7 Motion of a critically damped system

Case II: $\lambda^2 - \omega^2 = 0$ The system is said to be **critically damped** because any slight decrease in the damping force would result in oscillatory motion. The general solution of (11) is $x(t) = c_1e^{m_1t} + c_2te^{m_1t}$ or

$$x(t) = e^{-\lambda t}(c_1 + c_2t). \tag{14}$$

Some graphs of typical motion are given in Figure 5.1.7. Notice that the motion is quite similar to that of an overdamped system. It is also apparent from (14) that the mass can pass through the equilibrium position at most one time.

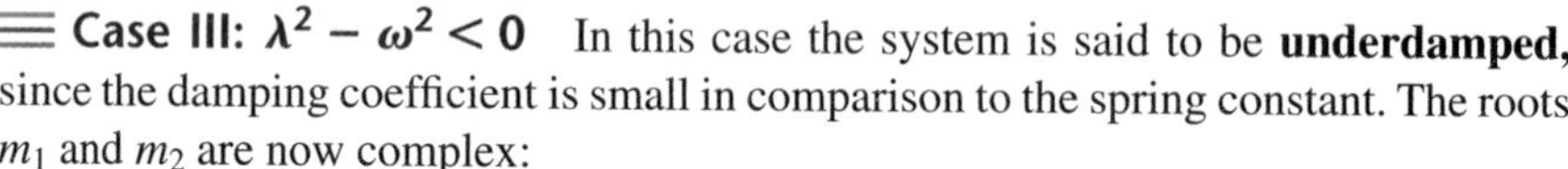

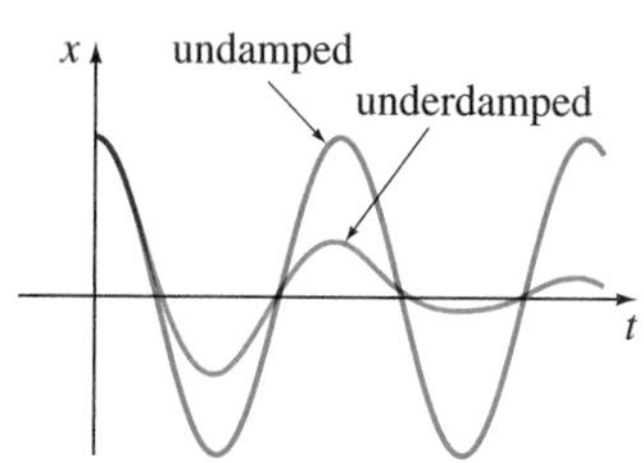

FIGURE 5.1.8 Motion of an underdamped system

Case III: $\lambda^2 - \omega^2 < 0$ In this case the system is said to be **underdamped,** since the damping coefficient is small in comparison to the spring constant. The roots m_1 and m_2 are now complex:

$$m_1 = -\lambda + \sqrt{\omega^2 - \lambda^2}\,i, \qquad m_2 = -\lambda - \sqrt{\omega^2 - \lambda^2}\,i.$$

Thus the general solution of equation (11) is

$$x(t) = e^{-\lambda t}\left(c_1 \cos\sqrt{\omega^2 - \lambda^2}\,t + c_2 \sin\sqrt{\omega^2 - \lambda^2}\,t\right). \tag{15}$$

As indicated in Figure 5.1.8, the motion described by (15) is oscillatory; but because of the coefficient $e^{-\lambda t}$, the amplitudes of vibration $\to 0$ as $t \to \infty$.

EXAMPLE 3 Overdamped Motion

It is readily verified that the solution of the initial-value problem

$$\frac{d^2x}{dt^2} + 5\frac{dx}{dt} + 4x = 0, \quad x(0) = 1, \quad x'(0) = 1$$

is

$$x(t) = \frac{5}{3}e^{-t} - \frac{2}{3}e^{-4t}. \tag{16}$$

The problem can be interpreted as representing the overdamped motion of a mass on a spring. The mass is initially released from a position 1 unit *below* the equilibrium position with a *downward* velocity of 1 ft/s.

To graph $x(t)$, we find the value of t for which the function has an extremum—that is, the value of time for which the first derivative (velocity) is zero. Differentiating (16) gives $x'(t) = -\frac{5}{3}e^{-t} + \frac{8}{3}e^{-4t}$, so $x'(t) = 0$ implies that

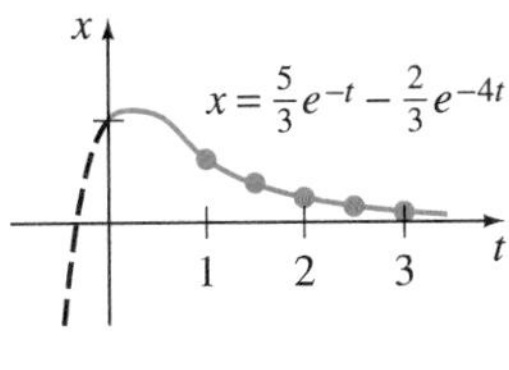

(a)

t	$x(t)$
1	0.601
1.5	0.370
2	0.225
2.5	0.137
3	0.083

(b)

FIGURE 5.1.9 Overdamped system in Example 3

$e^{3t} = \frac{8}{5}$ or $t = \frac{1}{3}\ln\frac{8}{5} = 0.157$. It follows from the first derivative test, as well as our physical intuition, that $x(0.157) = 1.069$ ft is actually a maximum. In other words, the mass attains an extreme displacement of 1.069 feet below the equilibrium position.

We should also check to see whether the graph crosses the t-axis—that is, whether the mass passes through the equilibrium position. This cannot happen in this instance because the equation $x(t) = 0$, or $e^{3t} = \frac{2}{5}$, has the physically irrelevant solution $t = \frac{1}{3}\ln\frac{2}{5} = -0.305$.

The graph of $x(t)$, along with some other pertinent data, is given in Figure 5.1.9. ≡

EXAMPLE 4 Critically Damped Motion

A mass weighing 8 pounds stretches a spring 2 feet. Assuming that a damping force numerically equal to 2 times the instantaneous velocity acts on the system, determine the equation of motion if the mass is initially released from the equilibrium position with an upward velocity of 3 ft/s.

SOLUTION From Hooke's law we see that $8 = k(2)$ gives $k = 4$ lb/ft and that $W = mg$ gives $m = \frac{8}{32} = \frac{1}{4}$ slug. The differential equation of motion is then

$$\frac{1}{4}\frac{d^2x}{dt^2} = -4x - 2\frac{dx}{dt} \quad \text{or} \quad \frac{d^2x}{dt^2} + 8\frac{dx}{dt} + 16x = 0. \tag{17}$$

The auxiliary equation for (17) is $m^2 + 8m + 16 = (m + 4)^2 = 0$, so $m_1 = m_2 = -4$. Hence the system is critically damped, and

$$x(t) = c_1e^{-4t} + c_2te^{-4t}. \tag{18}$$

Applying the initial conditions $x(0) = 0$ and $x'(0) = -3$, we find, in turn, that $c_1 = 0$ and $c_2 = -3$. Thus the equation of motion is

$$x(t) = -3te^{-4t}. \tag{19}$$

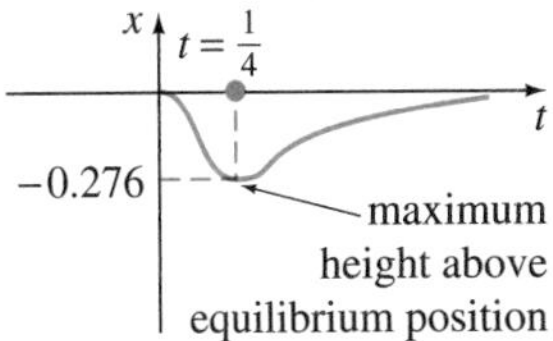

FIGURE 5.1.10 Critically damped system in Example 4

To graph $x(t)$, we proceed as in Example 3. From $x'(t) = -3e^{-4t}(1 - 4t)$ we see that $x'(t) = 0$ when $t = \frac{1}{4}$. The corresponding extreme displacement is $x\left(\frac{1}{4}\right) = -3\left(\frac{1}{4}\right)e^{-1} = -0.276$ ft. As shown in Figure 5.1.10, we interpret this value to mean that the mass reaches a maximum height of 0.276 foot above the equilibrium position. ≡

EXAMPLE 5 Underdamped Motion

A mass weighing 16 pounds is attached to a 5-foot-long spring. At equilibrium the spring measures 8.2 feet. If the mass is initially released from rest at a point 2 feet above the equilibrium position, find the displacements $x(t)$ if it is further known that the surrounding medium offers a resistance numerically equal to the instantaneous velocity.

SOLUTION The elongation of the spring after the mass is attached is $8.2 - 5 = 3.2$ ft, so it follows from Hooke's law that $16 = k(3.2)$ or $k = 5$ lb/ft. In addition, $m = \frac{16}{32} = \frac{1}{2}$ slug, so the differential equation is given by

$$\frac{1}{2}\frac{d^2x}{dt^2} = -5x - \frac{dx}{dt} \quad \text{or} \quad \frac{d^2x}{dt^2} + 2\frac{dx}{dt} + 10x = 0. \tag{20}$$

Proceeding, we find that the roots of $m^2 + 2m + 10 = 0$ are $m_1 = -1 + 3i$ and $m_2 = -1 - 3i$, which then implies that the system is underdamped, and

$$x(t) = e^{-t}(c_1 \cos 3t + c_2 \sin 3t). \tag{21}$$

Finally, the initial conditions $x(0) = -2$ and $x'(0) = 0$ yield $c_1 = -2$ and $c_2 = -\frac{2}{3}$, so the equation of motion is

$$x(t) = e^{-t}\left(-2\cos 3t - \frac{2}{3}\sin 3t\right). \tag{22}$$

Alternative Form of $x(t)$ In a manner identical to the procedure used on page 195, we can write any solution

$$x(t) = e^{-\lambda t}\left(c_1 \cos \sqrt{\omega^2 - \lambda^2}\,t + c_2 \sin \sqrt{\omega^2 - \lambda^2}\,t\right)$$

in the alternative form

$$x(t) = Ae^{-\lambda t}\sin\left(\sqrt{\omega^2 - \lambda^2}\,t + \phi\right), \tag{23}$$

where $A = \sqrt{c_1^2 + c_2^2}$ and the phase angle ϕ is determined from the equations

$$\sin\phi = \frac{c_1}{A}, \qquad \cos\phi = \frac{c_2}{A}, \qquad \tan\phi = \frac{c_1}{c_2}.$$

The coefficient $Ae^{-\lambda t}$ is sometimes called the **damped amplitude** of vibrations. Because (23) is not a periodic function, the number $2\pi/\sqrt{\omega^2 - \lambda^2}$ is called the **quasi period** and $\sqrt{\omega^2 - \lambda^2}/2\pi$ is the **quasi frequency.** The quasi period is the time interval between two successive maxima of $x(t)$. You should verify, for the equation of motion in Example 5, that $A = 2\sqrt{10}/3$ and $\phi = 4.391$. Therefore an equivalent form of (22) is

$$x(t) = \frac{2\sqrt{10}}{3}e^{-t}\sin(3t + 4.391).$$

5.1.3 SPRING/MASS SYSTEMS: DRIVEN MOTION

DE of Driven Motion with Damping Suppose we now take into consideration an external force $f(t)$ acting on a vibrating mass on a spring. For example, $f(t)$ could represent a driving force causing an oscillatory vertical motion of the support of the spring. See Figure 5.1.11. The inclusion of $f(t)$ in the formulation of Newton's second law gives the differential equation of **driven** or **forced motion:**

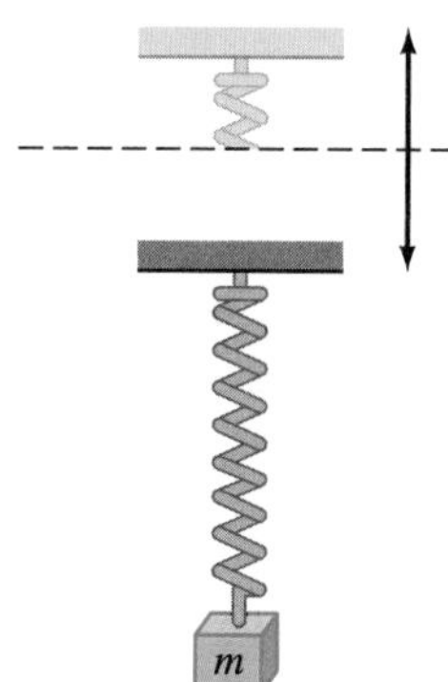

FIGURE 5.1.11 Oscillatory vertical motion of the support

$$m\frac{d^2x}{dt^2} = -kx - \beta\frac{dx}{dt} + f(t). \tag{24}$$

Dividing (24) by m gives

$$\frac{d^2x}{dt^2} + 2\lambda\frac{dx}{dt} + \omega^2 x = F(t), \tag{25}$$

where $F(t) = f(t)/m$ and, as in the preceding section, $2\lambda = \beta/m$, $\omega^2 = k/m$. To solve the latter nonhomogeneous equation, we can use either the method of undetermined coefficients or variation of parameters.

EXAMPLE 6 **Interpretation of an Initial-Value Problem**

Interpret and solve the initial-value problem

$$\frac{1}{5}\frac{d^2x}{dt^2} + 1.2\frac{dx}{dt} + 2x = 5\cos 4t, \quad x(0) = \frac{1}{2}, \quad x'(0) = 0. \tag{26}$$

SOLUTION We can interpret the problem to represent a vibrational system consisting of a mass ($m = \frac{1}{5}$ slug or kilogram) attached to a spring ($k = 2$ lb/ft or N/m).

The mass is initially released from rest $\frac{1}{2}$ unit (foot or meter) below the equilibrium position. The motion is damped ($\beta = 1.2$) and is being driven by an external periodic ($T = \pi/2$ s) force beginning at $t = 0$. Intuitively, we would expect that even with damping, the system would remain in motion until such time as the forcing function was "turned off," in which case the amplitudes would diminish. However, as the problem is given, $f(t) = 5 \cos 4t$ will remain "on" forever.

We first multiply the differential equation in (26) by 5 and solve

$$\frac{dx^2}{dt^2} + 6\frac{dx}{dt} + 10x = 0$$

by the usual methods. Because $m_1 = -3 + i$, $m_2 = -3 - i$, it follows that $x_c(t) = e^{-3t}(c_1 \cos t + c_2 \sin t)$. Using the method of undetermined coefficients, we assume a particular solution of the form $x_p(t) = A \cos 4t + B \sin 4t$. Differentiating $x_p(t)$ and substituting into the DE gives

$$x_p'' + 6x_p' + 10x_p = (-6A + 24B) \cos 4t + (-24A - 6B) \sin 4t = 25 \cos 4t.$$

The resulting system of equations

$$-6A + 24B = 25, \qquad -24A - 6B = 0$$

yields $A = -\frac{25}{102}$ and $B = \frac{50}{51}$. It follows that

$$x(t) = e^{-3t}(c_1 \cos t + c_2 \sin t) - \frac{25}{102} \cos 4t + \frac{50}{51} \sin 4t. \tag{27}$$

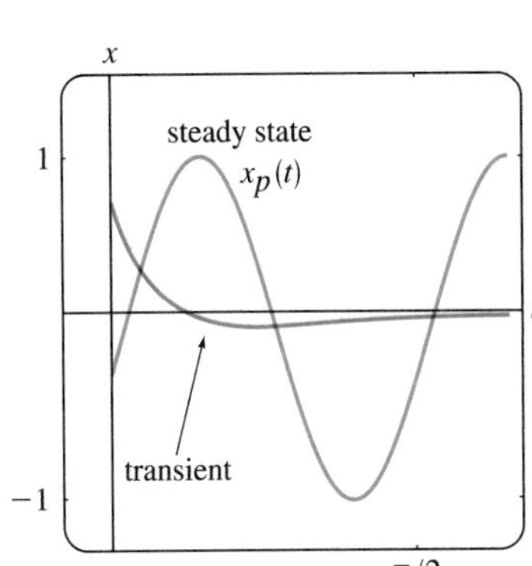

(a)

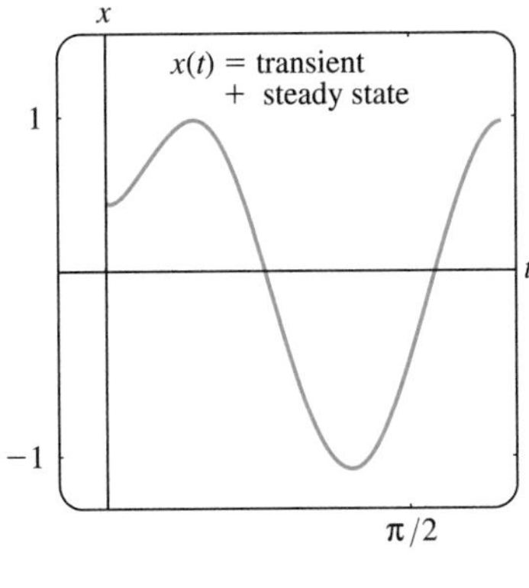

(b)

FIGURE 5.1.12 Graph of solution in (28) of Example 6

When we set $t = 0$ in the above equation, we obtain $c_1 = \frac{38}{51}$. By differentiating the expression and then setting $t = 0$, we also find that $c_2 = -\frac{86}{51}$. Therefore the equation of motion is

$$x(t) = e^{-3t}\left(\frac{38}{51} \cos t - \frac{86}{51} \sin t\right) - \frac{25}{102} \cos 4t + \frac{50}{51} \sin 4t. \tag{28}$$

Transient and Steady-State Terms When F is a periodic function, such as $F(t) = F_0 \sin \gamma t$ or $F(t) = F_0 \cos \gamma t$, the general solution of (25) for $\lambda > 0$ is the sum of a nonperiodic function $x_c(t)$ and a periodic function $x_p(t)$. Moreover, $x_c(t)$ dies off as time increases—that is, $\lim_{t\to\infty} x_c(t) = 0$. Thus for large values of time, the displacements of the mass are closely approximated by the particular solution $x_p(t)$. The complementary function $x_c(t)$ is said to be a **transient term** or **transient solution,** and the function $x_p(t)$, the part of the solution that remains after an interval of time, is called a **steady-state term** or **steady-state solution.** Note therefore that the effect of the initial conditions on a spring/mass system driven by F is transient. In the particular solution (28), $e^{-3t}\left(\frac{38}{51} \cos t - \frac{86}{51} \sin t\right)$ is a transient term, and $x_p(t) = -\frac{25}{102} \cos 4t + \frac{50}{51} \sin 4t$ is a steady-state term. The graphs of these two terms and the solution (28) are given in Figures 5.1.12(a) and 5.1.12(b), respectively.

EXAMPLE 7 Transient/Steady-State Solutions

The solution of the initial-value problem

$$\frac{d^2x}{dt^2} + 2\frac{dx}{dt} + 2x = 4 \cos t + 2 \sin t, \quad x(0) = 0, \quad x'(0) = x_1,$$

where x_1 is constant, is given by

$$x(t) = \underbrace{(x_1 - 2)\, e^{-t} \sin t}_{\text{transient}} + \underbrace{2 \sin t}_{\text{steady-state}}.$$

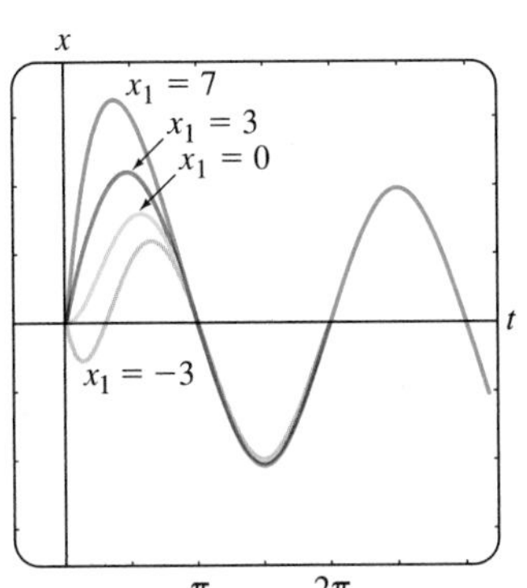

FIGURE 5.1.13 Graph of solution in Example 7 for various initial velocities x_1

Solution curves for selected values of the initial velocity x_1 are shown in Figure 5.1.13. The graphs show that the influence of the transient term is negligible for about $t > 3\pi/2$.

DE of Driven Motion without Damping With a periodic impressed force and no damping force, there is no transient term in the solution of a problem. Also, we shall see that a periodic impressed force with a frequency near or the same as the frequency of free undamped vibrations can cause a severe problem in any oscillatory mechanical system.

EXAMPLE 8 Undamped Forced Motion

Solve the initial-value problem

$$\frac{d^2x}{dt^2} + \omega^2 x = F_0 \sin \gamma t, \quad x(0) = 0, \quad x'(0) = 0, \tag{29}$$

where F_0 is a constant and $\gamma \neq \omega$.

SOLUTION The complementary function is $x_c(t) = c_1 \cos \omega t + c_2 \sin \omega t$. To obtain a particular solution, we assume $x_p(t) = A \cos \gamma t + B \sin \gamma t$ so that

$$x_p'' + \omega^2 x_p = A(\omega^2 - \gamma^2) \cos \gamma t + B(\omega^2 - \gamma^2) \sin \gamma t = F_0 \sin \gamma t.$$

Equating coefficients immediately gives $A = 0$ and $B = F_0/(\omega^2 - \gamma^2)$. Therefore

$$x_p(t) = \frac{F_0}{\omega^2 - \gamma^2} \sin \gamma t.$$

Applying the given initial conditions to the general solution

$$x(t) = c_1 \cos \omega t + c_2 \sin \omega t + \frac{F_0}{\omega^2 - \gamma^2} \sin \gamma t$$

yields $c_1 = 0$ and $c_2 = -\gamma F_0/\omega(\omega^2 - \gamma^2)$. Thus the solution is

$$x(t) = \frac{F_0}{\omega(\omega^2 - \gamma^2)}(-\gamma \sin \omega t + \omega \sin \gamma t), \qquad \gamma \neq \omega. \tag{30}$$

Pure Resonance Although equation (30) is not defined for $\gamma = \omega$, it is interesting to observe that its limiting value as $\gamma \to \omega$ can be obtained by applying L'Hôpital's Rule. This limiting process is analogous to "tuning in" the frequency of the driving force $(\gamma/2\pi)$ to the frequency of free vibrations $(\omega/2\pi)$. Intuitively, we expect that over a length of time we should be able to substantially increase the amplitudes of vibration. For $\gamma = \omega$ we define the solution to be

$$\begin{aligned} x(t) = \lim_{\gamma \to \omega} F_0 \frac{-\gamma \sin \omega t + \omega \sin \gamma t}{\omega(\omega^2 - \gamma^2)} &= F_0 \lim_{\gamma \to \omega} \frac{\dfrac{d}{d\gamma}(-\gamma \sin \omega t + \omega \sin \gamma t)}{\dfrac{d}{d\gamma}(\omega^3 - \omega\gamma^2)} \\ &= F_0 \lim_{\gamma \to \omega} \frac{-\sin \omega t + \omega t \cos \gamma t}{-2\omega\gamma} \\ &= F_0 \frac{-\sin \omega t + \omega t \cos \omega t}{-2\omega^2} \\ &= \frac{F_0}{2\omega^2} \sin \omega t - \frac{F_0}{2\omega} t \cos \omega t. \end{aligned} \tag{31}$$

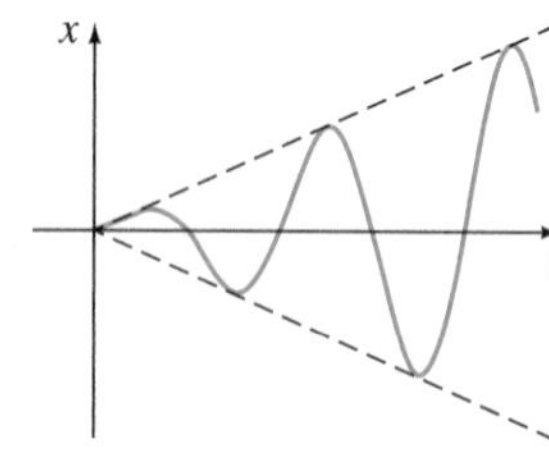

FIGURE 5.1.14 Pure resonance

As suspected, when $t \to \infty$, the displacements become large; in fact, $|x(t_n)| \to \infty$ when $t_n = n\pi/\omega, n = 1, 2, \ldots$. The phenomenon that we have just described is known as **pure resonance.** The graph given in Figure 5.1.14 shows typical motion in this case.

In conclusion it should be noted that there is no actual need to use a limiting process on (30) to obtain the solution for $\gamma = \omega$. Alternatively, equation (31) follows by solving the initial-value problem

$$\frac{d^2x}{dt^2} + \omega^2 x = F_0 \sin \omega t, \quad x(0) = 0, \quad x'(0) = 0$$

directly by conventional methods.

If the displacements of a spring/mass system were actually described by a function such as (31), the system would necessarily fail. Large oscillations of the mass would eventually force the spring beyond its elastic limit. One might argue too that the resonating model presented in Figure 5.1.14 is completely unrealistic because it ignores the retarding effects of ever-present damping forces. Although it is true that pure resonance cannot occur when the smallest amount of damping is taken into consideration, large and equally destructive amplitudes of vibration (although bounded as $t \to \infty$) can occur. See Problem 43 in Exercises 5.1.

5.1.4 SERIES CIRCUIT ANALOGUE

≡ ***LRC*-Series Circuits** As was mentioned in the introduction to this chapter, many different physical systems can be described by a linear second-order differential equation similar to the differential equation of forced motion with damping:

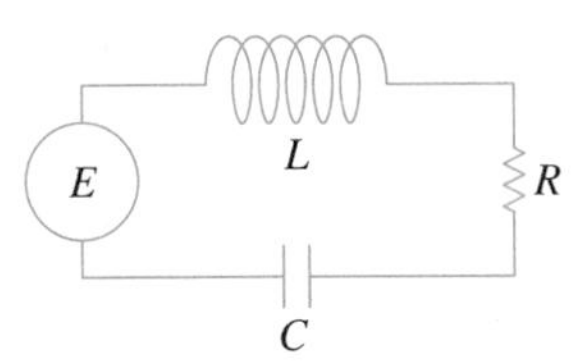

FIGURE 5.1.15 *LRC*-series circuit

$$m\frac{d^2x}{dt^2} + \beta\frac{dx}{dt} + kx = f(t). \tag{32}$$

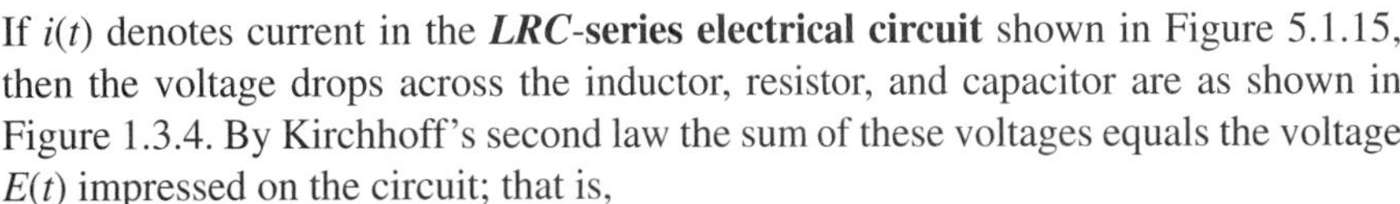

If $i(t)$ denotes current in the ***LRC*-series electrical circuit** shown in Figure 5.1.15, then the voltage drops across the inductor, resistor, and capacitor are as shown in Figure 1.3.4. By Kirchhoff's second law the sum of these voltages equals the voltage $E(t)$ impressed on the circuit; that is,

$$L\frac{di}{dt} + Ri + \frac{1}{C}q = E(t). \tag{33}$$

But the charge $q(t)$ on the capacitor is related to the current $i(t)$ by $i = dq/dt$, so (33) becomes the linear second-order differential equation

$$L\frac{d^2q}{dt^2} + R\frac{dq}{dt} + \frac{1}{C}q = E(t). \tag{34}$$

The nomenclature used in the analysis of circuits is similar to that used to describe spring/mass systems.

If $E(t) = 0$, the **electrical vibrations** of the circuit are said to be **free.** Because the auxiliary equation for (34) is $Lm^2 + Rm + 1/C = 0$, there will be three forms of the solution with $R \neq 0$, depending on the value of the discriminant $R^2 - 4L/C$. We say that the circuit is

overdamped if $R^2 - 4L/C > 0$,

critically damped if $R^2 - 4L/C = 0$,

and **underdamped** if $R^2 - 4L/C < 0$.

In each of these three cases the general solution of (34) contains the factor $e^{-Rt/2L}$, so $q(t) \to 0$ as $t \to \infty$. In the underdamped case when $q(0) = q_0$, the charge on the capacitor oscillates as it decays; in other words, the capacitor is charging and discharging as $t \to \infty$. When $E(t) = 0$ and $R = 0$, the circuit is said to be undamped, and the electrical vibrations do not approach zero as t increases without bound; the response of the circuit is **simple harmonic.**

EXAMPLE 9 Underdamped Series Circuit

Find the charge $q(t)$ on the capacitor in an *LRC*-series circuit when $L = 0.25$ henry (h), $R = 10$ ohms (Ω), $C = 0.001$ farad (f), $E(t) = 0$, $q(0) = q_0$ coulombs (C), and $i(0) = 0$.

SOLUTION Since $1/C = 1000$, equation (34) becomes

$$\frac{1}{4}q'' + 10q' + 1000q = 0 \quad \text{or} \quad q'' + 40q' + 4000q = 0.$$

Solving this homogeneous equation in the usual manner, we find that the circuit is underdamped and $q(t) = e^{-20t}(c_1 \cos 60t + c_2 \sin 60t)$. Applying the initial conditions, we find $c_1 = q_0$ and $c_2 = \frac{1}{3} q_0$. Thus

$$q(t) = q_0 e^{-20t}\left(\cos 60t + \frac{1}{3}\sin 60t\right).$$

Using (23), we can write the foregoing solution as

$$q(t) = \frac{q_0\sqrt{10}}{3} e^{-20t} \sin(60t + 1.249).$$

≡

When there is an impressed voltage $E(t)$ on the circuit, the electrical vibrations are said to be **forced.** In the case when $R \neq 0$, the complementary function $q_c(t)$ of (34) is called a **transient solution.** If $E(t)$ is periodic or a constant, then the particular solution $q_p(t)$ of (34) is a **steady-state solution.**

EXAMPLE 10 Steady-State Current

Find the steady-state solution $q_p(t)$ and the **steady-state current** in an *LRC*-series circuit when the impressed voltage is $E(t) = E_0 \sin \gamma t$.

SOLUTION The steady-state solution $q_p(t)$ is a particular solution of the differential equation

$$L\frac{d^2q}{dt^2} + R\frac{dq}{dt} + \frac{1}{C}q = E_0 \sin \gamma t.$$

Using the method of undetermined coefficients, we assume a particular solution of the form $q_p(t) = A \sin \gamma t + B \cos \gamma t$. Substituting this expression into the differential equation, simplifying, and equating coefficients gives

$$A = \frac{E_0\left(L\gamma - \dfrac{1}{C\gamma}\right)}{-\gamma\left(L^2\gamma^2 - \dfrac{2L}{C} + \dfrac{1}{C^2\gamma^2} + R^2\right)}, \qquad B = \frac{E_0 R}{-\gamma\left(L^2\gamma^2 - \dfrac{2L}{C} + \dfrac{1}{C^2\gamma^2} + R^2\right)}.$$

It is convenient to express A and B in terms of some new symbols.

If $\quad X = L\gamma - \dfrac{1}{C\gamma}, \quad$ then $\quad X^2 = L^2\gamma^2 - \dfrac{2L}{C} + \dfrac{1}{C^2\gamma^2}.$

If $\quad Z = \sqrt{X^2 + R^2}, \quad$ then $\quad Z^2 = L^2\gamma^2 - \dfrac{2L}{C} + \dfrac{1}{C^2\gamma^2} + R^2.$

Therefore $A = E_0X/(-\gamma Z^2)$ and $B = E_0R/(-\gamma Z^2)$, so the steady-state charge is

$$q_p(t) = -\frac{E_0X}{\gamma Z^2}\sin \gamma t - \frac{E_0R}{\gamma Z^2}\cos \gamma t.$$

Now the steady-state current is given by $i_p(t) = q_p'(t)$:

$$i_p(t) = \frac{E_0}{Z}\left(\frac{R}{Z}\sin \gamma t - \frac{X}{Z}\cos \gamma t\right). \tag{35}$$

≡

The quantities $X = L\gamma - 1/C\gamma$ and $Z = \sqrt{X^2 + R^2}$ defined in Example 10 are called the **reactance** and **impedance,** respectively, of the circuit. Both the reactance and the impedance are measured in ohms.

EXERCISES 5.1

Answers to selected odd-numbered problems begin on page ANS-7.

5.1.1 SPRING/MASS SYSTEMS: FREE UNDAMPED MOTION

1. A mass weighing 4 pounds is attached to a spring whose spring constant is 16 lb/ft. What is the period of simple harmonic motion?

2. A 20-kilogram mass is attached to a spring. If the frequency of simple harmonic motion is $2/\pi$ cycles/s, what is the spring constant k? What is the frequency of simple harmonic motion if the original mass is replaced with an 80-kilogram mass?

3. A mass weighing 24 pounds, attached to the end of a spring, stretches it 4 inches. Initially, the mass is released from rest from a point 3 inches above the equilibrium position. Find the equation of motion.

4. Determine the equation of motion if the mass in Problem 3 is initially released from the equilibrium position with a downward velocity of 2 ft/s.

5. A mass weighing 20 pounds stretches a spring 6 inches. The mass is initially released from rest from a point 6 inches below the equilibrium position.

(a) Find the position of the mass at the times $t = \pi/12$, $\pi/8$, $\pi/6$, $\pi/4$, and $9\pi/32$ s.

(b) What is the velocity of the mass when $t = 3\pi/16$ s? In which direction is the mass heading at this instant?

(c) At what times does the mass pass through the equilibrium position?

6. A force of 400 newtons stretches a spring 2 meters. A mass of 50 kilograms is attached to the end of the spring and is initially released from the equilibrium position with an upward velocity of 10 m/s. Find the equation of motion.

7. Another spring whose constant is 20 N/m is suspended from the same rigid support but parallel to the spring/mass system in Problem 6. A mass of 20 kilograms is attached to the second spring, and both masses are initially released from the equilibrium position with an upward velocity of 10 m/s.

(a) Which mass exhibits the greater amplitude of motion?

(b) Which mass is moving faster at $t = \pi/4$ s? At $\pi/2$ s?

(c) At what times are the two masses in the same position? Where are the masses at these times? In which directions are the masses moving?

8. A mass weighing 32 pounds stretches a spring 2 feet. Determine the amplitude and period of motion if the mass is initially released from a point 1 foot above the equilibrium position with an upward velocity of 2 ft/s. How many complete cycles will the mass have completed at the end of 4π seconds?

9. A mass weighing 8 pounds is attached to a spring. When set in motion, the spring/mass system exhibits simple harmonic motion.

(a) Determine the equation of motion if the spring constant is 1 lb/ft and the mass is initially released from a point 6 inches below the equilibrium position with a downward velocity of $\frac{3}{2}$ ft/s.

(b) Express the equation of motion in the form given in (6).

(c) Express the equation of motion in the form given in (6′).

10. A mass weighing 10 pounds stretches a spring $\frac{1}{4}$ foot. This mass is removed and replaced with a mass of 1.6 slugs, which is initially released from a point $\frac{1}{3}$ foot above the equilibrium position with a downward velocity of $\frac{5}{4}$ ft/s.

(a) Express the equation of motion in the form given in (6).

(b) Express the equation of motion in the form given in (6′)

(c) Use one of the solutions obtained in parts (a) and (b) to determine the times the mass attains a displacement below the equilibrium position numerically equal to $\frac{1}{2}$ the amplitude of motion.

11. A mass weighing 64 pounds stretches a spring 0.32 foot. The mass is initially released from a point 8 inches above the equilibrium position with a downward velocity of 5 ft/s.

(a) Find the equation of motion.

(b) What are the amplitude and period of motion?

(c) How many complete cycles will the mass have completed at the end of 3π seconds?

(d) At what time does the mass pass through the equilibrium position heading downward for the second time?

(e) At what times does the mass attain its extreme displacements on either side of the equilibrium position?

(f) What is the position of the mass at $t = 3$ s?

(g) What is the instantaneous velocity at $t = 3$ s?

(h) What is the acceleration at $t = 3$ s?

(i) What is the instantaneous velocity at the times when the mass passes through the equilibrium position?

(j) At what times is the mass 5 inches below the equilibrium position?

(k) At what times is the mass 5 inches below the equilibrium position heading in the upward direction?

12. A mass of 1 slug is suspended from a spring whose spring constant is 9 lb/ft. The mass is initially released from a point 1 foot above the equilibrium position with an upward velocity of $\sqrt{3}$ ft/s. Find the times at which the mass is heading downward at a velocity of 3 ft/s.

13. Under some circumstances when two parallel springs, with constants k_1 and k_2, support a single mass, the **effective spring constant** of the system is given by $k = 4k_1k_2/(k_1 + k_2)$. A mass weighing 20 pounds stretches one spring 6 inches and another spring 2 inches. The springs are attached to a common rigid support and then to a metal plate. As shown in Figure 5.1.16, the mass is attached to the center of the plate in the double-spring arrangement. Determine the effective spring constant of this system. Find the equation of motion if the mass is initially released from the equilibrium position with a downward velocity of 2 ft/s.

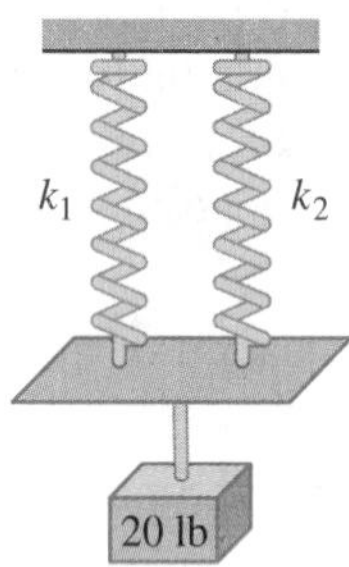

FIGURE 5.1.16 Double-spring system in Problem 13

14. A certain mass stretches one spring $\frac{1}{3}$ foot and another spring $\frac{1}{2}$ foot. The two springs are attached to a common rigid support in the manner described in Problem 13 and Figure 5.1.16. The first mass is set aside, a mass weighing 8 pounds is attached to the double-spring arrangement, and the system is set in motion. If the period of motion is $\pi/15$ second, determine how much the first mass weighs.

15. A model of a spring/mass system is $4x'' + e^{-0.1t}x = 0$. By inspection of the differential equation only, discuss the behavior of the system over a long period of time.

16. A model of a spring/mass system is $4x'' + tx = 0$. By inspection of the differential equation only, discuss the behavior of the system over a long period of time.

5.1.2 SPRING/MASS SYSTEMS: FREE DAMPED MOTION

In Problems 17–20 the given figure represents the graph of an equation of motion for a damped spring/mass system. Use the graph to determine

(a) whether the initial displacement is above or below the equilibrium position and

(b) whether the mass is initially released from rest, heading downward, or heading upward.

17.

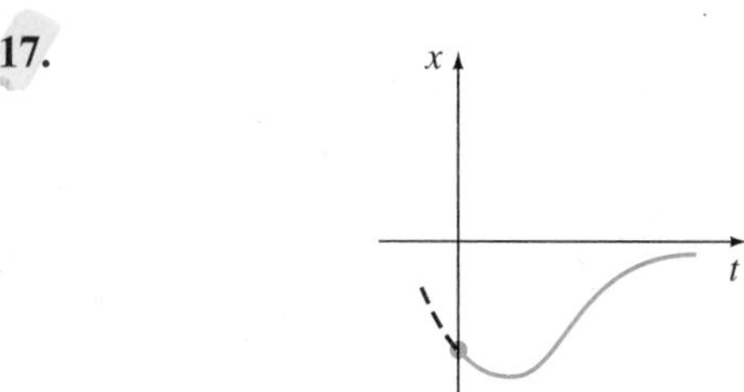

FIGURE 5.1.17 Graph for Problem 17

18.

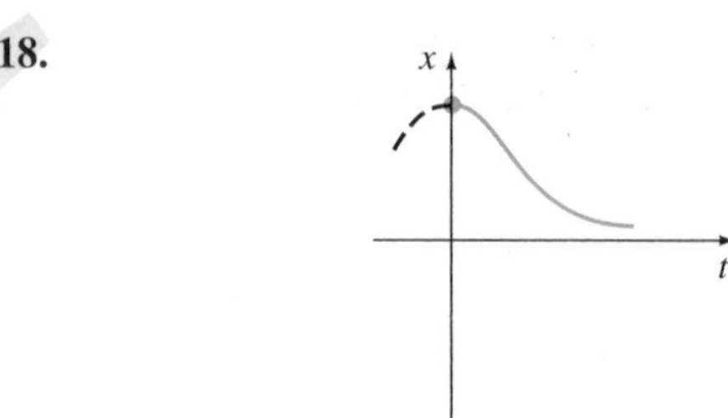

FIGURE 5.1.18 Graph for Problem 18

19.

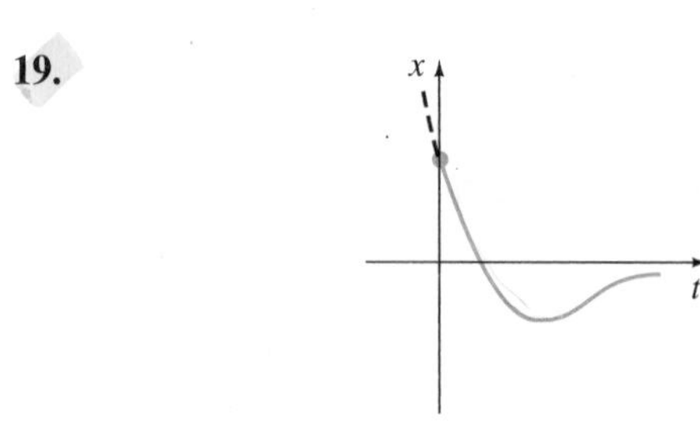

FIGURE 5.1.19 Graph for Problem 19

20.

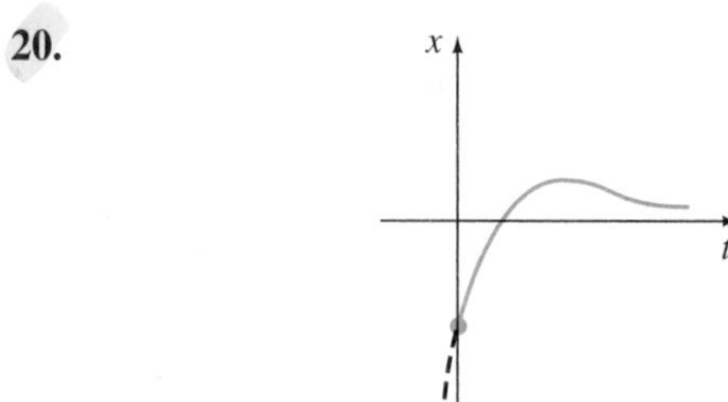

FIGURE 5.1.20 Graph for Problem 20

21. A mass weighing 4 pounds is attached to a spring whose constant is 2 lb/ft. The medium offers a damping force that is numerically equal to the instantaneous velocity. The mass is initially released from a point 1 foot above the equilibrium position with a downward velocity of 8 ft/s. Determine the time at which the mass passes through the equilibrium position. Find the time at which the mass attains its extreme displacement from the equilibrium position. What is the position of the mass at this instant?

22. A 4-foot spring measures 8 feet long after a mass weighing 8 pounds is attached to it. The medium through which the mass moves offers a damping force numerically equal to $\sqrt{2}$ times the instantaneous velocity. Find the equation of motion if the mass is initially released from the equilibrium position with a downward velocity of 5 ft/s. Find the time at which the mass attains its extreme displacement from the equilibrium position. What is the position of the mass at this instant?

23. A 1-kilogram mass is attached to a spring whose constant is 16 N/m, and the entire system is then submerged in a liquid that imparts a damping force numerically equal to 10 times the instantaneous velocity. Determine the equations of motion if

(a) the mass is initially released from rest from a point 1 meter below the equilibrium position, and then

(b) the mass is initially released from a point 1 meter below the equilibrium position with an upward velocity of 12 m/s.

24. In parts (a) and (b) of Problem 23 determine whether the mass passes through the equilibrium position. In each case find the time at which the mass attains its extreme displacement from the equilibrium position. What is the position of the mass at this instant?

25. A force of 2 pounds stretches a spring 1 foot. A mass weighing 3.2 pounds is attached to the spring, and the system is then immersed in a medium that offers a damping force that is numerically equal to 0.4 times the instantaneous velocity.

(a) Find the equation of motion if the mass is initially released from rest from a point 1 foot above the equilibrium position.

(b) Express the equation of motion in the form given in (23).

(c) Find the first time at which the mass passes through the equilibrium position heading upward.

26. After a mass weighing 10 pounds is attached to a 5-foot spring, the spring measures 7 feet. This mass is removed and replaced with another mass that weighs 8 pounds. The entire system is placed in a medium that offers a damping force that is numerically equal to the instantaneous velocity.

(a) Find the equation of motion if the mass is initially released from a point $\frac{1}{2}$ foot below the equilibrium position with a downward velocity of 1 ft/s.

(b) Express the equation of motion in the form given in (23).

(c) Find the times at which the mass passes through the equilibrium position heading downward.

(d) Graph the equation of motion.

27. A mass weighing 10 pounds stretches a spring 2 feet. The mass is attached to a dashpot device that offers a damping force numerically equal to β ($\beta > 0$) times the instantaneous velocity. Determine the values of the damping constant β so that the subsequent motion is **(a)** overdamped, **(b)** critically damped, and **(c)** underdamped.

28. A mass weighing 24 pounds stretches a spring 4 feet. The subsequent motion takes place in medium that offers a damping force numerically equal to β ($\beta > 0$) times the instantaneous velocity. If the mass is initially released from the equilibrium position with an upward velocity of 2 ft/s, show that when $\beta > 3\sqrt{2}$ the equation of motion is

$$x(t) = \frac{-3}{\sqrt{\beta^2 - 18}} e^{-2\beta t/3} \sinh \frac{2}{3} \sqrt{\beta^2 - 18}\,t.$$

5.1.3 SPRING/MASS SYSTEMS: DRIVEN MOTION

29. A mass weighing 16 pounds stretches a spring $\frac{8}{3}$ feet. The mass is initially released from rest from a point 2 feet below the equilibrium position, and the subsequent motion takes place in a medium that offers a damping force that is numerically equal to $\frac{1}{2}$ the instantaneous velocity. Find the equation of motion if the mass is driven by an external force equal to $f(t) = 10 \cos 3t$.

30. A mass of 1 slug is attached to a spring whose constant is 5 lb/ft. Initially, the mass is released 1 foot below the equilibrium position with a downward velocity of 5 ft/s, and the subsequent motion takes place in a medium that offers a damping force that is numerically equal to 2 times the instantaneous velocity.

(a) Find the equation of motion if the mass is driven by an external force equal to $f(t) = 12 \cos 2t + 3 \sin 2t$.

(b) Graph the transient and steady-state solutions on the same coordinate axes.

(c) Graph the equation of motion.

31. A mass of 1 slug, when attached to a spring, stretches it 2 feet and then comes to rest in the equilibrium position. Starting at $t = 0$, an external force equal to $f(t) = 8 \sin 4t$ is applied to the system. Find the equation of motion if the surrounding medium offers a damping force that is numerically equal to 8 times the instantaneous velocity.

32. In Problem 31 determine the equation of motion if the external force is $f(t) = e^{-t} \sin 4t$. Analyze the displacements for $t \to \infty$.

33. When a mass of 2 kilograms is attached to a spring whose constant is 32 N/m, it comes to rest in the equilibrium position. Starting at $t = 0$, a force equal to $f(t) = 68e^{-2t} \cos 4t$ is applied to the system. Find the equation of motion in the absence of damping.

34. In Problem 33 write the equation of motion in the form $x(t) = A\sin(\omega t + \phi) + Be^{-2t}\sin(4t + \theta)$. What is the amplitude of vibrations after a very long time?

35. A mass m is attached to the end of a spring whose constant is k. After the mass reaches equilibrium, its support begins to oscillate vertically about a horizontal line L according to a formula $h(t)$. The value of h represents the distance in feet measured from L. See Figure 5.1.21.

(a) Determine the differential equation of motion if the entire system moves through a medium offering a damping force that is numerically equal to $\beta(dx/dt)$.

(b) Solve the differential equation in part (a) if the spring is stretched 4 feet by a mass weighing 16 pounds and $\beta = 2$, $h(t) = 5 \cos t$, $x(0) = x'(0) = 0$.

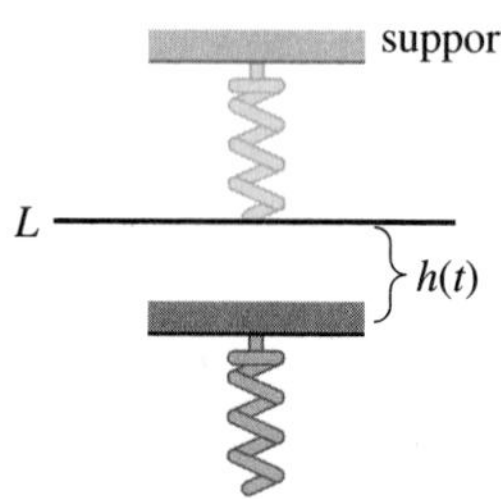

FIGURE 5.1.21 Oscillating support in Problem 35

36. A mass of 100 grams is attached to a spring whose constant is 1600 dynes/cm. After the mass reaches equilibrium, its support oscillates according to the formula $h(t) = \sin 8t$, where h represents displacement from its original position. See Problem 35 and Figure 5.1.21.

(a) In the absence of damping, determine the equation of motion if the mass starts from rest from the equilibrium position.

(b) At what times does the mass pass through the equilibrium position?

(c) At what times does the mass attain its extreme displacements?

(d) What are the maximum and minimum displacements?

(e) Graph the equation of motion.

In Problems 37 and 38 solve the given initial-value problem.

37. $\dfrac{d^2x}{dt^2} + 4x = -5 \sin 2t + 3 \cos 2t,$

$x(0) = -1, \quad x'(0) = 1$

38. $\dfrac{d^2x}{dt^2} + 9x = 5 \sin 3t, \quad x(0) = 2, \quad x'(0) = 0$

39. (a) Show that the solution of the initial-value problem

$$\frac{d^2x}{dt^2} + \omega^2 x = F_0 \cos \gamma t, \quad x(0) = 0, \quad x'(0) = 0$$

is $$x(t) = \frac{F_0}{\omega^2 - \gamma^2}(\cos \gamma t - \cos \omega t).$$

(b) Evaluate $\displaystyle\lim_{\gamma \to \omega} \frac{F_0}{\omega^2 - \gamma^2}(\cos \gamma t - \cos \omega t)$.

40. Compare the result obtained in part (b) of Problem 39 with the solution obtained using variation of parameters when the external force is $F_0 \cos \omega t$.

41. (a) Show that $x(t)$ given in part (a) of Problem 39 can be written in the form

$$x(t) = \frac{-2F_0}{\omega^2 - \gamma^2} \sin \frac{1}{2}(\gamma - \omega)t \sin \frac{1}{2}(\gamma + \omega)t.$$

(b) If we define $\varepsilon = \frac{1}{2}(\gamma - \omega)$, show that when ε is small an approximate solution is

$$x(t) = \frac{F_0}{2\varepsilon\gamma} \sin \varepsilon t \sin \gamma t.$$

When ε is small, the frequency $\gamma/2\pi$ of the impressed force is close to the frequency $\omega/2\pi$ of free vibrations. When this occurs, the motion is as indicated in Figure 5.1.22. Oscillations of this kind are called **beats** and are due to the fact that the frequency of $\sin \varepsilon t$ is quite small in comparison to the frequency of $\sin \gamma t$. The dashed curves, or envelope of the graph of $x(t)$, are obtained from the graphs of $\pm(F_0/2\varepsilon\gamma) \sin \varepsilon t$. Use a graphing utility with various values of F_0, ε, and γ to verify the graph in Figure 5.1.22.

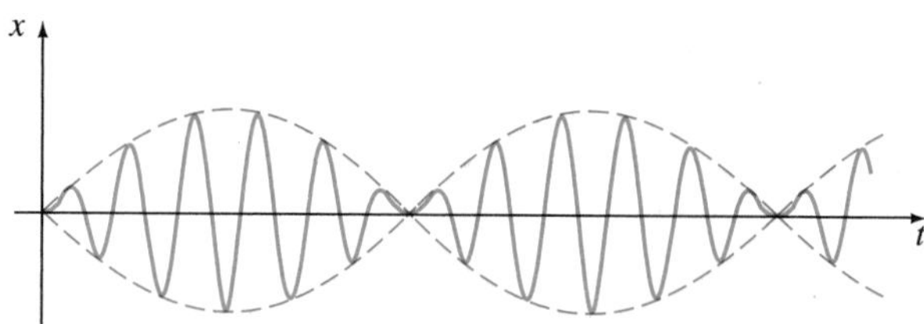

FIGURE 5.1.22 Beats phenomenon in Problem 41

Computer Lab Assignments

42. Can there be beats when a damping force is added to the model in part (a) of Problem 39? Defend your position with graphs obtained either from the explicit solution of the problem

$$\frac{d^2x}{dt^2} + 2\lambda\frac{dx}{dt} + \omega^2 x = F_0 \cos \gamma t, \quad x(0) = 0, \quad x'(0) = 0$$

or from solution curves obtained using a numerical solver.

43. (a) Show that the general solution of

$$\frac{d^2x}{dt^2} + 2\lambda\frac{dx}{dt} + \omega^2 x = F_0 \sin \gamma t$$

is

$$x(t) = Ae^{-\lambda t}\sin\left(\sqrt{\omega^2 - \lambda^2}\,t + \phi\right) + \frac{F_0}{\sqrt{(\omega^2 - \gamma^2)^2 + 4\lambda^2\gamma^2}}\sin(\gamma t + \theta),$$

where $A = \sqrt{c_1^2 + c_2^2}$ and the phase angles ϕ and θ are, respectively, defined by $\sin\phi = c_1/A$, $\cos\phi = c_2/A$ and

$$\sin\theta = \frac{-2\lambda\gamma}{\sqrt{(\omega^2 - \gamma^2)^2 + 4\lambda^2\gamma^2}},$$

$$\cos\theta = \frac{\omega^2 - \gamma^2}{\sqrt{(\omega^2 - \gamma^2)^2 + 4\lambda^2\gamma^2}}.$$

(b) The solution in part (a) has the form $x(t) = x_c(t) + x_p(t)$. Inspection shows that $x_c(t)$ is transient, and hence for large values of time, the solution is approximated by $x_p(t) = g(\gamma)\sin(\gamma t + \theta)$, where

$$g(\gamma) = \frac{F_0}{\sqrt{(\omega^2 - \gamma^2)^2 + 4\lambda^2\gamma^2}}.$$

Although the amplitude $g(\gamma)$ of $x_p(t)$ is bounded as $t \to \infty$, show that the maximum oscillations will occur at the value $\gamma_1 = \sqrt{\omega^2 - 2\lambda^2}$. What is the maximum value of g? The number $\sqrt{\omega^2 - 2\lambda^2}/2\pi$ is said to be the **resonance frequency** of the system.

(c) When $F_0 = 2$, $m = 1$, and $k = 4$, g becomes

$$g(\gamma) = \frac{2}{\sqrt{(4 - \gamma^2)^2 + \beta^2\gamma^2}}.$$

Construct a table of the values of γ_1 and $g(\gamma_1)$ corresponding to the damping coefficients $\beta = 2$, $\beta = 1$, $\beta = \frac{3}{4}$, $\beta = \frac{1}{2}$, and $\beta = \frac{1}{4}$. Use a graphing utility to obtain the graphs of g corresponding to these damping coefficients. Use the same coordinate axes. This family of graphs is called the **resonance curve** or **frequency response curve** of the system. What is γ_1 approaching as $\beta \to 0$? What is happening to the resonance curve as $\beta \to 0$?

44. Consider a driven undamped spring/mass system described by the initial-value problem

$$\frac{d^2x}{dt^2} + \omega^2 x = F_0\sin^n\gamma t, \quad x(0) = 0, \quad x'(0) = 0.$$

(a) For $n = 2$, discuss why there is a single frequency $\gamma_1/2\pi$ at which the system is in pure resonance.

(b) For $n = 3$, discuss why there are two frequencies $\gamma_1/2\pi$ and $\gamma_2/2\pi$ at which the system is in pure resonance.

(c) Suppose $\omega = 1$ and $F_0 = 1$. Use a numerical solver to obtain the graph of the solution of the initial-value problem for $n = 2$ and $\gamma = \gamma_1$ in part (a). Obtain the graph of the solution of the initial-value problem for $n = 3$ corresponding, in turn, to $\gamma = \gamma_1$ and $\gamma = \gamma_2$ in part (b).

5.1.4 SERIES CIRCUIT ANALOGUE

45. Find the charge on the capacitor in an *LRC*-series circuit at $t = 0.01$ s when $L = 0.05$ h, $R = 2\ \Omega$, $C = 0.01$ f, $E(t) = 0$ V, $q(0) = 5$ C, and $i(0) = 0$ A. Determine the first time at which the charge on the capacitor is equal to zero.

46. Find the charge on the capacitor in an *LRC*-series circuit when $L = \frac{1}{4}$ h, $R = 20\ \Omega$, $C = \frac{1}{300}$ f, $E(t) = 0$ V, $q(0) = 4$ C, and $i(0) = 0$ A. Is the charge on the capacitor ever equal to zero?

In Problems 47 and 48 find the charge on the capacitor and the current in the given *LRC*-series circuit. Find the maximum charge on the capacitor.

47. $L = \frac{5}{3}$ h, $R = 10\ \Omega$, $C = \frac{1}{30}$ f, $E(t) = 300$ V, $q(0) = 0$ C, $i(0) = 0$ A

48. $L = 1$ h, $R = 100\ \Omega$, $C = 0.0004$ f, $E(t) = 30$ V, $q(0) = 0$ C, $i(0) = 2$ A

49. Find the steady-state charge and the steady-state current in an *LRC*-series circuit when $L = 1$ h, $R = 2\ \Omega$, $C = 0.25$ f, and $E(t) = 50\cos t$ V.

50. Show that the amplitude of the steady-state current in the *LRC*-series circuit in Example 10 is given by E_0/Z, where Z is the impedance of the circuit.

51. Use Problem 50 to show that the steady-state current in an *LRC*-series circuit when $L = \frac{1}{2}$ h, $R = 20\ \Omega$, $C = 0.001$ f, and $E(t) = 100\sin 60t$ V, is given by $i_p(t) = 4.160\sin(60t - 0.588)$.

52. Find the steady-state current in an *LRC*-series circuit when $L = \frac{1}{2}$ h, $R = 20\ \Omega$, $C = 0.001$ f, and $E(t) = 100\sin 60t + 200\cos 40t$ V.

53. Find the charge on the capacitor in an *LRC*-series circuit when $L = \frac{1}{2}$ h, $R = 10\ \Omega$, $C = 0.01$ f, $E(t) = 150$ V, $q(0) = 1$ C, and $i(0) = 0$ A. What is the charge on the capacitor after a long time?

54. Show that if L, R, C, and E_0 are constant, then the amplitude of the steady-state current in Example 10 is a maximum when $\gamma = 1/\sqrt{LC}$. What is the maximum amplitude?

55. Show that if L, R, E_0, and γ are constant, then the amplitude of the steady-state current in Example 10 is a maximum when the capacitance is $C = 1/L\gamma^2$.

56. Find the charge on the capacitor and the current in an *LC*-series circuit when $L = 0.1$ h, $C = 0.1$ f, $E(t) = 100\sin\gamma t$ V, $q(0) = 0$ C, and $i(0) = 0$ A.

57. Find the charge on the capacitor and the current in an *LC*-series circuit when $E(t) = E_0\cos\gamma t$ V, $q(0) = q_0$ C, and $i(0) = i_0$ A.

58. In Problem 57 find the current when the circuit is in resonance.

5.2 LINEAR MODELS: BOUNDARY-VALUE PROBLEMS

REVIEW MATERIAL

- Section 4.1 (page 117)
- Problems 37–40 in Exercises 4.3
- Problems 37–40 in Exercises 4.4

INTRODUCTION The preceding section was devoted to systems in which a second-order mathematical model was accompanied by initial conditions—that is, side conditions that are specified on the unknown function and its first derivative at a single point. But often the mathematical description of a physical system demands that we solve a linear differential equation subject to boundary conditions—that is, conditions specified on the unknown function, or on one of its derivatives, or even on a linear combination of the unknown function and one of its derivatives at two (or more) different points.

Deflection of a Beam Many structures are constructed by using girders or beams, and these beams deflect or distort under their own weight or under the influence of some external force. As we shall now see, this deflection $y(x)$ is governed by a relatively simple linear fourth-order differential equation.

To begin, let us assume that a beam of length L is homogeneous and has uniform cross sections along its length. In the absence of any load on the beam (including its weight), a curve joining the centroids of all its cross sections is a straight line called the **axis of symmetry.** See Figure 5.2.1(a). If a load is applied to the beam in a vertical plane containing the axis of symmetry, the beam, as shown in Figure 5.2.1(b), undergoes a distortion, and the curve connecting the centroids of all cross sections is called the **deflection curve** or **elastic curve.** The deflection curve approximates the shape of the beam. Now suppose that the x-axis coincides with the axis of symmetry and that the deflection $y(x)$, measured from this axis, is positive if downward. In the theory of elasticity it is shown that the bending moment $M(x)$ at a point x along the beam is related to the load per unit length $w(x)$ by the equation

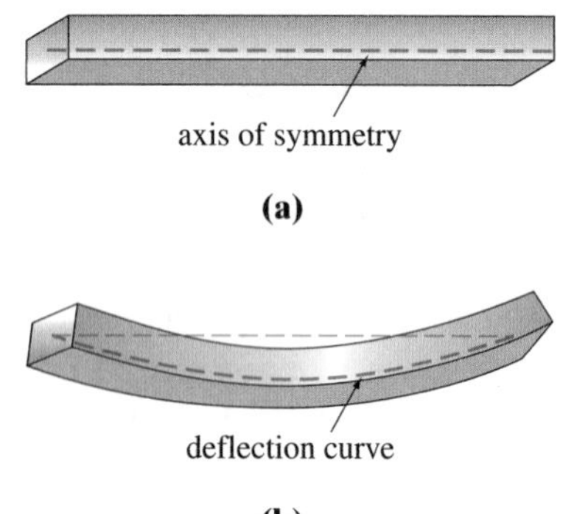

FIGURE 5.2.1 Deflection of a homogeneous beam

$$\frac{d^2M}{dx^2} = w(x). \tag{1}$$

In addition, the bending moment $M(x)$ is proportional to the curvature κ of the elastic curve

$$M(x) = EI\kappa, \tag{2}$$

where E and I are constants; E is Young's modulus of elasticity of the material of the beam, and I is the moment of inertia of a cross section of the beam (about an axis known as the neutral axis). The product EI is called the **flexural rigidity** of the beam.

Now, from calculus, curvature is given by $\kappa = y''/[1 + (y')^2]^{3/2}$. When the deflection $y(x)$ is small, the slope $y' \approx 0$, and so $[1 + (y')^2]^{3/2} \approx 1$. If we let $\kappa \approx y''$, equation (2) becomes $M = EI\,y''$. The second derivative of this last expression is

$$\frac{d^2M}{dx^2} = EI\frac{d^2}{dx^2}y'' = EI\frac{d^4y}{dx^4}. \tag{3}$$

Using the given result in (1) to replace d^2M/dx^2 in (3), we see that the deflection $y(x)$ satisfies the fourth-order differential equation

$$EI\frac{d^4y}{dx^4} = w(x). \tag{4}$$

(a) embedded at both ends

(b) cantilever beam: embedded at the left end, free at the right end

(c) simply supported at both ends

FIGURE 5.2.2 Beams with various end conditions

TABLE 5.2.1

Ends of the Beam	Boundary Conditions
embedded	$y = 0,\quad y' = 0$
free	$y'' = 0,\quad y''' = 0$
simply supported or hinged	$y = 0,\quad y'' = 0$

Boundary conditions associated with equation (4) depend on how the ends of the beam are supported. A cantilever beam is **embedded** or **clamped** at one end and **free** at the other. A diving board, an outstretched arm, an airplane wing, and a balcony are common examples of such beams, but even trees, flagpoles, skyscrapers, and the George Washington Monument can act as cantilever beams because they are embedded at one end and are subject to the bending force of the wind. For a cantilever beam the deflection $y(x)$ must satisfy the following two conditions at the embedded end $x = 0$:

- $y(0) = 0$ because there is no deflection, and
- $y'(0) = 0$ because the deflection curve is tangent to the x-axis (in other words, the slope of the deflection curve is zero at this point).

At $x = L$ the free-end conditions are

- $y''(L) = 0$ because the bending moment is zero, and
- $y'''(L) = 0$ because the shear force is zero.

The function $F(x) = dM/dx = EI\, d^3y/dx^3$ is called the shear force. If an end of a beam is **simply supported** or **hinged** (also called **pin supported** and **fulcrum supported**) then we must have $y = 0$ and $y'' = 0$ at that end. Table 5.2.1 summarizes the boundary conditions that are associated with (4). See Figure 5.2.2.

EXAMPLE 1 An Embedded Beam

A beam of length L is embedded at both ends. Find the deflection of the beam if a constant load w_0 is uniformly distributed along its length—that is, $w(x) = w_0$, $0 < x < L$.

SOLUTION From (4) we see that the deflection $y(x)$ satisfies

$$EI\frac{d^4y}{dx^4} = w_0.$$

Because the beam is embedded at both its left end ($x = 0$) and its right end ($x = L$), there is no vertical deflection and the line of deflection is horizontal at these points. Thus the boundary conditions are

$$y(0) = 0, \qquad y'(0) = 0, \qquad y(L) = 0, \qquad y'(L) = 0.$$

We can solve the nonhomogeneous differential equation in the usual manner (find y_c by observing that $m = 0$ is root of multiplicity four of the auxiliary equation $m^4 = 0$ and then find a particular solution y_p by undetermined coefficients), or we can simply integrate the equation $d^4y/dx^4 = w_0/EI$ four times in succession. Either way, we find the general solution of the equation $y = y_c + y_p$ to be

$$y(x) = c_1 + c_2x + c_3x^2 + c_4x^3 + \frac{w_0}{24EI}x^4.$$

Now the conditions $y(0) = 0$ and $y'(0) = 0$ give, in turn, $c_1 = 0$ and $c_2 = 0$, whereas the remaining conditions $y(L) = 0$ and $y'(L) = 0$ applied to $y(x) = c_3x^2 + c_4x^3 + \frac{w_0}{24EI}x^4$ yield the simultaneous equations

$$c_3L^2 + c_4L^3 + \frac{w_0}{24EI}L^4 = 0$$

$$2c_3L + 3c_4L^2 + \frac{w_0}{6EI}L^3 = 0.$$

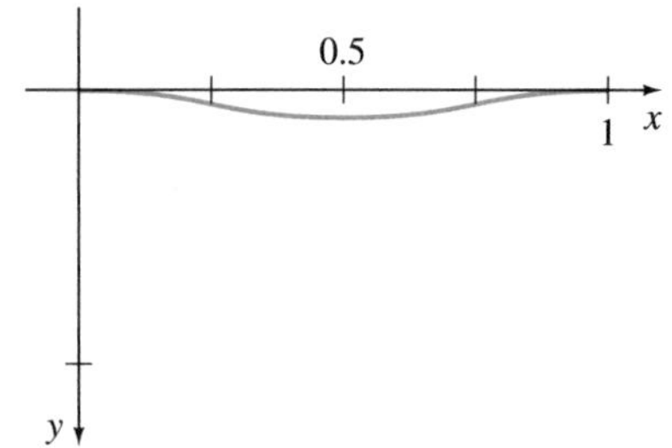

FIGURE 5.2.3 Deflection curve for BVP in Example 1

Solving this system gives $c_3 = w_0L^2/24EI$ and $c_4 = -w_0L/12EI$. Thus the deflection is

$$y(x) = \frac{w_0L^2}{24EI}x^2 - \frac{w_0L}{12EI}x^3 + \frac{w_0}{24EI}x^4$$

or $y(x) = \dfrac{w_0}{24EI}x^2(x - L)^2$. By choosing $w_0 = 24EI$, and $L = 1$, we obtain the deflection curve in Figure 5.2.3. ≡

≡ Eigenvalues and Eigenfunctions Many applied problems demand that we solve a two-point boundary-value problem (BVP) involving a linear differential equation that contains a parameter λ. We seek the values of λ for which the boundary-value problem has *nontrivial,* that is, *nonzero,* solutions.

EXAMPLE 2 Nontrivial Solutions of a BVP

Solve the boundary-value problem

$$y'' + \lambda y = 0, \quad y(0) = 0, \quad y(L) = 0.$$

SOLUTION We shall consider three cases: $\lambda = 0$, $\lambda < 0$, and $\lambda > 0$.

Case I: For $\lambda = 0$ the solution of $y'' = 0$ is $y = c_1x + c_2$. The conditions $y(0) = 0$ and $y(L) = 0$ applied to this solution imply, in turn, $c_2 = 0$ and $c_1 = 0$. Hence for $\lambda = 0$ the only solution of the boundary-value problem is the trivial solution $y = 0$.

Note that we use hyperbolic functions here. Reread "Two Equations Worth Knowing" on pages 134–135.

Case II: For $\lambda < 0$ it is convenient to write $\lambda = -\alpha^2$, where α denotes a positive number. With this notation the roots of the auxiliary equation $m^2 - \alpha^2 = 0$ are $m_1 = \alpha$ and $m_2 = -\alpha$. Since the interval on which we are working is finite, we choose to write the general solution of $y'' - \alpha^2y = 0$ as $y = c_1 \cosh \alpha x + c_2 \sinh \alpha x$. Now $y(0)$ is

$$y(0) = c_1 \cosh 0 + c_2 \sinh 0 = c_1 \cdot 1 + c_2 \cdot 0 = c_1,$$

and so $y(0) = 0$ implies that $c_1 = 0$. Thus $y = c_2 \sinh \alpha x$. The second condition, $y(L) = 0$, demands that $c_2 \sinh \alpha L = 0$. For $\alpha \neq 0$, $\sinh \alpha L \neq 0$; consequently, we are forced to choose $c_2 = 0$. Again the only solution of the BVP is the trivial solution $y = 0$.

Case III: For $\lambda > 0$ we write $\lambda = \alpha^2$, where α is a positive number. Because the auxiliary equation $m^2 + \alpha^2 = 0$ has complex roots $m_1 = i\alpha$ and $m_2 = -i\alpha$, the general solution of $y'' + \alpha^2y = 0$ is $y = c_1 \cos \alpha x + c_2 \sin \alpha x$. As before, $y(0) = 0$ yields $c_1 = 0$, and so $y = c_2 \sin \alpha x$. Now the last condition $y(L) = 0$, or

$$c_2 \sin \alpha L = 0,$$

is satisfied by choosing $c_2 = 0$. But this means that $y = 0$. If we require $c_2 \neq 0$, then $\sin \alpha L = 0$ is satisfied whenever αL is an integer multiple of π.

$$\alpha L = n\pi \quad \text{or} \quad \alpha = \frac{n\pi}{L} \quad \text{or} \quad \lambda_n = \alpha_n^2 = \left(\frac{n\pi}{L}\right)^2, \quad n = 1, 2, 3, \ldots.$$

Therefore for any real nonzero c_2, $y = c_2 \sin(n\pi x/L)$ is a solution of the problem for each n. Because the differential equation is homogeneous, any constant multiple of a solution is also a solution, so we may, if desired, simply take $c_2 = 1$. In other words, for each number in the sequence

$$\lambda_1 = \frac{\pi^2}{L^2}, \quad \lambda_2 = \frac{4\pi^2}{L^2}, \quad \lambda_3 = \frac{9\pi^2}{L^2}, \ldots,$$

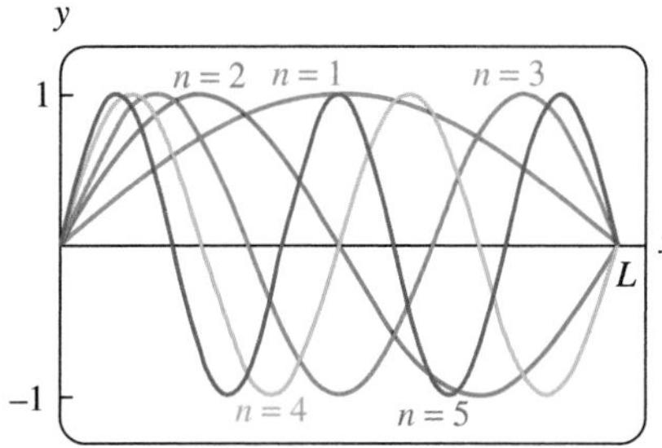

FIGURE 5.2.4 Graphs of eigenfunctions $y_n = \sin(n\pi x/L)$, for $n = 1, 2, 3, 4, 5$

the corresponding function in the sequence

$$y_1 = \sin\frac{\pi}{L}x, \quad y_2 = \sin\frac{2\pi}{L}x, \quad y_3 = \sin\frac{3\pi}{L}x, \ldots$$

is a nontrivial solution of the problem $y'' + \lambda_n y = 0, y(0) = 0, y(L) = 0$ for $n = 1, 2, 3, \ldots$, respectively. ≡

The numbers $\lambda_n = n^2\pi^2/L^2$, $n = 1, 2, 3, \ldots$ for which the boundary-value problem in Example 2 possesses nontrivial solutions are known as **eigenvalues.** The nontrivial solutions that depend on these values of λ_n, $y_n = c_2 \sin(n\pi x/L)$ or simply $y_n = \sin(n\pi x/L)$, are called **eigenfunctions.** The graphs of the eigenfunctions for $n = 1, 2, 3, 4, 5$ are shown in Figure 5.2.4. Note that each graph passes through the two points $(0, 0)$ and $(0, L)$.

EXAMPLE 3 Example 2 Revisited

It follows from Example 2 and the preceding disucussion that the boundary-value problem

$$y'' + 5y = 0, \ y(0) = 0, y(L) = 0$$

possesses only the trivial solution $y = 0$ because 5 is *not* an eigenvalue. ≡

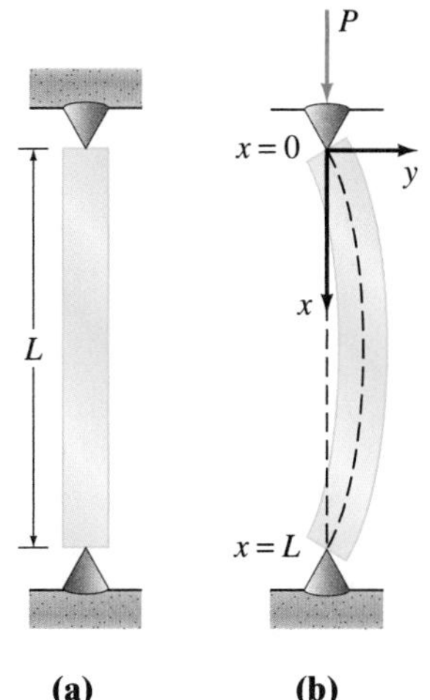

FIGURE 5.2.5 Elastic column buckling under a compressive force

≡ Buckling of a Thin Vertical Column

In the eighteenth century Leonhard Euler was one of the first mathematicians to study an eigenvalue problem in analyzing how a thin elastic column buckles under a compressive axial force.

Consider a long, slender vertical column of uniform cross section and length L. Let $y(x)$ denote the deflection of the column when a constant vertical compressive force, or load, P is applied to its top, as shown in Figure 5.2.5. By comparing bending moments at any point along the column, we obtain

$$EI\frac{d^2y}{dx^2} = -Py \quad \text{or} \quad EI\frac{d^2y}{dx^2} + Py = 0, \tag{5}$$

where E is Young's modulus of elasticity and I is the moment of inertia of a cross section about a vertical line through its centroid.

EXAMPLE 4 The Euler Load

Find the deflection of a thin vertical homogeneous column of length L subjected to a constant axial load P if the column is hinged at both ends.

SOLUTION The boundary-value problem to be solved is

$$EI\frac{d^2y}{dx^2} + Py = 0, \quad y(0) = 0, \quad y(L) = 0.$$

First note that $y = 0$ is a perfectly good solution of this problem. This solution has a simple intuitive interpretation: If the load P is not great enough, there is no deflection. The question then is this: For what values of P will the column bend? In mathematical terms: For what values of P does the given boundary-value problem possess nontrivial solutions?

By writing $\lambda = P/EI$, we see that

$$y'' + \lambda y = 0, \quad y(0) = 0, \quad y(L) = 0$$

is identical to the problem in Example 2. From Case III of that discussion we see that the deflections are $y_n(x) = c_2 \sin(n\pi x/L)$ corresponding to the eigenvalues $\lambda_n = P_n/EI = n^2\pi^2/L^2$, $n = 1, 2, 3, \ldots$. Physically, this means that the column will buckle or deflect only when the compressive force is one of the values $P_n = n^2\pi^2 EI/L^2$, $n = 1, 2, 3, \ldots$. These different forces are called **critical**

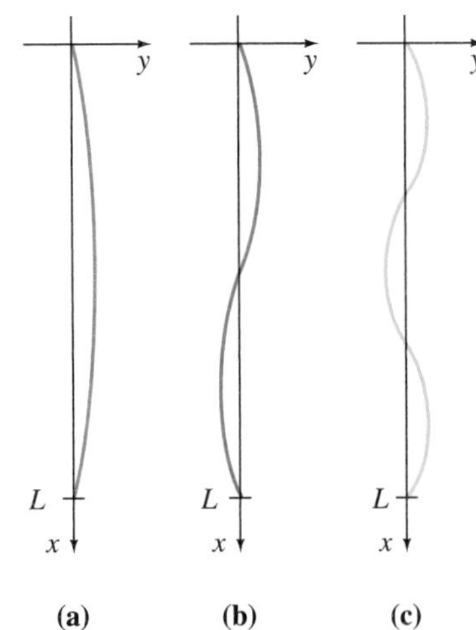

FIGURE 5.2.6 Deflection curves corresponding to compressive forces P_1, P_2, P_3

loads. The deflection corresponding to the smallest critical load $P_1 = \pi^2 EI/L^2$, called the **Euler load,** is $y_1(x) = c_2 \sin(\pi x/L)$ and is known as the **first buckling mode.**

The deflection curves in Example 4 corresponding to $n = 1$, $n = 2$, and $n = 3$ are shown in Figure 5.2.6. Note that if the original column has some sort of physical restraint put on it at $x = L/2$, then the smallest critical load will be $P_2 = 4\pi^2 EI/L^2$, and the deflection curve will be as shown in Figure 5.2.6(b). If restraints are put on the column at $x = L/3$ and at $x = 2L/3$, then the column will not buckle until the critical load $P_3 = 9\pi^2 EI/L^2$ is applied, and the deflection curve will be as shown in Figure 5.2.6(c). See Problem 23 in Exercises 5.2.

Rotating String The simple linear second-order differential equation

$$y'' + \lambda y = 0 \qquad (6)$$

occurs again and again as a mathematical model. In Section 5.1 we saw (6) in the forms $d^2x/dt^2 + (k/m)x = 0$ and $d^2q/dt^2 + (1/LC)q = 0$ as models for, respectively, the simple harmonic motion of a spring/mass system and the simple harmonic response of a series circuit. It is apparent when the model for the deflection of a thin column in (5) is written as $d^2y/dx^2 + (P/EI)y = 0$ that it is the same as (6). We encounter the basic equation (6) one more time in this section: as a model that defines the deflection curve or the shape $y(x)$ assumed by a rotating string. The physical situation is analogous to when two people hold a jump rope and twirl it in a synchronous manner. See Figures 5.2.7(a) and 5.2.7(b).

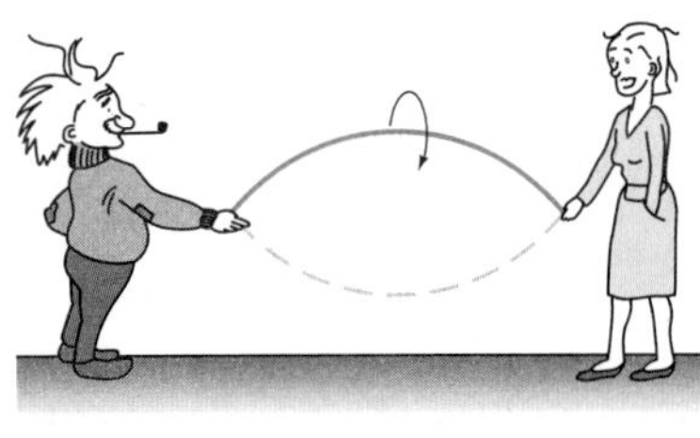
(a)

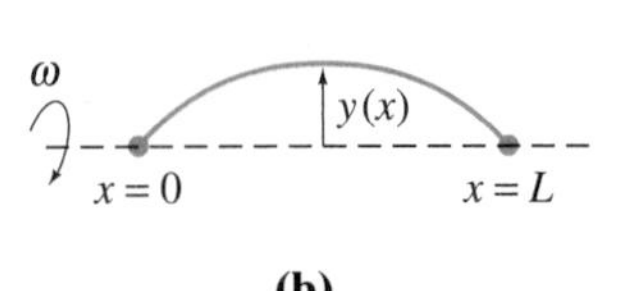

(b)

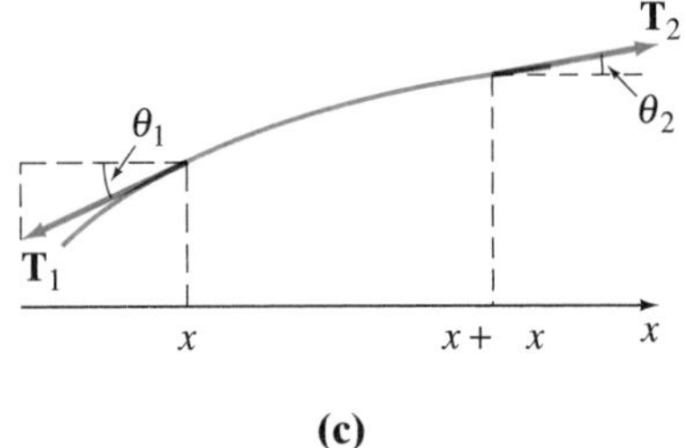

(c)

FIGURE 5.2.7 Rotating string and forces acting on it

Suppose a string of length L with constant linear density ρ (mass per unit length) is stretched along the x-axis and fixed at $x = 0$ and $x = L$. Suppose the string is then rotated about that axis at a constant angular speed ω. Consider a portion of the string on the interval $[x, x + \Delta x]$, where Δx is small. If the magnitude T of the tension $\mathbf{T}$, acting tangential to the string, is constant along the string, then the desired differential equation can be obtained by equating two different formulations of the net force acting on the string on the interval $[x, x + \Delta x]$. First, we see from Figure 5.2.7(c) that the net vertical force is

$$F = T\sin\theta_2 - T\sin\theta_1. \qquad (7)$$

When angles θ_1 and θ_2 (measured in radians) are small, we have $\sin\theta_2 \approx \tan\theta_2$ and $\sin\theta_1 \approx \tan\theta_1$. Moreover, since $\tan\theta_2$ and $\tan\theta_1$ are, in turn, slopes of the lines containing the vectors $\mathbf{T}_2$ and $\mathbf{T}_1$, we can also write

$$\tan\theta_2 = y'(x + \Delta x) \qquad \text{and} \qquad \tan\theta_1 = y'(x).$$

Thus (7) becomes

$$F \approx T[y'(x + \Delta x) - y'(x)]. \qquad (8)$$

Second, we can obtain a different form of this same net force using Newton's second law, $F = ma$. Here the mass of the string on the interval is $m = \rho\,\Delta x$; the centripetal acceleration of a body rotating with angular speed ω in a circle of radius r is $a = r\omega^2$. With Δx small we take $r = y$. Thus the net vertical force is also approximated by

$$F \approx -(\rho\,\Delta x)y\omega^2, \qquad (9)$$

where the minus sign comes from the fact that the acceleration points in the direction opposite to the positive y-direction. Now by equating (8) and (9), we have

$$T[y'(x + \Delta x) - y'(x)] = -(\rho\Delta x)y\omega^2 \qquad \text{or} \qquad T\,\underbrace{\frac{y'(x + \Delta x) - y'(x)}{\Delta x}}_{\text{difference quotient}} + \rho\omega^2 y = 0. \qquad (10)$$

For Δx close to zero the difference quotient in (10) is approximately the second derivative d^2y/dx^2. Finally, we arrive at the model

$$T\frac{d^2y}{dx^2} + \rho\omega^2 y = 0. \qquad (11)$$

Since the string is anchored at its ends $x = 0$ and $x = L$, we expect that the solution $y(x)$ of equation (11) should also satisfy the boundary conditions $y(0) = 0$ and $y(L) = 0$.

REMARKS

(*i*) Eigenvalues are not always easily found, as they were in Example 2; you might have to approximate roots of equations such as $\tan x = -x$ or $\cos x \cosh x = 1$. See Problems 34–38 in Exercises 5.2.

(*ii*) Boundary conditions applied to a general solution of a linear differential equation can lead to a homogeneous algebraic system of linear equations in which the unknowns are the coefficients c_i in the general solution. A homogeneous algebraic system of linear equations is always consistent because it possesses at least a trivial solution. But a homogeneous system of n linear equations in n unknowns has a nontrivial solution if and only if the determinant of the coefficients equals zero. You might need to use this last fact in Problems 19 and 20 in Exercises 5.2.

EXERCISES 5.2

Answers to selected odd-numbered problems begin on page ANS-8.

Deflection of a Beam

In Problems 1–5 solve equation (4) subject to the appropriate boundary conditions. The beam is of length L, and w_0 is a constant.

1. **(a)** The beam is embedded at its left end and free at its right end, and $w(x) = w_0$, $0 < x < L$.

(b) Use a graphing utility to graph the deflection curve when $w_0 = 24EI$ and $L = 1$.

2. **(a)** The beam is simply supported at both ends, and $w(x) = w_0$, $0 < x < L$.

(b) Use a graphing utility to graph the deflection curve when $w_0 = 24EI$ and $L = 1$.

3. **(a)** The beam is embedded at its left end and simply supported at its right end, and $w(x) = w_0$, $0 < x < L$.

(b) Use a graphing utility to graph the deflection curve when $w_0 = 48EI$ and $L = 1$.

4. **(a)** The beam is embedded at its left end and simply supported at its right end, and $w(x) = w_0 \sin(\pi x/L)$, $0 < x < L$.

(b) Use a graphing utility to graph the deflection curve when $w_0 = 2\pi^3 EI$ and $L = 1$.

(c) Use a root-finding application of a CAS (or a graphic calculator) to approximate the point in the graph in part (b) at which the maximum deflection occurs. What is the maximum deflection?

5. **(a)** The beam is simply supported at both ends, and $w(x) = w_0 x$, $0 < x < L$.

(b) Use a graphing utility to graph the deflection curve when $w_0 = 36EI$ and $L = 1$.

(c) Use a root-finding application of a CAS (or a graphic calculator) to approximate the point in the graph in part (b) at which the maximum deflection occurs. What is the maximum deflection?

6. **(a)** Find the maximum deflection of the cantilever beam in Problem 1.

(b) How does the maximum deflection of a beam that is half as long compare with the value in part (a)?

(c) Find the maximum deflection of the simply supported beam in Problem 2.

(d) How does the maximum deflection of the simply supported beam in part (c) compare with the value of maximum deflection of the embedded beam in Example 1?

7. A cantilever beam of length L is embedded at its right end, and a horizontal tensile force of P pounds is applied to its free left end. When the origin is taken at its free end, as shown in Figure 5.2.8, the deflection $y(x)$ of the beam can be shown to satisfy the differential equation

$$EIy'' = Py - w(x)\frac{x}{2}.$$

Find the deflection of the cantilever beam if $w(x) = w_0 x$, $0 < x < L$, and $y(0) = 0$, $y'(L) = 0$.

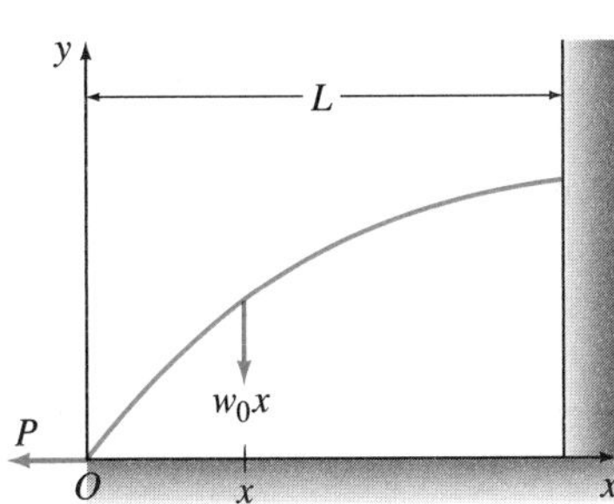

FIGURE 5.2.8 Deflection of cantilever beam in Problem 7

8. When a compressive instead of a tensile force is applied at the free end of the beam in Problem 7, the differential equation of the deflection is

$$EIy'' = -Py - w(x)\frac{x}{2}.$$

Solve this equation if $w(x) = w_0 x$, $0 < x < L$, and $y(0) = 0$, $y'(L) = 0$.

Eigenvalues and Eigenfunctions

In Problems 9–18 find the eigenvalues and eigenfunctions for the given boundary-value problem.

9. $y'' + \lambda y = 0, \quad y(0) = 0, \quad y(\pi) = 0$

10. $y'' + \lambda y = 0, \quad y(0) = 0, \quad y(\pi/4) = 0$

11. $y'' + \lambda y = 0, \quad y'(0) = 0, \quad y(L) = 0$

12. $y'' + \lambda y = 0, \quad y(0) = 0, \quad y'(\pi/2) = 0$

13. $y'' + \lambda y = 0, \quad y'(0) = 0, \quad y'(\pi) = 0$

14. $y'' + \lambda y = 0, \quad y(-\pi) = 0, \quad y(\pi) = 0$

15. $y'' + 2y' + (\lambda + 1)y = 0, \quad y(0) = 0, \quad y(5) = 0$

16. $y'' + (\lambda + 1)y = 0, \quad y'(0) = 0, \quad y'(1) = 0$

17. $x^2y'' + xy' + \lambda y = 0, \quad y(1) = 0, \quad y(e^{\pi}) = 0$

18. $x^2y'' + xy' + \lambda y = 0, \quad y'(e^{-1}) = 0, \quad y(1) = 0$

In Problems 19 and 20 find the eigenvalues and eigenfunctions for the given boundary-value problem. Consider only the case $\lambda = \alpha^4$, $\alpha > 0$.

19. $y^{(4)} - \lambda y = 0, \quad y(0) = 0, \quad y''(0) = 0, \quad y(1) = 0, \quad y''(1) = 0$

20. $y^{(4)} - \lambda y = 0, \quad y'(0) = 0, \quad y'''(0) = 0, \quad y(\pi) = 0, \quad y''(\pi) = 0$

Buckling of a Thin Column

21. Consider Figure 5.2.6. Where should physical restraints be placed on the column if we want the critical load to be P_4? Sketch the deflection curve corresponding to this load.

22. The critical loads of thin columns depend on the end conditions of the column. The value of the Euler load P_1 in Example 4 was derived under the assumption that the column was hinged at both ends. Suppose that a thin vertical homogeneous column is embedded at its base ($x = 0$) and free at its top ($x = L$) and that a constant axial load P is applied to its free end. This load either causes a small deflection δ as shown in Figure 5.2.9 or does not cause such a deflection. In either case the differential equation for the deflection $y(x)$ is

$$EI\frac{d^2y}{dx^2} + Py = P\delta.$$

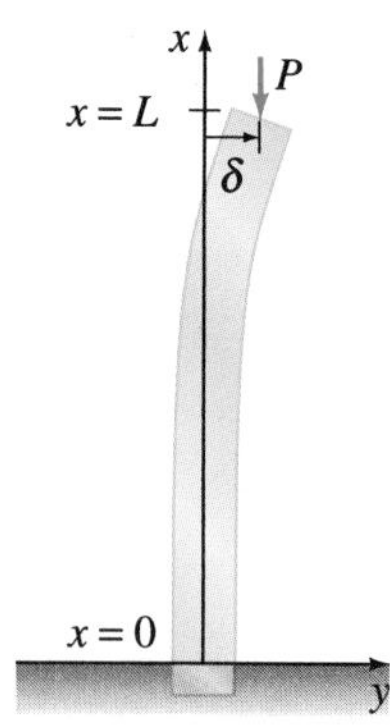

FIGURE 5.2.9 Deflection of vertical column in Problem 22

(a) What is the predicted deflection when $\delta = 0$?

(b) When $\delta \neq 0$, show that the Euler load for this column is one-fourth of the Euler load for the hinged column in Example 4.

23. As was mentioned in Problem 22, the differential equation (5) that governs the deflection $y(x)$ of a thin elastic column subject to a constant compressive axial force P is valid only when the ends of the column are hinged. In general, the differential equation governing the deflection of the column is given by

$$\frac{d^2}{dx^2}\left(EI\frac{d^2y}{dx^2}\right) + P\frac{d^2y}{dx^2} = 0.$$

Assume that the column is uniform (EI is a constant) and that the ends of the column are hinged. Show that the solution of this fourth-order differential equation subject to the boundary conditions $y(0) = 0$, $y''(0) = 0$, $y(L) = 0$, $y''(L) = 0$ is equivalent to the analysis in Example 4.

24. Suppose that a uniform thin elastic column is hinged at the end $x = 0$ and embedded at the end $x = L$.

(a) Use the fourth-order differential equation given in Problem 23 to find the eigenvalues λ_n, the critical loads P_n, the Euler load P_1, and the deflections $y_n(x)$.

(b) Use a graphing utility to graph the first buckling mode.

Rotating String

25. Consider the boundary-value problem introduced in the construction of the mathematical model for the shape of a rotating string:

$$T\frac{d^2y}{dx^2} + \rho\omega^2 y = 0, \quad y(0) = 0, \quad y(L) = 0.$$

For constant T and ρ, define the critical speeds of angular rotation ω_n as the values of ω for which the boundary-value problem has nontrivial solutions. Find the critical speeds ω_n and the corresponding deflections $y_n(x)$.

26. When the magnitude of tension T is not constant, then a model for the deflection curve or shape $y(x)$ assumed by a rotating string is given by

$$\frac{d}{dx}\left[T(x)\frac{dy}{dx}\right] + \rho\omega^2 y = 0.$$

Suppose that $1 < x < e$ and that $T(x) = x^2$.

(a) If $y(1) = 0$, $y(e) = 0$, and $\rho\omega^2 > 0.25$, show that the critical speeds of angular rotation are $\omega_n = \frac{1}{2}\sqrt{(4n^2\pi^2 + 1)/\rho}$ and the corresponding deflections are

$$y_n(x) = c_2 x^{-1/2}\sin(n\pi \ln x), \quad n = 1, 2, 3, \ldots.$$

(b) Use a graphing utility to graph the deflection curves on the interval $[1, e]$ for $n = 1, 2, 3$. Choose $c_2 = 1$.

Miscellaneous Boundary-Value Problems

27. Temperature in a Sphere Consider two concentric spheres of radius $r = a$ and $r = b$, $a < b$. See Figure 5.2.10. The temperature $u(r)$ in the region between the spheres is determined from the boundary-value problem

$$r\frac{d^2u}{dr^2} + 2\frac{du}{dr} = 0, \quad u(a) = u_0, \quad u(b) = u_1,$$

where u_0 and u_1 are constants. Solve for $u(r)$.

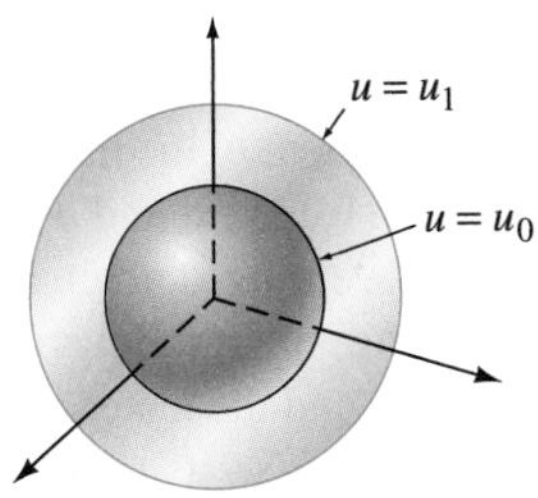

FIGURE 5.2.10 Concentric spheres in Problem 27

28. Temperature in a Ring The temperature $u(r)$ in the circular ring shown in Figure 5.2.11 is determined from the boundary-value problem

$$r\frac{d^2u}{dr^2} + \frac{du}{dr} = 0, \quad u(a) = u_0, \quad u(b) = u_1,$$

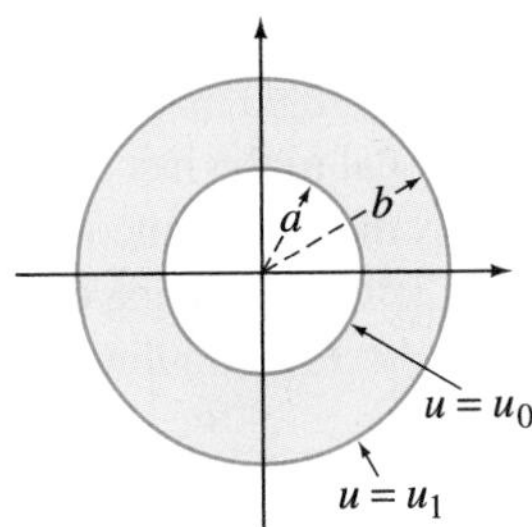

FIGURE 5.2.11 Circular ring in Problem 28

where u_0 and u_1 are constants. Show that

$$u(r) = \frac{u_0 \ln(r/b) - u_1 \ln(r/a)}{\ln(a/b)}.$$

Discussion Problems

29. Simple Harmonic Motion The model $mx'' + kx = 0$ for simple harmonic motion, discussed in Section 5.1, can be related to Example 2 of this section.

Consider a free undamped spring/mass system for which the spring constant is, say, $k = 10$ lb/ft. Determine those masses m_n that can be attached to the spring so that when each mass is released at the equilibrium position at $t = 0$ with a nonzero velocity v_0, it will then pass through the equilibrium position at $t = 1$ second. How many times will each mass m_n pass through the equilibrium position in the time interval $0 < t < 1$?

30. Damped Motion Assume that the model for the spring/mass system in Problem 29 is replaced by

$$mx'' + 2x' + kx = 0.$$

In other words, the system is free but is subjected to damping numerically equal to 2 times the instantaneous velocity. With the same initial conditions and spring constant as in Problem 29, investigate whether a mass m can be found that will pass through the equilibrium position at $t = 1$ second.

In Problems 31 and 32 determine whether it is possible to find values y_0 and y_1 (Problem 31) and values of $L > 0$ (Problem 32) so that the given boundary-value problem has **(a)** precisely one nontrivial solution, **(b)** more than one solution, **(c)** no solution, **(d)** the trivial solution.

31. $y'' + 16y = 0, \quad y(0) = y_0, y(\pi/2) = y_1$

32. $y'' + 16y = 0, \quad y(0) = 1, y(L) = 1$

33. Consider the boundary-value problem

$$y'' + \lambda y = 0, \quad y(-\pi) = y(\pi), \quad y'(-\pi) = y'(\pi).$$

(a) The type of boundary conditions specified are called **periodic boundary conditions.** Give a geometric interpretation of these conditions.

(b) Find the eigenvalues and eigenfunctions of the problem.

(c) Use a graphing utility to graph some of the eigenfunctions. Verify your geometric interpretation of the boundary conditions given in part (a).

34. Show that the eigenvalues and eigenfunctions of the boundary-value problem

$$y'' + \lambda y = 0, \quad y(0) = 0, \quad y(1) + y'(1) = 0$$

are $\lambda_n = \alpha_n^2$ and $y_n = \sin \alpha_n x$, respectively, where α_n, $n = 1, 2, 3, \ldots$ are the consecutive positive roots of the equation $\tan \alpha = -\alpha$.

Computer Lab Assignments

35. Use a CAS to plot graphs to convince yourself that the equation $\tan\alpha = -\alpha$ in Problem 34 has an infinite number of roots. Explain why the negative roots of the equation can be ignored. Explain why $\lambda = 0$ is not an eigenvalue even though $\alpha = 0$ is an obvious solution of the equation $\tan\alpha = -\alpha$.

36. Use a root-finding application of a CAS to approximate the first four eigenvalues λ_1, λ_2, λ_3, and λ_4 for the BVP in Problem 34.

In Problems 37 and 38 find the eigenvalues and eigenfunctions of the given boundary-value problem. Use a CAS to approximate the first four eigenvalues λ_1, λ_2, λ_3, and λ_4.

37. $y'' + \lambda y = 0,\quad y(0) = 0,\quad y(1) - \frac{1}{2}y'(1) = 0$

38. $y^{(4)} - \lambda y = 0$, $y(0) = 0$, $y'(0) = 0$, $y(1) = 0$, $y'(1) = 0$ [*Hint*: Consider only $\lambda = \alpha^4$, $\alpha > 0$.]

5.3 NONLINEAR MODELS

REVIEW MATERIAL

- Section 4.10

INTRODUCTION In this section we examine some nonlinear higher-order mathematical models. We are able to solve some of these models using the substitution method (leading to reduction of the order of the DE) introduced on page 186. In some cases in which the model cannot be solved, we show how a nonlinear DE can be replaced by a linear DE through a process called *linearization.*

Nonlinear Springs The mathematical model in (1) of Section 5.1 has the form

$$m\frac{d^2x}{dt^2} + F(x) = 0, \tag{1}$$

where $F(x) = kx$. Because x denotes the displacement of the mass from its equilibrium position, $F(x) = kx$ is Hooke's law—that is, the force exerted by the spring that tends to restore the mass to the equilibrium position. A spring acting under a linear restoring force $F(x) = kx$ is naturally referred to as a **linear spring.** But springs are seldom perfectly linear. Depending on how it is constructed and the material that is used, a spring can range from "mushy," or soft, to "stiff," or hard, so its restorative force may vary from something below to something above that given by the linear law. In the case of free motion, if we assume that a nonaging spring has some nonlinear characteristics, then it might be reasonable to assume that the restorative force of a spring—that is, $F(x)$ in (1)—is proportional to, say, the cube of the displacement x of the mass beyond its equilibrium position or that $F(x)$ is a linear combination of powers of the displacement such as that given by the nonlinear function $F(x) = kx + k_1x^3$. A spring whose mathematical model incorporates a nonlinear restorative force, such as

$$m\frac{d^2x}{dt^2} + kx^3 = 0 \quad \text{or} \quad m\frac{d^2x}{dt^2} + kx + k_1x^3 = 0, \tag{2}$$

is called a **nonlinear spring.** In addition, we examined mathematical models in which damping imparted to the motion was proportional to the instantaneous velocity dx/dt and the restoring force of a spring was given by the linear function $F(x) = kx$. But these were simply assumptions; in more realistic situations damping could be proportional to some power of the instantaneous velocity dx/dt. The nonlinear differential equation

$$m\frac{d^2x}{dt^2} + \left|\frac{dx}{dt}\right|\frac{dx}{dt} + kx = 0 \tag{3}$$

is one model of a free spring/mass system in which the damping force is proportional to the square of the velocity. One can then envision other kinds of models: linear damping and nonlinear restoring force, nonlinear damping and nonlinear restoring force, and so on. The point is that nonlinear characteristics of a physical system lead to a mathematical model that is nonlinear.

Notice in (2) that both $F(x) = kx^3$ and $F(x) = kx + k_1x^3$ are odd functions of x. To see why a polynomial function containing only odd powers of x provides a reasonable model for the restoring force, let us express F as a power series centered at the equilibrium position $x = 0$:

$$F(x) = c_0 + c_1x + c_2x^2 + c_3x^3 + \cdots.$$

When the displacements x are small, the values of x^n are negligible for n sufficiently large. If we truncate the power series with, say, the fourth term, then $F(x) = c_0 + c_1x + c_2x^2 + c_3x^3$. For the force at $x > 0$,

$$F(x) = c_0 + c_1x + c_2x^2 + c_3x^3,$$

and for the force at $-x < 0$,

$$F(-x) = c_0 - c_1x + c_2x^2 - c_3x^3$$

to have the same magnitude but act in the opposite direction, we must have $F(-x) = -F(x)$. Because this means that F is an odd function, we must have $c_0 = 0$ and $c_2 = 0$, and so $F(x) = c_1x + c_3x^3$. Had we used only the first two terms in the series, the same argument yields the linear function $F(x) = c_1x$. A restoring force with mixed powers, such as $F(x) = c_1x + c_2x^2$, and the corresponding vibrations are said to be **unsymmetrical**. In the next discussion we shall write $c_1 = k$ and $c_3 = k_1$.

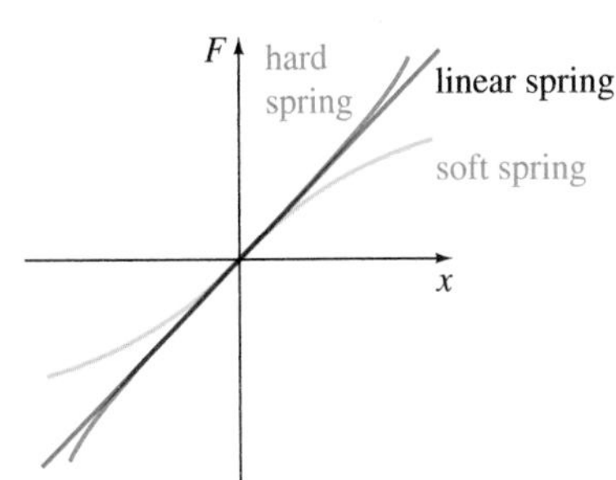

FIGURE 5.3.1 Hard and soft springs

Hard and Soft Springs Let us take a closer look at the equation in (1) in the case in which the restoring force is given by $F(x) = kx + k_1x^3$, $k > 0$. The spring is said to be **hard** if $k_1 > 0$ and **soft** if $k_1 < 0$. Graphs of three types of restoring forces are illustrated in Figure 5.3.1. The next example illustrates these two special cases of the differential equation $md^2x/dt^2 + kx + k_1x^3 = 0$, $m > 0$, $k > 0$.

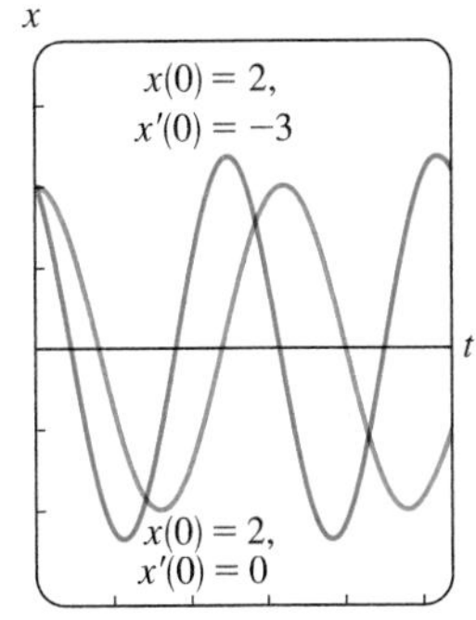

(a) hard spring

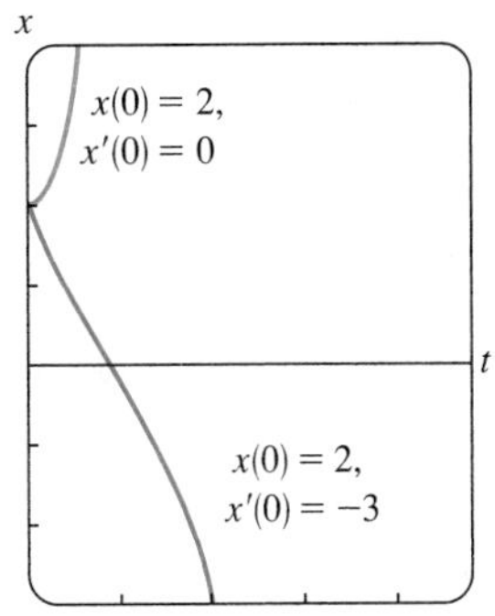

(b) soft spring

FIGURE 5.3.2 Numerical solution curves

EXAMPLE 1 Comparison of Hard and Soft Springs

The differential equations

$$\frac{d^2x}{dt^2} + x + x^3 = 0 \tag{4}$$

and

$$\frac{d^2x}{dt^2} + x - x^3 = 0 \tag{5}$$

are special cases of the second equation in (2) and are models of a hard spring and a soft spring, respectively. Figure 5.3.2(a) shows two solutions of (4) and Figure 5.3.2(b) shows two solutions of (5) obtained from a numerical solver. The curves shown in red are solutions that satisfy the initial conditions $x(0) = 2$, $x'(0) = -3$; the two curves in blue are solutions that satisfy $x(0) = 2$, $x'(0) = 0$. These solution curves certainly suggest that the motion of a mass on the hard spring is oscillatory, whereas motion of a mass on the soft spring appears to be nonoscillatory. But we must be careful about drawing conclusions based on a couple of numerical solution curves. A more complete picture of the nature of the solutions of both of these equations can be obtained from the qualitative analysis discussed in Chapter 10. ≡

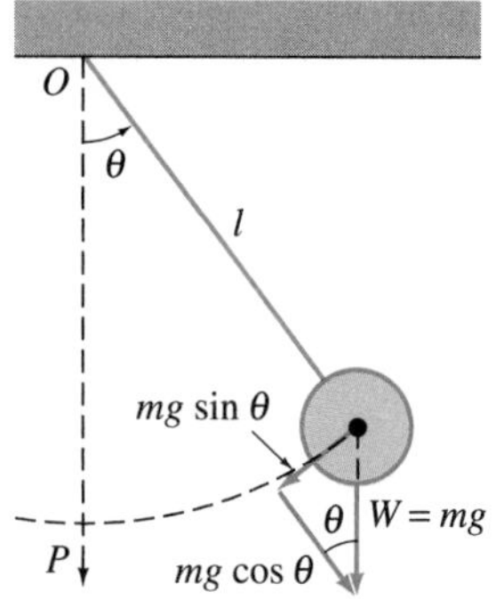

FIGURE 5.3.3 Simple pendulum

Nonlinear Pendulum Any object that swings back and forth is called a **physical pendulum.** The **simple pendulum** is a special case of the physical pendulum and consists of a rod of length l to which a mass m is attached at one end. In describing the motion of a simple pendulum in a vertical plane, we make the simplifying assumptions that the mass of the rod is negligible and that no external damping or driving forces act on the system. The displacement angle θ of the pendulum, measured from the vertical as shown in Figure 5.3.3, is considered positive when measured to the right of OP and negative to the left of OP. Now recall the arc s of a circle of radius l is related to the central angle θ by the formula $s = l\theta$. Hence angular acceleration is

$$a = \frac{d^2s}{dt^2} = l\,\frac{d^2\theta}{dt^2}.$$

From Newton's second law we then have

$$F = ma = ml\,\frac{d^2\theta}{dt^2}.$$

From Figure 5.3.3 we see that the magnitude of the tangential component of the force due to the weight W is $mg \sin \theta$. In direction this force is $-mg \sin \theta$ because it points to the left for $\theta > 0$ and to the right for $\theta < 0$. We equate the two different versions of the tangential force to obtain $ml\, d^2\theta/dt^2 = -mg \sin \theta$, or

$$\frac{d^2\theta}{dt^2} + \frac{g}{l}\sin\theta = 0. \tag{6}$$

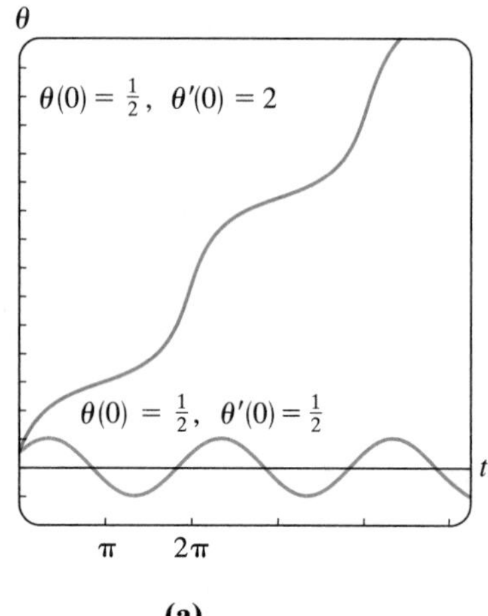

(a)

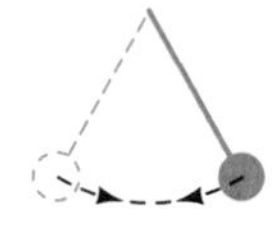

(b) $\theta(0) = \frac{1}{2}$, $\theta'(0) = \frac{1}{2}$

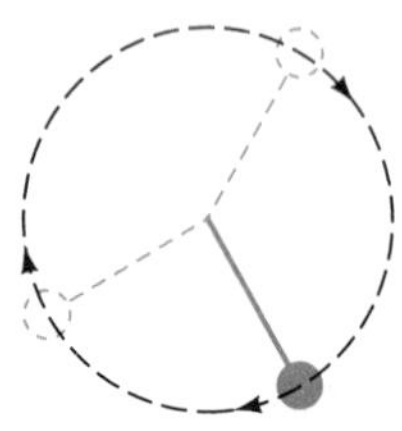

(c) $\theta(0) = \frac{1}{2}$, $\theta'(0) = 2$

FIGURE 5.3.4 In Example 2, oscillating pendulum in (b); whirling pendulum in (c)

Linearization Because of the presence of $\sin\theta$, the model in (6) is nonlinear. In an attempt to understand the behavior of the solutions of nonlinear higher-order differential equations, one sometimes tries to simplify the problem by replacing nonlinear terms by certain approximations. For example, the Maclaurin series for $\sin\theta$ is given by

$$\sin\theta = \theta - \frac{\theta^3}{3!} + \frac{\theta^5}{5!} - \cdots$$

so if we use the approximation $\sin\theta \approx \theta - \theta^3/6$, equation (6) becomes

$$\frac{d^2\theta}{dt^2} + \frac{g}{l}\theta - \frac{g}{6l}\theta^3 = 0.$$

Observe that this last equation is the same as the second nonlinear equation in (2) with $m = 1$, $k = g/l$, and $k_1 = -g/6l$. However, if we assume that the displacements θ are small enough to justify using the replacement $\sin\theta \approx \theta$, then (6) becomes

$$\frac{d^2\theta}{dt^2} + \frac{g}{l}\theta = 0. \tag{7}$$

See Problem 25 in Exercises 5.3. If we set $\omega^2 = g/l$, we recognize (7) as the differential equation (2) of Section 5.1 that is a model for the free undamped vibrations of a linear spring/mass system. In other words, (7) is again the basic linear equation $y'' + \lambda y = 0$ discussed on page 212 of Section 5.2. As a consequence we say that equation (7) is a **linearization** of equation (6). Because the general solution of (7) is $\theta(t) = c_1 \cos \omega t + c_2 \sin \omega t$, this linearization suggests that for initial conditions amenable to small oscillations the motion of the pendulum described by (6) will be periodic.

EXAMPLE 2 Two Initial-Value Problems

The graphs in Figure 5.3.4(a) were obtained with the aid of a numerical solver and represent approximate or numerical solution curves of (6) when $\omega^2 = 1$. The blue curve depicts the solution of (6) that satisfies the initial conditions $\theta(0) = \frac{1}{2}$, $\theta'(0) = \frac{1}{2}$, whereas the red curve is the solution of (6) that satisfies

$\theta(0) = \frac{1}{2}$, $\theta'(0) = 2$. The blue curve represents a periodic solution—the pendulum *oscillating* back and forth as shown in Figure 5.3.4(b) with an apparent amplitude $A \leq 1$. The red curve shows that θ increases without bound as time increases—the pendulum, starting from the same initial displacement, is given an initial velocity of magnitude great enough to send it over the top; in other words, the pendulum is *whirling* about its pivot as shown in Figure 5.3.4(c). In the absence of damping, the motion in each case is continued indefinitely. ≡

≡ **Telephone Wires** The first-order differential equation $dy/dx = W/T_1$ is equation (16) of Section 1.3. This differential equation, established with the aid of Figure 1.3.8 on page 26, serves as a mathematical model for the shape of a flexible cable suspended between two vertical supports when the cable is carrying a vertical load. In Section 2.2 we solved this simple DE under the assumption that the vertical load carried by the cables of a suspension bridge was the weight of a horizontal roadbed distributed evenly along the x-axis. With $W = \rho x$, ρ the weight per unit length of the roadbed, the shape of each cable between the vertical supports turned out to be parabolic. We are now in a position to determine the shape of a uniform flexible cable hanging only under its own weight, such as a wire strung between two telephone posts. The vertical load is now the wire itself, and so if ρ is the linear density of the wire (measured, say, in pounds per feet) and s is the length of the segment P_1P_2 in Figure 1.3.8 then $W = \rho s$. Hence

$$\frac{dy}{dx} = \frac{\rho s}{T_1}. \tag{8}$$

Since the arc length between points P_1 and P_2 is given by

$$s = \int_0^x \sqrt{1 + \left(\frac{dy}{dx}\right)^2}\, dx, \tag{9}$$

it follows from the fundamental theorem of calculus that the derivative of (9) is

$$\frac{ds}{dx} = \sqrt{1 + \left(\frac{dy}{dx}\right)^2}. \tag{10}$$

Differentiating (8) with respect to x and using (10) lead to the second-order equation

$$\frac{d^2y}{dx^2} = \frac{\rho}{T_1}\frac{ds}{dx} \quad \text{or} \quad \frac{d^2y}{dx^2} = \frac{\rho}{T_1}\sqrt{1 + \left(\frac{dy}{dx}\right)^2}. \tag{11}$$

In the example that follows we solve (11) and show that the curve assumed by the suspended cable is a **catenary.** Before proceeding, observe that the nonlinear second-order differential equation (11) is one of those equations having the form $F(x, y', y'') = 0$ discussed in Section 4.10. Recall that we have a chance of solving an equation of this type by reducing the order of the equation by means of the substitution $u = y'$.

EXAMPLE 3 A Solution of (11)

From the position of the y-axis in Figure 1.3.8 it is apparent that initial conditions associated with the second differential equation in (11) are $y(0) = a$ and $y'(0) = 0$. If we substitute $u = y'$, then the equation in (11) becomes $\dfrac{du}{dx} = \dfrac{\rho}{T_1}\sqrt{1 + u^2}$. Separating variables, we find that

$$\int \frac{du}{\sqrt{1 + u^2}} = \frac{\rho}{T_1}\int dx \quad \text{gives} \quad \sinh^{-1}u = \frac{\rho}{T_1}x + c_1.$$

Now, $y'(0) = 0$ is equivalent to $u(0) = 0$. Since $\sinh^{-1} 0 = 0$, $c_1 = 0$, so $u = \sinh(\rho x/T_1)$. Finally, by integrating both sides of

$$\frac{dy}{dx} = \sinh\frac{\rho}{T_1}x, \qquad \text{we get} \qquad y = \frac{T_1}{\rho}\cosh\frac{\rho}{T_1}x + c_2.$$

Using $y(0) = a$, $\cosh 0 = 1$, the last equation implies that $c_2 = a - T_1/\rho$. Thus we see that the shape of the hanging wire is given by $y = (T_1/\rho)\cosh(\rho x/T_1) + a - T_1/\rho$. ≡

In Example 3, had we been clever enough at the start to choose $a = T_1/\rho$, then the solution of the problem would have been simply the hyperbolic cosine $y = (T_1/\rho)\cosh(\rho x/T_1)$.

Rocket Motion In (12) of Section 1.3 we saw that the differential equation of a free-falling body of mass m near the surface of the Earth is given by

$$m\frac{d^2s}{dt^2} = -mg \qquad \text{or simply} \qquad \frac{d^2s}{dt^2} = -g,$$

where s represents the distance from the surface of the Earth to the object and the positive direction is considered to be upward. In other words, the underlying assumption here is that the distance s to the object is small when compared with the radius R of the Earth; put yet another way, the distance y from the center of the Earth to the object is approximately the same as R. If, on the other hand, the distance y to the object, such as a rocket or a space probe, is large when compared to R, then we combine Newton's second law of motion and his universal law of gravitation to derive a differential equation in the variable y.

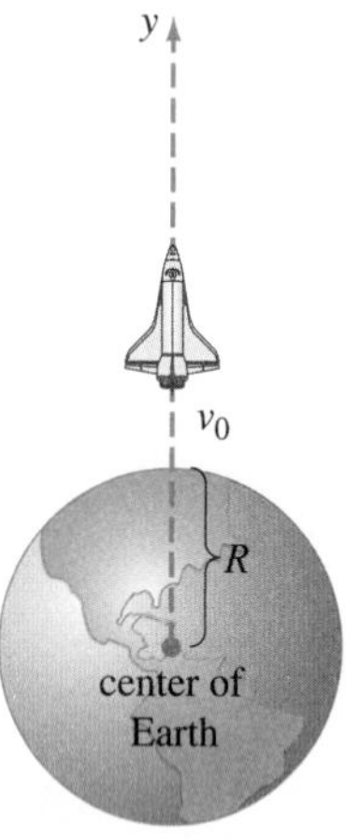

FIGURE 5.3.5 Distance to rocket is large compared to R.

Suppose a rocket is launched vertically upward from the ground as shown in Figure 5.3.5. If the positive direction is upward and air resistance is ignored, then the differential equation of motion after fuel burnout is

$$m\frac{d^2y}{dt^2} = -k\frac{Mm}{y^2} \qquad \text{or} \qquad \frac{d^2y}{dt^2} = -k\frac{M}{y^2}, \tag{12}$$

where k is a constant of proportionality, y is the distance from the center of the Earth to the rocket, M is the mass of the Earth, and m is the mass of the rocket. To determine the constant k, we use the fact that when $y = R$, $kMm/R^2 = mg$ or $k = gR^2/M$. Thus the last equation in (12) becomes

$$\frac{d^2y}{dt^2} = -g\frac{R^2}{y^2}. \tag{13}$$

See Problem 14 in Exercises 5.3.

Variable Mass Notice in the preceding discussion that we described the motion of the rocket after it has burned all its fuel, when presumably its mass m is constant. Of course, during its powered ascent the total mass of the rocket varies as its fuel is being expended. We saw in (17) of Exercises 1.3 that the second law of motion, as originally advanced by Newton, states that when a body of mass m moves through a force field with velocity v, the time rate of change of the momentum mv of the body is equal to applied or net force F acting on the body:

$$F = \frac{d}{dt}(mv). \tag{14}$$

If m is constant, then (14) yields the more familiar form $F = m\,dv/dt = ma$, where a is acceleration. We use the form of Newton's second law given in (14) in the next example, in which the mass m of the body is variable.

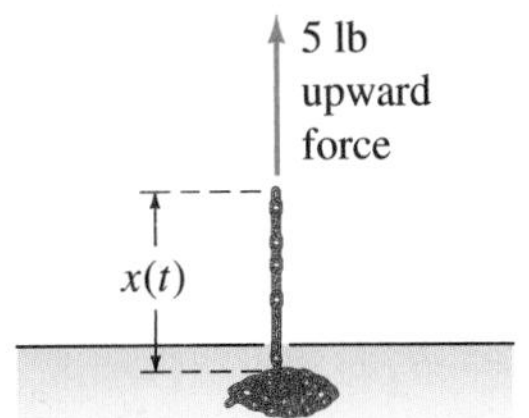

FIGURE 5.3.6 Chain pulled upward by a constant force in Example 4

EXAMPLE 4 Chain Pulled Upward by a Constant Force

A uniform 10-foot-long chain is coiled loosely on the ground. One end of the chain is pulled vertically upward by means of constant force of 5 pounds. The chain weighs 1 pound per foot. Determine the height of the end above ground level at time t. See Figure 5.3.6.

SOLUTION Let us suppose that $x = x(t)$ denotes the height of the end of the chain in the air at time t, $v = dx/dt$, and the positive direction is upward. For the portion of the chain that is in the air at time t we have the following variable quantities:

$$\begin{aligned} &\textit{weight:} && W = (x \text{ ft}) \cdot (1 \text{ lb/ft}) = x, \\ &\textit{mass:} && m = W/g = x/32, \\ &\textit{net force:} && F = 5 - W = 5 - x. \end{aligned}$$

Thus from (14) we have

$$\frac{d}{dt}\left(\frac{x}{32}v\right) = 5 - x \quad \text{or} \quad \overbrace{x\frac{dv}{dt} + v\frac{dx}{dt}}^{\text{Product Rule}} = 160 - 32x. \tag{15}$$

Because $v = dx/dt$, the last equation becomes

$$x\frac{d^2x}{dt^2} + \left(\frac{dx}{dt}\right)^2 + 32x = 160. \tag{16}$$

The nonlinear second-order differential equation (16) has the form $F(x, x', x'') = 0$, which is the second of the two forms considered in Section 4.10 that can possibly be solved by reduction of order. To solve (16), we revert back to (15) and use $v = x'$ along with the Chain Rule. From $\dfrac{dv}{dt} = \dfrac{dv}{dx}\dfrac{dx}{dt} = v\dfrac{dv}{dx}$ the second equation in (15) can be rewritten as

$$xv\frac{dv}{dx} + v^2 = 160 - 32x. \tag{17}$$

On inspection (17) might appear intractable, since it cannot be characterized as any of the first-order equations that were solved in Chapter 2. However, by rewriting (17) in differential form $M(x, v)dx + N(x, v)dv = 0$, we observe that although the equation

$$(v^2 + 32x - 160)dx + xv\,dv = 0 \tag{18}$$

is not exact, it can be transformed into an exact equation by multiplying it by an integrating factor. From $(M_v - N_x)/N = 1/x$ we see from (13) of Section 2.4 that an integrating factor is $e^{\int dx/x} = e^{\ln x} = x$. When (18) is multiplied by $\mu(x) = x$, the resulting equation is exact (verify). By identifying $\partial f/\partial x = xv^2 + 32x^2 - 160x$, $\partial f/\partial v = x^2v$ and then proceeding as in Section 2.4, we obtain

$$\frac{1}{2}x^2v^2 + \frac{32}{3}x^3 - 80x^2 = c_1. \tag{19}$$

Since we have assumed that all of the chain is on the floor initially, we have $x(0) = 0$. This last condition applied to (19) yields $c_1 = 0$. By solving the algebraic equation $\frac{1}{2}x^2v^2 + \frac{32}{3}x^3 - 80x^2 = 0$ for $v = dx/dt > 0$, we get another first-order differential equation,

$$\frac{dx}{dt} = \sqrt{160 - \frac{64}{3}x}.$$

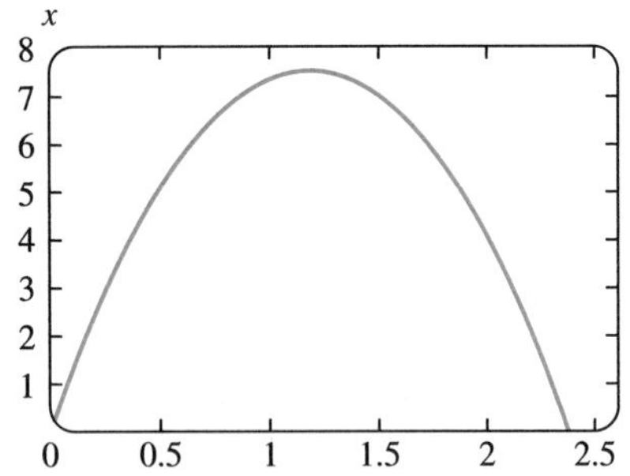

FIGURE 5.3.7 Graph of (21) in Example 4

The last equation can be solved by separation of variables. You should verify that

$$-\frac{3}{32}\left(160 - \frac{64}{3}x\right)^{1/2} = t + c_2. \tag{20}$$

This time the initial condition $x(0) = 0$ implies that $c_2 = -3\sqrt{10}/8$. Finally, by squaring both sides of (20) and solving for x, we arrive at the desired result,

$$x(t) = \frac{15}{2} - \frac{15}{2}\left(1 - \frac{4\sqrt{10}}{15}t\right)^2. \tag{21}$$

The graph of (21) given in Figure 5.3.7 should not, on physical grounds, be taken at face value. See Problem 15 in Exercises 5.3. ≡

EXERCISES 5.3

Answers to selected odd-numbered problems begin on page ANS-9.

To the Instructor In addition to Problems 24 and 25, all or portions of Problems 1–6, 8–13, 15, 20, and 21 could serve as Computer Lab Assignments.

Nonlinear Springs

In Problems 1–4 the given differential equation is model of an undamped spring/mass system in which the restoring force $F(x)$ in (1) is nonlinear. For each equation use a numerical solver to plot the solution curves that satisfy the given initial conditions. If the solutions appear to be periodic use the solution curve to estimate the period T of oscillations.

1. $\dfrac{d^2x}{dt^2} + x^3 = 0,$

$x(0) = 1, x'(0) = 1; \quad x(0) = \frac{1}{2}, x'(0) = -1$

2. $\dfrac{d^2x}{dt^2} + 4x - 16x^3 = 0,$

$x(0) = 1, x'(0) = 1; \quad x(0) = -2, x'(0) = 2$

3. $\dfrac{d^2x}{dt^2} + 2x - x^2 = 0,$

$x(0) = 1, x'(0) = 1; \quad x(0) = \frac{3}{2}, x'(0) = -1$

4. $\dfrac{d^2x}{dt^2} + xe^{0.01x} = 0,$

$x(0) = 1, x'(0) = 1; \quad x(0) = 3, x'(0) = -1$

5. In Problem 3, suppose the mass is released from the initial position $x(0) = 1$ with an initial velocity $x'(0) = x_1$. Use a numerical solver to estimate the smallest value of $|x_1|$ at which the motion of the mass is nonperiodic.

6. In Problem 3, suppose the mass is released from an initial position $x(0) = x_0$ with the initial velocity $x'(0) = 1$. Use a numerical solver to estimate an interval $a \le x_0 \le b$ for which the motion is oscillatory.

7. Find a linearization of the differential equation in Problem 4.

8. Consider the model of an undamped nonlinear spring/mass system given by $x'' + 8x - 6x^3 + x^5 = 0$. Use a numerical solver to discuss the nature of the oscillations of the system corresponding to the initial conditions:

$x(0) = 1, x'(0) = 1; \quad x(0) = -2, x'(0) = \frac{1}{2};$

$x(0) = \sqrt{2}, x'(0) = 1; \quad x(0) = 2, x'(0) = \frac{1}{2};$

$x(0) = 2, x'(0) = 0; \quad x(0) = -\sqrt{2}, x'(0) = -1.$

In Problems 9 and 10 the given differential equation is a model of a damped nonlinear spring/mass system. Predict the behavior of each system as $t \to \infty$. For each equation use a numerical solver to obtain the solution curves satisfying the given initial conditions.

9. $\dfrac{d^2x}{dt^2} + \dfrac{dx}{dt} + x + x^3 = 0,$

$x(0) = -3, x'(0) = 4; \quad x(0) = 0, x'(0) = -8$

10. $\dfrac{d^2x}{dt^2} + \dfrac{dx}{dt} + x - x^3 = 0,$

$x(0) = 0, x'(0) = \frac{3}{2}; \quad x(0) = -1, x'(0) = 1$

11. The model $mx'' + kx + k_1x^3 = F_0\cos \omega t$ of an undamped periodically driven spring/mass system is called **Duffing's differential equation.** Consider the initial-value problem $x'' + x + k_1x^3 = 5 \cos t$, $x(0) = 1$, $x'(0) = 0$. Use a numerical solver to investigate the behavior of the system for values of $k_1 > 0$ ranging from $k_1 = 0.01$ to $k_1 = 100$. State your conclusions.

12. **(a)** Find values of $k_1 < 0$ for which the system in Problem 11 is oscillatory.

(b) Consider the initial-value problem

$$x'' + x + k_1x^3 = \cos \tfrac{3}{2}t, \quad x(0) = 0, \quad x'(0) = 0.$$

Find values for $k_1 < 0$ for which the system is oscillatory.

Nonlinear Pendulum

13. Consider the model of the free damped nonlinear pendulum given by

$$\frac{d^2\theta}{dt^2} + 2\lambda\frac{d\theta}{dt} + \omega^2 \sin\theta = 0.$$

Use a numerical solver to investigate whether the motion in the two cases $\lambda^2 - \omega^2 > 0$ and $\lambda^2 - \omega^2 < 0$ corresponds, respectively, to the overdamped and underdamped cases discussed in Section 5.1 for spring/mass systems. For $\lambda^2 - \omega^2 > 0$, use $\lambda = 2, \omega = 1, \theta(0) = 1$, and $\theta'(0) = 2$. For $\lambda^2 - \omega^2 < 0$, use $\lambda = \frac{1}{3}, \omega = 1, \theta(0) = -2$, and $\theta'(0) = 4$.

Rocket Motion

14. **(a)** Use the substitution $v = dy/dt$ to solve (13) for v in terms of y. Assuming that the velocity of the rocket at burnout is $v = v_0$ and $y \approx R$ at that instant, show that the approximate value of the constant c of integration is $c = -gR + \frac{1}{2}v_0^2$.

(b) Use the solution for v in part (a) to show that the escape velocity of the rocket is given by $v_0 = \sqrt{2gR}$. [*Hint*: Take $y \to \infty$ and assume $v > 0$ for all time t.]

(c) The result in part (b) holds for any body in the Solar System. Use the values $g = 32$ ft/s^2 and $R = 4000$ mi to show that the escape velocity from the Earth is (approximately) $v_0 = 25{,}000$ mi/h.

(d) Find the escape velocity from the Moon if the acceleration of gravity is $0.165g$ and $R = 1080$ mi.

Variable Mass

15. **(a)** In Example 4, how much of the chain would you intuitively expect the constant 5-pound force to be able to lift?

(b) What is the initial velocity of the chain?

(c) Why is the time interval corresponding to $x(t) \geq 0$ given in Figure 5.3.7 not the interval I of definition of the solution (21)? Determine the interval I. How much chain is actually lifted? Explain any difference between this answer and your prediction in part (a).

(d) Why would you expect $x(t)$ to be a periodic solution?

16. A uniform chain of length L, measured in feet, is held vertically so that the lower end just touches the floor. The chain weighs 2 lb/ft. The upper end that is held is released from rest at $t = 0$ and the chain falls straight down. If $x(t)$ denotes the length of the chain on the floor at time t, air resistance is ignored, and the positive direction is taken to be downward, then

$$(L - x)\frac{d^2x}{dt^2} - \left(\frac{dx}{dt}\right)^2 = Lg.$$

(a) Solve for v in terms of x. Solve for x in terms of t. Express v in terms of t.

(b) Determine how long it takes for the chain to fall completely to the ground.

(c) What velocity does the model in part (a) predict for the upper end of the chain as it hits the ground?

Miscellaneous Mathematical Models

17. Pursuit Curve In a naval exercise a ship S_1 is pursued by a submarine S_2 as shown in Figure 5.3.8. Ship S_1 departs point (0, 0) at $t = 0$ and proceeds along a straight-line course (the y-axis) at a constant speed v_1. The submarine S_2 keeps ship S_1 in visual contact, indicated by the straight dashed line L in the figure, while traveling at a constant speed v_2 along a curve C. Assume that ship S_2 starts at the point $(a, 0)$, $a > 0$, at $t = 0$ and that L is tangent to C.

(a) Determine a mathematical model that describes the curve C.

(b) Find an explicit solution of the differential equation. For convenience define $r = v_1/v_2$.

(c) Determine whether the paths of S_1 and S_2 will ever intersect by considering the cases $r > 1, r < 1$, and $r = 1$.

[*Hint*: $\dfrac{dt}{dx} = \dfrac{dt}{ds}\dfrac{ds}{dx}$, where s is arc length measured along C.]

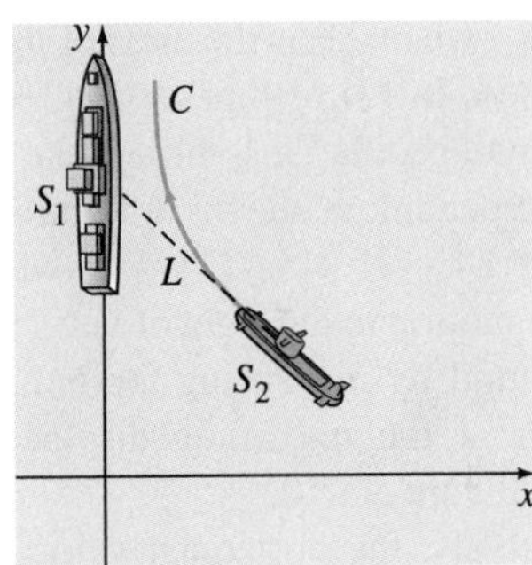

FIGURE 5.3.8 Pursuit curve in Problem 17

18. Pursuit Curve In another naval exercise a destroyer S_1 pursues a submerged submarine S_2. Suppose that S_1 at (9, 0) on the x-axis detects S_2 at (0, 0) and that S_2 simultaneously detects S_1. The captain of the destroyer S_1 assumes that the submarine will take immediate evasive action and conjectures that its likely new course is the straight line indicated in Figure 5.3.9. When S_1 is at (3, 0), it changes from its straight-line course toward the origin to a pursuit curve C. Assume that the speed of the destroyer is, at all times, a constant 30 mi/h and that the submarine's speed is a constant 15 mi/h.

(a) Explain why the captain waits until S_1 reaches (3, 0) before ordering a course change to C.

(b) Using polar coordinates, find an equation $r = f(\theta)$ for the curve C.

(c) Let T denote the time, measured from the initial detection, at which the destroyer intercepts the submarine. Find an upper bound for T.

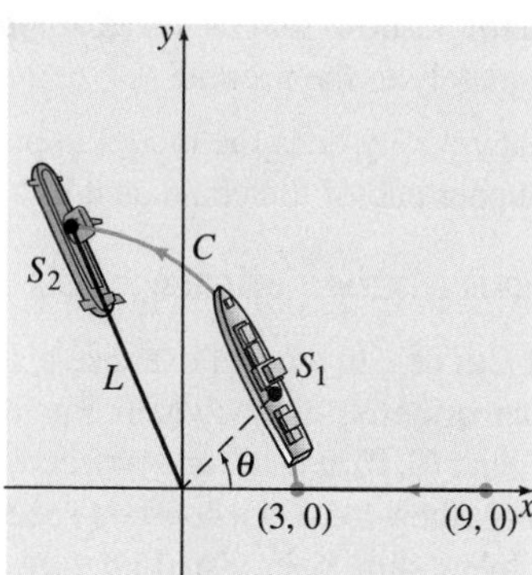

FIGURE 5.3.9 Pursuit curve in Problem 18

19. The Ballistic Pendulum Historically, in order to maintain quality control over munitions (bullets) produced by an assembly line, the manufacturer would use a **ballistic pendulum** to determine the muzzle velocity of a gun, that is, the speed of a bullet as it leaves the barrel. Invented in 1742 by the English engineer Benjamin Robins, the ballistic pendulum is simply a plane pendulum consisting of a rod of negligible mass to which a block of wood of mass m_w is attached. The system is set in motion by the impact of a bullet which is moving horizontally at the unknown velocity v_b; at the time of the impact, which we take as $t = 0$, the combined mass is $m_w + m_b$, where m_b is the mass of the bullet imbedded in the wood. In (7) of this section, we saw that in the case of small oscillations, the angular displacement $\theta(t)$ of a plane pendulum shown in Figure 5.3.3 is given by the linear DE $\theta'' + (g/l)\theta = 0$, where $\theta > 0$ corresponds to motion to the right of vertical. The velocity v_b can be found by measuring the height h of the mass $m_w + m_b$ at the maximum displacement angle θ_{max} shown in Figure 5.3.10.

Intuitively, the horizontal velocity V of the combined mass (wood plus bullet) after impact is only a fraction of the velocity v_b of the bullet, that is,

$$V = \left(\frac{m_b}{m_w + m_b}\right)v_b.$$

Now recall, a distance s traveled by a particle moving along a circular path is related to the radius l and central angle θ by the formula $s = l\theta$. By differentiating the last formula with respect to time t, it follows that the angular velocity ω of the mass and its linear velocity v are related by $v = l\omega$. Thus the initial angular velocity ω_0 at the time t at which the bullet impacts the wood block is related to V by $V = l\omega_0$ or

$$\omega_0 = \left(\frac{m_b}{m_w + m_b}\right)\frac{v_b}{l}.$$

(a) Solve the initial-value problem

$$\frac{d^2\theta}{dt^2} + \frac{g}{l}\theta = 0, \; \theta(0) = 0, \; \theta'(0) = \omega_0.$$

(b) Use the result from part (a) to show that

$$v_b = \left(\frac{m_w + m_b}{m_b}\right)\sqrt{lg}\,\theta_{max}.$$

(c) Use Figure 5.3.10 to express $\cos\theta_{max}$ in terms of l and h. Then use the first two terms of the Maclaurin series for $\cos\theta$ to express θ_{max} in terms of l and h. Finally, show that v_b is given (approximately) by

$$v_b = \left(\frac{m_w + m_b}{m_b}\right)\sqrt{2gh}.$$

(d) Use the result in part (c) to find v_b and $m_b = 5$ g, $m_w = 1$ kg, and $h = 6$ cm.

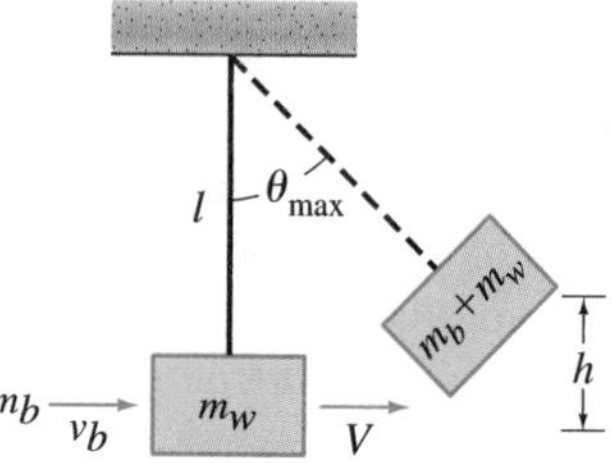

FIGURE 5.3.10 Ballistic pendulum in Problem 19

20. Relief Supplies As shown in Figure 5.3.11, a plane flying horizontally at a constant speed v_0 drops a relief supply pack to a person on the ground. Assume the origin is the point where the supply pack is released and that the positive x-axis points forward and that positive y-axis points downward. Under the assumption that the horizontal and vertical components of the air resistance are proportional to $(dx/dt)^2$ and $(dy/dt)^2$, respectively, and if the position of the supply pack is given by $\mathbf{r}(t) = x(t)\mathbf{i} + y(t)\mathbf{j}$, then its velocity is $\mathbf{v}(t) = (dx/dt)\mathbf{i} + (dy/dt)\mathbf{j}$. Equating components in the vector form of Newton's second law of motion,

$$m\frac{d\mathbf{v}}{dt} = mg - k\left[\left(\frac{dx}{dt}\right)^2\mathbf{i} + \left(\frac{dy}{dt}\right)^2\mathbf{j}\right]$$

gives

$$m\frac{d^2x}{dt^2} = mg - k\left(\frac{dx}{dt}\right)^2, \quad x(0) = 0, \; x'(0) = v_0$$

$$m\frac{d^2y}{dt^2} = mg - k\left(\frac{dy}{dt}\right)^2, \quad y(0) = 0, \; y'(0) = 0.$$

(a) Solve both of the foregoing initial-value problems by means of the substitutions $u = dx/dt$, $w = dy/dt$, and separation of variables. [*Hint*: See the *Remarks* at the end of Section 3.2.]

(b) Suppose the plane files at an altitude of 1000 ft and that its constant speed is 300 mi/h. Assume that the constant of proportionality for air resistance is $k = 0.0053$ and that the supply pack weighs 256 lb. Use a root-finding application of a CAS or a graphic calculator to determine the horizontal distance the pack travels, measured from its point of release to the point where it hits the ground.

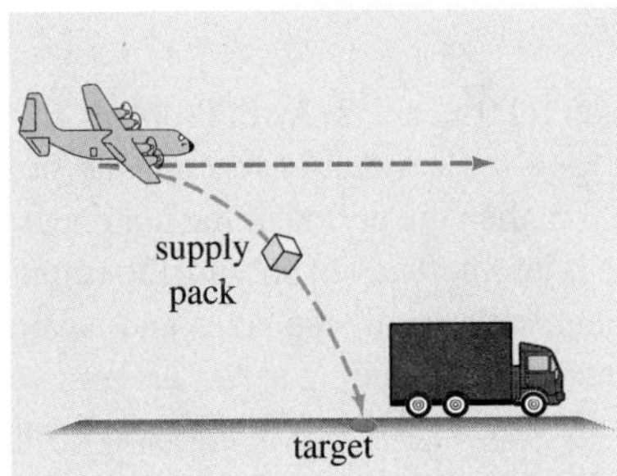

FIGURE 5.3.11 Airplane drop in Problem 20

Discussion Problems

21. Discuss why the damping term in equation (3) is written as

$$\beta\left|\frac{dx}{dt}\right|\frac{dx}{dt} \text{ instead of } \beta\left(\frac{dx}{dt}\right)^2.$$

22. **(a)** Experiment with a calculator to find an interval $0 \le \theta < \theta_1$, where θ is measured in radians, for which you think $\sin\theta \approx \theta$ is a fairly good estimate. Then use a graphing utility to plot the graphs of $y = x$ and $y = \sin x$ on the same coordinate axes for $0 \le x \le \pi/2$. Do the graphs confirm your observations with the calculator?

(b) Use a numerical solver to plot the solution curves of the initial-value problems

$$\frac{d^2\theta}{dt^2} + \sin\theta = 0, \quad \theta(0) = \theta_0, \quad \theta'(0) = 0$$

and $$\frac{d^2\theta}{dt^2} + \theta = 0, \quad \theta(0) = \theta_0, \quad \theta'(0) = 0$$

for several values of θ_0 in the interval $0 \le \theta < \theta_1$ found in part (a). Then plot solution curves of the initial-value problems for several values of θ_0 for which $\theta_0 > \theta_1$.

23. Pendulum Motion on the Moon Does a pendulum of length l oscillate faster on the Earth or on the Moon?

(a) Take $l = 3$ and $g = 32$ for the acceleration of gravity on Earth. Use a numerical solver to generate a numerical solution curve for the nonlinear model (6) subject to the initial conditions $\theta(0) = 1$, $\theta'(0) = 2$. Repeat using the same values but use $0.165g$ for the acceleration of gravity on the Moon.

(b) From the graphs in part (a), determine which pendulum oscillates faster. Which pendulum has the greater amplitude of motion?

24. Pendulum Motion on the Moon-Continued Repeat the two parts of Problem 23 this time using the linear model (7).

Computer Lab Assignments

25. Consider the initial-value problem

$$\frac{d^2\theta}{dt^2} + \sin\theta = 0, \quad \theta(0) = \frac{\pi}{12}, \quad \theta'(0) = -\frac{1}{3}$$

for a nonlinear pendulum. Since we cannot solve the differential equation, we can find no explicit solution of this problem. But suppose we wish to determine the first time $t_1 > 0$ for which the pendulum in Figure 5.3.3, starting from its initial position to the right, reaches the position OP—that is, the first positive root of $\theta(t) = 0$. In this problem and the next we examine several ways to proceed.

(a) Approximate t_1 by solving the linear problem

$$\frac{d^2\theta}{dt^2} + \theta = 0, \quad \theta(0) = \frac{\pi}{12}, \quad \theta'(0) = -\frac{1}{3}.$$

(b) Use the method illustrated in Example 3 of Section 4.10 to find the first four nonzero terms of a Taylor series solution $\theta(t)$ centered at 0 for the nonlinear initial-value problem. Give the exact values of all coefficients.

(c) Use the first two terms of the Taylor series in part (b) to approximate t_1.

(d) Use the first three terms of the Taylor series in part (b) to approximate t_1.

(e) Use a root-finding application of a CAS or a graphic calculator and the first four terms of the Taylor series in part (b) to approximate t_1.

(f) In this part of the problem you are led through the commands in *Mathematica* that enable you to approximate the root t_1. The procedure is easily modified so that any root of $\theta(t) = 0$ can be approximated. *(If you do not have Mathematica, adapt the given procedure by finding the corresponding syntax for the CAS you have on hand.)* Precisely reproduce and then, in turn, execute each line in the given sequence of commands.

```
sol = NDSolve[{y''[t] + Sin[y[t]] == 0,
        y[0] == Pi/12, y'[0] == -1/3},
        y, {t, 0, 5}]//Flatten
solution = y[t]/.sol
Clear[y]
y[t_]: = Evaluate[solution]
y[t]
gr1 = Plot[y[t], {t, 0, 5}]
root = FindRoot[y[t] == 0, {t, 1}]
```

(g) Appropriately modify the syntax in part (f) and find the next two positive roots of $\theta(t) = 0$.

26. Consider a pendulum that is released from rest from an initial displacement of θ_0 radians. Solving the linear model (7) subject to the initial conditions $\theta(0) = \theta_0$, $\theta'(0) = 0$ gives $\theta(t) = \theta_0 \cos\sqrt{g/l}\,t$. The period of oscillations predicted by this model is given by the familiar formula $T = 2\pi/\sqrt{g/l} = 2\pi\sqrt{l/g}$. The interesting thing about this formula for T is that it does not depend on the magnitude of the initial displacement θ_0. In other words, the linear model predicts that the time it would take the pendulum to swing from an initial displacement of, say, $\theta_0 = \pi/2$ $(= 90°)$ to $-\pi/2$ and back again would be exactly the same as the time it would take to cycle from, say, $\theta_0 = \pi/360$ $(= 0.5°)$ to $-\pi/360$. This is intuitively unreasonable; the actual period must depend on θ_0.

If we assume that $g = 32$ ft/s^2 and $l = 32$ ft, then the period of oscillation of the linear model is $T = 2\pi$ s. Let us compare this last number with the period predicted by the nonlinear model when $\theta_0 = \pi/4$. Using a numerical solver that is capable of generating hard data, approximate the solution of

$$\frac{d^2\theta}{dt^2} + \sin\theta = 0, \quad \theta(0) = \frac{\pi}{4}, \quad \theta'(0) = 0$$

on the interval $0 \le t \le 2$. As in Problem 25, if t_1 denotes the first time the pendulum reaches the position OP in Figure 5.3.3, then the period of the nonlinear pendulum is $4t_1$. Here is another way of solving the equation $\theta(t) = 0$. Experiment with small step sizes and advance the time, starting at $t = 0$ and ending at $t = 2$. From your hard data observe the time t_1 when $\theta(t)$ changes, for the first time, from positive to negative. Use the value t_1 to determine the true value of the period of the nonlinear pendulum. Compute the percentage relative error in the period estimated by $T = 2\pi$.

CHAPTER 5 IN REVIEW

Answers to selected odd-numbered problems begin on page ANS-9.

Answer Problems 1–8 without referring back to the text. Fill in the blank or answer true/false.

1. If a mass weighing 10 pounds stretches a spring 2.5 feet, a mass weighing 32 pounds will stretch it ________ feet.

2. The period of simple harmonic motion of mass weighing 8 pounds attached to a spring whose constant is 6.25 lb/ft is ________ seconds.

3. The differential equation of a spring/mass system is $x'' + 16x = 0$. If the mass is initially released from a point 1 meter above the equilibrium position with a downward velocity of 3 m/s, the amplitude of vibrations is ________ meters.

4. Pure resonance cannot take place in the presence of a damping force. ________

5. In the presence of a damping force, the displacements of a mass on a spring will always approach zero as $t \to \infty$. ________

6. A mass on a spring whose motion is critically damped can possibly pass through the equilibrium position twice. ________

7. At critical damping any increase in damping will result in an ________ system.

8. If simple harmonic motion is described by $x = (\sqrt{2}/2)\sin(2t + \phi)$, the phase angle ϕ is ______ when the initial conditions are $x(0) = -\frac{1}{2}$ and $x'(0) = 1$.

In Problems 9 and 10 the eigenvalues and eigenfunctions of the boundary-value problem $y'' + \lambda y = 0$, $y'(0) = 0$, $y'(\pi) = 0$ are $\lambda_n = n^2$, $n = 0, 1, 2, \ldots$, and $y = \cos nx$, respectively. Fill in the blanks.

9. A solution of the BVP when $\lambda = 8$ is $y =$ ________ because ________.

10. A solution of the BVP when $\lambda = 36$ is $y =$ ________ because ________.

11. A free undamped spring/mass system oscillates with a period of 3 seconds. When 8 pounds are removed from the spring, the system has a period of 2 seconds. What was the weight of the original mass on the spring?

12. A mass weighing 12 pounds stretches a spring 2 feet. The mass is initially released from a point 1 foot below the equilibrium position with an upward velocity of 4 ft/s.
   (a) Find the equation of motion.
   (b) What are the amplitude, period, and frequency of the simple harmonic motion?
   (c) At what times does the mass return to the point 1 foot below the equilibrium position?
   (d) At what times does the mass pass through the equilibrium position moving upward? Moving downward?
   (e) What is the velocity of the mass at $t = 3\pi/16$ s?
   (f) At what times is the velocity zero?

13. A force of 2 pounds stretches a spring 1 foot. With one end held fixed, a mass weighing 8 pounds is attached to the other end. The system lies on a table that imparts a frictional force numerically equal to $\frac{3}{2}$ times the instantaneous velocity. Initially, the mass is displaced 4 inches above the equilibrium position and released from rest. Find the equation of motion if the motion takes place along a horizontal straight line that is taken as the x-axis.

14. A mass weighing 32 pounds stretches a spring 6 inches. The mass moves through a medium offering a damping force that is numerically equal to β times the instantaneous velocity. Determine the values of $\beta > 0$ for which the spring/mass system will exhibit oscillatory motion.

15. A spring with constant $k = 2$ is suspended in a liquid that offers a damping force numerically equal to 4 times the instantaneous velocity. If a mass m is suspended from the spring, determine the values of m for which the subsequent free motion is nonoscillatory.

16. The vertical motion of a mass attached to a spring is described by the IVP $\frac{1}{4}x'' + x' + x = 0$, $x(0) = 4$, $x'(0) = 2$. Determine the maximum vertical displacement of the mass.

17. A mass weighing 4 pounds stretches a spring 18 inches. A periodic force equal to $f(t) = \cos \gamma t + \sin \gamma t$ is impressed on the system starting at $t = 0$. In the absence of a damping force, for what value of γ will the system be in a state of pure resonance?

18. Find a particular solution for $x'' + 2\lambda x' + \omega^2 x = A$, where A is a constant force.

19. A mass weighing 4 pounds is suspended from a spring whose constant is 3 lb/ft. The entire system is immersed in a fluid offering a damping force numerically equal to the instantaneous velocity. Beginning at $t = 0$, an external force equal to $f(t) = e^{-t}$ is impressed on the system. Determine the equation of motion if the mass is initially released from rest at a point 2 feet below the equilibrium position.

20. **(a)** Two springs are attached in series as shown in Figure 5.R.1. If the mass of each spring is ignored, show that the effective spring constant k of the system is defined by $1/k = 1/k_1 + 1/k_2$.

(b) A mass weighing W pounds stretches a spring $\frac{1}{2}$ foot and stretches a different spring $\frac{1}{4}$ foot. The two springs are attached, and the mass is then attached to the double spring as shown in Figure 5.R.1. Assume that the motion is free and that there is no damping force present. Determine the equation of motion if the mass is initially released at a point 1 foot below the equilibrium position with a downward velocity of $\frac{2}{3}$ ft/s.

(c) Show that the maximum speed of the mass is $\frac{2}{3}\sqrt{3g + 1}$.

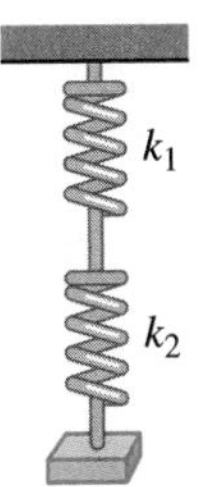

FIGURE 5.R.1 Attached springs in Problem 20

21. A series circuit contains an inductance of $L = 1$ h, a capacitance of $C = 10^{-4}$ f, and an electromotive force of $E(t) = 100 \sin 50t$ V. Initially, the charge q and current i are zero.

(a) Determine the charge $q(t)$.

(b) Determine the current $i(t)$.

(c) Find the times for which the charge on the capacitor is zero.

22. **(a)** Show that the current $i(t)$ in an LRC-series circuit satisfies $L\dfrac{d^2i}{dt^2} + R\dfrac{di}{dt} + \dfrac{1}{C}i = E'(t)$, where $E'(t)$ denotes the derivative of $E(t)$.

(b) Two initial conditions $i(0)$ and $i'(0)$ can be specified for the DE in part (a). If $i(0) = i_0$ and $q(0) = q_0$, what is $i'(0)$?

23. Consider the boundary-value problem

$$y'' + \lambda y = 0, \quad y(0) = y(2\pi), \quad y'(0) = y'(2\pi).$$

Show that except for the case $\lambda = 0$, there are two independent eigenfunctions corresponding to each eigenvalue.

24. A bead is constrained to slide along a frictionless rod of length L. The rod is rotating in a vertical plane with a constant angular velocity ω about a pivot P fixed at the midpoint of the rod, but the design of the pivot allows the bead to move along the entire length of the rod. Let $r(t)$ denote the position of the bead relative to this rotating coordinate system as shown in Figure 5.R.2. To apply Newton's second law of motion to this rotating frame of reference, it is necessary to use the fact that the net force acting on the bead is the sum of the real forces (in this case, the force due to gravity) and the inertial forces (coriolis, transverse, and centrifugal). The mathematics is a little complicated, so we just give the resulting differential equation for r:

$$m\frac{d^2r}{dt^2} = m\omega^2 r - mg \sin \omega t.$$

(a) Solve the foregoing DE subject to the initial conditions $r(0) = r_0$, $r'(0) = v_0$.

(b) Determine the initial conditions for which the bead exhibits simple harmonic motion. What is the minimum length L of the rod for which it can accommodate simple harmonic motion of the bead?

(c) For initial conditions other than those obtained in part (b), the bead must eventually fly off the rod. Explain using the solution $r(t)$ in part (a).

(d) Suppose $\omega = 1$ rad/s. Use a graphing utility to graph the solution $r(t)$ for the initial conditions $r(0) = 0$, $r'(0) = v_0$, where v_0 is 0, 10, 15, 16, 16.1, and 17.

(e) Suppose the length of the rod is $L = 40$ ft. For each pair of initial conditions in part (d), use a root-finding application to find the total time that the bead stays on the rod.

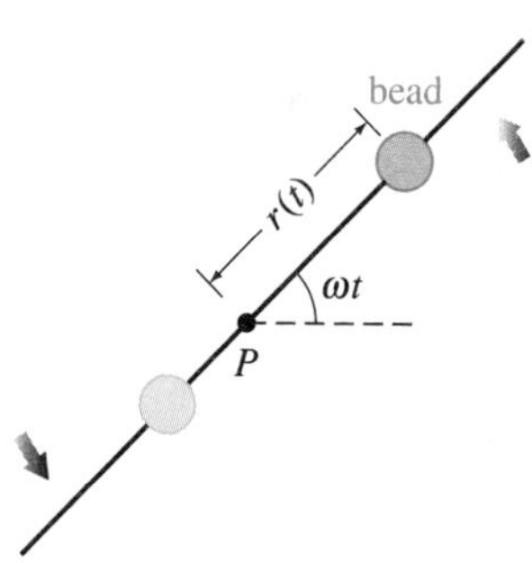

FIGURE 5.R.2 Rotating rod in Problem 24

25. Suppose a mass m lying on a flat dry frictionless surface is attached to the free end of a spring whose constant is k. In Figure 5.R.3(a) the mass is shown at the equilibrium position $x = 0$, that is, the spring is neither stretched nor compressed. As shown in Figure 5.R.3(b), the displacement $x(t)$ of the mass to the right of the equilibrium position is positive and negative to the left. Determine a differential equation for the displacement $x(t)$ of the freely sliding mass. Discuss the difference between the derivation of this DE and the analysis leading to (1) of Section 5.1.

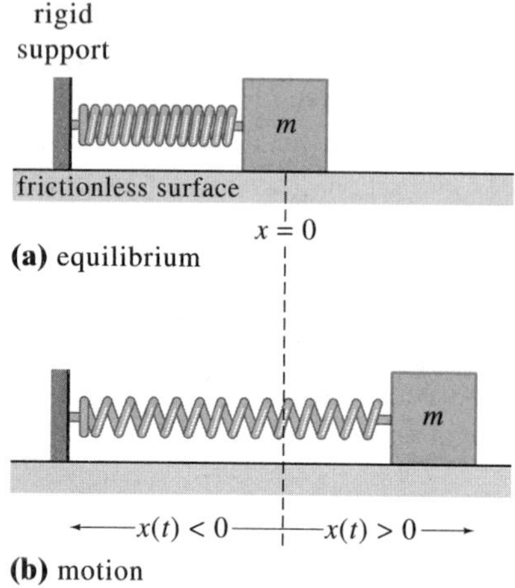

FIGURE 5.R.3 Sliding spring/mass system in Problem 25

26. Suppose the mass m on the flat, dry, frictionless surface in Problem 25 is attached to two springs as shown Figure 5.R.4. If the spring constants are k_1 and k_2, determine a differential equation for the displacement $x(t)$ of the freely sliding mass.

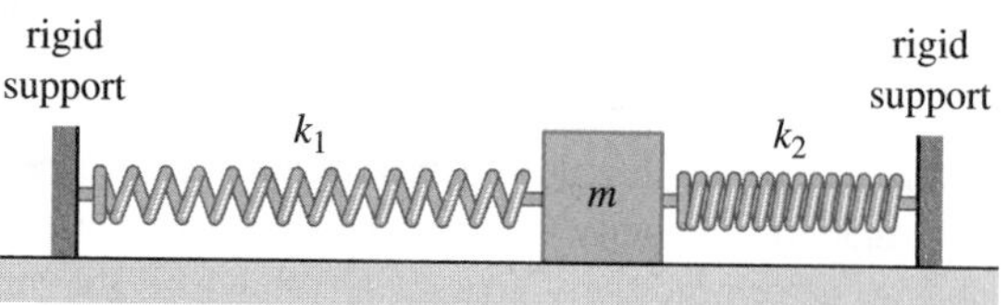

FIGURE 5.R.4 Double spring system in Problem 26

27. Suppose the mass m in the spring/mass system in Problem 25 slides over a dry surface whose coefficient of sliding friction is $\mu > 0$. If the retarding force of kinetic friction has the constant magnitude $f_k = \mu mg$, where mg is the weight of the mass, and acts opposite to the direction of motion, then it is known as **coulomb friction.** By using the **signum function**

$$\operatorname{sgn}(x') = \begin{cases} -1, & x' < 0 \text{ (motion to left)} \\ 1, & x' > 0 \text{ (motion to right)} \end{cases}$$

determine a piecewise-defined differential equation for the displacement $x(t)$ of the damped sliding mass.

28. For simplicity, let us assume in Problem 27 that $m = 1$, $k = 1$, and $f_k = 1$.

(a) Find the displacement $x(t)$ of the mass if it is released from rest from a point 5.5 units to the right of the equilibrium position, that is, the initial conditions are $x(0) = 5.5$, $x'(0) = 0$. When released, intuitively the motion of the mass will be to the left. Give a time interval $[0, t_1]$ over which this solution is defined. Where is the mass at time t_1?

(b) For $t > t_1$ assume that the motion is now to the right. Using initial conditions at t_1, find $x(t)$ and give a time interval $[t_1, t_2]$ over which this solution is defined. Where is the mass at time t_2?

(c) For $t > t_2$ assume that the motion is now to the left. Using initial conditions at t_2, find $x(t)$ and give a time interval $[t_2, t_3]$ over which this solution is defined. Where is the mass at time t_3?

(d) Using initial conditions at t_3, show that the model predicts that there is no furthere motion for $t > t_3$.

(e) Graph the displacement $x(t)$ on the interval $[0, t_3]$.

6 Series Solutions of Linear Equations

Up to this point in our study of differential equations we have primarily solved linear equations of order two (or higher) that have constant coefficients. The only exception was the Cauchy-Euler equation in Section 4.7. In applications, higher-order linear equations with variable coefficients are just as important as, if not more than, differential equations with constant coefficients. As pointed out in Section 4.7, even a simple linear second-order equation with variable coefficients such as $y'' + xy = 0$ does not possess solutions that are elementary functions. But this is not to say that we *can't* find two linearly independent solutions of $y'' + xy = 0$; we can. In Sections 6.2 and 6.4 we shall see that the functions that are solutions of this equation are defined by infinite series.

In this chapter we shall study two infinite-series methods for finding solutions of homogeneous linear second-order DEs $a_2(x)y'' + a_1(x)y' + a_0(x)y = 0$, where the variable coefficients $a_2(x)$, $a_1(x)$, and $a_0(x)$ are, for the most part, simple polynomial functions.

6.1 REVIEW OF POWER SERIES

REVIEW MATERIAL

- Infinite series of constants, *p*-series, harmonic series, alternating harmonic series, geometric series, tests for convergence especially the ratio test
- Power series, Taylor series, Maclaurin series (See any calculus text)

INTRODUCTION In Section 4.3 we saw that solving a homogeneous linear DE with constant coefficients was essentially a problem in algebra. By finding the roots of the auxiliary equation, we could write a general solution of the DE as a linear combination of the elementary functions $e^{\alpha x}$, $x^k e^{\alpha x}$, $x^k e^{\alpha x}\cos\beta x$, and $x^k e^{\alpha x}\sin\beta x$. But as was pointed out in the introduction to Section 4.7, most linear higher-order DEs with variable coefficients cannot be solved in terms of elementary functions. A usual course of action for equations of this sort is to assume a solution in the form of an infinite series and proceed in a manner similar to the method of undetermined coefficients (Section 4.4). In Section 6.2 we consider linear second-order DEs with variable coefficients that possess solutions in the form of a power series, and so it is appropriate that we begin this chapter with a review of that topic.

Power Series

Recall from calculus that **power series** in $x - a$ is an infinite series of the form

The index of summation need not start at $n = 0$. ▶

$$\sum_{n=0}^{\infty} c_n(x-a)^n = c_0 + c_1(x-a) + c_2(x-a)^2 + \cdots.$$

Such a series also said to be a **power series centered at *a*.** For example, the power series $\sum_{n=0}^{\infty}(x+1)^n$ is centered at $a = -1$. In the next section we will be concerned principally with power series in x, in other words, power series that are centered at $a = 0$. For example,

$$\sum_{n=0}^{\infty} 2^n x^n = 1 + 2x + 4x^2 + \cdots$$

is a power series in x.

Important Facts

The following bulleted list summarizes some important facts about power series $\sum_{n=0}^{\infty} c_n(x-a)^n$.

- **Convergence** A power series is **convergent** at a specified value of x if its sequence of partial sums $\{S_N(x)\}$ converges, that is, $\lim_{N\to\infty} S_N(x) = \lim_{N\to\infty} \sum_{n=0}^{N} c_n(x-a)^n$ exists. If the limit does not exist at x, then the series is said to be **divergent.**
- **Interval of Convergence** Every power series has an **interval of convergence.** The interval of convergence is the set of *all* real numbers x for which the series converges. The center of the interval of convergence is the center a of the series.
- **Radius of Convergence** The radius R of the interval of convergence of a power series is called its **radius of convergence.** If $R > 0$, then a power series converges for $|x - a| < R$ and diverges for $|x - a| > R$. If the series converges only at its center a, then $R = 0$. If the series converges for all x, then we write $R = \infty$. Recall, the absolute-value inequality $|x - a| < R$ is equivalent to the simultaneous inequality $a - R < x < a + R$. A power series may or may not converge at the endpoints $a - R$ and $a + R$ of this interval.
- **Absolute Convergence** Within its interval of convergence a power series **converges absolutely.** In other words, if x is in the interval of convergence and is not an endpoint of the interval, then the series of absolute values $\sum_{n=0}^{\infty} |c_n(x-a)^n|$ converges. See Figure 6.1.1.

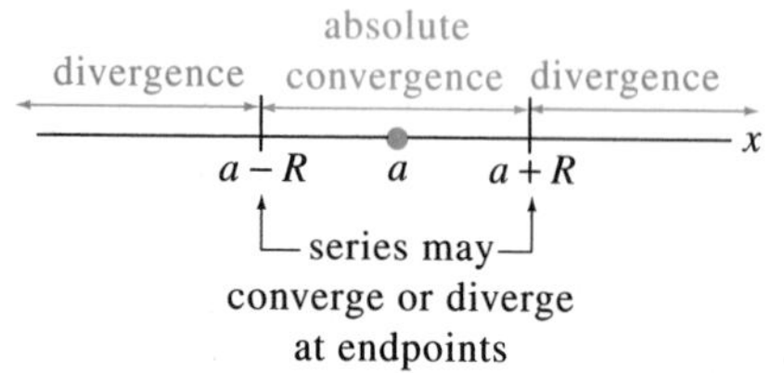

FIGURE 6.1.1 Absolute convergence within the interval of convergence and divergence outside of this interval

- **Ratio Test** Convergence of power series can often be determined by the **ratio test.** Suppose $c_n \neq 0$ for all n in $\Sigma_{n=0}^{\infty} c_n(x-a)^n$, and that

$$\lim_{n\to\infty}\left|\frac{c_{n+1}(x-a)^{n+1}}{c_n(x-a)^n}\right| = |x-a|\lim_{n\to\infty}\left|\frac{c_{n+1}}{c_n}\right| = L.$$

If $L < 1$, the series converges absolutely; if $L > 1$ the series diverges; and if $L = 1$ the test is inconclusive. The ratio test is always inconclusive at an endpoint $a \pm R$.

EXAMPLE 1 Interval of Convergence

Find the interval and radius of convergence for $\sum_{n=1}^{\infty}\frac{(x-3)^n}{2^n n}$.

SOLUTION The ratio test gives

$$\lim_{n\to\infty}\left|\frac{\dfrac{(x-3)^{n+1}}{2^{n+1}(n+1)}}{\dfrac{(x-3)^n}{2^n n}}\right| = |x-3|\lim_{n\to\infty}\frac{n+1}{2n} = \frac{1}{2}|x-3|.$$

The series converges absolutely for $\frac{1}{2}|x-3| < 1$ or $|x-3| < 2$ or $1 < x < 5$. This last inequality defines the *open* interval of convergence. The series diverges for $|x-3| > 2$, that is, for $x > 5$ or $x < 1$. At the left endpoint $x = 1$ of the open interval of convergence, the series of constants $\Sigma_{n=1}^{\infty}((-1)^n/n)$ is convergent by the alternating series test. At the right endpoint $x = 5$, the series $\Sigma_{n=1}^{\infty}(1/n)$ is the divergent harmonic series. The interval of convergence of the series is $[1, 5)$, and the radius of convergence is $R = 2$. ≡

- **A Power Series Defines a Function** A power series defines a function, that is, $f(x) = \Sigma_{n=0}^{\infty} c_n(x-a)^n$ whose domain is the interval of convergence of the series. If the radius of convergence is $R > 0$ or $R = \infty$, then f is continuous, differentiable, and integrable on the intervals $(a-R, a+R)$ or $(-\infty, \infty)$, respectively. Moreover, $f'(x)$ and $\int f(x)\,dx$ can be found by term-by-term differentiation and integration. Convergence at an endpoint may be either lost by differentiation or gained through integration. If

$$y = \sum_{n=1}^{\infty} c_n x^n = c_0 + c_1x + c_2x^2 + c_3x^3 + \cdots$$

is a power series in x, then the first two derivatives are $y' = \Sigma_{n=0}^{\infty} nx^{n-1}$ and $y'' = \Sigma_{n=0}^{\infty} n(n-1)x^{n-2}$. Notice that the first term in the first derivative and the first two terms in the second derivative are zero. We omit these zero terms and write

$$y' = \sum_{n=1}^{\infty} c_n n x^{n-1} = c_1 + 2c_2x + 3c_3x^2 + 4c_4x^3 + \cdots$$
$$y'' = \sum_{n=2}^{\infty} c_n n(n-1)x^{n-2} = 2c_2 + 6c_3x + 12c_4x^2 + \cdots. \tag{1}$$

Be sure you understand the two results given in (1); especially note where the index of summation starts in each series. These results are important and will be used in all examples in the next section.

- **Identity Property** If $\Sigma_{n=0}^{\infty} c_n(x-a)^n = 0$, $R > 0$, for all numbers x in some open interval, then $c_n = 0$ for all n.
- **Analytic at a Point** A function f is said to be **analytic at a point *a*** if it can be represented by a power series in $x - a$ with either a positive or an infinite radius of convergence. In calculus it is seen that infinitely

differentiable functions such as e^x, $\sin x$, $\cos x$, $e^x \ln(1 + x)$, and so on, can be represented by Taylor series

$$\sum_{n=0}^{\infty} \frac{f^{(n)}(a)}{n!}(x - a)^n = f(a) + \frac{f'(a)}{1!}(x - a) + \frac{f''(a)}{1!}(x - a)^2 + \cdots$$

or by a Maclaurin series

$$\sum_{n=0}^{\infty} \frac{f^{(n)}(0)}{n!}x^n = f(0) + \frac{f'(0)}{1!}x + \frac{f''(0)}{1!}x^2 + \cdots.$$

You might remember some of the following Maclaurin series representations.

Maclaurin Series	Interval of Convergence	
$e^x = 1 + \frac{x}{1!} + \frac{x^2}{2!} + \frac{x^3}{3!} + \cdots = \sum_{n=0}^{\infty} \frac{1}{n!}x^n$	$(-\infty, \infty)$	
$\cos x = 1 - \frac{x^2}{2!} + \frac{x^4}{4!} - \frac{x^6}{6!} + \cdots = \sum_{n=0}^{\infty} \frac{(-1)^n}{(2n)!}x^{2n}$	$(-\infty, \infty)$	
$\sin x = x - \frac{x^3}{3!} + \frac{x^5}{5!} - \frac{x^7}{7!} + \cdots = \sum_{n=0}^{\infty} \frac{(-1)^n}{(2n+1)!}x^{2n+1}$	$(-\infty, \infty)$	
$\tan^{-1} x = x - \frac{x^3}{3} + \frac{x^5}{5} - \frac{x^7}{7} + \cdots = \sum_{n=0}^{\infty} \frac{(-1)^n}{2n+1}x^{2n+1}$	$[-1, 1]$	(2)
$\cosh x = 1 + \frac{x^2}{2!} + \frac{x^4}{4!} + \frac{x^6}{6!} + \cdots = \sum_{n=0}^{\infty} \frac{1}{(2n)!}x^{2n}$	$(-\infty, \infty)$	
$\sinh x = x + \frac{x^3}{3!} + \frac{x^5}{5!} + \frac{x^7}{7!} + \cdots = \sum_{n=0}^{\infty} \frac{1}{(2n+1)!}x^{2n+1}$	$(-\infty, \infty)$	
$\ln(1 + x) = x - \frac{x^2}{2} + \frac{x^3}{3} - \frac{x^4}{4} + \cdots = \sum_{n=1}^{\infty} \frac{(-1)^{n+1}}{n}x^n$	$(-1, 1]$	
$\frac{1}{1-x} = 1 + x + x^2 + x^3 + \cdots = \sum_{n=0}^{\infty} x^n$	$(-1, 1)$	

These results can be used to obtain power series representations of other functions. For example, if we wish to find the Maclaurin series representation of, say, e^{x^2} we need only replace x in the Maclaurin series for e^x:

$$e^{x^2} = 1 + \frac{x^2}{1!} + \frac{x^4}{2!} + \frac{x^6}{3!} + \cdots = \sum_{n=0}^{\infty} \frac{1}{n!}x^{2n}.$$

Similarly, to obtain a Taylor series representation of $\ln x$ centered at $a = 1$ we replace x by $x - 1$ in the Maclaurin series for $\ln(1 + x)$:

$$\ln x = \ln(1 + (x - 1)) = (x - 1) - \frac{(x-1)^2}{2} + \frac{(x-1)^3}{3} - \frac{(x-1)^4}{4} + \cdots = \sum_{n=1}^{\infty} \frac{(-1)^{n+1}}{n}(x - 1)^n.$$

You can also verify that the interval of convergence is (0, 2] by using the ratio test.

The interval of convergence for the power series representation of e^{x^2} is the same as that of e^x, that is, $(-\infty, \infty)$. But the interval of convergence of the Taylor series of $\ln x$ is now $(0, 2]$; this interval is $(-1, 1]$ shifted 1 unit to the right.

- **Arithmetic of Power Series** Power series can be combined through the operations of addition, multiplication, and division. The procedures for powers series are similar to the way in which two polynomials are added, multiplied, and divided—that is, we add coefficients of like powers of x, use the distributive law and collect like terms, and perform long division.

EXAMPLE 2 Multiplication of Power Series

Find a power series representation of $e^x \sin x$.

SOLUTION We use the power series for e^x and $\sin x$:

$$\begin{aligned} e^x \sin x &= \left(1 + x + \frac{x^2}{2} + \frac{x^3}{6} + \frac{x^4}{24} + \cdots\right)\left(x - \frac{x^3}{6} + \frac{x^5}{120} - \frac{x^7}{5040} + \cdots\right) \\ &= (1)x + (1)x^2 + \left(-\frac{1}{6} + \frac{1}{2}\right)x^3 + \left(-\frac{1}{6} + \frac{1}{6}\right)x^4 + \left(\frac{1}{120} - \frac{1}{12} + \frac{1}{24}\right)x^5 + \cdots \\ &= x + x^2 + \frac{x^3}{3} - \frac{x^5}{30} + \cdots. \end{aligned}$$

Since the power series of e^x and $\sin x$ both converge on $(-\infty, \infty)$, the product series converges on the same interval. Problems involving multiplication or division of power series can be done with minimal fuss using a computer algebra system. ≡

≡ Shifting the Summation Index For the three remaining sections of this chapter, it is crucial that you become adept at simplifying the sum of two or more power series, each series expressed in summation notation, to an expression with a single Σ. As the next example illustrates, combining two or more summations as a single summation often requires a reindexing, that is, a shift in the index of summation.

EXAMPLE 3 Addition of Power Series

Write

$$\sum_{n=2}^{\infty} n(n-1)c_n x^{n-2} + \sum_{n=0}^{\infty} c_n x^{n+1}$$

as one power series.

SOLUTION In order to add the two series given in summation notation, it is necessary that both indices of summation start with the same number and that the powers of x in each series be "in phase," in other words, if one series starts with a multiple of, say, x to the first power, then we want the other series to start with the same power. Note that in the given problem, the first series starts with x^0 whereas the second series starts with x^1. By writing the first term of the first series outside of the summation notation,

$$\sum_{n=2}^{\infty} n(n-1)c_n x^{n-2} + \sum_{n=0}^{\infty} c_n x^{n+1} = 2 \cdot 1 c_2 x^0 + \underbrace{\sum_{n=3}^{\infty} n(n-1)c_n x^{n-2}}_{\text{series starts with } x \text{ for } n=3} + \underbrace{\sum_{n=0}^{\infty} c_n x^{n+1}}_{\text{series starts with } x \text{ for } n=0} \tag{3}$$

we see that both series on the right side start with the same power of x, namely, x^1. Now to get the same summation index we are inspired by the exponents of x; we let $k = n - 2$ in the first series and at the same time let $k = n + 1$ in the second series. For $n = 3$ in $k = n - 2$ we get $k = 1$, and for $n = 0$ in $k = n + 1$ we get $k = 1$, and so the right-hand side of (3) becomes

$$2c_2 + \sum_{k=1}^{\infty} (k+2)(k+1)c_{k+2}x^k + \sum_{k=1}^{\infty} c_{k-1}x^k. \tag{4}$$

(The powers x^k are marked "same" and the starting indices $k=1$ are marked "same".)

Remember the summation index is a "dummy" variable; the fact that $k = n - 2$ in one case and $k = n + 1$ in the other should cause no confusion if you keep in mind that it is the *value* of the summation index that is important. In both cases k takes on the same successive values $k = 1, 2, 3, \ldots$ when n takes on the values $n = 2, 3, 4, \ldots$ for $k = n - 1$ and $n = 0, 1, 2, \ldots$ for $k = n + 1$. We are now in a position to add the series in (4) term-by-term:

$$\sum_{n=2}^{\infty} n(n-1)c_n x^{n-2} + \sum_{n=0}^{\infty} c_n x^{n+1} = 2c_2 + \sum_{k=1}^{\infty} [(k+2)(k+1)c_{k+2} + c_{k-1}]x^k. \quad (5)$$

If you are not totally convinced of the result in (5), then write out a few terms on both sides of the equality.

A Preview The point of this section is to remind you of the salient facts about power series so that you are comfortable using power series in the next section to find solutions of linear second-order DEs. In the last example in this section we tie up many of the concepts just discussed; it also gives a preview of the method that will used in Section 6.2. We purposely keep the example simple by solving a linear first-order equation. Also suspend, for the sake of illustration, the fact that you already know how to solve the given equation by the integrating-factor method in Section 2.3.

EXAMPLE 4 A Power Series Solution

Find a power series solution $y = \sum_{n=0}^{\infty} c_n x^n$ of the differential equation $y' + y = 0$.

SOLUTION We break down the solution into a sequence of steps.

(*i*) First calculate the derivative of the assumed solution:

$$y' = \sum_{n=1}^{\infty} c_n n x^{n-1} \leftarrow \text{see the first line in (1)}$$

(*ii*) Then substitute y and y' into the given DE:

$$y' + y = \sum_{n=1}^{\infty} c_n n x^{n-1} + \sum_{n=0}^{\infty} c_n x^n.$$

(*iii*) Now shift the indices of summation. When the indices of summation have the same starting point and the powers of x agree, combine the summations:

$$\begin{aligned} y' + y &= \underbrace{\sum_{n=1}^{\infty} c_n n x^{n-1}}_{k=n-1} + \underbrace{\sum_{n=0}^{\infty} c_n x^n}_{k=n} \\ &= \sum_{k=0}^{\infty} c_{k+1}(k+1)x^k + \sum_{k=0}^{\infty} c_k x^k \\ &= \sum_{k=0}^{\infty} [c_{k+1}(k+1) + c_k]x^k. \end{aligned}$$

(*iv*) Because we want $y' + y = 0$ for all x in some interval,

$$\sum_{k=0}^{\infty} [c_{k+1}(k+1) + c_k]x^k = 0$$

is an identity and so we must have $c_{k+1}(k+1) + c_k = 0$, or

$$c_{k+1} = -\frac{1}{k+1}c_k, \quad k = 0, 1, 2, \ldots.$$

(*v*) By letting k take on successive integer values starting with $k = 0$, we find

$$c_1 = -\frac{1}{1}c_0 = -c_0$$
$$c_2 = -\frac{1}{2}c_1 = -\frac{1}{2}(-c_0) = \frac{1}{2}c_0$$
$$c_3 = -\frac{1}{3}c_2 = -\frac{1}{3}\left(\frac{1}{2}c_0\right) = -\frac{1}{3 \cdot 2}c_0$$
$$c_4 = -\frac{1}{4}c_2 = -\frac{1}{4}\left(-\frac{1}{3 \cdot 2}c_0\right) = \frac{1}{4 \cdot 3 \cdot 2}c_0$$

and so on, where c_0 is arbitrary.

(*vi*) Using the original assumed solution and the results in part (*v*) we obtain a formal power series solution

$$\begin{aligned} y &= c_0 + c_1x + c_2x^2 + c_3x^3 + c_4x^4 + \cdots \\ &= c_0 - c_0x + \frac{1}{2}c_0x^2 - c_0\frac{1}{3 \cdot 2}x^3 + c_0\frac{1}{4 \cdot 3 \cdot 2}x^4 - \cdots \\ &= c_0\left[1 - x + \frac{1}{2}x^2 - \frac{1}{3 \cdot 2}x^3 + \frac{1}{4 \cdot 3 \cdot 2}x^4 - \cdots\right]. \end{aligned}$$

It should be fairly obvious that the pattern of the coefficients in part (*v*) is $c_k = c_0(-1)^k/k!$, $k = 0, 1, 2, \ldots$. so that in summation notation we can write

If desired we could switch back to n as the index of summation. ▶

$$y = c_0\sum_{k=0}^{\infty}\frac{(-1)^k}{k!}x^k. \tag{8}$$ ≡

From the first power series representation in (2) the solution in (8) is recognized as $y = c_0e^{-x}$. Had you used the method of Section 2.3, you would have found that $y = ce^{-x}$ is a solution of $y' + y = 0$ on the interval $(-\infty, \infty)$. This interval is also the interval of convergence of the power series in (8).

EXERCISES 6.1

Answers to selected odd-numbered problems begin on page ANS-9.

In Problems 1–10 find the interval and radius of convergence for the given power series.

1. $\displaystyle\sum_{n=1}^{\infty}\frac{(-1)^n}{n}x^n$
2. $\displaystyle\sum_{n=1}^{\infty}\frac{1}{n^2}x^n$
3. $\displaystyle\sum_{n=1}^{\infty}\frac{2^n}{n}x^n$
4. $\displaystyle\sum_{n=0}^{\infty}\frac{5^n}{n!}x^n$
5. $\displaystyle\sum_{k=1}^{\infty}\frac{(-1)^k}{10^k}(x-5)^k$
6. $\displaystyle\sum_{k=0}^{\infty}k!(x-1)^k$
7. $\displaystyle\sum_{k=1}^{\infty}\frac{1}{k^2+k}(3x-1)^k$
8. $\displaystyle\sum_{k=0}^{\infty}3^{-k}(4x-5)^k$
9. $\displaystyle\sum_{k=1}^{\infty}\frac{2^{5k}}{5^{2k}}\left(\frac{x}{3}\right)^k$
10. $\displaystyle\sum_{n=0}^{\infty}\frac{(-1)^n}{9^n}x^{2n+1}$

In Problems 11–16 use an appropriate series in (2) to find the Maclaurin series of the given function. Write your answer in summation notation.

11. $e^{-x/2}$
12. xe^{3x}
13. $\dfrac{1}{2+x}$
14. $\dfrac{x}{1+x^2}$
15. $\ln(1-x)$
16. $\sin x^2$

In Problems 17 and 18 use an appropriate series in (2) to find the Taylor series of the given function centered at the indicated value of a. Write your answer in summation notation.

17. $\sin x$, $a = 2\pi$ [*Hint*: Use periodicity.]
18. $\ln x$; $a = 2$ [*Hint*: $x = 2[1 + (x-2)/2]$]

In Problems 19 and 20 the given function is analytic at $a = 0$. Use appropriate series in (2) and multiplication to find the first four nonzero terms of the Maclaurin series of the given function.

19. $\sin x \cos x$
20. $e^{-x}\cos x$

In Problems 21 and 22 the given function is analytic at $a = 0$. Use appropriate series in (2) and long division to find the first four nonzero terms of the Maclaurin series of the given function.

21. $\sec x$
22. $\tan x$

In Problems 23 and 24 use a substitution to shift the summation index so that the general term of given power series involves x^k.

23. $\sum_{n=1}^{\infty} nc_n x^{n+2}$

24. $\sum_{n=3}^{\infty} (2n - 1)c_n x^{n-3}$

In Problems 25–30 proceed as in Example 3 to rewrite the given expression using a single power series whose general term involves x^k.

25. $\sum_{n=1}^{\infty} nc_n x^{n-1} - \sum_{n=0}^{\infty} c_n x^n$

26. $\sum_{n=1}^{\infty} nc_n x^{n-1} + 3\sum_{n=0}^{\infty} c_n x^{n+2}$

27. $\sum_{n=1}^{\infty} 2nc_n x^{n-1} + \sum_{n=0}^{\infty} 6c_n x^{n+1}$

28. $\sum_{n=2}^{\infty} n(n - 1)c_n x^{n-2} + \sum_{n=0}^{\infty} c_n x^{n+2}$

29. $\sum_{n=2}^{\infty} n(n - 1)c_n x^{n-2} - 2\sum_{n=1}^{\infty} nc_n x^n + \sum_{n=0}^{\infty} c_n x^n$

30. $\sum_{n=2}^{\infty} n(n - 1)c_n x^n + 2\sum_{n=2}^{\infty} n(n - 1)c_n x^{n-2} + 3\sum_{n=1}^{\infty} nc_n x^n$

In Problems 31–34 verify by direct substitution that the given power series is a solution of the indicated differential equation. [*Hint*: For a power x^{2n+1} let $k = n + 1$.]

31. $y = \sum_{n=0}^{\infty} \frac{(-1)^n}{n!} x^{2n}, \quad y' + 2xy = 0$

32. $y = \sum_{n=0}^{\infty} (-1)^n x^{2n}, \quad (1 + x^2)y' + 2xy = 0$

33. $y = \sum_{n=1}^{\infty} \frac{(-1)^{n+1}}{n} x^n, \quad (x + 1)y'' + y' = 0$

34. $y = \sum_{n=0}^{\infty} \frac{(-1)^n}{2^{2n}(n!)^2} x^{2n}, \quad xy'' + y' + xy = 0$

In Problems 35–38 proceed as in Example 4 and find a power series solution $y = \sum_{n=0}^{\infty} c_n x^n$ of the given linear first-order differential equation.

35. $y' - 5y = 0$

36. $4y' + y = 0$

37. $y' = xy$

38. $(1 + x)y' + y = 0$

Discussion Problems

39. In Problem 19, find an easier way than multiplying two power series to obtain the Maclaurin series representation of $\sin x \cos x$.

40. In Problem 21, what do you think is the interval of convergence for the Maclaurin series of $\sec x$?

6.2 SOLUTIONS ABOUT ORDINARY POINTS

REVIEW MATERIAL

- Power series, analytic at a point, shifting the index of summation in Section 6.1

INTRODUCTION At the end of the last section we illustrated how to obtain a power series solution of a linear first-order differential equation. In this section we turn to the more important problem of finding power series solutions of linear second-order equations. More to the point, we are going to find solutions of linear second-order equations in the form of power series whose center is a number x_0 that is an **ordinary point** of the DE. We begin with the definition of an ordinary point.

≡ A Definition If we divide the homogeneous linear second-order differential equation

$$a_2(x)y'' + a_1(x)y' + a_0(x)y = 0 \tag{1}$$

by the lead coefficient $a_2(x)$ we obtain the standard form

$$y'' + P(x)y' + Q(x)y = 0. \tag{2}$$

We have the following definition.

DEFINITION 6.2.1 Ordinary and Singular Points

A point $x = x_0$ is said to be an **ordinary point** of the differential of the differential equation (1) if both coefficients $P(x)$ and $Q(x)$ in the standard form (2) are analytic at x_0. A point that is *not* an ordinary point of (1) is said to be a **singular point** of the DE.

EXAMPLE 1 Ordinary Points

(a) A homogeneous linear second-order differential equation with constant coefficients, such as

$$y'' + y = 0 \quad \text{and} \quad y'' + 3y' + 2y = 0,$$

can have no singular points. In other words, every finite value* of x is an ordinary point of such equations.
(b) Every finite value of x is an ordinary point of the differential equation

$$y'' + e^x y' + (\sin x)y = 0.$$

Specifically $x = 0$ is an ordinary point of the DE, because we have already seen in (2) of Section 6.1 that both e^x and $\sin x$ are analytic at this point. ≡

The negation of the second sentence in Definition 6.2.1 stipulates that if *at least one* of the coefficient functions $P(x)$ and $Q(x)$ in (2) fails to be analytic at x_0, then x_0 is a singular point.

EXAMPLE 2 Singular Points

(a) The differential equation

$$y'' + xy' + (\ln x)y = 0$$

is already in standard form. The coefficient functions are

$$P(x) = x \quad \text{and} \quad Q(x) = \ln x.$$

Now $P(x) = x$ is analytic at every real number, and $Q(x) = \ln x$ is analytic at every *positive* real number. However, since $Q(x) = \ln x$ is discontinuous at $x = 0$ it cannot be represented by a power series in x, that is, a power series centered at 0. We conclude that $x = 0$ is a singular point of the DE.

(b) By putting $xy'' + y' + xy = 0$ in the standard form

$$y'' + \frac{1}{x}y' + y = 0,$$

we see that $P(x) = 1/x$ fails to be analytic at $x = 0$. Hence $x = 0$ is a singular point of the equation. ≡

≡ **Polynomial Coefficients** We will primarily be interested in the case when the coefficients $a_2(x)$, $a_1(x)$, and $a_0(x)$ in (1) are polynomial functions with no common factors. A polynomial function is analytic at any value of x, and a rational function is analytic except at points where its denominator is zero. Thus, in (2) both coefficients

$$P(x) = \frac{a_1(x)}{a_2(x)} \quad \text{and} \quad Q(x) = \frac{a_0(x)}{a_2(x)}$$

*For our purposes, ordinary points and singular points will always be finite points. It is possible for an ODE to have, say, a singular point at infinity.

are analytic except at those numbers for which $a_2(x) = 0$. It follows, then, that

A number $x = x_0$ is an ordinary point of (1) *if $a_2(x_0) \neq 0$, whereas $x = x_0$ is a singular point of* (1) *if $a_2(x_0) = 0$.*

EXAMPLE 3 Ordinary and Singular Points

(a) The only singular points of the differential equation

$$(x^2 - 1)y'' + 2xy' + 6y = 0$$

are the solutions of $x^2 - 1 = 0$ or $x = \pm 1$. All other values of x are ordinary points.

(b) Inspection of the Cauchy-Euler

$$\overset{\downarrow a_2(x) = x^2 = 0 \text{ at } x = 0}{x^2y'' + y = 0}$$

shows that it has a singular point at $x = 0$. All other values of x are ordinary points.

(c) Singular points need not be real numbers. The equation

$$(x^2 + 1)y'' + xy' - y = 0$$

has singular points at the solutions of $x^2 + 1 = 0$—namely, $x = \pm i$. All other values of x, real or complex, are ordinary points. ≡

We state the following theorem about the existence of power series solutions without proof.

THEOREM 6.2.1 Existence of Power Series Solutions

If $x = x_0$ is an ordinary point of the differential equation (1), we can always find two linearly independent solutions in the form of a power series centered at x_0, that is,

$$y = \sum_{n=0}^{\infty} c_n(x - x_0)^n.$$

A power series solution converges at least on some interval defined by $|x - x_0| < R$, where R is the distance from x_0 to the closest singular point.

A solution of the form $y = \sum_{n=0}^{\infty} c_n(x - x_0)^n$ is said to be a **solution about the ordinary point x_0.** The distance R in Theorem 6.2.1 is the *minimum* value or *lower bound* for the radius of convergence.

EXAMPLE 4 Minimum Radius of Convergence

Find the minimum radius of convergence of a power series solution of the second-order differential equation

$$(x^2 - 2x + 5)y'' + xy' - y = 0$$

(a) about the ordinary point $x = 0$, **(b)** about the ordinary point $x = -1$.

SOLUTION By the quadratic formula we see from $x^2 - 2x + 5 = 0$ that the singular points of the given differential equation are the complex numbers $1 \pm 2i$.

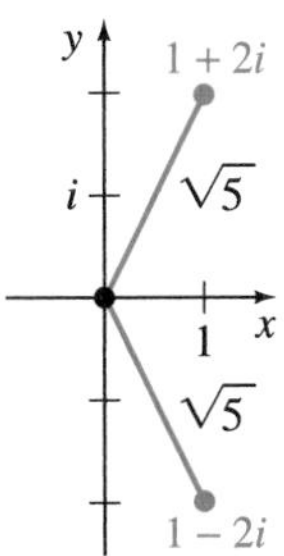

FIGURE 6.2.1 Distance from singular points to the ordinary point 0 in Example 4

(a) Because $x = 0$ is an ordinary point of the DE, Theorem 6.2.1 guarantees that we can find two power series solutions centered at 0. That is, solutions that look like $y = \sum_{n=0}^{\infty} c_n x^n$ and, moreover, we know without actually finding these solutions that each series must converge *at least* for $|x| < \sqrt{5}$, where $R = \sqrt{5}$ is the distance in the complex plane from either of the numbers $1 + 2i$ (the point $(1, 2)$) or $1 - 2i$ (the point $(1, -2)$) to the ordinary point 0 (the point $(0, 0)$). See Figure 6.2.1.

(b) Because $x = -1$ is an ordinary point of the DE, Theorem 6.2.1 guarantees that we can find two power series solutions that look like $y = \sum_{n=0}^{\infty} c_n (x + 1)^n$. Each of power series converges at least for $|x + 1| < 2\sqrt{2}$ since the distance from each of the singular points to -1 (the point $(-1, 0)$) is $R = \sqrt{8} = 2\sqrt{2}$. ≡

In part (a) of Example 4, *one* of the two power series solutions centered at 0 of the differential equation is valid on an interval much larger than $(-\sqrt{5}, \sqrt{5})$; in actual fact this solution is valid on the interval $(-\infty, \infty)$ because it can be shown that one of the two solutions about 0 reduces to a polynomial.

≡ **Note** In the examples that follow as well as in the problems of Exercises 6.2 we will, for the sake of simplicity, find only power series solutions about the ordinary point $x = 0$. If it is necessary to find a power series solutions of an ODE about an ordinary point $x_0 \neq 0$, we can simply make the change of variable $t = x - x_0$ in the equation (this translates $x = x_0$ to $t = 0$), find solutions of the new equation of the form $y = \sum_{n=0}^{\infty} c_n t^n$, and then resubstitute $t = x - x_0$.

≡ **Finding a Power Series Solution** Finding a power series solution of a homogeneous linear second-order ODE has been accurately described as "the method of undetermined *series* coefficients" since the procedure is quite analogous to what we did in Section 4.4. In case you did not work through Example 4 of Section 6.1 here, in brief, is the idea. Substitute $y = \sum_{n=0}^{\infty} c_n x^n$ into the differential equation, combine series as we did in Example 3 of Section 6.1, and then equate the all coefficients to the right-hand side of the equation to determine the coefficients c_n. But because the right-hand side is zero, the last step requires, by the *identity property* in the bulleted list in Section 6.1, that all coefficients of x must be equated to zero. No, this does *not* mean that all coefficients *are* zero; this would not make sense, after all Theorem 6.2.1 guarantees that we can find two solutions. We will see in Example 5 how the single assumption that $y = \sum_{n=0}^{\infty} c_n x^n = c_0 + c_1 x + c_2 x^2 + \cdots$ leads to two sets of coefficients so that we have two distinct power series $y_1(x)$ and $y_2(x)$, both expanded about the ordinary point $x = 0$. The general solution of the differential equation is $y = C_1 y_1(x) + C_2 y_2(x)$; indeed, it can be shown that $C_1 = c_0$ and $C_2 = c_1$.

EXAMPLE 5 Power Series Solutions

Before working through this example, we recommend that you reread Example 4 of Section 6.1. ▶

Solve $y'' + xy = 0$.

SOLUTION Since there are no singular points, Theorem 6.2.1 guarantees two power series solutions centered at 0 that converge for $|x| < \infty$. Substituting $y = \sum_{n=0}^{\infty} c_n x^n$ and the second derivative $y'' = \sum_{n=2}^{\infty} n(n - 1)c_n x^{n-2}$ (see (1) in Section 6.1) into the differential equation give

$$y'' + xy = \sum_{n=2}^{\infty} c_n n(n - 1)x^{n-2} + x\sum_{n=0}^{\infty} c_n x^n = \sum_{n=2}^{\infty} c_n n(n - 1)x^{n-2} + \sum_{n=0}^{\infty} c_n x^{n+1}. \quad (3)$$

We have already added the last two series on the right-hand side of the equality in (3) by shifting the summation index. From the result given in (5) of Section 6.1

$$y'' + xy = 2c_2 + \sum_{k=1}^{\infty} [(k + 1)(k + 2)c_{k+2} + c_{k-1}]x^k = 0. \quad (4)$$

At this point we invoke the identity property. Since (4) is identically zero, it is necessary that the coefficient of each power of x be set equal to zero—that is, $2c_2 = 0$ (it is the coefficient of x^0), and

$$(k+1)(k+2)c_{k+2} + c_{k-1} = 0, \qquad k = 1, 2, 3, \ldots. \tag{5}$$

Now $2c_2 = 0$ obviously dictates that $c_2 = 0$. But the expression in (5), called a **recurrence relation,** determines the c_k in such a manner that we can choose a certain subset of the set of coefficients to be *nonzero*. Since $(k+1)(k+2) \neq 0$ for all values of k, we can solve (5) for c_{k+2} in terms of c_{k-1}:

$$c_{k+2} = -\frac{c_{k-1}}{(k+1)(k+2)}, \qquad k = 1, 2, 3, \ldots. \tag{6}$$

This relation generates consecutive coefficients of the assumed solution one at a time as we let k take on the successive integers indicated in (6):

$$k = 1, \qquad c_3 = -\frac{c_0}{2 \cdot 3}$$

$$k = 2, \qquad c_4 = -\frac{c_1}{3 \cdot 4}$$

$$k = 3, \qquad c_5 = -\frac{c_2}{4 \cdot 5} = 0 \qquad \leftarrow c_2 \text{ is zero}$$

$$k = 4, \qquad c_6 = -\frac{c_3}{5 \cdot 6} = \frac{1}{2 \cdot 3 \cdot 5 \cdot 6} c_0$$

$$k = 5, \qquad c_7 = -\frac{c_4}{6 \cdot 7} = \frac{1}{3 \cdot 4 \cdot 6 \cdot 7} c_1$$

$$k = 6, \qquad c_8 = -\frac{c_5}{7 \cdot 8} = 0 \qquad \leftarrow c_5 \text{ is zero}$$

$$k = 7, \qquad c_9 = -\frac{c_6}{8 \cdot 9} = -\frac{1}{2 \cdot 3 \cdot 5 \cdot 6 \cdot 8 \cdot 9} c_0$$

$$k = 8, \qquad c_{10} = -\frac{c_7}{9 \cdot 10} = -\frac{1}{3 \cdot 4 \cdot 6 \cdot 7 \cdot 9 \cdot 10} c_1$$

$$k = 9, \qquad c_{11} = -\frac{c_8}{10 \cdot 11} = 0 \qquad \leftarrow c_8 \text{ is zero}$$

and so on. Now substituting the coefficients just obtained into the original assumption

$$y = c_0 + c_1 x + c_2 x^2 + c_3 x^3 + c_4 x^4 + c_5 x^5 + c_6 x^6 + c_7 x^7 + c_8 x^8 + c_9 x^9 + c_{10} x^{10} + c_{11} x^{11} + \cdots,$$

we get

$$y = c_0 + c_1 x + 0 - \frac{c_0}{2 \cdot 3} x^3 - \frac{c_1}{3 \cdot 4} x^4 + 0 + \frac{c_0}{2 \cdot 3 \cdot 5 \cdot 6} x^6$$
$$+ \frac{c_1}{3 \cdot 4 \cdot 6 \cdot 7} x^7 + 0 - \frac{c_0}{2 \cdot 3 \cdot 5 \cdot 6 \cdot 8 \cdot 9} x^9 - \frac{c_1}{3 \cdot 4 \cdot 6 \cdot 7 \cdot 9 \cdot 10} x^{10} + 0 + \cdots.$$

After grouping the terms containing c_0 and the terms containing c_1, we obtain $y = c_0 y_1(x) + c_1 y_2(x)$, where

$$y_1(x) = 1 - \frac{1}{2 \cdot 3} x^3 + \frac{1}{2 \cdot 3 \cdot 5 \cdot 6} x^6 - \frac{1}{2 \cdot 3 \cdot 5 \cdot 6 \cdot 8 \cdot 9} x^9 + \cdots = 1 + \sum_{k=1}^{\infty} \frac{(-1)^k}{2 \cdot 3 \cdots (3k-1)(3k)} x^{3k}$$

$$y_2(x) = x - \frac{1}{3 \cdot 4} x^4 + \frac{1}{3 \cdot 4 \cdot 6 \cdot 7} x^7 - \frac{1}{3 \cdot 4 \cdot 6 \cdot 7 \cdot 9 \cdot 10} x^{10} + \cdots = x + \sum_{k=1}^{\infty} \frac{(-1)^k}{3 \cdot 4 \cdots (3k)(3k+1)} x^{3k+1}.$$

Because the recursive use of (6) leaves c_0 and c_1 completely undetermined, they can be chosen arbitrarily. As was mentioned prior to this example, the linear combination $y = c_0 y_1(x) + c_1 y_2(x)$ actually represents the general solution of the differential equation. Although we know from Theorem 6.2.1 that each series solution converges for $|x| < \infty$, that is, on the interval $(-\infty, \infty)$. This fact can also be verified by the ratio test. ≡

The differential equation in Example 5 is called **Airy's equation** and is named after the English mathematician and astronomer **George Biddel Airy** (1801–1892). Airy's differential equation is encountered in the study of diffraction of light, diffraction of radio waves around the surface of the Earth, aerodynamics, and the deflection of a uniform thin vertical column that bends under its own weight. Other common forms of Airy's equation are $y'' - xy = 0$ and $y'' + \alpha^2 xy = 0$. See Problem 41 in Exercises 6.4 for an application of the last equation.

EXAMPLE 6 Power Series Solution

Solve $(x^2 + 1)y'' + xy' - y = 0$.

SOLUTION As we have already seen on page 240, the given differential equation has singular points at $x = \pm i$, and so a power series solution centered at 0 will converge at least for $|x| < 1$, where 1 is the distance in the complex plane from 0 to either i or $-i$. The assumption $y = \sum_{n=0}^{\infty} c_n x^n$ and its first two derivatives lead to

$$(x^2 + 1)\sum_{n=2}^{\infty} n(n-1)c_n x^{n-2} + x\sum_{n=1}^{\infty} nc_n x^{n-1} - \sum_{n=0}^{\infty} c_n x^n$$

$$= \sum_{n=2}^{\infty} n(n-1)c_n x^n + \sum_{n=2}^{\infty} n(n-1)c_n x^{n-2} + \sum_{n=1}^{\infty} nc_n x^n - \sum_{n=0}^{\infty} c_n x^n$$

$$= 2c_2 x^0 - c_0 x^0 + 6c_3 x + c_1 x - c_1 x + \underbrace{\sum_{n=2}^{\infty} n(n-1)c_n x^n}_{k=n}$$

$$+ \underbrace{\sum_{n=4}^{\infty} n(n-1)c_n x^{n-2}}_{k=n-2} + \underbrace{\sum_{n=2}^{\infty} nc_n x^n}_{k=n} - \underbrace{\sum_{n=2}^{\infty} c_n x^n}_{k=n}$$

$$= 2c_2 - c_0 + 6c_3 x + \sum_{k=2}^{\infty} [k(k-1)c_k + (k+2)(k+1)c_{k+2} + kc_k - c_k]x^k$$

$$= 2c_2 - c_0 + 6c_3 x + \sum_{k=2}^{\infty} [(k+1)(k-1)c_k + (k+2)(k+1)c_{k+2}]x^k = 0.$$

From this identity we conclude that $2c_2 - c_0 = 0$, $6c_3 = 0$, and

$$(k+1)(k-1)c_k + (k+2)(k+1)c_{k+2} = 0.$$

Thus

$$c_2 = \frac{1}{2}c_0$$

$$c_3 = 0$$

$$c_{k+2} = \frac{1-k}{k+2}c_k, \qquad k = 2, 3, 4, \ldots .$$

Substituting $k = 2, 3, 4, \ldots$ into the last formula gives

$$c_4 = -\frac{1}{4}c_2 = -\frac{1}{2 \cdot 4}c_0 = -\frac{1}{2^2 2!}c_0$$

$$c_5 = -\frac{2}{5}c_3 = 0 \qquad \leftarrow c_3 \text{ is zero}$$

$$c_6 = -\frac{3}{6}c_4 = \frac{3}{2\cdot 4\cdot 6}c_0 = \frac{1\cdot 3}{2^3 3!}c_0$$

$$c_7 = -\frac{4}{7}c_5 = 0 \qquad \leftarrow c_5 \text{ is zero}$$

$$c_8 = -\frac{5}{8}c_6 = -\frac{3\cdot 5}{2\cdot 4\cdot 6\cdot 8}c_0 = -\frac{1\cdot 3\cdot 5}{2^4 4!}c_0$$

$$c_9 = -\frac{6}{9}c_7 = 0, \qquad \leftarrow c_7 \text{ is zero}$$

$$c_{10} = -\frac{7}{10}c_8 = \frac{3\cdot 5\cdot 7}{2\cdot 4\cdot 6\cdot 8\cdot 10}c_0 = \frac{1\cdot 3\cdot 5\cdot 7}{2^5 5!}c_0,$$

and so on. Therefore

$$\begin{aligned} y &= c_0 + c_1x + c_2x^2 + c_3x^3 + c_4x^4 + c_5x^5 + c_6x^6 + c_7x^7 + c_8x^8 + c_9x^9 + c_{10}x^{10} + \cdots \\ &= c_0\left[1 + \frac{1}{2}x^2 - \frac{1}{2^2 2!}x^4 + \frac{1\cdot 3}{2^3 3!}x^6 - \frac{1\cdot 3\cdot 5}{2^4 4!}x^8 + \frac{1\cdot 3\cdot 5\cdot 7}{2^5 5!}x^{10} - \cdots\right] + c_1x \\ &= c_0y_1(x) + c_1y_2(x). \end{aligned}$$

The solutions are the polynomial $y_2(x) = x$ and the power series

$$y_1(x) = 1 + \frac{1}{2}x^2 + \sum_{n=2}^{\infty}(-1)^{n-1}\frac{1\cdot 3\cdot 5\cdots(2n-3)}{2^n n!}x^{2n}, \qquad |x| < 1. \quad \equiv$$

EXAMPLE 7 Three-Term Recurrence Relation

If we seek a power series solution $y = \sum_{n=0}^{\infty} c_nx^n$ for the differential equation

$$y'' - (1 + x)y = 0,$$

we obtain $c_2 = \frac{1}{2}c_0$ and the three-term recurrence relation

$$c_{k+2} = \frac{c_k + c_{k-1}}{(k+1)(k+2)}, \qquad k = 1, 2, 3, \ldots.$$

It follows from these two results that all coefficients c_n, for $n \geq 3$, are expressed in terms of *both* c_0 and c_1. To simplify life, we can first choose $c_0 \neq 0$, $c_1 = 0$; this yields coefficients for one solution expressed entirely in terms of c_0. Next, if we choose $c_0 = 0$, $c_1 \neq 0$, then coefficients for the other solution are expressed in terms of c_1. Using $c_2 = \frac{1}{2}c_0$ in both cases, the recurrence relation for $k = 1, 2, 3, \ldots$ gives

$c_0 \neq 0, c_1 = 0$	$c_0 = 0, c_1 \neq 0$
$c_2 = \frac{1}{2}c_0$	$c_2 = \frac{1}{2}c_0 = 0$
$c_3 = \frac{c_1 + c_0}{2\cdot 3} = \frac{c_0}{2\cdot 3} = \frac{c_0}{6}$	$c_3 = \frac{c_1 + c_0}{2\cdot 3} = \frac{c_1}{2\cdot 3} = \frac{c_1}{6}$
$c_4 = \frac{c_2 + c_1}{3\cdot 4} = \frac{c_0}{2\cdot 3\cdot 4} = \frac{c_0}{24}$	$c_4 = \frac{c_2 + c_1}{3\cdot 4} = \frac{c_1}{3\cdot 4} = \frac{c_1}{12}$
$c_5 = \frac{c_3 + c_2}{4\cdot 5} = \frac{c_0}{4\cdot 5}\left[\frac{1}{6} + \frac{1}{2}\right] = \frac{c_0}{30}$	$c_5 = \frac{c_3 + c_2}{4\cdot 5} = \frac{c_1}{4\cdot 5\cdot 6} = \frac{c_1}{120}$

and so on. Finally, we see that the general solution of the equation is $y = c_0y_1(x) + c_1y_2(x)$, where

$$y_1(x) = 1 + \frac{1}{2}x^2 + \frac{1}{6}x^3 + \frac{1}{24}x^4 + \frac{1}{30}x^5 + \cdots$$

and

$$y_2(x) = x + \frac{1}{6}x^3 + \frac{1}{12}x^4 + \frac{1}{120}x^5 + \cdots.$$

Each series converges for all finite values of x. ≡

≡ Nonpolynomial Coefficients The next example illustrates how to find a power series solution about the ordinary point $x_0 = 0$ of a differential equation when its coefficients are not polynomials. In this example we see an application of the multiplication of two power series.

EXAMPLE 8 DE with Nonpolynomial Coefficients

Solve $y'' + (\cos x)y = 0$.

SOLUTION We see that $x = 0$ is an ordinary point of the equation because, as we have already seen, $\cos x$ is analytic at that point. Using the Maclaurin series for $\cos x$ given in (2) of Section 6.1, along with the usual assumption $y = \sum_{n=0}^{\infty} c_nx^n$ and the results in (1) of Section 6.1 we find

$$\begin{aligned} y'' + (\cos x)y &= \sum_{n=2}^{\infty} n(n-1)c_nx^{n-2} + \left(1 - \frac{x^2}{2!} + \frac{x^4}{4!} - \frac{x^6}{6!} + \cdots\right)\sum_{n=0}^{\infty} c_nx^n \\ &= 2c_2 + 6c_3x + 12c_4x^2 + 20c_5x^3 + \cdots + \left(1 - \frac{x^2}{2!} + \frac{x^4}{4!} + \cdots\right)(c_0 + c_1x + c_2x^2 + c_3x^3 + \cdots) \\ &= 2c_2 + c_0 + (6c_3 + c_1)x + \left(12c_4 + c_2 - \frac{1}{2}c_0\right)x^2 + \left(20c_5 + c_3 - \frac{1}{2}c_1\right)x^3 + \cdots = 0. \end{aligned}$$

It follows that

$$2c_2 + c_0 = 0, \quad 6c_3 + c_1 = 0, \quad 12c_4 + c_2 - \frac{1}{2}c_0 = 0, \quad 20c_5 + c_3 - \frac{1}{2}c_1 = 0,$$

and so on. This gives $c_2 = -\frac{1}{2}c_0$, $c_3 = -\frac{1}{6}c_1$, $c_4 = \frac{1}{12}c_0$, $c_5 = \frac{1}{30}c_1, \ldots$. By grouping terms, we arrive at the general solution $y = c_0y_1(x) + c_1y_2(x)$, where

$$y_1(x) = 1 - \frac{1}{2}x^2 + \frac{1}{12}x^4 - \cdots \quad \text{and} \quad y_2(x) = x - \frac{1}{6}x^3 + \frac{1}{30}x^5 - \cdots.$$

Because the differential equation has no finite singular points, both power series converge for $|x| < \infty$. ≡

≡ Solution Curves The approximate graph of a power series solution $y(x) = \sum_{n=0}^{\infty} c_nx^n$ can be obtained in several ways. We can always resort to graphing the terms in the sequence of partial sums of the series—in other words, the graphs of the polynomials $S_N(x) = \sum_{n=0}^{N} c_nx^n$. For large values of N, $S_N(x)$ should give us an indication of the behavior of $y(x)$ near the ordinary point $x = 0$. We can also obtain an approximate or numerical solution curve by using a solver as we did in Section 4.10. For example, if you carefully scrutinize the series solutions of Airy's equation in

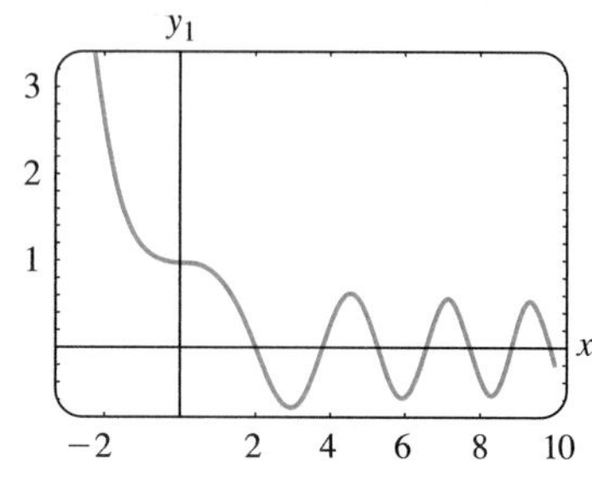

(a) plot of $y_1(x)$

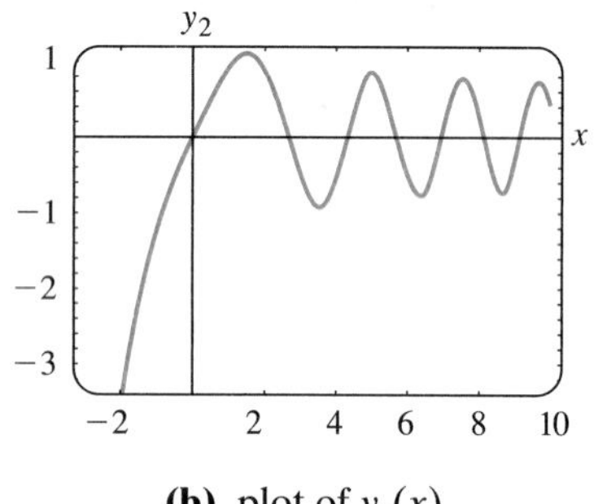

(b) plot of $y_2(x)$

FIGURE 6.2.2 Numerical solution curves for Airy's DE

Example 5, you should see that $y_1(x)$ and $y_2(x)$ are, in turn, the solutions of the initial-value problems

$$\begin{aligned} y'' + xy &= 0, \quad y(0) = 1, \quad y'(0) = 0, \\ y'' + xy &= 0, \quad y(0) = 0, \quad y'(0) = 1. \end{aligned} \tag{11}$$

The specified initial conditions "pick out" the solutions $y_1(x)$ and $y_2(x)$ from $y = c_0y_1(x) + c_1y_2(x)$, since it should be apparent from our basic series assumption $y = \sum_{n=0}^{\infty} c_nx^n$ that $y(0) = c_0$ and $y'(0) = c_1$. Now if your numerical solver requires a system of equations, the substitution $y' = u$ in $y'' + xy = 0$ gives $y'' = u' = -xy$, and so a system of two first-order equations equivalent to Airy's equation is

$$\begin{aligned} y' &= u \\ u' &= -xy. \end{aligned} \tag{12}$$

Initial conditions for the system in (12) are the two sets of initial conditions in (11) rewritten as $y(0) = 1$, $u(0) = 0$, and $y(0) = 0$, $u(0) = 1$. The graphs of $y_1(x)$ and $y_2(x)$ shown in Figure 6.2.2 were obtained with the aid of a numerical solver. The fact that the numerical solution curves appear to be oscillatory is consistent with the fact that Airy's equation appeared in Section 5.1 (page 197) in the form $mx'' + ktx = 0$ as a model of a spring whose "spring constant" $K(t) = kt$ increases with time.

REMARKS

(*i*) In the problems that follow, do not expect to be able to write a solution in terms of summation notation in each case. Even though we can generate as many terms as desired in a series solution $y = \sum_{n=0}^{\infty} c_nx^n$ either through the use of a recurrence relation or, as in Example 8, by multiplication, it might not be possible to deduce any general term for the coefficients c_n. We might have to settle, as we did in Examples 7 and 8, for just writing out the first few terms of the series.

(*ii*) A point x_0 is an ordinary point of a *nonhomogeneous* linear second-order DE $y'' + P(x)y' + Q(x)y = f(x)$ if $P(x)$, $Q(x)$, and $f(x)$ are analytic at x_0. Moreover, Theorem 6.2.1 extends to such DEs; in other words, we can find power series solutions $y = \sum_{n=0}^{\infty} c_n(x - x_0)^n$ of nonhomogeneous linear DEs in the same manner as in Examples 5–8. See Problem 26 in Exercises 6.2.

EXERCISES 6.2

Answers to selected odd-numbered problems begin on page ANS-9.

In Problems 1 and 2 without actually solving the given differential equation, find the minimum radius of convergence of power series solutions about the ordinary point $x = 0$. About the ordinary point $x = 1$.

1. $(x^2 - 25)y'' + 2xy' + y = 0$

2. $(x^2 - 2x + 10)y'' + xy' - 4y = 0$

In Problems 3–6 find two power series solutions of the given differential equation about the ordinary point $x = 0$. Compare the series solutions with the solutions of the differential equations obtained using the method of Section 4.3. Try to explain any differences between the two forms of the solutions.

3. $y'' + y = 0$ **4.** $y'' - y = 0$

5. $y'' - y' = 0$ **6.** $y'' + 2y' = 0$

In Problems 7–18 find two power series solutions of the given differential equation about the ordinary point $x = 0$.

7. $y'' - xy = 0$ **8.** $y'' + x^2y = 0$

9. $y'' - 2xy' + y = 0$ **10.** $y'' - xy' + 2y = 0$

11. $y'' + x^2y' + xy = 0$ **12.** $y'' + 2xy' + 2y = 0$

13. $(x - 1)y'' + y' = 0$ **14.** $(x + 2)y'' + xy' - y = 0$

15. $y'' - (x + 1)y' - y = 0$

16. $(x^2 + 1)y'' - 6y = 0$

17. $(x^2 + 2)y'' + 3xy' - y = 0$

18. $(x^2 - 1)y'' + xy' - y = 0$

In Problems 19–22 use the power series method to solve the given initial-value problem.

19. $(x - 1)y'' - xy' + y = 0,\ y(0) = -2,\ y'(0) = 6$

20. $(x + 1)y'' - (2 - x)y' + y = 0,\ y(0) = 2,\ y'(0) = -1$

21. $y'' - 2xy' + 8y = 0,\ y(0) = 3,\ y'(0) = 0$

22. $(x^2 + 1)y'' + 2xy' = 0,\ y(0) = 0,\ y'(0) = 1$

In Problems 23 and 24 use the procedure in Example 8 to find two power series solutions of the given differential equation about the ordinary point $x = 0$.

23. $y'' + (\sin x)y = 0$

24. $y'' + e^x y' - y = 0$

Discussion Problems

25. Without actually solving the differential equation $(\cos x)y'' + y' + 5y = 0$, find the minimum radius of convergence of power series solutions about the ordinary point $x = 0$. About the ordinary point $x = 1$.

26. How can the power series method be used to solve the *nonhomogeneous* equation $y'' - xy = 1$ about the ordinary point $x = 0$? Of $y'' - 4xy' - 4y = e^x$? Carry out your ideas by solving both DEs.

27. Is $x = 0$ an ordinary or a singular point of the differential equation $xy'' + (\sin x)y = 0$? Defend your answer with sound mathematics. [*Hint*: Use the Maclaurin series of $\sin x$ and then examine $(\sin x)/x$.

28. Is $x = 0$ an ordinary point of the differential equation $y'' + 5xy' + \sqrt{x}y = 0$?

Computer Lab Assignments

29. **(a)** Find two power series solutions for $y'' + xy' + y = 0$ and express the solutions $y_1(x)$ and $y_2(x)$ in terms of summation notation.

(b) Use a CAS to graph the partial sums $S_N(x)$ for $y_1(x)$. Use $N = 2, 3, 5, 6, 8, 10$. Repeat using the partial sums $S_N(x)$ for $y_2(x)$.

(c) Compare the graphs obtained in part (b) with the curve obtained by using a numerical solver. Use the initial-conditions $y_1(0) = 1,\ y_1'(0) = 0$, and $y_2(0) = 0,\ y_2'(0) = 1$.

(d) Reexamine the solution $y_1(x)$ in part (a). Express this series as an elementary function. Then use (5) of Section 4.2 to find a second solution of the equation. Verify that this second solution is the same as the power series solution $y_2(x)$.

30. **(a)** Find one more nonzero term for each of the solutions $y_1(x)$ and $y_2(x)$ in Example 8.

(b) Find a series solution $y(x)$ of the initial-value problem $y'' + (\cos x)y = 0,\ y(0) = 1,\ y'(0) = 1$.

(c) Use a CAS to graph the partial sums $S_N(x)$ for the solution $y(x)$ in part (b). Use $N = 2, 3, 4, 5, 6, 7$.

(d) Compare the graphs obtained in part (c) with the curve obtained using a numerical solver for the initial-value problem in part (b).

6.3 SOLUTIONS ABOUT SINGULAR POINTS

REVIEW MATERIAL

- Section 4.2 (especially (5) of that section)
- The definition of a singular point in Definition 6.2.1

INTRODUCTION The two differential equations

$$y'' + xy = 0 \quad \text{and} \quad xy'' + y = 0$$

are similar only in that they are both examples of simple linear second-order DEs with variable coefficients. That is all they have in common. Since $x = 0$ is an *ordinary point* of $y'' + xy = 0$, we saw in Section 6.2 that there was no problem in finding two distinct power series solutions centered at that point. In contrast, because $x = 0$ is a *singular point* of $xy'' + y = 0$, finding two infinite series—notice that we did not say *power series*—solutions of the equation about that point becomes a more difficult task.

The solution method that is discussed in this section does not always yield two infinite series solutions. When only one solution is found, we can use the formula given in (5) of Section 4.2 to find a second solution.

A Definition A singular point x_0 of a linear differential equation

$$a_2(x)y'' + a_1(x)y' + a_0(x)y = 0 \tag{1}$$

is further classified as either regular or irregular. The classification again depends on the functions P and Q in the standard form

$$y'' + P(x)y' + Q(x)y = 0. \tag{2}$$

DEFINITION 6.3.1 Regular and Irregular Singular Points

A singular point $x = x_0$ is said to be a **regular singular point** of the differential equation (1) if the functions $p(x) = (x - x_0)\,P(x)$ and $q(x) = (x - x_0)^2Q(x)$ are both analytic at x_0. A singular point that is not regular is said to be an **irregular singular point** of the equation.

The second sentence in Definition 6.3.1 indicates that if one or both of the functions $p(x) = (x - x_0)\,P(x)$ and $q(x) = (x - x_0)^2Q(x)$ fail to be analytic at x_0, then x_0 is an irregular singular point.

Polynomial Coefficients As in Section 6.2, we are mainly interested in linear equations (1) where the coefficients $a_2(x)$, $a_1(x)$, and $a_0(x)$ are polynomials with no common factors. We have already seen that if $a_2(x_0) = 0$, then $x = x_0$ is a singular point of (1), since at least one of the rational functions $P(x) = a_1(x)/a_2(x)$ and $Q(x) = a_0(x)/a_2(x)$ in the standard form (2) fails to be analytic at that point. But since $a_2(x)$ is a polynomial and x_0 is one of its zeros, it follows from the Factor Theorem of algebra that $x - x_0$ is a factor of $a_2(x)$. This means that after $a_1(x)/a_2(x)$ and $a_0(x)/a_2(x)$ are reduced to lowest terms, the factor $x - x_0$ must remain, to some positive integer power, in one or both denominators. Now suppose that $x = x_0$ is a singular point of (1) but both the functions defined by the products $p(x) = (x - x_0)\,P(x)$ and $q(x) = (x - x_0)^2Q(x)$ are analytic at x_0. We are led to the conclusion that multiplying $P(x)$ by $x - x_0$ and $Q(x)$ by $(x - x_0)^2$ has the effect (through cancellation) that $x - x_0$ no longer appears in either denominator. We can now determine whether x_0 is regular by a quick visual check of denominators:

If $x - x_0$ appears at most to the first power in the denominator of $P(x)$ and at most to the second power in the denominator of $Q(x)$, then $x = x_0$ is a regular singular point.

Moreover, observe that if $x = x_0$ is a regular singular point and we multiply (2) by $(x - x_0)^2$, then the original DE can be put into the form

$$(x - x_0)^2y'' + (x - x_0)p(x)y' + q(x)y = 0, \tag{3}$$

where p and q are analytic at $x = x_0$.

EXAMPLE 1 Classification of Singular Points

It should be clear that $x = 2$ and $x = -2$ are singular points of

$$(x^2 - 4)^2y'' + 3(x - 2)y' + 5y = 0.$$

After dividing the equation by $(x^2 - 4)^2 = (x - 2)^2(x + 2)^2$ and reducing the coefficients to lowest terms, we find that

$$P(x) = \frac{3}{(x - 2)(x + 2)^2} \quad \text{and} \quad Q(x) = \frac{5}{(x - 2)^2(x + 2)^2}.$$

We now test $P(x)$ and $Q(x)$ at each singular point.

For $x = 2$ to be a regular singular point, the factor $x - 2$ can appear at most to the first power in the denominator of $P(x)$ and at most to the second power in the denominator of $Q(x)$. A check of the denominators of $P(x)$ and $Q(x)$ shows that both these conditions are satisfied, so $x = 2$ is a regular singular point. Alternatively, we are led to the same conclusion by noting that both rational functions

$$p(x) = (x - 2)P(x) = \frac{3}{(x + 2)^2} \quad \text{and} \quad q(x) = (x - 2)^2Q(x) = \frac{5}{(x + 2)^2}$$

are analytic at $x = 2$.

Now since the factor $x - (-2) = x + 2$ appears to the second power in the denominator of $P(x)$, we can conclude immediately that $x = -2$ is an irregular singular point of the equation. This also follows from the fact that

$$p(x) = (x + 2)P(x) = \frac{3}{(x - 2)(x + 2)}$$

is not analytic at $x = -2$. ≡

In Example 1, notice that since $x = 2$ is a regular singular point, the original equation can be written as

$$(x - 2)^2y'' + (x - 2)\underbrace{\frac{3}{(x + 2)^2}}_{p(x) \text{ analytic at } x = 2}y' + \underbrace{\frac{5}{(x + 2)^2}}_{q(x) \text{ analytic at } x = 2}y = 0.$$

As another example, we can see that $x = 0$ is an irregular singular point of $x^3y'' - 2xy' + 8y = 0$ by inspection of the denominators of $P(x) = -2/x^2$ and $Q(x) = 8/x^3$. On the other hand, $x = 0$ is a regular singular point of $xy'' - 2xy' + 8y = 0$, since $x - 0$ and $(x - 0)^2$ do not even appear in the respective denominators of $P(x) = -2$ and $Q(x) = 8/x$. For a singular point $x = x_0$ any nonnegative power of $x - x_0$ less than one (namely, zero) and any nonnegative power less than two (namely, zero and one) in the denominators of $P(x)$ and $Q(x)$, respectively, imply that x_0 is a regular singular point. A singular point can be a complex number. You should verify that $x = 3i$ and $x = -3i$ are two regular singular points of $(x^2 + 9)y'' - 3xy' + (1 - x)y = 0$.

≡ Note Any second-order Cauchy-Euler equation $ax^2y'' + bxy' + cy = 0$, where a, b, and c are real constants, has a regular singular point at $x = 0$. You should verify that two solutions of the Cauchy-Euler equation $x^2y'' - 3xy' + 4y = 0$ on the interval $(0, \infty)$ are $y_1 = x^2$ and $y_2 = x^2 \ln x$. If we attempted to find a power series solution about the regular singular point $x = 0$ (namely, $y = \sum_{n=0}^{\infty} c_nx^n$), we would succeed in obtaining only the polynomial solution $y_1 = x^2$. The fact that we would not obtain the second solution is not surprising because $\ln x$ (and consequently $y_2 = x^2 \ln x$) is not analytic at $x = 0$—that is, y_2 does not possess a Taylor series expansion centered at $x = 0$.

≡ Method of Frobenius To solve a differential equation (1) about a regular singular point, we employ the following theorem due to the eminent German mathematician **Ferdinand Georg Frobenius** (1849–1917).

THEOREM 6.3.1 Frobenius' Theorem

If $x = x_0$ is a regular singular point of the differential equation (1), then there exists at least one solution of the form

$$y = (x - x_0)^r\sum_{n=0}^{\infty} c_n(x - x_0)^n = \sum_{n=0}^{\infty} c_n(x - x_0)^{n+r}, \tag{4}$$

where the number r is a constant to be determined. The series will converge at least on some interval $0 < x - x_0 < R$.

Notice the words *at least* in the first sentence of Theorem 6.3.1. This means that in contrast to Theorem 6.2.1, Theorem 6.3.1 gives us no assurance that *two* series solutions of the type indicated in (4) can be found. The **method of Frobenius,** finding series solutions about a regular singular point x_0, is similar to the power-series method in the preceding section in that we substitute $y = \sum_{n=0}^{\infty} c_n(x - x_0)^{n+r}$ into the given differential equation and determine the unknown coefficients c_n by a recurrence relation. However, we have an additional task in this procedure: Before determining the coefficients, we must find the unknown exponent r. If r is found to be a number that is not a nonnegative integer, then the corresponding solution $y = \sum_{n=0}^{\infty} c_n(x - x_0)^{n+r}$ is not a power series.

As we did in the discussion of solutions about ordinary points, we shall always assume, for the sake of simplicity in solving differential equations, that the regular singular point is $x = 0$.

EXAMPLE 2 Two Series Solutions

Because $x = 0$ is a regular singular point of the differential equation

$$3xy'' + y' - y = 0, \tag{5}$$

we try to find a solution of the form $y = \sum_{n=0}^{\infty} c_n x^{n+r}$. Now

$$y' = \sum_{n=0}^{\infty} (n + r)c_n x^{n+r-1} \qquad \text{and} \qquad y'' = \sum_{n=0}^{\infty} (n + r)(n + r - 1)c_n x^{n+r-2},$$

so

$$\begin{aligned} 3xy'' + y' - y &= 3\sum_{n=0}^{\infty} (n + r)(n + r - 1)c_n x^{n+r-1} + \sum_{n=0}^{\infty} (n + r)c_n x^{n+r-1} - \sum_{n=0}^{\infty} c_n x^{n+r} \\ &= \sum_{n=0}^{\infty} (n + r)(3n + 3r - 2)c_n x^{n+r-1} - \sum_{n=0}^{\infty} c_n x^{n+r} \\ &= x^r\left[r(3r - 2)c_0 x^{-1} + \underbrace{\sum_{n=1}^{\infty} (n + r)(3n + 3r - 2)c_n x^{n-1}}_{k=n-1} - \underbrace{\sum_{n=0}^{\infty} c_n x^n}_{k=n}\right] \\ &= x^r\left[r(3r - 2)c_0 x^{-1} + \sum_{k=0}^{\infty} [(k + r + 1)(3k + 3r + 1)c_{k+1} - c_k]x^k\right] = 0, \end{aligned}$$

which implies that $\qquad r(3r - 2)c_0 = 0$

and $\qquad (k + r + 1)(3k + 3r + 1)c_{k+1} - c_k = 0, \qquad k = 0, 1, 2, \ldots.$

Because nothing is gained by taking $c_0 = 0$, we must then have

$$r(3r - 2) = 0 \tag{6}$$

and

$$c_{k+1} = \frac{c_k}{(k + r + 1)(3k + 3r + 1)}, \qquad k = 0, 1, 2, \ldots. \tag{7}$$

When substituted in (7), the two values of r that satisfy the quadratic equation (6), $r_1 = \frac{2}{3}$ and $r_2 = 0$, give two different recurrence relations:

$$r_1 = \tfrac{2}{3}, \qquad c_{k+1} = \frac{c_k}{(3k + 5)(k + 1)}, \qquad k = 0, 1, 2, \ldots \tag{8}$$

$$r_2 = 0, \qquad c_{k+1} = \frac{c_k}{(k + 1)(3k + 1)}, \qquad k = 0, 1, 2, \ldots. \tag{9}$$

From (8) we find

$$c_1 = \frac{c_0}{5 \cdot 1}$$

$$c_2 = \frac{c_1}{8 \cdot 2} = \frac{c_0}{2!5 \cdot 8}$$

$$c_3 = \frac{c_2}{11 \cdot 3} = \frac{c_0}{3!5 \cdot 8 \cdot 11}$$

$$c_4 = \frac{c_3}{14 \cdot 4} = \frac{c_0}{4!5 \cdot 8 \cdot 11 \cdot 14}$$

$$\vdots$$

$$c_n = \frac{c_0}{n!5 \cdot 8 \cdot 11 \cdots (3n+2)}.$$

From (9) we find

$$c_1 = \frac{c_0}{1 \cdot 1}$$

$$c_2 = \frac{c_1}{2 \cdot 4} = \frac{c_0}{2!1 \cdot 4}$$

$$c_3 = \frac{c_2}{3 \cdot 7} = \frac{c_0}{3!1 \cdot 4 \cdot 7}$$

$$c_4 = \frac{c_3}{4 \cdot 10} = \frac{c_0}{4!1 \cdot 4 \cdot 7 \cdot 10}$$

$$\vdots$$

$$c_n = \frac{c_0}{n!1 \cdot 4 \cdot 7 \cdots (3n-2)}.$$

Here we encounter something that did not happen when we obtained solutions about an ordinary point; we have what looks to be two different sets of coefficients, but each set contains the *same* multiple c_0. If we omit this term, the series solutions are

$$y_1(x) = x^{2/3}\left[1 + \sum_{n=1}^{\infty} \frac{1}{n!5 \cdot 8 \cdot 11 \cdots (3n+2)} x^n\right] \tag{10}$$

$$y_2(x) = x^0\left[1 + \sum_{n=1}^{\infty} \frac{1}{n!1 \cdot 4 \cdot 7 \cdots (3n-2)} x^n\right]. \tag{11}$$

By the ratio test it can be demonstrated that both (10) and (11) converge for all values of x—that is, $|x| < \infty$. Also, it should be apparent from the form of these solutions that neither series is a constant multiple of the other, and therefore $y_1(x)$ and $y_2(x)$ are linearly independent on the entire x-axis. Hence by the superposition principle, $y = C_1y_1(x) + C_2y_2(x)$ is another solution of (5). On any interval that does not contain the origin, such as $(0, \infty)$, this linear combination represents the general solution of the differential equation. ≡

≡ Indicial Equation Equation (6) is called the **indicial equation** of the problem, and the values $r_1 = \frac{2}{3}$ and $r_2 = 0$ are called the **indicial roots, or exponents,** of the singularity $x = 0$. In general, after substituting $y = \sum_{n=0}^{\infty} c_n x^{n+r}$ into the given differential equation and simplifying, the indicial equation is a quadratic equation in r that results from equating the *total coefficient of the lowest power of x to zero.* We solve for the two values of r and substitute these values into a recurrence relation such as (7). Theorem 6.3.1 guarantees that at least one solution of the assumed series form can be found.

It is possible to obtain the indicial equation in advance of substituting $y = \sum_{n=0}^{\infty} c_n x^{n+r}$ into the differential equation. If $x = 0$ is a regular singular point of (1), then by Definition 6.3.1 both functions $p(x) = xP(x)$ and $q(x) = x^2Q(x)$, where P and Q are defined by the standard form (2), are analytic at $x = 0$; that is, the power series expansions

$$p(x) = xP(x) = a_0 + a_1x + a_2x^2 + \cdots \quad \text{and} \quad q(x) = x^2Q(x) = b_0 + b_1x + b_2x^2 + \cdots \tag{12}$$

are valid on intervals that have a positive radius of convergence. By multiplying (2) by x^2, we get the form given in (3):

$$x^2y'' + x[xP(x)]y' + [x^2Q(x)]y = 0. \tag{13}$$

After substituting $y = \sum_{n=0}^{\infty} c_n x^{n+r}$ and the two series in (12) into (13) and carrying out the multiplication of series, we find the general indicial equation to be

$$r(r - 1) + a_0 r + b_0 = 0, \tag{14}$$

where a_0 and b_0 are as defined in (12). See Problems 13 and 14 in Exercises 6.3.

EXAMPLE 3 Two Series Solutions

Solve $2xy'' + (1 + x)y' + y = 0$.

SOLUTION Substituting $y = \sum_{n=0}^{\infty} c_n x^{n+r}$ gives

$$\begin{aligned} 2xy'' + (1 + x)y' + y &= 2\sum_{n=0}^{\infty}(n + r)(n + r - 1)c_n x^{n+r-1} + \sum_{n=0}^{\infty}(n + r)c_n x^{n+r-1} \\ &\quad + \sum_{n=0}^{\infty}(n + r)c_n x^{n+r} + \sum_{n=0}^{\infty} c_n x^{n+r} \\ &= \sum_{n=0}^{\infty}(n + r)(2n + 2r - 1)c_n x^{n+r-1} + \sum_{n=0}^{\infty}(n + r + 1)c_n x^{n+r} \\ &= x^r\Bigg[r(2r - 1)c_0 x^{-1} + \underbrace{\sum_{n=1}^{\infty}(n + r)(2n + 2r - 1)c_n x^{n-1}}_{k=n-1} + \underbrace{\sum_{n=0}^{\infty}(n + r + 1)c_n x^n}_{k=n}\Bigg] \\ &= x^r\Bigg[r(2r - 1)c_0 x^{-1} + \sum_{k=0}^{\infty}[(k + r + 1)(2k + 2r + 1)c_{k+1} + (k + r + 1)c_k]x^k\Bigg], \end{aligned}$$

which implies that

$$r(2r - 1) = 0 \tag{15}$$

and

$$(k + r + 1)(2k + 2r + 1)c_{k+1} + (k + r + 1)c_k = 0, \tag{16}$$

$k = 0, 1, 2, \ldots$. From (15) we see that the indicial roots are $r_1 = \frac{1}{2}$ and $r_2 = 0$.

For $r_1 = \frac{1}{2}$ we can divide by $k + \frac{3}{2}$ in (16) to obtain

$$c_{k+1} = \frac{-c_k}{2(k + 1)}, \qquad k = 0, 1, 2, \ldots, \tag{17}$$

whereas for $r_2 = 0$, (16) becomes

$$c_{k+1} = \frac{-c_k}{2k + 1}, \qquad k = 0, 1, 2, \ldots. \tag{18}$$

From (17) we find	From (18) we find
$c_1 = \dfrac{-c_0}{2 \cdot 1}$	$c_1 = \dfrac{-c_0}{1}$
$c_2 = \dfrac{-c_1}{2 \cdot 2} = \dfrac{c_0}{2^2 \cdot 2!}$	$c_2 = \dfrac{-c_1}{3} = \dfrac{c_0}{1 \cdot 3}$
$c_3 = \dfrac{-c_2}{2 \cdot 3} = \dfrac{-c_0}{2^3 \cdot 3!}$	$c_3 = \dfrac{-c_2}{5} = \dfrac{-c_0}{1 \cdot 3 \cdot 5}$
$c_4 = \dfrac{-c_3}{2 \cdot 4} = \dfrac{c_0}{2^4 \cdot 4!}$	$c_4 = \dfrac{-c_3}{7} = \dfrac{c_0}{1 \cdot 3 \cdot 5 \cdot 7}$
$\vdots$	$\vdots$
$c_n = \dfrac{(-1)^n c_0}{2^n n!}.$	$c_n = \dfrac{(-1)^n c_0}{1 \cdot 3 \cdot 5 \cdot 7 \cdots (2n - 1)}.$

Thus for the indicial root $r_1 = \frac{1}{2}$ we obtain the solution

$$y_1(x) = x^{1/2}\left[1 + \sum_{n=1}^{\infty} \frac{(-1)^n}{2^n n!} x^n\right] = \sum_{n=0}^{\infty} \frac{(-1)^n}{2^n n!} x^{n+1/2},$$

where we have again omitted c_0. The series converges for $x \geq 0$; as given, the series is not defined for negative values of x because of the presence of $x^{1/2}$. For $r_2 = 0$ a second solution is

$$y_2(x) = 1 + \sum_{n=1}^{\infty} \frac{(-1)^n}{1 \cdot 3 \cdot 5 \cdot 7 \cdots (2n-1)} x^n, \qquad |x| < \infty.$$

On the interval $(0, \infty)$ the general solution is $y = C_1 y_1(x) + C_2 y_2(x)$. ≡

EXAMPLE 4 Only One Series Solution

Solve $xy'' + y = 0$.

SOLUTION From $xP(x) = 0$, $x^2Q(x) = x$ and the fact that 0 and x are their own power series centered at 0 we conclude that $a_0 = 0$ and $b_0 = 0$, so from (14) the indicial equation is $r(r-1) = 0$. You should verify that the two recurrence relations corresponding to the indicial roots $r_1 = 1$ and $r_2 = 0$ yield exactly the same set of coefficients. In other words, in this case the method of Frobenius produces only a single series solution

$$y_1(x) = \sum_{n=0}^{\infty} \frac{(-1)^n}{n!(n+1)!} x^{n+1} = x - \frac{1}{2}x^2 + \frac{1}{12}x^3 - \frac{1}{144}x^4 + \cdots.$$ ≡

≡ **Three Cases** For the sake of discussion let us again suppose that $x = 0$ is a regular singular point of equation (1) and that the indicial roots r_1 and r_2 of the singularity are real. When using the method of Frobenius, we distinguish three cases corresponding to the nature of the indicial roots r_1 and r_2. In the first two cases the symbol r_1 denotes the largest of two distinct roots, that is, $r_1 > r_2$. In the last case $r_1 = r_2$.

≡ **Case I:** If r_1 and r_2 are distinct and the difference $r_1 - r_2$ is not a positive integer, then there exist two linearly independent solutions of equation (1) of the form

$$y_1(x) = \sum_{n=0}^{\infty} c_n x^{n+r_1}, \quad c_0 \neq 0, \qquad y_2(x) = \sum_{n=0}^{\infty} b_n x^{n+r_2}, \quad b_0 \neq 0.$$

This is the case illustrated in Examples 2 and 3.

Next we assume that the difference of the roots is N, where N is a positive integer. In this case the second solution *may* contain a logarithm.

≡ **Case II:** If r_1 and r_2 are distinct and the difference $r_1 - r_2$ is a positive integer, then there exist two linearly independent solutions of equation (1) of the form

$$y_1(x) = \sum_{n=0}^{\infty} c_n x^{n+r_1}, \qquad c_0 \neq 0, \tag{19}$$

$$y_2(x) = Cy_1(x)\ln x + \sum_{n=0}^{\infty} b_n x^{n+r_2}, \qquad b_0 \neq 0, \tag{20}$$

where C is a constant that could be zero.

Finally, in the last case, the case when $r_1 = r_2$, a second solution will *always* contain a logarithm. The situation is analogous to the solution of a Cauchy-Euler equation when the roots of the auxiliary equation are equal.

Case III: If r_1 and r_2 are equal, then there always exist two linearly independent solutions of equation (1) of the form

$$y_1(x) = \sum_{n=0}^{\infty} c_n x^{n+r_1}, \qquad c_0 \neq 0, \tag{21}$$

$$y_2(x) = y_1(x) \ln x + \sum_{n=1}^{\infty} b_n x^{n+r_1}. \tag{22}$$

Finding a Second Solution When the difference $r_1 - r_2$ is a positive integer (Case II), we *may* or *may not* be able to find two solutions having the form $y = \sum_{n=0}^{\infty} c_n x^{n+r}$. This is something that we do not know in advance but is determined after we have found the indicial roots and have carefully examined the recurrence relation that defines the coefficients c_n. We just may be lucky enough to find two solutions that involve only powers of x, that is, $y_1(x) = \sum_{n=0}^{\infty} c_n x^{n+r_1}$ (equation (19)) and $y_2(x) = \sum_{n=0}^{\infty} b_n x^{n+r_2}$ (equation (20) with $C = 0$). See Problem 31 in Exercises 6.3. On the other hand, in Example 4 we see that the difference of the indicial roots is a positive integer ($r_1 - r_2 = 1$) and the method of Frobenius failed to give a second series solution. In this situation equation (20), with $C \neq 0$, indicates what the second solution looks like. Finally, when the difference $r_1 - r_2$ is a zero (Case III), the method of Frobenius fails to give a second series solution; the second solution (22) always contains a logarithm and can be shown to be equivalent to (20) with $C = 1$. One way to obtain the second solution with the logarithmic term is to use the fact that

$$y_2(x) = y_1(x) \int \frac{e^{-\int P(x)dx}}{y_1^2(x)}\, dx \tag{23}$$

is also a solution of $y'' + P(x)y' + Q(x)y = 0$ whenever $y_1(x)$ is a known solution. We illustrate how to use (23) in the next example.

EXAMPLE 5 Example 4 Revisited Using a CAS

Find the general solution of $xy'' + y = 0$.

SOLUTION From the known solution given in Example 4,

$$y_1(x) = x - \frac{1}{2}x^2 + \frac{1}{12}x^3 - \frac{1}{144}x^4 + \cdots,$$

we can construct a second solution $y_2(x)$ using formula (23). Those with the time, energy, and patience can carry out the drudgery of squaring a series, long division, and integration of the quotient by hand. But all these operations can be done with relative ease with the help of a CAS. We give the results:

$$y_2(x) = y_1(x) \int \frac{e^{-\int 0 dx}}{[y_1(x)]^2}\, dx = y_1(x) \int \frac{dx}{\left[x - \frac{1}{2}x^2 + \frac{1}{12}x^3 - \frac{1}{144}x^4 + \cdots\right]^2}$$

$$= y_1(x) \int \frac{dx}{\left[x^2 - x^3 + \frac{5}{12}x^4 - \frac{7}{72}x^5 + \cdots\right]} \qquad \leftarrow \text{after squaring}$$

$$= y_1(x) \int \left[\frac{1}{x^2} + \frac{1}{x} + \frac{7}{12} + \frac{19}{72}x + \cdots\right] dx \qquad \leftarrow \text{after long division}$$

$$= y_1(x) \left[-\frac{1}{x} + \ln x + \frac{7}{12}x + \frac{19}{144}x^2 + \cdots\right] \qquad \leftarrow \text{after integrating}$$

$$= y_1(x) \ln x + y_1(x) \left[-\frac{1}{x} + \frac{7}{12}x + \frac{19}{144}x^2 + \cdots\right],$$

$$\text{or} \qquad y_2(x) = y_1(x) \ln x + \left[-1 - \frac{1}{2}x + \frac{1}{2}x^2 + \cdots\right]. \qquad \leftarrow \text{after multiplying out}$$

On the interval $(0, \infty)$ the general solution is $y = C_1y_1(x) + C_2y_2(x)$. ≡

Note that the final form of y_2 in Example 5 matches (20) with $C = 1$; the series in the brackets corresponds to the summation in (20) with $r_2 = 0$.

REMARKS

(i) The three different forms of a linear second-order differential equation in (1), (2), and (3) were used to discuss various theoretical concepts. But on a practical level, when it comes to actually solving a differential equation using the method of Frobenius, it is advisable to work with the form of the DE given in (1).

(ii) When the difference of indicial roots $r_1 - r_2$ is a positive integer $(r_1 > r_2)$, it sometimes pays to iterate the recurrence relation using the smaller root r_2 first. See Problems 31 and 32 in Exercises 6.3.

(iii) Because an indicial root r is a solution of a quadratic equation, it could be complex. We shall not, however, investigate this case.

(iv) If $x = 0$ is an irregular singular point, then we might not be able to find *any* solution of the DE of form $y = \sum_{n=0}^{\infty} c_n x^{n+r}$.

EXERCISES 6.3

Answers to selected odd-numbered problems begin on page ANS-10.

In Problems 1–10 determine the singular points of the given differential equation. Classify each singular point as regular or irregular.

1. $x^3y'' + 4x^2y' + 3y = 0$

2. $x(x + 3)^2y'' - y = 0$

3. $(x^2 - 9)^2y'' + (x + 3)y' + 2y = 0$

4. $y'' - \frac{1}{x}y' + \frac{1}{(x - 1)^3}y = 0$

5. $(x^3 + 4x)y'' - 2xy' + 6y = 0$

6. $x^2(x - 5)^2y'' + 4xy' + (x^2 - 25)y = 0$

7. $(x^2 + x - 6)y'' + (x + 3)y' + (x - 2)y = 0$

8. $x(x^2 + 1)^2y'' + y = 0$

9. $x^3(x^2 - 25)(x - 2)^2y'' + 3x(x - 2)y' + 7(x + 5)y = 0$

10. $(x^3 - 2x^2 + 3x)^2y'' + x(x - 3)^2y' - (x + 1)y = 0$

In Problems 11 and 12 put the given differential equation into form (3) for each regular singular point of the equation. Identify the functions $p(x)$ and $q(x)$.

11. $(x^2 - 1)y'' + 5(x + 1)y' + (x^2 - x)y = 0$

12. $xy'' + (x + 3)y' + 7x^2y = 0$

In Problems 13 and 14, $x = 0$ is a regular singular point of the given differential equation. Use the general form of the indicial equation in (14) to find the indicial roots of the singularity. Without solving, discuss the number of series solutions you would expect to find using the method of Frobenius.

13. $x^2y'' + \left(\frac{5}{3}x + x^2\right)y' - \frac{1}{3}y = 0$

14. $xy'' + y' + 10y = 0$

In Problems 15–24, $x = 0$ is a regular singular point of the given differential equation. Show that the indicial roots of the singularity do not differ by an integer. Use the method of Frobenius to obtain two linearly independent series solutions about $x = 0$. Form the general solution on $(0, \infty)$.

15. $2xy'' - y' + 2y = 0$

16. $2xy'' + 5y' + xy = 0$

17. $4xy'' + \frac{1}{2}y' + y = 0$

18. $2x^2y'' - xy' + (x^2 + 1)y = 0$

19. $3xy'' + (2 - x)y' - y = 0$

20. $x^2y'' - \left(x - \frac{2}{9}\right)y = 0$

21. $2xy'' - (3 + 2x)y' + y = 0$

22. $x^2y'' + xy' + \left(x^2 - \frac{4}{9}\right)y = 0$

23. $9x^2y'' + 9x^2y' + 2y = 0$

24. $2x^2y'' + 3xy' + (2x - 1)y = 0$

In Problems 25–30, $x = 0$ is a regular singular point of the given differential equation. Show that the indicial roots of the singularity differ by an integer. Use the method of Frobenius to obtain at least one series solution about $x = 0$. Use (23) where necessary and a CAS, if instructed, to find a second solution. Form the general solution on $(0, \infty)$.

25. $xy'' + 2y' - xy = 0$

26. $x^2y'' + xy' + \left(x^2 - \frac{1}{4}\right)y = 0$

27. $xy'' - xy' + y = 0$

28. $y'' + \dfrac{3}{x}y' - 2y = 0$

29. $xy'' + (1 - x)y' - y = 0$

30. $xy'' + y' + y = 0$

In Problems 31 and 32, $x = 0$ is a regular singular point of the given differential equation. Show that the indicial roots of the singularity differ by an integer. Use the recurrence relation found by the method of Frobenius first with the larger root r_1. How many solutions did you find? Next use the recurrence relation with the smaller root r_2. How many solutions did you find?

31. $xy'' + (x - 6)y' - 3y = 0$

32. $x(x - 1)y'' + 3y' - 2y = 0$

33. (a) The differential equation $x^4y'' + \lambda y = 0$ has an irregular singular point at $x = 0$. Show that the substitution $t = 1/x$ yields the DE

$$\frac{d^2y}{dt^2} + \frac{2}{t}\frac{dy}{dt} + \lambda y = 0,$$

which now has a regular singular point at $t = 0$.

(b) Use the method of this section to find two series solutions of the second equation in part (a) about the regular singular point $t = 0$.

(c) Express each series solution of the original equation in terms of elementary functions.

Mathematical Model

34. **Buckling of a Tapered Column** In Example 4 of Section 5.2 we saw that when a constant vertical compressive force or load P was applied to a thin column of uniform cross section, the deflection $y(x)$ was a solution of the boundary-value problem

$$EI\frac{d^2y}{dx^2} + Py = 0, \quad y(0) = 0, \quad y(L) = 0. \tag{24}$$

The assumption here is that the column is hinged at both ends. The column will buckle or deflect only when the compressive force is a critical load P_n.

(a) In this problem let us assume that the column is of length L, is hinged at both ends, has circular cross sections, and is tapered as shown in Figure 6.3.1(a). If the column, a truncated cone, has a linear taper $y = cx$ as shown in cross section in Figure 6.3.1(b), the moment of inertia of a cross section with respect to an axis perpendicular to the xy-plane is $I = \frac{1}{4}\pi r^4$, where $r = y$ and $y = cx$. Hence we can write $I(x) = I_0(x/b)^4$, where $I_0 = I(b) = \frac{1}{4}\pi(cb)^4$. Substituting $I(x)$ into the differential equation in (24), we see that the deflection in this case is determined from the BVP

$$x^4\frac{d^2y}{dx^2} + \lambda y = 0, \quad y(a) = 0, \quad y(b) = 0,$$

where $\lambda = Pb^4/EI_0$. Use the results of Problem 33 to find the critical loads P_n for the tapered column. Use an appropriate identity to express the buckling modes $y_n(x)$ as a single function.

(b) Use a CAS to plot the graph of the first buckling mode $y_1(x)$ corresponding to the Euler load P_1 when $b = 11$ and $a = 1$.

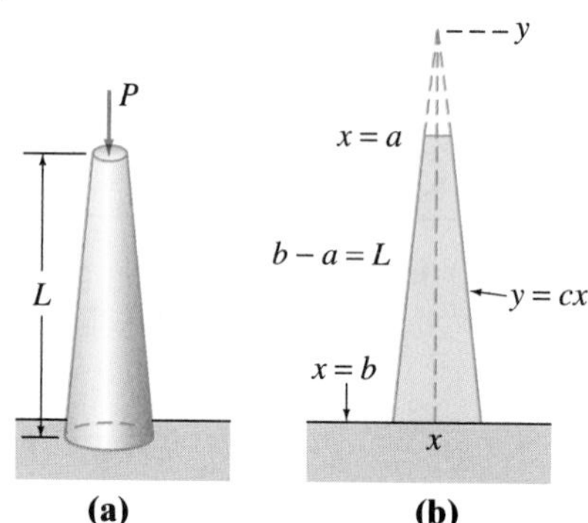

FIGURE 6.3.1 Tapered column in Problem 34

Discussion Problems

35. Discuss how you would define a regular singular point for the linear third-order differential equation

$$a_3(x)y''' + a_2(x)y'' + a_1(x)y' + a_0(x)y = 0.$$

36. Each of the differential equations

$$x^3y'' + y = 0 \quad \text{and} \quad x^2y'' + (3x - 1)y' + y = 0$$

has an irregular singular point at $x = 0$. Determine whether the method of Frobenius yields a series solution of each differential equation about $x = 0$. Discuss and explain your findings.

37. We have seen that $x = 0$ is a regular singular point of any Cauchy-Euler equation $ax^2y'' + bxy' + cy = 0$. Are the indicial equation (14) for a Cauchy-Euler equation and its auxiliary equation related? Discuss.

6.4 SPECIAL FUNCTIONS

REVIEW MATERIAL

- Sections 6.2 and 6.3

INTRODUCTION In the *Remarks* at the end of Section 2.3 we mentioned the branch of mathematics called **special functions**. Perhaps a better title for this field of applied mathematics might be *named functions* because many of the functions studied bear proper names: Bessel functions, Legendre functions, Airy functions, Chebyshev polynomials, Hermite polynomials, Jacobi polynomials, Laguerre polynomials, Gauss' hypergeometric function, Mathieu functions, and so on. Historically, special functions were often the by-product of necessity: Someone needed a solution of a very specialized differential equation that arose from an attempt to solve a physical problem. In effect, a special function was determined or defined by the differential equation and many properties of the function could be discerned from the series form of the solution.

In this section we use the methods of Sections 6.2 and 6.3 to find solutions of two differential equations

$$x^2y'' + xy' + (x^2 - \nu^2)y = 0 \tag{1}$$

$$(1 - x^2)y'' - 2xy' + n(n + 1)y = 0 \tag{2}$$

that arise in advanced studies of applied mathematics, physics, and engineering. They are called, respectively, **Bessel's equation of order** ν, named after the German mathematician and astronomer **Friedrich Wilhelm Bessel** (1784–1846), and **Legendre's equation of order** n, named after the French mathematician **Adrien-Marie Legendre** (1752–1833). When we solve (1) we shall assume that $\nu \geq 0$, whereas in (2) we shall consider only the case when n in a nonnegative integer.

Solution of Bessel's Equation

Because $x = 0$ is a regular singular point of Bessel's equation, we know that there exists at least one solution of the form $y = \sum_{n=0}^{\infty} c_n x^{n+r}$. Substituting the last expression into (1) gives

$$\begin{aligned} x^2y'' + xy' + (x^2 - \nu^2)y &= \sum_{n=0}^{\infty} c_n(n + r)(n + r - 1)x^{n+r} + \sum_{n=0}^{\infty} c_n(n + r)x^{n+r} + \sum_{n=0}^{\infty} c_n x^{n+r+2} - \nu^2 \sum_{n=0}^{\infty} c_n x^{n+r} \\ &= c_0(r^2 - r + r - \nu^2)x^r + x^r \sum_{n=1}^{\infty} c_n[(n + r)(n + r - 1) + (n + r) - \nu^2]x^n + x^r \sum_{n=0}^{\infty} c_n x^{n+2} \\ &= c_0(r^2 - \nu^2)x^r + x^r \sum_{n=1}^{\infty} c_n[(n + r)^2 - \nu^2]x^n + x^r \sum_{n=0}^{\infty} c_n x^{n+2}. \end{aligned} \tag{3}$$

From (3) we see that the indicial equation is $r^2 - \nu^2 = 0$, so the indicial roots are $r_1 = \nu$ and $r_2 = -\nu$. When $r_1 = \nu$, (3) becomes

$$\begin{aligned} &x^\nu \sum_{n=1}^{\infty} c_n n(n + 2\nu)x^n + x^\nu \sum_{n=0}^{\infty} c_n x^{n+2} \\ &\qquad = x^\nu \Bigg[(1 + 2\nu)c_1 x + \underbrace{\sum_{n=2}^{\infty} c_n n(n + 2\nu)x^n}_{k = n - 2} + \underbrace{\sum_{n=0}^{\infty} c_n x^{n+2}}_{k = n}\Bigg] \\ &\qquad = x^\nu \Bigg[(1 + 2\nu)c_1 x + \sum_{k=0}^{\infty} [(k + 2)(k + 2 + 2\nu)c_{k+2} + c_k]x^{k+2}\Bigg] = 0. \end{aligned}$$

Therefore by the usual argument we can write $(1 + 2\nu)c_1 = 0$ and

$$(k + 2)(k + 2 + 2\nu)c_{k+2} + c_k = 0$$

or

$$c_{k+2} = \frac{-c_k}{(k + 2)(k + 2 + 2\nu)}, \qquad k = 0, 1, 2, \ldots. \tag{4}$$

The choice $c_1 = 0$ in (4) implies that $c_3 = c_5 = c_7 = \cdots = 0$, so for $k = 0, 2, 4, \ldots$ we find, after letting $k + 2 = 2n$, $n = 1, 2, 3, \ldots$, that

$$c_{2n} = -\frac{c_{2n-2}}{2^2 n(n + \nu)}. \tag{5}$$

Thus

$$\begin{aligned}
c_2 &= -\frac{c_0}{2^2 \cdot 1 \cdot (1 + \nu)} \\
c_4 &= -\frac{c_2}{2^2 \cdot 2(2 + \nu)} = \frac{c_0}{2^4 \cdot 1 \cdot 2(1 + \nu)(2 + \nu)} \\
c_6 &= -\frac{c_4}{2^2 \cdot 3(3 + \nu)} = -\frac{c_0}{2^6 \cdot 1 \cdot 2 \cdot 3(1 + \nu)(2 + \nu)(3 + \nu)} \\
&\ \vdots \\
c_{2n} &= \frac{(-1)^n c_0}{2^{2n} n!(1 + \nu)(2 + \nu) \cdots (n + \nu)}, \qquad n = 1, 2, 3, \ldots.
\end{aligned} \tag{6}$$

It is standard practice to choose c_0 to be a specific value, namely,

$$c_0 = \frac{1}{2^\nu \Gamma(1 + \nu)},$$

where $\Gamma(1 + \nu)$ is the gamma function. See Appendix I. Since this latter function possesses the convenient property $\Gamma(1 + \alpha) = \alpha\Gamma(\alpha)$, we can reduce the indicated product in the denominator of (6) to one term. For example,

$$\begin{aligned}
\Gamma(1 + \nu + 1) &= (1 + \nu)\Gamma(1 + \nu) \\
\Gamma(1 + \nu + 2) &= (2 + \nu)\Gamma(2 + \nu) = (2 + \nu)(1 + \nu)\Gamma(1 + \nu).
\end{aligned}$$

Hence we can write (6) as

$$c_{2n} = \frac{(-1)^n}{2^{2n+\nu} n!(1 + \nu)(2 + \nu) \cdots (n + \nu)\Gamma(1 + \nu)} = \frac{(-1)^n}{2^{2n+\nu} n!\Gamma(1 + \nu + n)}$$

for $n = 0, 1, 2, \ldots$.

Bessel Functions of the First Kind Using the coefficients c_{2n} just obtained and $r = \nu$, a series solution of (1) is $y = \sum_{n=0}^{\infty} c_{2n} x^{2n+\nu}$. This solution is usually denoted by $J_\nu(x)$:

$$J_\nu(x) = \sum_{n=0}^{\infty} \frac{(-1)^n}{n!\Gamma(1 + \nu + n)} \left(\frac{x}{2}\right)^{2n+\nu}. \tag{7}$$

If $\nu \geq 0$, the series converges at least on the interval $[0, \infty)$. Also, for the second exponent $r_2 = -\nu$ we obtain, in exactly the same manner,

$$J_{-\nu}(x) = \sum_{n=0}^{\infty} \frac{(-1)^n}{n!\Gamma(1 - \nu + n)} \left(\frac{x}{2}\right)^{2n-\nu}. \tag{8}$$

The functions $J_\nu(x)$ and $J_{-\nu}(x)$ are called **Bessel functions of the first kind** of order ν and $-\nu$, respectively. Depending on the value of ν, (8) may contain negative powers of x and hence converges on $(0, \infty)$.*

*When we replace x by $|x|$, the series given in (7) and (8) converge for $0 < |x| < \infty$.

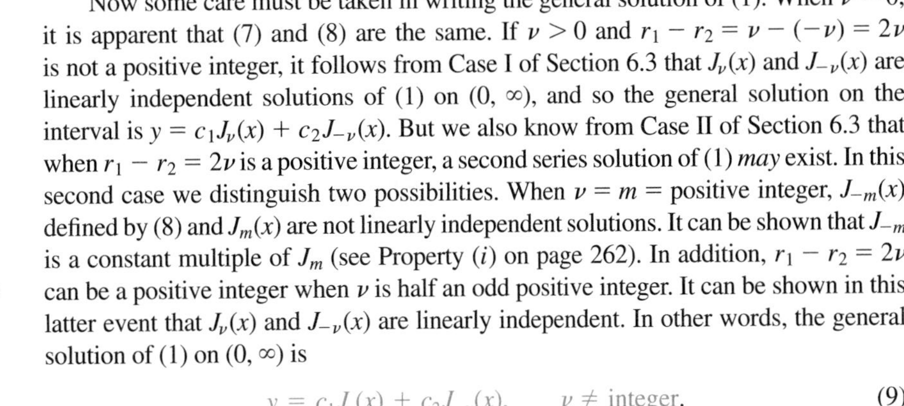

Now some care must be taken in writing the general solution of (1). When $\nu = 0$, it is apparent that (7) and (8) are the same. If $\nu > 0$ and $r_1 - r_2 = \nu - (-\nu) = 2\nu$ is not a positive integer, it follows from Case I of Section 6.3 that $J_\nu(x)$ and $J_{-\nu}(x)$ are linearly independent solutions of (1) on $(0, \infty)$, and so the general solution on the interval is $y = c_1 J_\nu(x) + c_2 J_{-\nu}(x)$. But we also know from Case II of Section 6.3 that when $r_1 - r_2 = 2\nu$ is a positive integer, a second series solution of (1) *may* exist. In this second case we distinguish two possibilities. When $\nu = m$ = positive integer, $J_{-m}(x)$ defined by (8) and $J_m(x)$ are not linearly independent solutions. It can be shown that J_{-m} is a constant multiple of J_m (see Property (*i*) on page 262). In addition, $r_1 - r_2 = 2\nu$ can be a positive integer when ν is half an odd positive integer. It can be shown in this latter event that $J_\nu(x)$ and $J_{-\nu}(x)$ are linearly independent. In other words, the general solution of (1) on $(0, \infty)$ is

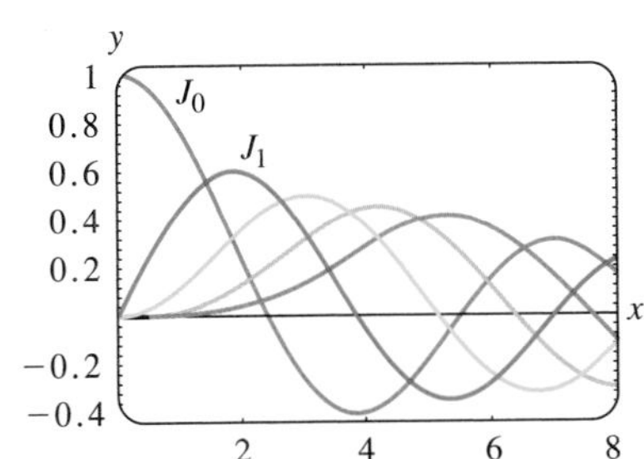

FIGURE 6.4.1 Bessel functions of the first kind for $n = 0, 1, 2, 3, 4$

$$y = c_1 J_\nu(x) + c_2 J_{-\nu}(x), \qquad \nu \neq \text{integer}. \tag{9}$$

The graphs of $y = J_0(x)$ and $y = J_1(x)$ are given in Figure 6.4.1.

EXAMPLE 1 Bessel's Equation of Order $\frac{1}{2}$

By identifying $\nu^2 = \frac{1}{4}$ and $\nu = \frac{1}{2}$, we can see from (9) that the general solution of the equation $x^2y'' + xy' + \left(x^2 - \frac{1}{4}\right)y = 0$ on $(0, \infty)$ is $y = c_1 J_{1/2}(x) + c_2 J_{-1/2}(x)$. ≡

≡ Bessel Functions of the Second Kind If $\nu \neq$ integer, the function defined by the linear combination

$$Y_\nu(x) = \frac{\cos \nu\pi J_\nu(x) - J_{-\nu}(x)}{\sin \nu\pi} \tag{10}$$

and the function $J_\nu(x)$ are linearly independent solutions of (1). Thus another form of the general solution of (1) is $y = c_1 J_\nu(x) + c_2 Y_\nu(x)$, provided that $\nu \neq$ integer. As $\nu \to m$, m an integer, (10) has the indeterminate form $0/0$. However, it can be shown by L'Hôpital's Rule that $\lim_{\nu\to m} Y_\nu(x)$ exists. Moreover, the function

$$Y_m(x) = \lim_{\nu\to m} Y_\nu(x)$$

and $J_m(x)$ are linearly independent solutions of $x^2y'' + xy' + (x^2 - m^2)y = 0$. Hence for *any* value of ν the general solution of (1) on $(0, \infty)$ can be written as

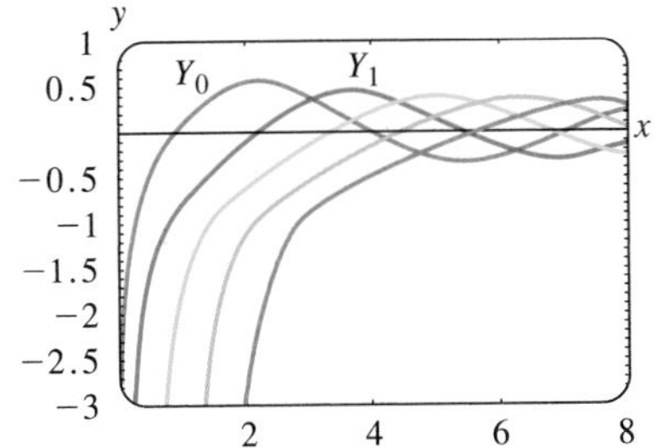

FIGURE 6.4.2 Bessel functions of the second kind for $n = 0, 1, 2, 3, 4$

$$y = c_1 J_\nu(x) + c_2 Y_\nu(x). \tag{11}$$

$Y_\nu(x)$ is called the **Bessel function of the second kind** of order ν. Figure 6.4.2 shows the graphs of $Y_0(x)$ and $Y_1(x)$.

EXAMPLE 2 Bessel's Equation of Order 3

By identifying $\nu^2 = 9$ and $\nu = 3$, we see from (11) that the general solution of the equation $x^2y'' + xy' + (x^2 - 9)y = 0$ on $(0, \infty)$ is $y = c_1 J_3(x) + c_2 Y_3(x)$. ≡

≡ DEs Solvable in Terms of Bessel Functions Sometimes it is possible to transform a differential equation into equation (1) by means of a change of variable. We can then express the solution of the original equation in terms of Bessel functions. For example, if we let $t = \alpha x$, $\alpha > 0$, in

$$x^2y'' + xy' + (\alpha^2x^2 - \nu^2)y = 0, \tag{12}$$

then by the Chain Rule,

$$\frac{dy}{dx} = \frac{dy}{dt}\frac{dt}{dx} = \alpha\frac{dy}{dt} \quad \text{and} \quad \frac{d^2y}{dx^2} = \frac{d}{dt}\left(\frac{dy}{dx}\right)\frac{dt}{dx} = \alpha^2\frac{d^2y}{dt^2}.$$

Accordingly, (12) becomes

$$\left(\frac{t}{\alpha}\right)^2 \alpha^2 \frac{d^2y}{dt^2} + \left(\frac{t}{\alpha}\right)\alpha \frac{dy}{dt} + (t^2 - \nu^2)y = 0 \qquad \text{or} \qquad t^2 \frac{d^2y}{dt^2} + t\frac{dy}{dt} + (t^2 - \nu^2)y = 0.$$

The last equation is Bessel's equation of order ν with solution $y = c_1J_\nu(t) + c_2Y_\nu(t)$. By resubstituting $t = \alpha x$ in the last expression, we find that the general solution of (12) is

$$y = c_1J_\nu(\alpha x) + c_2Y_\nu(\alpha x). \tag{13}$$

Equation (12), called the **parametric Bessel equation of order ν**, and its general solution (13) are very important in the study of certain boundary-value problems involving partial differential equations that are expressed in cylindrical coordinates.

Modified Bessel Functions Another equation that bears a resemblance to (1) is the **modified Bessel equation of order ν**,

$$x^2y'' + xy' - (x^2 + \nu^2)y = 0. \tag{14}$$

This DE can be solved in the manner just illustrated for (12). This time if we let $t = ix$, where $i^2 = -1$, then (14) becomes

$$t^2 \frac{d^2y}{dt^2} + t\frac{dy}{dt} + (t^2 - \nu^2)y = 0.$$

Because solutions of the last DE are $J_\nu(t)$ and $Y_\nu(t)$, *complex-valued* solutions of (14) are $J_\nu(ix)$ and $Y_\nu(ix)$. A real-valued solution, called the **modified Bessel function of the first kind** of order ν, is defined in terms of $J_\nu(ix)$:

$$I_\nu(x) = i^{-\nu}J_\nu(ix). \tag{15}$$

See Problem 21 in Exercises 6.4.

Analogous to (10), the **modified Bessel function of the second kind** of order $\nu \neq$ integer is defined to be

$$K_\nu(x) = \frac{\pi}{2}\frac{I_{-\nu}(x) - I_\nu(x)}{\sin \nu\pi}, \tag{16}$$

and for integer $\nu = n$,

$$K_n(x) = \lim_{\nu \to n} K_\nu(x).$$

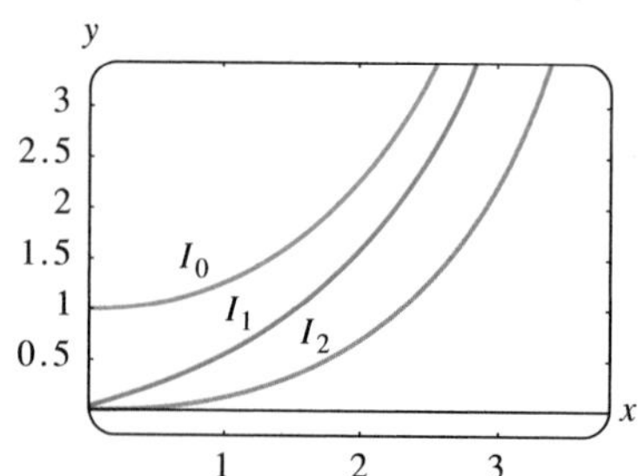

FIGURE 6.4.3 Modified Bessel functions of the first kind for $n = 0, 1, 2$

Because I_ν and K_ν are linearly independent on the interval $(0, \infty)$ for any value of ν, the general solution of (14) on that interval is

$$y = c_1I_\nu(x) + c_2K_\nu(x). \tag{17}$$

The graphs of $y = I_0(x)$, $y = I_1(x)$, and $y = I_2(x)$ are given in Figure 6.4.3 and the graphs of $y = K_0(x)$, $y = K_1(x)$, and $y = K_2(x)$ are given in Figure 6.4.4. Unlike the Bessel functions of the first and second kinds, the modified Bessel functions of the first and second kind are not oscillatory. Figures 6.4.3 and 6.4.4 also illustrate the fact that the modified Bessel functions $I_n(x)$ and $K_n(x)$, $n = 0, 1, 2, \ldots$ have no real zeros in the interval $(0, \infty)$. Also notice that the modified Bessel functions of the second kind $K_n(x)$ like the Bessel functions of the second kind $Y_n(x)$ become unbounded as $x \to 0^+$.

A change of variable in (14) gives us the **parametric form** of the modified Bessel equation of order ν:

$$x^2y'' + xy' - (\alpha^2x^2 + \nu^2)y = 0.$$

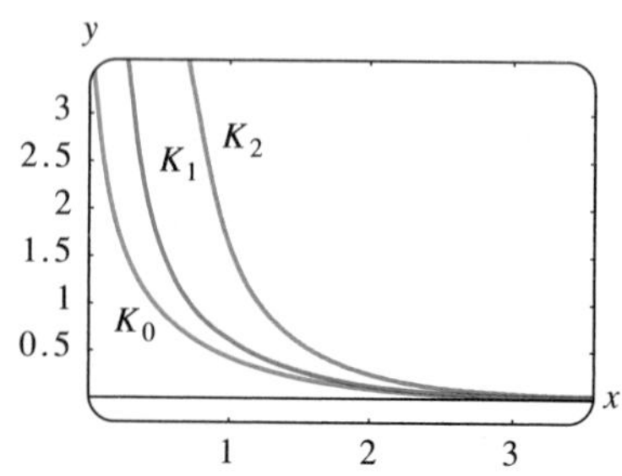

FIGURE 6.4.4 Modified Bessel functions of the second kind for $n = 0, 1, 2$

The general solution of the last equation on the interval $(0, \infty)$ is

$$y = c_1I_\nu(\alpha x) + c_2K_\nu(\alpha x).$$

Yet another equation, important because many DEs fit into its form by appropriate choices of the parameters, is

$$y'' + \frac{1-2a}{x}y' + \left(b^2c^2x^{2c-2} + \frac{a^2 - p^2c^2}{x^2}\right)y = 0, \qquad p \geq 0. \tag{18}$$

Although we shall not supply the details, the general solution of (18),

$$y = x^a\left[c_1J_p(bx^c) + c_2Y_p(bx^c)\right], \tag{19}$$

can be found by means of a change in both the independent and the dependent variables: $z = bx^c$, $y(x) = \left(\frac{z}{b}\right)^{a/c} w(z)$. If p is not an integer, then Y_p in (19) can be replaced by J_{-p}.

EXAMPLE 3 Using (18)

Find the general solution of $xy'' + 3y' + 9y = 0$ on $(0, \infty)$.

SOLUTION By writing the given DE as

$$y'' + \frac{3}{x}y' + \frac{9}{x}y = 0,$$

we can make the following identifications with (18):

$$1 - 2a = 3, \qquad b^2c^2 = 9, \qquad 2c - 2 = -1, \qquad \text{and} \qquad a^2 - p^2c^2 = 0.$$

The first and third equations imply that $a = -1$ and $c = \frac{1}{2}$. With these values the second and fourth equations are satisfied by taking $b = 6$ and $p = 2$. From (19) we find that the general solution of the given DE on the interval $(0, \infty)$ is $y = x^{-1}[c_1J_2(6x^{1/2}) + c_2Y_2(6x^{1/2})]$. ≡

EXAMPLE 4 The Aging Spring Revisited

Recall that in Section 5.1 we saw that one mathematical model for the free undamped motion of a mass on an aging spring is given by $mx'' + ke^{-\alpha t}x = 0$, $\alpha > 0$. We are now in a position to find the general solution of the equation. It is left as a problem to show that the change of variables $s = \frac{2}{\alpha}\sqrt{\frac{k}{m}}e^{-\alpha t/2}$ transforms the differential equation of the aging spring into

$$s^2\frac{d^2x}{ds^2} + s\frac{dx}{ds} + s^2x = 0.$$

The last equation is recognized as (1) with $\nu = 0$ and where the symbols x and s play the roles of y and x, respectively. The general solution of the new equation is $x = c_1J_0(s) + c_2Y_0(s)$. If we resubstitute s, then the general solution of $mx'' + ke^{-\alpha t}x = 0$ is seen to be

$$x(t) = c_1J_0\left(\frac{2}{\alpha}\sqrt{\frac{k}{m}}e^{-\alpha t/2}\right) + c_2Y_0\left(\frac{2}{\alpha}\sqrt{\frac{k}{m}}e^{-\alpha t/2}\right).$$

See Problems 33 and 39 in Exercises 6.4. ≡

The other model that was discussed in Section 5.1 of a spring whose characteristics change with time was $mx'' + ktx = 0$. By dividing through by m, we see that the equation $x'' + \frac{k}{m}tx = 0$ is Airy's equation $y'' + \alpha^2xy = 0$. See Example 5 in Section 6.2. The general solution of Airy's differential equation can also be written in terms of Bessel functions. See Problems 34, 35, and 40 in Exercises 6.4.

Properties We list below a few of the more useful properties of Bessel functions of order m, $m = 0, 1, 2, \ldots$:

$$(i)\ J_{-m}(x) = (-1)^m J_m(x), \qquad (ii)\ J_m(-x) = (-1)^m J_m(x),$$

$$(iii)\ J_m(0) = \begin{cases} 0, & m > 0 \\ 1, & m = 0, \end{cases} \qquad (iv)\ \lim_{x \to 0^+} Y_m(x) = -\infty.$$

Note that Property (ii) indicates that $J_m(x)$ is an even function if m is an even integer and an odd function if m is an odd integer. The graphs of $Y_0(x)$ and $Y_1(x)$ in Figure 6.4.2 illustrate Property (iv), namely, $Y_m(x)$ is unbounded at the origin. This last fact is not obvious from (10). The solutions of the Bessel equation of order 0 can be obtained by using the solutions $y_1(x)$ in (21) and $y_2(x)$ in (22) of Section 6.3. It can be shown that (21) of Section 6.3 is $y_1(x) = J_0(x)$, whereas (22) of that section is

$$y_2(x) = J_0(x)\ln x - \sum_{k=1}^{\infty} \frac{(-1)^k}{(k!)^2}\left(1 + \frac{1}{2} + \cdots + \frac{1}{k}\right)\left(\frac{x}{2}\right)^{2k}.$$

The Bessel function of the second kind of order 0, $Y_0(x)$, is then defined to be the linear combination $Y_0(x) = \dfrac{2}{\pi}(\gamma - \ln 2)y_1(x) + \dfrac{2}{\pi}y_2(x)$ for $x > 0$. That is,

$$Y_0(x) = \frac{2}{\pi}J_0(x)\left[\gamma + \ln\frac{x}{2}\right] - \frac{2}{\pi}\sum_{k=1}^{\infty} \frac{(-1)^k}{(k!)^2}\left(1 + \frac{1}{2} + \cdots + \frac{1}{k}\right)\left(\frac{x}{2}\right)^{2k},$$

where $\gamma = 0.57721566\ldots$ is **Euler's constant.** Because of the presence of the logarithmic term, it is apparent that $Y_0(x)$ is discontinuous at $x = 0$.

Numerical Values The first five nonnegative zeros of $J_0(x)$, $J_1(x)$, $Y_0(x)$, and $Y_1(x)$ are given in Table 6.4.1. Some additional function values of these four functions are given in Table 6.4.2.

TABLE 6.4.1 Zeros of J_0, J_1, Y_0, and Y_1

$J_0(x)$	$J_1(x)$	$Y_0(x)$	$Y_1(x)$
2.4048	0.0000	0.8936	2.1971
5.5201	3.8317	3.9577	5.4297
8.6537	7.0156	7.0861	8.5960
11.7915	10.1735	10.2223	11.7492
14.9309	13.3237	13.3611	14.8974

TABLE 6.4.2 Numerical Values of J_0, J_1, Y_0, and Y_1

x	$J_0(x)$	$J_1(x)$	$Y_0(x)$	$Y_1(x)$
0	1.0000	0.0000	—	—
1	0.7652	0.4401	0.0883	−0.7812
2	0.2239	0.5767	0.5104	−0.1070
3	−0.2601	0.3391	0.3769	0.3247
4	−0.3971	−0.0660	−0.0169	0.3979
5	−0.1776	−0.3276	−0.3085	0.1479
6	0.1506	−0.2767	−0.2882	−0.1750
7	0.3001	−0.0047	−0.0259	−0.3027
8	0.1717	0.2346	0.2235	−0.1581
9	−0.0903	0.2453	0.2499	0.1043
10	−0.2459	0.0435	0.0557	0.2490
11	−0.1712	−0.1768	−0.1688	0.1637
12	0.0477	−0.2234	−0.2252	−0.0571
13	0.2069	−0.0703	−0.0782	−0.2101
14	0.1711	0.1334	0.1272	−0.1666
15	−0.0142	0.2051	0.2055	0.0211

Differential Recurrence Relation Recurrence formulas that relate Bessel functions of different orders are important in theory and in applications. In the next example we derive a **differential recurrence relation.**

EXAMPLE 5 Derivation Using the Series Definition

Derive the formula $xJ'_\nu(x) = \nu J_\nu(x) - xJ_{\nu+1}(x)$.

SOLUTION It follows from (7) that

$$\begin{aligned}
xJ'_\nu(x) &= \sum_{n=0}^{\infty} \frac{(-1)^n(2n+\nu)}{n!\Gamma(1+\nu+n)}\left(\frac{x}{2}\right)^{2n+\nu} \\
&= \nu \sum_{n=0}^{\infty} \frac{(-1)^n}{n!\Gamma(1+\nu+n)}\left(\frac{x}{2}\right)^{2n+\nu} + 2\sum_{n=0}^{\infty} \frac{(-1)^n n}{n!\Gamma(1+\nu+n)}\left(\frac{x}{2}\right)^{2n+\nu} \\
&= \nu J_\nu(x) + x\underbrace{\sum_{n=1}^{\infty} \frac{(-1)^n}{(n-1)!\Gamma(1+\nu+n)}\left(\frac{x}{2}\right)^{2n+\nu-1}}_{k=n-1} \\
&= \nu J_\nu(x) - x\sum_{k=0}^{\infty} \frac{(-1)^k}{k!\Gamma(2+\nu+k)}\left(\frac{x}{2}\right)^{2k+\nu+1} = \nu J_\nu(x) - xJ_{\nu+1}(x).
\end{aligned}$$

≡

The result in Example 5 can be written in an alternative form. Dividing $xJ'_\nu(x) - \nu J_\nu(x) = -xJ_{\nu+1}(x)$ by x gives

$$J'_\nu(x) - \frac{\nu}{x}J_\nu(x) = -J_{\nu+1}(x).$$

This last expression is recognized as a linear first-order differential equation in $J_\nu(x)$. Multiplying both sides of the equality by the integrating factor $x^{-\nu}$ then yields

$$\frac{d}{dx}[x^{-\nu}J_\nu(x)] = -x^{-\nu}J_{\nu+1}(x). \tag{20}$$

It can be shown in a similar manner that

$$\frac{d}{dx}[x^{\nu}J_\nu(x)] = x^{\nu}J_{\nu-1}(x). \tag{21}$$

See Problem 27 in Exercises 6.4. The differential recurrence relations (20) and (21) are also valid for the Bessel function of the second kind $Y_\nu(x)$. Observe that when $\nu = 0$, it follows from (20) that

$$J'_0(x) = -J_1(x) \qquad \text{and} \qquad Y'_0(x) = -Y_1(x). \tag{22}$$

An application of these results is given in Problem 39 of Exercises 6.4.

≡ Bessel Functions of Half-Integral Order When the order is half an odd integer, that is, $\pm\frac{1}{2}, \pm\frac{3}{2}, \pm\frac{5}{2}, \ldots$, Bessel functions of the first and second kinds can be expressed in terms of the elementary functions $\sin x$, $\cos x$, and powers of x. Let's consider the case when $\nu = \frac{1}{2}$. From (7)

$$J_{1/2}(x) = \sum_{n=0}^{\infty} \frac{(-1)^n}{n!\Gamma\left(1+\frac{1}{2}+n\right)}\left(\frac{x}{2}\right)^{2n+1/2}.$$

In view of the property $\Gamma(1+\alpha)=\alpha\Gamma(\alpha)$ and the fact that $\Gamma(\frac{1}{2})=\sqrt{\pi}$ the values of $\Gamma(1+\frac{1}{2}+n)$ for $n=0$, $n=1$, $n=2$, and $n=3$ are, respectively,

$$\Gamma\left(\tfrac{3}{2}\right)=\Gamma\left(1+\tfrac{1}{2}\right)=\tfrac{1}{2}\Gamma\left(\tfrac{1}{2}\right)=\tfrac{1}{2}\sqrt{\pi}$$

$$\Gamma\left(\tfrac{5}{2}\right)=\Gamma\left(1+\tfrac{3}{2}\right)=\tfrac{3}{2}\Gamma\left(\tfrac{3}{2}\right)=\frac{3}{2^2}\sqrt{\pi}$$

$$\Gamma\left(\tfrac{7}{2}\right)=\Gamma\left(1+\tfrac{5}{2}\right)=\tfrac{5}{2}\Gamma\left(\tfrac{5}{2}\right)=\frac{5\cdot 3}{2^3}\sqrt{\pi}=\frac{5\cdot 4\cdot 3\cdot 2\cdot 1}{2^3 4\cdot 2}\sqrt{\pi}=\frac{5!}{2^5 2!}\sqrt{\pi}$$

$$\Gamma\left(\tfrac{9}{2}\right)=\Gamma\left(1+\tfrac{7}{2}\right)=\tfrac{7}{2}\Gamma\left(\tfrac{7}{2}\right)=\frac{7\cdot 5}{2^6\cdot 2!}\sqrt{\pi}=\frac{7\cdot 6\cdot 5!}{2^6\cdot 6\cdot 2!}\sqrt{\pi}=\frac{7!}{2^7 3!}\sqrt{\pi}.$$

In general,
$$\Gamma\left(1+\tfrac{1}{2}+n\right)=\frac{(2n+1)!}{2^{2n+1}n!}\sqrt{\pi}.$$

Hence
$$J_{1/2}(x)=\sum_{n=0}^{\infty}\frac{(-1)^n}{n!\dfrac{(2n+1)!}{2^{2n+1}n!}\sqrt{\pi}}\left(\frac{x}{2}\right)^{2n+1/2}=\sqrt{\frac{2}{\pi x}}\sum_{n=0}^{\infty}\frac{(-1)^n}{(2n+1)!}x^{2n+1}.$$

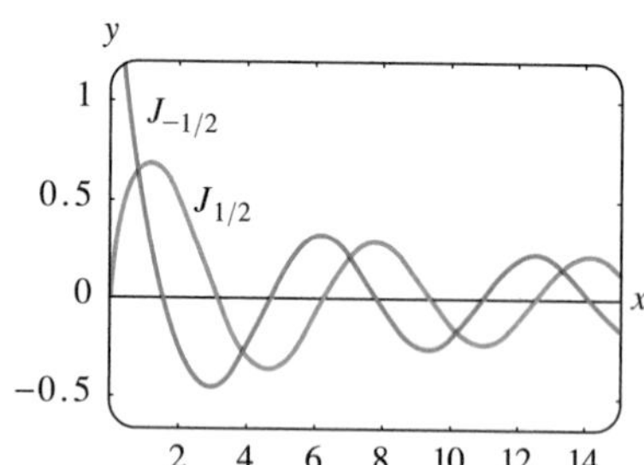

FIGURE 6.4.5 Bessel functions of order $\frac{1}{2}$ (blue) and order $-\frac{1}{2}$ (red)

From (2) of Section 6.1 you should recognize that the infinite series in the last line is the Maclaurin series for $\sin x$, and so we have shown that

$$J_{1/2}(x)=\sqrt{\frac{2}{\pi x}}\sin x. \tag{23}$$

We leave it as an exercise to show that

$$J_{-1/2}(x)=\sqrt{\frac{2}{\pi x}}\cos x. \tag{24}$$

See Figure 6.4.5 and Problems 31, 32, and 38 in Exercises 6.4.

If n is an integer, then $\nu=n+\frac{1}{2}$ is half an odd integer. Because $\cos(n+\frac{1}{2})\pi=0$ and $\sin(n+\frac{1}{2})\pi=\cos n\pi=(-1)^n$, we see from (10) that $Y_{n+1/2}(x)=(-1)^{n+1}J_{-(n+1/2)}(x)$. For $n=0$ and $n=-1$ we have, in turn, $Y_{1/2}(x)=-J_{-1/2}(x)$ and $Y_{-1/2}(x)=J_{1/2}(x)$. In view of (23) and (24) these results are the same as

$$Y_{1/2}(x)=-\sqrt{\frac{2}{\pi x}}\cos x \tag{25}$$

and

$$Y_{-1/2}(x)=\sqrt{\frac{2}{\pi x}}\sin x. \tag{26}$$

Spherical Bessel Functions Bessel functions of half-integral order are used to define two more important functions:

$$j_n(x)=\sqrt{\frac{\pi}{2x}}J_{n+1/2}(x)\quad\text{and}\quad y_n(x)=\sqrt{\frac{\pi}{2x}}Y_{n+1/2}(x). \tag{27}$$

The function $j_n(x)$ is called the **spherical Bessel function of the first kind** and $y_n(x)$ is the **spherical Bessel function of the second kind.** For example, for $n=0$ the expressions in (27) become

$$j_0(x)=\sqrt{\frac{\pi}{2x}}J_{1/2}(x)=\sqrt{\frac{\pi}{2x}}\sqrt{\frac{2}{\pi x}}\sin x=\frac{\sin x}{x}$$

and

$$y_0(x)=\sqrt{\frac{\pi}{2x}}Y_{1/2}(x)=-\sqrt{\frac{\pi}{2x}}\sqrt{\frac{2}{\pi x}}\cos x=-\frac{\cos x}{x}.$$

It is apparent from (27) and Figure 6.4.2 for $n \geq 0$ the spherical Bessel of the second kind $y_n(x)$ becomes unbounded as $x \to 0^+$.

Spherical Bessel functions arise in the solution of a special partial differential equation expressed in spherical coordinates. See Problem 54 in Exercises 6.4 and Problem 13 in Exercises 13.3.

Solution of Legendre's Equation Since $x = 0$ is an ordinary point of Legendre's equation (2), we substitute the series $y = \sum_{k=0}^{\infty} c_k x^k$, shift summation indices, and combine series to get

$$(1 - x^2)y'' - 2xy' + n(n + 1)y = [n(n + 1)c_0 + 2c_2] + [(n - 1)(n + 2)c_1 + 6c_3]x$$
$$+ \sum_{j=2}^{\infty} [(j + 2)(j + 1)c_{j+2} + (n - j)(n + j + 1)c_j]x^j = 0$$

which implies that

$$\begin{aligned} n(n + 1)c_0 + 2c_2 &= 0 \\ (n - 1)(n + 2)c_1 + 6c_3 &= 0 \\ (j + 2)(j + 1)c_{j+2} + (n - j)(n + j + 1)c_j &= 0 \end{aligned}$$

or

$$\begin{aligned} c_2 &= -\frac{n(n + 1)}{2!}c_0 \\ c_3 &= -\frac{(n - 1)(n + 2)}{3!}c_1 \\ c_{j+2} &= -\frac{(n - j)(n + j + 1)}{(j + 2)(j + 1)}c_j, \qquad j = 2, 3, 4, \ldots. \end{aligned} \tag{28}$$

If we let j take on the values 2, 3, 4, . . . , the recurrence relation (28) yields

$$\begin{aligned} c_4 &= -\frac{(n - 2)(n + 3)}{4 \cdot 3}c_2 = \frac{(n - 2)n(n + 1)(n + 3)}{4!}c_0 \\ c_5 &= -\frac{(n - 3)(n + 4)}{5 \cdot 4}c_3 = \frac{(n - 3)(n - 1)(n + 2)(n + 4)}{5!}c_1 \\ c_6 &= -\frac{(n - 4)(n + 5)}{6 \cdot 5}c_4 = -\frac{(n - 4)(n - 2)n(n + 1)(n + 3)(n + 5)}{6!}c_0 \\ c_7 &= -\frac{(n - 5)(n + 6)}{7 \cdot 6}c_5 = -\frac{(n - 5)(n - 3)(n - 1)(n + 2)(n + 4)(n + 6)}{7!}c_1 \end{aligned}$$

and so on. Thus for at least $|x| < 1$ we obtain two linearly independent power series solutions:

$$\begin{aligned} y_1(x) &= c_0\Big[1 - \frac{n(n + 1)}{2!}x^2 + \frac{(n - 2)n(n + 1)(n + 3)}{4!}x^4 \\ &\qquad - \frac{(n - 4)(n - 2)n(n + 1)(n + 3)(n + 5)}{6!}x^6 + \cdots\Big] \\ y_2(x) &= c_1\Big[x - \frac{(n - 1)(n + 2)}{3!}x^3 + \frac{(n - 3)(n - 1)(n + 2)(n + 4)}{5!}x^5 \\ &\qquad - \frac{(n - 5)(n - 3)(n - 1)(n + 2)(n + 4)(n + 6)}{7!}x^7 + \cdots\Big]. \end{aligned} \tag{29}$$

Notice that if n is an even integer, the first series terminates, whereas $y_2(x)$ is an infinite series. For example, if $n = 4$, then

$$y_1(x) = c_0\left[1 - \frac{4 \cdot 5}{2!}x^2 + \frac{2 \cdot 4 \cdot 5 \cdot 7}{4!}x^4\right] = c_0\left[1 - 10x^2 + \frac{35}{3}x^4\right].$$

Similarly, when n is an odd integer, the series for $y_2(x)$ terminates with x^n; that is, *when n is a nonnegative integer, we obtain an nth-degree polynomial solution* of Legendre's equation.

Because we know that a constant multiple of a solution of Legendre's equation is also a solution, it is traditional to choose specific values for c_0 or c_1, depending on whether n is an even or odd positive integer, respectively. For $n = 0$ we choose $c_0 = 1$, and for $n = 2, 4, 6, \ldots$

$$c_0 = (-1)^{n/2} \frac{1 \cdot 3 \cdots (n-1)}{2 \cdot 4 \cdots n},$$

whereas for $n = 1$ we choose $c_1 = 1$, and for $n = 3, 5, 7, \ldots$

$$c_1 = (-1)^{(n-1)/2} \frac{1 \cdot 3 \cdots n}{2 \cdot 4 \cdots (n-1)}.$$

For example, when $n = 4$, we have

$$y_1(x) = (-1)^{4/2} \frac{1 \cdot 3}{2 \cdot 4}\left[1 - 10x^2 + \frac{35}{3}x^4\right] = \frac{1}{8}(35x^4 - 30x^2 + 3).$$

Legendre Polynomials These specific nth-degree polynomial solutions are called **Legendre polynomials** and are denoted by $P_n(x)$. From the series for $y_1(x)$ and $y_2(x)$ and from the above choices of c_0 and c_1 we find that the first several Legendre polynomials are

$$\begin{aligned} &P_0(x) = 1, && P_1(x) = x, \\ &P_2(x) = \frac{1}{2}(3x^2 - 1), && P_3(x) = \frac{1}{2}(5x^3 - 3x), \\ &P_4(x) = \frac{1}{8}(35x^4 - 30x^2 + 3), && P_5(x) = \frac{1}{8}(63x^5 - 70x^3 + 15x). \end{aligned} \tag{30}$$

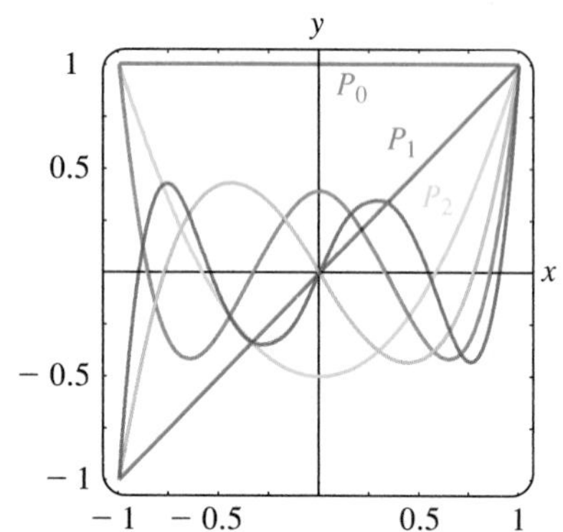

FIGURE 6.4.6 Legendre polynomials for $n = 0, 1, 2, 3, 4, 5$

Remember, $P_0(x)$, $P_1(x)$, $P_2(x)$, $P_3(x)$, ... are, in turn, particular solutions of the differential equations

$$\begin{aligned} n = 0: &\quad (1 - x^2)y'' - 2xy' = 0, \\ n = 1: &\quad (1 - x^2)y'' - 2xy' + 2y = 0, \\ n = 2: &\quad (1 - x^2)y'' - 2xy' + 6y = 0, \\ n = 3: &\quad (1 - x^2)y'' - 2xy' + 12y = 0, \\ \vdots &\qquad \vdots \end{aligned} \tag{31}$$

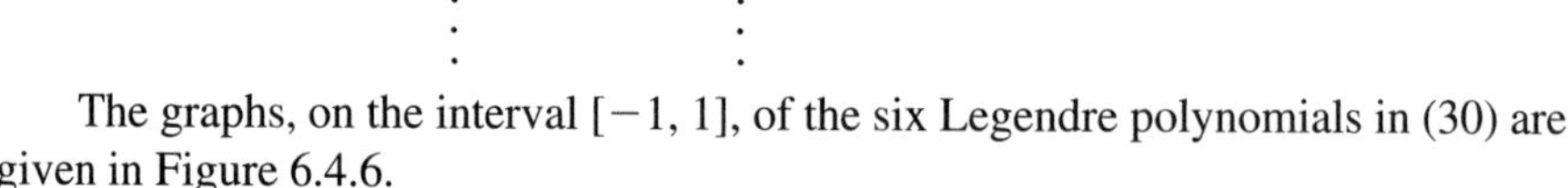

The graphs, on the interval $[-1, 1]$, of the six Legendre polynomials in (30) are given in Figure 6.4.6.

Properties You are encouraged to verify the following properties using the Legendre polynomials in (30).

$$(i)\ P_n(-x) = (-1)^n P_n(x)$$

$$(ii)\ P_n(1) = 1 \qquad (iii)\ P_n(-1) = (-1)^n$$

$$(iv)\ P_n(0) = 0, \quad n \text{ odd} \qquad (v)\ P_n'(0) = 0, \quad n \text{ even}$$

Property (i) indicates, as is apparent in Figure 6.4.6, that $P_n(x)$ is an even or odd function according to whether n is even or odd.

Recurrence Relation Recurrence relations that relate Legendre polynomials of different degrees are also important in some aspects of their applications. We state, without proof, the three-term recurrence relation

$$(k + 1)P_{k+1}(x) - (2k + 1)xP_k(x) + kP_{k-1}(x) = 0, \tag{32}$$

which is valid for $k = 1, 2, 3, \ldots$. In (30) we listed the first six Legendre polynomials. If, say, we wish to find $P_6(x)$, we can use (32) with $k = 5$. This relation expresses $P_6(x)$ in terms of the known $P_4(x)$ and $P_5(x)$. See Problem 45 in Exercises 6.4.

Another formula, although not a recurrence relation, can generate the Legendre polynomials by differentiation. **Rodrigues' formula** for these polynomials is

$$P_n(x) = \frac{1}{2^n n!}\frac{d^n}{dx^n}(x^2 - 1)^n, \qquad n = 0, 1, 2, \ldots . \tag{33}$$

See Problem 48 in Exercises 6.4.

REMARKS

Although we have assumed that the parameter n in Legendre's differential equation $(1 - x^2)y'' - 2xy' + n(n + 1)y = 0$, represented a nonnegative integer, in a more general setting n can represent any real number. Any solution of Legendre's equation is called a **Legendre function.** If n is *not* a nonnegative integer, then both Legendre functions $y_1(x)$ and $y_2(x)$ given in (29) are infinite series convergent on the open interval $(-1, 1)$ and divergent (unbounded) at $x = \pm 1$. If n is a nonnegative integer, then as we have just seen one of the Legendre functions in (29) is a polynomial and the other is an infinite series convergent for $-1 < x < 1$. You should be aware of the fact that Legendre's equation possesses solutions that are bounded on the *closed* interval $[-1, 1]$ only in the case when $n = 0, 1, 2, \ldots$. More to the point, the only Legendre functions that are bounded on the closed interval $[-1, 1]$ are the Legendre polynomials $P_n(x)$ or constant multiples of these polynomials. See Problem 47 in Exercises 6.4 and Problem 24 in Chapter 6 in Review.

EXERCISES 6.4

Answers to selected odd-numbered problems begin on page ANS-11.

Bessel's Equation

In Problems 1–6 use (1) to find the general solution of the given differential equation on $(0, \infty)$.

1. $x^2y'' + xy' + \left(x^2 - \frac{1}{9}\right)y = 0$

2. $x^2y'' + xy' + (x^2 - 1)y = 0$

3. $4x^2y'' + 4xy' + (4x^2 - 25)y = 0$

4. $16x^2y'' + 16xy' + (16x^2 - 1)y = 0$

5. $xy'' + y' + xy = 0$

6. $\dfrac{d}{dx}[xy'] + \left(x - \dfrac{4}{x}\right)y = 0$

In Problems 7–10 use (12) to find the general solution of the given differential equation on $(0, \infty)$.

7. $x^2y'' + xy' + (9x^2 - 4)y = 0$

8. $x^2y'' + xy' + \left(36x^2 - \frac{1}{4}\right)y = 0$

9. $x^2y'' + xy' + \left(25x^2 - \frac{4}{9}\right)y = 0$

10. $x^2y'' + xy' + (2x^2 - 64)y = 0$

In Problems 11 and 12 use the indicated change of variable to find the general solution of the given differential equation on $(0, \infty)$.

11. $x^2y'' + 2xy' + \alpha^2x^2y = 0; \quad y = x^{-1/2}v(x)$

12. $x^2y'' + \left(\alpha^2x^2 - \nu^2 + \frac{1}{4}\right)y = 0; \quad y = \sqrt{x}\,v(x)$

In Problems 13–20 use (18) to find the general solution of the given differential equation on $(0, \infty)$.

13. $xy'' + 2y' + 4y = 0$

14. $xy'' + 3y' + xy = 0$

15. $xy'' - y' + xy = 0$

16. $xy'' - 5y' + xy = 0$

17. $x^2y'' + (x^2 - 2)y = 0$

18. $4x^2y'' + (16x^2 + 1)y = 0$

19. $xy'' + 3y' + x^3y = 0$

20. $9x^2y'' + 9xy' + (x^6 - 36)y = 0$

21. Use the series in (7) to verify that $I_\nu(x) = i^{-\nu} J_\nu(ix)$ is a real function.

22. Assume that b in equation (18) can be pure imaginary, that is, $b = \beta i, \beta > 0, i^2 = -1$. Use this assumption to express the general solution of the given differential equation in terms the modified Bessel functions I_n and K_n.

(a) $y'' - x^2y = 0$
(b) $xy'' + y' - 7x^3y = 0$

In Problems 23–26 first use (18) to express the general solution of the given differential equation in terms of Bessel functions. Then use (23) and (24) to express the general solution in terms of elementary functions.

23. $y'' + y = 0$

24. $x^2y'' + 4xy' + (x^2 + 2)y = 0$

25. $16x^2y'' + 32xy' + (x^4 - 12)y = 0$

26. $4x^2y'' - 4xy' + (16x^2 + 3)y = 0$

27. (a) Proceed as in Example 5 to show that

$$xJ'_\nu(x) = -\nu J_\nu(x) + xJ_{\nu-1}(x).$$

[*Hint*: Write $2n + \nu = 2(n + \nu) - \nu$.]

(b) Use the result in part (a) to derive (21).

28. Use the formula obtained in Example 5 along with part (a) of Problem 27 to derive the recurrence relation

$$2\nu J_\nu(x) = xJ_{\nu+1}(x) + xJ_{\nu-1}(x).$$

In Problems 29 and 30 use (20) or (21) to obtain the given result.

29. $\int_0^x rJ_0(r)dr = xJ_1(x)$ **30.** $J'_0(x) = J_{-1}(x) = -J_1(x)$

31. Proceed as on page 264 to derive the elementary form of $J_{-1/2}(x)$ given in (24).

32. Use the recurrence relation in Problem 28 along with (23) and (24) to express $J_{3/2}(x)$, $J_{-3/2}(x)$, $J_{5/2}(x)$ and $J_{-5/2}(x)$ in terms of $\sin x$, $\cos x$, and powers of x.

33. Use the change of variables $s = \dfrac{2}{\alpha}\sqrt{\dfrac{k}{m}}e^{-\alpha t/2}$ to show that the differential equation of the aging spring $mx'' + ke^{-\alpha t}x = 0, \alpha > 0$, becomes

$$s^2\frac{d^2x}{ds^2} + s\frac{dx}{ds} + s^2x = 0.$$

34. Show that $y = x^{1/2}w\left(\frac{2}{3}\alpha x^{3/2}\right)$ is a solution of Airy's differential equation $y'' + \alpha^2xy = 0, x > 0$, whenever w is a solution of Bessel's equation of order $\frac{1}{3}$, that is, $t^2w'' + tw' + \left(t^2 - \frac{1}{9}\right)w = 0, t > 0$. [*Hint*: After differentiating, substituting, and simplifying, then let $t = \frac{2}{3}\alpha x^{3/2}$.]

35. (a) Use the result of Problem 34 to express the general solution of Airy's differential equation for $x > 0$ in terms of Bessel functions.

(b) Verify the results in part (a) using (18).

36. Use the Table 6.4.1 to find the first three positive eigenvalues and corresponding eigenfunctions of the boundary-value problem

$$xy'' + y' + \lambda xy = 0,$$

$$y(x), y'(x) \text{ bounded as } x \to 0^+, \quad y(2) = 0.$$

[*Hint*: By identifying $\lambda = \alpha^2$, the DE is the parametric Bessel equation of order zero.]

37. (a) Use (18) to show that the general solution of the differential equation $xy'' + \lambda y = 0$ on the interval $(0, \infty)$ is

$$y = c_1\sqrt{x}J_1(2\sqrt{\lambda x}) + c_2\sqrt{x}Y_1(2\sqrt{\lambda x}).$$

(b) Verify by direct substitution that $y = \sqrt{x}J_1(2\sqrt{x})$ is a particular solution of the DE in the case $\lambda = 1$.

Computer Lab Assignments

38. Use a CAS to graph $J_{3/2}(x)$, $J_{-3/2}(x)$, $J_{5/2}(x)$, and $J_{-5/2}(x)$.

39. (a) Use the general solution given in Example 4 to solve the IVP

$$4x'' + e^{-0.1t}x = 0, \quad x(0) = 1, \quad x'(0) = -\tfrac{1}{2}.$$

Also use $J'_0(x) = -J_1(x)$ and $Y'_0(x) = -Y_1(x)$ along with Table 6.4.1 or a CAS to evaluate coefficients.

(b) Use a CAS to graph the solution obtained in part (a) for $0 \le t \le \infty$.

40. (a) Use the general solution obtained in Problem 35 to solve the IVP

$$4x'' + tx = 0, \quad x(0.1) = 1, \quad x'(0.1) = -\tfrac{1}{2}.$$

Use a CAS to evaluate coefficients.

(b) Use a CAS to graph the solution obtained in part (a) for $0 \le t \le 200$.

41. Column Bending Under Its Own Weight A uniform thin column of length L, positioned vertically with one end embedded in the ground, will deflect, or bend away, from the vertical under the influence of its own weight when its length or height exceeds a certain critical value. It can be shown that the angular deflection $\theta(x)$ of the column from the vertical at a point $P(x)$ is a solution of the boundary-value problem:

$$EI\frac{d^2\theta}{dx^2} + \delta g(L - x)\theta = 0, \quad \theta(0) = 0, \quad \theta'(L) = 0,$$

where E is Young's modulus, I is the cross-sectional moment of inertia, δ is the constant linear density, and x is the distance along the column measured from its base. See Figure 6.4.7. The column will bend only for those values of L for which the boundary-value problem has a nontrivial solution.

(a) Restate the boundary-value problem by making the change of variables $t = L - x$. Then use the results of a problem earlier in this exercise set to express the general solution of the differential equation in terms of Bessel functions.

(b) Use the general solution found in part (a) to find a solution of the BVP and an equation which defines the critical length L, that is, the smallest value of L for which the column will start to bend.

(c) With the aid of a CAS, find the critical length L of a solid steel rod of radius $r = 0.05$ in., $\delta g = 0.28\,A$ lb/in., $E = 2.6 \times 10^7$ lb/in.^{2}, $A = \pi r^2$, and $I = \frac{1}{4}\pi r^4$.

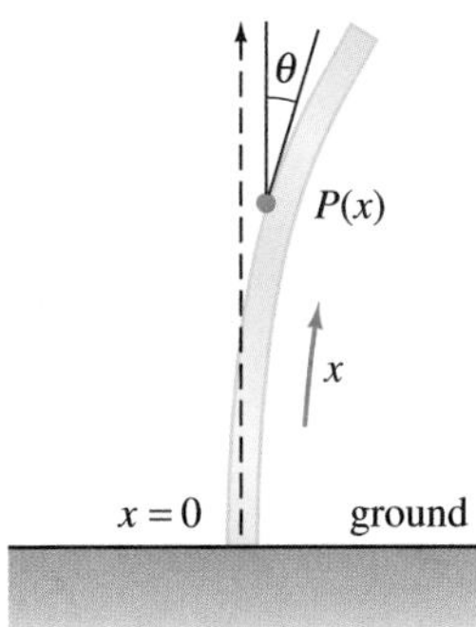

FIGURE 6.4.7 Beam in Problem 41

42. Buckling of a Thin Vertical Column In Example 4 of Section 5.2 we saw that when a constant vertical compressive force, or load, P was applied to a thin column of uniform cross section and hinged at both ends, the deflection $y(x)$ is a solution of the BVP:

$$EI\frac{d^2y}{dx^2} + Py = 0, \quad y(0) = 0, \quad y(L) = 0.$$

(a) If the bending stiffness factor EI is proportional to x, then $EI(x) = kx$, where k is a constant of proportionality. If $EI(L) = kL = M$ is the maximum stiffness factor, then $k = M/L$ and so $EI(x) = Mx/L$. Use the information in Problem 37 to find a solution of

$$M\frac{x}{L}\frac{d^2y}{dx^2} + Py = 0, \quad y(0) = 0, \quad y(L) = 0$$

if it is known that $\sqrt{x}\,Y_1(2\sqrt{\lambda x})$ is *not* zero at $x = 0$.

(b) Use Table 6.4.1 to find the Euler load P_1 for the column.

(c) Use a CAS to graph the first buckling mode $y_1(x)$ corresponding to the Euler load P_1. For simplicity assume that $c_1 = 1$ and $L = 1$.

43. Pendulum of Varying Length For the simple pendulum described on page 220 of Section 5.3, suppose that the rod holding the mass m at one end is replaced by a flexible wire or string and that the wire is strung over a pulley at the point of support O in Figure 5.3.3. In this manner, while it is in motion in a vertical plane, the mass m can be raised or lowered. In other words, the length $l(t)$ of the pendulum varies with time. Under the same assumptions leading to equation (6) in Section 5.3, it can be shown* that the differential equation for the displacement angle θ is now

$$l\theta'' + 2l'\theta' + g\sin\theta = 0.$$

(a) If l increases at constant rate v and if $l(0) = l_0$, show that a linearization of the foregoing DE is

$$(l_0 + vt)\theta'' + 2v\theta' + g\theta = 0. \tag{34}$$

(b) Make the change of variables $x = (l_0 + vt)/v$ and show that (34) becomes

$$\frac{d^2\theta}{dx^2} + \frac{2}{x}\frac{d\theta}{dx} + \frac{g}{vx}\theta = 0.$$

(c) Use part (b) and (18) to express the general solution of equation (34) in terms of Bessel functions.

(d) Use the general solution obtained in part (c) to solve the initial-value problem consisting of equation (34) and the initial conditions $\theta(0) = \theta_0$, $\theta'(0) = 0$. [*Hints*: To simplify calculations, use a further change of variable $u = \frac{2}{v}\sqrt{g(l_0 + vt)} = 2\sqrt{\frac{g}{v}}\,x^{1/2}$. Also, recall that (20) holds for both $J_1(u)$ and $Y_1(u)$. Finally, the identity

$$J_1(u)Y_2(u) - J_2(u)Y_1(u) = -\frac{2}{\pi u}$$

will be helpful.]

(problem continues on page 270)

*See *Mathematical Methods in Physical Sciences,* Mary Boas, John Wiley & Sons, Inc., 1966. Also see the article by Borelli, Coleman, and Hobson in *Mathematics Magazine,* vol. 58, no. 2, March 1985.

(e) Use a CAS to graph the solution $\theta(t)$ of the IVP in part (d) when $l_0 = 1$ ft, $\theta_0 = \frac{1}{10}$ radian, and $v = \frac{1}{60}$ ft/s. Experiment with the graph using different time intervals such as [0, 10], [0, 30], and so on.

(f) What do the graphs indicate about the displacement angle $\theta(t)$ as the length l of the wire increases with time?

Legendre's Equation

44. (a) Use the explicit solutions $y_1(x)$ and $y_2(x)$ of Legendre's equation given in (29) and the appropriate choice of c_0 and c_1 to find the Legendre polynomials $P_6(x)$ and $P_7(x)$.

(b) Write the differential equations for which $P_6(x)$ and $P_7(x)$ are particular solutions.

45. Use the recurrence relation (32) and $P_0(x) = 1$, $P_1(x) = x$, to generate the next six Legendre polynomials.

46. Show that the differential equation

$$\sin\theta\,\frac{d^2y}{d\theta^2} + \cos\theta\,\frac{dy}{d\theta} + n(n+1)(\sin\theta)y = 0$$

can be transformed into Legendre's equation by means of the substitution $x = \cos\theta$.

47. Find the first three positive values of λ for which the problem

$$(1 - x^2)y'' - 2xy' + \lambda y = 0,$$
$$y(0) = 0, \quad y(x),\, y'(x) \text{ bounded on } [-1,1]$$

has nontrivial solutions.

Computer Lab Assignments

48. For purposes of this problem ignore the list of Legendre polynomials given on page 266 and the graphs given in Figure 6.4.3. Use Rodrigues' formula (33) to generate the Legendre polynomials $P_1(x), P_2(x), \ldots, P_7(x)$. Use a CAS to carry out the differentiations and simplifications.

49. Use a CAS to graph $P_1(x), P_2(x), \ldots, P_7(x)$ on the interval $[-1, 1]$.

50. Use a root-finding application to find the zeros of $P_1(x), P_2(x), \ldots, P_7(x)$. If the Legendre polynomials are built-in functions of your CAS, find zeros of Legendre polynomials of higher degree. Form a conjecture about the location of the zeros of any Legendre polynomial $P_n(x)$, and then investigate to see whether it is true.

Miscellaneous Differential Equations

51. The differential equation

$$y'' - 2xy' + 2\alpha y = 0$$

is known as **Hermite's equation of order α** after the French mathematician **Charles Hermite** (1822–1901). Show that the general solution of the equation is $y(x) = c_0y_1(x) + c_1y_2(x)$, where

$$y_1(x) = 1 + \sum_{k=1}^{\infty}(-1)^k\frac{2^k\alpha(\alpha-2)\cdots(\alpha-2k+2)}{(2k)!}x^{2k}$$

$$y_2(x) = x + \sum_{k=1}^{\infty}(-1)^k\frac{2^k(\alpha-1)(\alpha-3)\cdots(\alpha-2k+1)}{(2k+1)!}x^{2k+1}$$

are power series solutions centered at the ordinary point 0.

52. (a) When $\alpha = n$ is a nonnegative integer, Hermite's differential equation always possesses a polynomial solution of degree n. Use $y_1(x)$, given in Problem 51, to find polynomial solutions for $n = 0$, $n = 2$, and $n = 4$. Then use $y_2(x)$ to find polynomial solutions for $n = 1$, $n = 3$, and $n = 5$.

(b) A **Hermite polynomial** $H_n(x)$ is defined to be the nth degree polynomial solution of Hermite's equation multiplied by an appropriate constant so that the coefficient of x^n in $H_n(x)$ is 2^n. Use the polynomial solutions in part (a) to show that the first six Hermite polynomials are

$$H_0(x) = 1$$
$$H_1(x) = 2x$$
$$H_2(x) = 4x^2 - 2$$
$$H_3(x) = 8x^3 - 12x$$
$$H_4(x) = 16x^4 - 48x^2 + 12$$
$$H_5(x) = 32x^5 - 160x^3 + 120x.$$

53. The differential equation

$$(1 - x^2)y'' - xy' + \alpha^2 y = 0,$$

where α is a parameter, is known as **Chebyshev's equation** after the Russian mathematician **Pafnuty Chebyshev** (1821–1894). When $\alpha = n$ is a nonnegative integer, Chebyshev's differential equation always possesses a polynomial solution of degree n. Find a fifth degree polynomial solution of this differential equation.

54. If n is an integer, use the substitution $R(x) = (\alpha x)^{-1/2}Z(x)$ to show that the general solution of the differential equation

$$x^2R'' + 2xR' + [\alpha^2x^2 - n(n+1)]R = 0$$

on the interval $(0, \infty)$ is $R(x) = c_1 j_n(\alpha x) + c_2 y_n(\alpha x)$, where $j_n(\alpha x)$ and $y_n(\alpha x)$ are the spherical Bessel functions of the first and second kind defined in (27).

CHAPTER 6 IN REVIEW

Answers to selected odd-numbered problems begin on page ANS-11.

In Problems 1 and 2 answer true or false without referring back to the text.

1. The general solution of $x^2y'' + xy' + (x^2 - 1)y = 0$ is $y = c_1J_1(x) + c_2J_{-1}(x)$. ________

2. Because $x = 0$ is an irregular singular point of $x^3y'' - xy' + y = 0$, the DE possesses no solution that is analytic at $x = 0$. ________

3. Both power series solutions of $y'' + \ln(x + 1)y' + y = 0$ centered at the ordinary point $x = 0$ are guaranteed to converge for all x in which *one* of the following intervals?

(a) $(-\infty, \infty)$ **(b)** $(-1, \infty)$
(c) $[-\frac{1}{2}, \frac{1}{2}]$ **(d)** $[-1, 1]$

4. $x = 0$ is an ordinary point of a certain linear differential equation. After the assumed solution $y = \sum_{n=0}^{\infty} c_nx^n$ is substituted into the DE, the following algebraic system is obtained by equating the coefficients of x^0, x^1, x^2, and x^3 to zero:

$$2c_2 + 2c_1 + c_0 = 0$$
$$6c_3 + 4c_2 + c_1 = 0$$
$$12c_4 + 6c_3 + c_2 - \tfrac{1}{3}c_1 = 0$$
$$20c_5 + 8c_4 + c_3 - \tfrac{2}{3}c_2 = 0.$$

Bearing in mind that c_0 and c_1 are arbitrary, write down the first five terms of two power series solutions of the differential equation.

5. Suppose the power series $\sum_{k=0}^{\infty} c_k(x - 4)^k$ is known to converge at -2 and diverge at 13. Discuss whether the series converges at -7, 0, 7, 10, and 11. Possible answers are *does, does not, might.*

6. Use the Maclaurin series for $\sin x$ and $\cos x$ along with long division to find the first three nonzero terms of a power series in x for the function $f(x) = \dfrac{\sin x}{\cos x}$.

In Problems 7 and 8 construct a linear second-order differential equation that has the given properties.

7. A regular singular point at $x = 1$ and an irregular singular point at $x = 0$

8. Regular singular points at $x = 1$ and at $x = -3$

In Problems 9–14 use an appropriate infinite series method about $x = 0$ to find two solutions of the given differential equation.

9. $2xy'' + y' + y = 0$

10. $y'' - xy' - y = 0$

11. $(x - 1)y'' + 3y = 0$

12. $y'' - x^2y' + xy = 0$

13. $xy'' - (x + 2)y' + 2y = 0$

14. $(\cos x)y'' + y = 0$

In Problems 15 and 16 solve the given initial-value problem.

15. $y'' + xy' + 2y = 0, \quad y(0) = 3, y'(0) = -2$

16. $(x + 2)y'' + 3y = 0, \quad y(0) = 0, y'(0) = 1$

17. Without actually solving the differential equation $(1 - 2\sin x)y'' + xy = 0$, find a lower bound for the radius of convergence of power series solutions about the ordinary point $x = 0$.

18. Even though $x = 0$ is an ordinary point of the differential equation, explain why it is not a good idea to try to find a solution of the IVP

$$y'' + xy' + y = 0, \quad y(1) = -6, \quad y'(1) = 3$$

of the form $y = \sum_{n=0}^{\infty} c_nx^n$. Using power series, find a better way to solve the problem.

In Problems 19 and 20 investigate whether $x = 0$ is an ordinary point, singular point, or irregular singular point of the given differential equation. [*Hint*: Recall the Maclaurin series for $\cos x$ and e^x.]

19. $xy'' + (1 - \cos x)y' + x^2y = 0$

20. $(e^x - 1 - x)y'' + xy = 0$

21. Note that $x = 0$ is an ordinary point of the differential equation $y'' + x^2y' + 2xy = 5 - 2x + 10x^3$. Use the assumption $y = \sum_{n=0}^{\infty} c_nx^n$ to find the general solution $y = y_c + y_p$ that consists of three power series centered at $x = 0$.

22. The first-order differential equation $dy/dx = x^2 + y^2$ cannot be solved in terms of elementary functions. However, a solution can be expressed in terms of Bessel functions.

(a) Show that the substitution $y = -\dfrac{1}{u}\dfrac{du}{dx}$ leads to the equation $u'' + x^2u = 0$.

(b) Use (18) in Section 6.4 to find the general solution of $u'' + x^2u = 0$.

(c) Use (20) and (21) in Section 6.4 in the forms

$$J'_\nu(x) = \frac{\nu}{x}J_\nu(x) - J_{\nu+1}(x)$$

and

$$J'_\nu(x) = -\frac{\nu}{x}J_\nu(x) + J_{\nu-1}(x)$$

as an aid to show that a one-parameter family of solutions of $dy/dx = x^2 + y^2$ is given by

$$y = x\,\frac{J_{3/4}\left(\frac{1}{2}x^2\right) - cJ_{-3/4}\left(\frac{1}{2}x^2\right)}{cJ_{1/4}\left(\frac{1}{2}x^2\right) + J_{-1/4}\left(\frac{1}{2}x^2\right)}.$$

23. **(a)** Use (10) of Section 6.4 and Problem 32 of Exercises 6.4 to show that

$$Y_{3/2}(x) = -\sqrt{\frac{2}{\pi x}}\left(\frac{\cos x}{x} + \sin x\right)$$

(b) Use (15) of Section 6.4 to show that

$$I_{1/2}(x) = \sqrt{\frac{2}{\pi x}}\sinh x \quad \text{and} \quad I_{-1/2}(x) = \sqrt{\frac{2}{\pi x}}\cosh x.$$

(c) Use (16) of Section 6.4 and part (b) to show that

$$K_{1/2}(x) = \sqrt{\frac{\pi}{2x}}\,e^{-x}.$$

24. **(a)** From (30) and (31) of Section 6.4 we know that when $n = 0$, Legendre's differential equation $(1 - x^2)y'' - 2xy' = 0$ has the polynomial solution $y = P_0(x) = 1$. Use (5) of Section 4.2 to show that a second Legendre function satisfying the DE for $-1 < x < 1$ is

$$y = \frac{1}{2}\ln\left(\frac{1+x}{1-x}\right).$$

(b) We also know from (30) and (31) of Section 6.4 that when $n = 1$, Legendre's differential equation $(1 - x^2)y'' - 2xy' + 2y = 0$ possesses the polynomial solution $y = P_1(x) = x$. Use (5) of Section 4.2 to show that a second Legendre function satisfying the DE for $-1 < x < 1$ is

$$y = \frac{x}{2}\ln\left(\frac{1+x}{1-x}\right) - 1.$$

(c) Use a graphing utility to graph the logarithmic Legendre functions given in parts (a) and (b).

25. **(a)** Use binomial series to formally show that

$$(1 - 2xt + t^2)^{-1/2} = \sum_{n=0}^{\infty} P_n(x)t^n.$$

(b) Use the result obtained in part (a) to show that $P_n(1) = 1$ and $P_n(-1) = (-1)^n$. See Properties (*ii*) and (*iii*) on page 266.

7 The Laplace Transform

In the linear mathematical models for a physical system such as a spring/mass system or a series electrical circuit, the right-hand member, or input, of the differential equations

$$m\frac{d^2x}{dt^2} + \beta\frac{dx}{dt} + kx = f(t) \qquad \text{or} \qquad L\frac{d^2q}{dt^2} + R\frac{dq}{dt} + \frac{1}{C}q = E(t)$$

is a driving function and represents either an external force $f(t)$ or an impressed voltage $E(t)$. In Section 5.1 we considered problems in which the functions f and E were continuous. However, discontinuous driving functions are not uncommon. For example, the impressed voltage on a circuit could be piecewise continuous and periodic, such as the "sawtooth" function shown on the left. Solving the differential equation of the circuit in this case is difficult using the techniques of Chapter 4. The Laplace transform studied in this chapter is an invaluable tool that simplifies the solution of problems such as these.

7.1 DEFINITION OF THE LAPLACE TRANSFORM

REVIEW MATERIAL

- Improper integrals with infinite limits of integration
- Integration by parts and partial fraction decomposition

INTRODUCTION In elementary calculus you learned that differentiation and integration are *transforms*; this means, roughly speaking, that these operations transform a function into another function. For example, the function $f(x) = x^2$ is transformed, in turn, into a linear function and a family of cubic polynomial functions by the operations of differentiation and integration:

$$\frac{d}{dx}x^2 = 2x \qquad \text{and} \qquad \int x^2\,dx = \frac{1}{3}x^3 + c.$$

Moreover, these two transforms possess the **linearity property** that the transform of a linear combination of functions is a linear combination of the transforms. For α and β constants

$$\frac{d}{dx}[\alpha f(x) + \beta g(x)] = \alpha f'(x) + \beta g'(x)$$

and

$$\int [\alpha f(x) + \beta g(x)]\,dx = \alpha \int f(x)\,dx + \beta \int g(x)\,dx$$

provided that each derivative and integral exists. In this section we will examine a special type of integral transform called the **Laplace transform.** In addition to possessing the linearity property the Laplace transform has many other interesting properties that make it very useful in solving linear initial-value problems.

Integral Transform If $f(x, y)$ is a function of two variables, then a definite integral of f with respect to one of the variables leads to a function of the other variable. For example, by holding y constant, we see that $\int_1^2 2xy^2\,dx = 3y^2$. Similarly, a definite integral such as $\int_a^b K(s, t)\,f(t)\,dt$ transforms a function f of the variable t into a function F of the variable s. We are particularly interested in an **integral transform,** where the interval of integration is the unbounded interval $[0, \infty)$. If $f(t)$ is defined for $t \geq 0$, then the improper integral $\int_0^\infty K(s, t)\,f(t)\,dt$ is defined as a limit:

$$\int_0^\infty K(s, t)\,f(t)\,dt = \lim_{b\to\infty} \int_0^b K(s, t)\,f(t)\,dt. \tag{1}$$

If the limit in (1) exists, then we say that the integral exists or is **convergent;** if the limit does not exist, the integral does not exist and is **divergent.** The limit in (1) will, in general, exist for only certain values of the variable s.

We will assume throughout that s is a real variable.

A Definition The function $K(s, t)$ in (1) is called the **kernel** of the transform. The choice $K(s, t) = e^{-st}$ as the kernel gives us an especially important integral transform.

DEFINITION 7.1.1 Laplace Transform

Let f be a function defined for $t \geq 0$. Then the integral

$$\mathscr{L}\{f(t)\} = \int_0^\infty e^{-st} f(t)\,dt \tag{2}$$

is said to be the **Laplace transform** of f, provided that the integral converges.

The Laplace transform is named in honor of the French mathematician and astronomer **Pierre-Simon Marquis de Laplace** (1749–1827).

When the defining integral (2) converges, the result is a function of s. In general discussion we shall use a lowercase letter to denote the function being transformed and the corresponding capital letter to denote its Laplace transform—for example,

$$\mathscr{L}\{f(t)\} = F(s), \qquad \mathscr{L}\{g(t)\} = G(s), \qquad \mathscr{L}\{y(t)\} = Y(s).$$

As the next four examples show, the domain of the function $F(s)$ depends on the function $f(t)$.

EXAMPLE 1 Applying Definition 7.1.1

Evaluate $\mathscr{L}\{1\}$.

SOLUTION From (2),

$$\mathscr{L}\{1\} = \int_0^\infty e^{-st}(1)\,dt = \lim_{b\to\infty}\int_0^b e^{-st}\,dt$$
$$= \lim_{b\to\infty}\frac{-e^{-st}}{s}\bigg|_0^b = \lim_{b\to\infty}\frac{-e^{-sb}+1}{s} = \frac{1}{s}$$

provided that $s > 0$. In other words, when $s > 0$, the exponent $-sb$ is negative, and $e^{-sb} \to 0$ as $b \to \infty$. The integral diverges for $s < 0$. ≡

The use of the limit sign becomes somewhat tedious, so we shall adopt the notation $|_0^\infty$ as a shorthand for writing $\lim_{b\to\infty}(\ )|_0^b$. For example,

$$\mathscr{L}\{1\} = \int_0^\infty e^{-st}(1)\,dt = \frac{-e^{-st}}{s}\bigg|_0^\infty = \frac{1}{s}, \qquad s > 0.$$

At the upper limit, it is understood that we mean $e^{-st} \to 0$ as $t \to \infty$ for $s > 0$.

EXAMPLE 2 Applying Definition 7.1.1

Evaluate $\mathscr{L}\{t\}$.

SOLUTION From Definition 7.1.1 we have $\mathscr{L}\{t\} = \int_0^\infty e^{-st}\,t\,dt$. Integrating by parts and using $\lim_{t\to\infty} te^{-st} = 0$, $s > 0$, along with the result from Example 1, we obtain

$$\mathscr{L}\{t\} = \frac{-te^{-st}}{s}\bigg|_0^\infty + \frac{1}{s}\int_0^\infty e^{-st}\,dt = \frac{1}{s}\mathscr{L}\{1\} = \frac{1}{s}\left(\frac{1}{s}\right) = \frac{1}{s^2}.$$ ≡

EXAMPLE 3 Applying Definition 7.1.1

Evaluate **(a)** $\mathscr{L}\{e^{-3t}\}$ **(b)** $\mathscr{L}\{e^{5t}\}$

SOLUTION In each case we use Definition 7.1.1.

(a)
$$\mathscr{L}\{e^{-3t}\} = \int_0^\infty e^{-3t}e^{-st}\,dt = \int_0^\infty e^{-(s+3)t}\,dt$$
$$= \frac{-e^{-(s+3)t}}{s+3}\bigg|_0^\infty$$
$$= \frac{1}{s+3}.$$

The last result is valid for $s > -3$ because in order to have $\lim_{t\to\infty} e^{-(s+3)t} = 0$ we must require that $s + 3 > 0$ or $s > -3$.

(b)
$$\mathscr{L}\{e^{5t}\} = \int_0^\infty e^{5t}e^{-st}\,dt = \int_0^\infty e^{-(s-5)t}\,dt$$
$$= \frac{-e^{-(s-5)t}}{s-5}\Bigg|_0^\infty$$
$$= \frac{1}{s-5}.$$

In contrast to part (a), this result is valid for $s > 5$ because $\lim_{t\to\infty} e^{-(s-5)t} = 0$ demands $s - 5 > 0$ or $s > 5$.

EXAMPLE 4 Applying Definition 7.1.1

Evaluate $\mathscr{L}\{\sin 2t\}$.

SOLUTION From Definition 7.1.1 and two applications of integration by parts we obtain

$$\mathscr{L}\{\sin 2t\} = \int_0^\infty e^{-st}\sin 2t\,dt = \frac{-e^{-st}\sin 2t}{s}\Bigg|_0^\infty + \frac{2}{s}\int_0^\infty e^{-st}\cos 2t\,dt$$
$$= \frac{2}{s}\int_0^\infty e^{-st}\cos 2t\,dt, \qquad s > 0$$

$\lim_{t\to\infty} e^{-st}\cos 2t = 0,\ s > 0$ ↓ … Laplace transform of $\sin 2t$ ↓

$$= \frac{2}{s}\left[\frac{-e^{-st}\cos 2t}{s}\Bigg|_0^\infty - \frac{2}{s}\int_0^\infty e^{-st}\sin 2t\,dt\right]$$
$$= \frac{2}{s^2} - \frac{4}{s^2}\mathscr{L}\{\sin 2t\}.$$

At this point we have an equation with $\mathscr{L}\{\sin 2t\}$ on both sides of the equality. Solving for that quantity yields the result

$$\mathscr{L}\{\sin 2t\} = \frac{2}{s^2+4}, \qquad s > 0.$$

$\mathscr{L}$ Is a Linear Transform For a linear combination of functions we can write

$$\int_0^\infty e^{-st}[\alpha f(t) + \beta g(t)]\,dt = \alpha\int_0^\infty e^{-st}f(t)\,dt + \beta\int_0^\infty e^{-st}g(t)\,dt$$

whenever both integrals converge for $s > c$. Hence it follows that

$$\mathscr{L}\{\alpha f(t) + \beta g(t)\} = \alpha\mathscr{L}\{f(t)\} + \beta\mathscr{L}\{g(t)\} = \alpha F(s) + \beta G(s). \qquad (3)$$

Because of the property given in (3), $\mathscr{L}$ is said to be a **linear transform.**

EXAMPLE 5 Linearity of the Laplace Transform

In this example we use the results of the preceding examples to illustrate the linearity of the Laplace transform.

(a) From Examples 1 and 2 we have for $s > 0$,

$$\mathscr{L}\{1 + 5t\} = \mathscr{L}\{1\} + 5\mathscr{L}\{t\} = \frac{1}{s} + \frac{5}{s^2}.$$

(b) From Examples 3 and 4 we have for $s > 5$,

$$\mathscr{L}\left\{4e^{5t} - 10\sin 2t\right\} = 4\mathscr{L}\left\{e^{5t}\right\} - 10\mathscr{L}\{\sin 2t\} = \frac{4}{s-5} - \frac{20}{s^2+4}.$$

(c) From Examples 1, 2, and 3 we have for $s > 0$,

$$\mathscr{L}\{20e^{-3t} + 7t - 9\} = 20\mathscr{L}\{e^{-3t}\} + 7\mathscr{L}\{t\} - 9\mathscr{L}\{1\}$$
$$= \frac{20}{s+3} + \frac{7}{s^2} - \frac{9}{s}.$$

We state the generalization of some of the preceding examples by means of the next theorem. From this point on we shall also refrain from stating any restrictions on s; it is understood that s is sufficiently restricted to guarantee the convergence of the appropriate Laplace transform.

THEOREM 7.1.1 Transforms of Some Basic Functions

(a) $\mathscr{L}\{1\} = \dfrac{1}{s}$

(b) $\mathscr{L}\{t^n\} = \dfrac{n!}{s^{n+1}}, \quad n = 1, 2, 3, \ldots$ **(c)** $\mathscr{L}\{e^{at}\} = \dfrac{1}{s-a}$

(d) $\mathscr{L}\{\sin kt\} = \dfrac{k}{s^2 + k^2}$ **(e)** $\mathscr{L}\{\cos kt\} = \dfrac{s}{s^2 + k^2}$

(f) $\mathscr{L}\{\sinh kt\} = \dfrac{k}{s^2 - k^2}$ **(g)** $\mathscr{L}\{\cosh kt\} = \dfrac{s}{s^2 - k^2}$

This result in (b) of Theorem 7.1.1 can be formally justified for n a positive integer using intergration by parts to first show that

$$\mathscr{L}\{t^n\} = \frac{n}{s}\,\mathscr{L}\{t^{n-1}\}.$$

Then for $n = 1, 2$, and 3, we have, respectively,

$$\mathscr{L}\{t\} = \frac{1}{s}\cdot\mathscr{L}\{1\} = \frac{1}{s}\cdot\frac{1}{s} = \frac{1}{s^2}$$

$$\mathscr{L}\{t^2\} = \frac{2}{s}\cdot\mathscr{L}\{t\} = \frac{2}{s}\cdot\frac{1}{s^2} = \frac{2\cdot 1}{s^3}$$

$$\mathscr{L}\{t^3\} = \frac{3}{s}\cdot\mathscr{L}\{t^2\} = \frac{3}{s}\cdot\frac{2\cdot 1}{s^3} = \frac{3\cdot 2\cdot 1}{s^4}$$

If we carry on in this manner, you should be convinced that

$$\mathscr{L}\{t^n\} = \frac{n\cdots 3\cdot 2\cdot 1}{s^{n+1}} = \frac{n!}{s^{n+1}}.$$

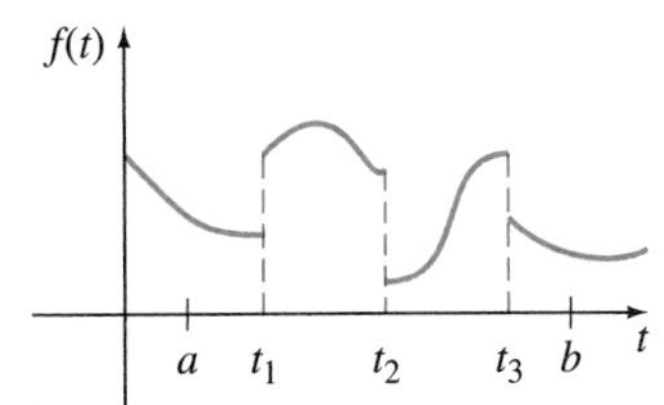

FIGURE 7.1.1 Piecewise continuous function

Sufficient Conditions for Existence of $\mathscr{L}\{f(t)\}$ The integral that defines the Laplace transform does not have to converge. For example, neither $\mathscr{L}\{1/t\}$ nor $\mathscr{L}\{e^{t^2}\}$ exists. Sufficient conditions guaranteeing the existence of $\mathscr{L}\{f(t)\}$ are that f be piecewise continuous on $[0, \infty)$ and that f be of exponential order for $t > T$. Recall that a function f is **piecewise continuous** on $[0, \infty)$ if, in any interval $0 \le a \le t \le b$, there are at most a finite number of points t_k, $k = 1, 2, \ldots, n$ $(t_{k-1} < t_k)$ at which f has finite discontinuities and is continuous on each open interval (t_{k-1}, t_k). See Figure 7.1.1. The concept of **exponential order** is defined in the following manner.

DEFINITION 7.1.2 Exponential Order

A function f is said to be of **exponential order** if there exist constants c, $M > 0$, and $T > 0$ such that $|f(t)| \le Me^{ct}$ for all $t > T$.

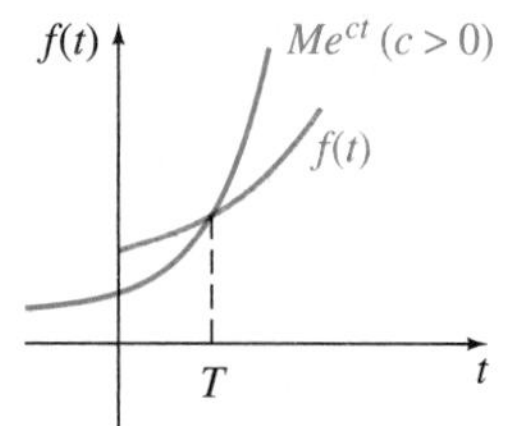

FIGURE 7.1.2 f is of exponential order

If f is an *increasing* function, then the condition $|f(t)| \leq Me^{ct}$, $t > T$, simply states that the graph of f on the interval (T, ∞) does not grow faster than the graph of the exponential function Me^{ct}, where c is a positive constant. See Figure 7.1.2. The functions $f(t) = t$, $f(t) = e^{-t}$, and $f(t) = 2\cos t$ are all of exponential order because for $c = 1$, $M = 1$, $T = 0$ we have, respectively, for $t > 0$

$$|t| \leq e^t, \qquad |e^{-t}| \leq e^t, \qquad \text{and} \qquad |2\cos t| \leq 2e^t.$$

A comparison of the graphs on the interval $[0, \infty)$ is given in Figure 7.1.3.

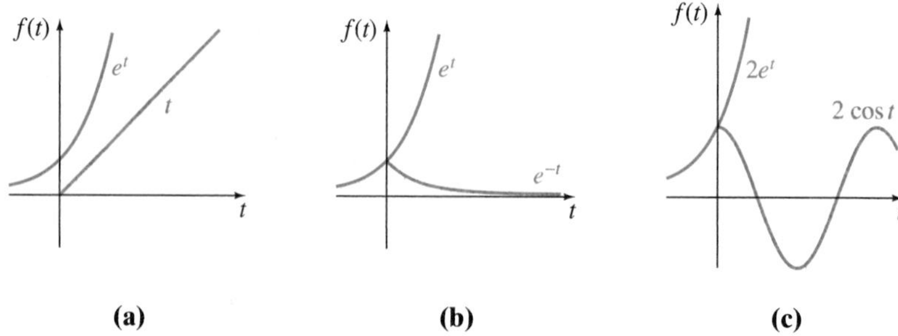

FIGURE 7.1.3 Three functions of exponential order

A positive integral power of t is always of exponential order, since, for $c > 0$,

$$|t^n| \leq Me^{ct} \quad \text{or} \quad \left|\frac{t^n}{e^{ct}}\right| \leq M \quad \text{for } t > T$$

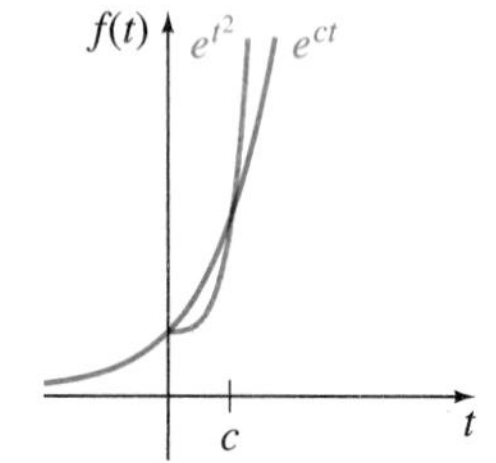

FIGURE 7.1.4 e^{t^2} is not of exponential order

is equivalent to showing that $\lim_{t\to\infty} t^n/e^{ct}$ is finite for $n = 1, 2, 3, \ldots$. The result follows from n applications of L'Hôpital's rule. A function such as $f(t) = e^{t^2}$ is not of exponential order since, as shown in Figure 7.1.4, e^{t^2} grows faster than any positive linear power of e for $t > c > 0$. This can also be seen from

$$\left|\frac{e^{t^2}}{e^{ct}}\right| = e^{t^2 - ct} = e^{t(t-c)} \longrightarrow \infty$$

as $t \to \infty$.

THEOREM 7.1.2 Sufficient Conditions for Existence

If f is piecewise continuous on $[0, \infty)$ and of exponential order, then $\mathscr{L}\{f(t)\}$ exists for $s > c$.

PROOF By the additive interval property of definite integrals we can write

$$\mathscr{L}\{f(t)\} = \int_0^T e^{-st} f(t)\,dt + \int_T^\infty e^{-st} f(t)\,dt = I_1 + I_2.$$

The integral I_1 exists because it can be written as a sum of integrals over intervals on which $e^{-st} f(t)$ is continuous. Now since f is of exponential order, there exist constants c, $M > 0$, $T > 0$ so that $|f(t)| \leq Me^{ct}$ for $t > T$. We can then write

$$|I_2| \leq \int_T^\infty |e^{-st} f(t)|\,dt \leq M \int_T^\infty e^{-st} e^{ct}\,dt = M \int_T^\infty e^{-(s-c)t}\,dt = M\frac{e^{-(s-c)T}}{s - c}$$

for $s > c$. Since $\int_T^\infty Me^{-(s-c)t}\,dt$ converges, the integral $\int_T^\infty |e^{-st} f(t)|\,dt$ converges by the comparison test for improper integrals. This, in turn, implies that I_2 exists for $s > c$. The existence of I_1 and I_2 implies that $\mathscr{L}\{f(t)\} = \int_0^\infty e^{-st}f(t)\,dt$ exists for $s > c$. ≡

EXAMPLE 6 Transform of a Piecewise Continuous Function

Evaluate $\mathscr{L}\{f(t)\}$ where $f(t) = \begin{cases} 0, & 0 \le t < 3 \\ 2, & t \ge 3. \end{cases}$

SOLUTION The function f, shown in Figure 7.1.5, is piecewise continuous and of exponential order for $t > 0$. Since f is defined in two pieces, $\mathscr{L}\{f(t)\}$ is expressed as the sum of two integrals:

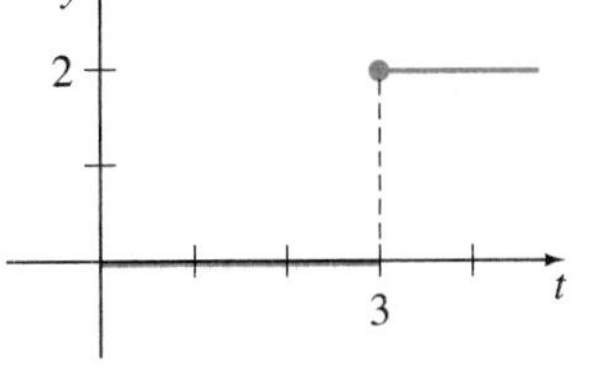

FIGURE 7.1.5 Piecewise continuous function in Example 6

$$\begin{aligned}\mathscr{L}\{f(t)\} = \int_0^\infty e^{-st}f(t)\,dt &= \int_0^3 e^{-st}(0)\,dt + \int_3^\infty e^{-st}(2)\,dt \\ &= 0 + \left.\frac{2e^{-st}}{-s}\right|_3^\infty \\ &= \frac{2e^{-3s}}{s}, \qquad s > 0.\end{aligned}$$

≡

We conclude this section with an additional bit of theory related to the types of functions of s that we will, generally, be working with. The next theorem indicates that not every arbitrary function of s is a Laplace transform of a piecewise continuous function of exponential order.

THEOREM 7.1.3 Behavior of $F(s)$ as $s \to \infty$

If f is piecewise continuous on $[0, \infty)$ and of exponential order and $F(s) = \mathscr{L}\{f(t)\}$, then $\lim_{s\to\infty} F(s) = 0$.

PROOF Since f is of exponential order, there exist constants γ, $M_1 > 0$, and $T > 0$ so that $|f(t)| \le M_1e^{\gamma t}$ for $t > T$. Also, since f is piecewise continuous for $0 \le t \le T$, it is necessarily bounded on the interval; that is, $|f(t)| \le M_2 = M_2e^{0t}$. If M denotes the maximum of the set $\{M_1, M_2\}$ and c denotes the maximum of $\{0, \gamma\}$, then

$$|F(s)| \le \int_0^\infty e^{-st}|f(t)|\,dt \le M\int_0^\infty e^{-st}e^{ct}\,dt = M\int_0^\infty e^{-(s-c)t}\,dt = \frac{M}{s-c}$$

for $s > c$. As $s \to \infty$, we have $|F(s)| \to 0$, and so $F(s) = \mathscr{L}\{f(t)\} \to 0$. ≡

REMARKS

(*i*) Throughout this chapter we shall be concerned primarily with functions that are both piecewise continuous and of exponential order. We note, however, that these two conditions are sufficient but not necessary for the existence of a Laplace transform. The function $f(t) = t^{-1/2}$ is not piecewise continuous on the interval $[0, \infty)$, but its Laplace transform exists. The function $f(t) = 2te^{t^2}\cos e^{t^2}$ is not of exponential order, but it can be shown that its Laplace transform exists. See Problems 43 and 54 in Exercises 7.1.

(*ii*) As a consequence of Theorem 7.1.3 we can say that functions of s such as $F_1(s) = 1$ and $F_2(s) = s/(s + 1)$ are not the Laplace transforms of piecewise continuous functions of exponential order, since $F_1(s) \not\to 0$ and $F_2(s) \not\to 0$ as $s \to \infty$. But you should not conclude from this that $F_1(s)$ and $F_2(s)$ are *not* Laplace transforms. There are other kinds of functions.

EXERCISES 7.1

Answers to selected odd-numbered problems begin on page ANS-11.

In Problems 1–18 use Definition 7.1.1 to find $\mathscr{L}\{f(t)\}$.

1. $f(t) = \begin{cases} -1, & 0 \le t < 1 \\ 1, & t \ge 1 \end{cases}$

2. $f(t) = \begin{cases} 4, & 0 \le t < 2 \\ 0, & t \ge 2 \end{cases}$

3. $f(t) = \begin{cases} t, & 0 \le t < 1 \\ 1, & t \ge 1 \end{cases}$

4. $f(t) = \begin{cases} 2t + 1, & 0 \le t < 1 \\ 0, & t \ge 1 \end{cases}$

5. $f(t) = \begin{cases} \sin t, & 0 \le t < \pi \\ 0, & t \ge \pi \end{cases}$

6. $f(t) = \begin{cases} 0, & 0 \le t < \pi/2 \\ \cos t, & t \ge \pi/2 \end{cases}$

7.

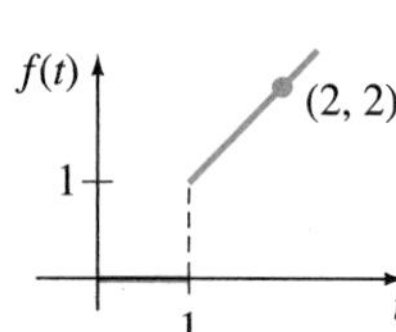

FIGURE 7.1.6 Graph for Problem 7

8.

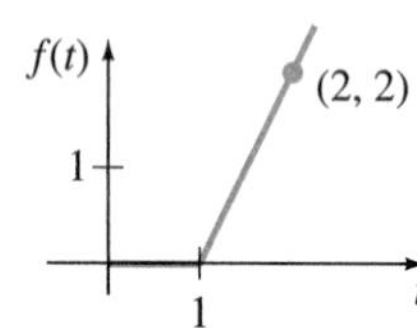

FIGURE 7.1.7 Graph for Problem 8

9.

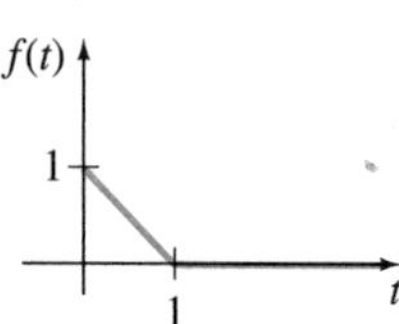

FIGURE 7.1.8 Graph for Problem 9

10.

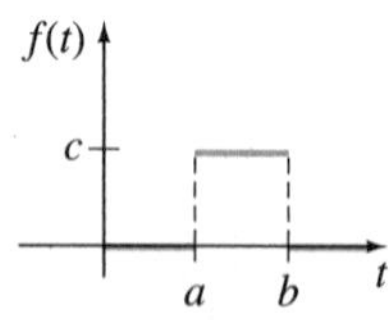

FIGURE 7.1.9 Graph for Problem 10

11. $f(t) = e^{t+7}$

12. $f(t) = e^{-2t-5}$

13. $f(t) = te^{4t}$

14. $f(t) = t^2e^{-2t}$

15. $f(t) = e^{-t} \sin t$

16. $f(t) = e^t \cos t$

17. $f(t) = t \cos t$

18. $f(t) = t \sin t$

In Problems 19–36 use Theorem 7.1.1 to find $\mathscr{L}\{f(t)\}$.

19. $f(t) = 2t^4$

20. $f(t) = t^5$

21. $f(t) = 4t - 10$

22. $f(t) = 7t + 3$

23. $f(t) = t^2 + 6t - 3$

24. $f(t) = -4t^2 + 16t + 9$

25. $f(t) = (t + 1)^3$

26. $f(t) = (2t - 1)^3$

27. $f(t) = 1 + e^{4t}$

28. $f(t) = t^2 - e^{-9t} + 5$

29. $f(t) = (1 + e^{2t})^2$

30. $f(t) = (e^t - e^{-t})^2$

31. $f(t) = 4t^2 - 5 \sin 3t$

32. $f(t) = \cos 5t + \sin 2t$

33. $f(t) = \sinh kt$

34. $f(t) = \cosh kt$

35. $f(t) = e^t \sinh t$

36. $f(t) = e^{-t} \cosh t$

In Problems 37–40 find $\mathscr{L}\{f(t)\}$ by first using a trigonometric identity.

37. $f(t) = \sin 2t \cos 2t$

38. $f(t) = \cos^2 t$

39. $f(t) = \sin(4t + 5)$

40. $f(t) = 10 \cos\left(t - \dfrac{\pi}{6}\right)$

41. We have encountered the **gamma function** $\Gamma(\alpha)$ in our study of Bessel functions in Section 6.4 (page 258). One definition of this function is given by the improper integral

$$\Gamma(\alpha) = \int_0^\infty t^{\alpha-1}e^{-t}\,dt, \quad \alpha > 0.$$

Use this definition to show that $\Gamma(\alpha + 1) = \alpha\Gamma(\alpha)$.

42. Use Problem 41 and a change of variables to obtain the generalization

$$\mathscr{L}\{t^\alpha\} = \frac{\Gamma(\alpha + 1)}{s^{\alpha+1}}, \quad \alpha > -1,$$

of the result in Theorem 7.1.1(b).

In Problems 43–46 use Problems 41 and 42 and the fact that $\Gamma(\frac{1}{2}) = \sqrt{\pi}$ to find the Laplace transform of the given function.

43. $f(t) = t^{-1/2}$

44. $f(t) = t^{1/2}$

45. $f(t) = t^{3/2}$

46. $f(t) = 2t^{1/2} + 8t^{5/2}$

Discussion Problems

47. Make up a function $F(t)$ that is of exponential order but where $f(t) = F'(t)$ is not of exponential order. Make up a function f that is not of exponential order but whose Laplace transform exists.

48. Suppose that $\mathscr{L}\{f_1(t)\} = F_1(s)$ for $s > c_1$ and that $\mathscr{L}\{f_2(t)\} = F_2(s)$ for $s > c_2$. When does

$$\mathscr{L}\{f_1(t) + f_2(t)\} = F_1(s) + F_2(s)?$$

49. Figure 7.1.4 suggests, but does not prove, that the function $f(t) = e^{t^2}$ is not of exponential order. How does the observation that $t^2 > \ln M + ct$, for $M > 0$ and t sufficiently large, show that $e^{t^2} > Me^{ct}$ for any c?

50. Use part (c) of Theorem 7.1.1 to show that $\mathscr{L}\{e^{(a+ib)t}\} = \dfrac{s - a + ib}{(s - a)^2 + b^2}$, where a and b are real and $i^2 = -1$. Show how Euler's formula (page 133) can then be used to deduce the results

$$\mathscr{L}\{e^{at}\cos bt\} = \frac{s-a}{(s-a)^2 + b^2}$$

$$\mathscr{L}\{e^{at}\sin bt\} = \frac{b}{(s-a)^2 + b^2}.$$

51. Under what conditions is a linear function $f(x) = mx + b$, $m \neq 0$, a linear transform?

52. Explain why the function

$$f(t) = \begin{cases} t, & 0 \le t < 2 \\ 4, & 2 < t < 5 \\ 1/(t-5), & t > 5 \end{cases}$$

is not piecewise continuous on $[0, \infty)$.

53. Show that the function $f(t) = 1/t^2$ does not possess a Laplace transform. [*Hint*: Write $\mathscr{L}\{1/t^2\}$ as two improper integrals:

$$\mathscr{L}\{1/t^2\} = \int_0^1 \frac{e^{-st}}{t^2}\,dt + \int_1^\infty \frac{e^{-st}}{t^2}\,dt = I_1 + I_2.$$

Show that I_1 diverges.]

54. Show that the Laplace transform $\mathscr{L}\{2te^{t^2}\cos e^{t^2}\}$ exists. [*Hint*: Start with integration by parts.]

55. If $\mathscr{L}\{f(t)\} = F(s)$ and $a > 0$ is a constant, show that

$$\mathscr{L}\{f(at)\} = \frac{1}{a}F\left(\frac{s}{a}\right).$$

This result is known as the **change of scale theorem.**

56. Use the given Laplace transform and the result in Problem 55 to find the indicated Laplace transform. Assume that a and k are positive constants.

(a) $\mathscr{L}\{e^t\} = \dfrac{1}{s-1}$; $\mathscr{L}\{e^{at}\}$

(b) $\mathscr{L}\{\sin t\} = \dfrac{1}{s^2+1}$; $\mathscr{L}\{\sin kt\}$

(c) $\mathscr{L}\{1 - \cos t\} = \dfrac{1}{s(s^2+1)}$; $\mathscr{L}\{1 - \cos kt\}$

(d) $\mathscr{L}\{\sin t \sinh t\} = \dfrac{2s}{s^4+4}$; $\mathscr{L}\{\sin kt \sinh kt\}$

7.2 INVERSE TRANSFORMS AND TRANSFORMS OF DERIVATIVES

REVIEW MATERIAL

- Partial fraction decomposition
- See the *Student Resource Manual*

INTRODUCTION In this section we take a few small steps into an investigation of how the Laplace transform can be used to solve certain types of equations for an unknown function. We begin the discussion with the concept of the inverse Laplace transform or, more precisely, the inverse of a Laplace transform $F(s)$. After some important preliminary background material on the Laplace transform of derivatives $f'(t), f''(t), \ldots$, we then illustrate how both the Laplace transform and the inverse Laplace transform come into play in solving some simple ordinary differential equations.

7.2.1 INVERSE TRANSFORMS

The Inverse Problem If $F(s)$ represents the Laplace transform of a function $f(t)$, that is, $\mathscr{L}\{f(t)\} = F(s)$, we then say $f(t)$ is the **inverse Laplace transform** of $F(s)$ and write $f(t) = \mathscr{L}^{-1}\{F(s)\}$. For example, from Examples 1, 2, and 3 of Section 7.1 we have, respectively,

Transform	Inverse Transform
$\mathscr{L}\{1\} = \dfrac{1}{s}$	$1 = \mathscr{L}^{-1}\left\{\dfrac{1}{s}\right\}$
$\mathscr{L}\{t\} = \dfrac{1}{s^2}$	$t = \mathscr{L}^{-1}\left\{\dfrac{1}{s^2}\right\}$
$\mathscr{L}\{e^{-3t}\} = \dfrac{1}{s+3}$	$e^{-3t} = \mathscr{L}^{-1}\left\{\dfrac{1}{s+3}\right\}$

We shall see shortly that in the application of the Laplace transform to equations we are not able to determine an unknown function $f(t)$ directly; rather, we are able to solve for the Laplace transform $F(s)$ of $f(t)$; but from that knowledge we ascertain f by computing $f(t) = \mathscr{L}^{-1}\{F(s)\}$. The idea is simply this: Suppose $F(s) = \dfrac{-2s + 6}{s^2 + 4}$ is a Laplace transform; find a function $f(t)$ such that $\mathscr{L}\{f(t)\} = F(s)$. We shall show how to solve this problem in Example 2.

For future reference the analogue of Theorem 7.1.1 for the inverse transform is presented as our next theorem.

THEOREM 7.2.1 Some Inverse Transforms

$$\textbf{(a)}\ 1 = \mathscr{L}^{-1}\left\{\frac{1}{s}\right\}$$

$$\textbf{(b)}\ t^n = \mathscr{L}^{-1}\left\{\frac{n!}{s^{n+1}}\right\}, \quad n = 1, 2, 3, \ldots \qquad \textbf{(c)}\ e^{at} = \mathscr{L}^{-1}\left\{\frac{1}{s - a}\right\}$$

$$\textbf{(d)}\ \sin kt = \mathscr{L}^{-1}\left\{\frac{k}{s^2 + k^2}\right\} \qquad \textbf{(e)}\ \cos kt = \mathscr{L}^{-1}\left\{\frac{s}{s^2 + k^2}\right\}$$

$$\textbf{(f)}\ \sinh kt = \mathscr{L}^{-1}\left\{\frac{k}{s^2 - k^2}\right\} \qquad \textbf{(g)}\ \cosh kt = \mathscr{L}^{-1}\left\{\frac{s}{s^2 - k^2}\right\}$$

In evaluating inverse transforms, it often happens that a function of s under consideration does not match *exactly* the form of a Laplace transform $F(s)$ given in a table. It may be necessary to "fix up" the function of s by multiplying and dividing by an appropriate constant.

EXAMPLE 1 Applying Theorem 7.2.1

Evaluate **(a)** $\mathscr{L}^{-1}\left\{\dfrac{1}{s^5}\right\}$ **(b)** $\mathscr{L}^{-1}\left\{\dfrac{1}{s^2 + 7}\right\}$.

SOLUTION **(a)** To match the form given in part (b) of Theorem 7.2.1, we identify $n + 1 = 5$ or $n = 4$ and then multiply and divide by 4!:

$$\mathscr{L}^{-1}\left\{\frac{1}{s^5}\right\} = \frac{1}{4!}\mathscr{L}^{-1}\left\{\frac{4!}{s^5}\right\} = \frac{1}{24}t^4.$$

(b) To match the form given in part (d) of Theorem 7.2.1, we identify $k^2 = 7$, so $k = \sqrt{7}$. We fix up the expression by multiplying and dividing by $\sqrt{7}$:

$$\mathscr{L}^{-1}\left\{\frac{1}{s^2 + 7}\right\} = \frac{1}{\sqrt{7}}\mathscr{L}^{-1}\left\{\frac{\sqrt{7}}{s^2 + 7}\right\} = \frac{1}{\sqrt{7}}\sin\sqrt{7}t.$$

≡

≡ $\mathscr{L}^{-1}$ is a Linear Transform

The inverse Laplace transform is also a linear transform; that is, for constants α and β

$$\mathscr{L}^{-1}\{\alpha F(s) + \beta G(s)\} = \alpha\mathscr{L}^{-1}\{F(s)\} + \beta\mathscr{L}^{-1}\{G(s)\}, \tag{1}$$

where F and G are the transforms of some functions f and g. Like (3) of Section 7.1, (1) extends to any finite linear combination of Laplace transforms.

EXAMPLE 2 Termwise Division and Linearity

Evaluate $\mathscr{L}^{-1}\left\{\dfrac{-2s+6}{s^2+4}\right\}$.

SOLUTION We first rewrite the given function of s as two expressions by means of termwise division and then use (1):

$$\mathscr{L}^{-1}\left\{\frac{-2s+6}{s^2+4}\right\} \overset{\text{termwise division}}{=} \mathscr{L}^{-1}\left\{\frac{-2s}{s^2+4}+\frac{6}{s^2+4}\right\} \overset{\text{linearity and fixing up constants}}{=} -2\,\mathscr{L}^{-1}\left\{\frac{s}{s^2+4}\right\}+\frac{6}{2}\mathscr{L}^{-1}\left\{\frac{2}{s^2+4}\right\} \tag{2}$$

$$= -2\cos 2t + 3\sin 2t. \qquad \leftarrow \text{parts (e) and (d) of Theorem 7.2.1 with } k=2$$

≡

≡ Partial Fractions Partial fractions play an important role in finding inverse Laplace transforms. The decomposition of a rational expression into component fractions can be done quickly by means of a single command on most computer algebra systems. Indeed, some CASs have packages that implement Laplace transform and inverse Laplace transform commands. But for those of you without access to such software, we will review in this and subsequent sections some of the basic algebra in the important cases in which the denominator of a Laplace transform $F(s)$ contains distinct linear factors, repeated linear factors, and quadratic polynomials with no real factors. Although we shall examine each of these cases as this chapter develops, it still might be a good idea for you to consult either a calculus text or a current precalculus text for a more comprehensive review of this theory.

The following example illustrates partial fraction decomposition in the case when the denominator of $F(s)$ is factorable into *distinct linear factors.*

EXAMPLE 3 Partial Fractions: Distinct Linear Factors

Evaluate $\mathscr{L}^{-1}\left\{\dfrac{s^2+6s+9}{(s-1)(s-2)(s+4)}\right\}$.

SOLUTION There exist unique real constants A, B, and C so that

$$\frac{s^2+6s+9}{(s-1)(s-2)(s+4)} = \frac{A}{s-1}+\frac{B}{s-2}+\frac{C}{s+4}$$

$$= \frac{A(s-2)(s+4)+B(s-1)(s+4)+C(s-1)(s-2)}{(s-1)(s-2)(s+4)}.$$

Since the denominators are identical, the numerators are identical:

$$s^2+6s+9 = A(s-2)(s+4)+B(s-1)(s+4)+C(s-1)(s-2). \tag{3}$$

By comparing coefficients of powers of s on both sides of the equality, we know that (3) is equivalent to a system of three equations in the three unknowns A, B, and C. However, there is a shortcut for determining these unknowns. If we set $s = 1$, $s = 2$, and $s = -4$ in (3), we obtain, respectively,

$$16 = A(-1)(5), \qquad 25 = B(1)(6), \qquad \text{and} \qquad 1 = C(-5)(-6),$$

and so $A = -\frac{16}{5}$, $B = \frac{25}{6}$, and $C = \frac{1}{30}$. Hence the partial fraction decomposition is

$$\frac{s^2+6s+9}{(s-1)(s-2)(s+4)} = -\frac{16/5}{s-1}+\frac{25/6}{s-2}+\frac{1/30}{s+4}, \tag{4}$$

and thus, from the linearity of $\mathscr{L}^{-1}$ and part (c) of Theorem 7.2.1,

$$\mathscr{L}^{-1}\left\{\frac{s^2+6s+9}{(s-1)(s-2)(s+4)}\right\} = -\frac{16}{5}\mathscr{L}^{-1}\left\{\frac{1}{s-1}\right\} + \frac{25}{6}\mathscr{L}^{-1}\left\{\frac{1}{s-2}\right\} + \frac{1}{30}\mathscr{L}^{-1}\left\{\frac{1}{s+4}\right\}$$

$$= -\frac{16}{5}e^t + \frac{25}{6}e^{2t} + \frac{1}{30}e^{-4t}. \qquad (5)$$

7.2.2 TRANSFORMS OF DERIVATIVES

Transform a Derivative As was pointed out in the introduction to this chapter, our immediate goal is to use the Laplace transform to solve differential equations. To that end we need to evaluate quantities such as $\mathscr{L}\{dy/dt\}$ and $\mathscr{L}\{d^2y/dt^2\}$. For example, if f' is continuous for $t \geq 0$, then integration by parts gives

$$\mathscr{L}\{f'(t)\} = \int_0^\infty e^{-st}f'(t)\,dt = e^{-st}f(t)\Big|_0^\infty + s\int_0^\infty e^{-st}f(t)\,dt$$

$$= -f(0) + s\mathscr{L}\{f(t)\}$$

or

$$\mathscr{L}\{f'(t)\} = sF(s) - f(0). \qquad (6)$$

Here we have assumed that $e^{-st}f(t) \to 0$ as $t \to \infty$. Similarly, with the aid of (6),

$$\mathscr{L}\{f''(t)\} = \int_0^\infty e^{-st}f''(t)\,dt = e^{-st}f'(t)\Big|_0^\infty + s\int_0^\infty e^{-st}f'(t)\,dt$$

$$= -f'(0) + s\mathscr{L}\{f'(t)\}$$

$$= s[sF(s) - f(0)] - f'(0) \qquad \leftarrow \text{from (6)}$$

or

$$\mathscr{L}\{f''(t)\} = s^2F(s) - sf(0) - f'(0). \qquad (7)$$

In like manner it can be shown that

$$\mathscr{L}\{f'''(t)\} = s^3F(s) - s^2f(0) - sf'(0) - f''(0). \qquad (8)$$

The recursive nature of the Laplace transform of the derivatives of a function f should be apparent from the results in (6), (7), and (8). The next theorem gives the Laplace transform of the nth derivative of f. The proof is omitted.

THEOREM 7.2.2 Transform of a Derivative

If $f, f', \ldots, f^{(n-1)}$ are continuous on $[0, \infty)$ and are of exponential order and if $f^{(n)}(t)$ is piecewise continuous on $[0, \infty)$, then

$$\mathscr{L}\{f^{(n)}(t)\} = s^nF(s) - s^{n-1}f(0) - s^{n-2}f'(0) - \cdots - f^{(n-1)}(0),$$

where $F(s) = \mathscr{L}\{f(t)\}$.

Solving Linear ODEs It is apparent from the general result given in Theorem 7.2.2 that $\mathscr{L}\{d^ny/dt^n\}$ depends on $Y(s) = \mathscr{L}\{y(t)\}$ and the $n - 1$ derivatives of $y(t)$ evaluated at $t = 0$. This property makes the Laplace transform ideally suited for solving linear initial-value problems in which the differential equation has *constant coefficients*. Such a differential equation is simply a linear combination of terms $y, y', y'', \ldots, y^{(n)}$:

$$a_n\frac{d^ny}{dt^n} + a_{n-1}\frac{d^{n-1}y}{dt^{n-1}} + \cdots + a_0y = g(t),$$

$$y(0) = y_0, y'(0) = y_1, \ldots, y^{(n-1)}(0) = y_{n-1},$$

where the a_i, $i = 0, 1, \ldots, n$ and $y_0, y_1, \ldots, y_{n-1}$ are constants. By the linearity property the Laplace transform of this linear combination is a linear combination of Laplace transforms:

$$a_n \mathscr{L}\left\{\frac{d^n y}{dt^n}\right\} + a_{n-1}\mathscr{L}\left\{\frac{d^{n-1}y}{dt^{n-1}}\right\} + \cdots + a_0 \mathscr{L}\{y\} = \mathscr{L}\{g(t)\}. \tag{9}$$

From Theorem 7.2.2, (9) becomes

$$\begin{aligned} &a_n[s^n Y(s) - s^{n-1}y(0) - \cdots - y^{(n-1)}(0)] \\ &\quad + a_{n-1}[s^{n-1}Y(s) - s^{n-2}y(0) - \cdots - y^{(n-2)}(0)] + \cdots + a_0 Y(s) = G(s), \end{aligned} \tag{10}$$

where $\mathscr{L}\{y(t)\} = Y(s)$ and $\mathscr{L}\{g(t)\} = G(s)$. In other words,

The Laplace transform of a linear differential equation with constant coefficients becomes an algebraic equation in $Y(s)$.

If we solve the general transformed equation (10) for the symbol $Y(s)$, we first obtain $P(s)Y(s) = Q(s) + G(s)$ and then write

$$Y(s) = \frac{Q(s)}{P(s)} + \frac{G(s)}{P(s)}, \tag{11}$$

where $P(s) = a_n s^n + a_{n-1}s^{n-1} + \cdots + a_0$, $Q(s)$ is a polynomial in s of degree less than or equal to $n - 1$ consisting of the various products of the coefficients a_i, $i = 1, \ldots, n$ and the prescribed initial conditions $y_0, y_1, \ldots, y_{n-1}$, and $G(s)$ is the Laplace transform of $g(t)$.* Typically, we put the two terms in (11) over the least common denominator and then decompose the expression into two or more partial fractions. Finally, the solution $y(t)$ of the original initial-value problem is $y(t) = \mathscr{L}^{-1}\{Y(s)\}$, where the inverse transform is done term by term.

The procedure is summarized in the diagram in Figure 7.2.1.

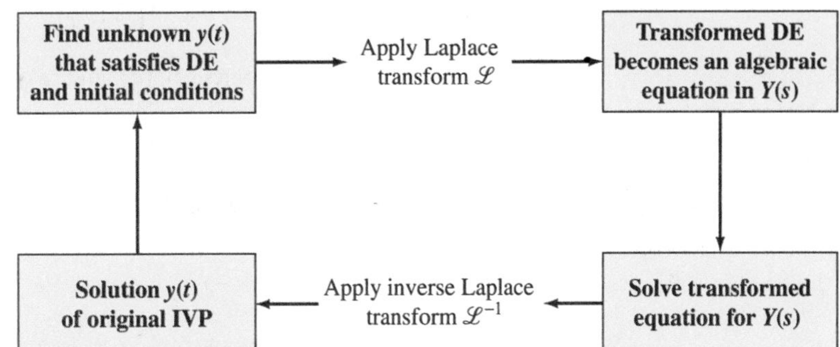

FIGURE 7.2.1 Steps in solving an IVP by the Laplace transform

The next example illustrates the foregoing method of solving DEs, as well as partial fraction decomposition in the case when the denominator of $Y(s)$ contains a *quadratic polynomial with no real factors.*

EXAMPLE 4 Solving a First-Order IVP

Use the Laplace transform to solve the initial-value problem

$$\frac{dy}{dt} + 3y = 13 \sin 2t, \quad y(0) = 6.$$

SOLUTION We first take the transform of each member of the differential equation:

$$\mathscr{L}\left\{\frac{dy}{dt}\right\} + 3\mathscr{L}\{y\} = 13\mathscr{L}\{\sin 2t\}. \tag{12}$$

*The polynomial $P(s)$ is the same as the nth-degree auxiliary polynomial in (12) in Section 4.3 with the usual symbol m replaced by s.

From (6), $\mathscr{L}\{dy/dt\} = sY(s) - y(0) = sY(s) - 6$, and from part (d) of Theorem 7.1.1, $\mathscr{L}\{\sin 2t\} = 2/(s^2 + 4)$, so (12) is the same as

$$sY(s) - 6 + 3Y(s) = \frac{26}{s^2 + 4} \quad \text{or} \quad (s + 3)Y(s) = 6 + \frac{26}{s^2 + 4}.$$

Solving the last equation for $Y(s)$, we get

$$Y(s) = \frac{6}{s + 3} + \frac{26}{(s + 3)(s^2 + 4)} = \frac{6s^2 + 50}{(s + 3)(s^2 + 4)}. \tag{13}$$

Since the quadratic polynomial $s^2 + 4$ does not factor using real numbers, its assumed numerator in the partial fraction decomposition is a linear polynomial in s:

$$\frac{6s^2 + 50}{(s + 3)(s^2 + 4)} = \frac{A}{s + 3} + \frac{Bs + C}{s^2 + 4}.$$

Putting the right-hand side of the equality over a common denominator and equating numerators gives $6s^2 + 50 = A(s^2 + 4) + (Bs + C)(s + 3)$. Setting $s = -3$ then immediately yields $A = 8$. Since the denominator has no more real zeros, we equate the coefficients of s^2 and s: $6 = A + B$ and $0 = 3B + C$. Using the value of A in the first equation gives $B = -2$, and then using this last value in the second equation gives $C = 6$. Thus

$$Y(s) = \frac{6s^2 + 50}{(s + 3)(s^2 + 4)} = \frac{8}{s + 3} + \frac{-2s + 6}{s^2 + 4}.$$

We are not quite finished because the last rational expression still has to be written as two fractions. This was done by termwise division in Example 2. From (2) of that example,

$$y(t) = 8\mathscr{L}^{-1}\left\{\frac{1}{s + 3}\right\} - 2\mathscr{L}^{-1}\left\{\frac{s}{s^2 + 4}\right\} + 3\mathscr{L}^{-1}\left\{\frac{2}{s^2 + 4}\right\}.$$

It follows from parts (c), (d), and (e) of Theorem 7.2.1 that the solution of the initial-value problem is $y(t) = 8e^{-3t} - 2\cos 2t + 3\sin 2t$. ≡

EXAMPLE 5 Solving a Second-Order IVP

Solve $y'' - 3y' + 2y = e^{-4t}, \quad y(0) = 1, \quad y'(0) = 5$.

SOLUTION Proceeding as in Example 4, we transform the DE. We take the sum of the transforms of each term, use (6) and (7), use the given initial conditions, use (c) of Theorem 7.1.1, and then solve for $Y(s)$:

$$\mathscr{L}\left\{\frac{d^2y}{dt^2}\right\} - 3\mathscr{L}\left\{\frac{dy}{dt}\right\} + 2\mathscr{L}\{y\} = \mathscr{L}\{e^{-4t}\}$$

$$s^2Y(s) - sy(0) - y'(0) - 3[sY(s) - y(0)] + 2Y(s) = \frac{1}{s + 4}$$

$$(s^2 - 3s + 2)Y(s) = s + 2 + \frac{1}{s + 4}$$

$$Y(s) = \frac{s + 2}{s^2 - 3s + 2} + \frac{1}{(s^2 - 3s + 2)(s + 4)} = \frac{s^2 + 6s + 9}{(s - 1)(s - 2)(s + 4)}. \tag{14}$$

The details of the partial fraction decomposition of $Y(s)$ have already been carried out in Example 3. In view of the results in (4) and (5) we have the solution of the initial-value problem

$$y(t) = \mathscr{L}^{-1}\{Y(s)\} = -\frac{16}{5}e^{t} + \frac{25}{6}e^{2t} + \frac{1}{30}e^{-4t}.$$ ≡

Examples 4 and 5 illustrate the basic procedure for using the Laplace transform to solve a linear initial-value problem, but these examples may appear to demonstrate a method that is not much better than the approach to such problems outlined in Sections 2.3 and 4.3–4.6. Don't draw any negative conclusions from only two examples. Yes, there is a lot of algebra inherent in the use of the Laplace transform, *but* observe that we do not have to use variation of parameters or worry about the cases and algebra in the method of undetermined coefficients. Moreover, since the method incorporates the prescribed initial conditions directly into the solution, there is no need for the separate operation of applying the initial conditions to the general solution $y = c_1y_1 + c_2y_2 + \cdots + c_ny_n + y_p$ of the DE to find specific constants in a particular solution of the IVP.

The Laplace transform has many operational properties. In the sections that follow we will examine some of these properties and see how they enable us to solve problems of greater complexity.

REMARKS

(*i*) The inverse Laplace transform of a function $F(s)$ may not be unique; in other words, it is possible that $\mathscr{L}\{f_1(t)\} = \mathscr{L}\{f_2(t)\}$ and yet $f_1 \neq f_2$. For our purposes this is not anything to be concerned about. If f_1 and f_2 are piecewise continuous on $[0, \infty)$ and of exponential order, then f_1 and f_2 are *essentially* the same. See Problem 44 in Exercises 7.2. However, if f_1 and f_2 are continuous on $[0, \infty)$ and $\mathscr{L}\{f_1(t)\} = \mathscr{L}\{f_2(t)\}$, then $f_1 = f_2$ on the interval.

(*ii*) This remark is for those of you who will be required to do partial fraction decompositions by hand. There is another way of determining the coefficients in a partial fraction decomposition in the special case when $\mathscr{L}\{f(t)\} = F(s)$ is a rational function of s and the denominator of F is a product of *distinct* linear factors. Let us illustrate by reexamining Example 3. Suppose we multiply both sides of the assumed decomposition

$$\frac{s^2 + 6s + 9}{(s-1)(s-2)(s+4)} = \frac{A}{s-1} + \frac{B}{s-2} + \frac{C}{s+4} \tag{15}$$

by, say, $s - 1$, simplify, and then set $s = 1$. Since the coefficients of B and C on the right-hand side of the equality are zero, we get

$$\left.\frac{s^2 + 6s + 9}{(s-2)(s+4)}\right|_{s=1} = A \quad \text{or} \quad A = -\frac{16}{5}.$$

Written another way,

$$\left.\frac{s^2 + 6s + 9}{\boxed{(s-1)}(s-2)(s+4)}\right|_{s=1} = -\frac{16}{5} = A,$$

where we have shaded, or *covered up*, the factor that canceled when the left-hand side was multiplied by $s - 1$. Now to obtain B and C, we simply evaluate the left-hand side of (15) while covering up, in turn, $s - 2$ and $s + 4$:

$$\left.\frac{s^2 + 6s + 9}{(s-1)\boxed{(s-2)}(s+4)}\right|_{s=2} = \frac{25}{6} = B$$

and
$$\left.\frac{s^2 + 6s + 9}{(s-1)(s-2)\boxed{(s+4)}}\right|_{s=-4} = \frac{1}{30} = C.$$

The desired decomposition (15) is given in (4). This special technique for determining coefficients is naturally known as the **cover-up method.**

(*iii*) In this remark we continue our introduction to the terminology of dynamical systems. Because of (9) and (10) the Laplace transform is well adapted to *linear* dynamical systems. The polynomial $P(s) = a_n s^n + a_{n-1}s^{n-1} + \cdots + a_0$ in (11) is the total coefficient of $Y(s)$ in (10) and is simply the left-hand side of the DE with the derivatives $d^k y/dt^k$ replaced by powers s^k, $k = 0, 1, \ldots, n$. It is usual practice to call the reciprocal of $P(s)$—namely, $W(s) = 1/P(s)$—the **transfer function** of the system and write (11) as

$$Y(s) = W(s)Q(s) + W(s)G(s). \tag{16}$$

In this manner we have separated, in an additive sense, the effects on the response that are due to the initial conditions (that is, $W(s)Q(s)$) from those due to the input function g (that is, $W(s)G(s)$). See (13) and (14). Hence the response $y(t)$ of the system is a superposition of two responses:

$$y(t) = \mathscr{L}^{-1}\{W(s)Q(s)\} + \mathscr{L}^{-1}\{W(s)G(s)\} = y_0(t) + y_1(t).$$

If the input is $g(t) = 0$, then the solution of the problem is $y_0(t) = \mathscr{L}^{-1}\{W(s)Q(s)\}$. This solution is called the **zero-input response** of the system. On the other hand, the function $y_1(t) = \mathscr{L}^{-1}\{W(s)G(s)\}$ is the output due to the input $g(t)$. Now if the initial state of the system is the zero state (all the initial conditions are zero), then $Q(s) = 0$, and so the only solution of the initial-value problem is $y_1(t)$. The latter solution is called the **zero-state response** of the system. Both $y_0(t)$ and $y_1(t)$ are particular solutions: $y_0(t)$ is a solution of the IVP consisting of the associated homogeneous equation with the given initial conditions, and $y_1(t)$ is a solution of the IVP consisting of the nonhomogeneous equation with zero initial conditions. In Example 5 we see from (14) that the transfer function is $W(s) = 1/(s^2 - 3s + 2)$, the zero-input response is

$$y_0(t) = \mathscr{L}^{-1}\left\{\frac{s+2}{(s-1)(s-2)}\right\} = -3e^t + 4e^{2t},$$

and the zero-state response is

$$y_1(t) = \mathscr{L}^{-1}\left\{\frac{1}{(s-1)(s-2)(s+4)}\right\} = -\frac{1}{5}e^t + \frac{1}{6}e^{2t} + \frac{1}{30}e^{-4t}.$$

Verify that the sum of $y_0(t)$ and $y_1(t)$ is the solution $y(t)$ in Example 5 and that $y_0(0) = 1$, $y_0'(0) = 5$, whereas $y_1(0) = 0$, $y_1'(0) = 0$.

EXERCISES 7.2

Answers to selected odd-numbered problems begin on page ANS-11.

7.2.1 INVERSE TRANSFORMS

In Problems 1–30 use appropriate algebra and Theorem 7.2.1 to find the given inverse Laplace transform.

1. $\mathscr{L}^{-1}\left\{\frac{1}{s^3}\right\}$

2. $\mathscr{L}^{-1}\left\{\frac{1}{s^4}\right\}$

3. $\mathscr{L}^{-1}\left\{\frac{1}{s^2} - \frac{48}{s^5}\right\}$

4. $\mathscr{L}^{-1}\left\{\left(\frac{2}{s} - \frac{1}{s^3}\right)^2\right\}$

5. $\mathscr{L}^{-1}\left\{\frac{(s+1)^3}{s^4}\right\}$

6. $\mathscr{L}^{-1}\left\{\frac{(s+2)^2}{s^3}\right\}$

7. $\mathscr{L}^{-1}\left\{\frac{1}{s^2} - \frac{1}{s} + \frac{1}{s-2}\right\}$

8. $\mathscr{L}^{-1}\left\{\frac{4}{s} + \frac{6}{s^5} - \frac{1}{s+8}\right\}$

9. $\mathscr{L}^{-1}\left\{\frac{1}{4s+1}\right\}$

10. $\mathscr{L}^{-1}\left\{\frac{1}{5s-2}\right\}$

11. $\mathscr{L}^{-1}\left\{\frac{5}{s^2+49}\right\}$

12. $\mathscr{L}^{-1}\left\{\frac{10s}{s^2+16}\right\}$

13. $\mathscr{L}^{-1}\left\{\frac{4s}{4s^2+1}\right\}$

14. $\mathscr{L}^{-1}\left\{\frac{1}{4s^2+1}\right\}$

15. $\mathscr{L}^{-1}\left\{\frac{2s-6}{s^2+9}\right\}$

16. $\mathscr{L}^{-1}\left\{\frac{s+1}{s^2+2}\right\}$

17. $\mathscr{L}^{-1}\left\{\frac{1}{s^2 + 3s}\right\}$

18. $\mathscr{L}^{-1}\left\{\frac{s+1}{s^2 - 4s}\right\}$

19. $\mathscr{L}^{-1}\left\{\frac{s}{s^2 + 2s - 3}\right\}$

20. $\mathscr{L}^{-1}\left\{\frac{1}{s^2 + s - 20}\right\}$

21. $\mathscr{L}^{-1}\left\{\frac{0.9s}{(s - 0.1)(s + 0.2)}\right\}$

22. $\mathscr{L}^{-1}\left\{\frac{s - 3}{(s - \sqrt{3})(s + \sqrt{3})}\right\}$

23. $\mathscr{L}^{-1}\left\{\frac{s}{(s - 2)(s - 3)(s - 6)}\right\}$

24. $\mathscr{L}^{-1}\left\{\frac{s^2 + 1}{s(s - 1)(s + 1)(s - 2)}\right\}$

25. $\mathscr{L}^{-1}\left\{\frac{1}{s^3 + 5s}\right\}$

26. $\mathscr{L}^{-1}\left\{\frac{s}{(s + 2)(s^2 + 4)}\right\}$

27. $\mathscr{L}^{-1}\left\{\frac{2s - 4}{(s^2 + s)(s^2 + 1)}\right\}$

28. $\mathscr{L}^{-1}\left\{\frac{1}{s^4 - 9}\right\}$

29. $\mathscr{L}^{-1}\left\{\frac{1}{(s^2 + 1)(s^2 + 4)}\right\}$

30. $\mathscr{L}^{-1}\left\{\frac{6s + 3}{s^4 + 5s^2 + 4}\right\}$

7.2.2 TRANSFORMS OF DERIVATIVES

In Problems 31–40 use the Laplace transform to solve the given initial-value problem.

31. $\frac{dy}{dt} - y = 1, \quad y(0) = 0$

32. $2\frac{dy}{dt} + y = 0, \quad y(0) = -3$

33. $y' + 6y = e^{4t}, \quad y(0) = 2$

34. $y' - y = 2\cos 5t, \quad y(0) = 0$

35. $y'' + 5y' + 4y = 0, \quad y(0) = 1, \quad y'(0) = 0$

36. $y'' - 4y' = 6e^{3t} - 3e^{-t}, \quad y(0) = 1, \quad y'(0) = -1$

37. $y'' + y = \sqrt{2}\sin\sqrt{2}t, \quad y(0) = 10, \quad y'(0) = 0$

38. $y'' + 9y = e^t, \quad y(0) = 0, \quad y'(0) = 0$

39. $2y''' + 3y'' - 3y' - 2y = e^{-t}, \quad y(0) = 0, \quad y'(0) = 0, \quad y''(0) = 1$

40. $y''' + 2y'' - y' - 2y = \sin 3t, \quad y(0) = 0, \quad y'(0) = 0, \quad y''(0) = 1$

The inverse forms of the results in Problem 50 in Exercises 7.1 are

$$\mathscr{L}^{-1}\left\{\frac{s - a}{(s - a)^2 + b^2}\right\} = e^{at}\cos bt$$

$$\mathscr{L}^{-1}\left\{\frac{b}{(s - a)^2 + b^2}\right\} = e^{at}\sin bt.$$

In Problems 41 and 42 use the Laplace transform and these inverses to solve the given initial-value problem.

41. $y' + y = e^{-3t}\cos 2t, \quad y(0) = 0$

42. $y'' - 2y' + 5y = 0, \quad y(0) = 1, \quad y'(0) = 3$

Discussion Problems

43. **(a)** With a slight change in notation the transform in (6) is the same as
$$\mathscr{L}\{f'(t)\} = s\mathscr{L}\{f(t)\} - f(0).$$
With $f(t) = te^{at}$, discuss how this result in conjunction with (c) of Theorem 7.1.1 can be used to evaluate $\mathscr{L}\{te^{at}\}$.

    (b) Proceed as in part (a), but this time discuss how to use (7) with $f(t) = t\sin kt$ in conjunction with (d) and (e) of Theorem 7.1.1 to evaluate $\mathscr{L}\{t\sin kt\}$.

44. Make up two functions f_1 and f_2 that have the same Laplace transform. Do not think profound thoughts.

45. Reread (*iii*) in the *Remarks* on page 288. Find the zero-input and the zero-state response for the IVP in Problem 36.

46. Suppose $f(t)$ is a function for which $f'(t)$ is piecewise continuous and of exponential order c. Use results in this section and Section 7.1 to justify
$$f(0) = \lim_{s\to\infty} sF(s),$$
where $F(s) = \mathscr{L}\{f(t)\}$. Verify this result with $f(t) = \cos kt$.

7.3 OPERATIONAL PROPERTIES I

REVIEW MATERIAL

- Keep practicing partial fraction decomposition
- Completion of the square

INTRODUCTION It is not convenient to use Definition 7.1.1 each time we wish to find the Laplace transform of a function $f(t)$. For example, the integration by parts involved in evaluating, say, $\mathscr{L}\{e^t t^2 \sin 3t\}$ is formidable, to say the least. In this section and the next we present several labor-saving operational properties of the Laplace transform that enable us to build up a more extensive list of transforms (see the table in Appendix III) without having to resort to the basic definition and integration.

7.3.1 TRANSLATION ON THE s-AXIS

A Translation Evaluating transforms such as $\mathscr{L}\{e^{5t}t^3\}$ and $\mathscr{L}\{e^{-2t}\cos 4t\}$ is straightforward provided that we know (and we do) $\mathscr{L}\{t^3\}$ and $\mathscr{L}\{\cos 4t\}$. In general, if we know the Laplace transform of a function f, $\mathscr{L}\{f(t)\} = F(s)$, it is possible to compute the Laplace transform of an exponential multiple of f, that is, $\mathscr{L}\{e^{at}f(t)\}$, with no additional effort other than *translating*, or *shifting*, the transform $F(s)$ to $F(s - a)$. This result is known as the **first translation theorem** or **first shifting theorem.**

THEOREM 7.3.1 First Translation Theorem

If $\mathscr{L}\{f(t)\} = F(s)$ and a is any real number, then

$$\mathscr{L}\{e^{at}f(t)\} = F(s - a).$$

PROOF The proof is immediate, since by Definition 7.1.1

$$\mathscr{L}\{e^{at}f(t)\} = \int_0^\infty e^{-st}e^{at}f(t)\,dt = \int_0^\infty e^{-(s-a)t}f(t)\,dt = F(s - a).$$

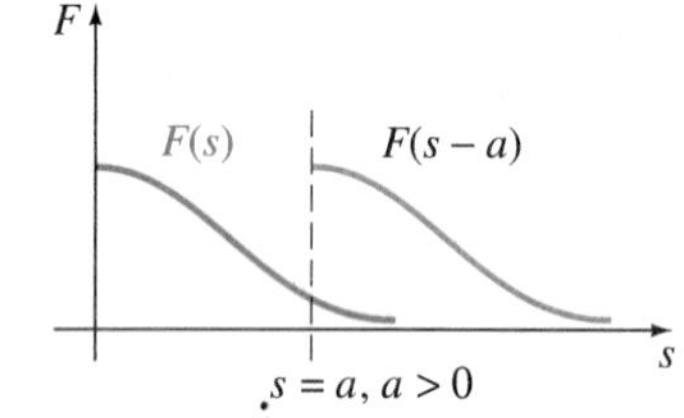

FIGURE 7.3.1 Shift on s-axis

If we consider s a real variable, then the graph of $F(s - a)$ is the graph of $F(s)$ shifted on the s-axis by the amount $|a|$. If $a > 0$, the graph of $F(s)$ is shifted a units to the right, whereas if $a < 0$, the graph is shifted $|a|$ units to the left. See Figure 7.3.1.

For emphasis it is sometimes useful to use the symbolism

$$\mathscr{L}\{e^{at}f(t)\} = \mathscr{L}\{f(t)\}|_{s \to s-a},$$

where $s \to s - a$ means that in the Laplace transform $F(s)$ of $f(t)$ we replace the symbol s wherever it appears by $s - a$.

EXAMPLE 1 **Using the First Translation Theorem**

Evaluate **(a)** $\mathscr{L}\{e^{5t}t^3\}$ **(b)** $\mathscr{L}\{e^{-2t}\cos 4t\}$.

SOLUTION The results follow from Theorems 7.1.1 and 7.3.1.

$$\text{(a)}\ \mathscr{L}\{e^{5t}t^3\} = \mathscr{L}\{t^3\}|_{s \to s-5} = \frac{3!}{s^4}\bigg|_{s \to s-5} = \frac{6}{(s-5)^4}$$

$$\text{(b)}\ \mathscr{L}\{e^{-2t}\cos 4t\} = \mathscr{L}\{\cos 4t\}|_{s \to s-(-2)} = \frac{s}{s^2 + 16}\bigg|_{s \to s+2} = \frac{s + 2}{(s + 2)^2 + 16}$$

Inverse Form of Theorem 7.3.1 To compute the inverse of $F(s - a)$, we must recognize $F(s)$, find $f(t)$ by taking the inverse Laplace transform of $F(s)$, and then multiply $f(t)$ by the exponential function e^{at}. This procedure can be summarized symbolically in the following manner:

$$\mathscr{L}^{-1}\{F(s - a)\} = \mathscr{L}^{-1}\{F(s)|_{s \to s-a}\} = e^{at}f(t), \tag{1}$$

where $f(t) = \mathscr{L}^{-1}\{F(s)\}$.

The first part of the next example illustrates partial fraction decomposition in the case when the denominator of $Y(s)$ contains *repeated linear factors.*

EXAMPLE 2 **Partial Fractions: Repeated Linear Factors**

Evaluate **(a)** $\mathscr{L}^{-1}\left\{\dfrac{2s+5}{(s-3)^2}\right\}$ **(b)** $\mathscr{L}^{-1}\left\{\dfrac{s/2+5/3}{s^2+4s+6}\right\}$.

SOLUTION **(a)** A repeated linear factor is a term $(s-a)^n$, where a is a real number and n is a positive integer ≥ 2. Recall that if $(s-a)^n$ appears in the denominator of a rational expression, then the assumed decomposition contains n partial fractions with constant numerators and denominators $s-a, (s-a)^2, \ldots, (s-a)^n$. Hence with $a=3$ and $n=2$ we write

$$\frac{2s+5}{(s-3)^2}=\frac{A}{s-3}+\frac{B}{(s-3)^2}.$$

By putting the two terms on the right-hand side over a common denominator, we obtain the numerator $2s+5=A(s-3)+B$, and this identity yields $A=2$ and $B=11$. Therefore

$$\frac{2s+5}{(s-3)^2}=\frac{2}{s-3}+\frac{11}{(s-3)^2} \tag{2}$$

and
$$\mathscr{L}^{-1}\left\{\frac{2s+5}{(s-3)^2}\right\}=2\mathscr{L}^{-1}\left\{\frac{1}{s-3}\right\}+11\mathscr{L}^{-1}\left\{\frac{1}{(s-3)^2}\right\}. \tag{3}$$

Now $1/(s-3)^2$ is $F(s)=1/s^2$ shifted three units to the right. Since $\mathscr{L}^{-1}\{1/s^2\}=t$, it follows from (1) that

$$\mathscr{L}^{-1}\left\{\frac{1}{(s-3)^2}\right\}=\mathscr{L}^{-1}\left\{\frac{1}{s^2}\bigg|_{s\to s-3}\right\}=e^{3t}t.$$

Finally, (3) is
$$\mathscr{L}^{-1}\left\{\frac{2s+5}{(s-3)^2}\right\}=2e^{3t}+11e^{3t}t. \tag{4}$$

(b) To start, observe that the quadratic polynomial s^2+4s+6 has no real zeros and so has no real linear factors. In this situation we *complete the square*:

$$\frac{s/2+5/3}{s^2+4s+6}=\frac{s/2+5/3}{(s+2)^2+2}. \tag{5}$$

Our goal here is to recognize the expression on the right-hand side as some Laplace transform $F(s)$ in which s has been replaced throughout by $s+2$. What we are trying to do is analogous to working part (b) of Example 1 backwards. The denominator in (5) is already in the correct form—that is, s^2+2 with s replaced by $s+2$. However, we must fix up the numerator by manipulating the constants: $\frac{1}{2}s+\frac{5}{3}=\frac{1}{2}(s+2)+\frac{5}{3}-\frac{2}{2}=\frac{1}{2}(s+2)+\frac{2}{3}$.

Now by termwise division, the linearity of $\mathscr{L}^{-1}$, parts (e) and (d) of Theorem 7.2.1, and finally (1),

$$\frac{s/2+5/3}{(s+2)^2+2}=\frac{\frac{1}{2}(s+2)+\frac{2}{3}}{(s+2)^2+2}=\frac{1}{2}\frac{s+2}{(s+2)^2+2}+\frac{2}{3}\frac{1}{(s+2)^2+2}$$

$$\mathscr{L}^{-1}\left\{\frac{s/2+5/3}{s^2+4s+6}\right\}=\frac{1}{2}\mathscr{L}^{-1}\left\{\frac{s+2}{(s+2)^2+2}\right\}+\frac{2}{3}\mathscr{L}^{-1}\left\{\frac{1}{(s+2)^2+2}\right\}$$

$$=\frac{1}{2}\mathscr{L}^{-1}\left\{\frac{s}{s^2+2}\bigg|_{s\to s+2}\right\}+\frac{2}{3\sqrt{2}}\mathscr{L}^{-1}\left\{\frac{\sqrt{2}}{s^2+2}\bigg|_{s\to s+2}\right\} \tag{6}$$

$$=\frac{1}{2}e^{-2t}\cos\sqrt{2}t+\frac{\sqrt{2}}{3}e^{-2t}\sin\sqrt{2}t. \tag{7}$$ ≡

EXAMPLE 3 An Initial-Value Problem

Solve $y'' - 6y' + 9y = t^2e^{3t}, \quad y(0) = 2, \quad y'(0) = 17.$

SOLUTION Before transforming the DE, note that its right-hand side is similar to the function in part (a) of Example 1. After using linearity, Theorem 7.3.1, and the initial conditions, we simplify and then solve for $Y(s) = \mathscr{L}\{f(t)\}$:

$$\mathscr{L}\{y''\} - 6\mathscr{L}\{y'\} + 9\mathscr{L}\{y\} = \mathscr{L}\{t^2e^{3t}\}$$

$$s^2Y(s) - sy(0) - y'(0) - 6[sY(s) - y(0)] + 9Y(s) = \frac{2}{(s-3)^3}$$

$$(s^2 - 6s + 9)Y(s) = 2s + 5 + \frac{2}{(s-3)^3}$$

$$(s-3)^2Y(s) = 2s + 5 + \frac{2}{(s-3)^3}$$

$$Y(s) = \frac{2s+5}{(s-3)^2} + \frac{2}{(s-3)^5}.$$

The first term on the right-hand side was already decomposed into individual partial fractions in (2) in part (a) of Example 2:

$$Y(s) = \frac{2}{s-3} + \frac{11}{(s-3)^2} + \frac{2}{(s-3)^5}.$$

Thus $$y(t) = 2\mathscr{L}^{-1}\left\{\frac{1}{s-3}\right\} + 11\mathscr{L}^{-1}\left\{\frac{1}{(s-3)^2}\right\} + \frac{2}{4!}\mathscr{L}^{-1}\left\{\frac{4!}{(s-3)^5}\right\}. \tag{8}$$

From the inverse form (1) of Theorem 7.3.1, the last two terms in (8) are

$$\mathscr{L}^{-1}\left\{\frac{1}{s^2}\bigg|_{s\to s-3}\right\} = te^{3t} \quad \text{and} \quad \mathscr{L}^{-1}\left\{\frac{4!}{s^5}\bigg|_{s\to s-3}\right\} = t^4e^{3t}.$$

Thus (8) is $y(t) = 2e^{3t} + 11te^{3t} + \frac{1}{12}t^4e^{3t}$. ≡

EXAMPLE 4 An Initial-Value Problem

Solve $y'' + 4y' + 6y = 1 + e^{-t}, \quad y(0) = 0, \quad y'(0) = 0.$

SOLUTION $$\mathscr{L}\{y''\} + 4\mathscr{L}\{y'\} + 6\mathscr{L}\{y\} = \mathscr{L}\{1\} + \mathscr{L}\{e^{-t}\}$$

$$s^2Y(s) - sy(0) - y'(0) + 4[sY(s) - y(0)] + 6Y(s) = \frac{1}{s} + \frac{1}{s+1}$$

$$(s^2 + 4s + 6)Y(s) = \frac{2s+1}{s(s+1)}$$

$$Y(s) = \frac{2s+1}{s(s+1)(s^2+4s+6)}$$

Since the quadratic term in the denominator does not factor into real linear factors, the partial fraction decomposition for $Y(s)$ is found to be

$$Y(s) = \frac{1/6}{s} + \frac{1/3}{s+1} - \frac{s/2 + 5/3}{s^2 + 4s + 6}.$$

Moreover, in preparation for taking the inverse transform we already manipulated the last term into the necessary form in part (b) of Example 2. So in view of the results in (6) and (7) we have the solution

$$y(t) = \frac{1}{6}\mathcal{L}^{-1}\left\{\frac{1}{s}\right\} + \frac{1}{3}\mathcal{L}^{-1}\left\{\frac{1}{s+1}\right\} - \frac{1}{2}\mathcal{L}^{-1}\left\{\frac{s+2}{(s+2)^2+2}\right\} - \frac{2}{3\sqrt{2}}\mathcal{L}^{-1}\left\{\frac{\sqrt{2}}{(s+2)^2+2}\right\}$$

$$= \frac{1}{6} + \frac{1}{3}e^{-t} - \frac{1}{2}e^{-2t}\cos\sqrt{2}t - \frac{\sqrt{2}}{3}e^{-2t}\sin\sqrt{2}t. \equiv$$

7.3.2 TRANSLATION ON THE *t*-AXIS

≡ Unit Step Function In engineering, one frequently encounters functions that are either "off" or "on." For example, an external force acting on a mechanical system or a voltage impressed on a circuit can be turned off after a period of time. It is convenient, then, to define a special function that is the number 0 (off) up to a certain time $t = a$ and then the number 1 (on) after that time. This function is called the **unit step function** or the **Heaviside function**, named after the English polymath **Oliver Heaviside** (1850–1925).

DEFINITION 7.3.1 Unit Step Function

The **unit step function** $\mathcal{U}(t-a)$ is defined to be

$$\mathcal{U}(t-a) = \begin{cases} 0, & 0 \le t < a \\ 1, & t \ge a. \end{cases}$$

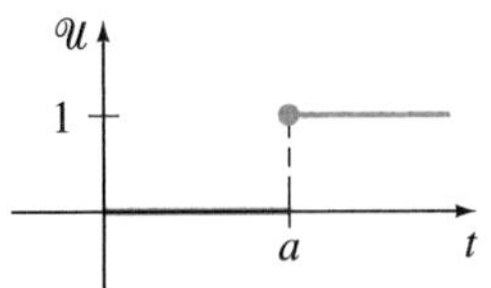

FIGURE 7.3.2 Graph of unit step function

Notice that we define $\mathcal{U}(t-a)$ only on the nonnegative t-axis, since this is all that we are concerned with in the study of the Laplace transform. In a broader sense $\mathcal{U}(t-a) = 0$ for $t < a$. The graph of $\mathcal{U}(t-a)$ is given in Figure 7.3.2. In the case when $a = 0$, we take $\mathcal{U}(t) = 1$ for $t \ge 0$.

When a function f defined for $t \ge 0$ is multiplied by $\mathcal{U}(t-a)$, the unit step function "turns off" a portion of the graph of that function. For example, consider the function $f(t) = 2t - 3$. To "turn off" the portion of the graph of f for $0 \le t < 1$, we simply form the product $(2t-3)\mathcal{U}(t-1)$. See Figure 7.3.3. In general, the graph of $f(t)\,\mathcal{U}(t-a)$ is 0 (off) for $0 \le t < a$ and is the portion of the graph of f (on) for $t \ge a$.

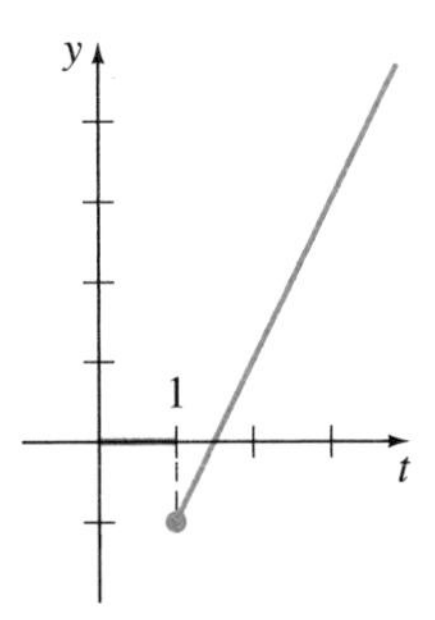

FIGURE 7.3.3 Function is $f(t) = (2t-3)\,\mathcal{U}(t-1)$

The unit step function can also be used to write piecewise-defined functions in a compact form. For example, if we consider $0 \le t < 2$, $2 \le t < 3$, and $t \ge 3$ and the corresponding values of $\mathcal{U}(t-2)$ and $\mathcal{U}(t-3)$, it should be apparent that the piecewise-defined function shown in Figure 7.3.4 is the same as $f(t) = 2 - 3\mathcal{U}(t-2) + \mathcal{U}(t-3)$. Also, a general piecewise-defined function of the type

$$f(t) = \begin{cases} g(t), & 0 \le t < a \\ h(t), & t \ge a \end{cases} \tag{9}$$

is the same as

$$f(t) = g(t) - g(t)\,\mathcal{U}(t-a) + h(t)\,\mathcal{U}(t-a). \tag{10}$$

Similarly, a function of the type

$$f(t) = \begin{cases} 0, & 0 \le t < a \\ g(t), & a \le t < b \\ 0, & t \ge b \end{cases} \tag{11}$$

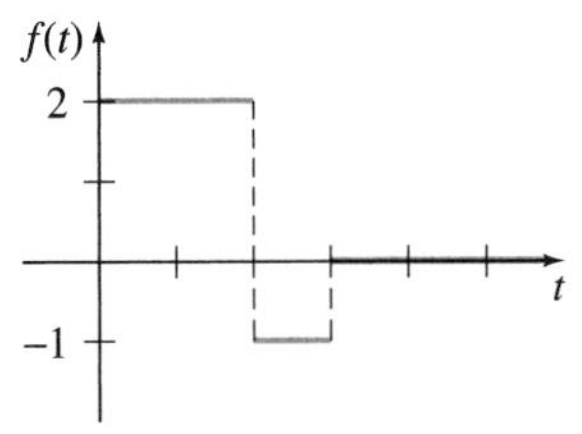

FIGURE 7.3.4 Function is $f(t) = 2 - 3\mathcal{U}(t-2) + \mathcal{U}(t-3)$

can be written

$$f(t) = g(t)[\mathcal{U}(t-a) - \mathcal{U}(t-b)]. \tag{12}$$

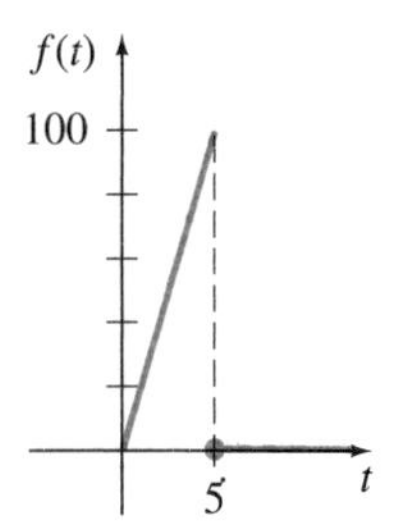

FIGURE 7.3.5 Function f in Example 5

EXAMPLE 5 A Piecewise-Defined Function

Express $f(t) = \begin{cases} 20t, & 0 \le t < 5 \\ 0, & t \ge 5 \end{cases}$ in terms of unit step functions. Graph.

SOLUTION The graph of f is given in Figure 7.3.5. Now from (9) and (10) with $a = 5$, $g(t) = 20t$, and $h(t) = 0$ we get $f(t) = 20t - 20t\,\mathscr{U}(t - 5)$. ≡

Consider a general function $y = f(t)$ defined for $t \ge 0$. The piecewise-defined function

$$f(t - a)\,\mathscr{U}(t - a) = \begin{cases} 0, & 0 \le t < a \\ f(t - a), & t \ge a \end{cases} \tag{13}$$

plays a significant role in the discussion that follows. As shown in Figure 7.3.6, for $a > 0$ the graph of the function $y = f(t - a)\,\mathscr{U}(t - a)$ coincides with the graph of $y = f(t - a)$ for $t \ge a$ (which is the *entire* graph of $y = f(t)$, $t \ge 0$ shifted a units to the right on the t-axis), but is identically zero for $0 \le t < a$.

We saw in Theorem 7.3.1 that an exponential multiple of $f(t)$ results in a translation of the transform $F(s)$ on the s-axis. As a consequence of the next theorem we see that whenever $F(s)$ is multiplied by an exponential function e^{-as}, $a > 0$, the inverse transform of the product $e^{-as}F(s)$ is the function f shifted along the t-axis in the manner illustrated in Figure 7.3.6(b). This result, presented next in its direct transform version, is called the **second translation theorem** or **second shifting theorem.**

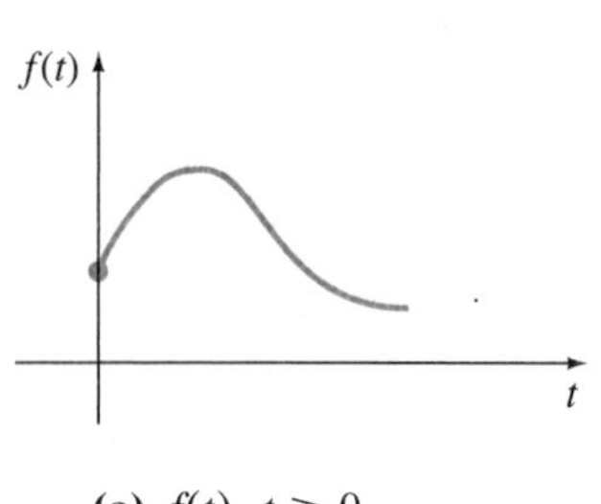

(a) $f(t)$, $t \ge 0$

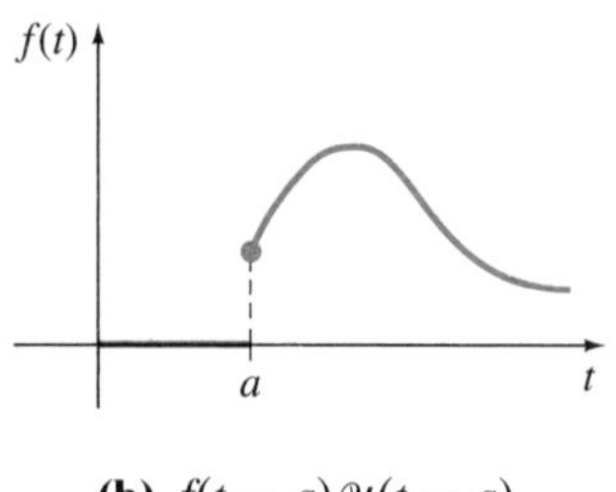

(b) $f(t - a)\mathscr{U}(t - a)$

FIGURE 7.3.6 Shift on t-axis

THEOREM 7.3.2 Second Translation Theorem

If $F(s) = \mathscr{L}\{f(t)\}$ and $a > 0$, then

$$\mathscr{L}\{f(t - a)\,\mathscr{U}(t - a)\} = e^{-as}F(s).$$

PROOF By the additive interval property of integrals,

$$\int_0^\infty e^{-st} f(t - a)\,\mathscr{U}(t - a)\,dt$$

can be written as two integrals:

$$\mathscr{L}\{f(t - a)\,\mathscr{U}(t - a)\} = \int_0^a e^{-st}f(t - a)\underbrace{\mathscr{U}(t - a)}_{\substack{\text{zero for}\\ 0 \le t < a}}\,dt + \int_a^\infty e^{-st}f(t - a)\underbrace{\mathscr{U}(t - a)}_{\substack{\text{one for}\\ t \ge a}}\,dt = \int_a^\infty e^{-st}f(t - a)\,dt.$$

Now if we let $v = t - a$, $dv = dt$ in the last integral, then

$$\mathscr{L}\{f(t - a)\,\mathscr{U}(t - a)\} = \int_0^\infty e^{-s(v+a)}f(v)\,dv = e^{-as}\int_0^\infty e^{-sv}f(v)\,dv = e^{-as}\mathscr{L}\{f(t)\}. \quad ≡$$

We often wish to find the Laplace transform of just a unit step function. This can be from either Definition 7.1.1 or Theorem 7.3.2. If we identify $f(t) = 1$ in Theorem 7.3.2, then $f(t - a) = 1$, $F(s) = \mathscr{L}\{1\} = 1/s$, and so

$$\mathscr{L}\{\mathscr{U}(t - a)\} = \frac{e^{-as}}{s}. \tag{14}$$

EXAMPLE 6 Figure 7.3.4 Revisited

Find the Laplace transform the function f in Figure 7.3.4.

SOLUTION We use f expressed in terms of the unit step function

$$f(t) = 2 - 3\mathscr{U}(t - 2) + \mathscr{U}(t - 3)$$

and the result given in (14):

$$\mathscr{L}\{f(t)\} = 2\mathscr{L}\{1\} - 3\mathscr{L}\{\mathscr{U}(t-2)\} + \mathscr{L}\{\mathscr{U}(t-3)\}$$
$$= \frac{2}{s} - 3\frac{e^{-2s}}{s} + \frac{e^{-3s}}{s}.$$ ≡

≡ **Inverse Form of Theorem 7.3.2** If $f(t) = \mathscr{L}^{-1}\{F(s)\}$, the inverse form of Theorem 7.3.2, $a > 0$, is

$$\mathscr{L}^{-1}\{e^{-as}F(s)\} = f(t-a)\,\mathscr{U}(t-a). \tag{15}$$

EXAMPLE 7 Using Formula (15)

Evaluate **(a)** $\mathscr{L}^{-1}\left\{\frac{1}{s-4}e^{-2s}\right\}$ **(b)** $\mathscr{L}^{-1}\left\{\frac{s}{s^2+9}e^{-\pi s/2}\right\}$.

SOLUTION **(a)** With the three identifications $a = 2$, $F(s) = 1/(s-4)$, and $\mathscr{L}^{-1}\{F(s)\} = e^{4t}$, we have from (15)

$$\mathscr{L}^{-1}\left\{\frac{1}{s-4}e^{-2s}\right\} = e^{4(t-2)}\,\mathscr{U}(t-2).$$

(b) With $a = \pi/2$, $F(s) = s/(s^2+9)$, and $\mathscr{L}^{-1}\{F(s)\} = \cos 3t$, (15) yields

$$\mathscr{L}^{-1}\left\{\frac{s}{s^2+9}e^{-\pi s/2}\right\} = \cos 3\left(t - \frac{\pi}{2}\right)\mathscr{U}\left(t - \frac{\pi}{2}\right).$$

The last expression can be simplified somewhat by using the addition formula for the cosine. Verify that the result is the same as $-\sin 3t\,\mathscr{U}\left(t - \frac{\pi}{2}\right)$. ≡

≡ **Alternative Form of Theorem 7.3.2** We are frequently confronted with the problem of finding the Laplace transform of a product of a function g and a unit step function $\mathscr{U}(t-a)$ where the function g lacks the precise shifted form $f(t-a)$ in Theorem 7.3.2. To find the Laplace transform of $g(t)\mathscr{U}(t-a)$, it is possible to fix up $g(t)$ into the required form $f(t-a)$ by algebraic manipulations. For example, if we wanted to use Theorem 7.3.2 to find the Laplace transform of $t^2\,\mathscr{U}(t-2)$, we would have to force $g(t) = t^2$ into the form $f(t-2)$. You should work through the details and verify that $t^2 = (t-2)^2 + 4(t-2) + 4$ is an identity. Therefore

$$\mathscr{L}\{t^2\,\mathscr{U}(t-2)\} = \mathscr{L}\{(t-2)^2\,\mathscr{U}(t-2) + 4(t-2)\,\mathscr{U}(t-2) + 4\mathscr{U}(t-2)\},$$

where each term on the right-hand side can now be evaluated by Theorem 7.3.2. But since these manipulations are time consuming and often not obvious, it is simpler to devise an alternative version of Theorem 7.3.2. Using Definition 7.1.1, the definition of $\mathscr{U}(t-a)$, and the substitution $u = t - a$, we obtain

$$\mathscr{L}\{g(t)\,\mathscr{U}(t-a)\} = \int_a^{\infty} e^{-st}g(t)\,dt = \int_0^{\infty} e^{-s(u+a)}\,g(u+a)\,du.$$

That is,

$$\mathscr{L}\{g(t)\mathscr{U}(t-a)\} = e^{-as}\,\mathscr{L}\{g(t+a)\}. \tag{16}$$

EXAMPLE 8 Second Translation Theorem—Alternative Form

Evaluate $\mathscr{L}\{\cos t\,\mathscr{U}(t-\pi)\}$.

SOLUTION With $g(t) = \cos t$ and $a = \pi$, then $g(t+\pi) = \cos(t+\pi) = -\cos t$ by the addition formula for the cosine function. Hence by (16),

$$\mathscr{L}\{\cos t\,\mathscr{U}(t-\pi)\} = -e^{-\pi s}\,\mathscr{L}\{\cos t\} = -\frac{s}{s^2+1}e^{-\pi s}.$$ ≡

EXAMPLE 9 An Initial-Value Problem

Solve $y' + y = f(t)$, $y(0) = 5$, where $f(t) = \begin{cases} 0, & 0 \le t < \pi \\ 3\cos t, & t \ge \pi. \end{cases}$

SOLUTION The function f can be written as $f(t) = 3\cos t\, \mathscr{U}(t - \pi)$, so by linearity, the results of Example 7, and the usual partial fractions, we have

$$\begin{aligned} \mathscr{L}\{y'\} + \mathscr{L}\{y\} &= 3\mathscr{L}\{\cos t\, \mathscr{U}(t - \pi)\} \\ sY(s) - y(0) + Y(s) &= -3\frac{s}{s^2 + 1}e^{-\pi s} \\ (s + 1)Y(s) &= 5 - \frac{3s}{s^2 + 1}e^{-\pi s} \end{aligned}$$

$$Y(s) = \frac{5}{s + 1} - \frac{3}{2}\left[-\frac{1}{s + 1}e^{-\pi s} + \frac{1}{s^2 + 1}e^{-\pi s} + \frac{s}{s^2 + 1}e^{-\pi s}\right]. \tag{17}$$

Now proceeding as we did in Example 7, it follows from (15) with $a = \pi$ that the inverses of the terms inside the brackets are

$$\mathscr{L}^{-1}\left\{\frac{1}{s + 1}e^{-\pi s}\right\} = e^{-(t - \pi)}\mathscr{U}(t - \pi), \qquad \mathscr{L}^{-1}\left\{\frac{1}{s^2 + 1}e^{-\pi s}\right\} = \sin(t - \pi)\,\mathscr{U}(t - \pi),$$

and
$$\mathscr{L}^{-1}\left\{\frac{s}{s^2 + 1}e^{-\pi s}\right\} = \cos(t - \pi)\,\mathscr{U}(t - \pi).$$

Thus the inverse of (17) is

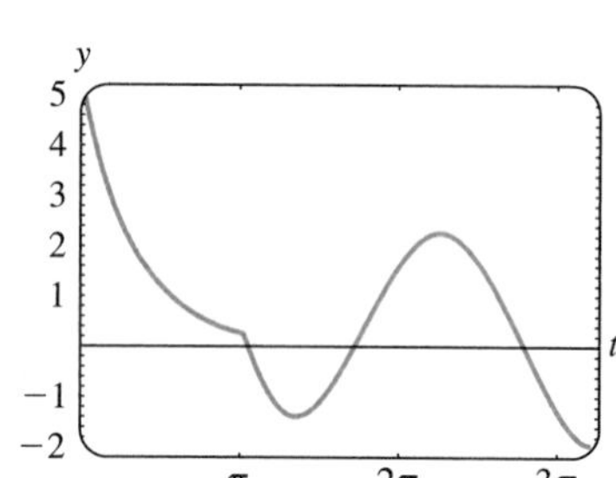

FIGURE 7.3.7 Graph of function (18) in Example 9

$$\begin{aligned} y(t) &= 5e^{-t} + \frac{3}{2}e^{-(t - \pi)}\mathscr{U}(t - \pi) - \frac{3}{2}\sin(t - \pi)\,\mathscr{U}(t - \pi) - \frac{3}{2}\cos(t - \pi)\,\mathscr{U}(t - \pi) \\ &= 5e^{-t} + \frac{3}{2}[e^{-(t - \pi)} + \sin t + \cos t]\,\mathscr{U}(t - \pi) \qquad \leftarrow \text{trigonometric identities} \end{aligned}$$

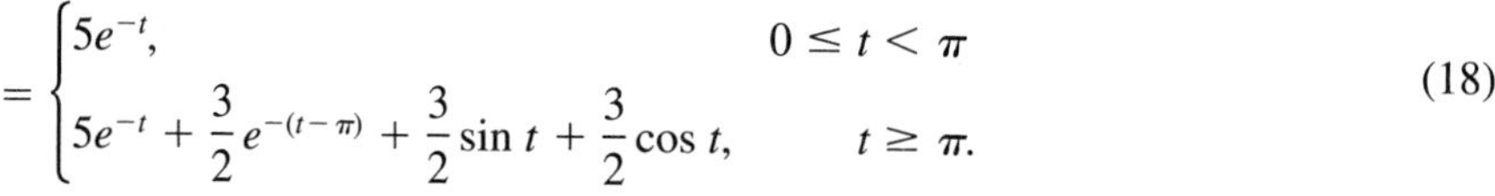

$$= \begin{cases} 5e^{-t}, & 0 \le t < \pi \\ 5e^{-t} + \frac{3}{2}e^{-(t - \pi)} + \frac{3}{2}\sin t + \frac{3}{2}\cos t, & t \ge \pi. \end{cases} \tag{18}$$

We obtained the graph of (18) shown in Figure 7.3.7 by using a graphing utility. ≡

≡ **Beams** In Section 5.2 we saw that the static deflection $y(x)$ of a uniform beam of length L carrying load $w(x)$ per unit length is found from the linear fourth-order differential equation

$$EI\frac{d^4y}{dx^4} = w(x), \tag{19}$$

where E is Young's modulus of elasticity and I is a moment of inertia of a cross section of the beam. The Laplace transform is particularly useful in solving (19) when $w(x)$ is piecewise-defined. However, to use the Laplace transform, we must tacitly assume that $y(x)$ and $w(x)$ are defined on $(0, \infty)$ rather than on $(0, L)$. Note, too, that the next example is a boundary-value problem rather than an initial-value problem.

EXAMPLE 10 A Boundary-Value Problem

A beam of length L is embedded at both ends, as shown in Figure 7.3.8. Find the deflection of the beam when the load is given by

$$w(x) = \begin{cases} w_0\left(1 - \frac{2}{L}x\right), & 0 < x < L/2 \\ 0, & L/2 < x < L. \end{cases}$$

FIGURE 7.3.8 Embedded beam with variable load in Example 10

SOLUTION Recall that because the beam is embedded at both ends, the boundary conditions are $y(0) = 0$, $y'(0) = 0$, $y(L) = 0$, $y'(L) = 0$. Now by (10) we can express $w(x)$ in terms of the unit step function:

$$w(x) = w_0\left(1 - \frac{2}{L}x\right) - w_0\left(1 - \frac{2}{L}x\right)\mathcal{U}\left(x - \frac{L}{2}\right)$$
$$= \frac{2w_0}{L}\left[\frac{L}{2} - x + \left(x - \frac{L}{2}\right)\mathcal{U}\left(x - \frac{L}{2}\right)\right].$$

Transforming (19) with respect to the variable x gives

$$EI\left(s^4Y(s) - s^3y(0) - s^2y'(0) - sy''(0) - y'''(0)\right) = \frac{2w_0}{L}\left[\frac{L/2}{s} - \frac{1}{s^2} + \frac{1}{s^2}e^{-Ls/2}\right]$$

or

$$s^4Y(s) - sy''(0) - y'''(0) = \frac{2w_0}{EIL}\left[\frac{L/2}{s} - \frac{1}{s^2} + \frac{1}{s^2}e^{-Ls/2}\right].$$

If we let $c_1 = y''(0)$ and $c_2 = y'''(0)$, then

$$Y(s) = \frac{c_1}{s^3} + \frac{c_2}{s^4} + \frac{2w_0}{EIL}\left[\frac{L/2}{s^5} - \frac{1}{s^6} + \frac{1}{s^6}e^{-Ls/2}\right],$$

and consequently

$$y(x) = \frac{c_1}{2!}\mathcal{L}^{-1}\left\{\frac{2!}{s^3}\right\} + \frac{c_2}{3!}\mathcal{L}^{-1}\left\{\frac{3!}{s^4}\right\} + \frac{2w_0}{EIL}\left[\frac{L/2}{4!}\mathcal{L}^{-1}\left\{\frac{4!}{s^5}\right\} - \frac{1}{5!}\mathcal{L}^{-1}\left\{\frac{5!}{s^6}\right\} + \frac{1}{5!}\mathcal{L}^{-1}\left\{\frac{5!}{s^6}e^{-Ls/2}\right\}\right]$$
$$= \frac{c_1}{2}x^2 + \frac{c_2}{6}x^3 + \frac{w_0}{60\,EIL}\left[\frac{5L}{2}x^4 - x^5 + \left(x - \frac{L}{2}\right)^5\mathcal{U}\left(x - \frac{L}{2}\right)\right].$$

Applying the conditions $y(L) = 0$ and $y'(L) = 0$ to the last result yields a system of equations for c_1 and c_2:

$$c_1\frac{L^2}{2} + c_2\frac{L^3}{6} + \frac{49w_0L^4}{1920EI} = 0$$
$$c_1 L + c_2\frac{L^2}{2} + \frac{85w_0L^3}{960EI} = 0.$$

Solving, we find $c_1 = 23w_0L^2/(960EI)$ and $c_2 = -9w_0L/(40EI)$. Thus the deflection is given by

$$y(x) = \frac{23w_0L^2}{1920EI}x^2 - \frac{3w_0L}{80EI}x^3 + \frac{w_0}{60EIL}\left[\frac{5L}{2}x^4 - x^5 + \left(x - \frac{L}{2}\right)^5\mathcal{U}\left(x - \frac{L}{2}\right)\right].$$ ≡

EXERCISES 7.3

Answers to selected odd-numbered problems begin on page ANS-12.

7.3.1 TRANSLATION ON THE s-AXIS

In Problems 1–20 find either $F(s)$ or $f(t)$, as indicated.

1. $\mathcal{L}\{te^{10t}\}$
2. $\mathcal{L}\{te^{-6t}\}$
3. $\mathcal{L}\{t^3e^{-2t}\}$
4. $\mathcal{L}\{t^{10}e^{-7t}\}$
5. $\mathcal{L}\{t(e^t + e^{2t})^2\}$
6. $\mathcal{L}\{e^{2t}(t - 1)^2\}$
7. $\mathcal{L}\{e^t \sin 3t\}$
8. $\mathcal{L}\{e^{-2t}\cos 4t\}$
9. $\mathcal{L}\{(1 - e^t + 3e^{-4t})\cos 5t\}$
10. $\mathcal{L}\left\{e^{3t}\left(9 - 4t + 10\sin\frac{t}{2}\right)\right\}$
11. $\mathcal{L}^{-1}\left\{\frac{1}{(s + 2)^3}\right\}$
12. $\mathcal{L}^{-1}\left\{\frac{1}{(s - 1)^4}\right\}$
13. $\mathcal{L}^{-1}\left\{\frac{1}{s^2 - 6s + 10}\right\}$
14. $\mathcal{L}^{-1}\left\{\frac{1}{s^2 + 2s + 5}\right\}$
15. $\mathcal{L}^{-1}\left\{\frac{s}{s^2 + 4s + 5}\right\}$
16. $\mathcal{L}^{-1}\left\{\frac{2s + 5}{s^2 + 6s + 34}\right\}$
17. $\mathcal{L}^{-1}\left\{\frac{s}{(s + 1)^2}\right\}$
18. $\mathcal{L}^{-1}\left\{\frac{5s}{(s - 2)^2}\right\}$
19. $\mathcal{L}^{-1}\left\{\frac{2s - 1}{s^2(s + 1)^3}\right\}$
20. $\mathcal{L}^{-1}\left\{\frac{(s + 1)^2}{(s + 2)^4}\right\}$

In Problems 21–30 use the Laplace transform to solve the given initial-value problem.

21. $y' + 4y = e^{-4t}$, $y(0) = 2$

22. $y' - y = 1 + te^t$, $y(0) = 0$

23. $y'' + 2y' + y = 0$, $y(0) = 1, y'(0) = 1$

24. $y'' - 4y' + 4y = t^3e^{2t}$, $y(0) = 0, y'(0) = 0$

25. $y'' - 6y' + 9y = t$, $y(0) = 0, y'(0) = 1$

26. $y'' - 4y' + 4y = t^3$, $y(0) = 1, y'(0) = 0$

27. $y'' - 6y' + 13y = 0$, $y(0) = 0, y'(0) = -3$

28. $2y'' + 20y' + 51y = 0$, $y(0) = 2, y'(0) = 0$

29. $y'' - y' = e^t \cos t$, $y(0) = 0, y'(0) = 0$

30. $y'' - 2y' + 5y = 1 + t$, $y(0) = 0, y'(0) = 4$

In Problems 31 and 32 use the Laplace transform and the procedure outlined in Example 10 to solve the given boundary-value problem.

31. $y'' + 2y' + y = 0$, $y'(0) = 2, y(1) = 2$

32. $y'' + 8y' + 20y = 0$, $y(0) = 0, y'(\pi) = 0$

33. A 4-pound weight stretches a spring 2 feet. The weight is released from rest 18 inches above the equilibrium position, and the resulting motion takes place in a medium offering a damping force numerically equal to $\frac{7}{8}$ times the instantaneous velocity. Use the Laplace transform to find the equation of motion $x(t)$.

34. Recall that the differential equation for the instantaneous charge $q(t)$ on the capacitor in an *LRC*-series circuit is given by

$$L\frac{d^2q}{dt^2} + R\frac{dq}{dt} + \frac{1}{C}q = E(t). \tag{20}$$

See Section 5.1. Use the Laplace transform to find $q(t)$ when $L = 1$ h, $R = 20\ \Omega$, $C = 0.005$ f, $E(t) = 150$ V, $t > 0$, $q(0) = 0$, and $i(0) = 0$. What is the current $i(t)$?

35. Consider a battery of constant voltage E_0 that charges the capacitor shown in Figure 7.3.9. Divide equation (20) by L and define $2\lambda = R/L$ and $\omega^2 = 1/LC$. Use the Laplace transform to show that the solution $q(t)$ of $q'' + 2\lambda q' + \omega^2 q = E_0/L$ subject to $q(0) = 0$, $i(0) = 0$ is

$$q(t) = \begin{cases} E_0C\left[1 - e^{-\lambda t}\left(\cosh\sqrt{\lambda^2 - \omega^2}\,t \right.\right. \\ \quad \left.\left. + \dfrac{\lambda}{\sqrt{\lambda^2 - \omega^2}}\sinh\sqrt{\lambda^2 - \omega^2}\,t\right)\right], & \lambda > \omega, \\ E_0C[1 - e^{-\lambda t}(1 + \lambda t)], & \lambda = \omega, \\ E_0C\left[1 - e^{-\lambda t}\left(\cos\sqrt{\omega^2 - \lambda^2}\,t \right.\right. \\ \quad \left.\left. + \dfrac{\lambda}{\sqrt{\omega^2 - \lambda^2}}\sin\sqrt{\omega^2 - \lambda^2}\,t\right)\right], & \lambda < \omega. \end{cases}$$

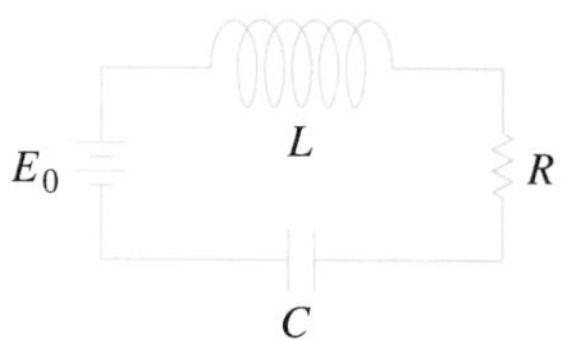

FIGURE 7.3.9 Series circuit in Problem 35

36. Use the Laplace transform to find the charge $q(t)$ in an *RC* series circuit when $q(0) = 0$ and $E(t) = E_0e^{-kt}$, $k > 0$. Consider two cases: $k \neq 1/RC$ and $k = 1/RC$.

7.3.2 TRANSLATION ON THE t-AXIS

In Problems 37–48 find either $F(s)$ or $f(t)$, as indicated.

37. $\mathscr{L}\{(t-1)\mathscr{U}(t-1)\}$

38. $\mathscr{L}\{e^{2-t}\mathscr{U}(t-2)\}$

39. $\mathscr{L}\{t\,\mathscr{U}(t-2)\}$

40. $\mathscr{L}\{(3t+1)\mathscr{U}(t-1)\}$

41. $\mathscr{L}\{\cos 2t\,\mathscr{U}(t-\pi)\}$

42. $\mathscr{L}\left\{\sin t\,\mathscr{U}\left(t-\frac{\pi}{2}\right)\right\}$

43. $\mathscr{L}^{-1}\left\{\dfrac{e^{-2s}}{s^3}\right\}$

44. $\mathscr{L}^{-1}\left\{\dfrac{(1+e^{-2s})^2}{s+2}\right\}$

45. $\mathscr{L}^{-1}\left\{\dfrac{e^{-\pi s}}{s^2+1}\right\}$

46. $\mathscr{L}^{-1}\left\{\dfrac{se^{-\pi s/2}}{s^2+4}\right\}$

47. $\mathscr{L}^{-1}\left\{\dfrac{e^{-s}}{s(s+1)}\right\}$

48. $\mathscr{L}^{-1}\left\{\dfrac{e^{-2s}}{s^2(s-1)}\right\}$

In Problems 49–54 match the given graph with one of the functions in (a)–(f). The graph of $f(t)$ is given in Figure 7.3.10.

(a) $f(t) - f(t)\,\mathscr{U}(t-a)$
(b) $f(t-b)\,\mathscr{U}(t-b)$
(c) $f(t)\,\mathscr{U}(t-a)$
(d) $f(t) - f(t)\,\mathscr{U}(t-b)$
(e) $f(t)\,\mathscr{U}(t-a) - f(t)\,\mathscr{U}(t-b)$
(f) $f(t-a)\,\mathscr{U}(t-a) - f(t-a)\,\mathscr{U}(t-b)$

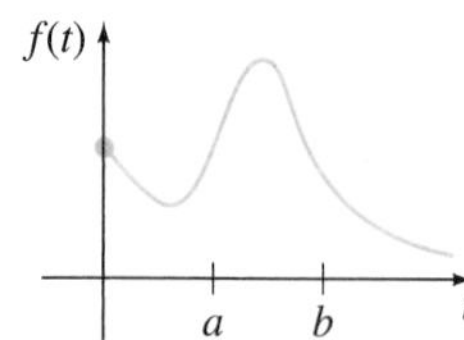

FIGURE 7.3.10 Graph for Problems 49–54

49.

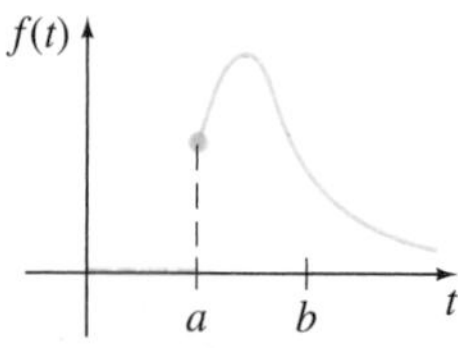

FIGURE 7.3.11 Graph for Problem 49

50.

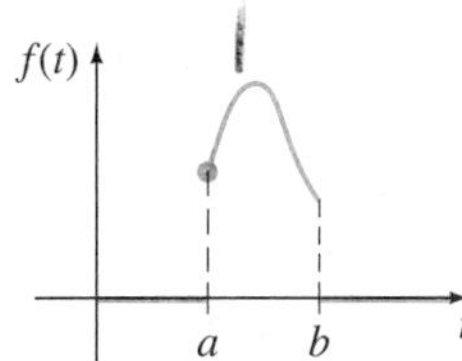

FIGURE 7.3.12 Graph for Problem 50

51.

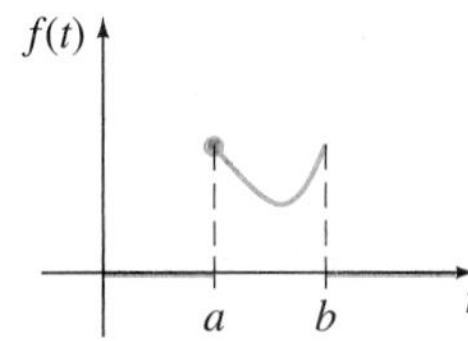

FIGURE 7.3.13 Graph for Problem 51

52.

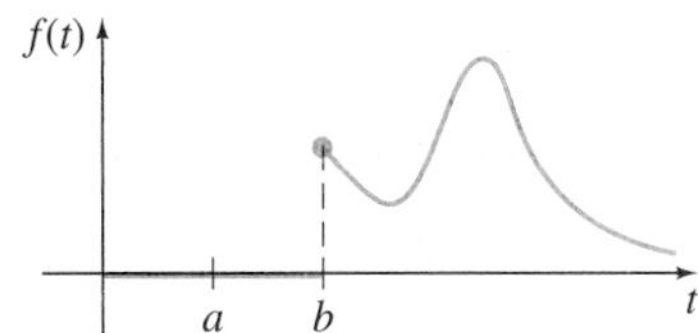

FIGURE 7.3.14 Graph for Problem 52

53.

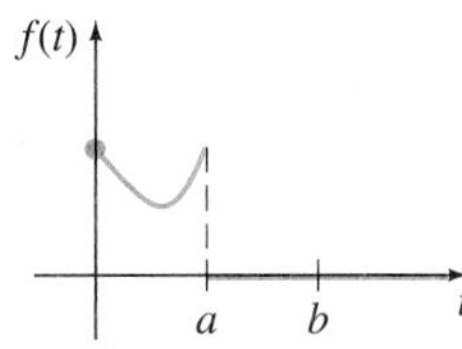

FIGURE 7.3.15 Graph for Problem 53

54.

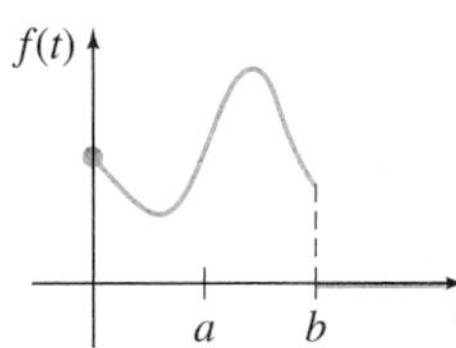

FIGURE 7.3.16 Graph for Problem 54

In Problems 55–62 write each function in terms of unit step functions. Find the Laplace transform of the given function.

55. $f(t) = \begin{cases} 2, & 0 \le t < 3 \\ -2, & t \ge 3 \end{cases}$

56. $f(t) = \begin{cases} 1, & 0 \le t < 4 \\ 0, & 4 \le t < 5 \\ 1, & t \ge 5 \end{cases}$

57. $f(t) = \begin{cases} 0, & 0 \le t < 1 \\ t^2, & t \ge 1 \end{cases}$

58. $f(t) = \begin{cases} 0, & 0 \le t < 3\pi/2 \\ \sin t, & t \ge 3\pi/2 \end{cases}$

59. $f(t) = \begin{cases} t, & 0 \le t < 2 \\ 0, & t \ge 2 \end{cases}$

60. $f(t) = \begin{cases} \sin t, & 0 \le t < 2\pi \\ 0, & t \ge 2\pi \end{cases}$

61.

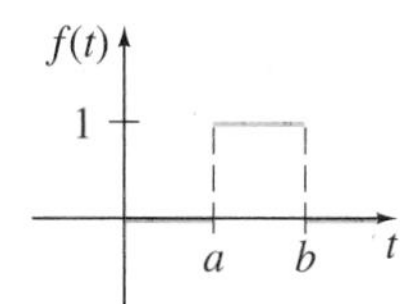

FIGURE 7.3.17 Graph for Problem 61

62.

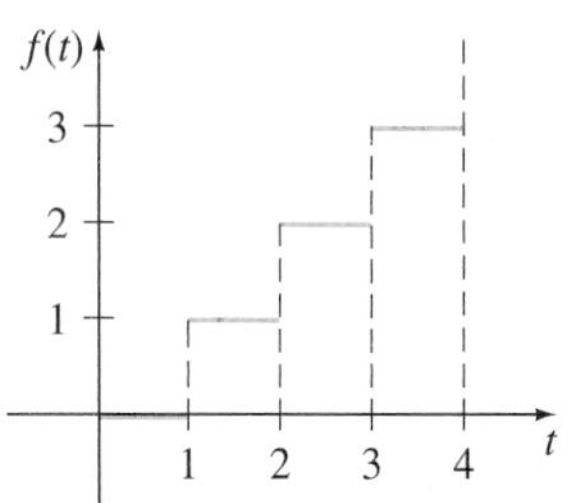

FIGURE 7.3.18 Graph for Problem 62

In Problems 63–70 use the Laplace transform to solve the given initial-value problem.

63. $y' + y = f(t), \quad y(0) = 0$, where $f(t) = \begin{cases} 0, & 0 \le t < 1 \\ 5, & t \ge 1 \end{cases}$

64. $y' + y = f(t), \quad y(0) = 0$, where

$$f(t) = \begin{cases} 1, & 0 \le t < 1 \\ -1, & t \ge 1 \end{cases}$$

65. $y' + 2y = f(t), \quad y(0) = 0$, where

$$f(t) = \begin{cases} t, & 0 \le t < 1 \\ 0, & t \ge 1 \end{cases}$$

66. $y'' + 4y = f(t), \quad y(0) = 0, y'(0) = -1$, where

$$f(t) = \begin{cases} 1, & 0 \le t < 1 \\ 0, & t \ge 1 \end{cases}$$

67. $y'' + 4y = \sin t\, \mathcal{U}(t - 2\pi), \quad y(0) = 1, y'(0) = 0$

68. $y'' - 5y' + 6y = \mathcal{U}(t - 1), \quad y(0) = 0, y'(0) = 1$

69. $y'' + y = f(t), \quad y(0) = 0, y'(0) = 1$, where

$$f(t) = \begin{cases} 0, & 0 \le t < \pi \\ 1, & \pi \le t < 2\pi \\ 0, & t \ge 2\pi \end{cases}$$

70. $y'' + 4y' + 3y = 1 - \mathcal{U}(t - 2) - \mathcal{U}(t - 4) + \mathcal{U}(t - 6)$, $y(0) = 0, y'(0) = 0$

71. Suppose a 32-pound weight stretches a spring 2 feet. If the weight is released from rest at the equilibrium position, find the equation of motion $x(t)$ if an impressed force $f(t) = 20t$ acts on the system for $0 \le t < 5$ and is then removed (see Example 5). Ignore any damping forces. Use a graphing utility to graph $x(t)$ on the interval [0, 10].

72. Solve Problem 71 if the impressed force $f(t) = \sin t$ acts on the system for $0 \le t < 2\pi$ and is then removed.

In Problems 73 and 74 use the Laplace transform to find the charge $q(t)$ on the capacitor in an RC-series circuit subject to the given conditions.

73. $q(0) = 0$, $R = 2.5\ \Omega$, $C = 0.08$ f, $E(t)$ given in Figure 7.3.19

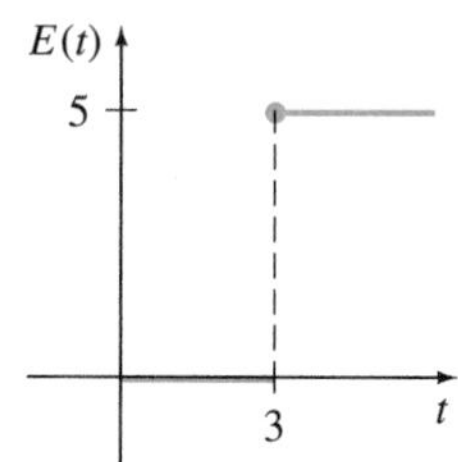

FIGURE 7.3.19 $E(t)$ in Problem 73

74. $q(0) = q_0$, $R = 10\ \Omega$, $C = 0.1$ f, $E(t)$ given in Figure 7.3.20

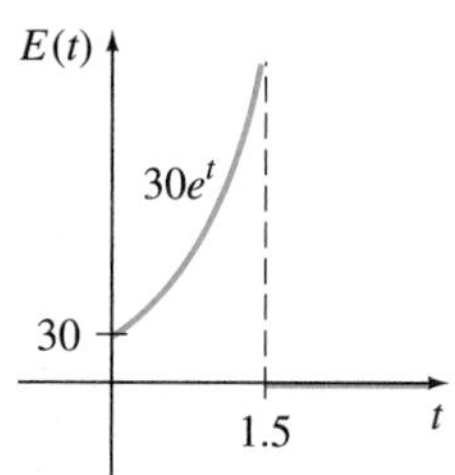

FIGURE 7.3.20 $E(t)$ in Problem 74

75. **(a)** Use the Laplace transform to find the current $i(t)$ in a single-loop LR-series circuit when $i(0) = 0$, $L = 1$ h, $R = 10\ \Omega$, and $E(t)$ is as given in Figure 7.3.21.

(b) Use a computer graphing program to graph $i(t)$ for $0 \le t \le 6$. Use the graph to estimate i_{max} and i_{min}, the maximum and minimum values of the current.

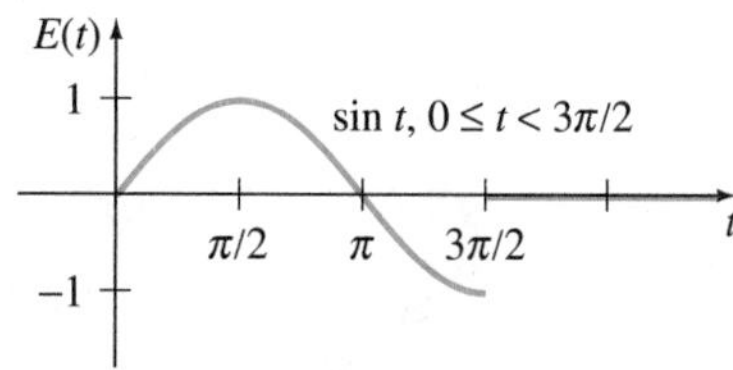

FIGURE 7.3.21 $E(t)$ in Problem 75

76. **(a)** Use the Laplace transform to find the charge $q(t)$ on the capacitor in an RC-series circuit when $q(0) = 0$, $R = 50\ \Omega$, $C = 0.01$ f, and $E(t)$ is as given in Figure 7.3.22.

(b) Assume that $E_0 = 100$ V. Use a computer graphing program to graph $q(t)$ for $0 \le t \le 6$. Use the graph to estimate q_{max}, the maximum value of the charge.

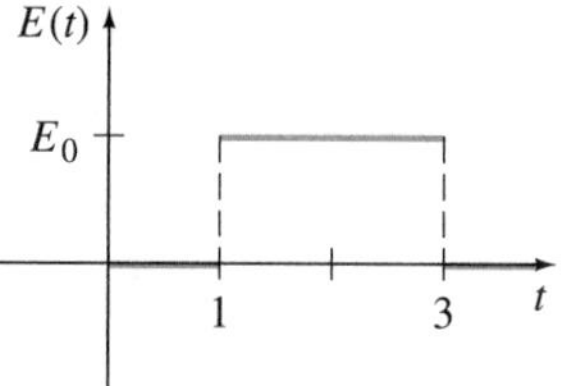

FIGURE 7.3.22 $E(t)$ in Problem 76

77. A cantilever beam is embedded at its left end and free at its right end. Use the Laplace transform to find the deflection $y(x)$ when the load is given by

$$w(x) = \begin{cases} w_0, & 0 < x < L/2 \\ 0, & L/2 \le x < L. \end{cases}$$

78. Solve Problem 77 when the load is given by

$$w(x) = \begin{cases} 0, & 0 < x < L/3 \\ w_0, & L/3 < x < 2L/3 \\ 0, & 2L/3 < x < L. \end{cases}$$

79. Find the deflection $y(x)$ of a cantilever beam embedded at its left end and free at its right end when the load is as given in Example 10.

80. A beam is embedded at its left end and simply supported at its right end. Find the deflection $y(x)$ when the load is as given in Problem 77.

Mathematical Model

81. **Cake Inside an Oven** Reread Example 4 in Section 3.1 on the cooling of a cake that is taken out of an oven.

(a) Devise a mathematical model for the temperature of a cake while it is *inside* the oven based on the following assumptions: At $t = 0$ the cake mixture is at the room temperature of 70°; the oven is not preheated, so at $t = 0$, when the cake mixture is placed into the oven, the temperature inside the oven is also 70°; the temperature of the oven increases linearly until $t = 4$ minutes, when the desired temperature of 300° is attained; the oven temperature is a constant 300° for $t \ge 4$.

(b) Use the Laplace transform to solve the initial-value problem in part (a).

Discussion Problems

82. Discuss how you would fix up each of the following functions so that Theorem 7.3.2 could be used directly to find the given Laplace transform. Check your answers using (16) of this section.

(a) $\mathscr{L}\{(2t + 1)\mathscr{U}(t - 1)\}$ **(b)** $\mathscr{L}\{e^t\,\mathscr{U}(t - 5)\}$

(c) $\mathscr{L}\{\cos t\,\mathscr{U}(t - \pi)\}$ **(d)** $\mathscr{L}\{(t^2 - 3t)\mathscr{U}(t - 2)\}$

83. (a) Assume that Theorem 7.3.1 holds when the symbol a is replaced by ki, where k is a real number and $i^2 = -1$. Show that $\mathscr{L}\{te^{kti}\}$ can be used to deduce

$$\mathscr{L}\{t\cos kt\} = \frac{s^2 - k^2}{(s^2 + k^2)^2}$$

$$\mathscr{L}\{t\sin kt\} = \frac{2ks}{(s^2 + k^2)^2}.$$

(b) Now use the Laplace transform to solve the initial-value problem $x'' + \omega^2 x = \cos \omega t$, $x(0) = 0$, $x'(0) = 0$.

7.4 OPERATIONAL PROPERTIES II

REVIEW MATERIAL

- Definition 7.1.1
- Theorems 7.3.1 and 7.3.2

INTRODUCTION In this section we develop several more operational properties of the Laplace transform. Specifically, we shall see how to find the transform of a function $f(t)$ that is multiplied by a monomial t^n, the transform of a special type of integral, and the transform of a periodic function. The last two transform properties allow us to solve some equations that we have not encountered up to this point: Volterra integral equations, integrodifferential equations, and ordinary differential equations in which the input function is a periodic piecewise-defined function.

7.4.1 DERIVATIVES OF A TRANSFORM

Multiplying a Function by t^n The Laplace transform of the product of a function $f(t)$ with t can be found by differentiating the Laplace transform of $f(t)$. To motivate this result, let us assume that $F(s) = \mathscr{L}\{f(t)\}$ exists and that it is possible to interchange the order of differentiation and integration. Then

$$\frac{d}{ds}F(s) = \frac{d}{ds}\int_0^\infty e^{-st}f(t)\,dt = \int_0^\infty \frac{\partial}{\partial s}[e^{-st}f(t)]\,dt = -\int_0^\infty e^{-st}tf(t)\,dt = -\mathscr{L}\{tf(t)\};$$

that is,
$$\mathscr{L}\{tf(t)\} = -\frac{d}{ds}\mathscr{L}\{f(t)\}.$$

We can use the last result to find the Laplace transform of $t^2f(t)$:

$$\mathscr{L}\{t^2f(t)\} = \mathscr{L}\{t\cdot tf(t)\} = -\frac{d}{ds}\mathscr{L}\{tf(t)\} = -\frac{d}{ds}\left(-\frac{d}{ds}\mathscr{L}\{f(t)\}\right) = \frac{d^2}{ds^2}\mathscr{L}\{f(t)\}.$$

The preceding two cases suggest the general result for $\mathscr{L}\{t^n f(t)\}$.

THEOREM 7.4.1 Derivatives of Transforms

If $F(s) = \mathscr{L}\{f(t)\}$ and $n = 1, 2, 3, \ldots$, then

$$\mathscr{L}\{t^nf(t)\} = (-1)^n\frac{d^n}{ds^n}F(s).$$

EXAMPLE 1 **Using Theorem 7.4.1**

Evaluate $\mathscr{L}\{t \sin kt\}$.

SOLUTION With $f(t) = \sin kt$, $F(s) = k/(s^2 + k^2)$, and $n = 1$, Theorem 7.4.1 gives

$$\mathscr{L}\{t \sin kt\} = -\frac{d}{ds}\mathscr{L}\{\sin kt\} = -\frac{d}{ds}\left(\frac{k}{s^2 + k^2}\right) = \frac{2ks}{(s^2 + k^2)^2}.$$

If we want to evaluate $\mathscr{L}\{t^2 \sin kt\}$ and $\mathscr{L}\{t^3 \sin kt\}$, all we need do, in turn, is take the negative of the derivative with respect to s of the result in Example 1 and then take the negative of the derivative with respect to s of $\mathscr{L}\{t^2 \sin kt\}$.

Note To find transforms of functions $t^n e^{at}$ we can use either Theorem 7.3.1 or Theorem 7.4.1. For example,

$$\text{Theorem 7.3.1:} \quad \mathscr{L}\{te^{3t}\} = \mathscr{L}\{t\}_{s \to s-3} = \left.\frac{1}{s^2}\right|_{s \to s-3} = \frac{1}{(s-3)^2}.$$

$$\text{Theorem 7.4.1:} \quad \mathscr{L}\{te^{3t}\} = -\frac{d}{ds}\mathscr{L}\{e^{3t}\} = -\frac{d}{ds}\frac{1}{s-3} = (s-3)^{-2} = \frac{1}{(s-3)^2}.$$

EXAMPLE 2 **An Initial-Value Problem**

Solve $x'' + 16x = \cos 4t$, $x(0) = 0$, $x'(0) = 1$.

SOLUTION The initial-value problem could describe the forced, undamped, and resonant motion of a mass on a spring. The mass starts with an initial velocity of 1 ft/s in the downward direction from the equilibrium position.

Transforming the differential equation gives

$$(s^2 + 16)X(s) = 1 + \frac{s}{s^2 + 16} \quad \text{or} \quad X(s) = \frac{1}{s^2 + 16} + \frac{s}{(s^2 + 16)^2}.$$

Now we just saw in Example 1 that

$$\mathscr{L}^{-1}\left\{\frac{2ks}{(s^2 + k^2)^2}\right\} = t \sin kt, \tag{1}$$

and so with the identification $k = 4$ in (1) and in part (d) of Theorem 7.2.1, we obtain

$$\begin{aligned} x(t) &= \frac{1}{4}\mathscr{L}^{-1}\left\{\frac{4}{s^2 + 16}\right\} + \frac{1}{8}\mathscr{L}^{-1}\left\{\frac{8s}{(s^2 + 16)^2}\right\} \\ &= \frac{1}{4}\sin 4t + \frac{1}{8}t \sin 4t. \end{aligned}$$

7.4.2 TRANSFORMS OF INTEGRALS

Convolution If functions f and g are piecewise continuous on the interval $[0, \infty)$, then a special product, denoted by $f * g$, is defined by the integral

$$f * g = \int_0^t f(\tau)\, g(t - \tau)\, d\tau \tag{2}$$

and is called the **convolution** of f and g. The convolution $f * g$ is a function of t. For example,

$$e^t * \sin t = \int_0^t e^\tau \sin(t - \tau)\, d\tau = \frac{1}{2}(-\sin t - \cos t + e^t). \tag{3}$$

It is left as an exercise to show that

$$\int_0^t f(\tau)\, g(t-\tau)\, d\tau = \int_0^t f(t-\tau)\, g(\tau)\, d\tau;$$

that is, $f * g = g * f$. This means that the convolution of two functions is commutative.

It is *not* true that the integral of a product of functions is the product of the integrals. However, it *is* true that the Laplace transform of the special product (2) is the product of the Laplace transform of f and g. This means that it is possible to find the Laplace transform of the convolution of two functions without actually evaluating the integral as we did in (3). The result that follows is known as the **convolution theorem.**

THEOREM 7.4.2 Convolution Theorem

If $f(t)$ and $g(t)$ are piecewise continuous on $[0, \infty)$ and of exponential order, then

$$\mathscr{L}\{f * g\} = \mathscr{L}\{f(t)\}\,\mathscr{L}\{g(t)\} = F(s)G(s).$$

PROOF Let $\quad F(s) = \mathscr{L}\{f(t)\} = \displaystyle\int_0^\infty e^{-s\tau} f(\tau)\, d\tau$

and $\quad G(s) = \mathscr{L}\{g(t)\} = \displaystyle\int_0^\infty e^{-s\beta} g(\beta)\, d\beta.$

Proceeding formally, we have

$$\begin{aligned} F(s)G(s) &= \left(\int_0^\infty e^{-s\tau} f(\tau)\, d\tau\right)\left(\int_0^\infty e^{-s\beta} g(\beta)\, d\beta\right) \\ &= \int_0^\infty \int_0^\infty e^{-s(\tau+\beta)} f(\tau) g(\beta)\, d\tau\, d\beta \\ &= \int_0^\infty f(\tau)\, d\tau \int_0^\infty e^{-s(\tau+\beta)} g(\beta)\, d\beta. \end{aligned}$$

Holding τ fixed, we let $t = \tau + \beta$, $dt = d\beta$, so that

$$F(s)G(s) = \int_0^\infty f(\tau)\, d\tau \int_\tau^\infty e^{-st} g(t-\tau)\, dt.$$

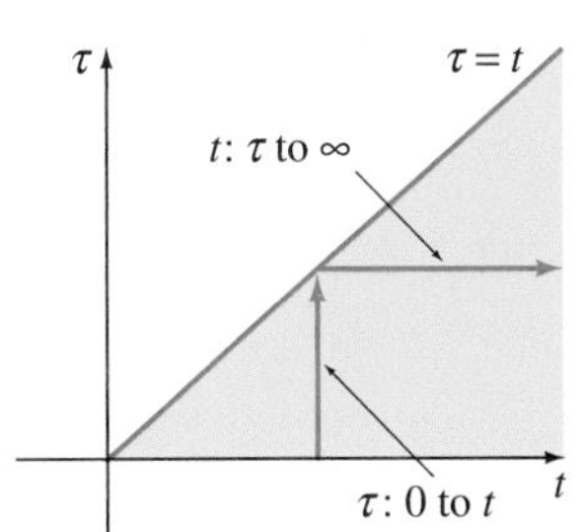

FIGURE 7.4.1 Changing order of integration from t first to τ first

In the $t\tau$-plane we are integrating over the shaded region in Figure 7.4.1. Since f and g are piecewise continuous on $[0, \infty)$ and of exponential order, it is possible to interchange the order of integration:

$$F(s)\,G(s) = \int_0^\infty e^{-st}\, dt \int_0^t f(\tau) g(t-\tau)\, d\tau = \int_0^\infty e^{-st}\left\{\int_0^t f(\tau)\, g(t-\tau)\, d\tau\right\} dt = \mathscr{L}\{f * g\}. \equiv$$

EXAMPLE 3 Transform of a Convolution

Evaluate $\mathscr{L}\left\{\displaystyle\int_0^t e^{\tau} \sin(t-\tau)\, d\tau\right\}.$

SOLUTION With $f(t) = e^t$ and $g(t) = \sin t$, the convolution theorem states that the Laplace transform of the convolution of f and g is the product of their Laplace transforms:

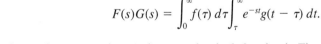

$$\mathscr{L}\left\{\int_0^t e^{\tau} \sin(t-\tau)\, d\tau\right\} = \mathscr{L}\{e^t\} \cdot \mathscr{L}\{\sin t\} = \frac{1}{s-1} \cdot \frac{1}{s^2+1} = \frac{1}{(s-1)(s^2+1)}. \equiv$$

Inverse Form of Theorem 7.4.2 The convolution theorem is sometimes useful in finding the inverse Laplace transform of the product of two Laplace transforms. From Theorem 7.4.2 we have

$$\mathscr{L}^{-1}\{F(s)G(s)\} = f * g. \tag{4}$$

Many of the results in the table of Laplace transforms in Appendix III can be derived using (4). For example, in the next example we obtain entry 25 of the table:

$$\mathscr{L}\{\sin kt - kt\cos kt\} = \frac{2k^3}{(s^2 + k^2)^2}. \tag{5}$$

EXAMPLE 4 Inverse Transform as a Convolution

Evaluate $\mathscr{L}^{-1}\left\{\dfrac{1}{(s^2 + k^2)^2}\right\}$.

SOLUTION Let $F(s) = G(s) = \dfrac{1}{s^2 + k^2}$ so that

$$f(t) = g(t) = \frac{1}{k}\mathscr{L}^{-1}\left\{\frac{k}{s^2 + k^2}\right\} = \frac{1}{k}\sin kt.$$

In this case (4) gives

$$\mathscr{L}^{-1}\left\{\frac{1}{(s^2 + k^2)^2}\right\} = \frac{1}{k^2}\int_0^t \sin k\tau \sin k(t - \tau)\, d\tau. \tag{6}$$

With the aid of the product-to-sum trigonometric identity

$$\sin A \sin B = \frac{1}{2}[\cos(A - B) - \cos(A + B)]$$

and the substitutions $A = k\tau$ and $B = k(t - \tau)$ we can carry out the integration in (6):

$$\begin{aligned}
\mathscr{L}^{-1}\left\{\frac{1}{(s^2 + k^2)^2}\right\} &= \frac{1}{2k^2}\int_0^t [\cos k(2\tau - t) - \cos kt]\, d\tau \\
&= \frac{1}{2k^2}\left[\frac{1}{2k}\sin k(2\tau - t) - \tau\cos kt\right]_0^t \\
&= \frac{\sin kt - kt\cos kt}{2k^3}.
\end{aligned}$$

Multiplying both sides by $2k^3$ gives the inverse form of (5). ≡

Transform of an Integral When $g(t) = 1$ and $\mathscr{L}\{g(t)\} = G(s) = 1/s$, the convolution theorem implies that the Laplace transform of the integral of f is

$$\mathscr{L}\left\{\int_0^t f(\tau)\, d\tau\right\} = \frac{F(s)}{s}. \tag{7}$$

The inverse form of (7),

$$\int_0^t f(\tau)\, d\tau = \mathscr{L}^{-1}\left\{\frac{F(s)}{s}\right\}, \tag{8}$$

can be used in lieu of partial fractions when s^n is a factor of the denominator and $f(t) = \mathscr{L}^{-1}\{F(s)\}$ is easy to integrate. For example, we know for $f(t) = \sin t$ that $F(s) = 1/(s^2 + 1)$, and so by (8)

$$\mathscr{L}^{-1}\left\{\frac{1}{s(s^2 + 1)}\right\} = \mathscr{L}^{-1}\left\{\frac{1/(s^2 + 1)}{s}\right\} = \int_0^t \sin\tau\, d\tau = 1 - \cos t$$

$$\mathscr{L}^{-1}\left\{\frac{1}{s^2(s^2 + 1)}\right\} = \mathscr{L}^{-1}\left\{\frac{1/s(s^2 + 1)}{s}\right\} = \int_0^t (1 - \cos\tau)\, d\tau = t - \sin t$$

$$\mathscr{L}^{-1}\left\{\frac{1}{s^3(s^2 + 1)}\right\} = \mathscr{L}^{-1}\left\{\frac{1/s^2(s^2 + 1)}{s}\right\} = \int_0^t (\tau - \sin\tau)\, d\tau = \tfrac{1}{2}t^2 - 1 + \cos t$$

and so on.

Volterra Integral Equation The convolution theorem and the result in (7) are useful in solving other types of equations in which an unknown function appears under an integral sign. In the next example we solve a **Volterra integral equation** for $f(t)$,

$$f(t) = g(t) + \int_0^t f(\tau)\, h(t - \tau)\, d\tau. \tag{9}$$

The functions $g(t)$ and $h(t)$ are known. Notice that the integral in (9) has the convolution form (2) with the symbol h playing the part of g.

EXAMPLE 5 An Integral Equation

Solve $f(t) = 3t^2 - e^{-t} - \int_0^t f(\tau)\, e^{t-\tau} d\tau$ for $f(t)$.

SOLUTION In the integral we identify $h(t - \tau) = e^{t-\tau}$ so that $h(t) = e^t$. We take the Laplace transform of each term; in particular, by Theorem 7.4.2 the transform of the integral is the product of $\mathscr{L}\{f(t)\} = F(s)$ and $\mathscr{L}\{e^t\} = 1/(s - 1)$:

$$F(s) = 3 \cdot \frac{2}{s^3} - \frac{1}{s+1} - F(s) \cdot \frac{1}{s-1}.$$

After solving the last equation for $F(s)$ and carrying out the partial fraction decomposition, we find

$$F(s) = \frac{6}{s^3} - \frac{6}{s^4} + \frac{1}{s} - \frac{2}{s+1}.$$

The inverse transform then gives

$$\begin{aligned} f(t) &= 3\mathscr{L}^{-1}\left\{\frac{2!}{s^3}\right\} - \mathscr{L}^{-1}\left\{\frac{3!}{s^4}\right\} + \mathscr{L}^{-1}\left\{\frac{1}{s}\right\} - 2\mathscr{L}^{-1}\left\{\frac{1}{s+1}\right\} \\ &= 3t^2 - t^3 + 1 - 2e^{-t}. \end{aligned}$$

Series Circuits In a single-loop or series circuit, Kirchhoff's second law states that the sum of the voltage drops across an inductor, resistor, and capacitor is equal to the impressed voltage $E(t)$. Now it is known that the voltage drops across an inductor, resistor, and capacitor are, respectively,

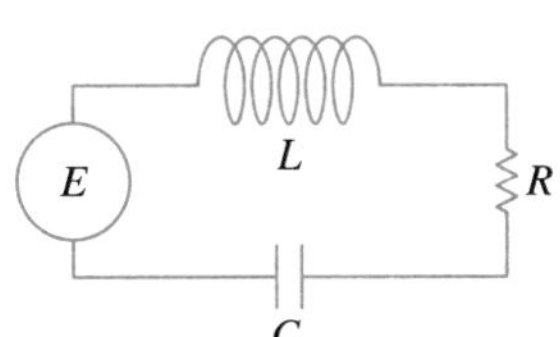

FIGURE 7.4.2 *LRC*-series circuit

$$L\frac{di}{dt}, \qquad Ri(t), \quad \text{and} \quad \frac{1}{C}\int_0^t i(\tau)\, d\tau,$$

where $i(t)$ is the current and L, R, and C are constants. It follows that the current in a circuit, such as that shown in Figure 7.4.2, is governed by the **integrodifferential equation**

$$L\frac{di}{dt} + Ri(t) + \frac{1}{C}\int_0^t i(\tau)\, d\tau = E(t). \tag{10}$$

EXAMPLE 6 An Integrodifferential Equation

Determine the current $i(t)$ in a single-loop *LRC*-series circuit when $L = 0.1$ h, $R = 2\ \Omega$, $C = 0.1$ f, $i(0) = 0$, and the impressed voltage is

$$E(t) = 120t - 120t\, \mathscr{U}(t - 1).$$

SOLUTION With the given data equation (10) becomes

$$0.1\frac{di}{dt} + 2i + 10\int_0^t i(\tau)\,d\tau = 120t - 120t\,\mathscr{U}(t-1).$$

Now by (7), $\mathscr{L}\{\int_0^t i(\tau)\,d\tau\} = I(s)/s$, where $I(s) = \mathscr{L}\{i(t)\}$. Thus the Laplace transform of the integrodifferential equation is

$$0.1sI(s) + 2I(s) + 10\frac{I(s)}{s} = 120\left[\frac{1}{s^2} - \frac{1}{s^2}e^{-s} - \frac{1}{s}e^{-s}\right]. \quad \leftarrow \text{by (16) of Section 7.3}$$

Multiplying this equation by $10s$, using $s^2 + 20s + 100 = (s+10)^2$, and then solving for $I(s)$ gives

$$I(s) = 1200\left[\frac{1}{s(s+10)^2} - \frac{1}{s(s+10)^2}e^{-s} - \frac{1}{(s+10)^2}e^{-s}\right].$$

By partial fractions,

$$I(s) = 1200\left[\frac{1/100}{s} - \frac{1/100}{s+10} - \frac{1/10}{(s+10)^2} - \frac{1/100}{s}e^{-s}\right.$$
$$\left. + \frac{1/100}{s+10}e^{-s} + \frac{1/10}{(s+10)^2}e^{-s} - \frac{1}{(s+10)^2}e^{-s}\right].$$

From the inverse form of the second translation theorem, (15) of Section 7.3, we finally obtain

$$i(t) = 12[1 - \mathscr{U}(t-1)] - 12[e^{-10t} - e^{-10(t-1)}\mathscr{U}(t-1)]$$
$$- 120te^{-10t} - 1080(t-1)e^{-10(t-1)}\mathscr{U}(t-1).$$

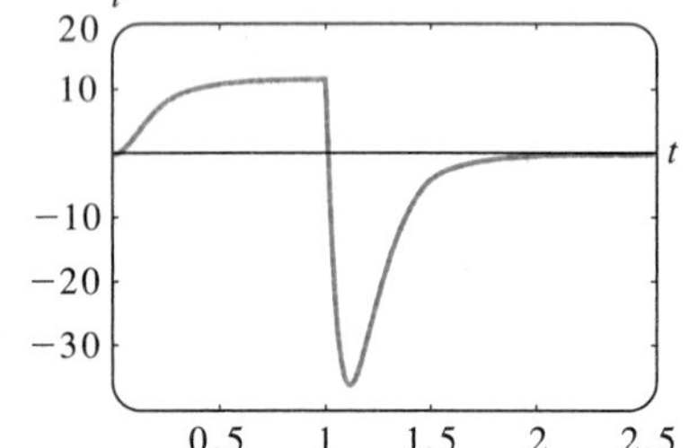

FIGURE 7.4.3 Graph of current $i(t)$ in Example 6

Written as a piecewise-defined function, the current is

$$i(t) = \begin{cases} 12 - 12e^{-10t} - 120te^{-10t}, & 0 \le t < 1 \\ -12e^{-10t} + 12e^{-10(t-1)} - 120te^{-10t} - 1080(t-1)e^{-10(t-1)}, & t \ge 1. \end{cases}$$

Using this last expression and a CAS, we graph $i(t)$ on each of the two intervals and then combine the graphs. Note in Figure 7.4.3 that even though the input $E(t)$ is discontinuous, the output or response $i(t)$ is a continuous function. ≡

Optional material if Section 4.8 was covered. ►

≡ Post Script—Green's Functions Redux By applying the Laplace transform to the initial-value problem

$$y'' + ay' + by = f(t), \quad y(0) = 0, y'(0) = 0,$$

where a and b are constants, we find that the transform of $y(t)$ is

$$Y(s) = \frac{F(s)}{s^2 + as + b},$$

where $F(s) = \mathscr{L}\{f(t)\}$. By rewriting the foregoing transform as the product

$$Y(s) = \frac{1}{s^2 + as + b}F(s)$$

we can use the inverse form of the convolution theorem (4) to write the solution of the IVP as

$$y(t) = \int_0^t g(t-\tau)f(\tau)d\tau, \tag{11}$$

where $\mathscr{L}^{-1}\left\{\dfrac{1}{s^2 + as + b}\right\} = g(t)$ and $\mathscr{L}^{-1}\{F(s)\} = f(t)$. On the other hand, we know from (10) of Section 4.8 that the solution of the IVP is also given by

$$y(t) = \int_0^t G(t, \tau) f(\tau)\, d\tau, \tag{12}$$

where $G(t, \tau)$ is the Green's function for the differential equation.

By comparing (11) and (12) we see that the Green's function for the differential equation is related to $\mathscr{L}^{-1}\left\{\dfrac{1}{s^2 + as + b}\right\} = g(t)$ by

$$G(t, \tau) = g(t - \tau). \tag{13}$$

For example, for the initial-value problem $y'' + 4y = f(t)$, $y(0) = 0$, $y'(0) = 0$ we find

$$\mathscr{L}^{-1}\left\{\frac{1}{s^2 + 4}\right\} = \tfrac{1}{2} \sin 2t = g(t).$$

In Example 4 of Section 4.8, the roles of the symbols x and t are played by t and τ in this discussion.

Thus from (13) we see that the Green's function for the DE $y'' + 4y = f(t)$ is $G(t, \tau) = g(t - \tau) = \frac{1}{2} \sin 2(t - \tau)$. See Example 4 in Section 4.8.

7.4.3 TRANSFORM OF A PERIODIC FUNCTION

Periodic Function If a periodic function has period T, $T > 0$, then $f(t + T) = f(t)$. The next theorem shows that the Laplace transform of a periodic function can be obtained by integration over one period.

THEOREM 7.4.3 Transform of a Periodic Function

If $f(t)$ is piecewise continuous on $[0, \infty)$, of exponential order, and periodic with period T, then

$$\mathscr{L}\{f(t)\} = \frac{1}{1 - e^{-sT}} \int_0^T e^{-st} f(t)\, dt.$$

PROOF Write the Laplace transform of f as two integrals:

$$\mathscr{L}\{f(t)\} = \int_0^T e^{-st} f(t)\, dt + \int_T^\infty e^{-st} f(t)\, dt.$$

When we let $t = u + T$, the last integral becomes

$$\int_T^\infty e^{-st} f(t)\, dt = \int_0^\infty e^{-s(u+T)} f(u + T)\, du = e^{-sT} \int_0^\infty e^{-su} f(u)\, du = e^{-sT} \mathscr{L}\{f(t)\}.$$

Therefore $$\mathscr{L}\{f(t)\} = \int_0^T e^{-st} f(t)\, dt + e^{-sT} \mathscr{L}\{f(t)\}.$$

Solving the equation in the last line for $\mathscr{L}\{f(t)\}$ proves the theorem. ≡

EXAMPLE 7 Transform of a Periodic Function

Find the Laplace transform of the periodic function shown in Figure 7.4.4.

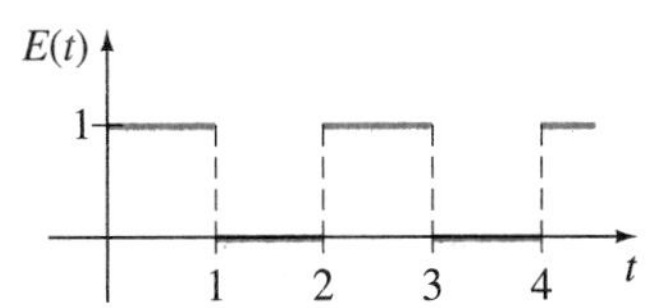

FIGURE 7.4.4 Square wave in Example 7

SOLUTION The function $E(t)$ is called a square wave and has period $T = 2$. For $0 \le t < 2$, $E(t)$ can be defined by

$$E(t) = \begin{cases} 1, & 0 \le t < 1 \\ 0, & 1 \le t < 2 \end{cases}$$

and outside the interval by $E(t+2) = E(t)$. Now from Theorem 7.4.3

$$\begin{aligned}\mathscr{L}\{E(t)\} &= \frac{1}{1-e^{-2s}}\int_0^2 e^{-st}E(t)\,dt = \frac{1}{1-e^{-2s}}\left[\int_0^1 e^{-st}\cdot 1\,dt + \int_1^2 e^{-st}\cdot 0\,dt\right] \\ &= \frac{1}{1-e^{-2s}}\,\frac{1-e^{-s}}{s} \qquad \leftarrow 1-e^{-2s} = (1+e^{-s})(1-e^{-s}) \\ &= \frac{1}{s(1+e^{-s})}. \end{aligned} \tag{14}$$

EXAMPLE 8 A Periodic Impressed Voltage

The differential equation for the current $i(t)$ in a single-loop LR-series circuit is

$$L\frac{di}{dt} + Ri = E(t). \tag{15}$$

Determine the current $i(t)$ when $i(0) = 0$ and $E(t)$ is the square wave function shown in Figure 7.4.4.

SOLUTION If we use the result in (14) of the preceding example, the Laplace transform of the DE is

$$LsI(s) + RI(s) = \frac{1}{s(1+e^{-s})} \qquad \text{or} \qquad I(s) = \frac{1/L}{s(s+R/L)}\cdot\frac{1}{1+e^{-s}}. \tag{16}$$

To find the inverse Laplace transform of the last function, we first make use of geometric series. With the identification $x = e^{-s}$, $s > 0$, the geometric series

$$\frac{1}{1+x} = 1 - x + x^2 - x^3 + \cdots \qquad \text{becomes} \qquad \frac{1}{1+e^{-s}} = 1 - e^{-s} + e^{-2s} - e^{-3s} + \cdots.$$

From

$$\frac{1}{s(s+R/L)} = \frac{L/R}{s} - \frac{L/R}{s+R/L}$$

we can then rewrite (16) as

$$\begin{aligned} I(s) &= \frac{1}{R}\left(\frac{1}{s} - \frac{1}{s+R/L}\right)(1 - e^{-s} + e^{-2s} - e^{-3s} + \cdots) \\ &= \frac{1}{R}\left(\frac{1}{s} - \frac{e^{-s}}{s} + \frac{e^{-2s}}{s} - \frac{e^{-3s}}{s} + \cdots\right) - \frac{1}{R}\left(\frac{1}{s+R/L} - \frac{1}{s+R/L}e^{-s} + \frac{e^{-2s}}{s+R/L} - \frac{e^{-3s}}{s+R/L} + \cdots\right). \end{aligned}$$

By applying the form of the second translation theorem to each term of both series, we obtain

$$\begin{aligned} i(t) &= \frac{1}{R}(1 - \mathscr{U}(t-1) + \mathscr{U}(t-2) - \mathscr{U}(t-3) + \cdots) \\ &\quad - \frac{1}{R}(e^{-Rt/L} - e^{-R(t-1)/L}\,\mathscr{U}(t-1) + e^{-R(t-2)/L}\,\mathscr{U}(t-2) - e^{-R(t-3)/L}\,\mathscr{U}(t-3) + \cdots) \end{aligned}$$

or, equivalently,

$$i(t) = \frac{1}{R}(1 - e^{-Rt/L}) + \frac{1}{R}\sum_{n=1}^{\infty}(-1)^n(1 - e^{-R(t-n)/L})\,\mathscr{U}(t-n).$$

To interpret the solution, let us assume for the sake of illustration that $R = 1$, $L = 1$, and $0 \le t < 4$. In this case

$$i(t) = 1 - e^{-t} - (1 - e^{t-1})\,\mathscr{U}(t-1) + (1 - e^{-(t-2)})\,\mathscr{U}(t-2) - (1 - e^{-(t-3)})\,\mathscr{U}(t-3);$$

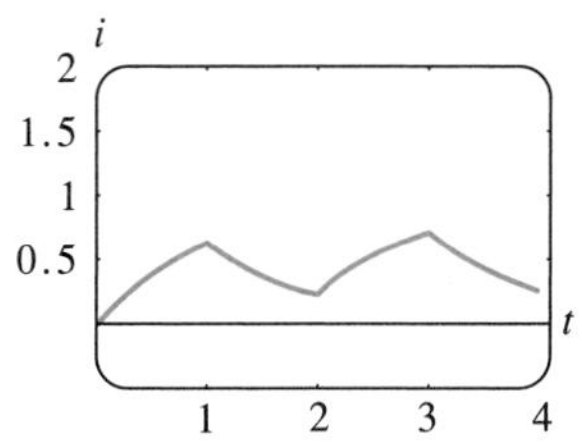

FIGURE 7.4.5 Graph of current $i(t)$ in Example 8

in other words,

$$i(t) = \begin{cases} 1 - e^{-t}, & 0 \le t < 1 \\ -e^{-t} + e^{-(t-1)}, & 1 \le t < 2 \\ 1 - e^{-t} + e^{-(t-1)} - e^{-(t-2)}, & 2 \le t < 3 \\ -e^{-t} + e^{-(t-1)} - e^{-(t-2)} + e^{-(t-3)}, & 3 \le t < 4. \end{cases}$$

The graph of $i(t)$ for $0 \le t < 4$, given in Figure 7.4.5, was obtained with the help of a CAS. ≡

EXERCISES 7.4

Answers to selected odd-numbered problems begin on page ANS-12.

7.4.1 DERIVATIVES OF A TRANSFORM

In Problems 1–8 use Theorem 7.4.1 to evaluate the given Laplace transform.

1. $\mathscr{L}\{te^{-10t}\}$ **2.** $\mathscr{L}\{t^3e^t\}$

3. $\mathscr{L}\{t\cos 2t\}$ **4.** $\mathscr{L}\{t\sinh 3t\}$

5. $\mathscr{L}\{t^2\sinh t\}$ **6.** $\mathscr{L}\{t^2\cos t\}$

7. $\mathscr{L}\{te^{2t}\sin 6t\}$ **8.** $\mathscr{L}\{te^{-3t}\cos 3t\}$

In Problems 9–14 use the Laplace transform to solve the given initial-value problem. Use the table of Laplace transforms in Appendix III as needed.

9. $y' + y = t\sin t, \quad y(0) = 0$

10. $y' - y = te^t\sin t, \quad y(0) = 0$

11. $y'' + 9y = \cos 3t, \quad y(0) = 2, \quad y'(0) = 5$

12. $y'' + y = \sin t, \quad y(0) = 1, \quad y'(0) = -1$

13. $y'' + 16y = f(t), \quad y(0) = 0, \quad y'(0) = 1$, where

$$f(t) = \begin{cases} \cos 4t, & 0 \le t < \pi \\ 0, & t \ge \pi \end{cases}$$

14. $y'' + y = f(t), \quad y(0) = 1, \quad y'(0) = 0$, where

$$f(t) = \begin{cases} 1, & 0 \le t < \pi/2 \\ \sin t, & t \ge \pi/2 \end{cases}$$

In Problems 15 and 16 use a graphing utility to graph the indicated solution.

15. $y(t)$ of Problem 13 for $0 \le t < 2\pi$

16. $y(t)$ of Problem 14 for $0 \le t < 3\pi$

In some instances the Laplace transform can be used to solve linear differential equations with variable monomial coefficients. In Problems 17 and 18 use Theorem 7.4.1 to reduce the given differential equation to a linear first-order DE in the transformed function $Y(s) = \mathscr{L}\{y(t)\}$. Solve the first-order DE for $Y(s)$ and then find $y(t) = \mathscr{L}^{-1}\{Y(s)\}$.

17. $ty'' - y' = 2t^2, \quad y(0) = 0$

18. $2y'' + ty' - 2y = 10, \quad y(0) = y'(0) = 0$

7.4.2 TRANSFORMS OF INTEGRALS

In Problems 19–30 use Theorem 7.4.2 to evaluate the given Laplace transform. Do not evaluate the integral before transforming.

19. $\mathscr{L}\{1 * t^3\}$ **20.** $\mathscr{L}\{t^2 * te^t\}$

21. $\mathscr{L}\{e^{-t} * e^t\cos t\}$ **22.** $\mathscr{L}\{e^{2t} * \sin t\}$

23. $\mathscr{L}\left\{\int_0^t e^\tau\,d\tau\right\}$ **24.** $\mathscr{L}\left\{\int_0^t \cos\tau\,d\tau\right\}$

25. $\mathscr{L}\left\{\int_0^t e^{-\tau}\cos\tau\,d\tau\right\}$ **26.** $\mathscr{L}\left\{\int_0^t \tau\sin\tau\,d\tau\right\}$

27. $\mathscr{L}\left\{\int_0^t \tau e^{t-\tau}\,d\tau\right\}$ **28.** $\mathscr{L}\left\{\int_0^t \sin\tau\cos(t-\tau)\,d\tau\right\}$

29. $\mathscr{L}\left\{t\int_0^t \sin\tau\,d\tau\right\}$ **30.** $\mathscr{L}\left\{t\int_0^t \tau e^{-\tau}\,d\tau\right\}$

In Problems 31–34 use (8) to evaluate the given inverse transform.

31. $\mathscr{L}^{-1}\left\{\dfrac{1}{s(s-1)}\right\}$ **32.** $\mathscr{L}^{-1}\left\{\dfrac{1}{s^2(s-1)}\right\}$

33. $\mathscr{L}^{-1}\left\{\dfrac{1}{s^3(s-1)}\right\}$ **34.** $\mathscr{L}^{-1}\left\{\dfrac{1}{s(s-a)^2}\right\}$

35. The table in Appendix III does not contain an entry for

$$\mathscr{L}^{-1}\left\{\frac{8k^3s}{(s^2+k^2)^3}\right\}.$$

(a) Use (4) along with the results in (5) to evaluate this inverse transform. Use a CAS as an aid in evaluating the convolution integral.

(b) Reexamine your answer to part (a). Could you have obtained the result in a different manner?

36. Use the Laplace transform and the results of Problem 35 to solve the initial-value problem

$$y'' + y = \sin t + t\sin t, \quad y(0) = 0, \quad y'(0) = 0.$$

Use a graphing utility to graph the solution.

In Problems 37–46 use the Laplace transform to solve the given integral equation or integrodifferential equation.

37. $f(t) + \int_0^t (t - \tau) f(\tau)\, d\tau = t$

38. $f(t) = 2t - 4\int_0^t \sin \tau f(t - \tau)\, d\tau$

39. $f(t) = te^t + \int_0^t \tau f(t - \tau)\, d\tau$

40. $f(t) + 2\int_0^t f(\tau) \cos (t - \tau)\, d\tau = 4e^{-t} + \sin t$

41. $f(t) + \int_0^t f(\tau)\, d\tau = 1$

42. $f(t) = \cos t + \int_0^t e^{-\tau} f(t - \tau)\, d\tau$

43. $f(t) = 1 + t - \frac{8}{3}\int_0^t (\tau - t)^3 f(\tau)\, d\tau$

44. $t - 2f(t) = \int_0^t (e^\tau - e^{-\tau}) f(t - \tau)\, d\tau$

45. $y'(t) = 1 - \sin t - \int_0^t y(\tau)\, d\tau, \quad y(0) = 0$

46. $\frac{dy}{dt} + 6y(t) + 9\int_0^t y(\tau)\, d\tau = 1, \quad y(0) = 0$

In Problems 47 and 48 solve equation (10) subject to $i(0) = 0$ with L, R, C, and $E(t)$ as given. Use a graphing utility to graph the solution for $0 \le t \le 3$.

47. $L = 0.1$ h, $R = 3\ \Omega$, $C = 0.05$ f,
$E(t) = 100[\mathcal{U}(t - 1) - \mathcal{U}(t - 2)]$

48. $L = 0.005$ h, $R = 1\ \Omega$, $C = 0.02$ f,
$E(t) = 100[t - (t - 1)\mathcal{U}(t - 1)]$

7.4.3 TRANSFORM OF A PERIODIC FUNCTION

In Problems 49–54 use Theorem 7.4.3 to find the Laplace transform of the given periodic function.

49.

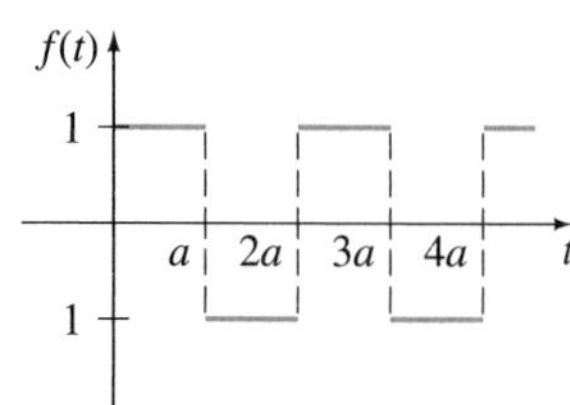

FIGURE 7.4.6 Graph for Problem 49

50.

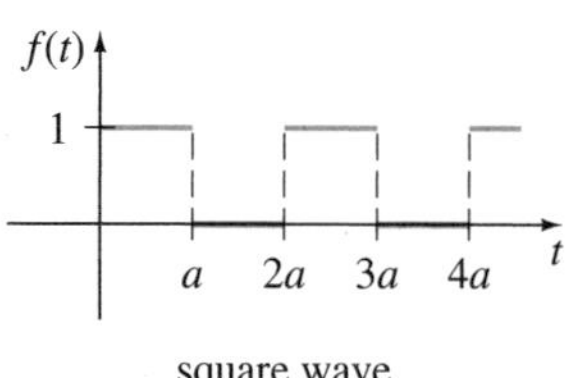

FIGURE 7.4.7 Graph for Problem 50

51.

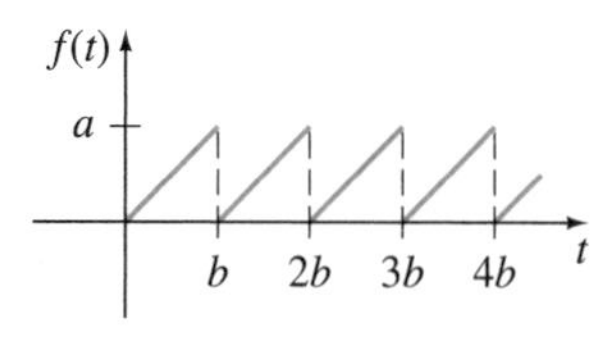

FIGURE 7.4.8 Graph for Problem 51

52.

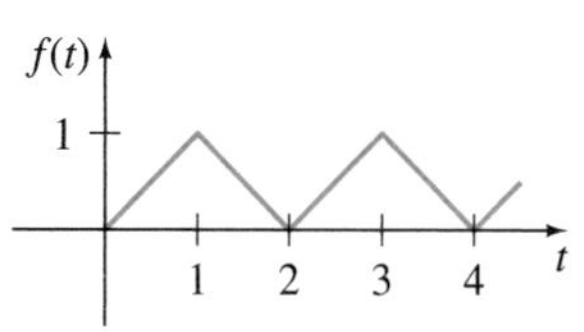

FIGURE 7.4.9 Graph for Problem 52

53.

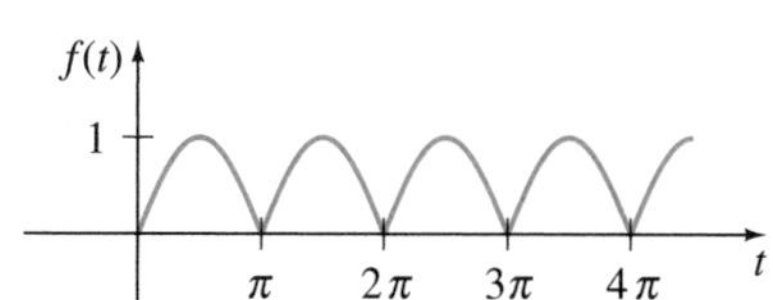

FIGURE 7.4.10 Graph for Problem 53

54.

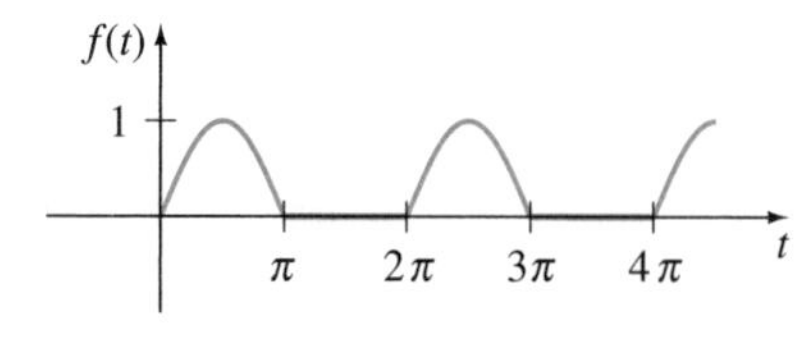

FIGURE 7.4.11 Graph for Problem 54

In Problems 55 and 56 solve equation (15) subject to $i(0) = 0$ with $E(t)$ as given. Use a graphing utility to graph the solution for $0 \le t < 4$ in the case when $L = 1$ and $R = 1$.

55. $E(t)$ is the meander function in Problem 49 with amplitude 1 and $a = 1$.

56. $E(t)$ is the sawtooth function in Problem 51 with amplitude 1 and $b = 1$.

In Problems 57 and 58 solve the model for a driven spring/mass system with damping

$$m\frac{d^2x}{dt^2} + \beta\frac{dx}{dt} + kx = f(t), \quad x(0) = 0, \quad x'(0) = 0,$$

where the driving function f is as specified. Use a graphing utility to graph $x(t)$ for the indicated values of t.

57. $m = \frac{1}{2}$, $\beta = 1$, $k = 5$, f is the meander function in Problem 49 with amplitude 10, and $a = \pi$, $0 \le t < 2\pi$.

58. $m = 1$, $\beta = 2$, $k = 1$, f is the square wave in Problem 50 with amplitude 5, and $a = \pi$, $0 \le t < 4\pi$.

Discussion Problems

59. Discuss how Theorem 7.4.1 can be used to find

$$\mathscr{L}^{-1}\left\{\ln\frac{s-3}{s+1}\right\}.$$

60. In Section 6.4 we saw that $ty'' + y' + ty = 0$ is Bessel's equation of order $\nu = 0$. In view of (22) of that section and Table 6.4.1 a solution of the initial-value problem $ty'' + y' + ty = 0$, $y(0) = 1$, $y'(0) = 0$, is $y = J_0(t)$. Use this result and the procedure outlined in the instructions to Problems 17 and 18 to show that

$$\mathscr{L}\{J_0(t)\} = \frac{1}{\sqrt{s^2+1}}.$$

[*Hint*: You might need to use Problem 46 in Exercises 7.2.]

61. (a) Laguerre's differential equation

$$ty'' + (1-t)y' + ny = 0$$

is known to possess polynomial solutions when n is a nonnegative integer. These solutions are naturally called **Laguerre polynomials** and are denoted by $L_n(t)$. Find $y = L_n(t)$, for $n = 0, 1, 2, 3, 4$ if it is known that $L_n(0) = 1$.

(b) Show that

$$\mathscr{L}\left\{\frac{e^t}{n!}\frac{d^n}{dt^n}t^n e^{-t}\right\} = Y(s),$$

where $Y(s) = \mathscr{L}\{y\}$ and $y = L_n(t)$ is a polynomial solution of the DE in part (a). Conclude that

$$L_n(t) = \frac{e^t}{n!}\frac{d^n}{dt^n}t^n e^{-t}, \qquad n = 0, 1, 2, \ldots.$$

This last relation for generating the Laguerre polynomials is the analogue of Rodrigues' formula for the Legendre polynomials. See (33) in Section 6.4.

62. The Laplace transform $\mathscr{L}\{e^{-t^2}\}$ exists, but without finding it solve the initial-value problem $y'' + y = e^{-t^2}$, $y(0) = 0$, $y'(0) = 0$.

63. Solve the integral equation

$$f(t) = e^t + e^t\int_0^t e^{-\tau}f(\tau)\,d\tau.$$

64. (a) Show that the square wave function $E(t)$ given in Figure 7.4.4 can be written

$$E(t) = \sum_{k=0}^{\infty}(-1)^k\,\mathscr{U}(t-k).$$

(b) Obtain (14) of this section by taking the Laplace transform of each term in the series in part (a).

65. Use the Laplace transform as an aide in evaluating the improper integral $\int_0^\infty te^{-2t}\sin 4t\,dt$.

66. If we assume that $\mathscr{L}\{f(t)/t\}$ exists and $\mathscr{L}\{f(t)\} = F(s)$, then

$$\mathscr{L}\left\{\frac{f(t)}{t}\right\} = \int_s^\infty F(u)du.$$

Use this result to find the Laplace transform of the given function. The symbols a and k are positive constants.

(a) $f(t) = \dfrac{\sin at}{t}$

(b) $f(t) = \dfrac{2(1-\cos kt)}{t}$

67. Transform of the Logarithm Because $f(t) = \ln t$ has an infinite discontinuity at $t = 0$ it might be assumed that $\mathscr{L}\{\ln t\}$ does not exist; however, this is incorrect. The point of this problem to guide you through the formal steps leading to the Laplace transform of $f(t) = \ln t$, $t > 0$.

(a) Use integration by parts to show that

$$\mathscr{L}\{\ln t\} = s\,\mathscr{L}\{t\ln t\} - \frac{1}{s}.$$

(b) If $\mathscr{L}\{\ln t\} = Y(s)$, use Theorem 7.4.1 with $n = 1$ to show that part (a) becomes

$$s\frac{dY}{ds} + Y = -\frac{1}{s}.$$

Find an explicit solution $Y(s)$ of the foregoing differential equation.

(c) Finally, the integral definition of **Euler's constant** (sometimes called the **Euler-Mascheroni constant**) is $\gamma = -\int_0^\infty e^{-t}\ln t\,dt$, where $\gamma = 0.5772156649\ldots$. Use $Y(1) = -\gamma$ in the solution in part (b) to show that

$$\mathscr{L}\{\ln t\} = -\frac{\gamma}{s} - \frac{\ln s}{s}, \quad s > 0.$$

Computer Lab Assignments

68. In this problem you are led through the commands in *Mathematica* that enable you to obtain the symbolic Laplace transform of a differential equation and the solution of the initial-value problem by finding the inverse transform. In *Mathematica* the Laplace transform of a function $y(t)$ is obtained using **LaplaceTransform [y[t], t, s]**. In line two of the syntax we replace **LaplaceTransform [y[t], t, s]** by the symbol **Y.** (*If you* *do not have* Mathematica, *then adapt the given procedure by finding the corresponding syntax for the CAS you have on hand.)*

Consider the initial-value problem

$$y'' + 6y' + 9y = t \sin t, \quad y(0) = 2, \quad y'(0) = -1.$$

Load the Laplace transform package. Precisely reproduce and then, in turn, execute each line in the following sequence of commands. Either copy the output by hand or print out the results.

```
diffequat = y''[t] + 6y'[t] + 9y[t] == t Sin[t]
transformdeq = LaplaceTransform [diffequat, t, s] /.
   {y[0] -> 2, y'[0] -> -1,
   LaplaceTransform [y[t], t, s] -> Y}
soln = Solve[transformdeq, Y]//Flatten
Y = Y/.soln
InverseLaplaceTransform[Y, s, t]
```

69. Appropriately modify the procedure of Problem 68 to find a solution of

$$y''' + 3y' - 4y = 0,$$
$$y(0) = 0, \quad y'(0) = 0, \quad y''(0) = 1.$$

70. The charge $q(t)$ on a capacitor in an LC-series circuit is given by

$$\frac{d^2q}{dt^2} + q = 1 - 4\mathcal{U}(t - \pi) + 6\mathcal{U}(t - 3\pi),$$
$$q(0) = 0, \quad q'(0) = 0.$$

Appropriately modify the procedure of Problem 68 to find $q(t)$. Graph your solution.

7.5 THE DIRAC DELTA FUNCTION

INTRODUCTION In the last paragraph on page 279, we indicated that as an immediate consequence of Theorem 7.1.3, $F(s) = 1$ cannot be the Laplace transform of a function f that is piecewise continuous on $[0, \infty)$ and of exponential order. In the discussion that follows we are going to introduce a function that is very different from the kinds that you have studied in previous courses. We shall see that there does indeed exist a function—or, more precisely, a *generalized function*—whose Laplace transform is $F(s) = 1$.

FIGURE 7.5.1 A golf club applies a force of large magnitude on the ball for a very short period of time

Unit Impulse

Mechanical systems are often acted on by an external force (or electromotive force in an electrical circuit) of large magnitude that acts only for a very short period of time. For example, a vibrating airplane wing could be struck by lightning, a mass on a spring could be given a sharp blow by a ball peen hammer, and a ball (baseball, golf ball, tennis ball) could be sent soaring when struck violently by some kind of club (baseball bat, golf club, tennis racket). See Figure 7.5.1. The graph of the piecewise-defined function

$$\delta_a(t - t_0) = \begin{cases} 0, & 0 \le t < t_0 - a \\ \dfrac{1}{2a}, & t_0 - a \le t < t_0 + a \\ 0, & t \ge t_0 + a, \end{cases} \tag{1}$$

$a > 0$, $t_0 > 0$, shown in Figure 7.5.2(a), could serve as a model for such a force. For a small value of a, $\delta_a(t - t_0)$ is essentially a constant function of large magnitude that is "on" for just a very short period of time, around t_0. The behavior of $\delta_a(t - t_0)$ as $a \to 0$ is illustrated in Figure 7.5.2(b). The function $\delta_a(t - t_0)$ is called a **unit impulse,** because it possesses the integration property $\int_0^\infty \delta_a(t - t_0)\, dt = 1$.

Dirac Delta Function

In practice it is convenient to work with another type of unit impulse, a "function" that approximates $\delta_a(t - t_0)$ and is defined by the limit

$$\delta(t - t_0) = \lim_{a \to 0} \delta_a(t - t_0). \tag{2}$$

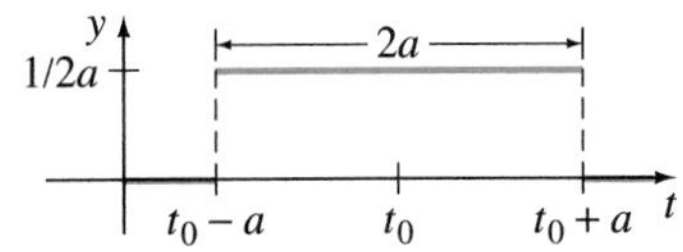

(a) graph of $\delta_a(t - t_0)$

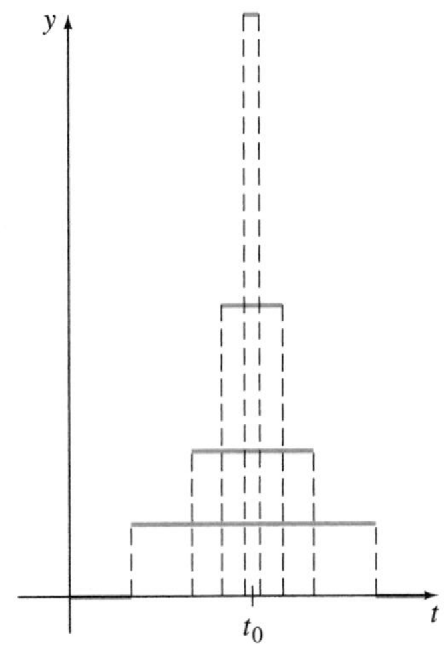

(b) behavior of δ_a as $a \to 0$

FIGURE 7.5.2 Unit impulse

The latter expression, which is not a function at all, can be characterized by the two properties

$$(i)\ \delta(t - t_0) = \begin{cases} \infty, & t = t_0 \\ 0, & t \neq t_0 \end{cases} \qquad \text{and} \qquad (ii) \int_0^\infty \delta(t - t_0)\, dt = 1.$$

The unit impulse $\delta(t - t_0)$ is called the **Dirac delta function.**

It is possible to obtain the Laplace transform of the Dirac delta function by the formal assumption that $\mathscr{L}\{\delta(t - t_0)\} = \lim_{a \to 0} \mathscr{L}\{\delta_a(t - t_0)\}$.

THEOREM 7.5.1 Transform of the Dirac Delta Function

For $t_0 > 0$,
$$\mathscr{L}\{\delta(t - t_0)\} = e^{-st_0}. \tag{3}$$

PROOF To begin, we can write $\delta_a(t - t_0)$ in terms of the unit step function by virtue of (11) and (12) of Section 7.3:

$$\delta_a(t - t_0) = \frac{1}{2a}[\mathscr{U}(t - (t_0 - a)) - \mathscr{U}(t - (t_0 + a))].$$

By linearity and (14) of Section 7.3 the Laplace transform of this last expression is

$$\mathscr{L}\{\delta_a(t - t_0)\} = \frac{1}{2a}\left[\frac{e^{-s(t_0 - a)}}{s} - \frac{e^{-s(t_0 + a)}}{s}\right] = e^{-st_0}\left(\frac{e^{sa} - e^{-sa}}{2sa}\right). \tag{4}$$

Since (4) has the indeterminate form 0/0 as $a \to 0$, we apply L'Hôpital's Rule:

$$\mathscr{L}\{\delta(t - t_0)\} = \lim_{a \to 0} \mathscr{L}\{\delta_a(t - t_0)\} = e^{-st_0} \lim_{a \to 0}\left(\frac{e^{sa} - e^{-sa}}{2sa}\right) = e^{-st_0}. \qquad \equiv$$

Now when $t_0 = 0$, it seems plausible to conclude from (3) that

$$\mathscr{L}\{\delta(t)\} = 1.$$

The last result emphasizes the fact that $\delta(t)$ is not the usual type of function that we have been considering, since we expect from Theorem 7.1.3 that $\mathscr{L}\{f(t)\} \to 0$ as $s \to \infty$.

EXAMPLE 1 Two Initial-Value Problems

Solve $y'' + y = 4\,\delta(t - 2\pi)$ subject to

(a) $y(0) = 1, \quad y'(0) = 0$ **(b)** $y(0) = 0, \quad y'(0) = 0$.

The two initial-value problems could serve as models for describing the motion of a mass on a spring moving in a medium in which damping is negligible. At $t = 2\pi$ the mass is given a sharp blow. In (a) the mass is released from rest 1 unit below the equilibrium position. In (b) the mass is at rest in the equilibrium position.

SOLUTION **(a)** From (3) the Laplace transform of the differential equation is

$$s^2Y(s) - s + Y(s) = 4e^{-2\pi s} \qquad \text{or} \qquad Y(s) = \frac{s}{s^2 + 1} + \frac{4e^{-2\pi s}}{s^2 + 1}.$$

Using the inverse form of the second translation theorem, we find

$$y(t) = \cos t + 4\sin(t - 2\pi)\,\mathscr{U}(t - 2\pi).$$

Since $\sin(t - 2\pi) = \sin t$, the foregoing solution can be written as

$$y(t) = \begin{cases} \cos t, & 0 \le t < 2\pi \\ \cos t + 4\sin t, & t \ge 2\pi. \end{cases} \tag{5}$$

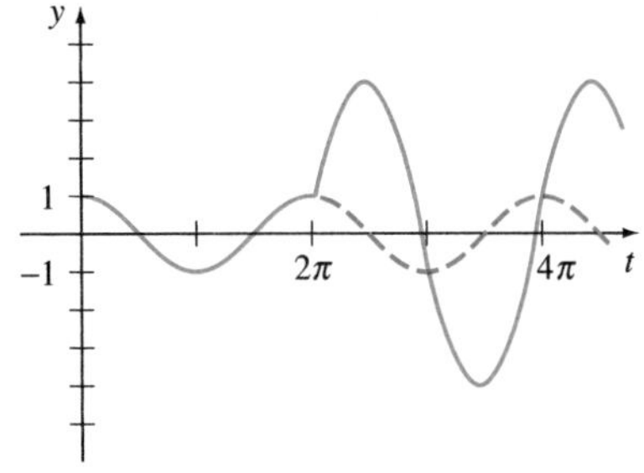

FIGURE 7.5.3 Mass is struck at $t = 2\pi$ in part (a) of Example 1

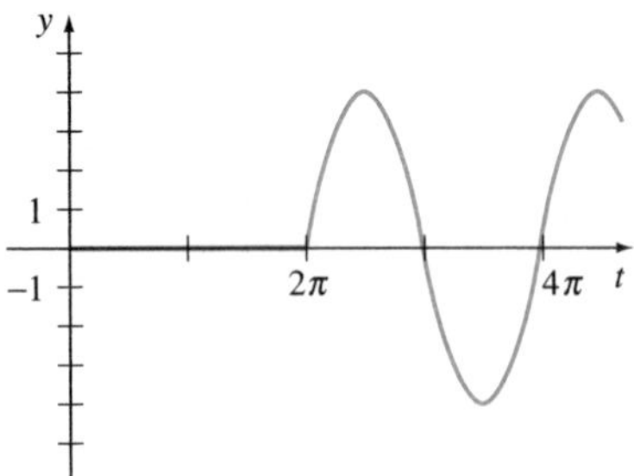

FIGURE 7.5.4 No motion until mass is struck at $t = 2\pi$ in part (b) of Example 1

In Figure 7.5.3 we see from the graph of (5) that the mass is exhibiting simple harmonic motion until it is struck at $t = 2\pi$. The influence of the unit impulse is to increase the amplitude of vibration to $\sqrt{17}$ for $t > 2\pi$.

(b) In this case the transform of the equation is simply

$$Y(s) = \frac{4e^{-2\pi s}}{s^2 + 1},$$

and so

$$\begin{aligned} y(t) &= 4 \sin(t - 2\pi)\,\mathscr{U}(t - 2\pi) \\ &= \begin{cases} 0, & 0 \le t < 2\pi \\ 4 \sin t, & t \ge 2\pi. \end{cases} \end{aligned} \tag{6}$$

The graph of (6) in Figure 7.5.4 shows, as we would expect from the initial conditions that the mass exhibits no motion until it is struck at $t = 2\pi$. ≡

REMARKS

(*i*) If $\delta(t - t_0)$ were a function in the usual sense, then property (*i*) on page 313 would imply $\int_0^\infty \delta(t - t_0)\,dt = 0$ rather than $\int_0^\infty \delta(t - t_0)\,dt = 1$. Because the Dirac delta function did not "behave" like an ordinary function, even though its users produced correct results, it was met initially with great scorn by mathematicians. However, in the 1940s Dirac's controversial function was put on a rigorous footing by the French mathematician Laurent Schwartz in his book *La Théorie de distribution*, and this, in turn, led to an entirely new branch of mathematics known as the **theory of distributions** or **generalized functions.** In this theory (2) is not an accepted definition of $\delta(t - t_0)$, nor does one speak of a function whose values are either ∞ or 0. Although we shall not pursue this topic any further, suffice it to say that the Dirac delta function is best characterized by its effect on other functions. If f is a continuous function, then

$$\int_0^\infty f(t)\,\delta(t - t_0)\,dt = f(t_0) \tag{7}$$

can be taken as the *definition* of $\delta(t - t_0)$. This result is known as the **sifting property,** since $\delta(t - t_0)$ has the effect of sifting the value $f(t_0)$ out of the set of values of f on $[0, \infty)$. Note that property (*ii*) (with $f(t) = 1$) and (3) (with $f(t) = e^{-st}$) are consistent with (7).

(*ii*) In (*iii*) in the *Remarks* at the end of Section 7.2 we indicated that the transfer function of a general linear nth-order differential equation with constant coefficients is $W(s) = 1/P(s)$, where $P(s) = a_n s^n + a_{n-1}s^{n-1} + \cdots + a_0$. The transfer function is the Laplace transform of function $w(t)$, called the **weight function** of a linear system. But $w(t)$ can also be characterized in terms of the discussion at hand. For simplicity let us consider a second-order linear system in which the input is a unit impulse at $t = 0$:

$$a_2 y'' + a_1 y' + a_0 y = \delta(t), \quad y(0) = 0, \quad y'(0) = 0.$$

Applying the Laplace transform and using $\mathscr{L}\{\delta(t)\} = 1$ shows that the transform of the response y in this case is the transfer function

$$Y(s) = \frac{1}{a_2 s^2 + a_1 s + a_0} = \frac{1}{P(s)} = W(s) \quad \text{and so} \quad y = \mathscr{L}^{-1}\left\{\frac{1}{P(s)}\right\} = w(t).$$

From this we can see, in general, that the weight function $y = w(t)$ of an nth-order linear system is the zero-state response of the system to a unit impulse. For this reason $w(t)$ is also called the **impulse response** of the system.

EXERCISES 7.5

Answers to selected odd-numbered problems begin on page ANS-13.

In Problems 1–12 use the Laplace transform to solve the given initial-value problem.

1. $y' - 3y = \delta(t - 2), \quad y(0) = 0$
2. $y' + y = \delta(t - 1), \quad y(0) = 2$
3. $y'' + y = \delta(t - 2\pi), \quad y(0) = 0, y'(0) = 1$
4. $y'' + 16y = \delta(t - 2\pi), \quad y(0) = 0, y'(0) = 0$
5. $y'' + y = \delta\left(t - \frac{1}{2}\pi\right) + \delta\left(t - \frac{3}{2}\pi\right),$ $y(0) = 0, y'(0) = 0$
6. $y'' + y = \delta(t - 2\pi) + \delta(t - 4\pi), \quad y(0) = 1, y'(0) = 0$
7. $y'' + 2y' = \delta(t - 1), \quad y(0) = 0, y'(0) = 1$
8. $y'' - 2y' = 1 + \delta(t - 2), \quad y(0) = 0, y'(0) = 1$
9. $y'' + 4y' + 5y = \delta(t - 2\pi), \quad y(0) = 0, y'(0) = 0$
10. $y'' + 2y' + y = \delta(t - 1), \quad y(0) = 0, y'(0) = 0$
11. $y'' + 4y' + 13y = \delta(t - \pi) + \delta(t - 3\pi),$ $y(0) = 1, y'(0) = 0$
12. $y'' - 7y' + 6y = e^t + \delta(t - 2) + \delta(t - 4),$ $y(0) = 0, y'(0) = 0$
13. A uniform beam of length L carries a concentrated load w_0 at $x = \frac{1}{2}L$. The beam is embedded at its left end and is free at its right end. Use the Laplace transform to determine the deflection $y(x)$ from
$$EI\frac{d^4y}{dx^4} = w_0\,\delta\left(x - \tfrac{1}{2}L\right),$$
where $y(0) = 0$, $y'(0) = 0$, $y''(L) = 0$, and $y'''(L) = 0$.
14. Solve the differential equation in Problem 13 subject to $y(0) = 0$, $y'(0) = 0$, $y(L) = 0$, $y'(L) = 0$. In this case the beam is embedded at both ends. See Figure 7.5.5.

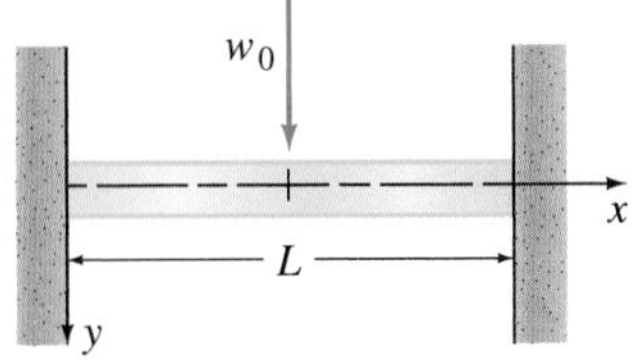

FIGURE 7.5.5 Beam in Problem 14

Discussion Problems

15. Someone tells you that the solutions of the two IVPs
$$y'' + 2y' + 10y = 0, \qquad y(0) = 0, \quad y'(0) = 1$$
$$y'' + 2y' + 10y = \delta(t), \qquad y(0) = 0, \quad y'(0) = 0$$
are exactly the same. Do you agree or disagree? Defend your answer.

7.6 SYSTEMS OF LINEAR DIFFERENTIAL EQUATIONS

REVIEW MATERIAL

- Solving systems of two equations in two unknowns

INTRODUCTION When initial conditions are specified, the Laplace transform of each equation in a system of linear differential equations with constant coefficients reduces the system of DEs to a set of simultaneous algebraic equations in the transformed functions. We solve the system of algebraic equations for each of the transformed functions and then find the inverse Laplace transforms in the usual manner.

Coupled Springs Two masses m_1 and m_2 are connected to two springs A and B of negligible mass having spring constants k_1 and k_2, respectively. In turn the two springs are attached as shown in Figure 7.6.1 on page 316. Let $x_1(t)$ and $x_2(t)$ denote the vertical displacements of the masses from their equilibrium positions. When the system is in motion, spring B is subject to both an elongation and a compression; hence its net elongation is $x_2 - x_1$. Therefore it follows from Hooke's law that springs A and B exert forces $-k_1x_1$ and $k_2(x_2 - x_1)$, respectively, on m_1. If no external force is impressed on the system and if no damping force is present, then the net force on m_1 is $-k_1x_1 + k_2(x_2 - x_1)$. By Newton's second law we can write

$$m_1\frac{d^2x_1}{dt^2} = -k_1x_1 + k_2(x_2 - x_1).$$

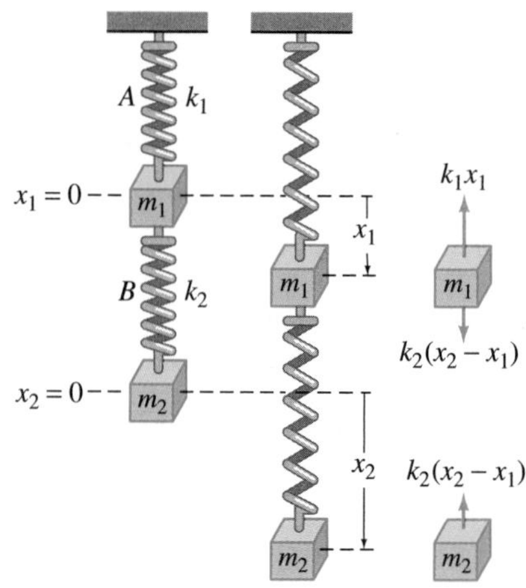

FIGURE 7.6.1 Coupled spring/mass system

Similarly, the net force exerted on mass m_2 is due solely to the net elongation of B; that is, $-k_2(x_2 - x_1)$. Hence we have

$$m_2\frac{d^2x_2}{dt^2} = -k_2(x_2 - x_1).$$

In other words, the motion of the coupled system is represented by the system of simultaneous second-order differential equations

$$\begin{aligned} m_1x_1'' &= -k_1x_1 + k_2(x_2 - x_1) \\ m_2x_2'' &= -k_2(x_2 - x_1). \end{aligned} \tag{1}$$

In the next example we solve (1) under the assumptions that $k_1 = 6$, $k_2 = 4$, $m_1 = 1$, $m_2 = 1$, and that the masses start from their equilibrium positions with opposite unit velocities.

EXAMPLE 1 **Coupled Springs**

Solve

$$\begin{aligned} x_1'' + 10x_1 \quad - 4x_2 &= 0 \\ -4x_1 + x_2'' + 4x_2 &= 0 \end{aligned} \tag{2}$$

subject to $x_1(0) = 0, x_1'(0) = 1, x_2(0) = 0, x_2'(0) = -1$.

SOLUTION The Laplace transform of each equation is

$$\begin{aligned} s^2X_1(s) - sx_1(0) - x_1'(0) + 10X_1(s) - 4X_2(s) &= 0 \\ -4X_1(s) + s^2X_2(s) - sx_2(0) - x_2'(0) + 4X_2(s) &= 0, \end{aligned}$$

where $X_1(s) = \mathscr{L}\{x_1(t)\}$ and $X_2(s) = \mathscr{L}\{x_2(t)\}$. The preceding system is the same as

$$\begin{aligned} (s^2 + 10)X_1(s) - \quad 4X_2(s) &= 1 \\ -4X_1(s) + (s^2 + 4)X_2(s) &= -1. \end{aligned} \tag{3}$$

Solving (3) for $X_1(s)$ and using partial fractions on the result yields

$$X_1(s) = \frac{s^2}{(s^2 + 2)(s^2 + 12)} = -\frac{1/5}{s^2 + 2} + \frac{6/5}{s^2 + 12},$$

and therefore

$$\begin{aligned} x_1(t) &= -\frac{1}{5\sqrt{2}}\mathscr{L}^{-1}\left\{\frac{\sqrt{2}}{s^2 + 2}\right\} + \frac{6}{5\sqrt{12}}\mathscr{L}^{-1}\left\{\frac{\sqrt{12}}{s^2 + 12}\right\} \\ &= -\frac{\sqrt{2}}{10}\sin\sqrt{2}t + \frac{\sqrt{3}}{5}\sin 2\sqrt{3}t. \end{aligned}$$

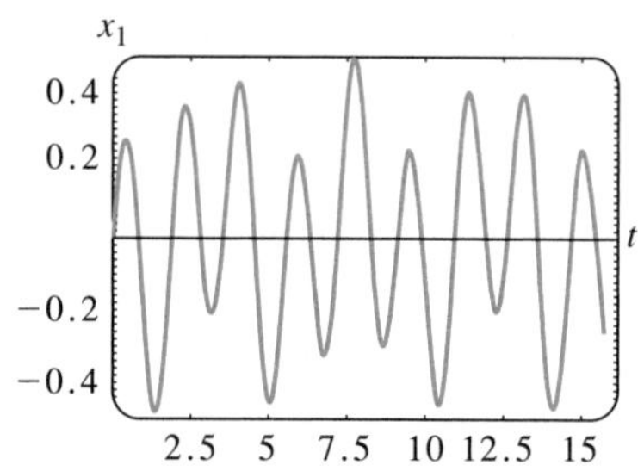

(a) plot of $x_1(t)$

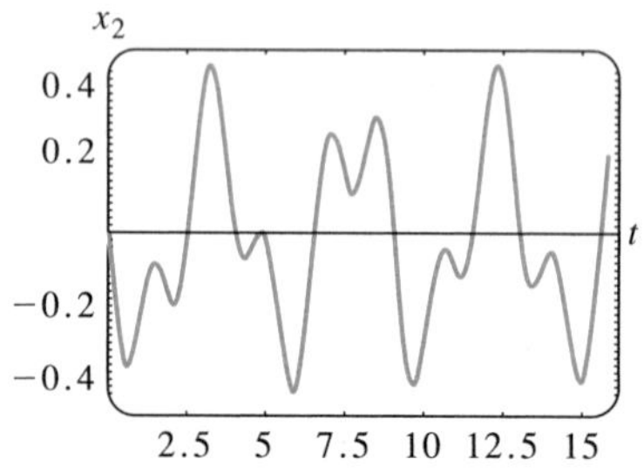

(b) plot of $x_2(t)$

FIGURE 7.6.2 Displacements of the two masses in Example 1

Substituting the expression for $X_1(s)$ into the first equation of (3) gives

$$X_2(s) = -\frac{s^2 + 6}{(s^2 + 2)(s^2 + 12)} = -\frac{2/5}{s^2 + 2} - \frac{3/5}{s^2 + 12}$$

and

$$\begin{aligned} x_2(t) &= -\frac{2}{5\sqrt{2}}\mathscr{L}^{-1}\left\{\frac{\sqrt{2}}{s^2 + 2}\right\} - \frac{3}{5\sqrt{12}}\mathscr{L}^{-1}\left\{\frac{\sqrt{12}}{s^2 + 12}\right\} \\ &= -\frac{\sqrt{2}}{5}\sin\sqrt{2}t - \frac{\sqrt{3}}{10}\sin 2\sqrt{3}t. \end{aligned}$$

Finally, the solution to the given system (2) is

$$\begin{aligned} x_1(t) &= -\frac{\sqrt{2}}{10}\sin\sqrt{2}t + \frac{\sqrt{3}}{5}\sin 2\sqrt{3}t \\ x_2(t) &= -\frac{\sqrt{2}}{5}\sin\sqrt{2}t - \frac{\sqrt{3}}{10}\sin 2\sqrt{3}t. \end{aligned} \tag{4}$$

The graphs of x_1 and x_2 in Figure 7.6.2 reveal the complicated oscillatory motion of each mass. ≡

≡ **Networks** In (18) of Section 3.3 we saw the currents $i_1(t)$ and $i_2(t)$ in the network shown in Figure 7.6.3, containing an inductor, a resistor, and a capacitor, were governed by the system of first-order differential equations

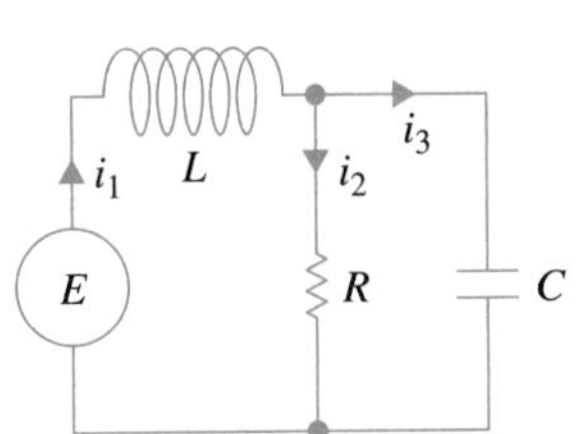

FIGURE 7.6.3 Electrical network

$$\begin{aligned} L\frac{di_1}{dt} + Ri_2 &= E(t) \\ RC\frac{di_2}{dt} + i_2 - i_1 &= 0. \end{aligned} \tag{5}$$

We solve this system by the Laplace transform in the next example.

EXAMPLE 2 An Electrical Network

Solve the system in (5) under the conditions $E(t) = 60$ V, $L = 1$ h, $R = 50$ Ω, $C = 10^{-4}$ f, and the currents i_1 and i_2 are initially zero.

SOLUTION We must solve

$$\begin{aligned} \frac{di_1}{dt} + 50i_2 &= 60 \\ 50(10^{-4})\frac{di_2}{dt} + i_2 - i_1 &= 0 \end{aligned}$$

subject to $i_1(0) = 0$, $i_2(0) = 0$.

Applying the Laplace transform to each equation of the system and simplifying gives

$$\begin{aligned} sI_1(s) + \quad 50I_2(s) &= \frac{60}{s} \\ -200I_1(s) + (s + 200)I_2(s) &= 0, \end{aligned}$$

where $I_1(s) = \mathscr{L}\{i_1(t)\}$ and $I_2(s) = \mathscr{L}\{i_2(t)\}$. Solving the system for I_1 and I_2 and decomposing the results into partial fractions gives

$$\begin{aligned} I_1(s) &= \frac{60s + 12{,}000}{s(s + 100)^2} = \frac{6/5}{s} - \frac{6/5}{s + 100} - \frac{60}{(s + 100)^2} \\ I_2(s) &= \frac{12{,}000}{s(s + 100)^2} = \frac{6/5}{s} - \frac{6/5}{s + 100} - \frac{120}{(s + 100)^2}. \end{aligned}$$

Taking the inverse Laplace transform, we find the currents to be

$$\begin{aligned} i_1(t) &= \frac{6}{5} - \frac{6}{5}e^{-100t} - 60te^{-100t} \\ i_2(t) &= \frac{6}{5} - \frac{6}{5}e^{-100t} - 120te^{-100t}. \end{aligned}$$

≡

Note that both $i_1(t)$ and $i_2(t)$ in Example 2 tend toward the value $E/R = \frac{6}{5}$ as $t \to \infty$. Furthermore, since the current through the capacitor is $i_3(t) = i_1(t) - i_2(t) = 60te^{-100t}$, we observe that $i_3(t) \to 0$ as $t \to \infty$.

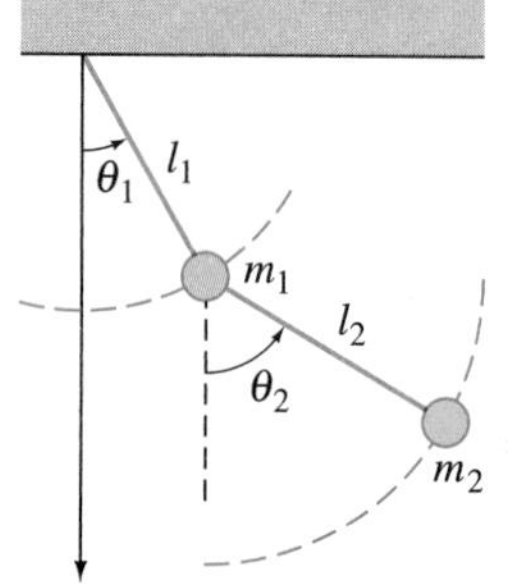

FIGURE 7.6.4 Double pendulum

Double Pendulum Consider the double-pendulum system consisting of a pendulum attached to a pendulum shown in Figure 7.6.4. We assume that the system oscillates in a vertical plane under the influence of gravity, that the mass of each rod is negligible, and that no damping forces act on the system. Figure 7.6.4 also shows that the displacement angle θ_1 is measured (in radians) from a vertical line extending downward from the pivot of the system and that θ_2 is measured from a vertical line extending downward from the center of mass m_1. The positive direction is to the right; the negative direction is to the left. As we might expect from the analysis leading to equation (6) of Section 5.3, the system of differential equations describing the motion is nonlinear:

$$\begin{aligned}(m_1 + m_2)l_1^2\theta_1'' + m_2l_1l_2\theta_2''\cos(\theta_1 - \theta_2) + m_2l_1l_2(\theta_2')^2\sin(\theta_1 - \theta_2) + (m_1 + m_2)l_1g\sin\theta_1 &= 0\\ m_2l_2^2\theta_2'' + m_2l_1l_2\theta_1''\cos(\theta_1 - \theta_2) - m_2l_1l_2(\theta_1')^2\sin(\theta_1 - \theta_2) + m_2l_2g\sin\theta_2 &= 0.\end{aligned} \tag{6}$$

But if the displacements $\theta_1(t)$ and $\theta_2(t)$ are assumed to be small, then the approximations $\cos(\theta_1 - \theta_2) \approx 1$, $\sin(\theta_1 - \theta_2) \approx 0$, $\sin\theta_1 \approx \theta_1$, $\sin\theta_2 \approx \theta_2$ enable us to replace system (6) by the linearization

$$\begin{aligned}(m_1 + m_2)l_1^2\theta_1'' + m_2l_1l_2\theta_2'' + (m_1 + m_2)l_1g\theta_1 &= 0\\ m_2l_2^2\theta_2'' + m_2l_1l_2\theta_1'' + m_2l_2g\theta_2 &= 0.\end{aligned} \tag{7}$$

EXAMPLE 3 Double Pendulum

It is left as an exercise to fill in the details of using the Laplace transform to solve system (7) when $m_1 = 3$, $m_2 = 1$, $l_1 = l_2 = 16$, $\theta_1(0) = 1$, $\theta_2(0) = -1$, $\theta_1'(0) = 0$, and $\theta_2'(0) = 0$. You should find that

$$\begin{aligned}\theta_1(t) &= \frac{1}{4}\cos\frac{2}{\sqrt{3}}t + \frac{3}{4}\cos 2t\\ \theta_2(t) &= \frac{1}{2}\cos\frac{2}{\sqrt{3}}t - \frac{3}{2}\cos 2t.\end{aligned} \tag{8}$$

With the aid of a CAS the positions of the two masses at $t = 0$ and at subsequent times are shown in Figure 7.6.5. See Problem 21 in Exercises 7.6.

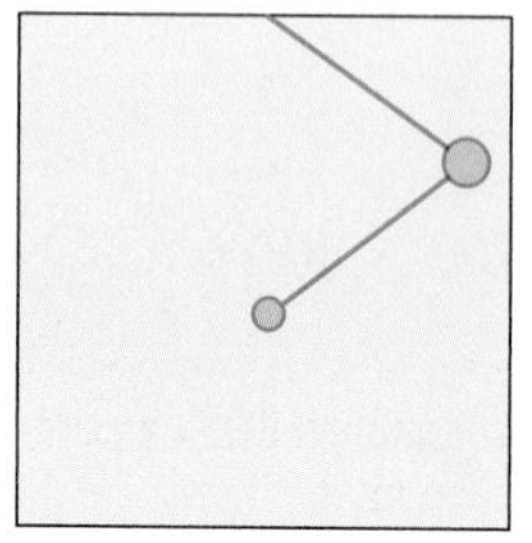

(a) $t = 0$

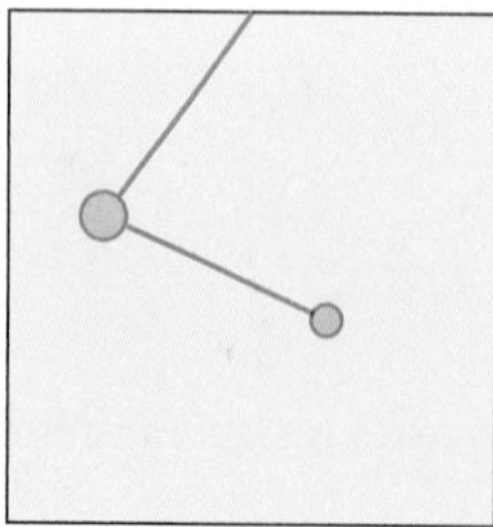

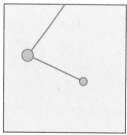

(b) $t = 1.4$

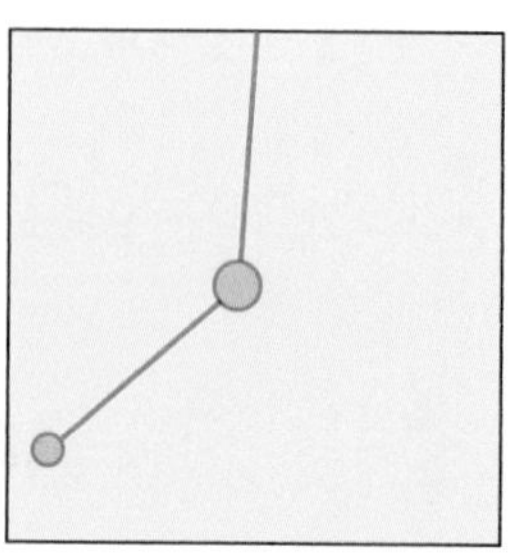

(c) $t = 2.5$

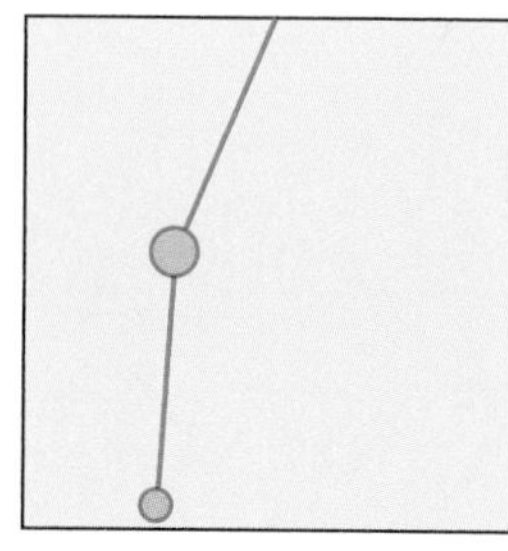

(d) $t = 8.5$

FIGURE 7.6.5 Positions of masses on double pendulum at various times in Example 3

EXERCISES 7.6

Answers to selected odd-numbered problems begin on page ANS-13.

In Problems 1–12 use the Laplace transform to solve the given system of differential equations.

1. $\dfrac{dx}{dt} = -x + y$

$\dfrac{dy}{dt} = 2x$

$x(0) = 0, \quad y(0) = 1$

2. $\dfrac{dx}{dt} = 2y + e^t$

$\dfrac{dy}{dt} = 8x - t$

$x(0) = 1, \quad y(0) = 1$

3. $\dfrac{dx}{dt} = x - 2y$

$\dfrac{dy}{dt} = 5x - y$

$x(0) = -1, \quad y(0) = 2$

4. $\dfrac{dx}{dt} + 3x + \dfrac{dy}{dt} = 1$

$\dfrac{dx}{dt} - x + \dfrac{dy}{dt} - y = e^t$

$x(0) = 0, \quad y(0) = 0$

5. $2\dfrac{dx}{dt} + \dfrac{dy}{dt} - 2x = 1$

$\dfrac{dx}{dt} + \dfrac{dy}{dt} - 3x - 3y = 2$

$x(0) = 0, \quad y(0) = 0$

6. $\dfrac{dx}{dt} + x - \dfrac{dy}{dt} + y = 0$

$\dfrac{dx}{dt} + \dfrac{dy}{dt} + 2y = 0$

$x(0) = 0, \quad y(0) = 1$

7. $\dfrac{d^2x}{dt^2} + x - y = 0$

$\dfrac{d^2y}{dt^2} + y - x = 0$

$x(0) = 0, \quad x'(0) = -2,$

$y(0) = 0, \quad y'(0) = 1$

8. $\dfrac{d^2x}{dt^2} + \dfrac{dx}{dt} + \dfrac{dy}{dt} = 0$

$\dfrac{d^2y}{dt^2} + \dfrac{dy}{dt} - 4\dfrac{dx}{dt} = 0$

$x(0) = 1, \quad x'(0) = 0,$

$y(0) = -1, \quad y'(0) = 5$

9. $\dfrac{d^2x}{dt^2} + \dfrac{d^2y}{dt^2} = t^2$

$\dfrac{d^2x}{dt^2} - \dfrac{d^2y}{dt^2} = 4t$

$x(0) = 8, \quad x'(0) = 0,$

$y(0) = 0, \quad y'(0) = 0$

10. $\dfrac{dx}{dt} - 4x + \dfrac{d^3y}{dt^3} = 6 \sin t$

$\dfrac{dx}{dt} + 2x - 2\dfrac{d^3y}{dt^3} = 0$

$x(0) = 0, \quad y(0) = 0,$

$y'(0) = 0, \quad y''(0) = 0$

11. $\dfrac{d^2x}{dt^2} + 3\dfrac{dy}{dt} + 3y = 0$

$\dfrac{d^2x}{dt^2} + 3y = te^{-t}$

$x(0) = 0, \quad x'(0) = 2, \quad y(0) = 0$

12. $\dfrac{dx}{dt} = 4x - 2y + 2\mathcal{U}(t-1)$

$\dfrac{dy}{dt} = 3x - y + \mathcal{U}(t-1)$

$x(0) = 0, \quad y(0) = \frac{1}{2}$

13. Solve system (1) when $k_1 = 3$, $k_2 = 2$, $m_1 = 1$, $m_2 = 1$ and $x_1(0) = 0$, $x_1'(0) = 1$, $x_2(0) = 1$, $x_2'(0) = 0$.

14. Derive the system of differential equations describing the straight-line vertical motion of the coupled springs shown in Figure 7.6.6. Use the Laplace transform to solve the system when $k_1 = 1$, $k_2 = 1$, $k_3 = 1$, $m_1 = 1$, $m_2 = 1$ and $x_1(0) = 0$, $x_1'(0) = -1$, $x_2(0) = 0$, $x_2'(0) = 1$.

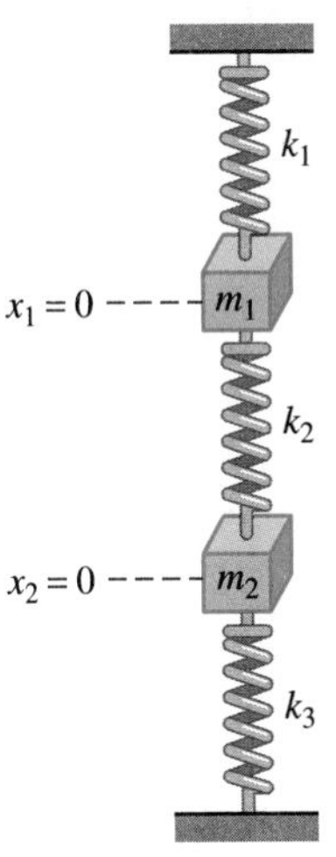

FIGURE 7.6.6 Coupled springs in Problem 14

15. **(a)** Show that the system of differential equations for the currents $i_2(t)$ and $i_3(t)$ in the electrical network shown in Figure 7.6.7 is

$$L_1\frac{di_2}{dt} + Ri_2 + Ri_3 = E(t)$$

$$L_2\frac{di_3}{dt} + Ri_2 + Ri_3 = E(t).$$

(b) Solve the system in part (a) if $R = 5\ \Omega$, $L_1 = 0.01$ h, $L_2 = 0.0125$ h, $E = 100$ V, $i_2(0) = 0$, and $i_3(0) = 0$.

(c) Determine the current $i_1(t)$.

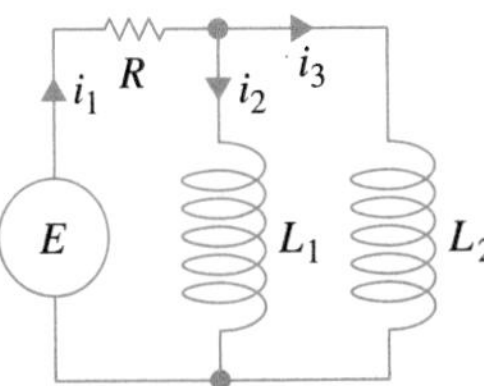

FIGURE 7.6.7 Network in Problem 15

16. **(a)** In Problem 12 in Exercises 3.3 you were asked to show that the currents $i_2(t)$ and $i_3(t)$ in the electrical network shown in Figure 7.6.8 satisfy

$$L\frac{di_2}{dt} + L\frac{di_3}{dt} + R_1 i_2 = E(t)$$

$$-R_1\frac{di_2}{dt} + R_2\frac{di_3}{dt} + \frac{1}{C}i_3 = 0.$$

Solve the system if $R_1 = 10\ \Omega$, $R_2 = 5\ \Omega$, $L = 1$ h, $C = 0.2$ f,

$$E(t) = \begin{cases} 120, & 0 \le t < 2 \\ 0, & t \ge 2, \end{cases}$$

$i_2(0) = 0$, and $i_3(0) = 0$.

(b) Determine the current $i_1(t)$.

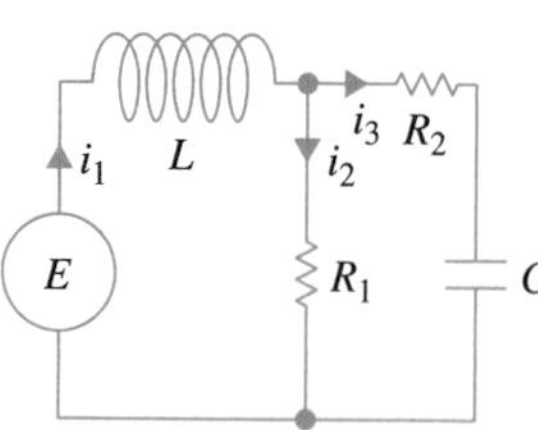

FIGURE 7.6.8 Network in Problem 16

17. Solve the system given in (17) of Section 3.3 when $R_1 = 6\ \Omega, R_2 = 5\ \Omega, L_1 = 1$ h, $L_2 = 1$ h, $E(t) = 50 \sin t$ V, $i_2(0) = 0$, and $i_3(0) = 0$.

18. Solve (5) when $E = 60$ V, $L = \frac{1}{2}$ h, $R = 50\ \Omega$, $C = 10^{-4}$ f, $i_1(0) = 0$, and $i_2(0) = 0$.

19. Solve (5) when $E = 60$ V, $L = 2$ h, $R = 50\ \Omega$, $C = 10^{-4}$ f, $i_1(0) = 0$, and $i_2(0) = 0$.

20. **(a)** Show that the system of differential equations for the charge on the capacitor $q(t)$ and the current $i_3(t)$ in the electrical network shown in Figure 7.6.9 is

$$R_1 \frac{dq}{dt} + \frac{1}{C} q + R_1 i_3 = E(t)$$

$$L \frac{di_3}{dt} + R_2 i_3 - \frac{1}{C} q = 0.$$

(b) Find the charge on the capacitor when $L = 1$ h, $R_1 = 1\ \Omega$, $R_2 = 1\ \Omega$, $C = 1$ f,

$$E(t) = \begin{cases} 0, & 0 < t < 1 \\ 50e^{-t}, & t \ge 1, \end{cases}$$

$i_3(0) = 0$, and $q(0) = 0$.

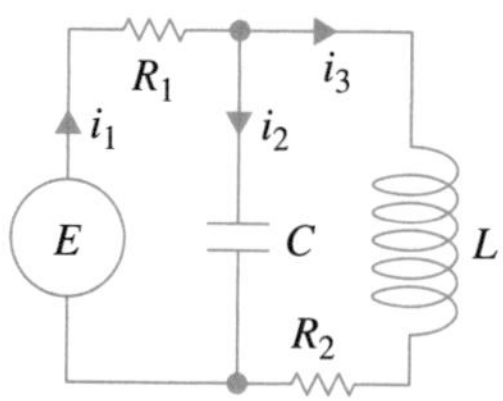

FIGURE 7.6.9 Network in Problem 20

Computer Lab Assignments

21. **(a)** Use the Laplace transform and the information given in Example 3 to obtain the solution (8) of the system given in (7).

(b) Use a graphing utility to graph $\theta_1(t)$ and $\theta_2(t)$ in the $t\theta$-plane. Which mass has extreme displacements of greater magnitude? Use the graphs to estimate the first time that each mass passes through its equilibrium position. Discuss whether the motion of the pendulums is periodic.

(c) Graph $\theta_1(t)$ and $\theta_2(t)$ in the $\theta_1\theta_2$-plane as parametric equations. The curve defined by these parametric equations is called a **Lissajous curve.**

(d) The positions of the masses at $t = 0$ are given in Figure 7.6.5(a). Note that we have used 1 radian $\approx 57.3°$. Use a calculator or a table application in a CAS to construct a table of values of the angles θ_1 and θ_2 for $t = 1, 2, \ldots, 10$ s. Then plot the positions of the two masses at these times.

(e) Use a CAS to find the first time that $\theta_1(t) = \theta_2(t)$ and compute the corresponding angular value. Plot the positions of the two masses at these times.

(f) Utilize the CAS to draw appropriate lines to simulate the pendulum rods, as in Figure 7.6.5. Use the animation capability of your CAS to make a "movie" of the motion of the double pendulum from $t = 0$ to $t = 10$ using a time increment of 0.1. [*Hint*: Express the coordinates $(x_1(t), y_1(t))$ and $(x_2(t), y_2(t))$ of the masses m_1 and m_2, respectively, in terms of $\theta_1(t)$ and $\theta_2(t)$.]

CHAPTER 7 IN REVIEW

Answers to selected odd-numbered problems begin on page ANS-13.

In Problems 1 and 2 use the definition of the Laplace transform to find $\mathscr{L}\{f(t)\}$.

1. $f(t) = \begin{cases} t, & 0 \le t < 1 \\ 2 - t, & t \ge 1 \end{cases}$

2. $f(t) = \begin{cases} 0, & 0 \le t < 2 \\ 1, & 2 \le t < 4 \\ 0, & t \ge 4 \end{cases}$

In Problems 3–24 fill in the blanks or answer true or false.

3. If f is not piecewise continuous on $[0, \infty)$, then $\mathscr{L}\{f(t)\}$ will not exist. _______

4. The function $f(t) = (e^t)^{10}$ is not of exponential order. _______

5. $F(s) = s^2/(s^2 + 4)$ is not the Laplace transform of a function that is piecewise continuous and of exponential order. _______

6. If $\mathscr{L}\{f(t)\} = F(s)$ and $\mathscr{L}\{g(t)\} = G(s)$, then $\mathscr{L}^{-1}\{F(s)G(s)\} = f(t)g(t)$. _______

7. $\mathscr{L}\{e^{-7t}\} =$ _______

8. $\mathscr{L}\{te^{-7t}\} =$ _______

9. $\mathscr{L}\{\sin 2t\} =$ _______

10. $\mathscr{L}\{e^{-3t}\sin 2t\} =$ _______

11. $\mathscr{L}\{t\sin 2t\} =$ _______

12. $\mathscr{L}\{\sin 2t\,\mathscr{U}(t-\pi)\} =$ _______

13. $\mathscr{L}^{-1}\left\{\dfrac{20}{s^6}\right\} =$ _______

14. $\mathscr{L}^{-1}\left\{\dfrac{1}{3s-1}\right\} =$ _______

15. $\mathscr{L}^{-1}\left\{\dfrac{1}{(s-5)^3}\right\} =$ _______

16. $\mathscr{L}^{-1}\left\{\dfrac{1}{s^2-5}\right\} =$ _______

17. $\mathscr{L}^{-1}\left\{\dfrac{s}{s^2-10s+29}\right\} =$ _______

18. $\mathscr{L}^{-1}\left\{\dfrac{e^{-5s}}{s^2}\right\} =$ _______

19. $\mathscr{L}^{-1}\left\{\dfrac{s+\pi}{s^2+\pi^2}e^{-s}\right\} =$ _______

20. $\mathscr{L}^{-1}\left\{\dfrac{1}{L^2s^2+n^2\pi^2}\right\} =$ _______

21. $\mathscr{L}\{e^{-5t}\}$ exists for $s >$ _______.

22. If $\mathscr{L}\{f(t)\} = F(s)$, then $\mathscr{L}\{te^{8t}f(t)\} =$ _______.

23. If $\mathscr{L}\{f(t)\} = F(s)$ and $k > 0$, then $\mathscr{L}\{e^{at}f(t-k)\mathscr{U}(t-k)\} =$ _______.

24. $\mathscr{L}\{\int_0^t e^{a\tau}f(\tau)\,d\tau\} =$ _______ whereas $\mathscr{L}\{e^{at}\int_0^t f(\tau)\,d\tau\} =$ _______.

In Problems 25–28 use the unit step function to find an equation for each graph in terms of the function $y = f(t)$, whose graph is given in Figure 7.R.1.

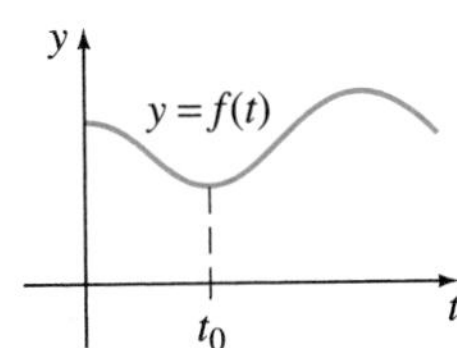

FIGURE 7.R.1 Graph for Problems 25–28

25.

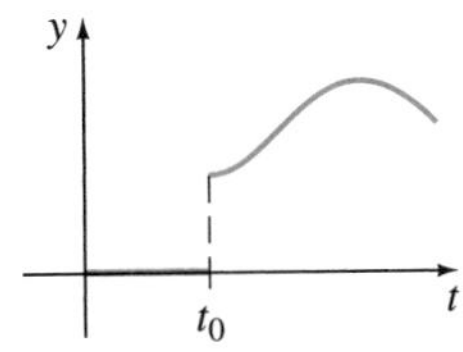

FIGURE 7.R.2 Graph for Problem 25

26.

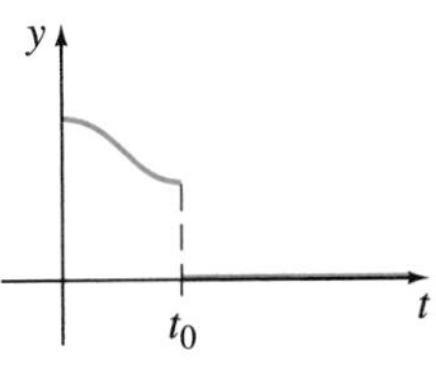

FIGURE 7.R.3 Graph for Problem 26

27.

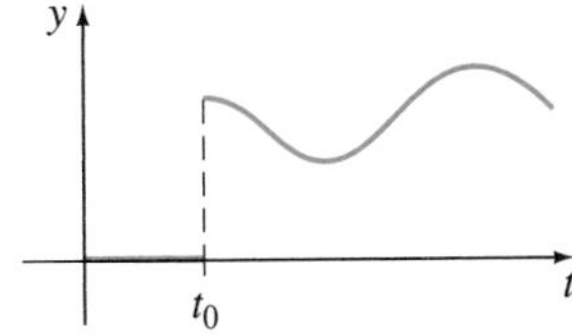

FIGURE 7.R.4 Graph for Problem 27

28.

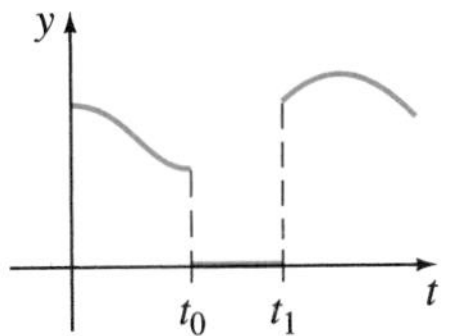

FIGURE 7.R.5 Graph for Problem 28

In Problems 29–32 express f in terms of unit step functions. Find $\mathscr{L}\{f(t)\}$ and $\mathscr{L}\{e^t f(t)\}$.

29.

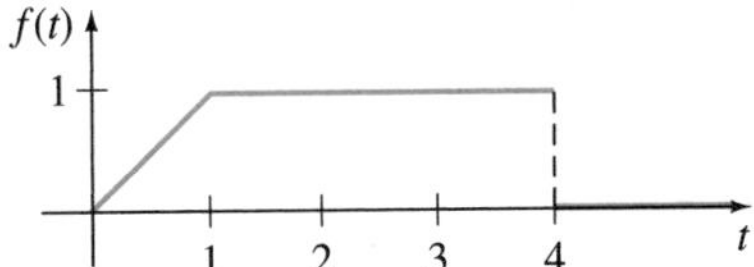

FIGURE 7.R.6 Graph for Problem 29

30.

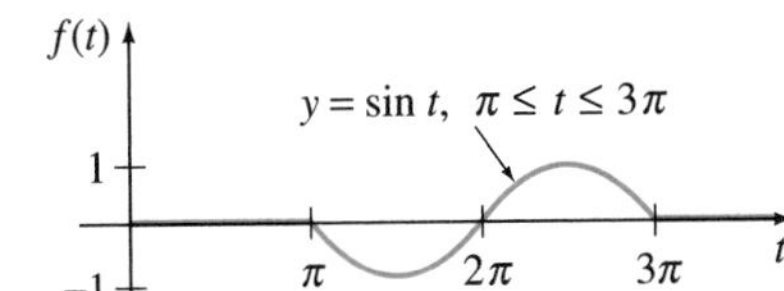

FIGURE 7.R.7 Graph for Problem 30

31.

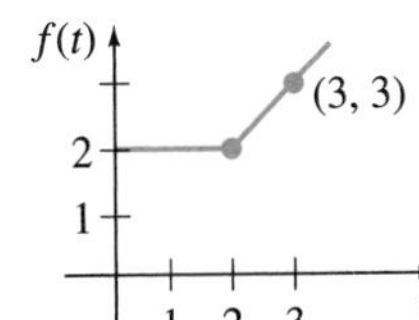

FIGURE 7.R.8 Graph for Problem 31

32.

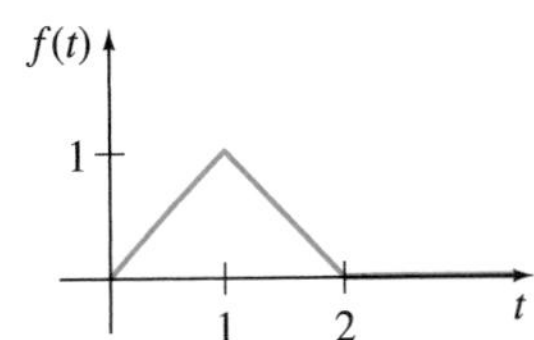

FIGURE 7.R.9 Graph for Problem 32

In Problems 33–40 use the Laplace transform to solve the given equation.

33. $y'' - 2y' + y = e^t, \quad y(0) = 0, y'(0) = 5$

34. $y'' - 8y' + 20y = te^t, \quad y(0) = 0, y'(0) = 0$

35. $y'' + 6y' + 5y = t - t\,\mathscr{U}(t - 2), \quad y(0) = 1, \ y'(0) = 0$

36. $y' - 5y = f(t)$, where

$$f(t) = \begin{cases} t^2, & 0 \le t < 1 \\ 0, & t \ge 1 \end{cases}, \quad y(0) = 1$$

37. $y' + 2y = f(t), \quad y(0) = 1$, where $f(t)$ is given in Figure 7.R.10.

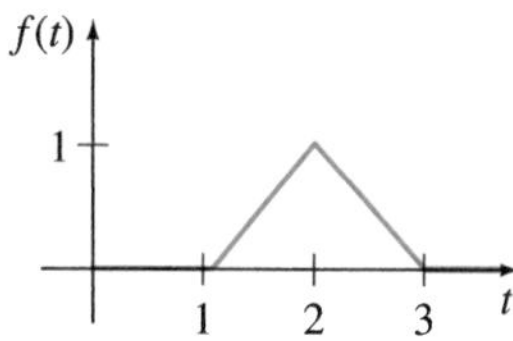

FIGURE 7.R.10 Graph for Problem 37

38. $y'' + 5y' + 4y = f(t), \quad y(0) = 0, y'(0) = 3$, where

$$f(t) = 12 \sum_{k=0}^{\infty} (-1)^k\, \mathscr{U}(t - k).$$

39. $y'(t) = \cos t + \displaystyle\int_0^t y(\tau)\cos(t - \tau)\,d\tau, \quad y(0) = 1$

40. $\displaystyle\int_0^t f(\tau) f(t - \tau)\,d\tau = 6t^3$

In Problems 41 and 42 use the Laplace transform to solve each system.

41. $x' + y = t$
$4x + y' = 0$
$x(0) = 1, \quad y(0) = 2$

42. $x'' + y'' = e^{2t}$
$2x' + y'' = -e^{2t}$
$x(0) = 0, \quad y(0) = 0,$
$x'(0) = 0, \quad y'(0) = 0$

43. The current $i(t)$ in an RC-series circuit can be determined from the integral equation

$$Ri + \frac{1}{C}\int_0^t i(\tau)\,d\tau = E(t),$$

where $E(t)$ is the impressed voltage. Determine $i(t)$ when $R = 10\ \Omega$, $C = 0.5$ f, and $E(t) = 2(t^2 + t)$.

44. A series circuit contains an inductor, a resistor, and a capacitor for which $L = \frac{1}{2}$ h, $R = 10\ \Omega$, and $C = 0.01$ f, respectively. The voltage

$$E(t) = \begin{cases} 10, & 0 \le t < 5 \\ 0, & t \ge 5 \end{cases}$$

is applied to the circuit. Determine the instantaneous charge $q(t)$ on the capacitor for $t > 0$ if $q(0) = 0$ and $q'(0) = 0$.

45. A uniform cantilever beam of length L is embedded at its left end ($x = 0$) and free at its right end. Find the deflection $y(x)$ if the load per unit length is given by

$$w(x) = \frac{2w_0}{L}\left[\frac{L}{2} - x + \left(x - \frac{L}{2}\right)\mathscr{U}\left(x - \frac{L}{2}\right)\right].$$

46. When a uniform beam is supported by an elastic foundation, the differential equation for its deflection $y(x)$ is

$$EI\frac{d^4y}{dx^4} + ky = w(x),$$

where k is the modulus of the foundation and $-ky$ is the restoring force of the foundation that acts in the direction opposite to that of the load $w(x)$. See Figure 7.R.11. For algebraic convenience suppose that the differential equation is written as

$$\frac{d^4y}{dx^4} + 4a^4y = \frac{w(x)}{EI},$$

where $a = (k/4EI)^{1/4}$. Assume $L = \pi$ and $a = 1$. Find the deflection $y(x)$ of a beam that is supported on an elastic foundation when

(a) the beam is simply supported at both ends and a constant load w_0 is uniformly distributed along its length,

(b) the beam is embedded at both ends and $w(x)$ is a concentrated load w_0 applied at $x = \pi/2$.

[*Hint*: In both parts of this problem use entries 35 and 36 in the table of Laplace transforms in Appendix III.]

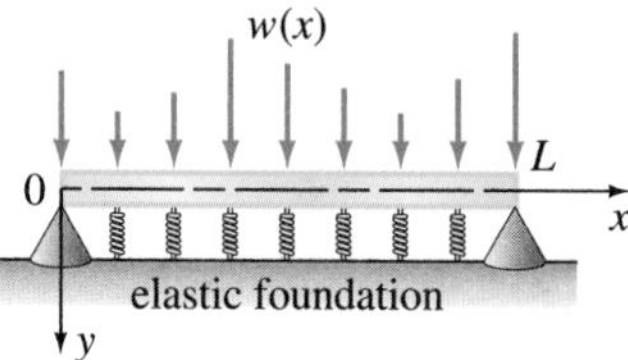

FIGURE 7.R.11 Beam on elastic foundation in Problem 46

47. **(a)** Suppose two identical pendulums are coupled by means of a spring with constant k. See Figure 7.R.12. Under the same assumptions made in the discussion preceding Example 3 in Section 7.6, it can be shown that when the displacement angles $\theta_1(t)$ and $\theta_2(t)$ are small, the system of linear differential equations describing the motion is

$$\theta_1'' + \frac{g}{l}\theta_1 = -\frac{k}{m}(\theta_1 - \theta_2)$$

$$\theta_2'' + \frac{g}{l}\theta_2 = \frac{k}{m}(\theta_1 - \theta_2).$$

Use the Laplace transform to solve the system when $\theta_1(0) = \theta_0, \theta_1'(0) = 0, \theta_2(0) = \psi_0, \theta_2'(0) = 0$, where θ_0 and ψ_0 constants. For convenience let $\omega^2 = g/l$, $K = k/m$.

(b) Use the solution in part (a) to discuss the motion of the coupled pendulums in the special case when the initial conditions are $\theta_1(0) = \theta_0$, $\theta_1'(0) = 0$, $\theta_2(0) = \theta_0$, $\theta_2'(0) = 0$. When the initial conditions are $\theta_1(0) = \theta_0$, $\theta_1'(0) = 0$, $\theta_2(0) = -\theta_0$, $\theta_2'(0) = 0$.

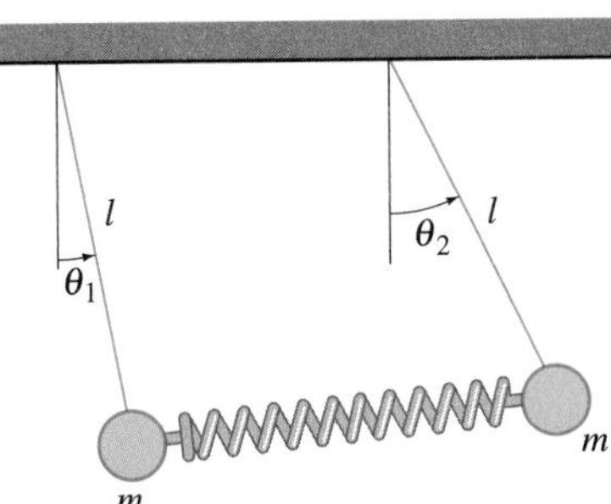

FIGURE 7.R.12 Coupled pendulums in Problem 47

48. Coulomb Friction Revisited In Problem 27 in Chapter 5 in Review we examined a spring/mass system in which a mass m slides over a dry horizontal surface whose coefficient of kinetic friction is a constant μ. The constant retarding force $f_k = \mu mg$ of the dry surface that acts opposite to the direction of motion is called Coulomb friction after the French physicist **Charles-Augustin de Coulomb** (1736–1806). You were asked to show that the piecewise-defined differential equation for the displacement $x(t)$ of the mass is given by

$$m\frac{d^2x}{dt^2} + kx = \begin{cases} f_k, & x' < 0 \text{ (motion to left)} \\ -f_k, & x' > 0 \text{ (motion to right)}. \end{cases}$$

(a) Suppose that the mass is released from rest from a point $x(0) = x_0 > 0$ and that there are no other external forces. Then the differential equations describing the motion of the mass m are

$$x'' + \omega^2 x = F, \quad 0 < t < T/2$$
$$x'' + \omega^2 x = -F, \quad T/2 < t < T$$
$$x'' + \omega^2 x = F, \quad T < t < 3T/2,$$

and so on, where $\omega^2 = k/m$, $F = f_k/m = \mu g$, $g = 32$, and $T = 2\pi/\omega$. Show that the times $0, T/2, T, 3T/2, \ldots$ correspond to $x'(t) = 0$.

(b) Explain why, in general, the initial displacement must satisfy $\omega^2|x_0| > F$.

(c) Explain why the interval $-F/\omega^2 \le x \le F/\omega^2$ is appropriately called the "dead zone" of the system.

(d) Use the Laplace transform and the concept of the meander function to solve for the displacement $x(t)$ for $t \ge 0$.

(e) Show that in the case $m = 1$, $k = 1$, $f_k = 1$, and $x_0 = 5.5$ that on the interval $[0, 2\pi)$ your solution agrees with parts (a) and (b) of Problem 28 in Chapter 5 in Review.

(f) Show that each successive oscillation is $2F/\omega^2$ shorter than the preceding one.

(g) Predict the long-term behavior of the system.

49. Range of a Projectile—No Air Resistance (a) A projectile, such as the canon ball shown in Figure 7.R.13, has weight $w = mg$ and initial velocity $\mathbf{v}_0$ that is tangent to its path of motion. If air resistance and all other forces except its weight are ignored, we saw in Problem 23 of Exercises 4.9 that motion of the projectile is describe by the system of linear differential equations

$$m\frac{d^2x}{dt^2} = 0$$
$$m\frac{d^2y}{dt^2} = -mg.$$

Use the Laplace transform to solve this system subject to the initial conditions $x(0) = 0$, $x'(0) = v_0 \cos\theta$, $y(0) = 0$, $y'(0) = v_0 \sin\theta$, where $v_0 = |\mathbf{v}_0|$ is constant and θ is the constant angle of elevation shown in Figure 7.R.13. The solutions $x(t)$ and $y(t)$ are parametric equations of the trajectory of the projectile.

(b) Use $x(t)$ in part (a) to eliminate the parameter t in $y(t)$. Use the resulting equation for y to show that the horizontal range R of the projectile is given by

$$R = \frac{v_0^2}{g}\sin 2\theta.$$

(c) From the formula in part (b), we see that R is a maximum when $\sin 2\theta = 1$ or when $\theta = \pi/4$. Show that the same range—less than the maximum—can be attained by firing the gun at either of two complementary angles θ and $\pi/2 - \theta$. The only difference is that the smaller angle results in a low trajectory whereas the larger angle gives a high trajectory.

(d) Suppose $g = 32$ ft/s^2, $\theta = 38°$, and $v_0 = 300$ ft/s. Use part (b) to find the horizontal range of the projectile. Find the time when the projectile hits the ground.

(e) Use the parametric equations $x(t)$ and $y(t)$ in part (a) along with the numerical data in part (d) to plot the ballistic curve of the projectile. Repeat with $\theta = 52°$ and $v_0 = 300$ ft/s. Superimpose both curves on the same coordinate system.

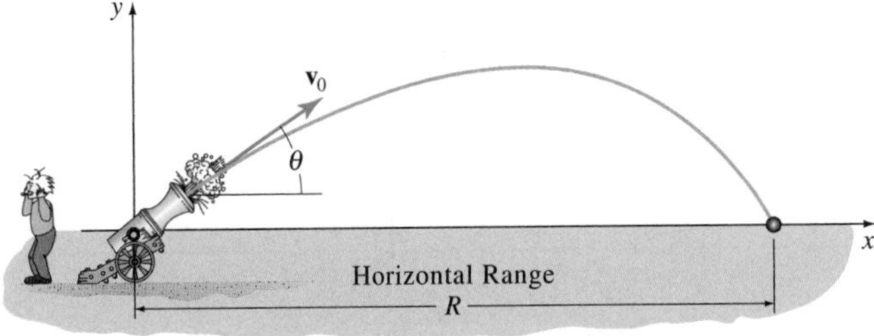

FIGURE 7.R.13 Projectile in Problem 49

50. Range of a Projectile—With Air Resistance **(a)** Now suppose that air resistance is a retarding force tangent to the path but acts opposite to the motion. If we take air resistance to be proportional to the velocity of the projectile, then we saw in Problem 24 of Exercises 4.9 that motion of the projectile is describe by the system of linear differential equations

$$m\frac{d^2x}{dt^2} = -\beta\frac{dx}{dt}$$

$$m\frac{d^2y}{dt^2} = -mg - \beta\frac{dy}{dt},$$

where $\beta > 0$. Use the Laplace transform to solve this system subject to the initial conditions $x(0) = 0$, $x'(0) = v_0 \cos\theta$, $y(0) = 0$, $y'(0) = v_0 \sin\theta$, where $v_0 = |\mathbf{v}_0|$ and θ are constant.

(b) Suppose $m = \frac{1}{4}$ slug, $g = 32$ ft/s^2, $\beta = 0.02$, $\theta = 38°$, and $v_0 = 300$ ft/s. Use a CAS to find the time when the projectile hits the ground and then compute its corresponding horizontal range.

(c) Repeat part (c) using the complementary angle $\theta = 52°$ and compare the range with that found in part (b). Does the property in part (c) of Problem 49 hold?

(d) Use the parametric equations $x(t)$ and $y(t)$ in part (a) along with the numerical data in part (b) to plot the ballistic curve of the projectile. Repeat with the same numerical data in part (b) but take $\theta = 52°$. Superimpose both curves on the same coordinate system. Compare these curves with those obtained in part (e) of Problem 49.

8 Systems of Linear First-Order Differential Equations

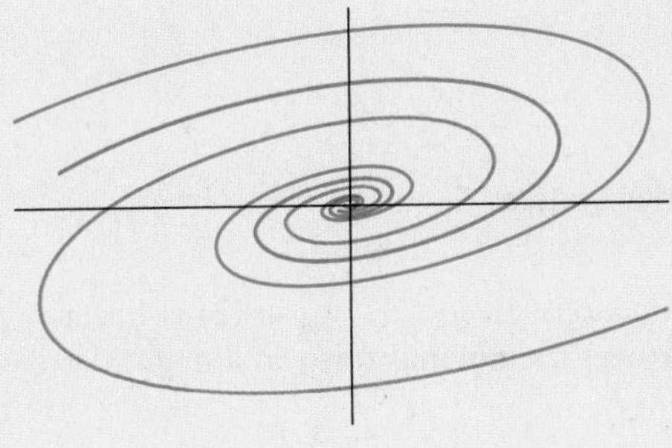

We encountered systems of ordinary differential equations in Sections 3.3, 4.9, and 7.6 and were able to solve some of these systems by means of either systematic elimination or by the Laplace transform. In this chapter we are going to concentrate only on *systems of linear first-order differential equations*. Although most of the systems that are considered could be solved using elimination or the Laplace transform, we are going to develop a general theory for these kinds of systems and in the case of systems with constant coefficients, a method of solution that utilizes some basic concepts from the algebra of matrices. We will see that this general theory and solution procedure is similar to that of linear higher-order differential equations considered in Chapter 4. This material is fundamental to the analysis of systems of nonlinear first-order equations in Chapter 10.

8.1 PRELIMINARY THEORY—LINEAR SYSTEMS

REVIEW MATERIAL

- Matrix notation and properties are used extensively throughout this chapter. It is imperative that you review either Appendix II or a linear algebra text if you unfamiliar with these concepts.

INTRODUCTION Recall that in Section 4.9 we illustrated how to solve systems of n linear differential equations in n unknowns of the form

$$\begin{aligned} P_{11}(D)x_1 + P_{12}(D)x_2 + \cdots + P_{1n}(D)x_n &= b_1(t) \\ P_{21}(D)x_1 + P_{22}(D)x_2 + \cdots + P_{2n}(D)x_n &= b_2(t) \\ &\vdots \\ P_{n1}(D)x_1 + P_{n2}(D)x_2 + \cdots + P_{nn}(D)x_n &= b_n(t), \end{aligned} \tag{1}$$

where the P_{ij} were polynomials of various degrees in the differential operator D. In this chapter we confine our study to systems of first-order DEs that are special cases of systems that have the normal form

$$\begin{aligned} \frac{dx_1}{dt} &= g_1(t, x_1, x_2, \ldots, x_n) \\ \frac{dx_2}{dt} &= g_2(t, x_1, x_2, \ldots, x_n) \\ &\vdots \\ \frac{dx_n}{dt} &= g_n(t, x_1, x_2, \ldots, x_n). \end{aligned} \tag{2}$$

A system such as (2) of n first-order equations is called a **first-order system.**

Linear Systems When each of the functions $g_1, g_2, \ldots, g_n$ in (2) is linear in the dependent variables $x_1, x_2, \ldots, x_n$, we get the **normal form** of a first-order system of linear equations:

$$\begin{aligned} \frac{dx_1}{dt} &= a_{11}(t)x_1 + a_{12}(t)x_2 + \cdots + a_{1n}(t)x_n + f_1(t) \\ \frac{dx_2}{dt} &= a_{21}(t)x_1 + a_{22}(t)x_2 + \cdots + a_{2n}(t)x_n + f_2(t) \\ &\vdots \\ \frac{dx_n}{dt} &= a_{n1}(t)x_1 + a_{n2}(t)x_2 + \cdots + a_{nn}(t)x_n + f_n(t). \end{aligned} \tag{3}$$

We refer to a system of the form given in (3) simply as a **linear system.** We assume that the coefficients a_{ij} as well as the functions f_i are continuous on a common interval I. When $f_i(t) = 0$, $i = 1, 2, \ldots, n$, the linear system (3) is said to be **homogeneous;** otherwise, it is **nonhomogeneous.**

Matrix Form of a Linear System If **X, A**(t), and **F**(t) denote the respective matrices

$$\mathbf{X} = \begin{pmatrix} x_1(t) \\ x_2(t) \\ \vdots \\ x_n(t) \end{pmatrix}, \quad \mathbf{A}(t) = \begin{pmatrix} a_{11}(t) & a_{12}(t) & \cdots & a_{1n}(t) \\ a_{21}(t) & a_{22}(t) & \cdots & a_{2n}(t) \\ \vdots & & & \vdots \\ a_{n1}(t) & a_{n2}(t) & \cdots & a_{nn}(t) \end{pmatrix}, \quad \mathbf{F}(t) = \begin{pmatrix} f_1(t) \\ f_2(t) \\ \vdots \\ f_n(t) \end{pmatrix},$$

then the system of linear first-order differential equations (3) can be written as

$$\frac{d}{dt}\begin{pmatrix} x_1 \\ x_2 \\ \vdots \\ x_n \end{pmatrix} = \begin{pmatrix} a_{11}(t) & a_{12}(t) & \cdots & a_{1n}(t) \\ a_{21}(t) & a_{22}(t) & \cdots & a_{2n}(t) \\ \vdots & & & \vdots \\ a_{n1}(t) & a_{n2}(t) & \cdots & a_{nn}(t) \end{pmatrix}\begin{pmatrix} x_1 \\ x_2 \\ \vdots \\ x_n \end{pmatrix} + \begin{pmatrix} f_1(t) \\ f_2(t) \\ \vdots \\ f_n(t) \end{pmatrix}$$

or simply

$$\mathbf{X}' = \mathbf{AX} + \mathbf{F}. \tag{4}$$

If the system is homogeneous, its matrix form is then

$$\mathbf{X}' = \mathbf{AX}. \tag{5}$$

EXAMPLE 1 Systems Written in Matrix Notation

(a) If $\mathbf{X} = \begin{pmatrix} x \\ y \end{pmatrix}$, then the matrix form of the homogeneous system

$$\begin{aligned} \frac{dx}{dt} &= 3x + 4y \\ \frac{dy}{dt} &= 5x - 7y \end{aligned} \quad \text{is} \quad \mathbf{X}' = \begin{pmatrix} 3 & 4 \\ 5 & -7 \end{pmatrix}\mathbf{X}.$$

(b) If $\mathbf{X} = \begin{pmatrix} x \\ y \\ z \end{pmatrix}$, then the matrix form of the nonhomogeneous system

$$\begin{aligned} \frac{dx}{dt} &= 6x + y + z + t \\ \frac{dy}{dt} &= 8x + 7y - z + 10t \\ \frac{dz}{dt} &= 2x + 9y - z + 6t \end{aligned} \quad \text{is} \quad \mathbf{X}' = \begin{pmatrix} 6 & 1 & 1 \\ 8 & 7 & -1 \\ 2 & 9 & -1 \end{pmatrix}\mathbf{X} + \begin{pmatrix} t \\ 10t \\ 6t \end{pmatrix}.$$

≡

DEFINITION 8.1.1 Solution Vector

A **solution vector** on an interval I is any column matrix

$$\mathbf{X} = \begin{pmatrix} x_1(t) \\ x_2(t) \\ \vdots \\ x_n(t) \end{pmatrix}$$

whose entries are differentiable functions satisfying the system (4) on the interval.

A solution vector of (4) is, of course, equivalent to n scalar equations $x_1 = \phi_1(t)$, $x_2 = \phi_2(t)$, . . . , $x_n = \phi_n(t)$ and can be interpreted geometrically as a set of parametric equations of a space curve. In the important case $n = 2$ the equations $x_1 = \phi_1(t)$, $x_2 = \phi_2(t)$ represent a curve in the x_1x_2-plane. It is common practice to call a curve in the plane a **trajectory** and to call the x_1x_2-plane the **phase plane.** We will come back to these concepts and illustrate them in the next section.

EXAMPLE 2 Verification of Solutions

Verify that on the interval $(-\infty, \infty)$

$$\mathbf{X}_1 = \begin{pmatrix} 1 \\ -1 \end{pmatrix} e^{-2t} = \begin{pmatrix} e^{-2t} \\ -e^{-2t} \end{pmatrix} \quad \text{and} \quad \mathbf{X}_2 = \begin{pmatrix} 3 \\ 5 \end{pmatrix} e^{6t} = \begin{pmatrix} 3e^{6t} \\ 5e^{6t} \end{pmatrix}$$

are solutions of

$$\mathbf{X}' = \begin{pmatrix} 1 & 3 \\ 5 & 3 \end{pmatrix} \mathbf{X}. \tag{6}$$

SOLUTION From $\mathbf{X}_1' = \begin{pmatrix} -2e^{-2t} \\ 2e^{-2t} \end{pmatrix}$ and $\mathbf{X}_2' = \begin{pmatrix} 18e^{6t} \\ 30e^{6t} \end{pmatrix}$ we see that

$$\mathbf{AX}_1 = \begin{pmatrix} 1 & 3 \\ 5 & 3 \end{pmatrix} \begin{pmatrix} e^{-2t} \\ -e^{-2t} \end{pmatrix} = \begin{pmatrix} e^{-2t} - 3e^{-2t} \\ 5e^{-2t} - 3e^{-2t} \end{pmatrix} = \begin{pmatrix} -2e^{-2t} \\ 2e^{-2t} \end{pmatrix} = \mathbf{X}_1',$$

and

$$\mathbf{AX}_2 = \begin{pmatrix} 1 & 3 \\ 5 & 3 \end{pmatrix} \begin{pmatrix} 3e^{6t} \\ 5e^{6t} \end{pmatrix} = \begin{pmatrix} 3e^{6t} + 15e^{6t} \\ 15e^{6t} + 15e^{6t} \end{pmatrix} = \begin{pmatrix} 18e^{6t} \\ 30e^{6t} \end{pmatrix} = \mathbf{X}_2'.$$

≡

Much of the theory of systems of n linear first-order differential equations is similar to that of linear nth-order differential equations.

Initial-Value Problem Let t_0 denote a point on an interval I and

$$\mathbf{X}(t_0) = \begin{pmatrix} x_1(t_0) \\ x_2(t_0) \\ \vdots \\ x_n(t_0) \end{pmatrix} \quad \text{and} \quad \mathbf{X}_0 = \begin{pmatrix} \gamma_1 \\ \gamma_2 \\ \vdots \\ \gamma_n \end{pmatrix},$$

where the γ_i, $i = 1, 2, \ldots, n$ are given constants. Then the problem

$$\begin{aligned} &\textit{Solve:} && \mathbf{X}' = \mathbf{A}(t)\mathbf{X} + \mathbf{F}(t) \\ &\textit{Subject to:} && \mathbf{X}(t_0) = \mathbf{X}_0 \end{aligned} \tag{7}$$

is an **initial-value problem** on the interval.

THEOREM 8.1.1 Existence of a Unique Solution

Let the entries of the matrices $\mathbf{A}(t)$ and $\mathbf{F}(t)$ be functions continuous on a common interval I that contains the point t_0. Then there exists a unique solution of the initial-value problem (7) on the interval.

≡ Homogeneous Systems In the next several definitions and theorems we are concerned only with homogeneous systems. Without stating it, we shall always assume that the a_{ij} and the f_i are continuous functions of t on some common interval I.

≡ Superposition Principle The following result is a **superposition principle** for solutions of linear systems.

THEOREM 8.1.2 Superposition Principle

Let $\mathbf{X}_1, \mathbf{X}_2, \ldots, \mathbf{X}_k$ be a set of solution vectors of the homogeneous system (5) on an interval I. Then the linear combination

$$\mathbf{X} = c_1\mathbf{X}_1 + c_2\mathbf{X}_2 + \cdots + c_k\mathbf{X}_k,$$

where the c_i, $i = 1, 2, \ldots, k$ are arbitrary constants, is also a solution on the interval.

It follows from Theorem 8.1.2 that a constant multiple of any solution vector of a homogeneous system of linear first-order differential equations is also a solution.

EXAMPLE 3 Using the Superposition Principle

You should practice by verifying that the two vectors

$$\mathbf{X}_1 = \begin{pmatrix} \cos t \\ -\frac{1}{2}\cos t + \frac{1}{2}\sin t \\ -\cos t - \sin t \end{pmatrix} \quad \text{and} \quad \mathbf{X}_2 = \begin{pmatrix} 0 \\ e^t \\ 0 \end{pmatrix}$$

are solutions of the system

$$\mathbf{X}' = \begin{pmatrix} 1 & 0 & 1 \\ 1 & 1 & 0 \\ -2 & 0 & -1 \end{pmatrix}\mathbf{X}. \tag{8}$$

By the superposition principle the linear combination

$$\mathbf{X} = c_1\mathbf{X}_1 + c_2\mathbf{X}_2 = c_1\begin{pmatrix} \cos t \\ -\frac{1}{2}\cos t + \frac{1}{2}\sin t \\ -\cos t - \sin t \end{pmatrix} + c_2\begin{pmatrix} 0 \\ e^t \\ 0 \end{pmatrix}$$

is yet another solution of the system. ≡

≡ Linear Dependence and Linear Independence We are primarily interested in linearly independent solutions of the homogeneous system (5).

DEFINITION 8.1.2 Linear Dependence/Independence

Let $\mathbf{X}_1, \mathbf{X}_2, \ldots, \mathbf{X}_k$ be a set of solution vectors of the homogeneous system (5) on an interval I. We say that the set is **linearly dependent** on the interval if there exist constants $c_1, c_2, \ldots, c_k$, not all zero, such that

$$c_1\mathbf{X}_1 + c_2\mathbf{X}_2 + \cdots + c_k\mathbf{X}_k = \mathbf{0}$$

for every t in the interval. If the set of vectors is not linearly dependent on the interval, it is said to be **linearly independent.**

The case when $k = 2$ should be clear; two solution vectors $\mathbf{X}_1$ and $\mathbf{X}_2$ are linearly dependent if one is a constant multiple of the other, and conversely. For $k > 2$ a set of solution vectors is linearly dependent if we can express at least one solution vector as a linear combination of the remaining vectors.

≡ Wronskian As in our earlier consideration of the theory of a single ordinary differential equation, we can introduce the concept of the **Wronskian** determinant as a test for linear independence. We state the following theorem without proof.

THEOREM 8.1.3 Criterion for Linearly Independent Solutions

Let
$$\mathbf{X}_1 = \begin{pmatrix} x_{11} \\ x_{21} \\ \vdots \\ x_{n1} \end{pmatrix}, \quad \mathbf{X}_2 = \begin{pmatrix} x_{12} \\ x_{22} \\ \vdots \\ x_{n2} \end{pmatrix}, \quad \ldots, \quad \mathbf{X}_n = \begin{pmatrix} x_{1n} \\ x_{2n} \\ \vdots \\ x_{nn} \end{pmatrix}$$

(continues on page 330)

be n solution vectors of the homogeneous system (5) on an interval I. Then the set of solution vectors is linearly independent on I if and only if the **Wronskian**

$$W(\mathbf{X}_1, \mathbf{X}_2, \ldots, \mathbf{X}_n) = \begin{vmatrix} x_{11} & x_{12} & \cdots & x_{1n} \\ x_{21} & x_{22} & \cdots & x_{2n} \\ \vdots & & & \vdots \\ x_{n1} & x_{n2} & \cdots & x_{nn} \end{vmatrix} \neq 0 \tag{9}$$

for every t in the interval.

It can be shown that if $\mathbf{X}_1, \mathbf{X}_2, \ldots, \mathbf{X}_n$ are solution vectors of (5), then for every t in I either $W(\mathbf{X}_1, \mathbf{X}_2, \ldots, \mathbf{X}_n) \neq 0$ or $W(\mathbf{X}_1, \mathbf{X}_2, \ldots, \mathbf{X}_n) = 0$. Thus if we can show that $W \neq 0$ for some t_0 in I, then $W \neq 0$ for every t, and hence the solutions are linearly independent on the interval.

Notice that, unlike our definition of the Wronskian in Section 4.1, here the definition of the determinant (9) does not involve differentiation.

EXAMPLE 4 Linearly Independent Solutions

In Example 2 we saw that $\mathbf{X}_1 = \begin{pmatrix} 1 \\ -1 \end{pmatrix} e^{-2t}$ and $\mathbf{X}_2 = \begin{pmatrix} 3 \\ 5 \end{pmatrix} e^{6t}$ are solutions of system (6). Clearly, $\mathbf{X}_1$ and $\mathbf{X}_2$ are linearly independent on the interval $(-\infty, \infty)$, since neither vector is a constant multiple of the other. In addition, we have

$$W(\mathbf{X}_1, \mathbf{X}_2) = \begin{vmatrix} e^{-2t} & 3e^{6t} \\ -e^{-2t} & 5e^{6t} \end{vmatrix} = 8e^{4t} \neq 0$$

for all real values of t. ≡

DEFINITION 8.1.3 Fundamental Set of Solutions

Any set $\mathbf{X}_1, \mathbf{X}_2, \ldots, \mathbf{X}_n$ of n linearly independent solution vectors of the homogeneous system (5) on an interval I is said to be a **fundamental set of solutions** on the interval.

THEOREM 8.1.4 Existence of a Fundamental Set

There exists a fundamental set of solutions for the homogeneous system (5) on an interval I.

The next two theorems are the linear system equivalents of Theorems 4.1.5 and 4.1.6.

THEOREM 8.1.5 General Solution—Homogeneous Systems

Let $\mathbf{X}_1, \mathbf{X}_2, \ldots, \mathbf{X}_n$ be a fundamental set of solutions of the homogeneous system (5) on an interval I. Then the **general solution** of the system on the interval is

$$\mathbf{X} = c_1\mathbf{X}_1 + c_2\mathbf{X}_2 + \cdots + c_n\mathbf{X}_n,$$

where the c_i, $i = 1, 2, \ldots, n$ are arbitrary constants.

EXAMPLE 5 General Solution of System (6)

From Example 2 we know that $\mathbf{X}_1 = \begin{pmatrix} 1 \\ -1 \end{pmatrix} e^{-2t}$ and $\mathbf{X}_2 = \begin{pmatrix} 3 \\ 5 \end{pmatrix} e^{6t}$ are linearly independent solutions of (6) on $(-\infty, \infty)$. Hence $\mathbf{X}_1$ and $\mathbf{X}_2$ form a fundamental set of solutions on the interval. The general solution of the system on the interval is then

$$\mathbf{X} = c_1\mathbf{X}_1 + c_2\mathbf{X}_2 = c_1\begin{pmatrix} 1 \\ -1 \end{pmatrix} e^{-2t} + c_2\begin{pmatrix} 3 \\ 5 \end{pmatrix} e^{6t}. \tag{10}$$

EXAMPLE 6 General Solution of System (8)

The vectors

$$\mathbf{X}_1 = \begin{pmatrix} \cos t \\ -\frac{1}{2}\cos t + \frac{1}{2}\sin t \\ -\cos t - \sin t \end{pmatrix}, \quad \mathbf{X}_2 = \begin{pmatrix} 0 \\ 1 \\ 0 \end{pmatrix} e^t, \quad \mathbf{X}_3 = \begin{pmatrix} \sin t \\ -\frac{1}{2}\sin t - \frac{1}{2}\cos t \\ -\sin t + \cos t \end{pmatrix}$$

are solutions of the system (8) in Example 3 (see Problem 16 in Exercises 8.1). Now

$$W(\mathbf{X}_1, \mathbf{X}_2, \mathbf{X}_3) = \begin{vmatrix} \cos t & 0 & \sin t \\ -\frac{1}{2}\cos t + \frac{1}{2}\sin t & e^t & -\frac{1}{2}\sin t - \frac{1}{2}\cos t \\ -\cos t - \sin t & 0 & -\sin t + \cos t \end{vmatrix} = e^t \neq 0$$

for all real values of t. We conclude that $\mathbf{X}_1$, $\mathbf{X}_2$, and $\mathbf{X}_3$ form a fundamental set of solutions on $(-\infty, \infty)$. Thus the general solution of the system on the interval is the linear combination $\mathbf{X} = c_1\mathbf{X}_1 + c_2\mathbf{X}_2 + c_3\mathbf{X}_3$; that is,

$$\mathbf{X} = c_1\begin{pmatrix} \cos t \\ -\frac{1}{2}\cos t + \frac{1}{2}\sin t \\ -\cos t - \sin t \end{pmatrix} + c_2\begin{pmatrix} 0 \\ 1 \\ 0 \end{pmatrix} e^t + c_3\begin{pmatrix} \sin t \\ -\frac{1}{2}\sin t - \frac{1}{2}\cos t \\ -\sin t + \cos t \end{pmatrix}.$$

Nonhomogeneous Systems For nonhomogeneous systems a **particular solution** $\mathbf{X}_p$ on an interval I is any vector, free of arbitrary parameters, whose entries are functions that satisfy the system (4).

THEOREM 8.1.6 General Solution—Nonhomogeneous Systems

Let $\mathbf{X}_p$ be a given solution of the nonhomogeneous system (4) on an interval I and let

$$\mathbf{X}_c = c_1\mathbf{X}_1 + c_2\mathbf{X}_2 + \cdots + c_n\mathbf{X}_n$$

denote the general solution on the same interval of the associated homogeneous system (5). Then the **general solution** of the nonhomogeneous system on the interval is

$$\mathbf{X} = \mathbf{X}_c + \mathbf{X}_p.$$

The general solution $\mathbf{X}_c$ of the associated homogeneous system (5) is called the **complementary function** of the nonhomogeneous system (4).

EXAMPLE 7 General Solution—Nonhomogeneous System

The vector $\mathbf{X}_p = \begin{pmatrix} 3t - 4 \\ -5t + 6 \end{pmatrix}$ is a particular solution of the nonhomogeneous system

$$\mathbf{X}' = \begin{pmatrix} 1 & 3 \\ 5 & 3 \end{pmatrix}\mathbf{X} + \begin{pmatrix} 12t - 11 \\ -3 \end{pmatrix} \tag{11}$$

on the interval $(-\infty, \infty)$. (Verify this.) The complementary function of (11) on the same interval, or the general solution of $\mathbf{X}' = \begin{pmatrix} 1 & 3 \\ 5 & 3 \end{pmatrix}\mathbf{X}$, was seen in (10) of Example 5 to be $\mathbf{X}_c = c_1\begin{pmatrix} 1 \\ -1 \end{pmatrix}e^{-2t} + c_2\begin{pmatrix} 3 \\ 5 \end{pmatrix}e^{6t}$. Hence by Theorem 8.1.6

$$\mathbf{X} = \mathbf{X}_c + \mathbf{X}_p = c_1\begin{pmatrix} 1 \\ -1 \end{pmatrix}e^{-2t} + c_2\begin{pmatrix} 3 \\ 5 \end{pmatrix}e^{6t} + \begin{pmatrix} 3t - 4 \\ -5t + 6 \end{pmatrix}$$

is the general solution of (11) on $(-\infty, \infty)$. ≡

EXERCISES 8.1

Answers to selected odd-numbered problems begin on page ANS-14.

In Problems 1–6 write the linear system in matrix form.

1. $\dfrac{dx}{dt} = 3x - 5y$

$\dfrac{dy}{dt} = 4x + 8y$

2. $\dfrac{dx}{dt} = 4x - 7y$

$\dfrac{dy}{dt} = 5x$

3. $\dfrac{dx}{dt} = -3x + 4y - 9z$

$\dfrac{dy}{dt} = 6x - y$

$\dfrac{dz}{dt} = 10x + 4y + 3z$

4. $\dfrac{dx}{dt} = x - y$

$\dfrac{dy}{dt} = x + 2z$

$\dfrac{dz}{dt} = -x + z$

5. $\dfrac{dx}{dt} = x - y + z + t - 1$

$\dfrac{dy}{dt} = 2x + y - z - 3t^2$

$\dfrac{dz}{dt} = x + y + z + t^2 - t + 2$

6. $\dfrac{dx}{dt} = -3x + 4y + e^{-t}\sin 2t$

$\dfrac{dy}{dt} = 5x + 9z + 4e^{-t}\cos 2t$

$\dfrac{dz}{dt} = y + 6z - e^{-t}$

In Problems 7–10 write the given system without the use of matrices.

7. $\mathbf{X}' = \begin{pmatrix} 4 & 2 \\ -1 & 3 \end{pmatrix}\mathbf{X} + \begin{pmatrix} 1 \\ -1 \end{pmatrix}e^t$

8. $\mathbf{X}' = \begin{pmatrix} 7 & 5 & -9 \\ 4 & 1 & 1 \\ 0 & -2 & 3 \end{pmatrix}\mathbf{X} + \begin{pmatrix} 0 \\ 2 \\ 1 \end{pmatrix}e^{5t} - \begin{pmatrix} 8 \\ 0 \\ 3 \end{pmatrix}e^{-2t}$

9. $\dfrac{d}{dt}\begin{pmatrix} x \\ y \\ z \end{pmatrix} = \begin{pmatrix} 1 & -1 & 2 \\ 3 & -4 & 1 \\ -2 & 5 & 6 \end{pmatrix}\begin{pmatrix} x \\ y \\ z \end{pmatrix} + \begin{pmatrix} 1 \\ 2 \\ 2 \end{pmatrix}e^{-t} - \begin{pmatrix} 3 \\ -1 \\ 1 \end{pmatrix}t$

10. $\dfrac{d}{dt}\begin{pmatrix} x \\ y \end{pmatrix} = \begin{pmatrix} 3 & -7 \\ 1 & 1 \end{pmatrix}\begin{pmatrix} x \\ y \end{pmatrix} + \begin{pmatrix} 4 \\ 8 \end{pmatrix}\sin t + \begin{pmatrix} t - 4 \\ 2t + 1 \end{pmatrix}e^{4t}$

In Problems 11–16 verify that the vector $\mathbf{X}$ is a solution of the given system.

11. $\dfrac{dx}{dt} = 3x - 4y$

$\dfrac{dy}{dt} = 4x - 7y$; $\mathbf{X} = \begin{pmatrix} 1 \\ 2 \end{pmatrix}e^{-5t}$

12. $\dfrac{dx}{dt} = -2x + 5y$

$\dfrac{dy}{dt} = -2x + 4y$; $\mathbf{X} = \begin{pmatrix} 5\cos t \\ 3\cos t - \sin t \end{pmatrix}e^t$

13. $\mathbf{X}' = \begin{pmatrix} -1 & \frac{1}{4} \\ 1 & -1 \end{pmatrix}\mathbf{X}$; $\mathbf{X} = \begin{pmatrix} -1 \\ 2 \end{pmatrix}e^{-3t/2}$

14. $\mathbf{X}' = \begin{pmatrix} 2 & 1 \\ -1 & 0 \end{pmatrix}\mathbf{X}$; $\mathbf{X} = \begin{pmatrix} 1 \\ 3 \end{pmatrix}e^t + \begin{pmatrix} 4 \\ -4 \end{pmatrix}te^t$

15. $\mathbf{X}' = \begin{pmatrix} 1 & 2 & 1 \\ 6 & -1 & 0 \\ -1 & -2 & -1 \end{pmatrix}\mathbf{X}; \quad \mathbf{X} = \begin{pmatrix} 1 \\ 6 \\ -13 \end{pmatrix}$

16. $\mathbf{X}' = \begin{pmatrix} 1 & 0 & 1 \\ 1 & 1 & 0 \\ -2 & 0 & -1 \end{pmatrix}\mathbf{X}; \quad \mathbf{X} = \begin{pmatrix} \sin t \\ -\frac{1}{2}\sin t - \frac{1}{2}\cos t \\ -\sin t + \cos t \end{pmatrix}$

In Problems 17–20 the given vectors are solutions of a system $\mathbf{X}' = \mathbf{AX}$. Determine whether the vectors form a fundamental set on the interval $(-\infty, \infty)$.

17. $\mathbf{X}_1 = \begin{pmatrix} 1 \\ 1 \end{pmatrix}e^{-2t}, \quad \mathbf{X}_2 = \begin{pmatrix} 1 \\ -1 \end{pmatrix}e^{-6t}$

18. $\mathbf{X}_1 = \begin{pmatrix} 1 \\ -1 \end{pmatrix}e^{t}, \quad \mathbf{X}_2 = \begin{pmatrix} 2 \\ 6 \end{pmatrix}e^{t} + \begin{pmatrix} 8 \\ -8 \end{pmatrix}te^{t}$

19. $\mathbf{X}_1 = \begin{pmatrix} 1 \\ -2 \\ 4 \end{pmatrix} + t\begin{pmatrix} 1 \\ 2 \\ 2 \end{pmatrix}, \quad \mathbf{X}_2 = \begin{pmatrix} 1 \\ -2 \\ 4 \end{pmatrix},$

$\mathbf{X}_3 = \begin{pmatrix} 3 \\ -6 \\ 12 \end{pmatrix} + t\begin{pmatrix} 2 \\ 4 \\ 4 \end{pmatrix}$

20. $\mathbf{X}_1 = \begin{pmatrix} 1 \\ 6 \\ -13 \end{pmatrix}, \quad \mathbf{X}_2 = \begin{pmatrix} 1 \\ -2 \\ -1 \end{pmatrix}e^{-4t}, \quad \mathbf{X}_3 = \begin{pmatrix} 2 \\ 3 \\ -2 \end{pmatrix}e^{3t}$

In Problems 21–24 verify that the vector $\mathbf{X}_p$ is a particular solution of the given system.

21. $\dfrac{dx}{dt} = x + 4y + 2t - 7$

$\dfrac{dy}{dt} = 3x + 2y - 4t - 18; \quad \mathbf{X}_p = \begin{pmatrix} 2 \\ -1 \end{pmatrix}t + \begin{pmatrix} 5 \\ 1 \end{pmatrix}$

22. $\mathbf{X}' = \begin{pmatrix} 2 & 1 \\ 1 & -1 \end{pmatrix}\mathbf{X} + \begin{pmatrix} -5 \\ 2 \end{pmatrix}; \quad \mathbf{X}_p = \begin{pmatrix} 1 \\ 3 \end{pmatrix}$

23. $\mathbf{X}' = \begin{pmatrix} 2 & 1 \\ 3 & 4 \end{pmatrix}\mathbf{X} - \begin{pmatrix} 1 \\ 7 \end{pmatrix}e^{t}; \quad \mathbf{X}_p = \begin{pmatrix} 1 \\ 1 \end{pmatrix}e^{t} + \begin{pmatrix} 1 \\ -1 \end{pmatrix}te^{t}$

24. $\mathbf{X}' = \begin{pmatrix} 1 & 2 & 3 \\ -4 & 2 & 0 \\ -6 & 1 & 0 \end{pmatrix}\mathbf{X} + \begin{pmatrix} -1 \\ 4 \\ 3 \end{pmatrix}\sin 3t; \quad \mathbf{X}_p = \begin{pmatrix} \sin 3t \\ 0 \\ \cos 3t \end{pmatrix}$

25. Prove that the general solution of

$$\mathbf{X}' = \begin{pmatrix} 0 & 6 & 0 \\ 1 & 0 & 1 \\ 1 & 1 & 0 \end{pmatrix}\mathbf{X}$$

on the interval $(-\infty, \infty)$ is

$$\mathbf{X} = c_1\begin{pmatrix} 6 \\ -1 \\ -5 \end{pmatrix}e^{-t} + c_2\begin{pmatrix} -3 \\ 1 \\ 1 \end{pmatrix}e^{-2t} + c_3\begin{pmatrix} 2 \\ 1 \\ 1 \end{pmatrix}e^{3t}.$$

26. Prove that the general solution of

$$\mathbf{X}' = \begin{pmatrix} -1 & -1 \\ -1 & 1 \end{pmatrix}\mathbf{X} + \begin{pmatrix} 1 \\ 1 \end{pmatrix}t^2 + \begin{pmatrix} 4 \\ -6 \end{pmatrix}t + \begin{pmatrix} -1 \\ 5 \end{pmatrix}$$

on the interval $(-\infty, \infty)$ is

$$\mathbf{X} = c_1\begin{pmatrix} 1 \\ -1 - \sqrt{2} \end{pmatrix}e^{\sqrt{2}t} + c_2\begin{pmatrix} 1 \\ -1 + \sqrt{2} \end{pmatrix}e^{-\sqrt{2}t} + \begin{pmatrix} 1 \\ 0 \end{pmatrix}t^2 + \begin{pmatrix} -2 \\ 4 \end{pmatrix}t + \begin{pmatrix} 1 \\ 0 \end{pmatrix}.$$

8.2 HOMOGENEOUS LINEAR SYSTEMS

REVIEW MATERIAL

- Section II.3 of Appendix II
- Also the *Student Resource Manual*

INTRODUCTION We saw in Example 5 of Section 8.1 that the general solution of the homogeneous system $\mathbf{X}' = \begin{pmatrix} 1 & 3 \\ 5 & 3 \end{pmatrix}\mathbf{X}$ is

$$\mathbf{X} = c_1\mathbf{X}_1 + c_2\mathbf{X}_2 = c_1\begin{pmatrix} 1 \\ -1 \end{pmatrix}e^{-2t} + c_2\begin{pmatrix} 3 \\ 5 \end{pmatrix}e^{6t}.$$

Because the solution vectors $\mathbf{X}_1$ and $\mathbf{X}_2$ have the form

$$\mathbf{X}_i = \begin{pmatrix} k_1 \\ k_2 \end{pmatrix}e^{\lambda_i t}, \qquad i = 1, 2,$$

(*continues on page 334*)

where k_1, k_2, λ_1, and λ_2 are constants, we are prompted to ask whether we can always find a solution of the form

$$\mathbf{X} = \begin{pmatrix} k_1 \\ k_2 \\ \vdots \\ k_n \end{pmatrix} e^{\lambda t} = \mathbf{K}e^{\lambda t} \tag{1}$$

for the general homogeneous linear first-order system

$$\mathbf{X}' = \mathbf{AX}, \tag{2}$$

where $\mathbf{A}$ is an $n \times n$ matrix of constants.

Eigenvalues and Eigenvectors If (1) is to be a solution vector of the homogeneous linear system (2), then $\mathbf{X}' = \mathbf{K}\lambda e^{\lambda t}$, so the system becomes $\mathbf{K}\lambda e^{\lambda t} = \mathbf{AK}e^{\lambda t}$. After dividing out $e^{\lambda t}$ and rearranging, we obtain $\mathbf{AK} = \lambda\mathbf{K}$ or $\mathbf{AK} - \lambda\mathbf{K} = \mathbf{0}$. Since $\mathbf{K} = \mathbf{IK}$, the last equation is the same as

$$(\mathbf{A} - \lambda\mathbf{I})\mathbf{K} = \mathbf{0}. \tag{3}$$

The matrix equation (3) is equivalent to the simultaneous algebraic equations

$$\begin{aligned} (a_{11} - \lambda)k_1 + \quad a_{12}k_2 + \cdots + \quad a_{1n}k_n &= 0 \\ a_{21}k_1 + (a_{22} - \lambda)k_2 + \cdots + \quad a_{2n}k_n &= 0 \\ \vdots \qquad\qquad\qquad & \vdots \\ a_{n1}k_1 + \quad a_{n2}k_2 + \cdots + (a_{nn} - \lambda)k_n &= 0. \end{aligned}$$

Thus to find a nontrivial solution $\mathbf{X}$ of (2), we must first find a nontrivial solution of the foregoing system; in other words, we must find a nontrivial vector $\mathbf{K}$ that satisfies (3). But for (3) to have solutions other than the obvious solution $k_1 = k_2 = \cdots = k_n = 0$, we must have

$$\det(\mathbf{A} - \lambda\mathbf{I}) = 0.$$

This polynomial equation in λ is called the **characteristic equation** of the matrix $\mathbf{A}$; its solutions are the **eigenvalues** of $\mathbf{A}$. A solution $\mathbf{K} \neq \mathbf{0}$ of (3) corresponding to an eigenvalue λ is called an **eigenvector** of $\mathbf{A}$. A solution of the homogeneous system (2) is then $\mathbf{X} = \mathbf{K}e^{\lambda t}$.

In the discussion that follows we examine three cases: real and distinct eigenvalues (that is, no eigenvalues are equal), repeated eigenvalues, and, finally, complex eigenvalues.

8.2.1 DISTINCT REAL EIGENVALUES

When the $n \times n$ matrix $\mathbf{A}$ possesses n distinct real eigenvalues $\lambda_1, \lambda_2, \ldots, \lambda_n$, then a set of n linearly independent eigenvectors $\mathbf{K}_1, \mathbf{K}_2, \ldots, \mathbf{K}_n$ can always be found, and

$$\mathbf{X}_1 = \mathbf{K}_1 e^{\lambda_1 t}, \quad \mathbf{X}_2 = \mathbf{K}_2 e^{\lambda_2 t}, \quad \ldots, \quad \mathbf{X}_n = \mathbf{K}_n e^{\lambda_n t}$$

is a fundamental set of solutions of (2) on the interval $(-\infty, \infty)$.

THEOREM 8.2.1 General Solution—Homogeneous Systems

Let $\lambda_1, \lambda_2, \ldots, \lambda_n$ be n distinct real eigenvalues of the coefficient matrix $\mathbf{A}$ of the homogeneous system (2) and let $\mathbf{K}_1, \mathbf{K}_2, \ldots, \mathbf{K}_n$ be the corresponding eigenvectors. Then the **general solution** of (2) on the interval $(-\infty, \infty)$ is given by

$$\mathbf{X} = c_1\mathbf{K}_1 e^{\lambda_1 t} + c_2\mathbf{K}_2 e^{\lambda_2 t} + \cdots + c_n\mathbf{K}_n e^{\lambda_n t}.$$

EXAMPLE 1 **Distinct Eigenvalues**

Solve
$$\frac{dx}{dt} = 2x + 3y$$
$$\frac{dy}{dt} = 2x + y. \tag{4}$$

SOLUTION We first find the eigenvalues and eigenvectors of the matrix of coefficients.

From the characteristic equation

$$\det(\mathbf{A} - \lambda\mathbf{I}) = \begin{vmatrix} 2-\lambda & 3 \\ 2 & 1-\lambda \end{vmatrix} = \lambda^2 - 3\lambda - 4 = (\lambda + 1)(\lambda - 4) = 0$$

we see that the eigenvalues are $\lambda_1 = -1$ and $\lambda_2 = 4$.

Now for $\lambda_1 = -1$, (3) is equivalent to

$$3k_1 + 3k_2 = 0$$
$$2k_1 + 2k_2 = 0.$$

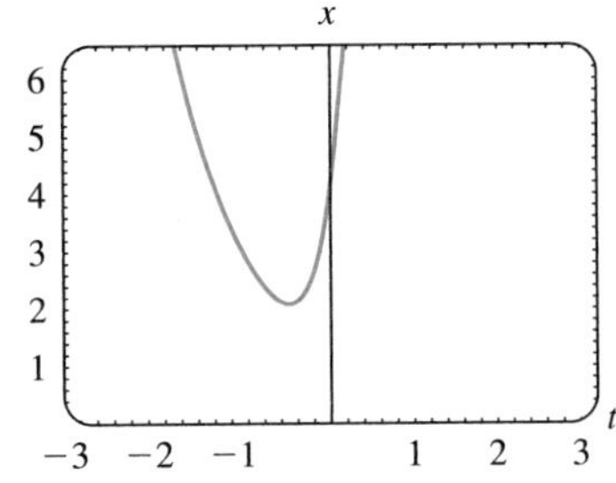

(a) graph of $x = e^{-t} + 3e^{4t}$

Thus $k_1 = -k_2$. When $k_2 = -1$, the related eigenvector is

$$\mathbf{K}_1 = \begin{pmatrix} 1 \\ -1 \end{pmatrix}.$$

For $\lambda_2 = 4$ we have
$$-2k_1 + 3k_2 = 0$$
$$2k_1 - 3k_2 = 0$$

so $k_1 = \frac{3}{2}k_2$; therefore with $k_2 = 2$ the corresponding eigenvector is

$$\mathbf{K}_2 = \begin{pmatrix} 3 \\ 2 \end{pmatrix}.$$

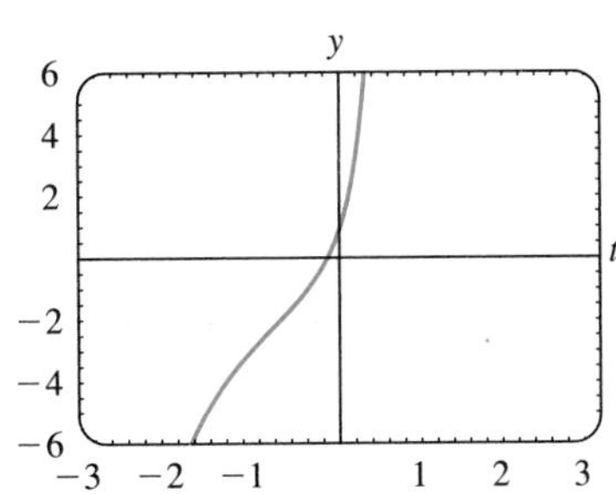

(b) graph of $y = -e^{-t} + 2e^{4t}$

Since the matrix of coefficients **A** is a 2 × 2 matrix and since we have found two linearly independent solutions of (4),

$$\mathbf{X}_1 = \begin{pmatrix} 1 \\ -1 \end{pmatrix} e^{-t} \quad \text{and} \quad \mathbf{X}_2 = \begin{pmatrix} 3 \\ 2 \end{pmatrix} e^{4t},$$

we conclude that the general solution of the system is

$$\mathbf{X} = c_1\mathbf{X}_1 + c_2\mathbf{X}_2 = c_1\begin{pmatrix} 1 \\ -1 \end{pmatrix} e^{-t} + c_2\begin{pmatrix} 3 \\ 2 \end{pmatrix} e^{4t}. \tag{5}$$

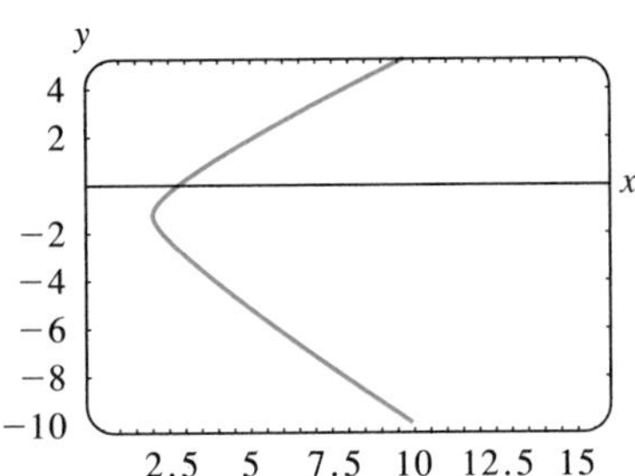

(c) trajectory defined by $x = e^{-t} + 3e^{4t}$, $y = -e^{-t} + 2e^{4t}$ in the phase plane

FIGURE 8.2.1 A solution from (5) yields three different curves in three different planes

Phase Portrait You should keep firmly in mind that writing a solution of a system of linear first-order differential equations in terms of matrices is simply an alternative to the method that we employed in Section 4.9, that is, listing the individual functions and the relationship between the constants. If we add the vectors on the right-hand side of (5) and then equate the entries with the corresponding entries in the vector on the left-hand side, we obtain the more familiar statement

$$x = c_1e^{-t} + 3c_2e^{4t}, \qquad y = -c_1e^{-t} + 2c_2e^{4t}.$$

As was pointed out in Section 8.1, we can interpret these equations as parametric equations of curves in the xy-plane or **phase plane.** Each curve, corresponding to specific choices for c_1 and c_2, is called a **trajectory.** For the choice of constants $c_1 = c_2 = 1$ in the solution (5) we see in Figure 8.2.1 the graph of $x(t)$ in the tx-plane, the graph of $y(t)$ in the ty-plane, and the trajectory consisting of the points

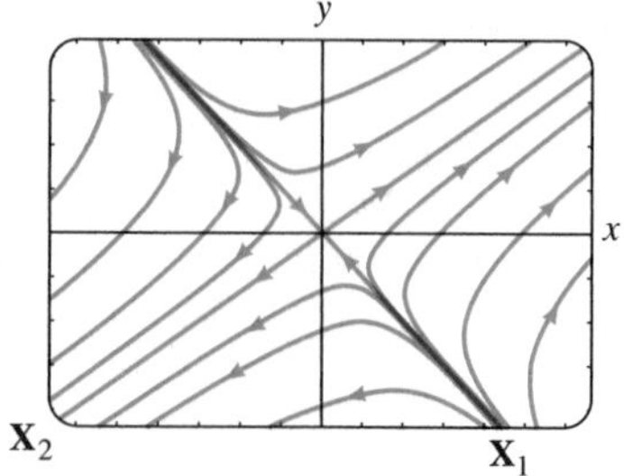

FIGURE 8.2.2 A phase portrait of system (4)

$(x(t), y(t))$ in the phase plane. A collection of representative trajectories in the phase plane, as shown in Figure 8.2.2, is said to be a **phase portrait** of the given linear system. What appears to be *two* red lines in Figure 8.2.2 are actually *four* red half-lines defined parametrically in the first, second, third, and fourth quadrants by the solutions $\mathbf{X}_2$, $-\mathbf{X}_1$, $-\mathbf{X}_2$, and $\mathbf{X}_1$, respectively. For example, the Cartesian equations $y = \frac{2}{3}x$, $x > 0$, and $y = -x$, $x > 0$, of the half-lines in the first and fourth quadrants were obtained by eliminating the parameter t in the solutions $x = 3e^{4t}$, $y = 2e^{4t}$, and $x = e^{-t}$, $y = -e^{-t}$, respectively. Moreover, each eigenvector can be visualized as a two-dimensional vector lying along one of these half-lines. The eigenvector $\mathbf{K}_2 = \begin{pmatrix} 3 \\ 2 \end{pmatrix}$ lies along $y = \frac{2}{3}x$ in the first quadrant, and $\mathbf{K}_1 = \begin{pmatrix} 1 \\ -1 \end{pmatrix}$ lies along $y = -x$ in the fourth quadrant. Each vector starts at the origin; $\mathbf{K}_2$ terminates at the point $(2, 3)$, and $\mathbf{K}_1$ terminates at $(1, -1)$.

The origin is not only a constant solution $x = 0$, $y = 0$ of every 2×2 homogeneous linear system $\mathbf{X}' = \mathbf{AX}$, but also an important point in the qualitative study of such systems. If we think in physical terms, the arrowheads on each trajectory in Figure 8.2.2 indicate the direction that a particle with coordinates $(x(t), y(t))$ on that trajectory at time t moves as time increases. Observe that the arrowheads, with the exception of only those on the half-lines in the second and fourth quadrants, indicate that a particle moves away from the origin as time t increases. If we imagine time ranging from $-\infty$ to ∞, then inspection of the solution $x = c_1e^{-t} + 3c_2e^{4t}$, $y = -c_1e^{-t} + 2c_2e^{4t}$, $c_1 \neq 0$, $c_2 \neq 0$ shows that a trajectory, or moving particle, "starts" asymptotic to one of the half-lines defined by $\mathbf{X}_1$ or $-\mathbf{X}_1$ (since e^{4t} is negligible for $t \rightarrow -\infty$) and "finishes" asymptotic to one of the half-lines defined by $\mathbf{X}_2$ and $-\mathbf{X}_2$ (since e^{-t} is negligible for $t \rightarrow \infty$).

We note in passing that Figure 8.2.2 represents a phase portrait that is typical of *all* 2×2 homogeneous linear systems $\mathbf{X}' = \mathbf{AX}$ with real eigenvalues of opposite signs. See Problem 17 in Exercises 8.2. Moreover, phase portraits in the two cases when distinct real eigenvalues have the same algebraic sign are typical of all such 2×2 linear systems; the only difference is that the arrowheads indicate that a particle moves away from the origin on any trajectory as $t \rightarrow \infty$ when both λ_1 and λ_2 are positive and moves toward the origin on any trajectory when both λ_1 and λ_2 are negative. Consequently, we call the origin a **repeller** in the case $\lambda_1 > 0$, $\lambda_2 > 0$ and an **attractor** in the case $\lambda_1 < 0$, $\lambda_2 < 0$. See Problem 18 in Exercises 8.2. The origin in Figure 8.2.2 is neither a repeller nor an attractor. Investigation of the remaining case when $\lambda = 0$ is an eigenvalue of a 2×2 homogeneous linear system is left as an exercise. See Problem 49 in Exercises 8.2.

EXAMPLE 2 Distinct Eigenvalues

Solve

$$\begin{aligned} \frac{dx}{dt} &= -4x + y + z \\ \frac{dy}{dt} &= x + 5y - z \\ \frac{dz}{dt} &= y - 3z. \end{aligned} \tag{6}$$

SOLUTION Using the cofactors of the third row, we find

$$\det(\mathbf{A} - \lambda\mathbf{I}) = \begin{vmatrix} -4-\lambda & 1 & 1 \\ 1 & 5-\lambda & -1 \\ 0 & 1 & -3-\lambda \end{vmatrix} = -(\lambda + 3)(\lambda + 4)(\lambda - 5) = 0,$$

and so the eigenvalues are $\lambda_1 = -3$, $\lambda_2 = -4$, and $\lambda_3 = 5$.

For $\lambda_1 = -3$ Gauss-Jordan elimination gives

$$(\mathbf{A} + 3\mathbf{I}|\mathbf{0}) = \left(\begin{array}{rrr|r} -1 & 1 & 1 & 0 \\ 1 & 8 & -1 & 0 \\ 0 & 1 & 0 & 0 \end{array}\right) \xrightarrow{\text{row operations}} \left(\begin{array}{rrr|r} 1 & 0 & -1 & 0 \\ 0 & 1 & 0 & 0 \\ 0 & 0 & 0 & 0 \end{array}\right).$$

Therefore $k_1 = k_3$ and $k_2 = 0$. The choice $k_3 = 1$ gives an eigenvector and corresponding solution vector

$$\mathbf{K}_1 = \begin{pmatrix} 1 \\ 0 \\ 1 \end{pmatrix}, \qquad \mathbf{X}_1 = \begin{pmatrix} 1 \\ 0 \\ 1 \end{pmatrix} e^{-3t}. \tag{7}$$

Similarly, for $\lambda_2 = -4$

$$(\mathbf{A} + 4\mathbf{I}|\mathbf{0}) = \left(\begin{array}{rrr|r} 0 & 1 & 1 & 0 \\ 1 & 9 & -1 & 0 \\ 0 & 1 & 1 & 0 \end{array}\right) \xrightarrow{\text{row operations}} \left(\begin{array}{rrr|r} 1 & 0 & -10 & 0 \\ 0 & 1 & 1 & 0 \\ 0 & 0 & 0 & 0 \end{array}\right)$$

implies that $k_1 = 10k_3$ and $k_2 = -k_3$. Choosing $k_3 = 1$, we get a second eigenvector and solution vector

$$\mathbf{K}_2 = \begin{pmatrix} 10 \\ -1 \\ 1 \end{pmatrix}, \qquad \mathbf{X}_2 = \begin{pmatrix} 10 \\ -1 \\ 1 \end{pmatrix} e^{-4t}. \tag{8}$$

Finally, when $\lambda_3 = 5$, the augmented matrices

$$(\mathbf{A} + 5\mathbf{I}|\mathbf{0}) = \left(\begin{array}{rrr|r} -9 & 1 & 1 & 0 \\ 1 & 0 & -1 & 0 \\ 0 & 1 & -8 & 0 \end{array}\right) \xrightarrow{\text{row operations}} \left(\begin{array}{rrr|r} 1 & 0 & -1 & 0 \\ 0 & 1 & -8 & 0 \\ 0 & 0 & 0 & 0 \end{array}\right)$$

yield

$$\mathbf{K}_3 = \begin{pmatrix} 1 \\ 8 \\ 1 \end{pmatrix}, \qquad \mathbf{X}_3 = \begin{pmatrix} 1 \\ 8 \\ 1 \end{pmatrix} e^{5t}. \tag{9}$$

The general solution of (6) is a linear combination of the solution vectors in (7), (8), and (9):

$$\mathbf{X} = c_1\begin{pmatrix} 1 \\ 0 \\ 1 \end{pmatrix} e^{-3t} + c_2\begin{pmatrix} 10 \\ -1 \\ 1 \end{pmatrix} e^{-4t} + c_3\begin{pmatrix} 1 \\ 8 \\ 1 \end{pmatrix} e^{5t}.$$ ≡

≡ **Use of Computers** Software packages such as MATLAB, *Mathematica, Maple,* and DERIVE can be real time savers in finding eigenvalues and eigenvectors of a matrix **A**.

8.2.2 REPEATED EIGENVALUES

Of course, not all of the n eigenvalues $\lambda_1, \lambda_2, \ldots, \lambda_n$ of an $n \times n$ matrix $\mathbf{A}$ need be distinct; that is, some of the eigenvalues may be repeated. For example, the characteristic equation of the coefficient matrix in the system

$$\mathbf{X}' = \begin{pmatrix} 3 & -18 \\ 2 & -9 \end{pmatrix}\mathbf{X} \tag{10}$$

is readily shown to be $(\lambda + 3)^2 = 0$, and therefore $\lambda_1 = \lambda_2 = -3$ is a root of *multiplicity two*. For this value we find the single eigenvector

$$\mathbf{K}_1 = \begin{pmatrix} 3 \\ 1 \end{pmatrix}, \quad \text{so} \quad \mathbf{X}_1 = \begin{pmatrix} 3 \\ 1 \end{pmatrix} e^{-3t} \tag{11}$$

is one solution of (10). But since we are obviously interested in forming the general solution of the system, we need to pursue the question of finding a second solution.

In general, if m is a positive integer and $(\lambda - \lambda_1)^m$ is a factor of the characteristic equation while $(\lambda - \lambda_1)^{m+1}$ is not a factor, then λ_1 is said to be an **eigenvalue of multiplicity *m*.** The next three examples illustrate the following cases:

(*i*) For some $n \times n$ matrices $\mathbf{A}$ it may be possible to find m linearly independent eigenvectors $\mathbf{K}_1, \mathbf{K}_2, \ldots, \mathbf{K}_m$ corresponding to an eigenvalue λ_1 of multiplicity $m \le n$. In this case the general solution of the system contains the linear combination

$$c_1\mathbf{K}_1e^{\lambda_1 t} + c_2\mathbf{K}_2e^{\lambda_1 t} + \cdots + c_m\mathbf{K}_me^{\lambda_1 t}.$$

(*ii*) If there is only one eigenvector corresponding to the eigenvalue λ_1 of multiplicity m, then m linearly independent solutions of the form

$$\begin{aligned} \mathbf{X}_1 &= \mathbf{K}_{11}e^{\lambda_1 t} \\ \mathbf{X}_2 &= \mathbf{K}_{21}te^{\lambda_1 t} + \mathbf{K}_{22}e^{\lambda_1 t} \\ &\ \vdots \\ \mathbf{X}_m &= \mathbf{K}_{m1}\frac{t^{m-1}}{(m-1)!}e^{\lambda_1 t} + \mathbf{K}_{m2}\frac{t^{m-2}}{(m-2)!}e^{\lambda_1 t} + \cdots + \mathbf{K}_{mm}e^{\lambda_1 t}, \end{aligned}$$

where $\mathbf{K}_{ij}$ are column vectors, can always be found.

Eigenvalue of Multiplicity Two We begin by considering eigenvalues of multiplicity two. In the first example we illustrate a matrix for which we can find two distinct eigenvectors corresponding to a double eigenvalue.

EXAMPLE 3 Repeated Eigenvalues

Solve $\mathbf{X}' = \begin{pmatrix} 1 & -2 & 2 \\ -2 & 1 & -2 \\ 2 & -2 & 1 \end{pmatrix}\mathbf{X}$.

SOLUTION Expanding the determinant in the characteristic equation

$$\det(\mathbf{A} - \lambda\mathbf{I}) = \begin{vmatrix} 1-\lambda & -2 & 2 \\ -2 & 1-\lambda & -2 \\ 2 & -2 & 1-\lambda \end{vmatrix} = 0$$

yields $-(\lambda + 1)^2(\lambda - 5) = 0$. We see that $\lambda_1 = \lambda_2 = -1$ and $\lambda_3 = 5$.

For $\lambda_1 = -1$ Gauss-Jordan elimination immediately gives

$$(\mathbf{A} + \mathbf{I}|\mathbf{0}) = \left(\begin{array}{ccc|c} 2 & -2 & 2 & 0 \\ -2 & 2 & -2 & 0 \\ 2 & -2 & 2 & 0 \end{array}\right) \xrightarrow{\text{row operations}} \left(\begin{array}{ccc|c} 1 & -1 & 1 & 0 \\ 0 & 0 & 0 & 0 \\ 0 & 0 & 0 & 0 \end{array}\right).$$

The first row of the last matrix means $k_1 - k_2 + k_3 = 0$ or $k_1 = k_2 - k_3$. The choices $k_2 = 1,\ k_3 = 0$ and $k_2 = 1,\ k_3 = 1$ yield, in turn, $k_1 = 1$ and $k_1 = 0$. Thus two eigenvectors corresponding to $\lambda_1 = -1$ are

$$\mathbf{K}_1 = \begin{pmatrix} 1 \\ 1 \\ 0 \end{pmatrix} \quad \text{and} \quad \mathbf{K}_2 = \begin{pmatrix} 0 \\ 1 \\ 1 \end{pmatrix}.$$

Since neither eigenvector is a constant multiple of the other, we have found two linearly independent solutions,

$$\mathbf{X}_1 = \begin{pmatrix} 1 \\ 1 \\ 0 \end{pmatrix} e^{-t} \quad \text{and} \quad \mathbf{X}_2 = \begin{pmatrix} 0 \\ 1 \\ 1 \end{pmatrix} e^{-t},$$

corresponding to the same eigenvalue. Lastly, for $\lambda_3 = 5$ the reduction

$$(\mathbf{A} + 5\mathbf{I}|\mathbf{0}) = \left(\begin{array}{rrr|r} -4 & -2 & 2 & 0 \\ -2 & -4 & -2 & 0 \\ 2 & -2 & -4 & 0 \end{array}\right) \xrightarrow[\text{operations}]{\text{row}} \left(\begin{array}{rrr|r} 1 & 0 & -1 & 0 \\ 0 & 1 & 1 & 0 \\ 0 & 0 & 0 & 0 \end{array}\right)$$

implies that $k_1 = k_3$ and $k_2 = -k_3$. Picking $k_3 = 1$ gives $k_1 = 1,\ k_2 = -1$; thus a third eigenvector is

$$\mathbf{K}_3 = \begin{pmatrix} 1 \\ -1 \\ 1 \end{pmatrix}.$$

We conclude that the general solution of the system is

$$\mathbf{X} = c_1 \begin{pmatrix} 1 \\ 1 \\ 0 \end{pmatrix} e^{-t} + c_2 \begin{pmatrix} 0 \\ 1 \\ 1 \end{pmatrix} e^{-t} + c_3 \begin{pmatrix} 1 \\ -1 \\ 1 \end{pmatrix} e^{5t}.$$

≡

The matrix of coefficients $\mathbf{A}$ in Example 3 is a special kind of matrix known as a symmetric matrix. An $n \times n$ matrix $\mathbf{A}$ is said to be **symmetric** if its transpose $\mathbf{A}^T$ (where the rows and columns are interchanged) is the same as $\mathbf{A}$—that is, if $\mathbf{A}^T = \mathbf{A}$. It can be proved that if the matrix $\mathbf{A}$ in the system $\mathbf{X}' = \mathbf{AX}$ is symmetric and has real entries, then we can always find n linearly independent eigenvectors $\mathbf{K}_1, \mathbf{K}_2, \ldots, \mathbf{K}_n$, and the general solution of such a system is as given in Theorem 8.2.1. As illustrated in Example 3, this result holds even when some of the eigenvalues are repeated.

≡ **Second Solution** Now suppose that λ_1 is an eigenvalue of multiplicity two and that there is only one eigenvector associated with this value. A second solution can be found of the form

$$\mathbf{X}_2 = \mathbf{K}te^{\lambda_1 t} + \mathbf{P}e^{\lambda_1 t}, \tag{12}$$

where

$$\mathbf{K} = \begin{pmatrix} k_1 \\ k_2 \\ \vdots \\ k_n \end{pmatrix} \quad \text{and} \quad \mathbf{P} = \begin{pmatrix} p_1 \\ p_2 \\ \vdots \\ p_n \end{pmatrix}.$$

To see this, we substitute (12) into the system $\mathbf{X}' = \mathbf{AX}$ and simplify:

$$(\mathbf{AK} - \lambda_1\mathbf{K})te^{\lambda_1 t} + (\mathbf{AP} - \lambda_1\mathbf{P} - \mathbf{K})e^{\lambda_1 t} = \mathbf{0}.$$

Since this last equation is to hold for all values of t, we must have

$$(\mathbf{A} - \lambda_1\mathbf{I})\mathbf{K} = \mathbf{0} \tag{13}$$

and

$$(\mathbf{A} - \lambda_1\mathbf{I})\mathbf{P} = \mathbf{K}. \tag{14}$$

Equation (13) simply states that $\mathbf{K}$ must be an eigenvector of $\mathbf{A}$ associated with λ_1. By solving (13), we find one solution $\mathbf{X}_1 = \mathbf{K}e^{\lambda_1 t}$. To find the second solution $\mathbf{X}_2$, we need only solve the additional system (14) for the vector $\mathbf{P}$.

EXAMPLE 4 Repeated Eigenvalues

Find the general solution of the system given in (10).

SOLUTION From (11) we know that $\lambda_1 = -3$ and that one solution is $\mathbf{X}_1 = \begin{pmatrix} 3 \\ 1 \end{pmatrix} e^{-3t}$. Identifying $\mathbf{K} = \begin{pmatrix} 3 \\ 1 \end{pmatrix}$ and $\mathbf{P} = \begin{pmatrix} p_1 \\ p_2 \end{pmatrix}$, we find from (14) that we must now solve

$$(\mathbf{A} + 3\mathbf{I})\mathbf{P} = \mathbf{K} \quad \text{or} \quad \begin{aligned} 6p_1 - 18p_2 &= 3 \\ 2p_1 - \ \ 6p_2 &= 1. \end{aligned}$$

Since this system is obviously equivalent to one equation, we have an infinite number of choices for p_1 and p_2. For example, by choosing $p_1 = 1$, we find $p_2 = \frac{1}{6}$. However, for simplicity we shall choose $p_1 = \frac{1}{2}$ so that $p_2 = 0$. Hence $\mathbf{P} = \begin{pmatrix} \frac{1}{2} \\ 0 \end{pmatrix}$. Thus from (12) we find $\mathbf{X}_2 = \begin{pmatrix} 3 \\ 1 \end{pmatrix} te^{-3t} + \begin{pmatrix} \frac{1}{2} \\ 0 \end{pmatrix} e^{-3t}$. The general solution of (10) is then $\mathbf{X} = c_1\mathbf{X}_1 + c_2\mathbf{X}_2$ or

$$\mathbf{X} = c_1\begin{pmatrix} 3 \\ 1 \end{pmatrix} e^{-3t} + c_2\left[\begin{pmatrix} 3 \\ 1 \end{pmatrix} te^{-3t} + \begin{pmatrix} \frac{1}{2} \\ 0 \end{pmatrix} e^{-3t}\right]. \equiv$$

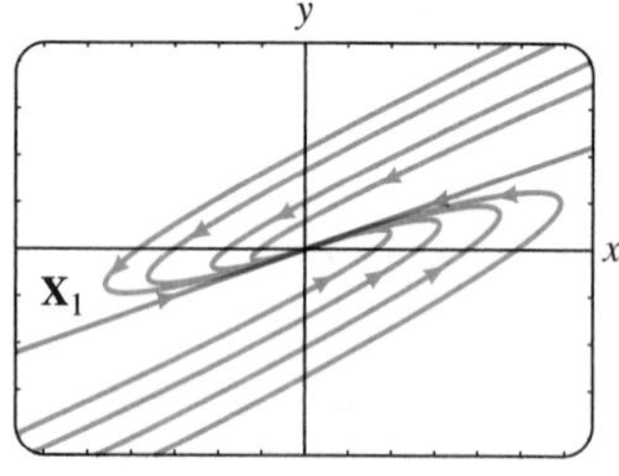

FIGURE 8.2.3 A phase portrait of system (10)

By assigning various values to c_1 and c_2 in the solution in Example 4, we can plot trajectories of the system in (10). A phase portrait of (10) is given in Figure 8.2.3. The solutions $\mathbf{X}_1$ and $-\mathbf{X}_1$ determine two half-lines $y = \frac{1}{3}x, x > 0$ and $y = \frac{1}{3}x, x < 0$, respectively, shown in red in the figure. Because the single eigenvalue is negative and $e^{-3t} \to 0$ as $t \to \infty$ on *every* trajectory, we have $(x(t), y(t)) \to (0, 0)$ as $t \to \infty$. This is why the arrowheads in Figure 8.2.3 indicate that a particle on any trajectory moves toward the origin as time increases and why the origin is an attractor in this case. Moreover, a moving particle or trajectory $x = 3c_1e^{-3t} + c_2(3te^{-3t} + \frac{1}{2}e^{-3t})$, $y = c_1e^{-3t} + c_2te^{-3t}$, $c_2 \neq 0$, approaches (0, 0) tangentially to one of the half-lines as $t \to \infty$. In contrast, when the repeated eigenvalue is positive, the situation is reversed and the origin is a repeller. See Problem 21 in Exercises 8.2. Analogous to Figure 8.2.2, Figure 8.2.3 is typical of *all* 2×2 homogeneous linear systems $\mathbf{X}' = \mathbf{AX}$ that have two repeated negative eigenvalues. See Problem 32 in Exercises 8.2.

≡ Eigenvalue of Multiplicity Three When the coefficient matrix $\mathbf{A}$ has only one eigenvector associated with an eigenvalue λ_1 of multiplicity three, we can find a second solution of the form (12) and a third solution of the form

$$\mathbf{X}_3 = \mathbf{K}\frac{t^2}{2}e^{\lambda_1 t} + \mathbf{P}te^{\lambda_1 t} + \mathbf{Q}e^{\lambda_1 t}, \tag{15}$$

where $\mathbf{K} = \begin{pmatrix} k_1 \\ k_2 \\ \vdots \\ k_n \end{pmatrix}$, $\mathbf{P} = \begin{pmatrix} p_1 \\ p_2 \\ \vdots \\ p_n \end{pmatrix}$, and $\mathbf{Q} = \begin{pmatrix} q_1 \\ q_2 \\ \vdots \\ q_n \end{pmatrix}$.

By substituting (15) into the system $\mathbf{X}' = \mathbf{AX}$, we find that the column vectors $\mathbf{K}$, $\mathbf{P}$, and $\mathbf{Q}$ must satisfy

$$(\mathbf{A} - \lambda_1\mathbf{I})\mathbf{K} = \mathbf{0} \tag{16}$$

$$(\mathbf{A} - \lambda_1\mathbf{I})\mathbf{P} = \mathbf{K} \tag{17}$$

and

$$(\mathbf{A} - \lambda_1\mathbf{I})\mathbf{Q} = \mathbf{P}. \tag{18}$$

Of course, the solutions of (16) and (17) can be used in forming the solutions $\mathbf{X}_1$ and $\mathbf{X}_2$.

EXAMPLE 5 Repeated Eigenvalues

Solve $\mathbf{X}' = \begin{pmatrix} 2 & 1 & 6 \\ 0 & 2 & 5 \\ 0 & 0 & 2 \end{pmatrix}\mathbf{X}$.

SOLUTION The characteristic equation $(\lambda - 2)^3 = 0$ shows that $\lambda_1 = 2$ is an eigenvalue of multiplicity three. By solving $(\mathbf{A} - 2\mathbf{I})\mathbf{K} = \mathbf{0}$, we find the single eigenvector

$$\mathbf{K} = \begin{pmatrix} 1 \\ 0 \\ 0 \end{pmatrix}.$$

We next solve the systems $(\mathbf{A} - 2\mathbf{I})\mathbf{P} = \mathbf{K}$ and $(\mathbf{A} - 2\mathbf{I})\mathbf{Q} = \mathbf{P}$ in succession and find that

$$\mathbf{P} = \begin{pmatrix} 0 \\ 1 \\ 0 \end{pmatrix} \quad \text{and} \quad \mathbf{Q} = \begin{pmatrix} 0 \\ -\frac{6}{5} \\ \frac{1}{5} \end{pmatrix}.$$

Using (12) and (15), we see that the general solution of the system is

$$\mathbf{X} = c_1\begin{pmatrix} 1 \\ 0 \\ 0 \end{pmatrix}e^{2t} + c_2\left[\begin{pmatrix} 1 \\ 0 \\ 0 \end{pmatrix}te^{2t} + \begin{pmatrix} 0 \\ 1 \\ 0 \end{pmatrix}e^{2t}\right] + c_3\left[\begin{pmatrix} 1 \\ 0 \\ 0 \end{pmatrix}\frac{t^2}{2}e^{2t} + \begin{pmatrix} 0 \\ 1 \\ 0 \end{pmatrix}te^{2t} + \begin{pmatrix} 0 \\ -\frac{6}{5} \\ \frac{1}{5} \end{pmatrix}e^{2t}\right]. \equiv$$

REMARKS

When an eigenvalue λ_1 has multiplicity m, either we can find m linearly independent eigenvectors or the number of corresponding eigenvectors is less than m. Hence the two cases listed on page 338 are not all the possibilities under which a repeated eigenvalue can occur. It can happen, say, that a 5×5 matrix has an eigenvalue of multiplicity five and there exist three corresponding linearly independent eigenvectors. See Problems 31 and 50 in Exercises 8.2.

8.2.3 COMPLEX EIGENVALUES

If $\lambda_1 = \alpha + \beta i$ and $\lambda_2 = \alpha - \beta i$, $\beta > 0$, $i^2 = -1$ are complex eigenvalues of the coefficient matrix **A**, we can then certainly expect their corresponding eigenvectors to also have complex entries.*

For example, the characteristic equation of the system

$$\begin{aligned}\frac{dx}{dt} &= 6x - y\\ \frac{dy}{dt} &= 5x + 4y\end{aligned} \tag{19}$$

is
$$\det(\mathbf{A} - \lambda\mathbf{I}) = \begin{vmatrix} 6-\lambda & -1 \\ 5 & 4-\lambda \end{vmatrix} = \lambda^2 - 10\lambda + 29 = 0.$$

From the quadratic formula we find $\lambda_1 = 5 + 2i$, $\lambda_2 = 5 - 2i$.

Now for $\lambda_1 = 5 + 2i$ we must solve

$$\begin{aligned}(1 - 2i)k_1 - \qquad\quad k_2 &= 0\\ 5k_1 - (1 + 2i)k_2 &= 0.\end{aligned}$$

Since $k_2 = (1 - 2i)k_1$,† the choice $k_1 = 1$ gives the following eigenvector and corresponding solution vector:

$$\mathbf{K}_1 = \begin{pmatrix} 1 \\ 1 - 2i \end{pmatrix}, \qquad \mathbf{X}_1 = \begin{pmatrix} 1 \\ 1 - 2i \end{pmatrix} e^{(5+2i)t}.$$

In like manner, for $\lambda_2 = 5 - 2i$ we find

$$\mathbf{K}_2 = \begin{pmatrix} 1 \\ 1 + 2i \end{pmatrix}, \qquad \mathbf{X}_2 = \begin{pmatrix} 1 \\ 1 + 2i \end{pmatrix} e^{(5-2i)t}.$$

We can verify by means of the Wronskian that these solution vectors are linearly independent, and so the general solution of (19) is

$$\mathbf{X} = c_1 \begin{pmatrix} 1 \\ 1 - 2i \end{pmatrix} e^{(5+2i)t} + c_2 \begin{pmatrix} 1 \\ 1 + 2i \end{pmatrix} e^{(5-2i)t}. \tag{20}$$

Note that the entries in $\mathbf{K}_2$ corresponding to λ_2 are the conjugates of the entries in $\mathbf{K}_1$ corresponding to λ_1. The conjugate of λ_1 is, of course, λ_2. We write this as $\lambda_2 = \overline{\lambda_1}$ and $\mathbf{K}_2 = \overline{\mathbf{K}_1}$. We have illustrated the following general result.

THEOREM 8.2.2 Solutions Corresponding to a Complex Eigenvalue

Let **A** be the coefficient matrix having real entries of the homogeneous system (2), and let $\mathbf{K}_1$ be an eigenvector corresponding to the complex eigenvalue $\lambda_1 = \alpha + i\beta$, α and β real. Then

$$\mathbf{K}_1 e^{\lambda_1 t} \qquad \text{and} \qquad \overline{\mathbf{K}_1} e^{\overline{\lambda_1} t}$$

are solutions of (2).

*When the characteristic equation has real coefficients, complex eigenvalues always appear in conjugate pairs.

†Note that the second equation is simply $(1 + 2i)$ times the first.

It is desirable and relatively easy to rewrite a solution such as (20) in terms of real functions. To this end we first use Euler's formula to write

$$e^{(5+2i)t} = e^{5t}e^{2ti} = e^{5t}(\cos 2t + i \sin 2t)$$
$$e^{(5-2i)t} = e^{5t}e^{-2ti} = e^{5t}(\cos 2t - i \sin 2t).$$

Then, after we multiply complex numbers, collect terms, and replace $c_1 + c_2$ by C_1 and $(c_1 - c_2)i$ by C_2, (20) becomes

$$\mathbf{X} = C_1\mathbf{X}_1 + C_2\mathbf{X}_2, \tag{21}$$

where
$$\mathbf{X}_1 = \left[\begin{pmatrix} 1 \\ 1 \end{pmatrix}\cos 2t - \begin{pmatrix} 0 \\ -2 \end{pmatrix}\sin 2t\right]e^{5t}$$

and
$$\mathbf{X}_2 = \left[\begin{pmatrix} 0 \\ -2 \end{pmatrix}\cos 2t + \begin{pmatrix} 1 \\ 1 \end{pmatrix}\sin 2t\right]e^{5t}.$$

It is now important to realize that the vectors $\mathbf{X}_1$ and $\mathbf{X}_2$ in (21) constitute a linearly independent set of *real* solutions of the original system. Consequently, we are justified in ignoring the relationship between C_1, C_2 and c_1, c_2, and we can regard C_1 and C_2 as completely arbitrary and real. In other words, the linear combination (21) is an alternative general solution of (19). Moreover, with the real form given in (21) we are able to obtain a phase portrait of the system in (19). From (21) we find $x(t)$ and $y(t)$ to be

$$x = C_1e^{5t}\cos 2t + C_2e^{5t}\sin 2t$$
$$y = (C_1 - 2C_2)e^{5t}\cos 2t + (2C_1 + C_2)e^{5t}\sin 2t.$$

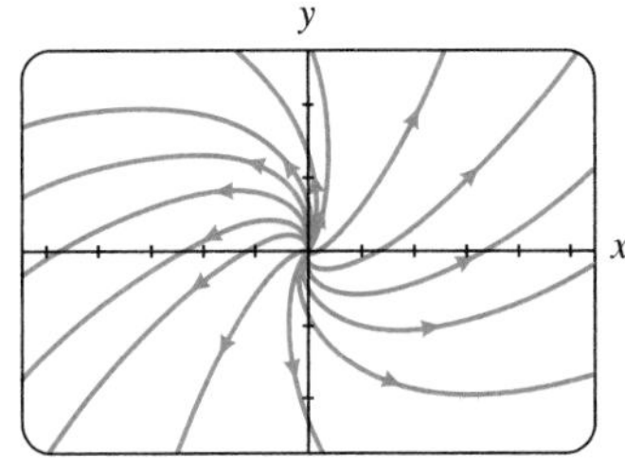

FIGURE 8.2.4 A phase portrait of system (19)

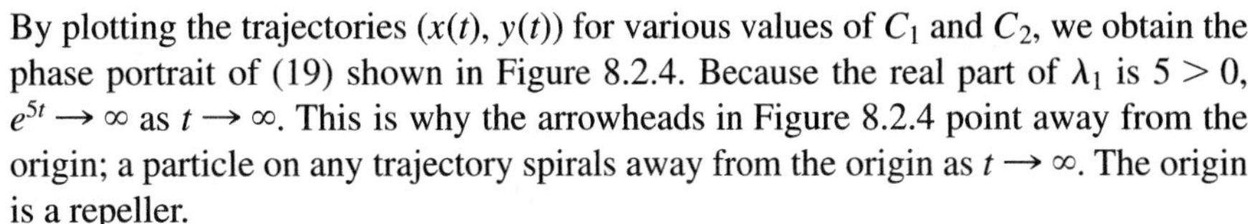

By plotting the trajectories $(x(t), y(t))$ for various values of C_1 and C_2, we obtain the phase portrait of (19) shown in Figure 8.2.4. Because the real part of λ_1 is $5 > 0$, $e^{5t} \to \infty$ as $t \to \infty$. This is why the arrowheads in Figure 8.2.4 point away from the origin; a particle on any trajectory spirals away from the origin as $t \to \infty$. The origin is a repeller.

The process by which we obtained the real solutions in (21) can be generalized. Let $\mathbf{K}_1$ be an eigenvector of the coefficient matrix $\mathbf{A}$ (with real entries) corresponding to the complex eigenvalue $\lambda_1 = \alpha + i\beta$. Then the solution vectors in Theorem 8.2.2 can be written as

$$\mathbf{K}_1e^{\lambda_1 t} = \mathbf{K}_1e^{\alpha t}e^{i\beta t} = \mathbf{K}_1e^{\alpha t}(\cos \beta t + i \sin \beta t)$$
$$\overline{\mathbf{K}}_1e^{\overline{\lambda}_1 t} = \overline{\mathbf{K}}_1e^{\alpha t}e^{-i\beta t} = \overline{\mathbf{K}}_1e^{\alpha t}(\cos \beta t - i \sin \beta t).$$

By the superposition principle, Theorem 8.1.2, the following vectors are also solutions:

$$\mathbf{X}_1 = \frac{1}{2}(\mathbf{K}_1e^{\lambda_1 t} + \overline{\mathbf{K}}_1e^{\overline{\lambda}_1 t}) = \frac{1}{2}(\mathbf{K}_1 + \overline{\mathbf{K}}_1)e^{\alpha t}\cos \beta t - \frac{i}{2}(-\mathbf{K}_1 + \overline{\mathbf{K}}_1)e^{\alpha t}\sin \beta t$$

$$\mathbf{X}_2 = \frac{i}{2}(-\mathbf{K}_1e^{\lambda_1 t} + \overline{\mathbf{K}}_1e^{\overline{\lambda}_1 t}) = \frac{i}{2}(-\mathbf{K}_1 + \overline{\mathbf{K}}_1)e^{\alpha t}\cos \beta t + \frac{1}{2}(\mathbf{K}_1 + \overline{\mathbf{K}}_1)e^{\alpha t}\sin \beta t.$$

Both $\frac{1}{2}(z + \overline{z}) = a$ and $\frac{1}{2}i(-z + \overline{z}) = b$ are *real* numbers for *any* complex number $z = a + ib$. Therefore, the entries in the column vectors $\frac{1}{2}(\mathbf{K}_1 + \overline{\mathbf{K}}_1)$ and $\frac{1}{2}i(-\mathbf{K}_1 + \overline{\mathbf{K}}_1)$ are real numbers. By defining

$$\mathbf{B}_1 = \frac{1}{2}(\mathbf{K}_1 + \overline{\mathbf{K}}_1) \quad \text{and} \quad \mathbf{B}_2 = \frac{i}{2}(-\mathbf{K}_1 + \overline{\mathbf{K}}_1), \tag{22}$$

we are led to the following theorem.

THEOREM 8.2.3 Real Solutions Corresponding to a Complex Eigenvalue

Let $\lambda_1 = \alpha + i\beta$ be a complex eigenvalue of the coefficient matrix **A** in the homogeneous system (2) and let $\mathbf{B}_1$ and $\mathbf{B}_2$ denote the column vectors defined in (22). Then

$$\begin{aligned}\mathbf{X}_1 &= [\mathbf{B}_1 \cos \beta t - \mathbf{B}_2 \sin \beta t]e^{\alpha t}\\ \mathbf{X}_2 &= [\mathbf{B}_2 \cos \beta t + \mathbf{B}_1 \sin \beta t]e^{\alpha t}\end{aligned} \tag{23}$$

are linearly independent solutions of (2) on $(-\infty, \infty)$.

The matrices $\mathbf{B}_1$ and $\mathbf{B}_2$ in (22) are often denoted by

$$\mathbf{B}_1 = \text{Re}(\mathbf{K}_1) \quad \text{and} \quad \mathbf{B}_2 = \text{Im}(\mathbf{K}_1) \tag{24}$$

since these vectors are, respectively, the *real* and *imaginary* parts of the eigenvector $\mathbf{K}_1$. For example, (21) follows from (23) with

$$\mathbf{K}_1 = \begin{pmatrix} 1 \\ 1 - 2i \end{pmatrix} = \begin{pmatrix} 1 \\ 1 \end{pmatrix} + i\begin{pmatrix} 0 \\ -2 \end{pmatrix},$$

$$\mathbf{B}_1 = \text{Re}(\mathbf{K}_1) = \begin{pmatrix} 1 \\ 1 \end{pmatrix} \quad \text{and} \quad \mathbf{B}_2 = \text{Im}(\mathbf{K}_1) = \begin{pmatrix} 0 \\ -2 \end{pmatrix}.$$

EXAMPLE 6 Complex Eigenvalues

Solve the initial-value problem

$$\mathbf{X}' = \begin{pmatrix} 2 & 8 \\ -1 & -2 \end{pmatrix}\mathbf{X}, \quad \mathbf{X}(0) = \begin{pmatrix} 2 \\ -1 \end{pmatrix}. \tag{25}$$

SOLUTION First we obtain the eigenvalues from

$$\det(\mathbf{A} - \lambda\mathbf{I}) = \begin{vmatrix} 2 - \lambda & 8 \\ -1 & -2 - \lambda \end{vmatrix} = \lambda^2 + 4 = 0.$$

The eigenvalues are $\lambda_1 = 2i$ and $\lambda_2 = \overline{\lambda_1} = -2i$. For λ_1 the system

$$\begin{aligned}(2 - 2i)\,k_1 + \qquad\quad 8k_2 &= 0\\ -k_1 + (-2 - 2i)k_2 &= 0\end{aligned}$$

gives $k_1 = -(2 + 2i)k_2$. By choosing $k_2 = -1$, we get

$$\mathbf{K}_1 = \begin{pmatrix} 2 + 2i \\ -1 \end{pmatrix} = \begin{pmatrix} 2 \\ -1 \end{pmatrix} + i\begin{pmatrix} 2 \\ 0 \end{pmatrix}.$$

Now from (24) we form

$$\mathbf{B}_1 = \text{Re}(\mathbf{K}_1) = \begin{pmatrix} 2 \\ -1 \end{pmatrix} \quad \text{and} \quad \mathbf{B}_2 = \text{Im}(\mathbf{K}_1) = \begin{pmatrix} 2 \\ 0 \end{pmatrix}.$$

Since $\alpha = 0$, it follows from (23) that the general solution of the system is

$$\begin{aligned}\mathbf{X} &= c_1\left[\begin{pmatrix} 2 \\ -1 \end{pmatrix}\cos 2t - \begin{pmatrix} 2 \\ 0 \end{pmatrix}\sin 2t\right] + c_2\left[\begin{pmatrix} 2 \\ 0 \end{pmatrix}\cos 2t + \begin{pmatrix} 2 \\ -1 \end{pmatrix}\sin 2t\right]\\ &= c_1\begin{pmatrix} 2\cos 2t - 2\sin 2t \\ -\cos 2t \end{pmatrix} + c_2\begin{pmatrix} 2\cos 2t + 2\sin 2t \\ -\sin 2t \end{pmatrix}.\end{aligned} \tag{26}$$

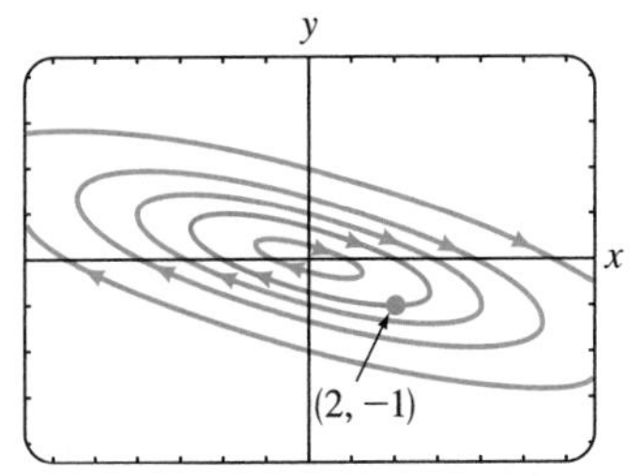

FIGURE 8.2.5 A phase portrait of (25) in Example 6

Some graphs of the curves or trajectories defined by solution (26) of the system are illustrated in the phase portrait in Figure 8.2.5. Now the initial condition $\mathbf{X}(0) = \begin{pmatrix} 2 \\ -1 \end{pmatrix}$ or, equivalently, $x(0) = 2$ and $y(0) = -1$ yields the algebraic system $2c_1 + 2c_2 = 2$, $-c_1 = -1$, whose solution is $c_1 = 1$, $c_2 = 0$. Thus the solution to the problem is $\mathbf{X} = \begin{pmatrix} 2\cos 2t - 2\sin 2t \\ -\cos 2t \end{pmatrix}$. The specific trajectory defined parametrically by the particular solution $x = 2\cos 2t - 2\sin 2t$, $y = -\cos 2t$ is the red curve in Figure 8.2.5. Note that this curve passes through $(2, -1)$. ≡

REMARKS

In this section we have examined exclusively homogeneous first-order systems of linear equations in normal form $\mathbf{X}' = \mathbf{AX}$. But often the mathematical model of a dynamical physical system is a homogeneous second-order system whose normal form is $\mathbf{X}'' = \mathbf{AX}$. For example, the model for the coupled springs in (1) of Section 7.6,

$$\begin{aligned} m_1x_1'' &= -k_1x_1 + k_2(x_2 - x_1) \\ m_2x_2'' &= -k_2(x_2 - x_1), \end{aligned} \tag{27}$$

can be written as

$$\mathbf{MX}'' = \mathbf{KX},$$

where

$$\mathbf{M} = \begin{pmatrix} m_1 & 0 \\ 0 & m_2 \end{pmatrix}, \quad \mathbf{K} = \begin{pmatrix} -k_1 - k_2 & k_2 \\ k_2 & -k_2 \end{pmatrix}, \quad \text{and} \quad \mathbf{X} = \begin{pmatrix} x_1(t) \\ x_2(t) \end{pmatrix}.$$

Since $\mathbf{M}$ is nonsingular, we can solve for $\mathbf{X}''$ as $\mathbf{X}'' = \mathbf{AX}$, where $\mathbf{A} = \mathbf{M}^{-1}\mathbf{K}$. Thus (27) is equivalent to

$$\mathbf{X}'' = \begin{pmatrix} -\dfrac{k_1}{m_1} - \dfrac{k_2}{m_1} & \dfrac{k_2}{m_1} \\ \dfrac{k_2}{m_2} & -\dfrac{k_2}{m_2} \end{pmatrix}\mathbf{X}. \tag{28}$$

The methods of this section can be used to solve such a system in two ways:

- First, the original system (27) can be transformed into a first-order system by means of substitutions. If we let $x_1' = x_3$ and $x_2' = x_4$, then $x_3' = x_1''$ and $x_4' = x_2''$ and so (27) is equivalent to a system of *four* linear first-order DEs:

$$\begin{aligned} x_1' &= x_3 \\ x_2' &= x_4 \\ x_3' &= -\left(\frac{k_1}{m_1} + \frac{k_2}{m_1}\right)x_1 + \frac{k_2}{m_1}x_2 \\ x_4' &= \frac{k_2}{m_2}x_1 - \frac{k_2}{m_2}x_2 \end{aligned} \quad \text{or} \quad \mathbf{X}' = \begin{pmatrix} 0 & 0 & 1 & 0 \\ 0 & 0 & 0 & 1 \\ -\dfrac{k_1}{m_1} - \dfrac{k_2}{m_1} & \dfrac{k_2}{m_1} & 0 & 0 \\ \dfrac{k_2}{m_2} & -\dfrac{k_2}{m_2} & 0 & 0 \end{pmatrix}\mathbf{X}. \tag{29}$$

By finding the eigenvalues and eigenvectors of the coefficient matrix $\mathbf{A}$ in (29), we see that the solution of this first-order system gives the complete state of the physical system—the positions of the masses relative to the equilibrium positions (x_1 and x_2) as well as the velocities of the masses (x_3 and x_4) at time t. See Problem 48(a) in Exercises 8.2.

- Second, because (27) describes free undamped motion, it can be argued that real-valued solutions of the second-order system (28) will have the form

$$\mathbf{X} = \mathbf{V}\cos\omega t \quad \text{and} \quad \mathbf{X} = \mathbf{V}\sin\omega t, \tag{30}$$

where $\mathbf{V}$ is a column matrix of constants. Substituting either of the functions in (30) into $\mathbf{X}'' = \mathbf{AX}$ yields $(\mathbf{A} + \omega^2\mathbf{I})\mathbf{V} = \mathbf{0}$. (Verify.) By identification with (3) of this section we conclude that $\lambda = -\omega^2$ represents an eigenvalue and $\mathbf{V}$ a corresponding eigenvector of $\mathbf{A}$. It can be shown that the eigenvalues $\lambda_i = -\omega_i^2$, $i = 1, 2$ of $\mathbf{A}$ are negative, and so $\omega_i = \sqrt{-\lambda_i}$ is a real number and represents a (circular) frequency of vibration (see (4) of Section 7.6). By superposition of solutions the general solution of (28) is then

$$\begin{aligned}\mathbf{X} &= c_1\mathbf{V}_1\cos\omega_1 t + c_2\mathbf{V}_1\sin\omega_1 t + c_3\mathbf{V}_2\cos\omega_2 t + c_4\mathbf{V}_2\sin\omega_2 t \\ &= (c_1\cos\omega_1 t + c_2\sin\omega_1 t)\mathbf{V}_1 + (c_3\cos\omega_2 t + c_4\sin\omega_2 t)\mathbf{V}_2,\end{aligned} \tag{31}$$

where $\mathbf{V}_1$ and $\mathbf{V}_2$ are, in turn, real eigenvectors of $\mathbf{A}$ corresponding to λ_1 and λ_2.

The result given in (31) generalizes. If $-\omega_1^2, -\omega_2^2, \ldots, -\omega_n^2$ are distinct negative eigenvalues and $\mathbf{V}_1, \mathbf{V}_2, \ldots, \mathbf{V}_n$ are corresponding real eigenvectors of the $n \times n$ coefficient matrix $\mathbf{A}$, then the homogeneous second-order system $\mathbf{X}'' = \mathbf{AX}$ has the general solution

$$\mathbf{X} = \sum_{i=1}^{n}(a_i\cos\omega_i t + b_i\sin\omega_i t)\mathbf{V}_i, \tag{32}$$

where a_i and b_i represent arbitrary constants. See Problem 48(b) in Exercises 8.2.

EXERCISES 8.2

Answers to selected odd-numbered problems begin on page ANS-14.

8.2.1 DISTINCT REAL EIGENVALUES

In Problems 1–12 find the general solution of the given system.

1. $\dfrac{dx}{dt} = x + 2y$

$\dfrac{dy}{dt} = 4x + 3y$

2. $\dfrac{dx}{dt} = 2x + 2y$

$\dfrac{dy}{dt} = x + 3y$

3. $\dfrac{dx}{dt} = -4x + 2y$

$\dfrac{dy}{dt} = -\dfrac{5}{2}x + 2y$

4. $\dfrac{dx}{dt} = -\dfrac{5}{2}x + 2y$

$\dfrac{dy}{dt} = \dfrac{3}{4}x - 2y$

5. $\mathbf{X}' = \begin{pmatrix} 10 & -5 \\ 8 & -12 \end{pmatrix}\mathbf{X}$

6. $\mathbf{X}' = \begin{pmatrix} -6 & 2 \\ -3 & 1 \end{pmatrix}\mathbf{X}$

7. $\dfrac{dx}{dt} = x + y - z$

$\dfrac{dy}{dt} = 2y$

$\dfrac{dz}{dt} = y - z$

8. $\dfrac{dx}{dt} = 2x - 7y$

$\dfrac{dy}{dt} = 5x + 10y + 4z$

$\dfrac{dz}{dt} = 5y + 2z$

9. $\mathbf{X}' = \begin{pmatrix} -1 & 1 & 0 \\ 1 & 2 & 1 \\ 0 & 3 & -1 \end{pmatrix}\mathbf{X}$

10. $\mathbf{X}' = \begin{pmatrix} 1 & 0 & 1 \\ 0 & 1 & 0 \\ 1 & 0 & 1 \end{pmatrix}\mathbf{X}$

11. $\mathbf{X}' = \begin{pmatrix} -1 & -1 & 0 \\ \frac{3}{4} & -\frac{3}{2} & 3 \\ \frac{1}{8} & \frac{1}{4} & -\frac{1}{2} \end{pmatrix}\mathbf{X}$

12. $\mathbf{X}' = \begin{pmatrix} -1 & 4 & 2 \\ 4 & -1 & -2 \\ 0 & 0 & 6 \end{pmatrix}\mathbf{X}$

In Problems 13 and 14 solve the given initial-value problem.

13. $\mathbf{X}' = \begin{pmatrix} \frac{1}{2} & 0 \\ 1 & -\frac{1}{2} \end{pmatrix}\mathbf{X}, \quad \mathbf{X}(0) = \begin{pmatrix} 3 \\ 5 \end{pmatrix}$

14. $\mathbf{X}' = \begin{pmatrix} 1 & 1 & 4 \\ 0 & 2 & 0 \\ 1 & 1 & 1 \end{pmatrix}\mathbf{X}, \quad \mathbf{X}(0) = \begin{pmatrix} 1 \\ 3 \\ 0 \end{pmatrix}$

Computer Lab Assignments

In Problems 15 and 16 use a CAS or linear algebra software as an aid in finding the general solution of the given system.

15. $\mathbf{X}' = \begin{pmatrix} 0.9 & 2.1 & 3.2 \\ 0.7 & 6.5 & 4.2 \\ 1.1 & 1.7 & 3.4 \end{pmatrix}\mathbf{X}$

16. $\mathbf{X}' = \begin{pmatrix} 1 & 0 & 2 & -1.8 & 0 \\ 0 & 5.1 & 0 & -1 & 3 \\ 1 & 2 & -3 & 0 & 0 \\ 0 & 1 & -3.1 & 4 & 0 \\ -2.8 & 0 & 0 & 1.5 & 1 \end{pmatrix}\mathbf{X}$

17. **(a)** Use computer software to obtain the phase portrait of the system in Problem 5. If possible, include arrowheads as in Figure 8.2.2. Also include four half-lines in your phase portrait.

(b) Obtain the Cartesian equations of each of the four half-lines in part (a).

(c) Draw the eigenvectors on your phase portrait of the system.

18. Find phase portraits for the systems in Problems 2 and 4. For each system find any half-line trajectories and include these lines in your phase portrait.

8.2.2 REPEATED EIGENVALUES

In Problems 19–28 find the general solution of the given system.

19. $\dfrac{dx}{dt} = 3x - y$, $\dfrac{dy}{dt} = 9x - 3y$

20. $\dfrac{dx}{dt} = -6x + 5y$, $\dfrac{dy}{dt} = -5x + 4y$

21. $\mathbf{X}' = \begin{pmatrix} -1 & 3 \\ -3 & 5 \end{pmatrix}\mathbf{X}$

22. $\mathbf{X}' = \begin{pmatrix} 12 & -9 \\ 4 & 0 \end{pmatrix}\mathbf{X}$

23. $\dfrac{dx}{dt} = 3x - y - z$, $\dfrac{dy}{dt} = x + y - z$, $\dfrac{dz}{dt} = x - y + z$

24. $\dfrac{dx}{dt} = 3x + 2y + 4z$, $\dfrac{dy}{dt} = 2x + 2z$, $\dfrac{dz}{dt} = 4x + 2y + 3z$

25. $\mathbf{X}' = \begin{pmatrix} 5 & -4 & 0 \\ 1 & 0 & 2 \\ 0 & 2 & 5 \end{pmatrix}\mathbf{X}$

26. $\mathbf{X}' = \begin{pmatrix} 1 & 0 & 0 \\ 0 & 3 & 1 \\ 0 & -1 & 1 \end{pmatrix}\mathbf{X}$

27. $\mathbf{X}' = \begin{pmatrix} 1 & 0 & 0 \\ 2 & 2 & -1 \\ 0 & 1 & 0 \end{pmatrix}\mathbf{X}$

28. $\mathbf{X}' = \begin{pmatrix} 4 & 1 & 0 \\ 0 & 4 & 1 \\ 0 & 0 & 4 \end{pmatrix}\mathbf{X}$

In Problems 29 and 30 solve the given initial-value problem.

29. $\mathbf{X}' = \begin{pmatrix} 2 & 4 \\ -1 & 6 \end{pmatrix}\mathbf{X}, \quad \mathbf{X}(0) = \begin{pmatrix} -1 \\ 6 \end{pmatrix}$

30. $\mathbf{X}' = \begin{pmatrix} 0 & 0 & 1 \\ 0 & 1 & 0 \\ 1 & 0 & 0 \end{pmatrix}\mathbf{X}, \quad \mathbf{X}(0) = \begin{pmatrix} 1 \\ 2 \\ 5 \end{pmatrix}$

31. Show that the 5×5 matrix

$$\mathbf{A} = \begin{pmatrix} 2 & 1 & 0 & 0 & 0 \\ 0 & 2 & 0 & 0 & 0 \\ 0 & 0 & 2 & 0 & 0 \\ 0 & 0 & 0 & 2 & 1 \\ 0 & 0 & 0 & 0 & 2 \end{pmatrix}$$

has an eigenvalue λ_1 of multiplicity 5. Show that three linearly independent eigenvectors corresponding to λ_1 can be found.

Computer Lab Assignments

32. Find phase portraits for the systems in Problems 20 and 21. For each system find any half-line trajectories and include these lines in your phase portrait.

8.2.3 COMPLEX EIGENVALUES

In Problems 33–44 find the general solution of the given system.

33. $\dfrac{dx}{dt} = 6x - y$, $\dfrac{dy}{dt} = 5x + 2y$

34. $\dfrac{dx}{dt} = x + y$, $\dfrac{dy}{dt} = -2x - y$

35. $\dfrac{dx}{dt} = 5x + y$, $\dfrac{dy}{dt} = -2x + 3y$

36. $\dfrac{dx}{dt} = 4x + 5y$, $\dfrac{dy}{dt} = -2x + 6y$

37. $\mathbf{X}' = \begin{pmatrix} 4 & -5 \\ 5 & -4 \end{pmatrix}\mathbf{X}$

38. $\mathbf{X}' = \begin{pmatrix} 1 & -8 \\ 1 & -3 \end{pmatrix}\mathbf{X}$

39. $\dfrac{dx}{dt} = z$, $\dfrac{dy}{dt} = -z$, $\dfrac{dz}{dt} = y$

40. $\dfrac{dx}{dt} = 2x + y + 2z$, $\dfrac{dy}{dt} = 3x + 6z$, $\dfrac{dz}{dt} = -4x - 3z$

41. $\mathbf{X}' = \begin{pmatrix} 1 & -1 & 2 \\ -1 & 1 & 0 \\ -1 & 0 & 1 \end{pmatrix}\mathbf{X}$

42. $\mathbf{X}' = \begin{pmatrix} 4 & 0 & 1 \\ 0 & 6 & 0 \\ -4 & 0 & 4 \end{pmatrix}\mathbf{X}$

43. $\mathbf{X}' = \begin{pmatrix} 2 & 5 & 1 \\ -5 & -6 & 4 \\ 0 & 0 & 2 \end{pmatrix}\mathbf{X}$ **44.** $\mathbf{X}' = \begin{pmatrix} 2 & 4 & 4 \\ -1 & -2 & 0 \\ -1 & 0 & -2 \end{pmatrix}\mathbf{X}$

In Problems 45 and 46 solve the given initial-value problem.

45. $\mathbf{X}' = \begin{pmatrix} 1 & -12 & -14 \\ 1 & 2 & -3 \\ 1 & 1 & -2 \end{pmatrix}\mathbf{X}, \quad \mathbf{X}(0) = \begin{pmatrix} 4 \\ 6 \\ -7 \end{pmatrix}$

46. $\mathbf{X}' = \begin{pmatrix} 6 & -1 \\ 5 & 4 \end{pmatrix}\mathbf{X}, \quad \mathbf{X}(0) = \begin{pmatrix} -2 \\ 8 \end{pmatrix}$

Computer Lab Assignments

47. Find phase portraits for the systems in Problems 36, 37, and 38.

48. **(a)** Solve (2) of Section 7.6 using the first method outlined in the *Remarks* (page 345)—that is, express (2) of Section 7.6 as a first-order system of four linear equations. Use a CAS or linear algebra software as an aid in finding eigenvalues and eigenvectors of a 4×4 matrix. Then apply the initial conditions to your general solution to obtain (4) of Section 7.6.

(b) Solve (2) of Section 7.6 using the second method outlined in the *Remarks*—that is, express (2) of Section 7.6 as a second-order system of two linear equations. Assume solutions of the form $\mathbf{X} = \mathbf{V} \sin \omega t$ and $\mathbf{X} = \mathbf{V} \cos \omega t$. Find the eigenvalues and eigenvectors of a 2×2 matrix. As in part (a), obtain (4) of Section 7.6.

Discussion Problems

49. Solve each of the following linear systems.

(a) $\mathbf{X}' = \begin{pmatrix} 1 & 1 \\ 1 & 1 \end{pmatrix}\mathbf{X}$ **(b)** $\mathbf{X}' = \begin{pmatrix} 1 & 1 \\ -1 & -1 \end{pmatrix}\mathbf{X}$

Find a phase portrait of each system. What is the geometric significance of the line $y = -x$ in each portrait?

50. Consider the 5×5 matrix given in Problem 31. Solve the system $\mathbf{X}' = \mathbf{AX}$ without the aid of matrix methods, but write the general solution using matrix notation. Use the general solution as a basis for a discussion of how the system can be solved using the matrix methods of this section. Carry out your ideas.

51. Obtain a Cartesian equation of the curve defined parametrically by the solution of the linear system in Example 6. Identify the curve passing through $(2, -1)$ in Figure 8.2.5. [*Hint*: Compute x^2, y^2, and xy.]

52. Examine your phase portraits in Problem 47. Under what conditions will the phase portrait of a 2×2 homogeneous linear system with complex eigenvalues consist of a family of closed curves? consist of a family of spirals? Under what conditions is the origin (0, 0) a repeller? An attractor?

8.3 NONHOMOGENEOUS LINEAR SYSTEMS

REVIEW MATERIAL

- Section 4.4 (Undetermined Coefficients)
- Section 4.6 (Variation of Parameters)

INTRODUCTION In Section 8.1 we saw that the general solution of a nonhomogeneous linear system $\mathbf{X}' = \mathbf{AX} + \mathbf{F}(t)$ on an interval I is $\mathbf{X} = \mathbf{X}_c + \mathbf{X}_p$, where $\mathbf{X}_c = c_1\mathbf{X}_1 + c_2\mathbf{X}_2 + \cdots + c_n\mathbf{X}_n$ is the **complementary function** or general solution of the associated homogeneous linear system $\mathbf{X}' = \mathbf{AX}$ and $\mathbf{X}_p$ is any **particular solution** of the nonhomogeneous system. In Section 8.2 we saw how to obtain $\mathbf{X}_c$ when the coefficient matrix $\mathbf{A}$ was an $n \times n$ matrix of constants. In the present section we consider two methods for obtaining $\mathbf{X}_p$.

The methods of **undetermined coefficients** and **variation of parameters** used in Chapter 4 to find particular solutions of nonhomogeneous linear ODEs can both be adapted to the solution of nonhomogeneous linear systems $\mathbf{X}' = \mathbf{AX} + \mathbf{F}(t)$. Of the two methods, variation of parameters is the more powerful technique. However, there are instances when the method of undetermined coefficients provides a quick means of finding a particular solution.

8.3.1 UNDETERMINED COEFFICIENTS

The Assumptions As in Section 4.4, the method of undetermined coefficients consists of making an educated guess about the form of a particular solution vector $\mathbf{X}_p$; the guess is motivated by the types of functions that make up the entries of the column matrix $\mathbf{F}(t)$. Not surprisingly, the matrix version of undetermined coefficients is applicable to $\mathbf{X}' = \mathbf{AX} + \mathbf{F}(t)$ only when the entries of $\mathbf{A}$ are constants and the entries of $\mathbf{F}(t)$ are constants, polynomials, exponential functions, sines and cosines, or finite sums and products of these functions.

EXAMPLE 1 Undetermined Coefficients

Solve the system $\mathbf{X}' = \begin{pmatrix} -1 & 2 \\ -1 & 1 \end{pmatrix}\mathbf{X} + \begin{pmatrix} -8 \\ 3 \end{pmatrix}$ on $(-\infty, \infty)$.

SOLUTION We first solve the associated homogeneous system

$$\mathbf{X}' = \begin{pmatrix} -1 & 2 \\ -1 & 1 \end{pmatrix}\mathbf{X}.$$

The characteristic equation of the coefficient matrix $\mathbf{A}$,

$$\det(\mathbf{A} - \lambda\mathbf{I}) = \begin{vmatrix} -1-\lambda & 2 \\ -1 & 1-\lambda \end{vmatrix} = \lambda^2 + 1 = 0,$$

yields the complex eigenvalues $\lambda_1 = i$ and $\lambda_2 = \overline{\lambda_1} = -i$. By the procedures of Section 8.2 we find

$$\mathbf{X}_c = c_1\begin{pmatrix} \cos t + \sin t \\ \cos t \end{pmatrix} + c_2\begin{pmatrix} \cos t - \sin t \\ -\sin t \end{pmatrix}.$$

Now since $\mathbf{F}(t)$ is a constant vector, we assume a constant particular solution vector $\mathbf{X}_p = \begin{pmatrix} a_1 \\ b_1 \end{pmatrix}$. Substituting this latter assumption into the original system and equating entries leads to

$$\begin{aligned} 0 &= -a_1 + 2b_1 - 8 \\ 0 &= -a_1 + b_1 + 3. \end{aligned}$$

Solving this algebraic system gives $a_1 = 14$ and $b_1 = 11$, and so a particular solution is $\mathbf{X}_p = \begin{pmatrix} 14 \\ 11 \end{pmatrix}$. The general solution of the original system of DEs on the interval $(-\infty, \infty)$ is then $\mathbf{X} = \mathbf{X}_c + \mathbf{X}_p$ or

$$\mathbf{X} = c_1\begin{pmatrix} \cos t + \sin t \\ \cos t \end{pmatrix} + c_2\begin{pmatrix} \cos t - \sin t \\ -\sin t \end{pmatrix} + \begin{pmatrix} 14 \\ 11 \end{pmatrix}.$$ ≡

EXAMPLE 2 Undetermined Coefficients

Solve the system $\mathbf{X}' = \begin{pmatrix} 6 & 1 \\ 4 & 3 \end{pmatrix}\mathbf{X} + \begin{pmatrix} 6t \\ -10t + 4 \end{pmatrix}$ on $(-\infty, \infty)$.

SOLUTION The eigenvalues and corresponding eigenvectors of the associated homogeneous system $\mathbf{X}' = \begin{pmatrix} 6 & 1 \\ 4 & 3 \end{pmatrix}\mathbf{X}$ are found to be $\lambda_1 = 2$, $\lambda_2 = 7$, $\mathbf{K}_1 = \begin{pmatrix} 1 \\ -4 \end{pmatrix}$, and $\mathbf{K}_2 = \begin{pmatrix} 1 \\ 1 \end{pmatrix}$. Hence the complementary function is

$$\mathbf{X}_c = c_1\begin{pmatrix} 1 \\ -4 \end{pmatrix}e^{2t} + c_2\begin{pmatrix} 1 \\ 1 \end{pmatrix}e^{7t}.$$

Now because $\mathbf{F}(t)$ can be written $\mathbf{F}(t) = \begin{pmatrix} 6 \\ -10 \end{pmatrix} t + \begin{pmatrix} 0 \\ 4 \end{pmatrix}$, we shall try to find a particular solution of the system that possesses the *same* form:

$$\mathbf{X}_p = \begin{pmatrix} a_2 \\ b_2 \end{pmatrix} t + \begin{pmatrix} a_1 \\ b_1 \end{pmatrix}.$$

Substituting this last assumption into the given system yields

$$\begin{pmatrix} a_2 \\ b_2 \end{pmatrix} = \begin{pmatrix} 6 & 1 \\ 4 & 3 \end{pmatrix} \left[\begin{pmatrix} a_2 \\ b_2 \end{pmatrix} t + \begin{pmatrix} a_1 \\ b_1 \end{pmatrix} \right] + \begin{pmatrix} 6 \\ -10 \end{pmatrix} t + \begin{pmatrix} 0 \\ 4 \end{pmatrix}$$

or

$$\begin{pmatrix} 0 \\ 0 \end{pmatrix} = \begin{pmatrix} (6a_2 + b_2 + 6)t + 6a_1 + b_1 - a_2 \\ (4a_2 + 3b_2 - 10)t + 4a_1 + 3b_1 - b_2 + 4 \end{pmatrix}.$$

From the last identity we obtain four algebraic equations in four unknowns

$$\begin{aligned} 6a_2 + b_2 + 6 &= 0 \\ 4a_2 + 3b_2 - 10 &= 0 \end{aligned} \qquad \text{and} \qquad \begin{aligned} 6a_1 + b_1 - a_2 &= 0 \\ 4a_1 + 3b_1 - b_2 + 4 &= 0. \end{aligned}$$

Solving the first two equations simultaneously yields $a_2 = -2$ and $b_2 = 6$. We then substitute these values into the last two equations and solve for a_1 and b_1. The results are $a_1 = -\frac{4}{7}$, $b_1 = \frac{10}{7}$. It follows, therefore, that a particular solution vector is

$$\mathbf{X}_p = \begin{pmatrix} -2 \\ 6 \end{pmatrix} t + \begin{pmatrix} -\frac{4}{7} \\ \frac{10}{7} \end{pmatrix}.$$

The general solution of the system on $(-\infty, \infty)$ is $\mathbf{X} = \mathbf{X}_c + \mathbf{X}_p$ or

$$\mathbf{X} = c_1 \begin{pmatrix} 1 \\ -4 \end{pmatrix} e^{2t} + c_2 \begin{pmatrix} 1 \\ 1 \end{pmatrix} e^{7t} + \begin{pmatrix} -2 \\ 6 \end{pmatrix} t + \begin{pmatrix} -\frac{4}{7} \\ \frac{10}{7} \end{pmatrix}.$$

≡

EXAMPLE 3 Form of $\mathbf{X}_p$

Determine the form of a particular solution vector $\mathbf{X}_p$ for the system

$$\begin{aligned} \frac{dx}{dt} &= 5x + 3y - 2e^{-t} + 1 \\ \frac{dy}{dt} &= -x + y + e^{-t} - 5t + 7. \end{aligned}$$

SOLUTION Because $\mathbf{F}(t)$ can be written in matrix terms as

$$\mathbf{F}(t) = \begin{pmatrix} -2 \\ 1 \end{pmatrix} e^{-t} + \begin{pmatrix} 0 \\ -5 \end{pmatrix} t + \begin{pmatrix} 1 \\ 7 \end{pmatrix}$$

a natural assumption for a particular solution would be

$$\mathbf{X}_p = \begin{pmatrix} a_3 \\ b_3 \end{pmatrix} e^{-t} + \begin{pmatrix} a_2 \\ b_2 \end{pmatrix} t + \begin{pmatrix} a_1 \\ b_1 \end{pmatrix}.$$

≡

REMARKS

The method of undetermined coefficients for linear systems is not as straightforward as the last three examples would seem to indicate. In Section 4.4 the form of a particular solution y_p was predicated on prior knowledge of the complementary function y_c. The same is true for the formation of $\mathbf{X}_p$. But there are further difficulties: The special rules governing the form of y_p in Section 4.4 do not *quite* carry to the formation of $\mathbf{X}_p$. For example, if $\mathbf{F}(t)$ is a constant vector, as in Example 1, and $\lambda = 0$ is an eigenvalue of multiplicity one, then $\mathbf{X}_c$ contains a constant vector. Under the Multiplication Rule on page 145 we would ordinarily try a particular solution of the form $\mathbf{X}_p = \begin{pmatrix} a_1 \\ b_1 \end{pmatrix} t$. This is not the proper assumption for linear systems; it should be $\mathbf{X}_p = \begin{pmatrix} a_2 \\ b_2 \end{pmatrix} t + \begin{pmatrix} a_1 \\ b_1 \end{pmatrix}$. Similarly, in Example 3, if we replace e^{-t} in $\mathbf{F}(t)$ by e^{2t} ($\lambda = 2$ is an eigenvalue), then the correct form of the particular solution vector is

$$\mathbf{X}_p = \begin{pmatrix} a_4 \\ b_4 \end{pmatrix} te^{2t} + \begin{pmatrix} a_3 \\ b_3 \end{pmatrix} e^{2t} + \begin{pmatrix} a_2 \\ b_2 \end{pmatrix} t + \begin{pmatrix} a_1 \\ b_1 \end{pmatrix}.$$

Rather than delving into these difficulties, we turn instead to the method of variation of parameters.

8.3.2 VARIATION OF PARAMETERS

A Fundamental Matrix If $\mathbf{X}_1, \mathbf{X}_2, \ldots, \mathbf{X}_n$ is a fundamental set of solutions of the homogeneous system $\mathbf{X}' = \mathbf{AX}$ on an interval I, then its general solution on the interval is the linear combination $\mathbf{X} = c_1\mathbf{X}_1 + c_2\mathbf{X}_2 + \cdots + c_n\mathbf{X}_n$ or

$$\mathbf{X} = c_1 \begin{pmatrix} x_{11} \\ x_{21} \\ \vdots \\ x_{n1} \end{pmatrix} + c_2 \begin{pmatrix} x_{12} \\ x_{22} \\ \vdots \\ x_{n2} \end{pmatrix} + \cdots + c_n \begin{pmatrix} x_{1n} \\ x_{2n} \\ \vdots \\ x_{nn} \end{pmatrix} = \begin{pmatrix} c_1x_{11} + c_2x_{12} + \cdots + c_nx_{1n} \\ c_1x_{21} + c_2x_{22} + \cdots + c_nx_{2n} \\ \vdots \\ c_1x_{n1} + c_2x_{n2} + \cdots + c_nx_{nn} \end{pmatrix}. \tag{1}$$

The last matrix in (1) is recognized as the product of an $n \times n$ matrix with an $n \times 1$ matrix. In other words, the general solution (1) can be written as the product

$$\mathbf{X} = \mathbf{\Phi}(t)\mathbf{C}, \tag{2}$$

where $\mathbf{C}$ is an $n \times 1$ column vector of arbitrary constants $c_1, c_2, \ldots, c_n$ and the $n \times n$ matrix, whose columns consist of the entries of the solution vectors of the system $\mathbf{X}' = \mathbf{AX}$,

$$\mathbf{\Phi}(t) = \begin{pmatrix} x_{11} & x_{12} & \cdots & x_{1n} \\ x_{21} & x_{22} & \cdots & x_{2n} \\ \vdots & & & \vdots \\ x_{n1} & x_{n2} & \cdots & x_{nn} \end{pmatrix},$$

is called a **fundamental matrix** of the system on the interval.

In the discussion that follows we need to use two properties of a fundamental matrix:

- A fundamental matrix $\mathbf{\Phi}(t)$ is nonsingular.
- If $\mathbf{\Phi}(t)$ is a fundamental matrix of the system $\mathbf{X}' = \mathbf{AX}$, then

$$\mathbf{\Phi}'(t) = \mathbf{A\Phi}(t). \tag{3}$$

A reexamination of (9) of Theorem 8.1.3 shows that det $\mathbf{\Phi}(t)$ is the same as the Wronskian $W(\mathbf{X}_1, \mathbf{X}_2, \ldots, \mathbf{X}_n)$. Hence the linear independence of the columns of $\mathbf{\Phi}(t)$ on the interval I guarantees that det $\mathbf{\Phi}(t) \neq 0$ for every t in the interval. Since $\mathbf{\Phi}(t)$ is nonsingular, the multiplicative inverse $\mathbf{\Phi}^{-1}(t)$ exists for every t in the interval. The result given in (3) follows immediately from the fact that every column of $\mathbf{\Phi}(t)$ is a solution vector of $\mathbf{X}' = \mathbf{AX}$.

Variation of Parameters Analogous to the procedure in Section 4.6 we ask whether it is possible to replace the matrix of constants $\mathbf{C}$ in (2) by a column matrix of functions

$$\mathbf{U}(t) = \begin{pmatrix} u_1(t) \\ u_2(t) \\ \vdots \\ u_n(t) \end{pmatrix} \quad \text{so} \quad \mathbf{X}_p = \mathbf{\Phi}(t)\mathbf{U}(t) \tag{4}$$

is a particular solution of the nonhomogeneous system

$$\mathbf{X}' = \mathbf{AX} + \mathbf{F}(t). \tag{5}$$

By the Product Rule the derivative of the last expression in (4) is

$$\mathbf{X}_p' = \mathbf{\Phi}(t)\mathbf{U}'(t) + \mathbf{\Phi}'(t)\mathbf{U}(t). \tag{6}$$

Note that the order of the products in (6) is very important. Since $\mathbf{U}(t)$ is a column matrix, the products $\mathbf{U}'(t)\mathbf{\Phi}(t)$ and $\mathbf{U}(t)\mathbf{\Phi}'(t)$ are not defined. Substituting (4) and (6) into (5) gives

$$\mathbf{\Phi}(t)\mathbf{U}'(t) + \mathbf{\Phi}'(t)\mathbf{U}(t) = \mathbf{A\Phi}(t)\mathbf{U}(t) + \mathbf{F}(t). \tag{7}$$

Now if we use (3) to replace $\mathbf{\Phi}'(t)$, (7) becomes

$$\mathbf{\Phi}(t)\mathbf{U}'(t) + \mathbf{A\Phi}(t)\mathbf{U}(t) = \mathbf{A\Phi}(t)\mathbf{U}(t) + \mathbf{F}(t)$$

or

$$\mathbf{\Phi}(t)\mathbf{U}'(t) = \mathbf{F}(t). \tag{8}$$

Multiplying both sides of equation (8) by $\mathbf{\Phi}^{-1}(t)$ gives

$$\mathbf{U}'(t) = \mathbf{\Phi}^{-1}(t)\mathbf{F}(t) \quad \text{and so} \quad \mathbf{U}(t) = \int \mathbf{\Phi}^{-1}(t)\mathbf{F}(t)\,dt.$$

Since $\mathbf{X}_p = \mathbf{\Phi}(t)\mathbf{U}(t)$, we conclude that a particular solution of (5) is

$$\mathbf{X}_p = \mathbf{\Phi}(t)\int \mathbf{\Phi}^{-1}(t)\mathbf{F}(t)\,dt. \tag{9}$$

To calculate the indefinite integral of the column matrix $\mathbf{\Phi}^{-1}(t)\mathbf{F}(t)$ in (9), we integrate each entry. Thus the general solution of the system (5) is $\mathbf{X} = \mathbf{X}_c + \mathbf{X}_p$ or

$$\mathbf{X} = \mathbf{\Phi}(t)\mathbf{C} + \mathbf{\Phi}(t)\int \mathbf{\Phi}^{-1}(t)\mathbf{F}(t)\,dt. \tag{10}$$

Note that it is not necessary to use a constant of integration in the evaluation of $\int \mathbf{\Phi}^{-1}(t)\mathbf{F}(t)\,dt$ for the same reasons stated in the discussion of variation of parameters in Section 4.6.

EXAMPLE 4 **Variation of Parameters**

Solve the system

$$\mathbf{X}' = \begin{pmatrix} -3 & 1 \\ 2 & -4 \end{pmatrix}\mathbf{X} + \begin{pmatrix} 3t \\ e^{-t} \end{pmatrix} \tag{11}$$

on $(-\infty, \infty)$.

SOLUTION We first solve the associated homogeneous system

$$\mathbf{X}' = \begin{pmatrix} -3 & 1 \\ 2 & -4 \end{pmatrix}\mathbf{X}. \tag{12}$$

The characteristic equation of the coefficient matrix is

$$\det(\mathbf{A} - \lambda\mathbf{I}) = \begin{vmatrix} -3-\lambda & 1 \\ 2 & -4-\lambda \end{vmatrix} = (\lambda + 2)(\lambda + 5) = 0,$$

so the eigenvalues are $\lambda_1 = -2$ and $\lambda_2 = -5$. By the usual method we find that the eigenvectors corresponding to λ_1 and λ_2 are, respectively, $\mathbf{K}_1 = \begin{pmatrix} 1 \\ 1 \end{pmatrix}$ and $\mathbf{K}_2 = \begin{pmatrix} 1 \\ -2 \end{pmatrix}$. The solution vectors of the homogeneous system (12) are then

$$\mathbf{X}_1 = \begin{pmatrix} 1 \\ 1 \end{pmatrix}e^{-2t} = \begin{pmatrix} e^{-2t} \\ e^{-2t} \end{pmatrix} \quad \text{and} \quad \mathbf{X}_2 = \begin{pmatrix} 1 \\ -2 \end{pmatrix}e^{-5t} = \begin{pmatrix} e^{-5t} \\ -2e^{-5t} \end{pmatrix}.$$

The entries in $\mathbf{X}_1$ form the first column of $\mathbf{\Phi}(t)$, and the entries in $\mathbf{X}_2$ form the second column of $\mathbf{\Phi}(t)$. Hence

$$\mathbf{\Phi}(t) = \begin{pmatrix} e^{-2t} & e^{-5t} \\ e^{-2t} & -2e^{-5t} \end{pmatrix} \quad \text{and} \quad \mathbf{\Phi}^{-1}(t) = \begin{pmatrix} \frac{2}{3}e^{2t} & \frac{1}{3}e^{2t} \\ \frac{1}{3}e^{5t} & -\frac{1}{3}e^{5t} \end{pmatrix}.$$

From (9) we obtain the particular solution

$$\begin{aligned}
\mathbf{X}_p = \mathbf{\Phi}(t)\int \mathbf{\Phi}^{-1}(t)\mathbf{F}(t)\,dt &= \begin{pmatrix} e^{-2t} & e^{-5t} \\ e^{-2t} & -2e^{-5t} \end{pmatrix} \int \begin{pmatrix} \frac{2}{3}e^{2t} & \frac{1}{3}e^{2t} \\ \frac{1}{3}e^{5t} & -\frac{1}{3}e^{5t} \end{pmatrix} \begin{pmatrix} 3t \\ e^{-t} \end{pmatrix} dt \\
&= \begin{pmatrix} e^{-2t} & e^{-5t} \\ e^{-2t} & -2e^{-5t} \end{pmatrix} \int \begin{pmatrix} 2te^{2t} + \frac{1}{3}e^{t} \\ te^{5t} - \frac{1}{3}e^{4t} \end{pmatrix} dt \\
&= \begin{pmatrix} e^{-2t} & e^{-5t} \\ e^{-2t} & -2e^{-5t} \end{pmatrix} \begin{pmatrix} te^{2t} - \frac{1}{2}e^{2t} + \frac{1}{3}e^{t} \\ \frac{1}{5}te^{5t} - \frac{1}{25}e^{5t} - \frac{1}{12}e^{4t} \end{pmatrix} \\
&= \begin{pmatrix} \frac{6}{5}t - \frac{27}{50} + \frac{1}{4}e^{-t} \\ \frac{3}{5}t - \frac{21}{50} + \frac{1}{2}e^{-t} \end{pmatrix}.
\end{aligned}$$

Hence from (10) the general solution of (11) on the interval is

$$\begin{aligned}
\mathbf{X} &= \begin{pmatrix} e^{-2t} & e^{-5t} \\ e^{-2t} & -2e^{-5t} \end{pmatrix}\begin{pmatrix} c_1 \\ c_2 \end{pmatrix} + \begin{pmatrix} \frac{6}{5}t - \frac{27}{50} + \frac{1}{4}e^{-t} \\ \frac{3}{5}t - \frac{21}{50} + \frac{1}{2}e^{-t} \end{pmatrix} \\
&= c_1\begin{pmatrix} 1 \\ 1 \end{pmatrix}e^{-2t} + c_2\begin{pmatrix} 1 \\ -2 \end{pmatrix}e^{-5t} + \begin{pmatrix} \frac{6}{5} \\ \frac{3}{5} \end{pmatrix}t - \begin{pmatrix} \frac{27}{50} \\ \frac{21}{50} \end{pmatrix} + \begin{pmatrix} \frac{1}{4} \\ \frac{1}{2} \end{pmatrix}e^{-t}.
\end{aligned}$$

≡

Initial-Value Problem The general solution of (5) on an interval can be written in the alternative manner

$$\mathbf{X} = \mathbf{\Phi}(t)\mathbf{C} + \mathbf{\Phi}(t)\int_{t_0}^{t} \mathbf{\Phi}^{-1}(s)\mathbf{F}(s)\,ds, \tag{13}$$

where t and t_0 are points in the interval. This last form is useful in solving (5) subject to an initial condition $\mathbf{X}(t_0) = \mathbf{X}_0$, because the limits of integration are chosen so that the particular solution vanishes at $t = t_0$. Substituting $t = t_0$ into (13) yields $\mathbf{X}_0 = \mathbf{\Phi}(t_0)\mathbf{C}$ from which we get $\mathbf{C} = \mathbf{\Phi}^{-1}(t_0)\mathbf{X}_0$. Substituting this last result into (13) gives the following solution of the initial-value problem:

$$\mathbf{X} = \mathbf{\Phi}(t)\mathbf{\Phi}^{-1}(t_0)\mathbf{X}_0 + \mathbf{\Phi}(t)\int_{t_0}^{t} \mathbf{\Phi}^{-1}(s)\mathbf{F}(s)\,ds. \tag{14}$$

EXERCISES 8.3

Answers to selected odd-numbered problems begin on page ANS-15.

8.3.1 UNDETERMINED COEFFICIENTS

In Problems 1–8 use the method of undetermined coefficients to solve the given system.

1. $\dfrac{dx}{dt} = 2x + 3y - 7$

$\dfrac{dy}{dt} = -x - 2y + 5$

2. $\dfrac{dx}{dt} = 5x + 9y + 2$

$\dfrac{dy}{dt} = -x + 11y + 6$

3. $\mathbf{X}' = \begin{pmatrix} 1 & 3 \\ 3 & 1 \end{pmatrix}\mathbf{X} + \begin{pmatrix} -2t^2 \\ t + 5 \end{pmatrix}$

4. $\mathbf{X}' = \begin{pmatrix} 1 & -4 \\ 4 & 1 \end{pmatrix}\mathbf{X} + \begin{pmatrix} 4t + 9e^{6t} \\ -t + e^{6t} \end{pmatrix}$

5. $\mathbf{X}' = \begin{pmatrix} 4 & \frac{1}{3} \\ 9 & 6 \end{pmatrix}\mathbf{X} + \begin{pmatrix} -3 \\ 10 \end{pmatrix}e^{t}$

6. $\mathbf{X}' = \begin{pmatrix} -1 & 5 \\ -1 & 1 \end{pmatrix}\mathbf{X} + \begin{pmatrix} \sin t \\ -2\cos t \end{pmatrix}$

7. $\mathbf{X}' = \begin{pmatrix} 1 & 1 & 1 \\ 0 & 2 & 3 \\ 0 & 0 & 5 \end{pmatrix}\mathbf{X} + \begin{pmatrix} 1 \\ -1 \\ 2 \end{pmatrix}e^{4t}$

8. $\mathbf{X}' = \begin{pmatrix} 0 & 0 & 5 \\ 0 & 5 & 0 \\ 5 & 0 & 0 \end{pmatrix}\mathbf{X} + \begin{pmatrix} 5 \\ -10 \\ 40 \end{pmatrix}$

9. Solve $\mathbf{X}' = \begin{pmatrix} -1 & -2 \\ 3 & 4 \end{pmatrix}\mathbf{X} + \begin{pmatrix} 3 \\ 3 \end{pmatrix}$ subject to

$\mathbf{X}(0) = \begin{pmatrix} -4 \\ 5 \end{pmatrix}.$

10. **(a)** The system of differential equations for the currents $i_2(t)$ and $i_3(t)$ in the electrical network shown in Figure 8.3.1 is

$$\frac{d}{dt}\begin{pmatrix} i_2 \\ i_3 \end{pmatrix} = \begin{pmatrix} -R_1/L_1 & -R_1/L_1 \\ -R_1/L_2 & -(R_1 + R_2)/L_2 \end{pmatrix}\begin{pmatrix} i_2 \\ i_3 \end{pmatrix} + \begin{pmatrix} E/L_1 \\ E/L_2 \end{pmatrix}.$$

Use the method of undetermined coefficients to solve the system if $R_1 = 2\ \Omega$, $R_2 = 3\ \Omega$, $L_1 = 1$ h, $L_2 = 1$ h, $E = 60$ V, $i_2(0) = 0$, and $i_3(0) = 0$.

(b) Determine the current $i_1(t)$.

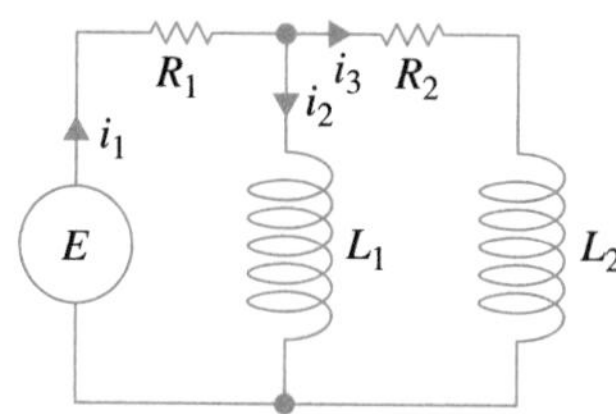

FIGURE 8.3.1 Network in Problem 10

8.3.2 VARIATION OF PARAMETERS

In Problems 11–30 use variation of parameters to solve the given system.

11. $\dfrac{dx}{dt} = 3x - 3y + 4$

$\dfrac{dy}{dt} = 2x - 2y - 1$

12. $\dfrac{dx}{dt} = 2x - y$

$\dfrac{dy}{dt} = 3x - 2y + 4t$

13. $\mathbf{X}' = \begin{pmatrix} 3 & -5 \\ \frac{3}{4} & -1 \end{pmatrix}\mathbf{X} + \begin{pmatrix} 1 \\ -1 \end{pmatrix}e^{t/2}$

14. $\mathbf{X}' = \begin{pmatrix} 2 & -1 \\ 4 & 2 \end{pmatrix}\mathbf{X} + \begin{pmatrix} \sin 2t \\ 2\cos 2t \end{pmatrix} e^{2t}$

15. $\mathbf{X}' = \begin{pmatrix} 0 & 2 \\ -1 & 3 \end{pmatrix}\mathbf{X} + \begin{pmatrix} 1 \\ -1 \end{pmatrix} e^{t}$

16. $\mathbf{X}' = \begin{pmatrix} 0 & 2 \\ -1 & 3 \end{pmatrix}\mathbf{X} + \begin{pmatrix} 2 \\ e^{-3t} \end{pmatrix}$

17. $\mathbf{X}' = \begin{pmatrix} 1 & 8 \\ 1 & -1 \end{pmatrix}\mathbf{X} + \begin{pmatrix} 12 \\ 12 \end{pmatrix} t$

18. $\mathbf{X}' = \begin{pmatrix} 1 & 8 \\ 1 & -1 \end{pmatrix}\mathbf{X} + \begin{pmatrix} e^{-t} \\ te^{t} \end{pmatrix}$

19. $\mathbf{X}' = \begin{pmatrix} 3 & 2 \\ -2 & -1 \end{pmatrix}\mathbf{X} + \begin{pmatrix} 2e^{-t} \\ e^{-t} \end{pmatrix}$

20. $\mathbf{X}' = \begin{pmatrix} 3 & 2 \\ -2 & -1 \end{pmatrix}\mathbf{X} + \begin{pmatrix} 1 \\ 1 \end{pmatrix}$

21. $\mathbf{X}' = \begin{pmatrix} 0 & -1 \\ 1 & 0 \end{pmatrix}\mathbf{X} + \begin{pmatrix} \sec t \\ 0 \end{pmatrix}$

22. $\mathbf{X}' = \begin{pmatrix} 1 & -1 \\ 1 & 1 \end{pmatrix}\mathbf{X} + \begin{pmatrix} 3 \\ 3 \end{pmatrix} e^{t}$

23. $\mathbf{X}' = \begin{pmatrix} 1 & -1 \\ 1 & 1 \end{pmatrix}\mathbf{X} + \begin{pmatrix} \cos t \\ \sin t \end{pmatrix} e^{t}$

24. $\mathbf{X}' = \begin{pmatrix} 2 & -2 \\ 8 & -6 \end{pmatrix}\mathbf{X} + \begin{pmatrix} 1 \\ 3 \end{pmatrix} \dfrac{e^{-2t}}{t}$

25. $\mathbf{X}' = \begin{pmatrix} 0 & 1 \\ -1 & 0 \end{pmatrix}\mathbf{X} + \begin{pmatrix} 0 \\ \sec t \tan t \end{pmatrix}$

26. $\mathbf{X}' = \begin{pmatrix} 0 & 1 \\ -1 & 0 \end{pmatrix}\mathbf{X} + \begin{pmatrix} 1 \\ \cot t \end{pmatrix}$

27. $\mathbf{X}' = \begin{pmatrix} 1 & 2 \\ -\frac{1}{2} & 1 \end{pmatrix}\mathbf{X} + \begin{pmatrix} \csc t \\ \sec t \end{pmatrix} e^{t}$

28. $\mathbf{X}' = \begin{pmatrix} 1 & -2 \\ 1 & -1 \end{pmatrix}\mathbf{X} + \begin{pmatrix} \tan t \\ 1 \end{pmatrix}$

29. $\mathbf{X}' = \begin{pmatrix} 1 & 1 & 0 \\ 1 & 1 & 0 \\ 0 & 0 & 3 \end{pmatrix}\mathbf{X} + \begin{pmatrix} e^{t} \\ e^{2t} \\ te^{3t} \end{pmatrix}$

30. $\mathbf{X}' = \begin{pmatrix} 3 & -1 & -1 \\ 1 & 1 & -1 \\ 1 & -1 & 1 \end{pmatrix}\mathbf{X} + \begin{pmatrix} 0 \\ t \\ 2e^{t} \end{pmatrix}$

In Problems 31 and 32 use (14) to solve the given initial-value problem.

31. $\mathbf{X}' = \begin{pmatrix} 3 & -1 \\ -1 & 3 \end{pmatrix}\mathbf{X} + \begin{pmatrix} 4e^{2t} \\ 4e^{4t} \end{pmatrix}, \quad \mathbf{X}(0) = \begin{pmatrix} 1 \\ 1 \end{pmatrix}$

32. $\mathbf{X}' = \begin{pmatrix} 1 & -1 \\ 1 & -1 \end{pmatrix}\mathbf{X} + \begin{pmatrix} 1/t \\ 1/t \end{pmatrix}, \quad \mathbf{X}(1) = \begin{pmatrix} 2 \\ -1 \end{pmatrix}$

33. The system of differential equations for the currents $i_1(t)$ and $i_2(t)$ in the electrical network shown in Figure 8.3.2 is

$$\frac{d}{dt}\begin{pmatrix} i_1 \\ i_2 \end{pmatrix} = \begin{pmatrix} -(R_1 + R_2)/L_2 & R_2/L_2 \\ R_2/L_1 & -R_2/L_1 \end{pmatrix}\begin{pmatrix} i_1 \\ i_2 \end{pmatrix} + \begin{pmatrix} E/L_2 \\ 0 \end{pmatrix}.$$

Use variation of parameters to solve the system if $R_1 = 8\ \Omega$, $R_2 = 3\ \Omega$, $L_1 = 1$ h, $L_2 = 1$ h, $E(t) = 100 \sin t$ V, $i_1(0) = 0$, and $i_2(0) = 0$.

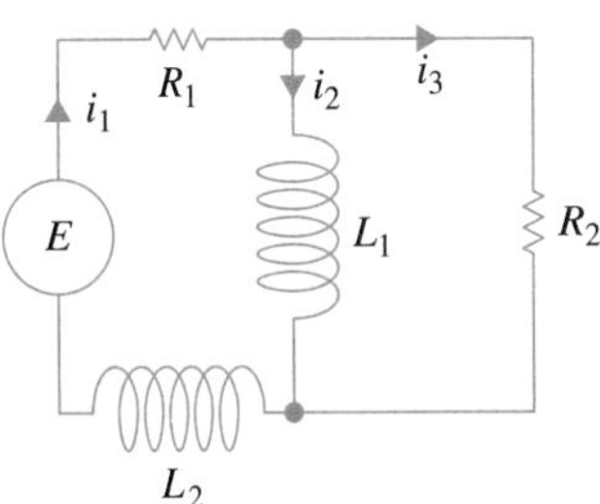

FIGURE 8.3.2 Network in Problem 33

Discussion Problems

34. If y_1 and y_2 are linearly independent solutions of the associated homogeneous DE for $y'' + P(x)y' + Q(x)y = f(x)$, show in the case of a nonhomogeneous linear second-order DE that (9) reduces to the form of variation of parameters discussed in Section 4.6.

Computer Lab Assignments

35. Solving a nonhomogeneous linear system $\mathbf{X}' = \mathbf{AX} + \mathbf{F}(t)$ by variation of parameters when $\mathbf{A}$ is a 3×3 (or larger) matrix is almost an impossible task to do by hand. Consider the system

$$\mathbf{X}' = \begin{pmatrix} 2 & -2 & 2 & 1 \\ -1 & 3 & 0 & 3 \\ 0 & 0 & 4 & -2 \\ 0 & 0 & 2 & -1 \end{pmatrix}\mathbf{X} + \begin{pmatrix} te^{t} \\ e^{-t} \\ e^{2t} \\ 1 \end{pmatrix}.$$

(a) Use a CAS or linear algebra software to find the eigenvalues and eigenvectors of the coefficient matrix.

(b) Form a fundamental matrix $\mathbf{\Phi}(t)$ and use the computer to find $\mathbf{\Phi}^{-1}(t)$.

(c) Use the computer to carry out the computations of: $\mathbf{\Phi}^{-1}(t)\mathbf{F}(t)$, $\int \mathbf{\Phi}^{-1}(t)\mathbf{F}(t)\,dt$, $\mathbf{\Phi}(t)\int \mathbf{\Phi}^{-1}(t)\mathbf{F}(t)\,dt$, $\mathbf{\Phi}(t)\mathbf{C}$, and $\mathbf{\Phi}(t)\mathbf{C} + \int \mathbf{\Phi}^{-1}(t)\mathbf{F}(t)\,dt$, where $\mathbf{C}$ is a column matrix of constants c_1, c_2, c_3, and c_4.

(d) Rewrite the computer output for the general solution of the system in the form $\mathbf{X} = \mathbf{X}_c + \mathbf{X}_p$, where $\mathbf{X}_c = c_1\mathbf{X}_1 + c_2\mathbf{X}_2 + c_3\mathbf{X}_3 + c_4\mathbf{X}_4$.

8.4 MATRIX EXPONENTIAL

REVIEW MATERIAL

- Appendix II.1 (Definitions II.10 and II.11)

INTRODUCTION Matrices can be used in an entirely different manner to solve a system of linear first-order differential equations. Recall that the simple linear first-order differential equation $x' = ax$, where a is constant, has the general solution $x = ce^{at}$, where c is a constant. It seems natural then to ask whether we can define a matrix exponential function $e^{\mathbf{A}t}$, where $\mathbf{A}$ is a matrix of constants, so that a solution of the linear system $\mathbf{X}' = \mathbf{AX}$ is $e^{\mathbf{A}t}$.

Homogeneous Systems We shall now see that it is possible to define a matrix exponential $e^{\mathbf{A}t}$ so that

$$\mathbf{X} = e^{\mathbf{A}t}\mathbf{C} \tag{1}$$

is a solution of the homogeneous system $\mathbf{X}' = \mathbf{AX}$. Here $\mathbf{A}$ is an $n \times n$ matrix of constants, and $\mathbf{C}$ is an $n \times 1$ column matrix of arbitrary constants. Note in (1) that the matrix $\mathbf{C}$ post multiplies $e^{\mathbf{A}t}$ because we want $e^{\mathbf{A}t}$ to be an $n \times n$ matrix. While the complete development of the meaning and theory of the matrix exponential would require a thorough knowledge of matrix algebra, one way of defining $e^{\mathbf{A}t}$ is inspired by the power series representation of the scalar exponential function e^{at}:

$$\begin{aligned} e^{at} &= 1 + at + \frac{(at)^2}{2!} + \cdots + \frac{(at)^k}{k!} + \cdots \\ &= 1 + at + a^2\frac{t^2}{2!} + \cdots + a^k\frac{t^k}{k!} + \cdots = \sum_{k=0}^{\infty} a^k \frac{t^k}{k!}. \end{aligned} \tag{2}$$

The series in (2) converges for all t. Using this series, with 1 replaced by the identity matrix $\mathbf{I}$ and the constant a replaced by an $n \times n$ matrix $\mathbf{A}$ of constants, we arrive at a definition for the $n \times n$ matrix $e^{\mathbf{A}t}$.

DEFINITION 8.4.1 Matrix Exponential

For any $n \times n$ matrix $\mathbf{A}$,

$$e^{\mathbf{A}t} = \mathbf{I} + \mathbf{A}t + \mathbf{A}^2\frac{t^2}{2!} + \cdots + \mathbf{A}^k\frac{t^k}{k!} + \cdots = \sum_{k=0}^{\infty} \mathbf{A}^k \frac{t^k}{k!}. \tag{3}$$

It can be shown that the series given in (3) converges to an $n \times n$ matrix for every value of t. Also, $\mathbf{A}^2 = \mathbf{AA}$, $\mathbf{A}^3 = \mathbf{A}(\mathbf{A}^2)$, and so on.

EXAMPLE 1 **Matrix Exponential Using (3)**

Compute $e^{\mathbf{A}t}$ for the matrix

$$\mathbf{A} = \begin{pmatrix} 2 & 0 \\ 0 & 3 \end{pmatrix}.$$

SOLUTION From the various powers

$$\mathbf{A}^2 = \begin{pmatrix} 2^2 & 0 \\ 0 & 3^2 \end{pmatrix}, \mathbf{A}^3 = \begin{pmatrix} 2^3 & 0 \\ 0 & 3^3 \end{pmatrix}, \mathbf{A}^4 = \begin{pmatrix} 2^4 & 0 \\ 0 & 3^4 \end{pmatrix}, \ldots, \mathbf{A}^n = \begin{pmatrix} 2^n & 0 \\ 0 & 3^n \end{pmatrix}, \ldots,$$

we see from (3) that

$$e^{\mathbf{A}t} = \mathbf{I} + \mathbf{A}t + \frac{\mathbf{A}^2}{2!}t^2 + \cdots$$

$$= \begin{pmatrix} 1 & 0 \\ 0 & 1 \end{pmatrix} + \begin{pmatrix} 2 & 0 \\ 0 & 3 \end{pmatrix}t + \begin{pmatrix} 2^2 & 0 \\ 0 & 3^2 \end{pmatrix}\frac{t^2}{2!} + \cdots + \begin{pmatrix} 2^n & 0 \\ 0 & 3^n \end{pmatrix}\frac{t^n}{n!} + \cdots$$

$$= \begin{pmatrix} 1 + 2t + 2^2\dfrac{t^2}{2!} + \cdots & 0 \\ 0 & 1 + 3t + 3^2\dfrac{t^2}{2!} + \cdots \end{pmatrix}.$$

In view of (2) and the identifications $a = 2$ and $a = 3$, the power series in the first and second rows of the last matrix represent, respectively, e^{2t} and e^{3t} and so we have

$$e^{\mathbf{A}t} = \begin{pmatrix} e^{2t} & 0 \\ 0 & e^{3t} \end{pmatrix}.$$

≡

The matrix in Example 1 is an example of a 2×2 diagonal matrix. In general, an $n \times n$ matrix $\mathbf{A}$ is a **diagonal matrix** if all its entries off the main diagonal are zero, that is,

$$\mathbf{A} = \begin{pmatrix} a_{11} & 0 & \cdots & 0 \\ 0 & a_{22} & \cdots & 0 \\ \vdots & \vdots & & \vdots \\ 0 & 0 & \cdots & a_{nn} \end{pmatrix}.$$

Hence if $\mathbf{A}$ is any $n \times n$ diagonal matrix it follows from Example 1 that

$$e^{\mathbf{A}t} = \begin{pmatrix} e^{a_{11}t} & 0 & \cdots & 0 \\ 0 & e^{a_{22}t} & \cdots & 0 \\ \vdots & \vdots & & \vdots \\ 0 & 0 & \cdots & e^{a_{nn}t} \end{pmatrix}.$$

Derivative of $e^{\mathbf{A}t}$ The derivative of the matrix exponential is analogous to the differentiation property of the scalar exponential $\frac{d}{dt}e^{at} = ae^{at}$. To justify

$$\frac{d}{dt}e^{\mathbf{A}t} = \mathbf{A}e^{\mathbf{A}t}, \tag{4}$$

we differentiate (3) term by term:

$$\frac{d}{dt}e^{\mathbf{A}t} = \frac{d}{dt}\left[\mathbf{I} + \mathbf{A}t + \mathbf{A}^2\frac{t^2}{2!} + \cdots + \mathbf{A}^k\frac{t^k}{k!} + \cdots\right] = \mathbf{A} + \mathbf{A}^2t + \frac{1}{2!}\mathbf{A}^3t^2 + \cdots$$

$$= \mathbf{A}\left[\mathbf{I} + \mathbf{A}t + \mathbf{A}^2\frac{t^2}{2!} + \cdots\right] = \mathbf{A}e^{\mathbf{A}t}.$$

Because of (4), we can now prove that (1) is a solution of $\mathbf{X}' = \mathbf{A}\mathbf{X}$ for every $n \times 1$ vector $\mathbf{C}$ of constants:

$$\mathbf{X}' = \frac{d}{dt}e^{\mathbf{A}t}\mathbf{C} = \mathbf{A}e^{\mathbf{A}t}\mathbf{C} = \mathbf{A}(e^{\mathbf{A}t}\mathbf{C}) = \mathbf{A}\mathbf{X}.$$

$e^{\mathbf{A}t}$ is a Fundamental Matrix If we denote the matrix exponential $e^{\mathbf{A}t}$ by the symbol $\mathbf{\Psi}(t)$, then (4) is equivalent to the matrix differential equation $\mathbf{\Psi}'(t) = \mathbf{A}\mathbf{\Psi}(t)$ (see (3) of Section 8.3). In addition, it follows immediately from

Definition 8.4.1 that $\mathbf{\Psi}(0) = e^{\mathbf{A}0} = \mathbf{I}$, and so det $\mathbf{\Psi}(0) \neq 0$. It turns out that these two properties are sufficient for us to conclude that $\mathbf{\Psi}(t)$ is a fundamental matrix of the system $\mathbf{X}' = \mathbf{AX}$.

Nonhomogeneous Systems We saw in (4) of Section 2.3 that the general solution of the single linear first-order differential equation $x' = ax + f(t)$, where a is a constant, can be expressed as

$$x = x_c + x_p = ce^{at} + e^{at}\int_{t_0}^{t} e^{-as}f(s)\,ds.$$

For a nonhomogeneous system of linear first-order differential equations it can be shown that the general solution of $\mathbf{X}' = \mathbf{AX} + \mathbf{F}(t)$, where $\mathbf{A}$ is an $n \times n$ matrix of constants, is

$$\mathbf{X} = \mathbf{X}_c + \mathbf{X}_p = e^{\mathbf{A}t}\mathbf{C} + e^{\mathbf{A}t}\int_{t_0}^{t} e^{-\mathbf{A}s}\mathbf{F}(s)\,ds. \tag{5}$$

Since the matrix exponential $e^{\mathbf{A}t}$ is a fundamental matrix, it is always nonsingular and $e^{-\mathbf{A}s} = (e^{\mathbf{A}s})^{-1}$. In practice, $e^{-\mathbf{A}s}$ can be obtained from $e^{\mathbf{A}t}$ by simply replacing t by $-s$.

Computation of $e^{\mathbf{A}t}$ The definition of $e^{\mathbf{A}t}$ given in (3) can, of course, always be used to compute $e^{\mathbf{A}t}$. However, the practical utility of (3) is limited by the fact that the entries in $e^{\mathbf{A}t}$ are power series in t. With a natural desire to work with simple and familiar things, we then try to recognize whether these series define a closed-form function. Fortunately, there are many alternative ways of computing $e^{\mathbf{A}t}$; the following discussion shows how the Laplace transform can be used.

Use of the Laplace Transform We saw in (5) that $\mathbf{X} = e^{\mathbf{A}t}$ is a solution of $\mathbf{X}' = \mathbf{AX}$. Indeed, since $e^{\mathbf{A}0} = \mathbf{I}$, $\mathbf{X} = e^{\mathbf{A}t}$ is a solution of the initial-value problem

$$\mathbf{X}' = \mathbf{AX}, \quad \mathbf{X}(0) = \mathbf{I}. \tag{6}$$

If $\mathbf{x}(s) = \mathscr{L}\{\mathbf{X}(t)\} = \mathscr{L}\{e^{\mathbf{A}t}\}$, then the Laplace transform of (6) is

$$s\mathbf{x}(s) - \mathbf{X}(0) = \mathbf{A}\mathbf{x}(s) \quad \text{or} \quad (s\mathbf{I} - \mathbf{A})\mathbf{x}(s) = \mathbf{I}.$$

Multiplying the last equation by $(s\mathbf{I} - \mathbf{A})^{-1}$ implies that $\mathbf{x}(s) = (s\mathbf{I} - \mathbf{A})^{-1}\mathbf{I} = (s\mathbf{I} - \mathbf{A})^{-1}$. In other words, $\mathscr{L}\{e^{\mathbf{A}t}\} = (s\mathbf{I} - \mathbf{A})^{-1}$ or

$$e^{\mathbf{A}t} = \mathscr{L}^{-1}\{(s\mathbf{I} - \mathbf{A})^{-1}\}. \tag{7}$$

EXAMPLE 2 **Matrix Exponential Using (7)**

Use the Laplace transform to compute $e^{\mathbf{A}t}$ for $\mathbf{A} = \begin{pmatrix} 1 & -1 \\ 2 & -2 \end{pmatrix}$.

SOLUTION First we compute the matrix $s\mathbf{I} - \mathbf{A}$ and find its inverse:

$$s\mathbf{I} - \mathbf{A} = \begin{pmatrix} s-1 & 1 \\ -2 & s+2 \end{pmatrix},$$

$$(s\mathbf{I} - \mathbf{A})^{-1} = \begin{pmatrix} s-1 & 1 \\ -2 & s+2 \end{pmatrix}^{-1} = \begin{pmatrix} \dfrac{s+2}{s(s+1)} & \dfrac{-1}{s(s+1)} \\[2ex] \dfrac{2}{s(s+1)} & \dfrac{s-1}{s(s+1)} \end{pmatrix}.$$

Then we decompose the entries of the last matrix into partial fractions:

$$(s\mathbf{I} - \mathbf{A})^{-1} = \begin{pmatrix} \frac{2}{s} - \frac{1}{s+1} & -\frac{1}{s} + \frac{1}{s+1} \\ \frac{2}{s} - \frac{2}{s+1} & -\frac{1}{s} + \frac{2}{s+1} \end{pmatrix}. \tag{8}$$

It follows from (7) that the inverse Laplace transform of (8) gives the desired result,

$$e^{\mathbf{A}t} = \begin{pmatrix} 2 - e^{-t} & -1 + e^{-t} \\ 2 - 2e^{-t} & -1 + 2e^{-t} \end{pmatrix}. \equiv$$

Use of Computers For those who are willing to momentarily trade understanding for speed of solution, $e^{\mathbf{A}t}$ can be computed with the aid of computer software. See Problems 27 and 28 in Exercises 8.4.

EXERCISES 8.4

Answers to selected odd-numbered problems begin on page ANS-16.

In Problems 1 and 2 use (3) to compute $e^{\mathbf{A}t}$ and $e^{-\mathbf{A}t}$.

1. $\mathbf{A} = \begin{pmatrix} 1 & 0 \\ 0 & 2 \end{pmatrix}$ **2.** $\mathbf{A} = \begin{pmatrix} 0 & 1 \\ 1 & 0 \end{pmatrix}$

In Problems 3 and 4 use (3) to compute $e^{\mathbf{A}t}$.

3. $\mathbf{A} = \begin{pmatrix} 1 & 1 & 1 \\ 1 & 1 & 1 \\ -2 & -2 & -2 \end{pmatrix}$

4. $\mathbf{A} = \begin{pmatrix} 0 & 0 & 0 \\ 3 & 0 & 0 \\ 5 & 1 & 0 \end{pmatrix}$

In Problems 5–8 use (1) to find the general solution of the given system.

5. $\mathbf{X}' = \begin{pmatrix} 1 & 0 \\ 0 & 2 \end{pmatrix}\mathbf{X}$ **6.** $\mathbf{X}' = \begin{pmatrix} 0 & 1 \\ 1 & 0 \end{pmatrix}\mathbf{X}$

7. $\mathbf{X}' = \begin{pmatrix} 1 & 1 & 1 \\ 1 & 1 & 1 \\ -2 & -2 & -2 \end{pmatrix}\mathbf{X}$ **8.** $\mathbf{X}' = \begin{pmatrix} 0 & 0 & 0 \\ 3 & 0 & 0 \\ 5 & 1 & 0 \end{pmatrix}\mathbf{X}$

In Problems 9–12 use (5) to find the general solution of the given system.

9. $\mathbf{X}' = \begin{pmatrix} 1 & 0 \\ 0 & 2 \end{pmatrix}\mathbf{X} + \begin{pmatrix} 3 \\ -1 \end{pmatrix}$

10. $\mathbf{X}' = \begin{pmatrix} 1 & 0 \\ 0 & 2 \end{pmatrix}\mathbf{X} + \begin{pmatrix} t \\ e^{4t} \end{pmatrix}$

11. $\mathbf{X}' = \begin{pmatrix} 0 & 1 \\ 1 & 0 \end{pmatrix}\mathbf{X} + \begin{pmatrix} 1 \\ 1 \end{pmatrix}$

12. $\mathbf{X}' = \begin{pmatrix} 0 & 1 \\ 1 & 0 \end{pmatrix}\mathbf{X} + \begin{pmatrix} \cosh t \\ \sinh t \end{pmatrix}$

13. Solve the system in Problem 7 subject to the initial condition

$$\mathbf{X}(0) = \begin{pmatrix} 1 \\ -4 \\ 6 \end{pmatrix}.$$

14. Solve the system in Problem 9 subject to the initial condition

$$\mathbf{X}(0) = \begin{pmatrix} 4 \\ 3 \end{pmatrix}.$$

In Problems 15–18 use the method of Example 2 to compute $e^{\mathbf{A}t}$ for the coefficient matrix. Use (1) to find the general solution of the given system.

15. $\mathbf{X}' = \begin{pmatrix} 4 & 3 \\ -4 & -4 \end{pmatrix}\mathbf{X}$ **16.** $\mathbf{X}' = \begin{pmatrix} 4 & -2 \\ 1 & 1 \end{pmatrix}\mathbf{X}$

17. $\mathbf{X}' = \begin{pmatrix} 5 & -9 \\ 1 & -1 \end{pmatrix}\mathbf{X}$ **18.** $\mathbf{X}' = \begin{pmatrix} 0 & 1 \\ -2 & -2 \end{pmatrix}\mathbf{X}$

Let $\mathbf{P}$ denote a matrix whose columns are eigenvectors $\mathbf{K}_1, \mathbf{K}_2, \ldots, \mathbf{K}_n$ corresponding to distinct eigenvalues $\lambda_1, \lambda_2, \ldots, \lambda_n$ of an $n \times n$ matrix $\mathbf{A}$. Then it can be shown that $\mathbf{A} = \mathbf{PDP}^{-1}$, where $\mathbf{D}$ is a diagonal matrix defined by

$$\mathbf{D} = \begin{pmatrix} \lambda_1 & 0 & \cdots & 0 \\ 0 & \lambda_2 & \cdots & 0 \\ \vdots & \vdots & & \vdots \\ 0 & 0 & \cdots & \lambda_n \end{pmatrix}. \tag{9}$$

In Problems 19 and 20 verify the foregoing result for the given matrix.

19. $\mathbf{A} = \begin{pmatrix} 2 & 1 \\ -3 & 6 \end{pmatrix}$ **20.** $\mathbf{A} = \begin{pmatrix} 2 & 1 \\ 1 & 2 \end{pmatrix}$

21. Suppose $\mathbf{A} = \mathbf{PDP}^{-1}$, where $\mathbf{D}$ is defined as in (9). Use (3) to show that $e^{\mathbf{A}t} = \mathbf{P}e^{\mathbf{D}t}\mathbf{P}^{-1}$.

22. If $\mathbf{D}$ is defined as in (9), then find $e^{\mathbf{D}t}$.

In Problems 23 and 24 use the results of Problems 19–22 to solve the given system.

23. $\mathbf{X}' = \begin{pmatrix} 2 & 1 \\ -3 & 6 \end{pmatrix}\mathbf{X}$

24. $\mathbf{X}' = \begin{pmatrix} 2 & 1 \\ 1 & 2 \end{pmatrix}\mathbf{X}$

Discussion Problems

25. Reread the discussion leading to the result given in (7). Does the matrix $s\mathbf{I} - \mathbf{A}$ always have an inverse? Discuss.

26. A matrix $\mathbf{A}$ is said to be **nilpotent** if there exists some positive integer m such that $\mathbf{A}^m = \mathbf{0}$. Verify that

$$\mathbf{A} = \begin{pmatrix} -1 & 1 & 1 \\ -1 & 0 & 1 \\ -1 & 1 & 1 \end{pmatrix}$$

is nilpotent. Discuss why it is relatively easy to compute $e^{\mathbf{A}t}$ when $\mathbf{A}$ is nilpotent. Compute $e^{\mathbf{A}t}$ and then use (1) to solve the system $\mathbf{X}' = \mathbf{AX}$.

Computer Lab Assignments

27. (a) Use (1) to find the general solution of $\mathbf{X}' = \begin{pmatrix} 4 & 2 \\ 3 & 3 \end{pmatrix}\mathbf{X}$. Use a CAS to find $e^{\mathbf{A}t}$. Then use the computer to find eigenvalues and eigenvectors of the coefficient matrix $\mathbf{A} = \begin{pmatrix} 4 & 2 \\ 3 & 3 \end{pmatrix}$ and form the general solution in the manner of Section 8.2. Finally, reconcile the two forms of the general solution of the system.

(b) Use (1) to find the general solution of $\mathbf{X}' = \begin{pmatrix} -3 & -1 \\ 2 & -1 \end{pmatrix}\mathbf{X}$. Use a CAS to find $e^{\mathbf{A}t}$. In the case of complex output, utilize the software to do the simplification; for example, in *Mathematica,* if **m = MatrixExp[A t]** has complex entries, then try the command **Simplify[ComplexExpand[m]].**

28. Use (1) to find the general solution of

$$\mathbf{X}' = \begin{pmatrix} -4 & 0 & 6 & 0 \\ 0 & -5 & 0 & -4 \\ -1 & 0 & 1 & 0 \\ 0 & 3 & 0 & 2 \end{pmatrix}\mathbf{X}.$$

Use MATLAB or a CAS to find $e^{\mathbf{A}t}$.

CHAPTER 8 IN REVIEW

Answers to selected odd-numbered problems begin on page ANS-16.

In Problems 1 and 2 fill in the blanks.

1. The vector $\mathbf{X} = k\begin{pmatrix} 4 \\ 5 \end{pmatrix}$ is a solution of

$$\mathbf{X}' = \begin{pmatrix} 1 & 4 \\ 2 & -1 \end{pmatrix}\mathbf{X} - \begin{pmatrix} 8 \\ 1 \end{pmatrix}$$

for $k =$ __________.

2. The vector $\mathbf{X} = c_1\begin{pmatrix} -1 \\ 1 \end{pmatrix}e^{-9t} + c_2\begin{pmatrix} 5 \\ 3 \end{pmatrix}e^{7t}$ is solution of the initial-value problem $\mathbf{X}' = \begin{pmatrix} 1 & 10 \\ 6 & -3 \end{pmatrix}\mathbf{X}$, $\mathbf{X}(0) = \begin{pmatrix} 2 \\ 0 \end{pmatrix}$ for $c_1 =$ __________ and $c_2 =$ __________.

3. Consider the linear system $\mathbf{X}' = \begin{pmatrix} 4 & 6 & 6 \\ 1 & 3 & 2 \\ -1 & -4 & -3 \end{pmatrix}\mathbf{X}$.

Without attempting to solve the system, determine which one of the vectors

$$\mathbf{K}_1 = \begin{pmatrix} 0 \\ 1 \\ 1 \end{pmatrix},\quad \mathbf{K}_2 = \begin{pmatrix} 1 \\ 1 \\ -1 \end{pmatrix},\quad \mathbf{K}_3 = \begin{pmatrix} 3 \\ 1 \\ -1 \end{pmatrix},\quad \mathbf{K}_4 = \begin{pmatrix} 6 \\ 2 \\ -5 \end{pmatrix}$$

is an eigenvector of the coefficient matrix. What is the solution of the system corresponding to this eigenvector?

4. Consider the linear system $\mathbf{X}' = \mathbf{AX}$ of two differential equations, where $\mathbf{A}$ is a real coefficient matrix. What is the general solution of the system if it is known that $\lambda_1 = 1 + 2i$ is an eigenvalue and $\mathbf{K}_1 = \begin{pmatrix} 1 \\ i \end{pmatrix}$ is a corresponding eigenvector?

In Problems 5–14 solve the given linear system.

5. $\dfrac{dx}{dt} = 2x + y$, $\quad \dfrac{dy}{dt} = -x$

6. $\dfrac{dx}{dt} = -4x + 2y$, $\quad \dfrac{dy}{dt} = 2x - 4y$

7. $\mathbf{X}' = \begin{pmatrix} 1 & 2 \\ -2 & 1 \end{pmatrix}\mathbf{X}$

8. $\mathbf{X}' = \begin{pmatrix} -2 & 5 \\ -2 & 4 \end{pmatrix}\mathbf{X}$

9. $\mathbf{X}' = \begin{pmatrix} 1 & -1 & 1 \\ 0 & 1 & 3 \\ 4 & 3 & 1 \end{pmatrix}\mathbf{X}$

10. $\mathbf{X}' = \begin{pmatrix} 0 & 2 & 1 \\ 1 & 1 & -2 \\ 2 & 2 & -1 \end{pmatrix}\mathbf{X}$

11. $\mathbf{X}' = \begin{pmatrix} 2 & 8 \\ 0 & 4 \end{pmatrix}\mathbf{X} + \begin{pmatrix} 2 \\ 16t \end{pmatrix}$

12. $\mathbf{X}' = \begin{pmatrix} 1 & 2 \\ -\frac{1}{2} & 1 \end{pmatrix}\mathbf{X} + \begin{pmatrix} 0 \\ e^t \tan t \end{pmatrix}$

13. $\mathbf{X}' = \begin{pmatrix} -1 & 1 \\ -2 & 1 \end{pmatrix}\mathbf{X} + \begin{pmatrix} 1 \\ \cot t \end{pmatrix}$

14. $\mathbf{X}' = \begin{pmatrix} 3 & 1 \\ -1 & 1 \end{pmatrix}\mathbf{X} + \begin{pmatrix} -2 \\ 1 \end{pmatrix}e^{2t}$

15. **(a)** Consider the linear system $\mathbf{X}' = \mathbf{AX}$ of three first-order differential equations, where the coefficient matrix is

$$\mathbf{A} = \begin{pmatrix} 5 & 3 & 3 \\ 3 & 5 & 3 \\ -5 & -5 & -3 \end{pmatrix}$$

and $\lambda = 2$ is known to be an eigenvalue of multiplicity two. Find two different solutions of the system corresponding to this eigenvalue without using a special formula (such as (12) of Section 8.2).

(b) Use the procedure of part (a) to solve

$$\mathbf{X}' = \begin{pmatrix} 1 & 1 & 1 \\ 1 & 1 & 1 \\ 1 & 1 & 1 \end{pmatrix}\mathbf{X}.$$

16. Verify that $\mathbf{X} = \begin{pmatrix} c_1 \\ c_2 \end{pmatrix}e^t$ is a solution of the linear system

$$\mathbf{X}' = \begin{pmatrix} 1 & 0 \\ 0 & 1 \end{pmatrix}\mathbf{X}$$

for arbitrary constants c_1 and c_2. By hand, draw a phase portrait of the system.

9 Numerical Solutions of Ordinary Differential Equations

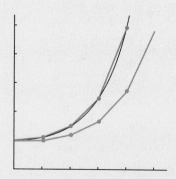

Even if it can be shown that a solution of a differential equation exists, we might not be able to exhibit it in explicit or implicit form. In many instances we have to be content with an approximation of the solution. If a solution exists, it represents a set of points in the Cartesian plane. In this chapter we continue to explore the basic idea introduced in Section 2.6, that is, using the differential equation to construct an algorithm to approximate the y-coordinates of points on the actual solution curve. Our concentration in this chapter is primarily on first-order initial-value problems $dy/dx = f(x, y)$, $y(x_0) = y_0$. We saw in Section 4.10 that numerical procedures developed for first-order DEs extend in a natural way to systems of first-order equations. Because of this extension, we are able to approximate solutions of a higher-order equation by rewriting it as a system of first-order DEs. Chapter 9 concludes with a method for approximating solutions of linear second-order boundary-value problems.

9.1 EULER METHODS AND ERROR ANALYSIS

REVIEW MATERIAL

- Section 2.6

INTRODUCTION In Chapter 2 we examined one of the simplest numerical methods for approximating solutions of first-order initial-value problems $y' = f(x, y)$, $y(x_0) = y_0$. Recall that the backbone of **Euler's method** is the formula

$$y_{n+1} = y_n + hf(x_n, y_n), \tag{1}$$

where f is the function obtained from the differential equation $y' = f(x, y)$. The recursive use of (1) for $n = 0, 1, 2, \ldots$ yields the y-coordinates $y_1, y_2, y_3, \ldots$ of points on successive "tangent lines" to the solution curve at $x_1, x_2, x_3, \ldots$ or $x_n = x_0 + nh$, where h is a constant and is the size of the step between x_n and x_{n+1}. The values $y_1, y_2, y_3, \ldots$ approximate the values of a solution $y(x)$ of the IVP at $x_1, x_2, x_3, \ldots$. But whatever advantage (1) has in its simplicity is lost in the crudeness of its approximations.

A Comparison

In Problem 4 in Exercises 2.6 you were asked to use Euler's method to obtain the approximate value of $y(1.5)$ for the solution of the initial-value problem $y' = 2xy$, $y(1) = 1$. You should have obtained the analytic solution $y = e^{x^2-1}$ and results similar to those given in Tables 9.1.1 and 9.1.2.

TABLE 9.1.1 Euler's Method with $h = 0.1$

x_n	y_n	Actual value	Abs. error	% Rel. error
1.00	1.0000	1.0000	0.0000	0.00
1.10	1.2000	1.2337	0.0337	2.73
1.20	1.4640	1.5527	0.0887	5.71
1.30	1.8154	1.9937	0.1784	8.95
1.40	2.2874	2.6117	0.3244	12.42
1.50	2.9278	3.4903	0.5625	16.12

TABLE 9.1.2 Euler's Method with $h = 0.05$

x_n	y_n	Actual value	Abs. error	% Rel. error
1.00	1.0000	1.0000	0.0000	0.00
1.05	1.1000	1.1079	0.0079	0.72
1.10	1.2155	1.2337	0.0182	1.47
1.15	1.3492	1.3806	0.0314	2.27
1.20	1.5044	1.5527	0.0483	3.11
1.25	1.6849	1.7551	0.0702	4.00
1.30	1.8955	1.9937	0.0982	4.93
1.35	2.1419	2.2762	0.1343	5.90
1.40	2.4311	2.6117	0.1806	6.92
1.45	2.7714	3.0117	0.2403	7.98
1.50	3.1733	3.4903	0.3171	9.08

In this case, with a step size $h = 0.1$ a 16% relative error in the calculation of the approximation to $y(1.5)$ is totally unacceptable. At the expense of doubling the number of calculations, some improvement in accuracy is obtained by halving the step size to $h = 0.05$.

Errors in Numerical Methods

In choosing and using a numerical method for the solution of an initial-value problem, we must be aware of the various sources of errors. For some kinds of computation the accumulation of errors might reduce the accuracy of an approximation to the point of making the computation useless. On the other hand, depending on the use to which a numerical solution may be put, extreme accuracy might not be worth the added expense and complication.

One source of error that is always present in calculations is **round-off error.** This error results from the fact that any calculator or computer can represent numbers using only a finite number of digits. Suppose, for the sake of illustration, that we have a calculator that uses base 10 arithmetic and carries four digits, so that $\frac{1}{3}$ is represented in the calculator as 0.3333 and $\frac{1}{9}$ is represented as 0.1111. If we use this calculator to compute $\left(x^2 - \frac{1}{9}\right)/\left(x - \frac{1}{3}\right)$ for $x = 0.3334$, we obtain

$$\frac{(0.3334)^2 - 0.1111}{0.3334 - 0.3333} = \frac{0.1112 - 0.1111}{0.3334 - 0.3333} = 1.$$

With the help of a little algebra, however, we see that

$$\frac{x^2 - \frac{1}{9}}{x - \frac{1}{3}} = \frac{\left(x - \frac{1}{3}\right)\left(x + \frac{1}{3}\right)}{x - \frac{1}{3}} = x + \frac{1}{3},$$

so when $x = 0.3334$, $\left(x^2 - \frac{1}{9}\right)/\left(x - \frac{1}{3}\right) \approx 0.3334 + 0.3333 = 0.6667$. This example shows that the effects of round-off error can be quite serious unless some care is taken. One way to reduce the effect of round-off error is to minimize the number of calculations. Another technique on a computer is to use double-precision arithmetic to check the results. In general, round-off error is unpredictable and difficult to analyze, and we will neglect it in the error analysis that follows. We will concentrate on investigating the error introduced by using a formula or algorithm to approximate the values of the solution.

Truncation Errors for Euler's Method In the sequence of values y_1, y_2, y_3, . . . generated from (1), usually the value of y_1 will not agree with the actual solution at x_1—namely, $y(x_1)$—because the algorithm gives only a straight-line approximation to the solution. See Figure 2.6.2. The error is called the **local truncation error, formula error,** or **discretization error.** It occurs at each step; that is, if we assume that y_n is accurate, then y_{n+1} will contain local truncation error.

To derive a formula for the local truncation error for Euler's method, we use Taylor's formula with remainder. If a function $y(x)$ possesses $k + 1$ derivatives that are continuous on an open interval containing a and x, then

$$y(x) = y(a) + y'(a)\frac{x - a}{1!} + \cdots + y^{(k)}(a)\frac{(x - a)^k}{k!} + y^{(k+1)}(c)\frac{(x - a)^{k+1}}{(k + 1)!},$$

where c is some point between a and x. Setting $k = 1$, $a = x_n$, and $x = x_{n+1} = x_n + h$, we get

$$y(x_{n+1}) = y(x_n) + y'(x_n)\frac{h}{1!} + y''(c)\frac{h^2}{2!}$$

or

$$y(x_{n+1}) = \underbrace{y_n + hf(x_n, y_n)}_{y_{n+1}} + y''(c)\frac{h^2}{2!}.$$

Euler's method (1) is the last formula without the last term; hence the local truncation error in y_{n+1} is

$$y''(c)\frac{h^2}{2!}, \qquad \text{where} \quad x_n < c < x_{n+1}.$$

The value of c is usually unknown (it exists theoretically), so the *exact* error cannot be calculated, but an upper bound on the absolute value of the error is $Mh^2/2!$, where $M = \max\limits_{x_n<x<x_{n+1}} |y''(x)|$.

In discussing errors that arise from the use of numerical methods, it is helpful to use the notation $O(h^n)$. To define this concept, we let $e(h)$ denote the error in a numerical calculation depending on h. Then $e(h)$ is said to be of order h^n, denoted by $O(h^n)$, if there exist a constant C and a positive integer n such that $|e(h)| \leq Ch^n$ for h sufficiently small. Thus the local truncation error for Euler's method is $O(h^2)$. We note that, in general, if $e(h)$ in a numerical method is of order h^n and h is halved, the new error is approximately $C(h/2)^n = Ch^n/2^n$; that is, the error is reduced by a factor of $1/2^n$.

EXAMPLE 1 Bound for Local Truncation Errors

Find a bound for the local truncation errors for Euler's method applied to $y' = 2xy$, $y(1) = 1$.

SOLUTION From the solution $y = e^{x^2-1}$ we get $y'' = (2 + 4x^2)e^{x^2-1}$, so the local truncation error is

$$y''(c)\frac{h^2}{2} = (2 + 4c^2)e^{(c^2-1)}\frac{h^2}{2},$$

where c is between x_n and $x_n + h$. In particular, for $h = 0.1$ we can get an upper bound on the local truncation error for y_1 by replacing c by 1.1:

$$[2 + (4)(1.1)^2]e^{((1.1)^2-1)}\frac{(0.1)^2}{2} = 0.0422.$$

From Table 9.1.1 we see that the error after the first step is 0.0337, less than the value given by the bound.

Similarly, we can get a bound for the local truncation error for any of the five steps given in Table 9.1.1 by replacing c by 1.5 (this value of c gives the largest value of $y''(c)$ for any of the steps and may be too generous for the first few steps). Doing this gives

$$[2 + (4)(1.5)^2]e^{((1.5)^2-1)}\frac{(0.1)^2}{2} = 0.1920 \tag{2}$$

as an upper bound for the local truncation error in each step. ≡

Note that if h is halved to 0.05 in Example 1, then the error bound is 0.0480, about one-fourth as much as shown in (2). This is expected because the local truncation error for Euler's method is $O(h^2)$.

In the above analysis we assumed that the value of y_n was exact in the calculation of y_{n+1}, but it is not because it contains local truncation errors from previous steps. The total error in y_{n+1} is an accumulation of the errors in each of the previous steps. This total error is called the **global truncation error.** A complete analysis of the global truncation error is beyond the scope of this text, but it can be shown that the global truncation error for Euler's method is $O(h)$.

We expect that, for Euler's method, if the step size is halved the error will be approximately halved as well. This is borne out in Tables 9.1.1 and 9.1.2 where the absolute error at $x = 1.50$ with $h = 0.1$ is 0.5625 and with $h = 0.05$ is 0.3171, approximately half as large.

In general it can be shown that if a method for the numerical solution of a differential equation has local truncation error $O(h^{\alpha+1})$, then the global truncation error is $O(h^{\alpha})$.

For the remainder of this section and in the subsequent sections we study methods that give significantly greater accuracy than does Euler's method.

≡ Improved Euler's Method The numerical method defined by the formula

$$y_{n+1} = y_n + h\frac{f(x_n, y_n) + f(x_{n+1}, y^*_{n+1})}{2}, \tag{3}$$

where

$$y^*_{n+1} = y_n + hf(x_n, y_n), \tag{4}$$

is commonly known as the **improved Euler's method.** To compute y_{n+1} for $n = 0, 1, 2, \ldots$ from (3), we must, at each step, first use Euler's method (4) to obtain an initial estimate y^*_{n+1}. For example, with $n = 0$, (4) gives $y^*_1 = y_0 + hf(x_0, y_0)$, and then, knowing this value, we use (3) to get $y_1 = y_0 + h\dfrac{f(x_0, y_0) + f(x_1, y^*_1)}{2}$, where

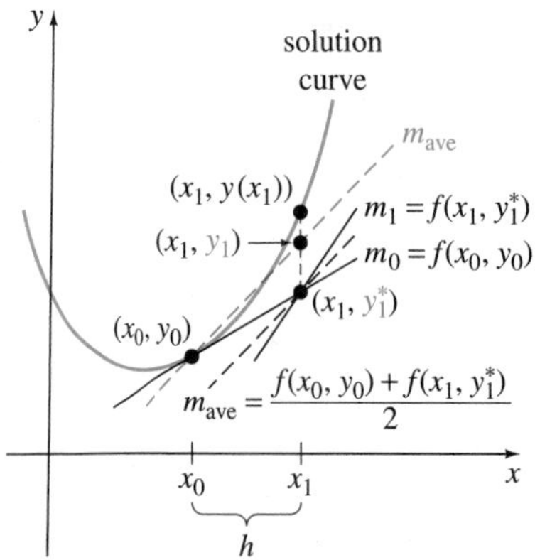

FIGURE 9.1.1 Slope of red dashed line is the average of m_0 and m_1

$x_1 = x_0 + h$. These equations can be readily visualized. In Figure 9.1.1, observe that $m_0 = f(x_0, y_0)$ and $m_1 = f(x_1, y_1^*)$ are slopes of the solid straight lines shown passing through the points (x_0, y_0) and (x_1, y_1^*), respectively. By taking an average of these slopes, that is, $m_{ave} = \dfrac{f(x_0, y_0) + f(x_1, y_1^*)}{2}$, we obtain the slope of the parallel dashed skew lines. With the first step, rather than advancing along the line through (x_0, y_0) with slope $f(x_0, y_0)$ to the point with y-coordinate y_1^* obtained by Euler's method, we advance instead along the red dashed line through (x_0, y_0) with slope m_{ave} until we reach x_1. It seems plausible from inspection of the figure that y_1 is an improvement over y_1^*.

In general, the improved Euler's method is an example of a **predictor-corrector method.** The value of y_{n+1}^* given by (4) predicts a value of $y(x_n)$, whereas the value of y_{n+1} defined by formula (3) corrects this estimate.

EXAMPLE 2 Improved Euler's Method

Use the improved Euler's method to obtain the approximate value of $y(1.5)$ for the solution of the initial-value problem $y' = 2xy$, $y(1) = 1$. Compare the results for $h = 0.1$ and $h = 0.05$.

SOLUTION With $x_0 = 1$, $y_0 = 1$, $f(x_n, y_n) = 2x_n y_n$, $n = 0$, and $h = 0.1$, we first compute (4):

$$y_1^* = y_0 + (0.1)(2x_0 y_0) = 1 + (0.1)2(1)(1) = 1.2.$$

We use this last value in (3) along with $x_1 = 1 + h = 1 + 0.1 = 1.1$:

$$y_1 = y_0 + (0.1)\frac{2x_0 y_0 + 2x_1 y_1^*}{2} = 1 + (0.1)\frac{2(1)(1) + 2(1.1)(1.2)}{2} = 1.232.$$

The comparative values of the calculations for $h = 0.1$ and $h = 0.05$ are given in Tables 9.1.3 and 9.1.4, respectively.

TABLE 9.1.3 Improved Euler's Method with $h = 0.1$

x_n	y_n	Actual value	Abs. error	% Rel. error
1.00	1.0000	1.0000	0.0000	0.00
1.10	1.2320	1.2337	0.0017	0.14
1.20	1.5479	1.5527	0.0048	0.31
1.30	1.9832	1.9937	0.0106	0.53
1.40	2.5908	2.6117	0.0209	0.80
1.50	3.4509	3.4904	0.0394	1.13

TABLE 9.1.4 Improved Euler's Method with $h = 0.05$

x_n	y_n	Actual value	Abs. error	% Rel. error
1.00	1.0000	1.0000	0.0000	0.00
1.05	1.1077	1.1079	0.0002	0.02
1.10	1.2332	1.2337	0.0004	0.04
1.15	1.3798	1.3806	0.0008	0.06
1.20	1.5514	1.5527	0.0013	0.08
1.25	1.7531	1.7551	0.0020	0.11
1.30	1.9909	1.9937	0.0029	0.14
1.35	2.2721	2.2762	0.0041	0.18
1.40	2.6060	2.6117	0.0057	0.22
1.45	3.0038	3.0117	0.0079	0.26
1.50	3.4795	3.4904	0.0108	0.31

≡

A brief word of caution is in order here. We cannot compute all the values of y_n^* first and then substitute these values into formula (3). In other words, we cannot use the data in Table 9.1.1 to help construct the values in Table 9.1.3. Why not?

≡ Truncation Errors for the Improved Euler's Method

The local truncation error for the improved Euler's method is $O(h^3)$. The derivation of this result is similar to the derivation of the local truncation error for Euler's method. Since the local truncation error for the improved Euler's method is $O(h^3)$, the global truncation error is $O(h^2)$. This can be seen in Example 2; when the step size is halved from $h = 0.1$ to $h = 0.05$, the absolute error at $x = 1.50$ is reduced from 0.0394 to 0.0108, a reduction of approximately $\left(\frac{1}{2}\right)^2 = \frac{1}{4}$.

EXERCISES 9.1

Answers to selected odd-numbered problems begin on page ANS-16.

In Problems 1–10 use the improved Euler's method to obtain a four-decimal approximation of the indicated value. First use $h = 0.1$ and then use $h = 0.05$.

1. $y' = 2x - 3y + 1, y(1) = 5;\quad y(1.5)$

2. $y' = 4x - 2y, y(0) = 2;\quad y(0.5)$

3. $y' = 1 + y^2, y(0) = 0;\quad y(0.5)$

4. $y' = x^2 + y^2, y(0) = 1;\quad y(0.5)$

5. $y' = e^{-y}, y(0) = 0;\quad y(0.5)$

6. $y' = x + y^2, y(0) = 0;\quad y(0.5)$

7. $y' = (x - y)^2, y(0) = 0.5;\quad y(0.5)$

8. $y' = xy + \sqrt{y}, y(0) = 1;\quad y(0.5)$

9. $y' = xy^2 - \dfrac{y}{x}, y(1) = 1;\quad y(1.5)$

10. $y' = y - y^2, y(0) = 0.5;\quad y(0.5)$

11. Consider the initial-value problem $y' = (x + y - 1)^2$, $y(0) = 2$. Use the improved Euler's method with $h = 0.1$ and $h = 0.05$ to obtain approximate values of the solution at $x = 0.5$. At each step compare the approximate value with the actual value of the analytic solution.

12. Although it might not be obvious from the differential equation, its solution could "behave badly" near a point x at which we wish to approximate $y(x)$. Numerical procedures may give widely differing results near this point. Let $y(x)$ be the solution of the initial-value problem $y' = x^2 + y^3, y(1) = 1$.

(a) Use a numerical solver to graph the solution on the interval [1, 1.4].

(b) Using the step size $h = 0.1$, compare the results obtained from Euler's method with the results from the improved Euler's method in the approximation of $y(1.4)$.

13. Consider the initial-value problem $y' = 2y$, $y(0) = 1$. The analytic solution is $y = e^{2x}$.

(a) Approximate $y(0.1)$ using one step and Euler's method.

(b) Find a bound for the local truncation error in y_1.

(c) Compare the error in y_1 with your error bound.

(d) Approximate $y(0.1)$ using two steps and Euler's method.

(e) Verify that the global truncation error for Euler's method is $O(h)$ by comparing the errors in parts (a) and (d).

14. Repeat Problem 13 using the improved Euler's method. Its global truncation error is $O(h^2)$.

15. Repeat Problem 13 using the initial-value problem $y' = x - 2y$, $y(0) = 1$. The analytic solution is

$$y = \tfrac{1}{2}x - \tfrac{1}{4} + \tfrac{5}{4}e^{-2x}.$$

16. Repeat Problem 15 using the improved Euler's method. Its global truncation error is $O(h^2)$.

17. Consider the initial-value problem $y' = 2x - 3y + 1$, $y(1) = 5$. The analytic solution is

$$y(x) = \tfrac{1}{9} + \tfrac{2}{3}x + \tfrac{38}{9}e^{-3(x-1)}.$$

(a) Find a formula involving c and h for the local truncation error in the nth step if Euler's method is used.

(b) Find a bound for the local truncation error in each step if $h = 0.1$ is used to approximate $y(1.5)$.

(c) Approximate $y(1.5)$ using $h = 0.1$ and $h = 0.05$ with Euler's method. See Problem 1 in Exercises 2.6.

(d) Calculate the errors in part (c) and verify that the global truncation error of Euler's method is $O(h)$.

18. Repeat Problem 17 using the improved Euler's method, which has a global truncation error $O(h^2)$. See Problem 1. You might need to keep more than four decimal places to see the effect of reducing the order of the error.

19. Repeat Problem 17 for the initial-value problem $y' = e^{-y}$, $y(0) = 0$. The analytic solution is $y(x) = \ln(x + 1)$. Approximate $y(0.5)$. See Problem 5 in Exercises 2.6.

20. Repeat Problem 19 using the improved Euler's method, which has global truncation error $O(h^2)$. See Problem 5. You might need to keep more than four decimal places to see the effect of reducing the order of error.

Discussion Problems

21. Answer the question "Why not?" that follows the three sentences after Example 2 on page 366.

9.2 RUNGE-KUTTA METHODS

REVIEW MATERIAL

- Section 2.6 (see page 78)

INTRODUCTION Probably one of the more popular as well as most accurate numerical procedures used in obtaining approximate solutions to a first-order initial-value problem $y' = f(x, y)$, $y(x_0) = y_0$ is the **fourth-order Runge-Kutta method.** As the name suggests, there are Runge-Kutta methods of different orders.

Runge-Kutta Methods Fundamentally, all Runge-Kutta methods are generalizations of the basic Euler formula (1) of Section 9.1 in that the slope function f is replaced by a weighted average of slopes over the interval $x_n \le x \le x_{n+1}$. That is,

$$y_{n+1} = y_n + h\,\overbrace{(w_1k_1 + w_2k_2 + \cdots + w_mk_m)}^{\text{weighted average}}. \tag{1}$$

Here the weights w_i, $i = 1, 2, \ldots, m$, are constants that generally satisfy $w_1 + w_2 + \cdots + w_m = 1$, and each k_i, $i = 1, 2, \ldots, m$, is the function f evaluated at a selected point (x, y) for which $x_n \le x \le x_{n+1}$. We shall see that the k_i are defined recursively. The number m is called the **order** of the method. Observe that by taking $m = 1$, $w_1 = 1$, and $k_1 = f(x_n, y_n)$, we get the familiar Euler formula $y_{n+1} = y_n + hf(x_n, y_n)$. Hence Euler's method is said to be a **first-order Runge-Kutta method.**

The average in (1) is not formed willy-nilly, but parameters are chosen so that (1) agrees with a Taylor polynomial of degree m. As we saw in the preceding section, if a function $y(x)$ possesses $k + 1$ derivatives that are continuous on an open interval containing a and x, then we can write

$$y(x) = y(a) + y'(a)\frac{x - a}{1!} + y''(a)\frac{(x - a)^2}{2!} + \cdots + y^{(k+1)}(c)\frac{(x - a)^{k+1}}{(k + 1)!},$$

where c is some number between a and x. If we replace a by x_n and x by $x_{n+1} = x_n + h$, then the foregoing formula becomes

$$y(x_{n+1}) = y(x_n + h) = y(x_n) + hy'(x_n) + \frac{h^2}{2!}y''(x_n) + \cdots + \frac{h^{k+1}}{(k + 1)!}y^{(k+1)}(c),$$

where c is now some number between x_n and x_{n+1}. When $y(x)$ is a solution of $y' = f(x, y)$ in the case $k = 1$ and the remainder $\frac{1}{2}h^2y''(c)$ is small, we see that a Taylor polynomial $y(x_{n+1}) = y(x_n) + hy'(x_n)$ of degree one agrees with the approximation formula of Euler's method

$$y_{n+1} = y_n + hy'_n = y_n + hf(x_n, y_n).$$

A Second-Order Runge-Kutta Method To further illustrate (1), we consider now a **second-order Runge-Kutta procedure.** This consists of finding constants or parameters w_1, w_2, α, and β so that the formula

$$y_{n+1} = y_n + h(w_1k_1 + w_2k_2), \tag{2}$$

where

$$k_1 = f(x_n, y_n)$$

$$k_2 = f(x_n + \alpha h, y_n + \beta hk_1),$$

agrees with a Taylor polynomial of degree two. For our purposes it suffices to say that this can be done whenever the constants satisfy

$$w_1 + w_2 = 1, \qquad w_2\alpha = \frac{1}{2}, \qquad \text{and} \qquad w_2\beta = \frac{1}{2}. \tag{3}$$

This is an algebraic system of three equations in four unknowns and has infinitely many solutions:

$$w_1 = 1 - w_2, \qquad \alpha = \frac{1}{2w_2}, \qquad \text{and} \qquad \beta = \frac{1}{2w_2}, \tag{4}$$

where $w_2 \neq 0$. For example, the choice $w_2 = \frac{1}{2}$ yields $w_1 = \frac{1}{2}$, $\alpha = 1$, and $\beta = 1$, and so (2) becomes

$$y_{n+1} = y_n + \frac{h}{2}(k_1 + k_2),$$

where $\qquad k_1 = f(x_n, y_n) \qquad \text{and} \qquad k_2 = f(x_n + h, y_n + hk_1).$

Since $x_n + h = x_{n+1}$ and $y_n + hk_1 = y_n + hf(x_n, y_n)$, the foregoing result is recognized to be the improved Euler's method that is summarized in (3) and (4) of Section 9.1.

In view of the fact that $w_2 \neq 0$ can be chosen arbitrarily in (4), there are many possible second-order Runge-Kutta methods. See Problem 2 in Exercises 9.2.

We shall skip any discussion of third-order methods in order to come to the principal point of discussion in this section.

≡ A Fourth-Order Runge-Kutta Method

A **fourth-order Runge-Kutta procedure** consists of finding parameters so that the formula

$$y_{n+1} = y_n + h(w_1k_1 + w_2k_2 + w_3k_3 + w_4k_4), \tag{5}$$

where
$$\begin{aligned} k_1 &= f(x_n, y_n) \\ k_2 &= f(x_n + \alpha_1 h, y_n + \beta_1 hk_1) \\ k_3 &= f(x_n + \alpha_2 h, y_n + \beta_2 hk_1 + \beta_3 hk_2) \\ k_4 &= f(x_n + \alpha_3 h, y_n + \beta_4 hk_1 + \beta_5 hk_2 + \beta_6 hk_3), \end{aligned}$$

agrees with a Taylor polynomial of degree four. This results in a system of 11 equations in 13 unknowns. The most commonly used set of values for the parameters yields the following result:

$$\begin{aligned} y_{n+1} &= y_n + \frac{h}{6}(k_1 + 2k_2 + 2k_3 + k_4), \\ k_1 &= f(x_n, y_n) \\ k_2 &= f\left(x_n + \tfrac{1}{2}h, y_n + \tfrac{1}{2}hk_1\right) \\ k_3 &= f\left(x_n + \tfrac{1}{2}h, y_n + \tfrac{1}{2}hk_2\right) \\ k_4 &= f(x_n + h, y_n + hk_3). \end{aligned} \tag{6}$$

While other fourth-order formulas are easily derived, the algorithm summarized in (6) is so widely used and recognized as a valuable computational tool it is often referred to as *the* fourth-order Runge-Kutta method or *the classical* Runge-Kutta method. It is (6) that we have in mind, hereafter, when we use the abbreviation the **RK4 method.**

You are advised to look carefully at the formulas in (6); note that k_2 depends on k_1, k_3 depends on k_2, and k_4 depends on k_3. Also, k_2 and k_3 involve approximations to the slope at the midpoint $x_n + \frac{1}{2}h$ of the interval defined by $x_n \leq x \leq x_{n+1}$.

EXAMPLE 1 RK4 Method

Use the RK4 method with $h = 0.1$ to obtain an approximation to $y(1.5)$ for the solution of $y' = 2xy$, $y(1) = 1$.

SOLUTION For the sake of illustration let us compute the case when $n = 0$. From (6) we find

$$\begin{aligned}
k_1 &= f(x_0, y_0) = 2x_0y_0 = 2\\
k_2 &= f\left(x_0 + \tfrac{1}{2}(0.1),\, y_0 + \tfrac{1}{2}(0.1)2\right)\\
&= 2\left(x_0 + \tfrac{1}{2}(0.1)\right)\left(y_0 + \tfrac{1}{2}(0.2)\right) = 2.31\\
k_3 &= f\left(x_0 + \tfrac{1}{2}(0.1),\, y_0 + \tfrac{1}{2}(0.1)2.31\right)\\
&= 2\left(x_0 + \tfrac{1}{2}(0.1)\right)\left(y_0 + \tfrac{1}{2}(0.231)\right) = 2.34255\\
k_4 &= f(x_0 + (0.1),\, y_0 + (0.1)2.34255)\\
&= 2(x_0 + 0.1)(y_0 + 0.234255) = 2.715361
\end{aligned}$$

and therefore

$$\begin{aligned}
y_1 &= y_0 + \frac{0.1}{6}(k_1 + 2k_2 + 2k_3 + k_4)\\
&= 1 + \frac{0.1}{6}(2 + 2(2.31) + 2(2.34255) + 2.715361) = 1.23367435.
\end{aligned}$$

The remaining calculations are summarized in Table 9.2.1, whose entries are rounded to four decimal places. ≡

TABLE 9.2.1 RK4 Method with $h = 0.1$

x_n	y_n	Actual value	Abs. error	% Rel. error
1.00	1.0000	1.0000	0.0000	0.00
1.10	1.2337	1.2337	0.0000	0.00
1.20	1.5527	1.5527	0.0000	0.00
1.30	1.9937	1.9937	0.0000	0.00
1.40	2.6116	2.6117	0.0001	0.00
1.50	3.4902	3.4904	0.0001	0.00

Inspection of Table 9.2.1 shows why the fourth-order Runge-Kutta method is so popular. If four-decimal-place accuracy is all that we desire, there is no need to use a smaller step size. Table 9.2.2 compares the results of applying Euler's, the improved Euler's, and the fourth-order Runge-Kutta methods to the initial-value problem $y' = 2xy$, $y(1) = 1$. (See Tables 9.1.1–9.1.4.)

TABLE 9.2.2 $y' = 2xy$, $y(1) = 1$

Comparison of numerical methods with $h = 0.1$					Comparison of numerical methods with $h = 0.05$				
x_n	Euler	Improved Euler	RK4	Actual value	x_n	Euler	Improved Euler	RK4	Actual value
1.00	1.0000	1.0000	1.0000	1.0000	1.00	1.0000	1.0000	1.0000	1.0000
1.10	1.2000	1.2320	1.2337	1.2337	1.05	1.1000	1.1077	1.1079	1.1079
1.20	1.4640	1.5479	1.5527	1.5527	1.10	1.2155	1.2332	1.2337	1.2337
1.30	1.8154	1.9832	1.9937	1.9937	1.15	1.3492	1.3798	1.3806	1.3806
1.40	2.2874	2.5908	2.6116	2.6117	1.20	1.5044	1.5514	1.5527	1.5527
1.50	2.9278	3.4509	3.4902	3.4904	1.25	1.6849	1.7531	1.7551	1.7551
					1.30	1.8955	1.9909	1.9937	1.9937
					1.35	2.1419	2.2721	2.2762	2.2762
					1.40	2.4311	2.6060	2.6117	2.6117
					1.45	2.7714	3.0038	3.0117	3.0117
					1.50	3.1733	3.4795	3.4903	3.4904

≡ Truncation Errors for the RK4 Method In Section 9.1 we saw that global truncation errors for Euler's method and for the improved Euler's method are, respectively, $O(h)$ and $O(h^2)$. Because the first equation in (6) agrees with a Taylor polynomial of degree four, the local truncation error for this method is $y^{(5)}(c)\, h^5/5!$ or $O(h^5)$, and the global truncation error is thus $O(h^4)$. It is now obvious why Euler's method, the improved Euler's method, and (6) are *first-*, *second-*, and *fourth-order* Runge-Kutta methods, respectively.

EXAMPLE 2 **Bound for Local Truncation Errors**

Find a bound for the local truncation errors for the RK4 method applied to $y' = 2xy$, $y(1) = 1$.

SOLUTION By computing the fifth derivative of the known solution $y(x) = e^{x^2-1}$, we get

$$y^{(5)}(c)\frac{h^5}{5!} = (120c + 160c^3 + 32c^5)e^{c^2-1}\frac{h^5}{5!}. \tag{7}$$

Thus with $c = 1.5$, (7) yields a bound of 0.00028 on the local truncation error for each of the five steps when $h = 0.1$. Note that in Table 9.2.1 the error in y_1 is much less than this bound.

Table 9.2.3 gives the approximations to the solution of the initial-value problem at $x = 1.5$ that are obtained from the RK4 method. By computing the value of the analytic solution at $x = 1.5$, we can find the error in these approximations. Because the method is so accurate, many decimal places must be used in the numerical solution to see the effect of halving the step size. Note that when h is halved, from $h = 0.1$ to $h = 0.05$, the error is divided by a factor of about $2^4 = 16$, as expected. ≡

TABLE 9.2.3 RK4 Method

h	Approx.	Error
0.1	3.49021064	$1.32321089 \times 10^{-4}$
0.05	3.49033382	$9.13776090 \times 10^{-6}$

≡ Adaptive Methods

We have seen that the accuracy of a numerical method for approximating solutions of differential equations can be improved by decreasing the step size h. Of course, this enhanced accuracy is usually obtained at a cost—namely, increased computation time and greater possibility of round-off error. In general, over the interval of approximation there may be subintervals where a relatively large step size suffices and other subintervals where a smaller step is necessary to keep the truncation error within a desired limit. Numerical methods that use a variable step size are called **adaptive methods.** One of the more popular of the adaptive routines is the **Runge-Kutta-Fehlberg method.** Because Fehlberg employed two Runge-Kutta methods of differing orders, a fourth- and a fifth-order method, this algorithm is frequently denoted as the **RKF45 method.***

*The Runge-Kutta method of order four used in RKF45 is *not* the same as that given in (6).

EXERCISES 9.2

Answers to selected odd-numbered problems begin on page ANS-17.

1. Use the RK4 method with $h = 0.1$ to approximate $y(0.5)$, where $y(x)$ is the solution of the initial-value problem $y' = (x + y - 1)^2$, $y(0) = 2$. Compare this approximate value with the actual value obtained in Problem 11 in Exercises 9.1.

2. Assume that $w_2 = \frac{3}{4}$ in (4). Use the resulting second-order Runge-Kutta method to approximate $y(0.5)$, where $y(x)$ is the solution of the initial-value problem in Problem 1. Compare this approximate value with the approximate value obtained in Problem 11 in Exercises 9.1.

In Problems 3–12 use the RK4 method with $h = 0.1$ to obtain a four-decimal approximation of the indicated value.

3. $y' = 2x - 3y + 1$, $y(1) = 5$; $y(1.5)$
4. $y' = 4x - 2y$, $y(0) = 2$; $y(0.5)$
5. $y' = 1 + y^2$, $y(0) = 0$; $y(0.5)$
6. $y' = x^2 + y^2$, $y(0) = 1$; $y(0.5)$
7. $y' = e^{-y}$, $y(0) = 0$; $y(0.5)$
8. $y' = x + y^2$, $y(0) = 0$; $y(0.5)$
9. $y' = (x - y)^2$, $y(0) = 0.5$; $y(0.5)$
10. $y' = xy + \sqrt{y}$, $y(0) = 1$; $y(0.5)$
11. $y' = xy^2 - \dfrac{y}{x}$, $y(1) = 1$; $y(1.5)$
12. $y' = y - y^2$, $y(0) = 0.5$; $y(0.5)$
13. If air resistance is proportional to the square of the instantaneous velocity, then the velocity v of a mass m dropped from a given height is determined from

$$m\frac{dv}{dt} = mg - kv^2, \qquad k > 0.$$

Let $v(0) = 0$, $k = 0.125$, $m = 5$ slugs, and $g = 32$ ft/s^2.

(a) Use the RK4 method with $h = 1$ to approximate the velocity $v(5)$.

(b) Use a numerical solver to graph the solution of the IVP on the interval [0, 6].

(c) Use separation of variables to solve the IVP and then find the actual value $v(5)$.

14. A mathematical model for the area A (in cm^2) that a colony of bacteria (*B. dendroides*) occupies is given by

$$\frac{dA}{dt} = A(2.128 - 0.0432A).^*$$

Suppose that the initial area is $0.24\ \text{cm}^2$.

(a) Use the RK4 method with $h = 0.5$ to complete the following table:

t (days)	1	2	3	4	5
A (observed)	2.78	13.53	36.30	47.50	49.40
A (approximated)					

(b) Use a numerical solver to graph the solution of the initial-value problem. Estimate the values $A(1)$, $A(2)$, $A(3)$, $A(4)$, and $A(5)$ from the graph.

(c) Use separation of variables to solve the initial-value problem and compute the actual values $A(1)$, $A(2)$, $A(3)$, $A(4)$, and $A(5)$.

15. Consider the initial-value problem $y' = x^2 + y^3$, $y(1) = 1$. See Problem 12 in Exercises 9.1.

(a) Compare the results obtained from using the RK4 method over the interval [1, 1.4] with step sizes $h = 0.1$ and $h = 0.05$.

(b) Use a numerical solver to graph the solution of the initial-value problem on the interval [1, 1.4].

16. Consider the initial-value problem $y' = 2y$, $y(0) = 1$. The analytic solution is $y(x) = e^{2x}$.

(a) Approximate $y(0.1)$ using one step and the RK4 method.

(b) Find a bound for the local truncation error in y_1.

(c) Compare the error in y_1 with your error bound.

(d) Approximate $y(0.1)$ using two steps and the RK4 method.

(e) Verify that the global truncation error for the RK4 method is $O(h^4)$ by comparing the errors in parts (a) and (d).

17. Repeat Problem 16 using the initial-value problem $y' = -2y + x$, $y(0) = 1$. The analytic solution is

$$y(x) = \tfrac{1}{2}x - \tfrac{1}{4} + \tfrac{5}{4}e^{-2x}.$$

18. Consider the initial-value problem $y' = 2x - 3y + 1$, $y(1) = 5$. The analytic solution is

$$y(x) = \tfrac{1}{9} + \tfrac{2}{3}x + \tfrac{38}{9}e^{-3(x-1)}.$$

(a) Find a formula involving c and h for the local truncation error in the nth step if the RK4 method is used.

(b) Find a bound for the local truncation error in each step if $h = 0.1$ is used to approximate $y(1.5)$.

(c) Approximate $y(1.5)$ using the RK4 method with $h = 0.1$ and $h = 0.05$. See Problem 3. You will need to carry more than six decimal places to see the effect of reducing the step size.

19. Repeat Problem 18 for the initial-value problem $y' = e^{-y}$, $y(0) = 0$. The analytic solution is $y(x) = \ln(x + 1)$. Approximate $y(0.5)$. See Problem 7.

Discussion Problems

20. A count of the number of evaluations of the function f used in solving the initial-value problem $y' = f(x, y)$, $y(x_0) = y_0$ is used as a measure of the computational complexity of a numerical method. Determine the number of evaluations of f required for each step of Euler's, the improved Euler's, and the RK4 methods. By considering some specific examples, compare the accuracy of these methods when used with comparable computational complexities.

Computer Lab Assignments

21. The RK4 method for solving an initial-value problem over an interval $[a, b]$ results in a finite set of points that are supposed to approximate points on the graph of the exact solution. To expand this set of discrete points to an approximate solution defined at all points on the interval $[a, b]$, we can use an **interpolating function.** This is a function, supported by most computer algebra systems, that agrees with the given data exactly and assumes a smooth transition between data points. These interpolating functions may be polynomials or sets of polynomials joined together smoothly. In *Mathematica* the command **y=Interpolation[data]** can be used to obtain an interpolating function through the points **data** $= \{\{x_0, y_0\}, \{x_1, y_1\}, \ldots, \{x_n, y_n\}\}$. The interpolating function **y[x]** can now be treated like any other function built into the computer algebra system.

(a) Find the analytic solution of the initial-value problem $y' = -y + 10 \sin 3x$; $y(0) = 0$ on the interval [0, 2]. Graph this solution and find its positive roots.

(b) Use the RK4 method with $h = 0.1$ to approximate a solution of the initial-value problem in part (a). Obtain an interpolating function and graph it. Find the positive roots of the interpolating function of the interval [0, 2].

*See V. A. Kostitzin, *Mathematical Biology* (London: Harrap, 1939).

9.3 MULTISTEP METHODS

REVIEW MATERIAL

- Sections 9.1 and 9.2

INTRODUCTION Euler's method, the improved Euler's method, and the Runge-Kutta methods are examples of **single-step** or **starting methods.** In these methods each successive value y_{n+1} is computed based only on information about the immediately preceding value y_n. On the other hand, **multistep** or **continuing methods** use the values from several computed steps to obtain the value of y_{n+1}. There are a large number of multistep method formulas for approximating solutions of DEs, but since it is not our intention to survey the vast field of numerical procedures, we will consider only one such method here.

Adams-Bashforth-Moulton Method The multistep method that is discussed in this section is called the fourth-order **Adams-Bashforth-Moulton method.** Like the improved Euler's method it is a predictor-corrector method—that is, one formula is used to predict a value y^*_{n+1}, which in turn is used to obtain a corrected value y_{n+1}. The predictor in this method is the Adams-Bashforth formula

$$\begin{aligned} y^*_{n+1} &= y_n + \frac{h}{24}(55y'_n - 59y'_{n-1} + 37y'_{n-2} - 9y'_{n-3}), \\ y'_n &= f(x_n, y_n) \\ y'_{n-1} &= f(x_{n-1}, y_{n-1}) \\ y'_{n-2} &= f(x_{n-2}, y_{n-2}) \\ y'_{n-3} &= f(x_{n-3}, y_{n-3}) \end{aligned} \tag{1}$$

for $n \geq 3$. The value of y^*_{n+1} is then substituted into the Adams-Moulton corrector

$$\begin{aligned} y_{n+1} &= y_n + \frac{h}{24}(9y'_{n+1} + 19y'_n - 5y'_{n-1} + y'_{n-2}) \\ y'_{n+1} &= f(x_{n+1}, y^*_{n+1}). \end{aligned} \tag{2}$$

Notice that formula (1) requires that we know the values of y_0, y_1, y_2, and y_3 to obtain y_4. The value of y_0 is, of course, the given initial condition. The local truncation error of the Adams-Bashforth-Moulton method is $O(h^5)$, the values of y_1, y_2, and y_3 are generally computed by a method with the same error property, such as the fourth-order Runge-Kutta method.

EXAMPLE 1 **Adams-Bashforth-Moulton Method**

Use the Adams-Bashforth-Moulton method with $h = 0.2$ to obtain an approximation to $y(0.8)$ for the solution of

$$y' = x + y - 1, \quad y(0) = 1.$$

SOLUTION With a step size of $h = 0.2$, $y(0.8)$ will be approximated by y_4. To get started, we use the RK4 method with $x_0 = 0$, $y_0 = 1$, and $h = 0.2$ to obtain

$$y_1 = 1.02140000, \quad y_2 = 1.09181796, \quad y_3 = 1.22210646.$$

Now with the identifications $x_0 = 0$, $x_1 = 0.2$, $x_2 = 0.4$, $x_3 = 0.6$, and $f(x, y) = x + y - 1$, we find

$$\begin{aligned} y'_0 &= f(x_0, y_0) = (0) + (1) - 1 = 0 \\ y'_1 &= f(x_1, y_1) = (0.2) + (1.02140000) - 1 = 0.22140000 \\ y'_2 &= f(x_2, y_2) = (0.4) + (1.09181796) - 1 = 0.49181796 \\ y'_3 &= f(x_3, y_3) = (0.6) + (1.22210646) - 1 = 0.82210646. \end{aligned}$$

With the foregoing values the predictor (1) then gives

$$y_4^* = y_3 + \frac{0.2}{24}(55y'_3 - 59y'_2 + 37y'_1 - 9y'_0) = 1.42535975.$$

To use the corrector (2), we first need

$$y'_4 = f(x_4, y_4^*) = 0.8 + 1.42535975 - 1 = 1.22535975.$$

Finally, (2) yields

$$y_4 = y_3 + \frac{0.2}{24}(9y'_4 + 19y'_3 - 5y'_2 + y'_1) = 1.42552788.$$

≡

You should verify that the actual value of $y(0.8)$ in Example 1 is $y(0.8) = 1.42554093$. See Problem 1 in Exercises 9.3.

≡ **Stability of Numerical Methods** An important consideration in using numerical methods to approximate the solution of an initial-value problem is the stability of the method. Simply stated, a numerical method is **stable** if small changes in the initial condition result in only small changes in the computed solution. A numerical method is said to be **unstable** if it is not stable. The reason that stability considerations are important is that in each step after the first step of a numerical technique we are essentially starting over again with a new initial-value problem, where the initial condition is the approximate solution value computed in the preceding step. Because of the presence of round-off error, this value will almost certainly vary at least slightly from the true value of the solution. Besides round-off error, another common source of error occurs in the initial condition itself; in physical applications the data are often obtained by imprecise measurements.

One possible method for detecting instability in the numerical solution of a specific initial-value problem is to compare the approximate solutions obtained when decreasing step sizes are used. If the numerical method is unstable, the error may actually increase with smaller step sizes. Another way of checking stability is to observe what happens to solutions when the initial condition is slightly perturbed (for example, change $y(0) = 1$ to $y(0) = 0.999$).

For a more detailed and precise discussion of stability, consult a numerical analysis text. In general, all of the methods that we have discussed in this chapter have good stability characteristics.

≡ **Advantages and Disadvantages of Multistep Methods** Many considerations enter into the choice of a method to solve a differential equation numerically. Single-step methods, particularly the RK4 method, are often chosen because of their accuracy and the fact that they are easy to program. However, a major drawback is that the right-hand side of the differential equation must be evaluated many times at each step. For instance, the RK4 method requires four function evaluations for each step. On the other hand, if the function evaluations in the previous step have been calculated and stored, a multistep method requires only one new function evaluation for each step. This can lead to great savings in time and expense.

As an example, solving $y' = f(x, y)$, $y(x_0) = y_0$ numerically using n steps by the fourth-order Runge-Kutta method requires $4n$ function evaluations. The Adams-Bashforth multistep method requires 16 function evaluations for the Runge-Kutta fourth-order starter and $n - 4$ for the n Adams-Bashforth steps, giving a total of $n + 12$ function evaluations for this method. In general the Adams-Bashforth multistep method requires slightly more than a quarter of the number of function evaluations required for the RK4 method. If the evaluation of $f(x, y)$ is complicated, the multistep method will be more efficient.

Another issue that is involved with multistep methods is how many times the Adams-Moulton corrector formula should be repeated in each step. Each time the corrector is used, another function evaluation is done, and so the accuracy is increased at the expense of losing an advantage of the multistep method. In practice, the corrector is calculated once, and if the value of y_{n+1} is changed by a large amount, the entire problem is restarted using a smaller step size. This is often the basis of the variable step size methods, whose discussion is beyond the scope of this text.

EXERCISES 9.3

Answers to selected odd-numbered problems begin on page ANS-17.

1. Find the analytic solution of the initial-value problem in Example 1. Compare the actual values of $y(0.2)$, $y(0.4)$, $y(0.6)$, and $y(0.8)$ with the approximations y_1, y_2, y_3, and y_4.

2. Write a computer program to implement the Adams-Bashforth-Moulton method.

In Problems 3 and 4 use the Adams-Bashforth-Moulton method to approximate $y(0.8)$, where $y(x)$ is the solution of the given initial-value problem. Use $h = 0.2$ and the RK4 method to compute y_1, y_2, and y_3.

3. $y' = 2x - 3y + 1, \quad y(0) = 1$

4. $y' = 4x - 2y, \quad y(0) = 2$

In Problems 5–8 use the Adams-Bashforth-Moulton method to approximate $y(1.0)$, where $y(x)$ is the solution of the given initial-value problem. First use $h = 0.2$ and then use $h = 0.1$. Use the RK4 method to compute y_1, y_2, and y_3.

5. $y' = 1 + y^2, \quad y(0) = 0$

6. $y' = y + \cos x, \quad y(0) = 1$

7. $y' = (x - y)^2, \quad y(0) = 0$

8. $y' = xy + \sqrt{y}, \quad y(0) = 1$

9.4 HIGHER-ORDER EQUATIONS AND SYSTEMS

REVIEW MATERIAL

- Section 1.1 (normal form of a second-order DE)
- Section 4.10 (second-order DE written as a system of first-order DEs)

INTRODUCTION So far, we have focused on numerical techniques that can be used to approximate the solution of a first-order initial-value problem $y' = f(x, y)$, $y(x_0) = y_0$. In order to approximate the solution of a second-order initial-value problem, we must express a second-order DE as a system of two first-order DEs. To do this, we begin by writing the second-order DE in normal form by solving for y'' in terms of x, y, and y'.

Second-Order IVPs A second-order initial-value problem

$$y'' = f(x, y, y'), \quad y(x_0) = y_0, \quad y'(x_0) = u_0 \tag{1}$$

can be expressed as an initial-value problem for a system of first-order differential equations. If we let $y' = u$, the differential equation in (1) becomes the system

$$\begin{aligned} y' &= u \\ u' &= f(x, y, u). \end{aligned} \tag{2}$$

Since $y'(x_0) = u(x_0)$, the corresponding initial conditions for (2) are then $y(x_0) = y_0$, $u(x_0) = u_0$. The system (2) can now be solved numerically by simply applying a particular numerical method to each first-order differential equation in the system. For example, **Euler's method** applied to the system (2) would be

$$\begin{aligned} y_{n+1} &= y_n + hu_n \\ u_{n+1} &= u_n + hf(x_n, y_n, u_n), \end{aligned} \tag{3}$$

whereas the **fourth-order Runge-Kutta method,** or **RK4 method,** would be

$$\begin{aligned} y_{n+1} &= y_n + \frac{h}{6}(m_1 + 2m_2 + 2m_3 + m_4) \\ u_{n+1} &= u_n + \frac{h}{6}(k_1 + 2k_2 + 2k_3 + k_4) \end{aligned} \tag{4}$$

where

$$\begin{aligned} m_1 &= u_n & k_1 &= f(x_n, y_n, u_n) \\ m_2 &= u_n + \tfrac{1}{2}hk_1 & k_2 &= f\left(x_n + \tfrac{1}{2}h, y_n + \tfrac{1}{2}hm_1, u_n + \tfrac{1}{2}hk_1\right) \\ m_3 &= u_n + \tfrac{1}{2}hk_2 & k_3 &= f\left(x_n + \tfrac{1}{2}h, y_n + \tfrac{1}{2}hm_2, u_n + \tfrac{1}{2}hk_2\right) \\ m_4 &= u_n + hk_3 & k_4 &= f(x_n + h, y_n + hm_3, u_n + hk_3). \end{aligned}$$

In general, we can express every nth-order differential equation $y^{(n)} = f(x, y, y', \ldots, y^{(n-1)})$ as a system of n first-order equations using the substitutions $y = u_1, y' = u_2, y'' = u_3, \ldots, y^{(n-1)} = u_n$.

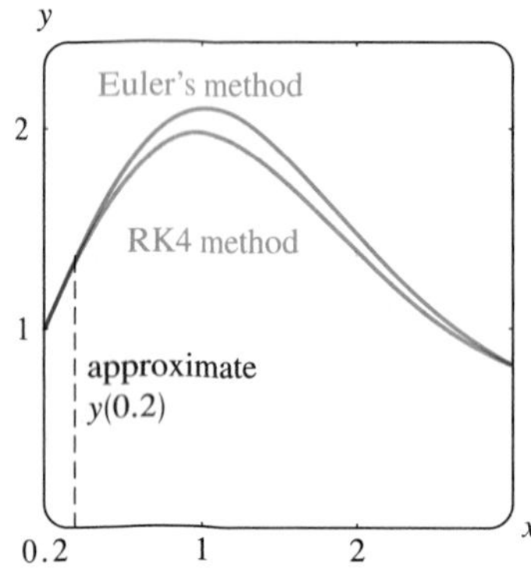

(a) Euler's method (red) and the RK4 method (blue)

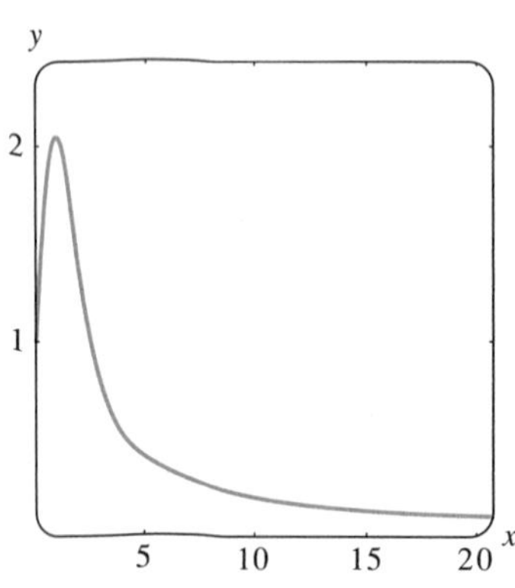

(b) RK4 method

FIGURE 9.4.1 Numerical solution curves generated by different methods

EXAMPLE 1 Euler's Method

Use Euler's method to obtain the approximate value of $y(0.2)$, where $y(x)$ is the solution of the initial-value problem

$$y'' + xy' + y = 0, \quad y(0) = 1, \quad y'(0) = 2. \tag{5}$$

SOLUTION In terms of the substitution $y' = u$, the equation is equivalent to the system

$$\begin{aligned} y' &= u \\ u' &= -xu - y. \end{aligned}$$

Thus from (3) we obtain

$$\begin{aligned} y_{n+1} &= y_n + hu_n \\ u_{n+1} &= u_n + h[-x_n u_n - y_n]. \end{aligned}$$

Using the step size $h = 0.1$ and $y_0 = 1$, $u_0 = 2$, we find

$$\begin{aligned} y_1 &= y_0 + (0.1)u_0 = 1 + (0.1)2 = 1.2 \\ u_1 &= u_0 + (0.1)\,[-x_0u_0 - y_0] = 2 + (0.1)[-(0)(2) - 1] = 1.9 \\ y_2 &= y_1 + (0.1)u_1 = 1.2 + (0.1)(1.9) = 1.39 \\ u_2 &= u_1 + (0.1)[-x_1u_1 - y_1] = 1.9 + (0.1)[-(0.1)(1.9) - 1.2] = 1.761. \end{aligned}$$

In other words, $y(0.2) \approx 1.39$ and $y'(0.2) \approx 1.761$. ≡

With the aid of the graphing feature of a numerical solver, in Figure 9.4.1(a) we compare the solution curve of (5) generated by Euler's method ($h = 0.1$) on the interval [0, 3] with the solution curve generated by the RK4 method ($h = 0.1$). From Figure 9.4.1(b) it appears that the solution $y(x)$ of (4) has the property that $y(x) \to 0$ and $x \to \infty$.

If desired, we can use the method of Section 6.2 to obtain two power series solutions of the differential equation in (5). But unless this method reveals that the DE possesses an elementary solution, we will still only be able to approximate $y(0.2)$ using a partial sum. Reinspection of the infinite series solutions of Airy's differential equation $y'' + xy = 0$, given on page 242, does not reveal the oscillatory behavior of the solutions $y_1(x)$ and $y_2(x)$ exhibited in the graphs in Figure 6.2.2. Those graphs were obtained from a numerical solver using the RK4 method with a step size of $h = 0.1$.

≡ Systems Reduced to First-Order Systems Using a procedure similar to that just discussed for second-order equations, we can often reduce a system of higher-order differential equations to a system of first-order equations by first solving for the highest-order derivative of each dependent variable and then making appropriate substitutions for the lower-order derivatives.

EXAMPLE 2 A System Rewritten as a First-Order System

Write
$$x'' - x' + 5x + 2y'' = e^t$$
$$-2x + y'' + 2y = 3t^2$$
as a system of first-order differential equations.

SOLUTION Write the system as
$$x'' + 2y'' = e^t - 5x + x'$$
$$y'' = 3t^2 + 2x - 2y$$
and then eliminate y'' by multiplying the second equation by 2 and subtracting. This gives
$$x'' = -9x + 4y + x' + e^t - 6t^2.$$
Since the second equation of the system already expresses the highest-order derivative of y in terms of the remaining functions, we are now in a position to introduce new variables. If we let $x' = u$ and $y' = v$, the expressions for x'' and y'' become, respectively,
$$u' = x'' = -9x + 4y + u + e^t - 6t^2$$
$$v' = y'' = 2x - 2y + 3t^2.$$
The original system can then be written in the form
$$x' = u$$
$$y' = v$$
$$u' = -9x + 4y + u + e^t - 6t^2$$
$$v' = 2x - 2y + 3t^2.$$
≡

It might not always be possible to carry out the reductions illustrated in Example 2.

≡ Numerical Solution of a System The solution of a system of the form
$$\frac{dx_1}{dt} = g_1(t, x_1, x_2, \ldots, x_n)$$
$$\frac{dx_2}{dt} = g_2(t, x_1, x_2, \ldots, x_n)$$
$$\vdots \qquad \qquad \vdots$$
$$\frac{dx_n}{dt} = g_n(t, x_1, x_2, \ldots, x_n)$$

can be approximated by a version of Euler's, the Runge-Kutta, or the Adams-Bashforth-Moulton method adapted to the system. For instance, the RK4 method applied to the system

$$\begin{aligned} x' &= f(t, x, y) \\ y' &= g(t, x, y) \\ x(t_0) = x_0, &\quad y(t_0) = y_0, \end{aligned} \tag{6}$$

looks like this:

$$\begin{aligned} x_{n+1} &= x_n + \frac{h}{6}(m_1 + 2m_2 + 2m_3 + m_4) \\ y_{n+1} &= y_n + \frac{h}{6}(k_1 + 2k_2 + 2k_3 + k_4), \end{aligned} \tag{7}$$

where

$$\begin{aligned} m_1 &= f(t_n, x_n, y_n) & k_1 &= g(t_n, x_n, y_n) \\ m_2 &= f\left(t_n + \tfrac{1}{2}h, x_n + \tfrac{1}{2}hm_1, y_n + \tfrac{1}{2}hk_1\right) & k_2 &= g\left(t_n + \tfrac{1}{2}h, x_n + \tfrac{1}{2}hm_1, y_n + \tfrac{1}{2}hk_1\right) \\ m_3 &= f\left(t_n + \tfrac{1}{2}h, x_n + \tfrac{1}{2}hm_2, y_n + \tfrac{1}{2}hk_2\right) & k_3 &= g\left(t_n + \tfrac{1}{2}h, x_n + \tfrac{1}{2}hm_2, y_n + \tfrac{1}{2}hk_2\right) \\ m_4 &= f(t_n + h, x_n + hm_3, y_n + hk_3) & k_4 &= g(t_n + h, x_n + hm_3, y_n + hk_3). \end{aligned} \tag{8}$$

EXAMPLE 3 RK4 Method

Consider the initial-value problem

$$\begin{aligned} x' &= 2x + 4y \\ y' &= -x + 6y \\ x(0) = -1, &\quad y(0) = 6. \end{aligned}$$

Use the RK4 method to approximate $x(0.6)$ and $y(0.6)$. Compare the results for $h = 0.2$ and $h = 0.1$.

SOLUTION We illustrate the computations of x_1 and y_1 with step size $h = 0.2$. With the identifications $f(t, x, y) = 2x + 4y$, $g(t, x, y) = -x + 6y$, $t_0 = 0$, $x_0 = -1$, and $y_0 = 6$ we see from (8) that

$$\begin{aligned} m_1 &= f(t_0, x_0, y_0) = f(0, -1, 6) = 2(-1) + 4(6) = 22 \\ k_1 &= g(t_0, x_0, y_0) = g(0, -1, 6) = -1(-1) + 6(6) = 37 \\ m_2 &= f\left(t_0 + \tfrac{1}{2}h, x_0 + \tfrac{1}{2}hm_1, y_0 + \tfrac{1}{2}hk_1\right) = f(0.1, 1.2, 9.7) = 41.2 \\ k_2 &= g\left(t_0 + \tfrac{1}{2}h, x_0 + \tfrac{1}{2}hm_1, y_0 + \tfrac{1}{2}hk_1\right) = g(0.1, 1.2, 9.7) = 57 \\ m_3 &= f\left(t_0 + \tfrac{1}{2}h, x_0 + \tfrac{1}{2}hm_2, y_0 + \tfrac{1}{2}hk_2\right) = f(0.1, 3.12, 11.7) = 53.04 \\ k_3 &= g\left(t_0 + \tfrac{1}{2}h, x_0 + \tfrac{1}{2}hm_2, y_0 + \tfrac{1}{2}hk_2\right) = g(0.1, 3.12, 11.7) = 67.08 \\ m_4 &= f(t_0 + h, x_0 + hm_3, y_0 + hk_3) = f(0.2, 9.608, 19.416) = 96.88 \\ k_4 &= g(t_0 + h, x_0 + hm_3, y_0 + hk_3) = g(0.2, 9.608, 19.416) = 106.888. \end{aligned}$$

TABLE 9.4.1 $h = 0.2$

t_n	x_n	y_n
0.00	−1.0000	6.0000
0.20	9.2453	19.0683
0.40	46.0327	55.1203
0.60	158.9430	150.8192

TABLE 9.4.2 $h = 0.1$

t_n	x_n	y_n
0.00	−1.0000	6.0000
0.10	2.3840	10.8883
0.20	9.3379	19.1332
0.30	22.5541	32.8539
0.40	46.5103	55.4420
0.50	88.5729	93.3006
0.60	160.7563	152.0025

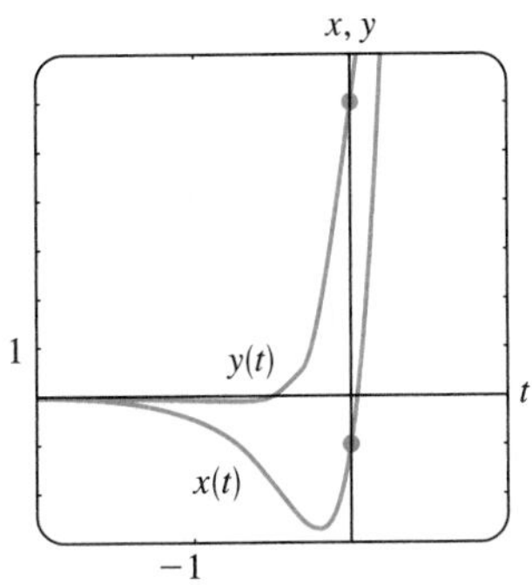

FIGURE 9.4.2 Numerical solution curves for IVP in Example 3

Therefore from (7) we get

$$x_1 = x_0 + \frac{0.2}{6}(m_1 + 2m_2 + 2m_3 + m_4)$$
$$= -1 + \frac{0.2}{6}(22 + 2(41.2) + 2(53.04) + 96.88) = 9.2453$$
$$y_1 = y_0 + \frac{0.2}{6}(k_1 + 2k_2 + 2k_3 + k_4)$$
$$= 6 + \frac{0.2}{6}(37 + 2(57) + 2(67.08) + 106.888) = 19.0683,$$

where, as usual, the computed values of x_1 and y_1 are rounded to four decimal places. These numbers give us the approximation $x_1 \approx x(0.2)$ and $y_1 \approx y(0.2)$. The subsequent values, obtained with the aid of a computer, are summarized in Tables 9.4.1 and 9.4.2. ≡

You should verify that the solution of the initial-value problem in Example 3 is given by $x(t) = (26t - 1)e^{4t}$, $y(t) = (13t + 6)e^{4t}$. From these equations we see that the actual values $x(0.6) = 160.9384$ and $y(0.6) = 152.1198$ compare favorably with the entries in the last line of Table 9.4.2. The graph of the solution in a neighborhood of $t = 0$ is shown in Figure 9.4.2; the graph was obtained from a numerical solver using the RK4 method with $h = 0.1$.

In conclusion, we state Euler's method for the general system (6):

$$x_{n+1} = x_n + hf(t_n, x_n, y_n)$$
$$y_{n+1} = y_n + hg(t_n, x_n, y_n).$$

EXERCISES 9.4

Answers to selected odd-numbered problems begin on page ANS-17.

1. Use Euler's method to approximate $y(0.2)$, where $y(x)$ is the solution of the initial-value problem
$$y'' - 4y' + 4y = 0, \quad y(0) = -2, \quad y'(0) = 1.$$
Use $h = 0.1$. Find the analytic solution of the problem, and compare the actual value of $y(0.2)$ with y_2.

2. Use Euler's method to approximate $y(1.2)$, where $y(x)$ is the solution of the initial-value problem
$$x^2y'' - 2xy' + 2y = 0, \quad y(1) = 4, \quad y'(1) = 9,$$
where $x > 0$. Use $h = 0.1$. Find the analytic solution of the problem, and compare the actual value of $y(1.2)$ with y_2.

In Problems 3 and 4 repeat the indicated problem using the RK4 method. First use $h = 0.2$ and then use $h = 0.1$.

3. Problem 1
4. Problem 2

5. Use the RK4 method to approximate $y(0.2)$, where $y(x)$ is the solution of the initial-value problem
$$y'' - 2y' + 2y = e^t \cos t, \quad y(0) = 1, \quad y'(0) = 2.$$
First use $h = 0.2$ and then use $h = 0.1$.

6. When $E = 100$ V, $R = 10\ \Omega$, and $L = 1$ h, the system of differential equations for the currents $i_1(t)$ and $i_3(t)$ in the electrical network given in Figure 9.4.3 is
$$\frac{di_1}{dt} = -20i_1 + 10i_3 + 100$$
$$\frac{di_3}{dt} = 10i_1 - 20i_3,$$
where $i_1(0) = 0$ and $i_3(0) = 0$. Use the RK4 method to approximate $i_1(t)$ and $i_3(t)$ at $t = 0.1, 0.2, 0.3, 0.4$, and 0.5. Use $h = 0.1$. Use a numerical solver to graph the solution for $0 \le t \le 5$. Use the graphs to predict the behavior of $i_1(t)$ and $i_3(t)$ as $t \to \infty$.

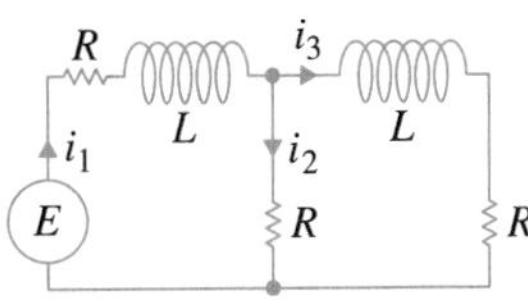

FIGURE 9.4.3 Network in Problem 6

In Problems 7–12 use the Runge-Kutta method to approximate $x(0.2)$ and $y(0.2)$. First use $h = 0.2$ and then use $h = 0.1$. Use a numerical solver and $h = 0.1$ to graph the solution in a neighborhood of $t = 0$.

7. $x' = 2x - y$
$y' = x$
$x(0) = 6, \quad y(0) = 2$

8. $x' = x + 2y$
$y' = 4x + 3y$
$x(0) = 1, \quad y(0) = 1$

9. $x' = -y + t$
$y' = x - t$
$x(0) = -3, \quad y(0) = 5$

10. $x' = 6x + y + 6t$
$y' = 4x + 3y - 10t + 4$
$x(0) = 0.5, \quad y(0) = 0.2$

11. $x' + 4x - y' = 7t$
$x' + y' - 2y = 3t$
$x(0) = 1, \quad y(0) = -2$

12. $x' + y' = 4t$
$-x' + y' + y = 6t^2 + 10$
$x(0) = 3, \quad y(0) = -1$

9.5 SECOND-ORDER BOUNDARY-VALUE PROBLEMS

REVIEW MATERIAL

- Section 4.1 (page 118)
- Exercises 4.3 (Problems 37–40)
- Exercises 4.4 (Problems 37–40)
- Section 5.2

INTRODUCTION We just saw in Section 9.4 how to approximate the solution of a *second-order initial-value problem*

$$y'' = f(x, y, y'), \quad y(x_0) = y_0, \quad y'(x_0) = u_0.$$

In this section we are going to examine two methods for approximating a solution of a *second-order boundary-value problem*

$$y'' = f(x, y, y'), \quad y(a) = \alpha, \quad y(b) = \beta.$$

Unlike the procedures that are used with second-order initial-value problems, the methods of second-order boundary-value problems do not require writing the second-order DE as a system of first-order DEs.

Finite Difference Approximations The Taylor series expansion, centered at a point a, of a function $y(x)$ is

$$y(x) = y(a) + y'(a)\frac{x - a}{1!} + y''(a)\frac{(x - a)^2}{2!} + y'''(a)\frac{(x - a)^3}{3!} + \cdots.$$

If we set $h = x - a$, then the preceding line is the same as

$$y(x) = y(a) + y'(a)\frac{h}{1!} + y''(a)\frac{h^2}{2!} + y'''(a)\frac{h^3}{3!} + \cdots.$$

For the subsequent discussion it is convenient then to rewrite this last expression in two alternative forms:

$$y(x + h) = y(x) + y'(x)h + y''(x)\frac{h^2}{2} + y'''(x)\frac{h^3}{6} + \cdots \tag{1}$$

and

$$y(x - h) = y(x) - y'(x)h + y''(x)\frac{h^2}{2} - y'''(x)\frac{h^3}{6} + \cdots. \tag{2}$$

If h is small, we can ignore terms involving $h^4, h^5, \ldots$ since these values are negligible. Indeed, if we ignore all terms involving h^2 and higher, then solving (1) and (2), in turn, for $y'(x)$ yields the following approximations for the first derivative:

$$y'(x) \approx \frac{1}{h}[y(x + h) - y(x)] \tag{3}$$

$$y'(x) \approx \frac{1}{h}[y(x) - y(x - h)]. \tag{4}$$

Subtracting (1) and (2) also gives

$$y'(x) \approx \frac{1}{2h}[y(x+h) - y(x-h)]. \tag{5}$$

On the other hand, if we ignore terms involving h^3 and higher, then by adding (1) and (2), we obtain an approximation for the second derivative $y''(x)$:

$$y''(x) \approx \frac{1}{h^2}[y(x+h) - 2y(x) + y(x-h)]. \tag{6}$$

The right-hand sides of (3), (4), (5), and (6) are called **difference quotients.** The expressions

$$y(x+h) - y(x), \quad y(x) - y(x-h), \quad y(x+h) - y(x-h),$$

and

$$y(x+h) - 2y(x) + y(x-h)$$

are called **finite differences.** Specifically, $y(x+h) - y(x)$ is called a **forward difference,** $y(x) - y(x-h)$ is a **backward difference,** and both $y(x+h) - y(x-h)$ and $y(x+h) - 2y(x) + y(x-h)$ are called **central differences.** The results given in (5) and (6) are referred to as **central difference approximations** for the derivatives y' and y''.

Finite Difference Method Consider now a linear second-order boundary-value problem

$$y'' + P(x)y' + Q(x)y = f(x), \qquad y(a) = \alpha, \quad y(b) = \beta. \tag{7}$$

Suppose $a = x_0 < x_1 < x_2 < \cdots < x_{n-1} < x_n = b$ represents a regular partition of the interval $[a, b]$, that is, $x_i = a + ih$, where $i = 0, 1, 2, \ldots, n$ and $h = (b - a)/n$. The points

$$x_1 = a + h, \qquad x_2 = a + 2h, \ldots, \qquad x_{n-1} = a + (n-1)h$$

are called **interior mesh points** of the interval $[a, b]$. If we let

$$y_i = y(x_i), \qquad P_i = P(x_i), \qquad Q_i = Q(x_i), \qquad \text{and} \qquad f_i = f(x_i)$$

and if y'' and y' in (7) are replaced by the central difference approximations (5) and (6), we get

$$\frac{y_{i+1} - 2y_i + y_{i-1}}{h^2} + P_i\frac{y_{i+1} - y_{i-1}}{2h} + Q_i y_i = f_i$$

or, after simplifying,

$$\left(1 + \frac{h}{2}P_i\right)y_{i+1} + (-2 + h^2Q_i)y_i + \left(1 - \frac{h}{2}P_i\right)y_{i-1} = h^2 f_i. \tag{8}$$

The last equation, known as a **finite difference equation,** is an approximation to the differential equation. It enables us to approximate the solution $y(x)$ of (7) at the interior mesh points $x_1, x_2, \ldots, x_{n-1}$ of the interval $[a, b]$. By letting i take on the values $1, 2, \ldots, n-1$ in (8), we obtain $n - 1$ equations in the $n - 1$ unknowns $y_1, y_2, \ldots, y_{n-1}$. Bear in mind that we know y_0 and y_n, since these are the prescribed boundary conditions $y_0 = y(x_0) = y(a) = \alpha$ and $y_n = y(x_n) = y(b) = \beta$.

In Example 1 we consider a boundary-value problem for which we can compare the approximate values that we find with the actual values of an explicit solution.

EXAMPLE 1 Using the Finite Difference Method

Use the difference equation (8) with $n = 4$ to approximate the solution of the boundary-value problem $y'' - 4y = 0$, $y(0) = 0$, $y(1) = 5$.

SOLUTION To use (8), we identify $P(x) = 0$, $Q(x) = -4$, $f(x) = 0$, and $h = (1 - 0)/4 = \frac{1}{4}$. Hence the difference equation is

$$y_{i+1} - 2.25y_i + y_{i-1} = 0. \tag{9}$$

Now the interior points are $x_1 = 0 + \frac{1}{4}, x_2 = 0 + \frac{2}{4}, x_3 = 0 + \frac{3}{4}$, so for $i = 1, 2,$ and 3, (9) yields the following system for the corresponding y_1, y_2, and y_3:

$$\begin{aligned} y_2 - 2.25y_1 + y_0 &= 0 \\ y_3 - 2.25y_2 + y_1 &= 0 \\ y_4 - 2.25y_3 + y_2 &= 0. \end{aligned}$$

With the boundary conditions $y_0 = 0$ and $y_4 = 5$ the foregoing system becomes

$$\begin{aligned} -2.25y_1 + y_2 \qquad\qquad &= 0 \\ y_1 - 2.25y_2 + y_3 &= 0 \\ y_2 - 2.25y_3 &= -5. \end{aligned}$$

Solving the system gives $y_1 = 0.7256$, $y_2 = 1.6327$, and $y_3 = 2.9479$.

Now the general solution of the given differential equation is $y = c_1 \cosh 2x + c_2 \sinh 2x$. The condition $y(0) = 0$ implies that $c_1 = 0$. The other boundary condition gives c_2. In this way we see that a solution of the boundary-value problem is $y(x) = (5 \sinh 2x)/\sinh 2$. Thus the actual values (rounded to four decimal places) of this solution at the interior points are as follows: $y(0.25) = 0.7184$, $y(0.5) = 1.6201$, and $y(0.75) = 2.9354$. ≡

The accuracy of the approximations in Example 1 can be improved by using a smaller value of h. Of course, the trade-off here is that a smaller value of h necessitates solving a larger system of equations. It is left as an exercise to show that with $h = \frac{1}{8}$, approximations to $y(0.25)$, $y(0.5)$, and $y(0.75)$ are 0.7202, 1.6233, and 2.9386, respectively. See Problem 11 in Exercises 9.5.

EXAMPLE 2 Using the Finite Difference Method

Use the difference equation (8) with $n = 10$ to approximate the solution of

$$y'' + 3y' + 2y = 4x^2, \qquad y(1) = 1, \quad y(2) = 6.$$

SOLUTION In this case we identify $P(x) = 3$, $Q(x) = 2$, $f(x) = 4x^2$, and $h = (2 - 1)/10 = 0.1$, and so (8) becomes

$$1.15y_{i+1} - 1.98y_i + 0.85y_{i-1} = 0.04x_i^2. \tag{10}$$

Now the interior points are $x_1 = 1.1$, $x_2 = 1.2$, $x_3 = 1.3$, $x_4 = 1.4$, $x_5 = 1.5$, $x_6 = 1.6$, $x_7 = 1.7$, $x_8 = 1.8$, and $x_9 = 1.9$. For $i = 1, 2, \ldots, 9$ and $y_0 = 1$, $y_{10} = 6$, (10) gives a system of nine equations and nine unknowns:

$$\begin{aligned} 1.15y_2 - 1.98y_1 \qquad\qquad &= -0.8016 \\ 1.15y_3 - 1.98y_2 + 0.85y_1 &= 0.0576 \\ 1.15y_4 - 1.98y_3 + 0.85y_2 &= 0.0676 \\ 1.15y_5 - 1.98y_4 + 0.85y_3 &= 0.0784 \end{aligned}$$

$$1.15y_6 - 1.98y_5 + 0.85y_4 = 0.0900$$
$$1.15y_7 - 1.98y_6 + 0.85y_5 = 0.1024$$
$$1.15y_8 - 1.98y_7 + 0.85y_6 = 0.1156$$
$$1.15y_9 - 1.98y_8 + 0.85y_7 = 0.1296$$
$$- 1.98y_9 + 0.85y_8 = -6.7556.$$

We can solve this large system using Gaussian elimination or, with relative ease, by means of a computer algebra system. The result is found to be $y_1 = 2.4047$, $y_2 = 3.4432$, $y_3 = 4.2010$, $y_4 = 4.7469$, $y_5 = 5.1359$, $y_6 = 5.4124$, $y_7 = 5.6117$, $y_8 = 5.7620$, and $y_9 = 5.8855$. ≡

≡ Shooting Method Another way of approximating a solution of a boundary-value problem $y'' = f(x, y, y')$, $y(a) = \alpha$, $y(b) = \beta$ is called the **shooting method.** The starting point in this method is the replacement of the second-order boundary-value problem by a second-order initial-value problem

$$y'' = f(x, y, y'), \quad y(a) = \alpha, \quad y'(a) = m_1. \tag{11}$$

The number m_1 in (11) is simply a guess for the unknown slope of the solution curve at the known point $(a, y(a))$. We then apply one of the step-by-step numerical techniques to the second-order equation in (11) to find an approximation β_1 for the value of $y(b)$. If β_1 agrees with the given value $y(b) = \beta$ to some preassigned tolerance, we stop; otherwise, the calculations are repeated, starting with a different guess $y'(a) = m_2$ to obtain a second approximation β_2 for $y(b)$. This method can be continued in a trial-and-error manner, or the subsequent slopes $m_3, m_4, \ldots$ can be adjusted in some systematic way; linear interpolation is particularly successful when the differential equation in (11) is linear. The procedure is analogous to shooting (the "aim" is the choice of the initial slope) at a target until the bull's-eye $y(b)$ is hit. See Problem 14 in Exercises 9.5.

Of course, underlying the use of these numerical methods is the assumption, which we know is not always warranted, that a solution of the boundary-value problem exists.

REMARKS

The approximation method using finite differences can be extended to boundary-value problems in which the first derivative is specified at a boundary—for example, a problem such as $y'' = f(x, y, y')$, $y'(a) = \alpha$, $y(b) = \beta$. See Problem 13 in Exercises 9.5.

EXERCISES 9.5

Answers to selected odd-numbered problems begin on page ANS-17.

In Problems 1–10 use the finite difference method and the indicated value of n to approximate the solution of the given boundary-value problem.

1. $y'' + 9y = 0, \quad y(0) = 4, y(2) = 1; \quad n = 4$
2. $y'' - y = x^2, \quad y(0) = 0, y(1) = 0; \quad n = 4$
3. $y'' + 2y' + y = 5x, \quad y(0) = 0, y(1) = 0; \quad n = 5$
4. $y'' - 10y' + 25y = 1, \quad y(0) = 1, y(1) = 0; \quad n = 5$
5. $y'' - 4y' + 4y = (x + 1)e^{2x},$ $y(0) = 3, y(1) = 0; \quad n = 6$
6. $y'' + 5y' = 4\sqrt{x}, \quad y(1) = 1, \quad y(2) = -1; \quad n = 6$
7. $x^2y'' + 3xy' + 3y = 0, \quad y(1) = 5, y(2) = 0; \quad n = 8$
8. $x^2y'' - xy' + y = \ln x, \quad y(1) = 0, y(2) = -2; \quad n = 8$
9. $y'' + (1 - x)y' + xy = x, \quad y(0) = 0, y(1) = 2; \quad n = 10$
10. $y'' + xy' + y = x, \quad y(0) = 1, y(1) = 0; \quad n = 10$
11. Rework Example 1 using $n = 8$.
12. The electrostatic potential u between two concentric spheres of radius $r = 1$ and $r = 4$ is determined from

$$\frac{d^2u}{dr^2} + \frac{2}{r}\frac{du}{dr} = 0, \quad u(1) = 50, \quad u(4) = 100.$$

Use the method of this section with $n = 6$ to approximate the solution of this boundary-value problem.

13. Consider the boundary-value problem $y'' + xy = 0$, $y'(0) = 1, y(1) = -1$.

(a) Find the difference equation corresponding to the differential equation. Show that for $i = 0, 1, 2, \ldots, n - 1$ the difference equation yields n equations in $n + 1$ unknows $y_{-1}, y_0, y_1, y_2, \ldots, y_{n-1}$. Here y_{-1} and y_0 are unknowns, since y_{-1} represents an approximation to y at the exterior point $x = -h$ and y_0 is not specified at $x = 0$.

(b) Use the central difference approximation (5) to show that $y_1 - y_{-1} = 2h$. Use this equation to eliminate y_{-1} from the system in part (a).

(c) Use $n = 5$ and the system of equations found in parts (a) and (b) to approximate the solution of the original boundary-value problem.

Computer Lab Assignments

14. Consider the boundary-value problem $y'' = y' - \sin(xy)$, $y(0) = 1, y(1) = 1.5$. Use the shooting method to approximate the solution of this problem. (The approximation can be obtained using a numerical technique—say, the RK4 method with $h = 0.1$; or, even better, if you have access to a CAS such as *Mathematica* or *Maple,* the **NDSolve** function can be used.)

CHAPTER 9 IN REVIEW

Answers to selected odd-numbered problems begin on page ANS-17.

In Problems 1–4 construct a table comparing the indicated values of $y(x)$ using Euler's method, the improved Euler's method, and the RK4 method. Compute to four rounded decimal places. First use $h = 0.1$ and then use $h = 0.05$.

1. $y' = 2 \ln xy, \quad y(1) = 2;$
$y(1.1), y(1.2), y(1.3), y(1.4), y(1.5)$

2. $y' = \sin x^2 + \cos y^2, \quad y(0) = 0;$
$y(0.1), y(0.2), y(0.3), y(0.4), y(0.5)$

3. $y' = \sqrt{x + y}, \quad y(0.5) = 0.5;$
$y(0.6), y(0.7), y(0.8), y(0.9), y(1.0)$

4. $y' = xy + y^2, \quad y(1) = 1;$
$y(1.1), y(1.2), y(1.3), y(1.4), y(1.5)$

5. Use Euler's method to approximate $y(0.2)$, where $y(x)$ is the solution of the initial-value problem $y'' - (2x + 1)y = 1, \ y(0) = 3, \ y'(0) = 1$. First use one step with $h = 0.2$ and then repeat the calculations using two steps with $h = 0.1$.

6. Use the Adams-Bashforth-Moulton method to approximate $y(0.4)$, where $y(x)$ is the solution of the initial-value problem $y' = 4x - 2y, \ y(0) = 2$. Use $h = 0.1$ and the RK4 method to compute y_1, y_2, and y_3.

7. Use Euler's method with $h = 0.1$ to approximate $x(0.2)$ and $y(0.2)$, where $x(t)$, $y(t)$ is the solution of the initial-value problem

$$\begin{aligned} x' &= x + y \\ y' &= x - y \end{aligned}$$
$$x(0) = 1, \qquad y(0) = 2.$$

8. Use the finite difference method with $n = 10$ to approximate the solution of the boundary-value problem $y'' + 6.55(1 + x)y = 1, y(0) = 0, y(1) = 0$.

Appendix I
Gamma Function

Euler's integral definition of the **gamma function** is

$$\Gamma(x) = \int_0^\infty t^{x-1}e^{-t}\,dt. \tag{1}$$

Convergence of the integral requires that $x - 1 > -1$ or $x > 0$. The recurrence relation

$$\Gamma(x + 1) = x\Gamma(x), \tag{2}$$

which we saw in Section 6.4, can be obtained from (1) with integration by parts. Now when $x = 1$, $\Gamma(1) = \int_0^\infty e^{-t}\,dt = 1$, and thus (2) gives

$$\Gamma(2) = 1\Gamma(1) = 1$$
$$\Gamma(3) = 2\Gamma(2) = 2 \cdot 1$$
$$\Gamma(4) = 3\Gamma(3) = 3 \cdot 2 \cdot 1,$$

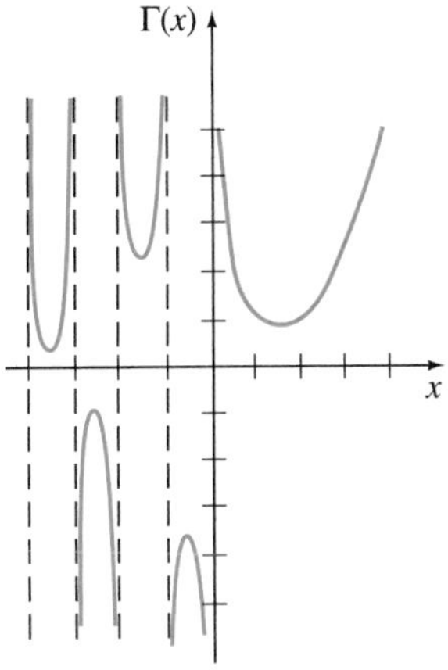

and so on. In this manner, it is seen that when n is a positive integer, $\Gamma(n + 1) = n!$. For this reason, the gamma function is often called the **generalized factorial function.**

Although the integral form (1) does not converge for $x < 0$, it can be shown by means of alternative definitions that the gamma function is defined for all real and complex numbers *except* $x = -n$, $n = 0, 1, 2, \ldots$. As a consequence, (2) is actually valid for $x \neq -n$. The graph of $\Gamma(x)$, considered as a function of a real variable x, is as given in Figure I.1. Observe that the nonpositive integers correspond to vertical asymptotes of the graph.

FIGURE I.1 Graph of $\Gamma(x)$ for x neither 0 nor a negative integer

In Problems 31 and 32 of Exercises 6.4 we used the fact that $\Gamma\left(\frac{1}{2}\right) = \sqrt{\pi}$. This result can be derived from (1) by setting $x = \frac{1}{2}$:

$$\Gamma\left(\tfrac{1}{2}\right) = \int_0^\infty t^{-1/2}e^{-t}\,dt. \tag{3}$$

When we let $t = u^2$, (3) can be written as $\Gamma\left(\frac{1}{2}\right) = 2\int_0^\infty e^{-u^2}\,du$. But $\int_0^\infty e^{-u^2}\,du = \int_0^\infty e^{-v^2}\,dv$, so

$$\left[\Gamma\left(\tfrac{1}{2}\right)\right]^2 = \left(2\int_0^\infty e^{-u^2}\,du\right)\left(2\int_0^\infty e^{-v^2}\,dv\right) = 4\int_0^\infty\int_0^\infty e^{-(u^2+v^2)}\,du\,dv.$$

Switching to polar coordinates $u = r\cos\theta$, $v = r\sin\theta$ enables us to evaluate the double integral:

$$4\int_0^\infty\int_0^\infty e^{-(u^2+v^2)}\,du\,dv = 4\int_0^{\pi/2}\int_0^\infty e^{-r^2}\,r\,dr\,d\theta = \pi.$$

Hence $$\left[\Gamma\left(\tfrac{1}{2}\right)\right]^2 = \pi \quad\text{or}\quad \Gamma\left(\tfrac{1}{2}\right) = \sqrt{\pi}. \tag{4}$$

EXAMPLE 1 Value of $\Gamma\left(-\frac{1}{2}\right)$

Evaluate $\Gamma\left(-\frac{1}{2}\right)$.

SOLUTION In view of (2) and (4), it follows that, with $x = -\frac{1}{2}$,

$$\Gamma\left(\tfrac{1}{2}\right) = -\tfrac{1}{2}\Gamma\left(-\tfrac{1}{2}\right).$$

Therefore $$\Gamma\left(-\tfrac{1}{2}\right) = -2\Gamma\left(\tfrac{1}{2}\right) = -2\sqrt{\pi}.$$ ≡

EXERCISES FOR APPENDIX I

Answers to selected odd-numbered problems begin on page ANS-31.

1. Evaluate.

   (a) $\Gamma(5)$ **(b)** $\Gamma(7)$

   (c) $\Gamma\left(-\frac{3}{2}\right)$ **(d)** $\Gamma\left(-\frac{5}{2}\right)$

2. Use (1) and the fact that $\Gamma\left(\frac{6}{5}\right) = 0.92$ to evaluate $\int_0^\infty x^5 e^{-x^5}\,dx$. [*Hint*: Let $t = x^5$.]

3. Use (1) and the fact that $\Gamma\left(\frac{5}{3}\right) = 0.89$ to evaluate $\int_0^\infty x^4 e^{-x^3}\,dx$.

4. Evaluate $\int_0^1 x^3\left(\ln\frac{1}{x}\right)^3 dx$. [*Hint*: Let $t = -\ln x$.]

5. Use the fact that $\Gamma(x) > \int_0^1 t^{x-1}e^{-t}\,dt$ to show that $\Gamma(x)$ is unbounded as $x \to 0^+$.

6. Use (1) to derive (2) for $x > 0$.

7. A definition of the gamma function due to Carl Friedrich Gauss that is valid for all real numbers, except $x = 0, -1, -2, \ldots$, is given by

$$\Gamma(x) = \lim_{n\to\infty} \frac{n!n^x}{x(x+1)(x+2)\cdots(x+n)}.$$

   Use this definition to show that $\Gamma(x+1) = x\Gamma(x)$.

Appendix II

Matrices

II.1 BASIC DEFINITIONS AND THEORY

DEFINITION II.1 Matrix

A **matrix A** is any rectangular array of numbers or functions:

$$\mathbf{A} = \begin{pmatrix} a_{11} & a_{12} & \cdots & a_{1n} \\ a_{21} & a_{22} & \cdots & a_{2n} \\ \vdots & & & \vdots \\ a_{m1} & a_{m2} & \cdots & a_{mn} \end{pmatrix}. \tag{1}$$

If a matrix has m rows and n columns, we say that its **size** is m by n (written $m \times n$). An $n \times n$ matrix is called a **square matrix** of order n.

The entry in the ith row and jth column of an $m \times n$ matrix $\mathbf{A}$ is written a_{ij}. An $m \times n$ matrix $\mathbf{A}$ is then abbreviated as $\mathbf{A} = (a_{ij})_{m\times n}$ or simply $\mathbf{A} = (a_{ij})$. A 1×1 matrix is sximply one constant or function.

DEFINITION II.2 Equality of Matrices

Two $m \times n$ matrices $\mathbf{A}$ and $\mathbf{B}$ are **equal** if $a_{ij} = b_{ij}$ for each i and j.

DEFINITION II.3 Column Matrix

A **column matrix X** is any matrix having n rows and one column:

$$\mathbf{X} = \begin{pmatrix} b_{11} \\ b_{21} \\ \vdots \\ b_{n1} \end{pmatrix} = (b_{i1})_{n\times 1}.$$

A column matrix is also called a **column vector** or simply a **vector.**

DEFINITION II.4 Multiples of Matrices

A **multiple** of a matrix $\mathbf{A}$ is defined to be

$$k\mathbf{A} = \begin{pmatrix} ka_{11} & ka_{12} & \cdots & ka_{1n} \\ ka_{21} & ka_{22} & \cdots & ka_{2n} \\ \vdots & & & \vdots \\ ka_{m1} & ka_{m2} & \cdots & ka_{mn} \end{pmatrix} = (ka_{ij})_{m\times n},$$

where k is a constant or a function.

EXAMPLE 1 Multiples of Matrices

(a) $5\begin{pmatrix} 2 & -3 \\ 4 & -1 \\ \frac{1}{5} & 6 \end{pmatrix} = \begin{pmatrix} 10 & -15 \\ 20 & -5 \\ 1 & 30 \end{pmatrix}$ **(b)** $e^t \begin{pmatrix} 1 \\ -2 \\ 4 \end{pmatrix} = \begin{pmatrix} e^t \\ -2e^t \\ 4e^t \end{pmatrix}$

We note in passing that, for any matrix **A**, the product $k\mathbf{A}$ is the same as $\mathbf{A}k$. For example,

$$e^{-3t}\begin{pmatrix} 2 \\ 5 \end{pmatrix} = \begin{pmatrix} 2e^{-3t} \\ 5e^{-3t} \end{pmatrix} = \begin{pmatrix} 2 \\ 5 \end{pmatrix} e^{-3t}.$$

DEFINITION II.5 Addition of Matrices

The **sum** of two $m \times n$ matrices **A** and **B** is defined to be the matrix

$$\mathbf{A} + \mathbf{B} = (a_{ij} + b_{ij})_{m \times n}.$$

In other words, when adding two matrices of the same size, we add the corresponding entries.

EXAMPLE 2 Matrix Addition

The sum of $\mathbf{A} = \begin{pmatrix} 2 & -1 & 3 \\ 0 & 4 & 6 \\ -6 & 10 & -5 \end{pmatrix}$ and $\mathbf{B} = \begin{pmatrix} 4 & 7 & -8 \\ 9 & 3 & 5 \\ 1 & -1 & 2 \end{pmatrix}$ is

$$\mathbf{A} + \mathbf{B} = \begin{pmatrix} 2+4 & -1+7 & 3+(-8) \\ 0+9 & 4+3 & 6+5 \\ -6+1 & 10+(-1) & -5+2 \end{pmatrix} = \begin{pmatrix} 6 & 6 & -5 \\ 9 & 7 & 11 \\ -5 & 9 & -3 \end{pmatrix}.$$

EXAMPLE 3 A Matrix Written as a Sum of Column Matrices

The single matrix $\begin{pmatrix} 3t^2 - 2e^t \\ t^2 + 7t \\ 5t \end{pmatrix}$ can be written as the sum of three column vectors:

$$\begin{pmatrix} 3t^2 - 2e^t \\ t^2 + 7t \\ 5t \end{pmatrix} = \begin{pmatrix} 3t^2 \\ t^2 \\ 0 \end{pmatrix} + \begin{pmatrix} 0 \\ 7t \\ 5t \end{pmatrix} + \begin{pmatrix} -2e^t \\ 0 \\ 0 \end{pmatrix} = \begin{pmatrix} 3 \\ 1 \\ 0 \end{pmatrix} t^2 + \begin{pmatrix} 0 \\ 7 \\ 5 \end{pmatrix} t + \begin{pmatrix} -2 \\ 0 \\ 0 \end{pmatrix} e^t.$$

The **difference** of two $m \times n$ matrices is defined in the usual manner: $\mathbf{A} - \mathbf{B} = \mathbf{A} + (-\mathbf{B})$, where $-\mathbf{B} = (-1)\mathbf{B}$.

DEFINITION II.6 Multiplication of Matrices

Let **A** be a matrix having m rows and n columns and **B** be a matrix having n rows and p columns. We define the **product AB** to be the $m \times p$ matrix

$$\mathbf{AB} = \begin{pmatrix} a_{11} & a_{12} & \cdots & a_{1n} \\ a_{21} & a_{22} & \cdots & a_{2n} \\ \vdots & & & \vdots \\ a_{m1} & a_{m2} & \cdots & a_{mn} \end{pmatrix} \begin{pmatrix} b_{11} & b_{12} & \cdots & b_{1p} \\ b_{21} & b_{22} & \cdots & b_{2p} \\ \vdots & & & \vdots \\ b_{n1} & b_{n2} & \cdots & b_{np} \end{pmatrix}$$

$$= \begin{pmatrix} a_{11}b_{11} + a_{12}b_{21} + \cdots + a_{1n}b_{n1} & \cdots & a_{11}b_{1p} + a_{12}b_{2p} + \cdots + a_{1n}b_{np} \\ a_{21}b_{11} + a_{22}b_{21} + \cdots + a_{2n}b_{n1} & \cdots & a_{21}b_{1p} + a_{22}b_{2p} + \cdots + a_{2n}b_{np} \\ \vdots & & \vdots \\ a_{m1}b_{11} + a_{m2}b_{21} + \cdots + a_{mn}b_{n1} & \cdots & a_{m1}b_{1p} + a_{m2}b_{2p} + \cdots + a_{mn}b_{np} \end{pmatrix}$$

$$= \left(\sum_{k=1}^{n} a_{ik}b_{kj} \right)_{m \times p}.$$

Note carefully in Definition II.6 that the product $\mathbf{AB} = \mathbf{C}$ is defined only when the number of columns in the matrix **A** is the same as the number of rows in **B**. The size of the product can be determined from

$$\mathbf{A}_{m \times n}\mathbf{B}_{n \times p} = \mathbf{C}_{m \times p}.$$

Also, you might recognize that the entries in, say, the ith row of the final matrix **AB** are formed by using the component definition of the inner, or dot, product of the ith row of **A** with each of the columns of **B**.

EXAMPLE 4 Multiplication of Matrices

(a) For $\mathbf{A} = \begin{pmatrix} 4 & 7 \\ 3 & 5 \end{pmatrix}$ and $\mathbf{B} = \begin{pmatrix} 9 & -2 \\ 6 & 8 \end{pmatrix}$,

$$\mathbf{AB} = \begin{pmatrix} 4 \cdot 9 + 7 \cdot 6 & 4 \cdot (-2) + 7 \cdot 8 \\ 3 \cdot 9 + 5 \cdot 6 & 3 \cdot (-2) + 5 \cdot 8 \end{pmatrix} = \begin{pmatrix} 78 & 48 \\ 57 & 34 \end{pmatrix}.$$

(b) For $\mathbf{A} = \begin{pmatrix} 5 & 8 \\ 1 & 0 \\ 2 & 7 \end{pmatrix}$ and $\mathbf{B} = \begin{pmatrix} -4 & -3 \\ 2 & 0 \end{pmatrix}$,

$$\mathbf{AB} = \begin{pmatrix} 5 \cdot (-4) + 8 \cdot 2 & 5 \cdot (-3) + 8 \cdot 0 \\ 1 \cdot (-4) + 0 \cdot 2 & 1 \cdot (-3) + 0 \cdot 0 \\ 2 \cdot (-4) + 7 \cdot 2 & 2 \cdot (-3) + 7 \cdot 0 \end{pmatrix} = \begin{pmatrix} -4 & -15 \\ -4 & -3 \\ 6 & -6 \end{pmatrix}.$$

≡

In general, *matrix multiplication is not commutative*; that is, $\mathbf{AB} \neq \mathbf{BA}$. Observe in part (a) of Example 4 that $\mathbf{BA} = \begin{pmatrix} 30 & 53 \\ 48 & 82 \end{pmatrix}$, whereas in part (b) the product **BA** is not defined, since Definition II.6 requires that the first matrix (in this case **B**) have the same number of columns as the second matrix has rows.

We are particularly interested in the product of a square matrix and a column vector.

EXAMPLE 5 **Multiplication of Matrices**

(a) $$\begin{pmatrix} 2 & -1 & 3 \\ 0 & 4 & 5 \\ 1 & -7 & 9 \end{pmatrix}\begin{pmatrix} -3 \\ 6 \\ 4 \end{pmatrix} = \begin{pmatrix} 2\cdot(-3) + (-1)\cdot 6 + 3\cdot 4 \\ 0\cdot(-3) + \quad 4\cdot 6 + 5\cdot 4 \\ 1\cdot(-3) + (-7)\cdot 6 + 9\cdot 4 \end{pmatrix} = \begin{pmatrix} 0 \\ 44 \\ -9 \end{pmatrix}$$

(b) $$\begin{pmatrix} -4 & 2 \\ 3 & 8 \end{pmatrix}\begin{pmatrix} x \\ y \end{pmatrix} = \begin{pmatrix} -4x + 2y \\ 3x + 8y \end{pmatrix}$$ ≡

≡ Multiplicative Identity For a given positive integer n the $n \times n$ matrix

$$\mathbf{I} = \begin{pmatrix} 1 & 0 & 0 & \cdots & 0 \\ 0 & 1 & 0 & \cdots & 0 \\ \vdots & & & & \vdots \\ 0 & 0 & 0 & \cdots & 1 \end{pmatrix}$$

is called the **multiplicative identity matrix.** It follows from Definition II.6 that for any $n \times n$ matrix **A**.

$$\mathbf{AI} = \mathbf{IA} = \mathbf{A}.$$

Also, it is readily verified that, if **X** is an $n \times 1$ column matrix, then $\mathbf{IX} = \mathbf{X}$.

≡ Zero Matrix A matrix consisting of all zero entries is called a **zero matrix** and is denoted by **0**. For example,

$$\mathbf{0} = \begin{pmatrix} 0 \\ 0 \end{pmatrix}, \qquad \mathbf{0} = \begin{pmatrix} 0 & 0 \\ 0 & 0 \end{pmatrix}, \qquad \mathbf{0} = \begin{pmatrix} 0 & 0 \\ 0 & 0 \\ 0 & 0 \end{pmatrix},$$

and so on. If **A** and **0** are $m \times n$ matrices, then

$$\mathbf{A} + \mathbf{0} = \mathbf{0} + \mathbf{A} = \mathbf{A}.$$

≡ Associative Law Although we shall not prove it, matrix multiplication is **associative.** If **A** is an $m \times p$ matrix, **B** a $p \times r$ matrix, and **C** an $r \times n$ matrix, then

$$\mathbf{A}(\mathbf{BC}) = (\mathbf{AB})\mathbf{C}$$

is an $m \times n$ matrix.

≡ Distributive Law If all products are defined, multiplication is **distributive** over addition:

$$\mathbf{A}(\mathbf{B} + \mathbf{C}) = \mathbf{AB} + \mathbf{AC} \quad \text{and} \quad (\mathbf{B} + \mathbf{C})\mathbf{A} = \mathbf{BA} + \mathbf{CA}.$$

≡ Determinant of a Matrix Associated with every *square* matrix **A** of constants is a number called the **determinant of the matrix,** which is denoted by det **A**.

EXAMPLE 6 **Determinant of a Square Matrix**

For $\mathbf{A} = \begin{pmatrix} 3 & 6 & 2 \\ 2 & 5 & 1 \\ -1 & 2 & 4 \end{pmatrix}$ we expand det **A** by cofactors of the first row:

$$\det \mathbf{A} = \begin{vmatrix} 3 & 6 & 2 \\ 2 & 5 & 1 \\ -1 & 2 & 4 \end{vmatrix} = 3\begin{vmatrix} 5 & 1 \\ 2 & 4 \end{vmatrix} - 6\begin{vmatrix} 2 & 1 \\ -1 & 4 \end{vmatrix} + 2\begin{vmatrix} 2 & 5 \\ -1 & 2 \end{vmatrix}$$

$$= 3(20 - 2) - 6(8 + 1) + 2(4 + 5) = 18.$$ ≡

It can be proved that a determinant det **A** can be expanded by cofactors using any row or column. If det **A** has a row (or a column) containing many zero entries, then wisdom dictates that we expand the determinant by that row (or column).

DEFINITION II.7 Transpose of a Matrix

The **transpose** of the $m \times n$ matrix (1) is the $n \times m$ matrix $\mathbf{A}^T$ given by

$$\mathbf{A}^T = \begin{pmatrix} a_{11} & a_{21} & \cdots & a_{m1} \\ a_{12} & a_{22} & \cdots & a_{m2} \\ \vdots & & & \vdots \\ a_{1n} & a_{2n} & \cdots & a_{mn} \end{pmatrix}.$$

In other words, the rows of a matrix **A** become the columns of its transpose $\mathbf{A}^T$.

EXAMPLE 7 Transpose of a Matrix

(a) The transpose of $\mathbf{A} = \begin{pmatrix} 3 & 6 & 2 \\ 2 & 5 & 1 \\ -1 & 2 & 4 \end{pmatrix}$ is $\mathbf{A}^T = \begin{pmatrix} 3 & 2 & -1 \\ 6 & 5 & 2 \\ 2 & 1 & 4 \end{pmatrix}$.

(b) If $\mathbf{X} = \begin{pmatrix} 5 \\ 0 \\ 3 \end{pmatrix}$, then $\mathbf{X}^T = (5 \quad 0 \quad 3)$. ≡

DEFINITION II.8 Multiplicative Inverse of a Matrix

Let **A** be an $n \times n$ matrix. If there exists an $n \times n$ matrix **B** such that

$$\mathbf{AB} = \mathbf{BA} = \mathbf{I},$$

where **I** is the multiplicative identity, then **B** is said to be the **multiplicative inverse of A** and is denoted by $\mathbf{B} = \mathbf{A}^{-1}$.

DEFINITION II.9 Nonsingular/Singular Matrices

Let **A** be an $n \times n$ matrix. If det $\mathbf{A} \neq 0$, then **A** is said to be **nonsingular.** If det $\mathbf{A} = 0$, then **A** is said to be **singular.**

The following theorem gives a necessary and sufficient condition for a square matrix to have a multiplicative inverse.

THEOREM II.1 Nonsingularity Implies A Has an Inverse

An $n \times n$ matrix **A** has a multiplicative inverse $\mathbf{A}^{-1}$ if and only if **A** is nonsingular.

The following theorem gives one way of finding the multiplicative inverse for a nonsingular matrix.

THEOREM II.2 A Formula for the Inverse of a Matrix

Let **A** be an $n \times n$ nonsingular matrix and let $C_{ij} = (-1)^{i+j}M_{ij}$, where M_{ij} is the determinant of the $(n-1) \times (n-1)$ matrix obtained by deleting the ith row and jth column from **A**. Then

$$\mathbf{A}^{-1} = \frac{1}{\det \mathbf{A}}(C_{ij})^T. \tag{2}$$

Each C_{ij} in Theorem II.2 is simply the **cofactor** (signed minor) of the corresponding entry a_{ij} in **A**. Note that the transpose is utilized in formula (2).

For future reference we observe in the case of a 2×2 nonsingular matrix

$$\mathbf{A} = \begin{pmatrix} a_{11} & a_{12} \\ a_{21} & a_{22} \end{pmatrix}$$

that $C_{11} = a_{22}$, $C_{12} = -a_{21}$, $C_{21} = -a_{12}$, and $C_{22} = a_{11}$. Thus

$$\mathbf{A}^{-1} = \frac{1}{\det \mathbf{A}}\begin{pmatrix} a_{22} & -a_{21} \\ -a_{12} & a_{11} \end{pmatrix}^T = \frac{1}{\det \mathbf{A}}\begin{pmatrix} a_{22} & -a_{12} \\ -a_{21} & a_{11} \end{pmatrix}. \tag{3}$$

For a 3×3 nonsingular matrix

$$\mathbf{A} = \begin{pmatrix} a_{11} & a_{12} & a_{13} \\ a_{21} & a_{22} & a_{23} \\ a_{31} & a_{32} & a_{33} \end{pmatrix},$$

$$C_{11} = \begin{vmatrix} a_{22} & a_{23} \\ a_{32} & a_{33} \end{vmatrix}, \quad C_{12} = -\begin{vmatrix} a_{21} & a_{23} \\ a_{31} & a_{33} \end{vmatrix}, \quad C_{13} = \begin{vmatrix} a_{21} & a_{22} \\ a_{31} & a_{32} \end{vmatrix},$$

and so on. Carrying out the transposition gives

$$\mathbf{A}^{-1} = \frac{1}{\det \mathbf{A}}\begin{pmatrix} C_{11} & C_{21} & C_{31} \\ C_{12} & C_{22} & C_{32} \\ C_{13} & C_{23} & C_{33} \end{pmatrix}. \tag{4}$$

EXAMPLE 8 Inverse of a 2 × 2 Matrix

Find the multiplicative inverse for $\mathbf{A} = \begin{pmatrix} 1 & 4 \\ 2 & 10 \end{pmatrix}$.

SOLUTION Since $\det \mathbf{A} = 10 - 8 = 2 \neq 0$, **A** is nonsingular. It follows from Theorem II.1 that $\mathbf{A}^{-1}$ exists. From (3) we find

$$\mathbf{A}^{-1} = \frac{1}{2}\begin{pmatrix} 10 & -4 \\ -2 & 1 \end{pmatrix} = \begin{pmatrix} 5 & -2 \\ -1 & \frac{1}{2} \end{pmatrix}.$$

≡

Not every square matrix has a multiplicative inverse. The matrix $\mathbf{A} = \begin{pmatrix} 2 & 2 \\ 3 & 3 \end{pmatrix}$ is singular, since $\det \mathbf{A} = 0$. Hence $\mathbf{A}^{-1}$ does not exist.

EXAMPLE 9 Inverse of a 3 × 3 Matrix

Find the multiplicative inverse for $\mathbf{A} = \begin{pmatrix} 2 & 2 & 0 \\ -2 & 1 & 1 \\ 3 & 0 & 1 \end{pmatrix}$.

SOLUTION Since det $\mathbf{A} = 12 \neq 0$, the given matrix is nonsingular. The cofactors corresponding to the entries in each row of det $\mathbf{A}$ are

$$C_{11} = \begin{vmatrix} 1 & 1 \\ 0 & 1 \end{vmatrix} = 1 \qquad C_{12} = -\begin{vmatrix} -2 & 1 \\ 3 & 1 \end{vmatrix} = 5 \qquad C_{13} = \begin{vmatrix} -2 & 1 \\ 3 & 0 \end{vmatrix} = -3$$

$$C_{21} = -\begin{vmatrix} 2 & 0 \\ 0 & 1 \end{vmatrix} = -2 \qquad C_{22} = \begin{vmatrix} 2 & 0 \\ 3 & 1 \end{vmatrix} = 2 \qquad C_{23} = -\begin{vmatrix} 2 & 2 \\ 3 & 0 \end{vmatrix} = 6$$

$$C_{31} = \begin{vmatrix} 2 & 0 \\ 1 & 1 \end{vmatrix} = 2 \qquad C_{32} = -\begin{vmatrix} 2 & 0 \\ -2 & 1 \end{vmatrix} = -2 \qquad C_{33} = \begin{vmatrix} 2 & 2 \\ -2 & 1 \end{vmatrix} = 6.$$

If follows from (4) that

$$\mathbf{A}^{-1} = \frac{1}{12}\begin{pmatrix} 1 & -2 & 2 \\ 5 & 2 & -2 \\ -3 & 6 & 6 \end{pmatrix} = \begin{pmatrix} \frac{1}{12} & -\frac{1}{6} & \frac{1}{6} \\ \frac{5}{12} & \frac{1}{6} & -\frac{1}{6} \\ -\frac{1}{4} & \frac{1}{2} & \frac{1}{2} \end{pmatrix}.$$

You are urged to verify that $\mathbf{A}^{-1}\mathbf{A} = \mathbf{A}\mathbf{A}^{-1} = \mathbf{I}$. ≡

Formula (2) presents obvious difficulties for nonsingular matrices larger than 3×3. For example, to apply (2) to a 4×4 matrix, we would have to calculate *sixteen* 3×3 determinants.* In the case of a large matrix there are more efficient ways of finding $\mathbf{A}^{-1}$. The curious reader is referred to any text in linear algebra.

Since our goal is to apply the concept of a matrix to systems of linear first-order differential equations, we need the following definitions.

DEFINITION II.10 Derivative of a Matrix of Functions

If $\mathbf{A}(t) = (a_{ij}(t))_{m \times n}$ is a matrix whose entries are functions differentiable on a common interval, then

$$\frac{d\mathbf{A}}{dt} = \left(\frac{d}{dt}a_{ij}\right)_{m \times n}.$$

DEFINITION II.11 Integral of a Matrix of Functions

If $\mathbf{A}(t) = (a_{ij}(t))_{m \times n}$ is a matrix whose entries are functions continuous on a common interval containing t and t_0, then

$$\int_{t_0}^{t} \mathbf{A}(s)\,ds = \left(\int_{t_0}^{t} a_{ij}(s)\,ds\right)_{m \times n}.$$

To differentiate (integrate) a matrix of functions, we simply differentiate (integrate) each entry. The derivative of a matrix is also denoted by $\mathbf{A}'(t)$.

EXAMPLE 10 Derivative/Integral of a Matrix

If $\quad \mathbf{X}(t) = \begin{pmatrix} \sin 2t \\ e^{3t} \\ 8t - 1 \end{pmatrix}, \quad$ then $\quad \mathbf{X}'(t) = \begin{pmatrix} \frac{d}{dt}\sin 2t \\ \frac{d}{dt}e^{3t} \\ \frac{d}{dt}(8t-1) \end{pmatrix} = \begin{pmatrix} 2\cos 2t \\ 3e^{3t} \\ 8 \end{pmatrix}$

*Strictly speaking, a determinant is a number, but it is sometimes convenient to refer to a determinant as if it were an array.

and $$\int_0^t \mathbf{X}(s)\,ds = \begin{pmatrix} \int_0^t \sin 2s\,ds \\ \int_0^t e^{3s}\,ds \\ \int_0^t (8s-1)\,ds \end{pmatrix} = \begin{pmatrix} -\frac{1}{2}\cos 2t + \frac{1}{2} \\ \frac{1}{3}e^{3t} - \frac{1}{3} \\ 4t^2 - t \end{pmatrix}.$$ ≡

II.2 GAUSSIAN AND GAUSS-JORDAN ELIMINATION

Matrices are an invaluable aid in solving algebraic systems of n linear equations in n variables or unknowns,

$$\begin{aligned} a_{11}x_1 + a_{12}x_2 + \cdots + a_{1n}x_n &= b_1 \\ a_{21}x_1 + a_{22}x_2 + \cdots + a_{2n}x_n &= b_2 \\ \vdots \qquad\qquad\quad & \quad\vdots \\ a_{n1}x_1 + a_{n2}x_2 + \cdots + a_{nn}x_n &= b_n. \end{aligned} \tag{5}$$

If $\mathbf{A}$ denotes the matrix of coefficients in (5), we know that Cramer's rule could be used to solve the system whenever $\det \mathbf{A} \neq 0$. However, that rule requires a herculean effort if $\mathbf{A}$ is larger than 3×3. The procedure that we shall now consider has the distinct advantage of being not only an efficient way of handling large systems, but also a means of solving consistent systems (5) in which $\det \mathbf{A} = 0$ and a means of solving m linear equations in n unknowns.

DEFINITION II.12 Augmented Matrix

The **augmented matrix** of the system (5) is the $n \times (n + 1)$ matrix

$$\left(\begin{array}{cccc|c} a_{11} & a_{12} & \cdots & a_{1n} & b_1 \\ a_{21} & a_{22} & \cdots & a_{2n} & b_2 \\ \vdots & & & & \vdots \\ a_{n1} & a_{n2} & \cdots & a_{nn} & b_n \end{array}\right).$$

If $\mathbf{B}$ is the column matrix of the b_i, $i = 1, 2, \ldots, n$, the augmented matrix of (5) is denoted by $(\mathbf{A}|\mathbf{B})$.

Elementary Row Operations

Recall from algebra that we can transform an algebraic system of equations into an equivalent system (that is, one having the same solution) by multiplying an equation by a nonzero constant, interchanging the positions of any two equations in a system, and adding a nonzero constant multiple of an equation to another equation. These operations on equations in a system are, in turn, equivalent to **elementary row operations** on an augmented matrix:

(i) Multiply a row by a nonzero constant.

(ii) Interchange any two rows.

(iii) Add a nonzero constant multiple of one row to any other row.

Elimination Methods

To solve a system such as (5) using an augmented matrix, we use either **Gaussian elimination** or the **Gauss-Jordan elimination method.** In the former method, we carry out a succession of elementary row operations until we arrive at an augmented matrix in **row-echelon form:**

(i) The first nonzero entry in a nonzero row is 1.

(ii) In consecutive nonzero rows the first entry 1 in the lower row appears to the right of the first 1 in the higher row.

(iii) Rows consisting of all 0's are at the bottom of the matrix.

In the Gauss-Jordan method the row operations are continued until we obtain an augmented matrix that is in **reduced row-echelon form.** A reduced row-echelon matrix has the same three properties listed above in addition to the following one:

(*iv*) A column containing a first entry 1 has 0's everywhere else.

EXAMPLE 11 Row-Echelon/Reduced Row-Echelon Form

(a) The augmented matrices

$$\left(\begin{array}{ccc|c} 1 & 5 & 0 & 2 \\ 0 & 1 & 0 & -1 \\ 0 & 0 & 0 & 0 \end{array}\right) \quad \text{and} \quad \left(\begin{array}{ccccc|c} 0 & 0 & 1 & -6 & 2 & 2 \\ 0 & 0 & 0 & 0 & 1 & 4 \end{array}\right)$$

are in row-echelon form. You should verify that the three criteria are satisfied.

(b) The augmented matrices

$$\left(\begin{array}{ccc|c} 1 & 0 & 0 & 7 \\ 0 & 1 & 0 & -1 \\ 0 & 0 & 0 & 0 \end{array}\right) \quad \text{and} \quad \left(\begin{array}{ccccc|c} 0 & 0 & 1 & -6 & 0 & -6 \\ 0 & 0 & 0 & 0 & 1 & 4 \end{array}\right)$$

are in reduced row-echelon form. Note that the remaining entries in the columns containing a leading entry 1 are all 0's. ≡

Note that in Gaussian elimination we stop once we have obtained *an* augmented matrix in row-echelon form. In other words, by using different sequences of row operations we may arrive at different row-echelon forms. This method then requires the use of back-substitution. In Gauss-Jordan elimination we stop when we have obtained *the* augmented matrix in reduced row-echelon form. Any sequence of row operations will lead to the same augmented matrix in reduced row-echelon form. This method does not require back-substitution; the solution of the system will be apparent by inspection of the final matrix. In terms of the equations of the original system, our goal in both methods is simply to make the coefficient of x_1 in the first equation* equal to 1 and then use multiples of that equation to eliminate x_1 from other equations. The process is repeated on the other variables.

To keep track of the row operations on an augmented matrix, we utilize the following notation:

Symbol	Meaning
R_{ij}	Interchange rows i and j
cR_i	Multiply the ith row by the nonzero constant c
$cR_i + R_j$	Multiply the ith row by c and add to the jth row

EXAMPLE 12 Solution by Elimination

Solve

$$\begin{aligned} 2x_1 + 6x_2 + x_3 &= 7 \\ x_1 + 2x_2 - x_3 &= -1 \\ 5x_1 + 7x_2 - 4x_3 &= 9 \end{aligned}$$

using **(a)** Gaussian elimination and **(b)** Gauss-Jordan elimination.

*We can always interchange equations so that the first equation contains the variable x_1.

SOLUTION **(a)** Using row operations on the augmented matrix of the system, we obtain

$$\left(\begin{array}{ccc|c} 2 & 6 & 1 & 7 \\ 1 & 2 & -1 & -1 \\ 5 & 7 & -4 & 9 \end{array}\right) \xrightarrow{R_{12}} \left(\begin{array}{ccc|c} 1 & 2 & -1 & -1 \\ 2 & 6 & 1 & 7 \\ 5 & 7 & -4 & 9 \end{array}\right) \xrightarrow[-5R_1 + R_3]{-2R_1 + R_2} \left(\begin{array}{ccc|c} 1 & 2 & -1 & -1 \\ 0 & 2 & 3 & 9 \\ 0 & -3 & 1 & 14 \end{array}\right)$$

$$\xrightarrow{\frac{1}{2}R_2} \left(\begin{array}{ccc|c} 1 & 2 & -1 & -1 \\ 0 & 1 & \frac{3}{2} & \frac{9}{2} \\ 0 & -3 & 1 & 14 \end{array}\right) \xrightarrow{3R_2 + R_3} \left(\begin{array}{ccc|c} 1 & 2 & -1 & -1 \\ 0 & 1 & \frac{3}{2} & \frac{9}{2} \\ 0 & 0 & \frac{11}{2} & \frac{55}{2} \end{array}\right) \xrightarrow{\frac{2}{11}R_3} \left(\begin{array}{ccc|c} 1 & 2 & -1 & -1 \\ 0 & 1 & \frac{3}{2} & \frac{9}{2} \\ 0 & 0 & 1 & 5 \end{array}\right).$$

The last matrix is in row-echelon form and represents the system

$$\begin{aligned} x_1 + 2x_2 - \phantom{\tfrac{3}{2}}x_3 &= -1 \\ x_2 + \frac{3}{2}x_3 &= \frac{9}{2} \\ x_3 &= 5. \end{aligned}$$

Substituting $x_3 = 5$ into the second equation then gives $x_2 = -3$. Substituting both these values back into the first equation finally yields $x_1 = 10$.

(b) We start with the last matrix above. Since the first entries in the second and third rows are 1's, we must, in turn, make the remaining entries in the second and third columns 0's:

$$\left(\begin{array}{ccc|c} 1 & 2 & -1 & -1 \\ 0 & 1 & \frac{3}{2} & \frac{9}{2} \\ 0 & 0 & 1 & 5 \end{array}\right) \xrightarrow{-2R_2 + R_1} \left(\begin{array}{ccc|c} 1 & 0 & -4 & -10 \\ 0 & 1 & \frac{3}{2} & \frac{9}{2} \\ 0 & 0 & 1 & 5 \end{array}\right) \xrightarrow[-\frac{3}{2}R_3 + R_2]{4R_3 + R_1} \left(\begin{array}{ccc|c} 1 & 0 & 0 & 10 \\ 0 & 1 & 0 & -3 \\ 0 & 0 & 1 & 5 \end{array}\right).$$

The last matrix is now in reduced row-echelon form. Because of what the matrix means in terms of equations, it is evident that the solution of the system is $x_1 = 10$, $x_2 = -3, x_3 = 5$. ≡

EXAMPLE 13 Gauss-Jordan Elimination

Solve

$$\begin{aligned} x + 3y - 2z &= -7 \\ 4x + y + 3z &= 5 \\ 2x - 5y + 7z &= 19. \end{aligned}$$

SOLUTION We solve the system using Gauss-Jordan elimination:

$$\left(\begin{array}{ccc|c} 1 & 3 & -2 & -7 \\ 4 & 1 & 3 & 5 \\ 2 & -5 & 7 & 19 \end{array}\right) \xrightarrow[-2R_1 + R_3]{-4R_1 + R_2} \left(\begin{array}{ccc|c} 1 & 3 & -2 & -7 \\ 0 & -11 & 11 & 33 \\ 0 & -11 & 11 & 33 \end{array}\right)$$

$$\xrightarrow[-\frac{1}{11}R_3]{-\frac{1}{11}R_2} \left(\begin{array}{ccc|c} 1 & 3 & -2 & -7 \\ 0 & 1 & -1 & -3 \\ 0 & 1 & -1 & -3 \end{array}\right) \xrightarrow[-R_2 + R_3]{-3R_2 + R_1} \left(\begin{array}{ccc|c} 1 & 0 & 1 & 1 \\ 0 & 1 & -1 & -3 \\ 0 & 0 & 0 & 0 \end{array}\right).$$

In this case, the last matrix in reduced row-echelon form implies that the original system of three equations in three unknowns is really equivalent to two equations in three unknowns. Since only z is common to both equations (the nonzero rows), we can assign its values arbitrarily. If we let $z = t$, where t represents any real number, then we see that the system has infinitely many solutions: $x = 2 - t$, $y = -3 + t$, $z = t$. Geometrically, these equations are the parametric equations for the line of intersection of the planes $x + 0y + z = 2$ and $0x + y - z = 3$. ≡

≡ Using Row Operations to Find an Inverse Because of the number of determinants that must be evaluated, formula (2) in Theorem II.2 is seldom used to find the inverse when the matrix **A** is large. In the case of 3×3 or larger matrices the method described in the next theorem is a particularly efficient means for finding $\mathbf{A}^{-1}$.

> **THEOREM II.3 Finding $\mathbf{A}^{-1}$ Using Elementary Row Operations**
>
> If an $n \times n$ matrix **A** can be transformed into the $n \times n$ identity **I** by a sequence of elementary row operations, then **A** is nonsingular. The same sequence of operations that transforms **A** into the identity **I** will also transform **I** into $\mathbf{A}^{-1}$.

It is convenient to carry out these row operations on **A** and **I** simultaneously by means of an $n \times 2n$ matrix obtained by augmenting **A** with the identity **I** as shown here:

$$(\mathbf{A} \mid \mathbf{I}) = \left(\begin{array}{cccc|cccc} a_{11} & a_{12} & \cdots & a_{1n} & 1 & 0 & \cdots & 0 \\ a_{21} & a_{22} & \cdots & a_{2n} & 1 & 0 & \cdots & 0 \\ \vdots & & & \vdots & \vdots & & & \vdots \\ a_{n1} & a_{n2} & \cdots & a_{nn} & 0 & 0 & \cdots & 1 \end{array}\right).$$

The procedure for finding $\mathbf{A}^{-1}$ is outlined in the following diagram:

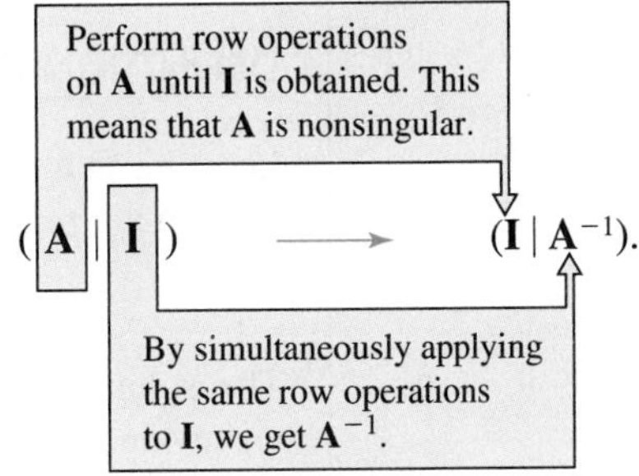
Perform row operations on A until I is obtained. This means that A is nonsingular.
(A | I) → (I | A⁻¹).
By simultaneously applying the same row operations to I, we get A⁻¹.

EXAMPLE 14 Inverse by Elementary Row Operations

Find the multiplicative inverse for $\mathbf{A} = \begin{pmatrix} 2 & 0 & 1 \\ -2 & 3 & 4 \\ -5 & 5 & 6 \end{pmatrix}$.

SOLUTION We shall use the same notation as we did when we reduced an augmented matrix to reduced row-echelon form:

$$\left(\begin{array}{ccc|ccc} 2 & 0 & 1 & 1 & 0 & 0 \\ -2 & 3 & 4 & 0 & 1 & 0 \\ -5 & 5 & 6 & 0 & 0 & 1 \end{array}\right) \xrightarrow{\frac{1}{2}R_1} \left(\begin{array}{ccc|ccc} 1 & 0 & \frac{1}{2} & \frac{1}{2} & 0 & 0 \\ -2 & 3 & 4 & 0 & 1 & 0 \\ -5 & 5 & 6 & 0 & 0 & 1 \end{array}\right) \xrightarrow[5R_1 + R_3]{2R_1 + R_2} \left(\begin{array}{ccc|ccc} 1 & 0 & \frac{1}{2} & \frac{1}{2} & 0 & 0 \\ 0 & 3 & 5 & 1 & 1 & 0 \\ 0 & 5 & \frac{17}{2} & \frac{5}{2} & 0 & 1 \end{array}\right)$$

$$\xrightarrow{\substack{\frac{1}{3}R_2\\ \frac{1}{5}R_3}} \left(\begin{array}{ccc|ccc} 1 & 0 & \frac{1}{2} & \frac{1}{2} & 0 & 0 \\ 0 & 1 & \frac{5}{3} & \frac{1}{3} & \frac{1}{3} & 0 \\ 0 & 1 & \frac{17}{10} & \frac{1}{2} & 0 & \frac{1}{5} \end{array}\right) \xrightarrow{-R_2 + R_3} \left(\begin{array}{ccc|ccc} 1 & 0 & \frac{1}{2} & \frac{1}{2} & 0 & 0 \\ 0 & 1 & \frac{5}{3} & \frac{1}{3} & \frac{1}{3} & 0 \\ 0 & 0 & \frac{1}{30} & \frac{1}{6} & -\frac{1}{3} & \frac{1}{5} \end{array}\right)$$

$$\xrightarrow{30R_3} \left(\begin{array}{ccc|ccc} 1 & 0 & \frac{1}{2} & \frac{1}{2} & 0 & 0 \\ 0 & 1 & \frac{5}{3} & \frac{1}{3} & \frac{1}{3} & 0 \\ 0 & 0 & 1 & 5 & -10 & 6 \end{array}\right) \xrightarrow{\substack{-\frac{1}{2}R_3 + R_1\\ -\frac{5}{3}R_3 + R_2}} \left(\begin{array}{ccc|ccc} 1 & 0 & 0 & -2 & 5 & -3 \\ 0 & 1 & 0 & -8 & 17 & -10 \\ 0 & 0 & 1 & 5 & -10 & 6 \end{array}\right).$$

Because **I** appears to the left of the vertical line, we conclude that the matrix to the right of the line is

$$\mathbf{A}^{-1} = \begin{pmatrix} -2 & 5 & -3 \\ -8 & 17 & -10 \\ 5 & -10 & 6 \end{pmatrix}.$$

≡

If row reduction of $(\mathbf{A}|\mathbf{I})$ leads to the situation

$$(\mathbf{A} \mid \mathbf{I}) \xrightarrow{\text{row operations}} (\mathbf{B} \mid \mathbf{C}),$$

where the matrix **B** contains a row of zeros, then necessarily **A** is singular. Since further reduction of **B** always yields another matrix with a row of zeros, we can never transform **A** into **I**.

II.3 THE EIGENVALUE PROBLEM

Gauss-Jordan elimination can be used to find the eigenvectors of a square matrix.

DEFINITION II.13 Eigenvalues and Eigenvectors

Let **A** be an $n \times n$ matrix. A number λ is said to be an **eigenvalue** of **A** if there exists a *nonzero* solution vector **K** of the linear system

$$\mathbf{AK} = \lambda\mathbf{K}. \tag{6}$$

The solution vector **K** is said to be an **eigenvector** corresponding to the eigenvalue λ.

The word *eigenvalue* is a combination of German and English terms adapted from the German word *eigenwert,* which, translated literally, is "proper value." Eigenvalues and eigenvectors are also called **characteristic values** and **characteristic vectors,** respectively.

EXAMPLE 15 Eigenvector of a Matrix

Verify that $\mathbf{K} = \begin{pmatrix} 1 \\ -1 \\ 1 \end{pmatrix}$ is an eigenvector of the matrix

$$\mathbf{A} = \begin{pmatrix} 0 & -1 & -3 \\ 2 & 3 & 3 \\ -2 & 1 & 1 \end{pmatrix}.$$

SOLUTION By carrying out the multiplication **AK**, we see that

$$\mathbf{AK} = \begin{pmatrix} 0 & -1 & -3 \\ 2 & 3 & 3 \\ -2 & 1 & 1 \end{pmatrix}\begin{pmatrix} 1 \\ -1 \\ 1 \end{pmatrix} = \begin{pmatrix} -2 \\ 2 \\ -2 \end{pmatrix} = (-2)\begin{pmatrix} 1 \\ -1 \\ 1 \end{pmatrix} = \underset{\text{eigenvalue}}{(-2)}\mathbf{K}.$$

We see from the preceding line and Definition II.13 that $\lambda = -2$ is an eigenvalue of **A**. ≡

Using properties of matrix algebra, we can write (6) in the alternative form

$$(\mathbf{A} - \lambda\mathbf{I})\mathbf{K} = \mathbf{0}, \tag{7}$$

where **I** is the multiplicative identity. If we let

$$\mathbf{K} = \begin{pmatrix} k_1 \\ k_2 \\ \vdots \\ k_n \end{pmatrix},$$

then (7) is the same as

$$\begin{aligned} (a_{11} - \lambda)k_1 + \quad a_{12}k_2 + \cdots + \quad a_{1n}k_n &= 0 \\ a_{21}k_1 + (a_{22} - \lambda)k_2 + \cdots + \quad a_{2n}k_n &= 0 \\ \vdots \qquad\qquad\qquad &\;\;\vdots \\ a_{n1}k_1 + \quad a_{n2}k_2 + \cdots + (a_{nn} - \lambda)k_n &= 0. \end{aligned} \tag{8}$$

Although an obvious solution of (8) is $k_1 = 0, k_2 = 0, \ldots, k_n = 0$, we are seeking only nontrivial solutions. It is known that a **homogeneous** system of n linear equations in n unknowns (that is, $b_i = 0$, $i = 1, 2, \ldots, n$ in (5)) has a nontrivial solution if and only if the determinant of the coefficient matrix is equal to zero. Thus to find a nonzero solution **K** for (7), we must have

$$\det(\mathbf{A} - \lambda\mathbf{I}) = 0. \tag{9}$$

Inspection of (8) shows that the expansion of $\det(\mathbf{A} - \lambda\mathbf{I})$ by cofactors results in an nth-degree polynomial in λ. The equation (9) is called the **characteristic equation** of **A**. Thus *the eigenvalues of* **A** *are the roots of the characteristic equation.* To find an eigenvector corresponding to an eigenvalue λ, we simply solve the system of equations $(\mathbf{A} - \lambda\mathbf{I})\mathbf{K} = \mathbf{0}$ by applying Gauss-Jordan elimination to the augmented matrix $(\mathbf{A} - \lambda\mathbf{I}|\mathbf{0})$.

EXAMPLE 16 Eigenvalues/Eigenvectors

Find the eigenvalues and eigenvectors of $\mathbf{A} = \begin{pmatrix} 1 & 2 & 1 \\ 6 & -1 & 0 \\ -1 & -2 & -1 \end{pmatrix}$.

SOLUTION To expand the determinant in the characteristic equation, we use the cofactors of the second row:

$$\det(\mathbf{A} - \lambda\mathbf{I}) = \begin{vmatrix} 1 - \lambda & 2 & 1 \\ 6 & -1 - \lambda & 0 \\ -1 & -2 & -1 - \lambda \end{vmatrix} = -\lambda^3 - \lambda^2 + 12\lambda = 0.$$

From $-\lambda^3 - \lambda^2 + 12\lambda = -\lambda(\lambda + 4)(\lambda - 3) = 0$ we see that the eigenvalues are $\lambda_1 = 0$, $\lambda_2 = -4$, and $\lambda_3 = 3$. To find the eigenvectors, we must now reduce $(\mathbf{A} - \lambda\mathbf{I}|\mathbf{0})$ three times corresponding to the three distinct eigenvalues.

For $\lambda_1 = 0$ we have

$$(\mathbf{A} - 0\mathbf{I} \mid \mathbf{0}) = \left(\begin{array}{rrr|r} 1 & 2 & 1 & 0 \\ 6 & -1 & 0 & 0 \\ -1 & -2 & -1 & 0 \end{array}\right) \xrightarrow[R_1 + R_3]{-6R_1 + R_2} \left(\begin{array}{rrr|r} 1 & 2 & 1 & 0 \\ 0 & -13 & -6 & 0 \\ 0 & 0 & 0 & 0 \end{array}\right)$$

$$\xrightarrow{-\frac{1}{13}R_2} \left(\begin{array}{rrr|r} 1 & 2 & 1 & 0 \\ 0 & 1 & \frac{6}{13} & 0 \\ 0 & 0 & 0 & 0 \end{array}\right) \xrightarrow{-2R_2 + R_1} \left(\begin{array}{rrr|r} 1 & 0 & \frac{1}{13} & 0 \\ 0 & 1 & \frac{6}{13} & 0 \\ 0 & 0 & 0 & 0 \end{array}\right).$$

Thus we see that $k_1 = -\frac{1}{13}k_3$ and $k_2 = -\frac{6}{13}k_3$. Choosing $k_3 = -13$, we get the eigenvector*

$$\mathbf{K}_1 = \begin{pmatrix} 1 \\ 6 \\ -13 \end{pmatrix}.$$

For $\lambda_2 = -4$,

$$(\mathbf{A} + 4\mathbf{I} \mid \mathbf{0}) = \left(\begin{array}{rrr|r} 5 & 2 & 1 & 0 \\ 6 & 3 & 0 & 0 \\ -1 & -2 & 3 & 0 \end{array}\right) \xrightarrow[R_{31}]{-R_3} \left(\begin{array}{rrr|r} 1 & 2 & -3 & 0 \\ 6 & 3 & 0 & 0 \\ 5 & 2 & 1 & 0 \end{array}\right)$$

$$\xrightarrow[-5R_1 + R_3]{-6R_1 + R_2} \left(\begin{array}{rrr|r} 1 & 2 & -3 & 0 \\ 0 & -9 & 18 & 0 \\ 0 & -8 & 16 & 0 \end{array}\right) \xrightarrow[-\frac{1}{8}R_3]{-\frac{1}{9}R_2} \left(\begin{array}{rrr|r} 1 & 2 & -3 & 0 \\ 0 & 1 & -2 & 0 \\ 0 & 1 & -2 & 0 \end{array}\right) \xrightarrow[-R_2 + R_3]{-2R_2 + R_1} \left(\begin{array}{rrr|r} 1 & 0 & 1 & 0 \\ 0 & 1 & -2 & 0 \\ 0 & 0 & 0 & 0 \end{array}\right)$$

implies that $k_1 = -k_3$ and $k_2 = 2k_3$. Choosing $k_3 = 1$ then yields the second eigenvector

$$\mathbf{K}_2 = \begin{pmatrix} -1 \\ 2 \\ 1 \end{pmatrix}.$$

Finally, for $\lambda_3 = 3$ Gauss-Jordan elimination gives

$$(\mathbf{A} - 3\mathbf{I} \mid \mathbf{0}) = \left(\begin{array}{rrr|r} -2 & 2 & 1 & 0 \\ 6 & -4 & 0 & 0 \\ -1 & -2 & -4 & 0 \end{array}\right) \xrightarrow[\text{operations}]{\text{row}} \left(\begin{array}{rrr|r} 1 & 0 & 1 & 0 \\ 0 & 1 & \frac{3}{2} & 0 \\ 0 & 0 & 0 & 0 \end{array}\right),$$

so $k_1 = -k_3$ and $k_2 = -\frac{3}{2}k_3$. The choice of $k_3 = -2$ leads to the third eigenvector:

$$\mathbf{K}_3 = \begin{pmatrix} 2 \\ 3 \\ -2 \end{pmatrix}.$$

≡

When an $n \times n$ matrix $\mathbf{A}$ possesses n distinct eigenvalues $\lambda_1, \lambda_2, \ldots, \lambda_n$, it can be proved that a set of n linearly independent† eigenvectors $\mathbf{K}_1, \mathbf{K}_2, \ldots, \mathbf{K}_n$ can be found. However, when the characteristic equation has repeated roots, it may not be possible to find n linearly independent eigenvectors for $\mathbf{A}$.

---

*Of course, k_3 could be chosen as any nonzero number. In other words, a nonzero constant multiple of an eigenvector is also an eigenvector.

†Linear independence of column vectors is defined in exactly the same manner as for functions.

EXAMPLE 17 **Eigenvalues/Eigenvectors**

Find the eigenvalues and eigenvectors of $\mathbf{A} = \begin{pmatrix} 3 & 4 \\ -1 & 7 \end{pmatrix}$.

SOLUTION From the characteristic equation

$$\det(\mathbf{A} - \lambda\mathbf{I}) = \begin{vmatrix} 3-\lambda & 4 \\ -1 & 7-\lambda \end{vmatrix} = (\lambda - 5)^2 = 0$$

we see that $\lambda_1 = \lambda_2 = 5$ is an eigenvalue of multiplicity two. In the case of a 2×2 matrix there is no need to use Gauss-Jordan elimination. To find the eigenvector(s) corresponding to $\lambda_1 = 5$, we resort to the system $(\mathbf{A} - 5\mathbf{I}|\mathbf{0})$ in its equivalent form

$$\begin{aligned} -2k_1 + 4k_2 &= 0 \\ -k_1 + 2k_2 &= 0. \end{aligned}$$

It is apparent from this system that $k_1 = 2k_2$. Thus if we choose $k_2 = 1$, we find the single eigenvector

$$\mathbf{K}_1 = \begin{pmatrix} 2 \\ 1 \end{pmatrix}.$$

≡

EXAMPLE 18 **Eigenvalues/Eigenvectors**

Find the eigenvalues and eigenvectors of $\mathbf{A} = \begin{pmatrix} 9 & 1 & 1 \\ 1 & 9 & 1 \\ 1 & 1 & 9 \end{pmatrix}$.

SOLUTION The characteristic equation

$$\det(\mathbf{A} - \lambda\mathbf{I}) = \begin{vmatrix} 9-\lambda & 1 & 1 \\ 1 & 9-\lambda & 1 \\ 1 & 1 & 9-\lambda \end{vmatrix} = -(\lambda - 11)(\lambda - 8)^2 = 0$$

shows that $\lambda_1 = 11$ and that $\lambda_2 = \lambda_3 = 8$ is an eigenvalue of multiplicity two.

For $\lambda_1 = 11$ Gauss-Jordan elimination gives

$$(\mathbf{A} - 11\mathbf{I}\,|\,\mathbf{0}) = \left(\begin{array}{ccc|c} -2 & 1 & 1 & 0 \\ 1 & -2 & 1 & 0 \\ 1 & 1 & -2 & 0 \end{array}\right) \xrightarrow{\text{row operations}} \left(\begin{array}{ccc|c} 1 & 0 & -1 & 0 \\ 0 & 1 & -1 & 0 \\ 0 & 0 & 0 & 0 \end{array}\right).$$

Hence $k_1 = k_3$ and $k_2 = k_3$. If $k_3 = 1$, then

$$\mathbf{K}_1 = \begin{pmatrix} 1 \\ 1 \\ 1 \end{pmatrix}.$$

Now for $\lambda_2 = 8$ we have

$$(\mathbf{A} - 8\mathbf{I}\,|\,\mathbf{0}) = \left(\begin{array}{ccc|c} 1 & 1 & 1 & 0 \\ 1 & 1 & 1 & 0 \\ 1 & 1 & 1 & 0 \end{array}\right) \xrightarrow{\text{row operations}} \left(\begin{array}{ccc|c} 1 & 1 & 1 & 0 \\ 0 & 0 & 0 & 0 \\ 0 & 0 & 0 & 0 \end{array}\right).$$

In the equation $k_1 + k_2 + k_3 = 0$, we are free to select two of the variables arbitrarily. Choosing, on the one hand, $k_2 = 1, k_3 = 0$ and, on the other, $k_2 = 0, k_3 = 1$, we obtain two linearly independent eigenvectors

$$\mathbf{K}_2 = \begin{pmatrix} -1 \\ 1 \\ 0 \end{pmatrix} \quad \text{and} \quad \mathbf{K}_3 = \begin{pmatrix} -1 \\ 0 \\ 1 \end{pmatrix}. \qquad \equiv$$

EXERCISES FOR APPENDIX II

Answers to selected odd-numbered problems begin on page ANS-31.

II.1 BASIC DEFINITIONS AND THEORY

1. If $\mathbf{A} = \begin{pmatrix} 4 & 5 \\ -6 & 9 \end{pmatrix}$ and $\mathbf{B} = \begin{pmatrix} -2 & 6 \\ 8 & -10 \end{pmatrix}$, find

(a) $\mathbf{A} + \mathbf{B}$ **(b)** $\mathbf{B} - \mathbf{A}$ **(c)** $2\mathbf{A} + 3\mathbf{B}$

2. If $\mathbf{A} = \begin{pmatrix} -2 & 0 \\ 4 & 1 \\ 7 & 3 \end{pmatrix}$ and $\mathbf{B} = \begin{pmatrix} 3 & -1 \\ 0 & 2 \\ -4 & -2 \end{pmatrix}$, find

(a) $\mathbf{A} - \mathbf{B}$ **(b)** $\mathbf{B} - \mathbf{A}$ **(c)** $2(\mathbf{A} + \mathbf{B})$

3. If $\mathbf{A} = \begin{pmatrix} 2 & -3 \\ -5 & 4 \end{pmatrix}$ and $\mathbf{B} = \begin{pmatrix} -1 & 6 \\ 3 & 2 \end{pmatrix}$, find

(a) $\mathbf{AB}$ **(b)** $\mathbf{BA}$ **(c)** $\mathbf{A}^2 = \mathbf{AA}$ **(d)** $\mathbf{B}^2 = \mathbf{BB}$

4. If $\mathbf{A} = \begin{pmatrix} 1 & 4 \\ 5 & 10 \\ 8 & 12 \end{pmatrix}$ and $\mathbf{B} = \begin{pmatrix} -4 & 6 & -3 \\ 1 & -3 & 2 \end{pmatrix}$, find

(a) $\mathbf{AB}$ **(b)** $\mathbf{BA}$

5. If $\mathbf{A} = \begin{pmatrix} 1 & -2 \\ -2 & 4 \end{pmatrix}$, $\mathbf{B} = \begin{pmatrix} 6 & 3 \\ 2 & 1 \end{pmatrix}$, and $\mathbf{C} = \begin{pmatrix} 0 & 2 \\ 3 & 4 \end{pmatrix}$, find

(a) $\mathbf{BC}$ **(b)** $\mathbf{A}(\mathbf{BC})$ **(c)** $\mathbf{C}(\mathbf{BA})$ **(d)** $\mathbf{A}(\mathbf{B} + \mathbf{C})$

6. If $\mathbf{A} = (5 \quad -6 \quad 7)$, $\mathbf{B} = \begin{pmatrix} 3 \\ 4 \\ -1 \end{pmatrix}$, and

$\mathbf{C} = \begin{pmatrix} 1 & 2 & 4 \\ 0 & 1 & -1 \\ 3 & 2 & 1 \end{pmatrix}$, find

(a) $\mathbf{AB}$ **(b)** $\mathbf{BA}$ **(c)** $(\mathbf{BA})\mathbf{C}$ **(d)** $(\mathbf{AB})\mathbf{C}$

7. If $\mathbf{A} = \begin{pmatrix} 4 \\ 8 \\ -10 \end{pmatrix}$ and $\mathbf{B} = (2 \quad 4 \quad 5)$, find

(a) $\mathbf{A}^T\mathbf{A}$ **(b)** $\mathbf{B}^T\mathbf{B}$ **(c)** $\mathbf{A} + \mathbf{B}^T$

8. If $\mathbf{A} = \begin{pmatrix} 1 & 2 \\ 2 & 4 \end{pmatrix}$ and $\mathbf{B} = \begin{pmatrix} -2 & 3 \\ 5 & 7 \end{pmatrix}$, find

(a) $\mathbf{A} + \mathbf{B}^T$ **(b)** $2\mathbf{A}^T - \mathbf{B}^T$ **(c)** $\mathbf{A}^T(\mathbf{A} - \mathbf{B})$

9. If $\mathbf{A} = \begin{pmatrix} 3 & 4 \\ 8 & 1 \end{pmatrix}$ and $\mathbf{B} = \begin{pmatrix} 5 & 10 \\ -2 & -5 \end{pmatrix}$, find

(a) $(\mathbf{AB})^T$ **(b)** $\mathbf{B}^T\mathbf{A}^T$

10. If $\mathbf{A} = \begin{pmatrix} 5 & 9 \\ -4 & 6 \end{pmatrix}$ and $\mathbf{B} = \begin{pmatrix} -3 & 11 \\ -7 & 2 \end{pmatrix}$, find

(a) $\mathbf{A}^T + \mathbf{B}^T$ **(b)** $(\mathbf{A} + \mathbf{B})^T$

In Problems 11–14 write the given sum as a single column matrix.

11. $4\begin{pmatrix} -1 \\ 2 \end{pmatrix} - 2\begin{pmatrix} 2 \\ 8 \end{pmatrix} + 3\begin{pmatrix} -2 \\ 3 \end{pmatrix}$

12. $3t\begin{pmatrix} 2 \\ t \\ -1 \end{pmatrix} + (t-1)\begin{pmatrix} -1 \\ -t \\ 3 \end{pmatrix} - 2\begin{pmatrix} 3t \\ 4 \\ -5t \end{pmatrix}$

13. $\begin{pmatrix} 2 & -3 \\ 1 & 4 \end{pmatrix}\begin{pmatrix} -2 \\ 5 \end{pmatrix} - \begin{pmatrix} -1 & 6 \\ -2 & 3 \end{pmatrix}\begin{pmatrix} -7 \\ 2 \end{pmatrix}$

14. $\begin{pmatrix} 1 & -3 & 4 \\ 2 & 5 & -1 \\ 0 & -4 & -2 \end{pmatrix}\begin{pmatrix} t \\ 2t-1 \\ -t \end{pmatrix} + \begin{pmatrix} -t \\ 1 \\ 4 \end{pmatrix} - \begin{pmatrix} 2 \\ 8 \\ -6 \end{pmatrix}$

In Problems 15–22 determine whether the given matrix is singular or nonsingular. If it is nonsingular, find $\mathbf{A}^{-1}$ using Theorem II.2.

15. $\mathbf{A} = \begin{pmatrix} -3 & 6 \\ -2 & 4 \end{pmatrix}$

16. $\mathbf{A} = \begin{pmatrix} 2 & 5 \\ 1 & 4 \end{pmatrix}$

17. $\mathbf{A} = \begin{pmatrix} 4 & 8 \\ -3 & -5 \end{pmatrix}$

18. $\mathbf{A} = \begin{pmatrix} 7 & 10 \\ 2 & 2 \end{pmatrix}$

19. $\mathbf{A} = \begin{pmatrix} 2 & 1 & 0 \\ -1 & 2 & 1 \\ 1 & 2 & 1 \end{pmatrix}$

20. $\mathbf{A} = \begin{pmatrix} 3 & 2 & 1 \\ 4 & 1 & 0 \\ -2 & 5 & -1 \end{pmatrix}$

21. $\mathbf{A} = \begin{pmatrix} 2 & 1 & 1 \\ 1 & -2 & -3 \\ 3 & 2 & 4 \end{pmatrix}$

22. $\mathbf{A} = \begin{pmatrix} 4 & 1 & -1 \\ 6 & 2 & -3 \\ -2 & -1 & 2 \end{pmatrix}$

In Problems 23 and 24 show that the given matrix is nonsingular for every real value of t. Find $\mathbf{A}^{-1}(t)$ using Theorem II.2.

23. $\mathbf{A}(t) = \begin{pmatrix} 2e^{-t} & e^{4t} \\ 4e^{-t} & 3e^{4t} \end{pmatrix}$

24. $\mathbf{A}(t) = \begin{pmatrix} 2e^{t}\sin t & -2e^{t}\cos t \\ e^{t}\cos t & e^{t}\sin t \end{pmatrix}$

In Problems 25–28 find $d\mathbf{X}/dt$.

25. $\mathbf{X} = \begin{pmatrix} 5e^{-t} \\ 2e^{-t} \\ -7e^{-t} \end{pmatrix}$

26. $\mathbf{X} = \begin{pmatrix} \frac{1}{2}\sin 2t - 4\cos 2t \\ -3\sin 2t + 5\cos 2t \end{pmatrix}$

27. $\mathbf{X} = 2\begin{pmatrix} 1 \\ -1 \end{pmatrix}e^{2t} + 4\begin{pmatrix} 2 \\ 1 \end{pmatrix}e^{-3t}$

28. $\mathbf{X} = \begin{pmatrix} 5te^{2t} \\ t\sin 3t \end{pmatrix}$

29. Let $\mathbf{A}(t) = \begin{pmatrix} e^{4t} & \cos \pi t \\ 2t & 3t^2 - 1 \end{pmatrix}$. Find

(a) $\dfrac{d\mathbf{A}}{dt}$ **(b)** $\displaystyle\int_0^2 \mathbf{A}(t)\,dt$ **(c)** $\displaystyle\int_0^t \mathbf{A}(s)\,ds$

30. Let $\mathbf{A}(t) = \begin{pmatrix} \dfrac{1}{t^2+1} & 3t \\ t^2 & t \end{pmatrix}$ and $\mathbf{B}(t) = \begin{pmatrix} 6t & 2 \\ 1/t & 4t \end{pmatrix}$. Find

(a) $\dfrac{d\mathbf{A}}{dt}$

(b) $\dfrac{d\mathbf{B}}{dt}$

(c) $\displaystyle\int_0^1 \mathbf{A}(t)\,dt$

(d) $\displaystyle\int_1^2 \mathbf{B}(t)\,dt$

(e) $\mathbf{A}(t)\mathbf{B}(t)$

(f) $\dfrac{d}{dt}\mathbf{A}(t)\mathbf{B}(t)$

(g) $\displaystyle\int_1^t \mathbf{A}(s)\mathbf{B}(s)\,ds$

II.2 GAUSSIAN AND GAUSS-JORDAN ELIMINATION

In Problems 31–38 solve the given system of equations by either Gaussian elimination or Gauss-Jordan elimination.

31.
$$\begin{aligned} x + y - 2z &= 14 \\ 2x - y + z &= 0 \\ 6x + 3y + 4z &= 1 \end{aligned}$$

32.
$$\begin{aligned} 5x - 2y + 4z &= 10 \\ x + y + z &= 9 \\ 4x - 3y + 3z &= 1 \end{aligned}$$

33.
$$\begin{aligned} y + z &= -5 \\ 5x + 4y - 16z &= -10 \\ x - y - 5z &= 7 \end{aligned}$$

34.
$$\begin{aligned} 3x + y + z &= 4 \\ 4x + 2y - z &= 7 \\ x + y - 3z &= 6 \end{aligned}$$

35.
$$\begin{aligned} 2x + y + z &= 4 \\ 10x - 2y + 2z &= -1 \\ 6x - 2y + 4z &= 8 \end{aligned}$$

36.
$$\begin{aligned} x + 2z &= 8 \\ x + 2y - 2z &= 4 \\ 2x + 5y - 6z &= 6 \end{aligned}$$

37.
$$\begin{aligned} x_1 + x_2 - x_3 - x_4 &= -1 \\ x_1 + x_2 + x_3 + x_4 &= 3 \\ x_1 - x_2 + x_3 - x_4 &= 3 \\ 4x_1 + x_2 - 2x_3 + x_4 &= 0 \end{aligned}$$

38.
$$\begin{aligned} 2x_1 + x_2 + x_3 &= 0 \\ x_1 + 3x_2 + x_3 &= 0 \\ 7x_1 + x_2 + 3x_3 &= 0 \end{aligned}$$

In Problems 39 and 40 use Gauss-Jordan elimination to demonstrate that the given system of equations has no solution.

39.
$$\begin{aligned} x + 2y + 4z &= 2 \\ 2x + 4y + 3z &= 1 \\ x + 2y - z &= 7 \end{aligned}$$

40.
$$\begin{aligned} x_1 + x_2 - x_3 + 3x_4 &= 1 \\ x_2 - x_3 - 4x_4 &= 0 \\ x_1 + 2x_2 - 2x_3 - x_4 &= 6 \\ 4x_1 + 7x_2 - 7x_3 &= 9 \end{aligned}$$

In Problems 41–46 use Theorem II.3 to find $\mathbf{A}^{-1}$ for the given matrix or show that no inverse exists.

41. $\mathbf{A} = \begin{pmatrix} 4 & 2 & 3 \\ 2 & 1 & 0 \\ -1 & -2 & 0 \end{pmatrix}$

42. $\mathbf{A} = \begin{pmatrix} 2 & 4 & -2 \\ 4 & 2 & -2 \\ 8 & 10 & -6 \end{pmatrix}$

43. $\mathbf{A} = \begin{pmatrix} -1 & 3 & 0 \\ 1 & -2 & 1 \\ 0 & 1 & 2 \end{pmatrix}$

44. $\mathbf{A} = \begin{pmatrix} 1 & 2 & 3 \\ 0 & 1 & 4 \\ 0 & 0 & 8 \end{pmatrix}$

45. $\mathbf{A} = \begin{pmatrix} 1 & 2 & 3 & 1 \\ -1 & 0 & 2 & 1 \\ 2 & 1 & -3 & 0 \\ 1 & 1 & 2 & 1 \end{pmatrix}$

46. $\mathbf{A} = \begin{pmatrix} 1 & 0 & 0 & 0 \\ 0 & 0 & 1 & 0 \\ 0 & 0 & 0 & 1 \\ 0 & 1 & 0 & 0 \end{pmatrix}$

II.3 THE EIGENVALUE PROBLEM

In Problems 47–54 find the eigenvalues and eigenvectors of the given matrix.

47. $\begin{pmatrix} -1 & 2 \\ -7 & 8 \end{pmatrix}$

48. $\begin{pmatrix} 2 & 1 \\ 2 & 1 \end{pmatrix}$

49. $\begin{pmatrix} -8 & -1 \\ 16 & 0 \end{pmatrix}$

50. $\begin{pmatrix} 1 & 1 \\ \frac{1}{4} & 1 \end{pmatrix}$

51. $\begin{pmatrix} 5 & -1 & 0 \\ 0 & -5 & 9 \\ 5 & -1 & 0 \end{pmatrix}$

52. $\begin{pmatrix} 3 & 0 & 0 \\ 0 & 2 & 0 \\ 4 & 0 & 1 \end{pmatrix}$

53. $\begin{pmatrix} 0 & 4 & 0 \\ -1 & -4 & 0 \\ 0 & 0 & -2 \end{pmatrix}$

54. $\begin{pmatrix} 1 & 6 & 0 \\ 0 & 2 & 1 \\ 0 & 1 & 2 \end{pmatrix}$

In Problems 55 and 56 show that the given matrix has complex eigenvalues. Find the eigenvectors of the matrix.

55. $\begin{pmatrix} -1 & 2 \\ -5 & 1 \end{pmatrix}$

56. $\begin{pmatrix} 2 & -1 & 0 \\ 5 & 2 & 4 \\ 0 & 1 & 2 \end{pmatrix}$

Miscellaneous Problems

57. If $\mathbf{A}(t)$ is a 2×2 matrix of differentiable functions and $\mathbf{X}(t)$ is a 2×1 column matrix of differentiable functions, prove the product rule

$$\frac{d}{dt}[\mathbf{A}(t)\mathbf{X}(t)] = \mathbf{A}(t)\mathbf{X}'(t) + \mathbf{A}'(t)\mathbf{X}(t).$$

58. Derive formula (3). [*Hint*: Find a matrix

$$\mathbf{B} = \begin{pmatrix} b_{11} & b_{12} \\ b_{21} & b_{22} \end{pmatrix}$$

for which $\mathbf{AB} = \mathbf{I}$. Solve for b_{11}, b_{12}, b_{21}, and b_{22}. Then show that $\mathbf{BA} = \mathbf{I}$.]

59. If $\mathbf{A}$ is nonsingular and $\mathbf{AB} = \mathbf{AC}$, show that $\mathbf{B} = \mathbf{C}$.

60. If $\mathbf{A}$ and $\mathbf{B}$ are nonsingular, show that $(\mathbf{AB})^{-1} = \mathbf{B}^{-1}\mathbf{A}^{-1}$.

61. Let $\mathbf{A}$ and $\mathbf{B}$ be $n \times n$ matrices. In general, is

$$(\mathbf{A} + \mathbf{B})^2 = \mathbf{A}^2 + 2\mathbf{AB} + \mathbf{B}^2?$$

62. A square matrix $\mathbf{A}$ is said to be a **diagonal matrix** if all its entries off the main diagonal are zero—that is, $a_{ij} = 0, i \neq j$. The entries a_{ii} on the main diagonal may or may not be zero. The multiplicative identity matrix $\mathbf{I}$ is an example of a diagonal matrix.

(a) Find the inverse of the 2×2 diagonal matrix

$$\mathbf{A} = \begin{pmatrix} a_{11} & 0 \\ 0 & a_{22} \end{pmatrix}$$

when $a_{11} \neq 0, a_{22} \neq 0$.

(b) Find the inverse of a 3×3 diagonal matrix $\mathbf{A}$ whose main diagonal entries a_{ii} are all nonzero.

(c) In general, what is the inverse of an $n \times n$ diagonal matrix $\mathbf{A}$ whose main diagonal entries a_{ii} are all nonzero?

Appendix III
Laplace Transforms

$f(t)$	$\mathscr{L}\{f(t)\} = F(s)$
1. 1	$\frac{1}{s}$
2. t	$\frac{1}{s^2}$
3. t^n	$\frac{n!}{s^{n+1}}$, n a positive integer
4. $t^{-1/2}$	$\sqrt{\frac{\pi}{s}}$
5. $t^{1/2}$	$\frac{\sqrt{\pi}}{2s^{3/2}}$
6. t^{α}	$\frac{\Gamma(\alpha + 1)}{s^{\alpha+1}}$, $\alpha > -1$
7. $\sin kt$	$\frac{k}{s^2 + k^2}$
8. $\cos kt$	$\frac{s}{s^2 + k^2}$
9. $\sin^2 kt$	$\frac{2k^2}{s(s^2 + 4k^2)}$
10. $\cos^2 kt$	$\frac{s^2 + 2k^2}{s(s^2 + 4k^2)}$
11. e^{at}	$\frac{1}{s - a}$
12. $\sinh kt$	$\frac{k}{s^2 - k^2}$
13. $\cosh kt$	$\frac{s}{s^2 - k^2}$
14. $\sinh^2 kt$	$\frac{2k^2}{s(s^2 - 4k^2)}$
15. $\cosh^2 kt$	$\frac{s^2 - 2k^2}{s(s^2 - 4k^2)}$
16. te^{at}	$\frac{1}{(s - a)^2}$
17. $t^n e^{at}$	$\frac{n!}{(s - a)^{n+1}}$, n a positive integer

$f(t)$	$\mathscr{L}\{f(t)\} = F(s)$
18. $e^{at} \sin kt$	$\dfrac{k}{(s-a)^2 + k^2}$
19. $e^{at} \cos kt$	$\dfrac{s-a}{(s-a)^2 + k^2}$
20. $e^{at} \sinh kt$	$\dfrac{k}{(s-a)^2 - k^2}$
21. $e^{at} \cosh kt$	$\dfrac{s-a}{(s-a)^2 - k^2}$
22. $t \sin kt$	$\dfrac{2ks}{(s^2 + k^2)^2}$
23. $t \cos kt$	$\dfrac{s^2 - k^2}{(s^2 + k^2)^2}$
24. $\sin kt + kt \cos kt$	$\dfrac{2ks^2}{(s^2 + k^2)^2}$
25. $\sin kt - kt \cos kt$	$\dfrac{2k^3}{(s^2 + k^2)^2}$
26. $t \sinh kt$	$\dfrac{2ks}{(s^2 - k^2)^2}$
27. $t \cosh kt$	$\dfrac{s^2 + k^2}{(s^2 - k^2)^2}$
28. $\dfrac{e^{at} - e^{bt}}{a - b}$	$\dfrac{1}{(s-a)(s-b)}$
29. $\dfrac{ae^{at} - be^{bt}}{a - b}$	$\dfrac{s}{(s-a)(s-b)}$
30. $1 - \cos kt$	$\dfrac{k^2}{s(s^2 + k^2)}$
31. $kt - \sin kt$	$\dfrac{k^3}{s^2(s^2 + k^2)}$
32. $\dfrac{a \sin bt - b \sin at}{ab(a^2 - b^2)}$	$\dfrac{1}{(s^2 + a^2)(s^2 + b^2)}$
33. $\dfrac{\cos bt - \cos at}{a^2 - b^2}$	$\dfrac{s}{(s^2 + a^2)(s^2 + b^2)}$
34. $\sin kt \sinh kt$	$\dfrac{2k^2 s}{s^4 + 4k^4}$
35. $\sin kt \cosh kt$	$\dfrac{k(s^2 + 2k^2)}{s^4 + 4k^4}$
36. $\cos kt \sinh kt$	$\dfrac{k(s^2 - 2k^2)}{s^4 + 4k^4}$
37. $\cos kt \cosh kt$	$\dfrac{s^3}{s^4 + 4k^4}$

$f(t)$	$\mathscr{L}\{f(t)\} = F(s)$
38. $J_0(kt)$	$\dfrac{1}{\sqrt{s^2 + k^2}}$
39. $\dfrac{e^{bt} - e^{at}}{t}$	$\ln \dfrac{s - a}{s - b}$
40. $\dfrac{2(1 - \cos kt)}{t}$	$\ln \dfrac{s^2 + k^2}{s^2}$
41. $\dfrac{2(1 - \cosh kt)}{t}$	$\ln \dfrac{s^2 - k^2}{s^2}$
42. $\dfrac{\sin at}{t}$	$\arctan\left(\dfrac{a}{s}\right)$
43. $\dfrac{\sin at \cos bt}{t}$	$\dfrac{1}{2}\arctan\dfrac{a + b}{s} + \dfrac{1}{2}\arctan\dfrac{a - b}{s}$
44. $\dfrac{1}{\sqrt{\pi t}} e^{-a^2/4t}$	$\dfrac{e^{-a\sqrt{s}}}{\sqrt{s}}$
45. $\dfrac{a}{2\sqrt{\pi t^3}} e^{-a^2/4t}$	$e^{-a\sqrt{s}}$
46. $\operatorname{erfc}\left(\dfrac{a}{2\sqrt{t}}\right)$	$\dfrac{e^{-a\sqrt{s}}}{s}$
47. $2\sqrt{\dfrac{t}{\pi}}\, e^{-a^2/4t} - a \operatorname{erfc}\left(\dfrac{a}{2\sqrt{t}}\right)$	$\dfrac{e^{-a\sqrt{s}}}{s\sqrt{s}}$
48. $e^{ab}e^{b^2 t} \operatorname{erfc}\left(b\sqrt{t} + \dfrac{a}{2\sqrt{t}}\right)$	$\dfrac{e^{-a\sqrt{s}}}{\sqrt{s}(\sqrt{s} + b)}$
49. $-e^{ab}e^{b^2 t} \operatorname{erfc}\left(b\sqrt{t} + \dfrac{a}{2\sqrt{t}}\right) + \operatorname{erfc}\left(\dfrac{a}{2\sqrt{t}}\right)$	$\dfrac{be^{-a\sqrt{s}}}{s(\sqrt{s} + b)}$
50. $e^{at} f(t)$	$F(s - a)$
51. $\mathscr{U}(t - a)$	$\dfrac{e^{-as}}{s}$
52. $f(t - a)\mathscr{U}(t - a)$	$e^{-as}F(s)$
53. $g(t)\mathscr{U}(t - a)$	$e^{-as}\mathscr{L}\{g(t + a)\}$
54. $f^{(n)}(t)$	$s^n F(s) - s^{(n-1)} f(0) - \cdots - f^{(n-1)}(0)$
55. $t^n f(t)$	$(-1)^n \dfrac{d^n}{ds^n} F(s)$
56. $\displaystyle\int_0^t f(\tau) g(t - \tau)\, d\tau$	$F(s)G(s)$
57. $\delta(t)$	1
58. $\delta(t - t_0)$	e^{-st_0}

ANSWERS FOR SELECTED ODD-NUMBERED PROBLEMS

EXERCISES 1.1 (PAGE 10)

1. linear, second order **3.** linear, fourth order
5. nonlinear, second order **7.** linear, third order
9. linear in x but nonlinear in y
15. domain of function is $[-2, \infty)$; largest interval of definition for solution is $(-2, \infty)$
17. domain of function is the set of real numbers except $x = 2$ and $x = -2$; largest intervals of definition for solution are $(-\infty, -2)$, $(-2, 2)$, or $(2, \infty)$
19. $X = \dfrac{e^t - 1}{e^t - 2}$ defined on $(-\infty, \ln 2)$ or on $(\ln 2, \infty)$
27. $m = -2$ **29.** $m = 2, m = 3$ **31.** $m = 0, m = -1$
33. $y = 2$ **35.** no constant solutions

EXERCISES 1.2 (PAGE 17)

1. $y = 1/(1 - 4e^{-x})$
3. $y = 1/(x^2 - 1)$; $(1, \infty)$
5. $y = 1/(x^2 + 1)$; $(-\infty, \infty)$
7. $x = -\cos t + 8 \sin t$
9. $x = \frac{\sqrt{3}}{4} \cos t + \frac{1}{4} \sin t$ **11.** $y = \frac{3}{2}e^x - \frac{1}{2}e^{-x}$
13. $y = 5e^{-x-1}$ **15.** $y = 0, y = x^3$
17. half-planes defined by either $y > 0$ or $y < 0$
19. half-planes defined by either $x > 0$ or $x < 0$
21. the regions defined by $y > 2$, $y < -2$, or $-2 < y < 2$
23. any region not containing $(0, 0)$
25. yes
27. no
29. **(a)** $y = cx$
(b) any rectangular region not touching the y-axis
(c) No, the function is not differentiable at $x = 0$.
31. **(b)** $y = 1/(1 - x)$ on $(-\infty, 1)$;
$y = -1/(x + 1)$ on $(-1, \infty)$;
(c) $y = 0$ on $(-\infty, \infty)$
39. $y = 3 \sin 2x$
41. $y = 0$
43. no solution

EXERCISES 1.3 (PAGE 28)

1. $\dfrac{dP}{dt} = kP + r$; $\dfrac{dP}{dt} = kP - r$
3. $\dfrac{dP}{dt} = k_1 P - k_2 P^2$
7. $\dfrac{dx}{dt} = kx(1000 - x)$
9. $\dfrac{dA}{dt} + \dfrac{1}{100}A = 0$; $A(0) = 50$
11. $\dfrac{dA}{dt} + \dfrac{7}{600 - t}A = 6$ **13.** $\dfrac{dh}{dt} = -\dfrac{c\pi}{450}\sqrt{h}$
15. $L\dfrac{di}{dt} + Ri = E(t)$ **17.** $m\dfrac{dv}{dt} = mg - kv^2$
19. $m\dfrac{d^2x}{dt^2} = -kx$
21. $m\dfrac{dv}{dt} + v\dfrac{dm}{dt} + kv = -mg + R$
23. $\dfrac{d^2r}{dt^2} + \dfrac{gR^2}{r^2} = 0$ **25.** $\dfrac{dA}{dt} = k(M - A), k > 0$
27. $\dfrac{dx}{dt} + kx = r, k > 0$ **29.** $\dfrac{dy}{dx} = \dfrac{-x + \sqrt{x^2 + y^2}}{y}$

CHAPTER 1 IN REVIEW (PAGE 33)

1. $\dfrac{dy}{dx} = 10y$ **3.** $y'' + k^2y = 0$
5. $y'' - 2y' + y = 0$ **7.** **(a)**, **(d)**
9. **(b)** **11.** **(b)**
13. $y = c_1$ and $y = c_2e^x$, c_1 and c_2 constants
15. $y' = x^2 + y^2$
17. **(a)** The domain is the set of all real numbers.
(b) either $(-\infty, 0)$ or $(0, \infty)$
19. For $x_0 = -1$ the interval is $(-\infty, 0)$, and for $x_0 = 2$ the interval is $(0, \infty)$.
21. **(c)** $y = \begin{cases} -x^2, & x < 0 \\ x^2, & x \geq 0 \end{cases}$ **23.** $(-\infty, \infty)$
25. $(0, \infty)$ **31.** $y = \frac{1}{2}e^{3x} - \frac{1}{2}e^{-x} - 2x$
33. $y = \frac{3}{2}e^{3x-3} + \frac{9}{2}e^{-x+1} - 2x$.
35. $y_0 = -3, y_1 = 0$
37. $\dfrac{dP}{dt} = k(P - 200 + 10t)$

EXERCISES 2.1 (PAGE 43)

21. 0 is asymptotically stable (attractor); 3 is unstable (repeller).
23. 2 is semi-stable.
25. -2 is unstable (repeller); 0 is semi-stable; 2 is asymptotically stable (attractor).
27. -1 is asymptotically stable (attractor); 0 is unstable (repeller).
39. $0 < P_0 < h/k$
41. $\sqrt{mg/k}$

EXERCISES 2.2 (PAGE 51)

1. $y = -\frac{1}{5}\cos 5x + c$
3. $y = \frac{1}{3}e^{-3x} + c$
5. $y = cx^4$
7. $-3e^{-2y} = 2e^{3x} + c$
9. $\frac{1}{3}x^3 \ln x - \frac{1}{9}x^3 = \frac{1}{2}y^2 + 2y + \ln|y| + c$
11. $4\cos y = 2x + \sin 2x + c$
13. $(e^x + 1)^{-2} + 2(e^y + 1)^{-1} = c$
15. $S = ce^{kr}$
17. $P = \dfrac{ce^t}{1 + ce^t}$
19. $(y + 3)^5 e^x = c(x + 4)^5 e^y$
21. $y = \sin\left(\frac{1}{2}x^2 + c\right)$
23. $x = \tan\left(4t - \frac{3}{4}\pi\right)$
25. $y = \dfrac{e^{-(1+1/x)}}{x}$
27. $y = \frac{1}{2}x + \frac{\sqrt{3}}{2}\sqrt{1 - x^2}$
29. $y = e^{\int_4^x e^{t^2} dt}$
31. $y = -\sqrt{x^2 + x - 1}$; $\left(-\infty, -\frac{1 + \sqrt{5}}{2}\right)$.
33. $y = -\ln(2 - e^x)$; $(-\infty, \ln 2)$
35. **(a)** $y = 2, y = -2, y = 2\dfrac{3 - e^{4x-1}}{3 + e^{4x-1}}$
37. $y = -1$ and $y = 1$ are singular solutions of Problem 21; $y = 0$ of Problem 22
39. $y = 1$
41. $y = 1 + \frac{1}{10}\tan\left(\frac{1}{10}x\right)$
45. $y = \tan x - \sec x + c$
47. $y = [-1 + c(1 + \sqrt{x})]^2$
49. $y = 2\sqrt{\sqrt{x}e^{\sqrt{x}} - e^{\sqrt{x}} + 4}$
57. $y(x) = (4h/L^2)x^2 + a$

EXERCISES 2.3 (PAGE 61)

1. $y = ce^{5x}, (-\infty, \infty)$
3. $y = \frac{1}{4}e^{3x} + ce^{-x}, (-\infty, \infty)$; ce^{-x} is transient
5. $y = \frac{1}{3} + ce^{-x^3}, (-\infty, \infty)$; ce^{-x^3} is transient
7. $y = x^{-1}\ln x + cx^{-1}, (0, \infty)$; solution is transient
9. $y = cx - x\cos x, (0, \infty)$
11. $y = \frac{1}{7}x^3 - \frac{1}{5}x + cx^{-4}, (0, \infty)$; cx^{-4} is transient
13. $y = \frac{1}{2}x^{-2}e^x + cx^{-2}e^{-x}, (0, \infty)$; $cx^{-2}e^{-x}$ is transient
15. $x = 2y^6 + cy^4, (0, \infty)$
17. $y = \sin x + c\cos x, (-\pi/2, \pi/2)$
19. $(x + 1)e^x y = x^2 + c, (-1, \infty)$; solution is transient
21. $(\sec\theta + \tan\theta)r = \theta - \cos\theta + c, (-\pi/2, \pi/2)$
23. $y = e^{-3x} + cx^{-1}e^{-3x}, (0, \infty)$; solution is transient
25. $y = -\frac{1}{5}x - \frac{1}{25} + \frac{76}{25}e^{5x}; (-\infty, \infty)$
27. $y = x^{-1}e^x + (2 - e)x^{-1}; (0, \infty)$
29. $i = \dfrac{E}{R} + \left(i_0 - \dfrac{E}{R}\right)e^{-Rt/L}; (-\infty, \infty)$
31. $y = 2x + 1 + 5/x; (0, \infty)$
33. $(x + 1)y = x\ln x - x + 21; (0, \infty)$
35. $y = -2 + 3e^{-\cos x}; (-\infty, \infty)$
37. $y = \begin{cases} \frac{1}{2}(1 - e^{-2x}), & 0 \le x \le 3 \\ \frac{1}{2}(e^6 - 1)e^{-2x}, & x > 3 \end{cases}$
39. $y = \begin{cases} \frac{1}{2} + \frac{3}{2}e^{-x^2}, & 0 \le x < 1 \\ \left(\frac{1}{2}e + \frac{3}{2}\right)e^{-x^2}, & x \ge 1 \end{cases}$
41. $y = \begin{cases} 2x - 1 + 4e^{-2x}, & 0 \le x \le 1 \\ 4x^2\ln x + (1 + 4e^{-2})x^2, & x > 1 \end{cases}$
43. $y = e^{x^2-1} + \frac{1}{2}\sqrt{\pi}e^{x^2}(\operatorname{erf}(x) - \operatorname{erf}(1))$
53. $E(t) = E_0 e^{-(t-4)/RC}$

EXERCISES 2.4 (PAGE 69)

1. $x^2 - x + \frac{3}{2}y^2 + 7y = c$
3. $\frac{5}{2}x^2 + 4xy - 2y^4 = c$
5. $x^2y^2 - 3x + 4y = c$
7. not exact
9. $xy^3 + y^2\cos x - \frac{1}{2}x^2 = c$
11. not exact
13. $xy - 2xe^x + 2e^x - 2x^3 = c$
15. $x^3y^3 - \tan^{-1} 3x = c$
17. $-\ln|\cos x| + \cos x \sin y = c$
19. $t^4y - 5t^3 - ty + y^3 = c$
21. $\frac{1}{3}x^3 + x^2y + xy^2 - y = \frac{4}{3}$
23. $4ty + t^2 - 5t + 3y^2 - y = 8$
25. $y^2\sin x - x^3y - x^2 + y\ln y - y = 0$
27. $k = 10$
29. $x^2y^2\cos x = c$
31. $x^2y^2 + x^3 = c$
33. $3x^2y^3 + y^4 = c$
35. $-2ye^{3x} + \frac{10}{3}e^{3x} + x = c$
37. $e^{y^2}(x^2 + 4) = 20$
39. **(c)** $y_1(x) = -x^2 - \sqrt{x^4 - x^3 + 4}$
$y_2(x) = -x^2 + \sqrt{x^4 - x^3 + 4}$
45. **(a)** $v(x) = 8\sqrt{\dfrac{x}{3} - \dfrac{9}{x^2}}$ **(b)** 12.7 ft/s

EXERCISES 2.5 (PAGE 74)

1. $y + x\ln|x| = cx$
3. $(x - y)\ln|x - y| = y + c(x - y)$
5. $x + y\ln|x| = cy$
7. $\ln(x^2 + y^2) + 2\tan^{-1}(y/x) = c$
9. $4x = y(\ln|y| - c)^2$
11. $y^3 + 3x^3\ln|x| = 8x^3$
13. $\ln|x| = e^{y/x} - 1$
15. $y^3 = 1 + cx^{-3}$
17. $y^{-3} = x + \frac{1}{3} + ce^{3x}$
19. $e^{t/y} = ct$
21. $y^{-3} = -\frac{9}{5}x^{-1} + \frac{49}{5}x^{-6}$
23. $y = -x - 1 + \tan(x + c)$

25. $2y - 2x + \sin 2(x + y) = c$
27. $4(y - 2x + 3) = (x + c)^2$
29. $-\cot(x + y) + \csc(x + y) = x + \sqrt{2} - 1$
35. (b) $y = \dfrac{2}{x} + \left(-\frac{1}{4}x + cx^{-3}\right)^{-1}$

EXERCISES 2.6 (PAGE 79)

1. $y_2 = 2.9800, \quad y_4 = 3.1151$
3. $y_{10} = 2.5937, \quad y_{20} = 2.6533; y = e^x$
5. $y_5 = 0.4198, \quad y_{10} = 0.4124$
7. $y_5 = 0.5639, \quad y_{10} = 0.5565$
9. $y_5 = 1.2194, \quad y_{10} = 1.2696$
13. Euler: $y_{10} = 3.8191, \quad y_{20} = 5.9363$
RK4: $y_{10} = 42.9931, \quad y_{20} = 84.0132$

CHAPTER 2 IN REVIEW (PAGE 80)

1. $-A/k$, a repeller for $k > 0$, an attractor for $k < 0$
3. true
5. $\dfrac{d^3y}{dx^3} = x\sin y$
7. true
9. $y = c_1e^{e^x}$
11. $\dfrac{dy}{dx} + (\sin x)y = x$
13. $\dfrac{dy}{dx} = (y - 1)^2(y - 3)^2$
15. semi-stable for n even and unstable for n odd; semi-stable for n even and asymptotically stable for n odd.
19. $2x + \sin 2x = 2\ln(y^2 + 1) + c$
21. $(6x + 1)y^3 = -3x^3 + c$
23. $Q = ct^{-1} + \frac{1}{25}t^4(-1 + 5\ln t)$
25. $y = \frac{1}{4} + c(x^2 + 4)^{-4}$
27. $y = \csc x, (\pi, 2\pi)$
29. (b) $y = \frac{1}{4}(x + 2\sqrt{y_0} - x_0)^2, (x_0 - 2\sqrt{y_0}, \infty)$

EXERCISES 3.1 (PAGE 90)

1. 7.9 yr; 10 yr
3. 760; approximately 11 persons/yr
5. 11 h
7. 136.5 h
9. $I(15) = 0.00098I_0$ or approximately 0.1% of I_0
11. 15,600 years
13. $T(1) = 36.67°$ F; approximately 3.06 min
15. approximately 82.1 s; approximately 145.7 s
17. 390°
19. about 1.6 hours prior to the discovery of the body
21. $A(t) = 200 - 170e^{-t/50}$
23. $A(t) = 1000 - 1000e^{-t/100}$
25. $A(t) = 1000 - 10t - \frac{1}{10}(100 - t)^2$; 100 min
27. 64.38 lb
29. $i(t) = \frac{3}{5} - \frac{3}{5}e^{-500t}$; $i \to \frac{3}{5}$ as $t \to \infty$
31. $q(t) = \frac{1}{100} - \frac{1}{100}e^{-50t}$; $i(t) = \frac{1}{2}e^{-50t}$
33. $i(t) = \begin{cases} 60 - 60e^{-t/10}, & 0 \le t \le 20 \\ 60(e^2 - 1)e^{-t/10}, & t > 20 \end{cases}$
35. (a) $v(t) = \dfrac{mg}{k} + \left(v_0 - \dfrac{mg}{k}\right)e^{-kt/m}$
(b) $v \to \dfrac{mg}{k}$ as $t \to \infty$
(c) $s(t) = \dfrac{mg}{k}t - \dfrac{m}{k}\left(v_0 - \dfrac{mg}{k}\right)e^{-kt/m} + \dfrac{m}{k}\left(v_0 - \dfrac{mg}{k}\right)$
39. (a) $v(t) = \dfrac{\rho g}{4k}\left(\dfrac{k}{\rho}t + r_0\right) - \dfrac{\rho g r_0}{4k}\left(\dfrac{r_0}{\dfrac{k}{\rho}t + r_0}\right)^3$
(c) $33\frac{1}{3}$ seconds
41. (a) $P(t) = P_0e^{(k_1 - k_2)t}$
43. (a) As $t \to \infty$, $x(t) \to r/k$.
(b) $x(t) = r/k - (r/k)e^{-kt}$; $(\ln 2)/k$
47. (c) 1.988 ft

EXERCISES 3.2 (PAGE 100)

1. (a) $N = 2000$
(b) $N(t) = \dfrac{2000\,e^t}{1999 + e^t}$; $N(10) = 1834$
3. 1,000,000; 5.29 mo
5. (b) $P(t) = \dfrac{4(P_0 - 1) - (P_0 - 4)e^{-3t}}{(P_0 - 1) - (P_0 - 4)e^{-3t}}$
(c) For $0 < P_0 < 1$, time of extinction is
$$t = -\frac{1}{3}\ln\frac{4(P_0 - 1)}{P_0 - 4}.$$
7. $P(t) = \dfrac{5}{2} + \dfrac{\sqrt{3}}{2}\tan\left[-\dfrac{\sqrt{3}}{2}t + \tan^{-1}\left(\dfrac{2P_0 - 5}{\sqrt{3}}\right)\right]$;
time of extinction is
$$t = \frac{2}{\sqrt{3}}\left[\tan^{-1}\frac{5}{\sqrt{3}} + \tan^{-1}\left(\frac{2P_0 - 5}{\sqrt{3}}\right)\right]$$
9. 29.3 g; $X \to 60$ as $t \to \infty$; 0 g of A and 30 g of B
11. (a) $h(t) = \left(\sqrt{H} - \dfrac{4A_h}{A_w}t\right)^2$; I is $0 \le t \le \sqrt{H}A_w/4A_h$
(b) $576\sqrt{10}$ s or 30.36 min
13. (a) approximately 858.65 s or 14.31 min
(b) 243 s or 4.05 min

15. **(a)** $v(t) = \sqrt{\frac{mg}{k}}\tanh\left(\sqrt{\frac{kg}{m}}t + c_1\right)$

where $c_1 = \tanh^{-1}\left(\sqrt{\frac{k}{mg}}v_0\right)$

(b) $\sqrt{\frac{mg}{k}}$

(c) $s(t) = \frac{m}{k}\ln\cosh\left(\sqrt{\frac{kg}{m}}t + c_1\right) + c_2,$

where $c_2 = -(m/k)\ln\cosh c_1$

17. **(a)** $m\frac{dv}{dt} = mg - kv^2 - \rho V,$

where ρ is the weight density of water

(b) $v(t) = \sqrt{\frac{mg - \rho V}{k}}\tanh\left(\frac{\sqrt{kmg - k\rho V}}{m}t + c_1\right)$

(c) $\sqrt{\frac{mg - \rho V}{k}}$

19. **(a)** $W = 0$ and $W = 2$
(b) $W(x) = 2\,\text{sech}^2(x - c_1)$
(c) $W(x) = 2\,\text{sech}^2 x$

21. **(a)** $P(t) = \frac{1}{(-0.001350t + 10^{-0.01})^{100}}$
(b) approximately 724 months
(b) approximately 12,839 and 28,630,966

EXERCISES 3.3 (PAGE 110)

1. $x(t) = x_0\,e^{-\lambda_1 t}$

$y(t) = \frac{x_0\lambda_1}{\lambda_2 - \lambda_1}(e^{-\lambda_1 t} - e^{-\lambda_2 t})$

$z(t) = x_0\left(1 - \frac{\lambda_2}{\lambda_2 - \lambda_1}e^{-\lambda_1 t} + \frac{\lambda_1}{\lambda_2 - \lambda_1}e^{-\lambda_2 t}\right)$

3. 5, 20, 147 days. The time when $y(t)$ and $z(t)$ are the same makes sense because most of A and half of B are gone, so half of C should have been formed.

5. $\frac{dx_1}{dt} = 6 - \frac{2}{25}x_1 + \frac{1}{50}x_2$

$\frac{dx_2}{dt} = \frac{2}{25}x_1 - \frac{2}{25}x_2$

7. **(a)** $\frac{dx_1}{dt} = 3\frac{x_2}{100 - t} - 2\frac{x_1}{100 + t}$

$\frac{dx_2}{dt} = 2\frac{x_1}{100 + t} - 3\frac{x_2}{100 - t}$

(b) $x_1(t) + x_2(t) = 150$; $x_2(30) \approx 47.4$ lb

13. $L_1\frac{di_2}{dt} + (R_1 + R_2)i_2 + R_1 i_3 = E(t)$

$L_2\frac{di_3}{dt} + R_1 i_2 + (R_1 + R_3)\,i_3 = E(t)$

15. $i(0) = i_0$, $s(0) = n - i_0$, $r(0) = 0$

CHAPTER 3 IN REVIEW (PAGE 113)

1. $dP/dt = 0.15P$
3. $P(45) = 8.99$ billion

5. $x = 10\ln\left(\frac{10 + \sqrt{100 - y^2}}{y}\right) - \sqrt{100 - y^2}$

7. **(a)** $\frac{BT_1 + T_2}{1 + B}, \frac{BT_1 + T_2}{1 + B}$

(b) $T(t) = \frac{BT_1 + T_2}{1 + B} + \frac{T_1 - T_2}{1 + B}e^{k(1+B)t}$

9. $i(t) = \begin{cases} 4t - \frac{1}{5}t^2, & 0 \le t < 10 \\ 20, & t \ge 10 \end{cases}$

11. $x(t) = \frac{\alpha c_1 e^{\alpha k_1 t}}{1 + c_1 e^{\alpha k_1 t}}$, $y(t) = c_2(1 + c_1 e^{\alpha k_1 t})^{k_2/k_1}$

13. $x = -y + 1 + c_2 e^{-y}$

15. **(a)** $K(t) = K_0 e^{-(\lambda_1 + \lambda_2)t}$,

$C(t) = \frac{\lambda_1}{\lambda_1 + \lambda_2}K_0\left[1 - e^{-(\lambda_1 + \lambda_2)t}\right],$

$A(t) = \frac{\lambda_2}{\lambda_1 + \lambda_2}K_0\left[1 - e^{-(\lambda_1 + \lambda_2)t}\right]$

(b) 1.3×10^9 years

(c) 89%, 11%

EXERCISES 4.1 (PAGE 127)

1. $y = \frac{1}{2}e^x - \frac{1}{2}e^{-x}$
3. $y = 3x - 4x\ln x$
9. $(-\infty, 2)$

11. **(a)** $y = \frac{e}{e^2 - 1}(e^x - e^{-x})$ **(b)** $y = \frac{\sinh x}{\sinh 1}$

13. **(a)** $y = e^x\cos x - e^x\sin x$
(b) no solution
(c) $y = e^x\cos x + e^{-\pi/2}e^x\sin x$
(d) $y = c_2 e^x\sin x$, where c_2 is arbitrary

15. dependent **17.** dependent
19. dependent **21.** independent

23. The functions satisfy the DE and are linearly independent on the interval since $W(e^{-3x}, e^{4x}) = 7e^x \ne 0$; $y = c_1 e^{-3x} + c_2 e^{4x}$.

25. The functions satisfy the DE and are linearly independent on the interval since $W(e^x\cos 2x, e^x\sin 2x) = 2e^{2x} \ne 0$; $y = c_1 e^x\cos 2x + c_2 e^x\sin 2x$.

27. The functions satisfy the DE and are linearly independent on the interval since $W(x^3, x^4) = x^6 \ne 0$; $y = c_1 x^3 + c_2 x^4$.

29. The functions satisfy the DE and are linearly independent on the interval since $W(x, x^{-2}, x^{-2}\ln x) = 9x^{-6} \ne 0$; $y = c_1 x + c_2 x^{-2} + c_3 x^{-2}\ln x$.

35. **(b)** $y_p = x^2 + 3x + 3e^{2x}$; $y_p = -2x^2 - 6x - \frac{1}{3}e^{2x}$

EXERCISES 4.2 (PAGE 131)

1. $y_2 = xe^{2x}$
3. $y_2 = \sin 4x$
5. $y_2 = \sinh x$
7. $y_2 = xe^{2x/3}$
9. $y_2 = x^4 \ln|x|$
11. $y_2 = 1$
13. $y_2 = x\cos(\ln x)$
15. $y_2 = x^2 + x + 2$
17. $y_2 = e^{2x}, y_p = -\frac{1}{2}$
19. $y_2 = e^{2x}, y_p = \frac{5}{2}e^{3x}$

EXERCISES 4.3 (PAGE 137)

1. $y = c_1 + c_2e^{-x/4}$
3. $y = c_1e^{3x} + c_2e^{-2x}$
5. $y = c_1e^{-4x} + c_2xe^{-4x}$
7. $y = c_1e^{2x/3} + c_2e^{-x/4}$
9. $y = c_1\cos 3x + c_2\sin 3x$
11. $y = e^{2x}(c_1\cos x + c_2\sin x)$
13. $y = e^{-x/3}\left(c_1\cos\frac{1}{3}\sqrt{2}\,x + c_2\sin\frac{1}{3}\sqrt{2}\,x\right)$
15. $y = c_1 + c_2e^{-x} + c_3e^{5x}$
17. $y = c_1e^{-x} + c_2e^{3x} + c_3xe^{3x}$
19. $u = c_1e^t + e^{-t}(c_2\cos t + c_3\sin t)$
21. $y = c_1e^{-x} + c_2xe^{-x} + c_3x^2e^{-x}$
23. $y = c_1 + c_2x + e^{-x/2}\left(c_3\cos\frac{1}{2}\sqrt{3}x + c_4\sin\frac{1}{2}\sqrt{3}x\right)$
25. $y = c_1\cos\frac{1}{2}\sqrt{3}x + c_2\sin\frac{1}{2}\sqrt{3}x$
$+ c_3x\cos\frac{1}{2}\sqrt{3}x + c_4x\sin\frac{1}{2}\sqrt{3}x$
27. $u = c_1e^r + c_2re^r + c_3e^{-r} + c_4re^{-r} + c_5e^{-5r}$
29. $y = 2\cos 4x - \frac{1}{2}\sin 4x$
31. $y = -\frac{1}{3}e^{-(t-1)} + \frac{1}{3}e^{5(t-1)}$
33. $y = 0$
35. $y = \frac{5}{36} - \frac{5}{36}e^{-6x} + \frac{1}{6}xe^{-6x}$
37. $y = e^{5x} - xe^{5x}$
39. $y = 0$
41. $y = \dfrac{1}{2}\left(1 - \dfrac{5}{\sqrt{3}}\right)e^{-\sqrt{3}x} + \dfrac{1}{2}\left(1 + \dfrac{5}{\sqrt{3}}\right)e^{\sqrt{3}x}$;
$y = \cosh\sqrt{3}x + \dfrac{5}{\sqrt{3}}\sinh\sqrt{3}x$
49. $y'' - 6y' + 5y = 0$
51. $y'' - 2y' = 0$
53. $y'' + 9y = 0$
55. $y'' + 2y' + 2y = 0$
57. $y''' - 8y'' = 0$

EXERCISES 4.4 (PAGE 147)

1. $y = c_1e^{-x} + c_2e^{-2x} + 3$
3. $y = c_1e^{5x} + c_2xe^{5x} + \frac{6}{5}x + \frac{3}{5}$
5. $y = c_1e^{-2x} + c_2xe^{-2x} + x^2 - 4x + \frac{7}{2}$
7. $y = c_1\cos\sqrt{3}x + c_2\sin\sqrt{3}x + \left(-4x^2 + 4x - \frac{4}{3}\right)e^{3x}$
9. $y = c_1 + c_2e^x + 3x$
11. $y = c_1e^{x/2} + c_2xe^{x/2} + 12 + \frac{1}{2}x^2e^{x/2}$
13. $y = c_1\cos 2x + c_2\sin 2x - \frac{3}{4}x\cos 2x$
15. $y = c_1\cos x + c_2\sin x - \frac{1}{2}x^2\cos x + \frac{1}{2}x\sin x$
17. $y = c_1e^x\cos 2x + c_2e^x\sin 2x + \frac{1}{4}xe^x\sin 2x$
19. $y = c_1e^{-x} + c_2xe^{-x} - \frac{1}{2}\cos x$
$+ \frac{12}{25}\sin 2x - \frac{9}{25}\cos 2x$
21. $y = c_1 + c_2x + c_3e^{6x} - \frac{1}{4}x^2 - \frac{6}{37}\cos x + \frac{1}{37}\sin x$
23. $y = c_1e^x + c_2xe^x + c_3x^2e^x - x - 3 - \frac{2}{3}x^3e^x$
25. $y = c_1\cos x + c_2\sin x + c_3x\cos x + c_4x\sin x$
$+ x^2 - 2x - 3$
27. $y = \sqrt{2}\sin 2x - \frac{1}{2}$
29. $y = -200 + 200e^{-x/5} - 3x^2 + 30x$
31. $y = -10e^{-2x}\cos x + 9e^{-2x}\sin x + 7e^{-4x}$
33. $x = \dfrac{F_0}{2\omega^2}\sin\omega t - \dfrac{F_0}{2\omega}t\cos\omega t$
35. $y = 11 - 11e^x + 9xe^x + 2x - 12x^2e^x + \frac{1}{2}e^{5x}$
37. $y = 6\cos x - 6(\cot 1)\sin x + x^2 - 1$
39. $y = \dfrac{-4\sin\sqrt{3}x}{\sin\sqrt{3} + \sqrt{3}\cos\sqrt{3}} + 2x$
41. $y = \begin{cases} \cos 2x + \frac{5}{6}\sin 2x + \frac{1}{3}\sin x, & 0 \le x \le \pi/2 \\ \frac{2}{3}\cos 2x + \frac{5}{6}\sin 2x, & x > \pi/2 \end{cases}$

EXERCISES 4.5 (PAGE 155)

1. $(3D - 2)(3D + 2)y = \sin x$
3. $(D - 6)(D + 2)y = x - 6$
5. $D(D + 5)^2y = e^x$
7. $(D - 1)(D - 2)(D + 5)y = xe^{-x}$
9. $D(D + 2)(D^2 - 2D + 4)y = 4$
15. D^4
17. $D(D - 2)$
19. $D^2 + 4$
21. $D^3(D^2 + 16)$
23. $(D + 1)(D - 1)^3$
25. $D(D^2 - 2D + 5)$
27. $1, x, x^2, x^3, x^4$
29. $e^{6x}, e^{-3x/2}$
31. $\cos\sqrt{5}x, \sin\sqrt{5}x$
33. $1, e^{5x}, xe^{5x}$
35. $y = c_1e^{-3x} + c_2e^{3x} - 6$
37. $y = c_1 + c_2e^{-x} + 3x$
39. $y = c_1e^{-2x} + c_2xe^{-2x} + \frac{1}{2}x + 1$
41. $y = c_1 + c_2x + c_3e^{-x} + \frac{2}{3}x^4 - \frac{8}{3}x^3 + 8x^2$
43. $y = c_1e^{-3x} + c_2e^{4x} + \frac{1}{7}xe^{4x}$
45. $y = c_1e^{-x} + c_2e^{3x} - e^x + 3$
47. $y = c_1\cos 5x + c_2\sin 5x + \frac{1}{4}\sin x$
49. $y = c_1e^{-3x} + c_2xe^{-3x} - \frac{1}{49}xe^{4x} + \frac{2}{343}e^{4x}$
51. $y = c_1e^{-x} + c_2e^x + \frac{1}{6}x^3e^x - \frac{1}{4}x^2e^x + \frac{1}{4}xe^x - 5$
53. $y = e^x(c_1\cos 2x + c_2\sin 2x) + \frac{1}{3}e^x\sin x$
55. $y = c_1\cos 5x + c_2\sin 5x - 2x\cos 5x$
57. $y = e^{-x/2}\left(c_1\cos\dfrac{\sqrt{3}}{2}x + c_2\sin\dfrac{\sqrt{3}}{2}x\right)$
$+ \sin x + 2\cos x - x\cos x$
59. $y = c_1 + c_2x + c_3e^{-8x} + \frac{11}{256}x^2 + \frac{7}{32}x^3 - \frac{1}{16}x^4$
61. $y = c_1e^x + c_2xe^x + c_3x^2e^x + \frac{1}{6}x^3e^x + x - 13$
63. $y = c_1 + c_2x + c_3e^x + c_4xe^x + \frac{1}{2}x^2e^x + \frac{1}{2}x^2$
65. $y = \frac{5}{8}e^{-8x} + \frac{5}{8}e^{8x} - \frac{1}{4}$
67. $y = -\frac{41}{125} + \frac{41}{125}e^{5x} - \frac{1}{10}x^2 + \frac{9}{25}x$
69. $y = -\pi\cos x - \frac{11}{3}\sin x - \frac{8}{3}\cos 2x + 2x\cos x$
71. $y = 2e^{2x}\cos 2x - \frac{3}{64}e^{2x}\sin 2x + \frac{1}{8}x^3 + \frac{3}{16}x^2 + \frac{3}{32}x$

EXERCISES 4.6 (PAGE 161)

1. $y = c_1 \cos x + c_2 \sin x + x \sin x + \cos x \ln|\cos x|$

3. $y = c_1 \cos x + c_2 \sin x - \frac{1}{2} x \cos x$

5. $y = c_1 \cos x + c_2 \sin x + \frac{1}{2} - \frac{1}{6} \cos 2x$

7. $y = c_1 e^x + c_2 e^{-x} + \frac{1}{2} x \sinh x$

9. $y = c_1 e^{2x} + c_2 e^{-2x} + \frac{1}{4}\left(e^{2x} \ln|x| - e^{-2x} \int_{x_0}^{x} \frac{e^{4t}}{t}\, dt\right)$, $x_0 > 0$

11. $y = c_1 e^{-x} + c_2 e^{-2x} + (e^{-x} + e^{-2x}) \ln(1 + e^x)$

13. $y = c_1 e^{-2x} + c_2 e^{-x} - e^{-2x} \sin e^x$

15. $y = c_1 e^{-t} + c_2 t e^{-t} + \frac{1}{2} t^2 e^{-t} \ln t - \frac{3}{4} t^2 e^{-t}$

17. $y = c_1 e^x \sin x + c_2 e^x \cos x + \frac{1}{3} x e^x \sin x + \frac{1}{3} e^x \cos x \ln|\cos x|$

19. $y = \frac{1}{4} e^{-x/2} + \frac{3}{4} e^{x/2} + \frac{1}{8} x^2 e^{x/2} - \frac{1}{4} x e^{x/2}$

21. $y = \frac{4}{9} e^{-4x} + \frac{25}{36} e^{2x} - \frac{1}{4} e^{-2x} + \frac{1}{9} e^{-x}$

23. $y = c_1 x^{-1/2} \cos x + c_2 x^{-1/2} \sin x + x^{-1/2}$

25. $y = c_1 + c_2 \cos x + c_3 \sin x - \ln|\cos x| - \sin x \ln|\sec x + \tan x|$

27. $y = c_1 e^x + c_2 e^{-x} + c_3 e^{2x} + \frac{1}{30} e^{4x}$

EXERCISES 4.7 (PAGE 168)

1. $y = c_1 x^{-1} + c_2 x^2$

3. $y = c_1 + c_2 \ln x$

5. $y = c_1 \cos(2 \ln x) + c_2 \sin(2 \ln x)$

7. $y = c_1 x^{(2-\sqrt{6})} + c_2 x^{(2+\sqrt{6})}$

9. $y = c_1 \cos\left(\frac{1}{5} \ln x\right) + c_2 \sin\left(\frac{1}{5} \ln x\right)$

11. $y = c_1 x^{-2} + c_2 x^{-2} \ln x$

13. $y = x^{-1/2}\left[c_1 \cos\left(\frac{1}{6}\sqrt{3} \ln x\right) + c_2 \sin\left(\frac{1}{6}\sqrt{3} \ln x\right)\right]$

15. $y = c_1 x^3 + c_2 \cos\left(\sqrt{2} \ln x\right) + c_3 \sin\left(\sqrt{2} \ln x\right)$

17. $y = c_1 + c_2 x + c_3 x^2 + c_4 x^{-3}$

19. $y = c_1 + c_2 x^5 + \frac{1}{5} x^5 \ln x$

21. $y = c_1 x + c_2 x \ln x + x(\ln x)^2$

23. $y = c_1 x^{-1} + c_2 x - \ln x$

25. $y = 2 - 2x^{-2}$

27. $y = \cos(\ln x) + 2 \sin(\ln x)$

29. $y = \frac{3}{4} - \ln x + \frac{1}{4} x^2$

31. $y = c_1 x^{-10} + c_2 x^2$

33. $y = c_1 x^{-1} + c_2 x^{-8} + \frac{1}{30} x^2$

35. $y = x^2[c_1 \cos(3 \ln x) + c_2 \sin(3 \ln x)] + \frac{4}{13} + \frac{3}{10} x$

37. $y = 2(-x)^{1/2} - 5(-x)^{1/2} \ln(-x), x < 0$

39. $y = c_1(x + 3)^2 + c_2(x + 3)^7$

41. $y = c_1 \cos[\ln(x + 2)] + c_2 \sin[\ln(x + 2)]$

EXERCISES 4.8 (PAGE 179)

1. $y_p(x) = \frac{1}{4} \int_{x_0}^{x} \sinh 4(x - t) f(t)\, dt$

3. $y_p(x) = \int_{x_0}^{x} (x - t) e^{-(x-t)} f(t)\, dt$

5. $y_p(x) = \frac{1}{3} \int_{x_0}^{x} \sin 3(x - t) f(t)\, dt$

7. $y = c_1 e^{-4x} + c_2 e^{4x} + \frac{1}{4} \int_{x_0}^{x} \sinh 4(x - t) t e^{-2t}\, dt$

9. $y = c_1 e^{-x} + c_2 x e^{-x} + \int_{x_0}^{x} (x - t) e^{-(x-t)} e^{-t}\, dt$

11. $y = c_1 \cos 3x + c_2 \sin 3x + \frac{1}{3} \int_{x_0}^{x} \sin 3(x - t)(t + \sin t)\, dt$

13. $y_p(x) = \frac{1}{4} x e^{2x} - \frac{1}{16} e^{2x} + \frac{1}{16} e^{-2x}$

15. $y_p(x) = \frac{1}{2} x^2 e^{5x}$

17. $y_p(x) = -\cos x + \frac{\pi}{2} \sin x - x \sin x - \cos x \ln|\sin x|$

19. $y = \frac{25}{16} e^{-2x} - \frac{9}{16} e^{2x} + \frac{1}{4} x e^{2x}$

21. $y = -e^{5x} + 6x e^{5x} + \frac{1}{2} x^2 e^{5x}$

23. $y = -x \sin x - \cos x \ln|\sin x|$

25. $y = (\cos 1 - 2)e^{-x} + (1 + \sin 1 - \cos 1)e^{-2x} - e^{-2x} \sin e^x$

27. $y = 4x - 2x^2 - x \ln x$

29. $y = \frac{46}{45} x^3 - \frac{1}{20} x^{-2} + \frac{1}{36} - \frac{1}{6} \ln x$

31. $y(x) = 5e^x + 3e^{-x} + y_p(x)$,

where $y_p(x) = \begin{cases} 1 - \cosh x, & x < 0 \\ -1 + \cosh x, & x \geq 0 \end{cases}$

33. $y = \cos x - \sin x + y_p(x)$,

where $y_p(x) = \begin{cases} 0, & x < 0 \\ 10 - 10 \cos x, & 0 \leq x \leq 3\pi \\ -20 \cos x, & x > 3\pi \end{cases}$

35. $y_p(x) = (x - 1) \int_{0}^{x} t f(t)\, dt + x \int_{x}^{1} (t - 1) f(t)\, dt$

37. $y_p(x) = \frac{1}{2} x^2 - \frac{1}{2} x$

39. $y_p(x) = \dfrac{\sin(x - 1)}{\sin 1} - \dfrac{\sin x}{\sin 1} + 1$

41. $y_p(x) = -e^x \cos x - e^x \sin x + e^x$

43. $y_p(x) = \frac{1}{2} (\ln x)^2 + \frac{1}{2} \ln x$

EXERCISES 4.9 (PAGE 184)

1. $x = c_1 e^t + c_2 t e^t$
$y = (c_1 - c_2) e^t + c_2 t e^t$

3. $x = c_1 \cos t + c_2 \sin t + t + 1$
$y = c_1 \sin t - c_2 \cos t + t - 1$

5. $x = \frac{1}{2}c_1 \sin t + \frac{1}{2}c_2 \cos t - 2c_3 \sin \sqrt{6}t - 2c_4 \cos \sqrt{6}t$

$y = c_1 \sin t + c_2 \cos t + c_3 \sin \sqrt{6}t + c_4 \cos \sqrt{6}t$

7. $x = c_1 e^{2t} + c_2 e^{-2t} + c_3 \sin 2t + c_4 \cos 2t + \frac{1}{5}e^t$

$y = c_1 e^{2t} + c_2 e^{-2t} - c_3 \sin 2t - c_4 \cos 2t - \frac{1}{5}e^t$

9. $x = c_1 - c_2 \cos t + c_3 \sin t + \frac{17}{15}e^{3t}$

$y = c_1 + c_2 \sin t + c_3 \cos t - \frac{4}{15}e^{3t}$

11. $x = c_1 e^t + c_2 e^{-t/2} \cos \frac{1}{2}\sqrt{3}t + c_3 e^{-t/2} \sin \frac{1}{2}\sqrt{3}t$

$y = \left(-\frac{3}{2}c_2 - \frac{1}{2}\sqrt{3}c_3\right)e^{-t/2} \cos \frac{1}{2}\sqrt{3}t$
$+ \left(\frac{1}{2}\sqrt{3}c_2 - \frac{3}{2}c_3\right)e^{-t/2} \sin \frac{1}{2}\sqrt{3}t$

13. $x = c_1 e^{4t} + \frac{4}{3}e^t$

$y = -\frac{3}{4}c_1 e^{4t} + c_2 + 5e^t$

15. $x = c_1 + c_2 t + c_3 e^t + c_4 e^{-t} - \frac{1}{2}t^2$

$y = (c_1 - c_2 + 2) + (c_2 + 1)t + c_4 e^{-t} - \frac{1}{2}t^2$

17. $x = c_1 e^t + c_2 e^{-t/2} \sin \frac{1}{2}\sqrt{3}t + c_3 e^{-t/2} \cos \frac{1}{2}\sqrt{3}t$

$y = c_1 e^t + \left(-\frac{1}{2}c_2 - \frac{1}{2}\sqrt{3}c_3\right)e^{-t/2} \sin \frac{1}{2}\sqrt{3}t$
$+ \left(\frac{1}{2}\sqrt{3}c_2 - \frac{1}{2}c_3\right)e^{-t/2} \cos \frac{1}{2}\sqrt{3}t$

$z = c_1 e^t + \left(-\frac{1}{2}c_2 + \frac{1}{2}\sqrt{3}c_3\right)e^{-t/2} \sin \frac{1}{2}\sqrt{3}t$
$+ \left(-\frac{1}{2}\sqrt{3}c_2 - \frac{1}{2}c_3\right)e^{-t/2} \cos \frac{1}{2}\sqrt{3}t$

19. $x = -6c_1 e^{-t} - 3c_2 e^{-2t} + 2c_3 e^{3t}$
$y = c_1 e^{-t} + c_2 e^{-2t} + c_3 e^{3t}$
$z = 5c_1 e^{-t} + c_2 e^{-2t} + c_3 e^{3t}$

21. $x = e^{-3t+3} - te^{-3t+3}$
$y = -e^{-3t+3} + 2te^{-3t+3}$

23. $mx'' = 0$
$my'' = -mg;$
$x = c_1 t + c_2$
$y = -\frac{1}{2}gt^2 + c_3 t + c_4$

EXERCISES 4.10 (PAGE 189)

3. $y = \ln|\cos(c_1 - x)| + c_2$

5. $y = \dfrac{1}{c_1^2}\ln|c_1 x + 1| - \dfrac{1}{c_1}x + c_2$

7. $\frac{1}{3}y^3 - c_1 y = x + c_2$

9. $y = \frac{2}{3}(x + 1)^{3/2} + \frac{4}{3}$

11. $y = \tan\left(\frac{1}{4}\pi - \frac{1}{2}x\right), -\frac{1}{2}\pi < x < \frac{3}{2}\pi$

13. $y = -\dfrac{1}{c_1}\sqrt{1 - c_1^2 x^2} + c_2$

15. $y = 1 + x + \frac{1}{2}x^2 + \frac{1}{2}x^3 + \frac{1}{6}x^4 + \frac{1}{10}x^5 + \cdots$

17. $y = 1 + x - \frac{1}{2}x^2 + \frac{2}{3}x^3 - \frac{1}{4}x^4 + \frac{7}{60}x^5 + \cdots$

19. $y = -\sqrt{1 - x^2}$

CHAPTER 4 IN REVIEW (PAGE 190)

1. $y = 0$

3. false

5. $y = c_1 \cos 5x + c_2 \sin 5x$

7. $x^2y'' - 3xy' + 4y = 0$

9. $y_p = x^2 + x - 2$

11. $(-\infty, 0); (0, \infty)$

13. $y = c_1 e^{3x} + c_2 e^{-5x} + c_3 x e^{-5x} + c_4 e^x + c_5 x e^x + c_6 x^2 e^x;$
$y = c_1 x^3 + c_2 x^{-5} + c_3 x^{-5} \ln x + c_4 x + c_5 x \ln x$
$+ c_6 x (\ln x)^2$

15. $y = c_1 e^{(1+\sqrt{3})x} + c_2 e^{(1-\sqrt{3})x}$

17. $y = c_1 + c_2 e^{-5x} + c_3 x e^{-5x}$

19. $y = c_1 e^{-x/3} + e^{-3x/2}\left(c_2 \cos \frac{1}{2}\sqrt{7}x + c_3 \sin \frac{1}{2}\sqrt{7}x\right)$

21. $y = e^{3x/2}\left(c_2 \cos \frac{1}{2}\sqrt{11}x + c_3 \sin \frac{1}{2}\sqrt{11}x\right) + \frac{4}{5}x^3 + \frac{36}{25}x^2$
$+ \frac{46}{125}x - \frac{222}{625}$

23. $y = c_1 + c_2 e^{2x} + c_3 e^{3x} + \frac{1}{5}\sin x - \frac{1}{5}\cos x + \frac{4}{3}x$

25. $y = e^x(c_1 \cos x + c_2 \sin x)$
$- e^x \cos x \ln|\sec x + \tan x|$

27. $y = c_1 x^{-1/3} + c_2 x^{1/2}$

29. $y = c_1 x^2 + c_2 x^3 + x^4 - x^2 \ln x$

31. **(a)** $y = c_1 \cos \omega x + c_2 \sin \omega x + A \cos \alpha x$
$+ B \sin \alpha x, \quad \omega \neq \alpha;$
$y = c_1 \cos \omega x + c_2 \sin \omega x + Ax \cos \omega x$
$+ Bx \sin \omega x, \quad \omega = \alpha$

(b) $y = c_1 e^{-\omega x} + c_2 e^{\omega x} + Ae^{\alpha x}, \omega \neq \alpha;$
$y = c_1 e^{-\omega x} + c_2 e^{\omega x} + Axe^{\omega x}, \omega = \alpha$

33. **(a)** $y = c_1 \cosh x + c_2 \sinh x + c_3 x \cosh x$
$+ c_4 x \sinh x$

(b) $y_p = Ax^2 \cosh x + Bx^2 \sinh x$

35. $y = e^{x-\pi} \cos x$

37. $y = \frac{13}{4}e^x - \frac{5}{4}e^{-x} - x - \frac{1}{2}\sin x$

39. $y = x^2 + 4$

43. $x = -c_1 e^t - \frac{3}{2}c_2 e^{2t} + \frac{5}{2}$
$y = c_1 e^t + c_2 e^{2t} - 3$

45. $x = c_1 e^t + c_2 e^{5t} + te^t$
$y = -c_1 e^t + 3c_2 e^{5t} - te^t + 2e^t$

EXERCISES 5.1 (PAGE 205)

1. $\dfrac{\sqrt{2}\,\pi}{8}$

3. $x(t) = -\frac{1}{4}\cos 4\sqrt{6}t$

5. **(a)** $x\left(\frac{\pi}{12}\right) = -\frac{1}{4}; x\left(\frac{\pi}{8}\right) = -\frac{1}{2}; x\left(\frac{\pi}{6}\right) = -\frac{1}{4};$

$x\left(\frac{\pi}{4}\right) = \frac{1}{2}; x\left(\frac{9\pi}{32}\right) = \frac{\sqrt{2}}{4}$

(b) 4 ft/s; downward

(c) $t = \dfrac{(2n + 1)\pi}{16}, n = 0, 1, 2, \ldots$

7. **(a)** the 20-kg mass
(b) the 20-kg mass; the 50-kg mass
(c) $t = n\pi, n = 0, 1, 2, \ldots$; at the equilibrium position; the 50-kg mass is moving upward whereas the 20-kg mass is moving upward when n is even and downward when n is odd.

9. **(a)** $x(t) = \frac{1}{2}\cos 2t + \frac{3}{4}\sin 2t$
(b) $x(t) = \frac{\sqrt{13}}{4}\sin(2t + 0.588)$
(c) $x(t) = \frac{\sqrt{13}}{4}\cos(2t - 0.983)$

11. **(a)** $x(t) = -\frac{2}{3}\cos 10t + \frac{1}{2}\sin 10t$
$= \frac{5}{6}\sin(10t - 0.927)$
(b) $\frac{5}{6}$ ft; $\frac{\pi}{5}$
(c) 15 cycles
(d) 0.721 s
(e) $\frac{(2n+1)\pi}{20} + 0.0927, n = 0, 1, 2, \ldots$
(f) $x(3) = -0.597$ ft
(g) $x'(3) = -5.814$ ft/s
(h) $x''(3) = 59.702$ ft/s^2
(i) $\pm 8\frac{1}{3}$ ft/s
(j) $0.1451 + \frac{n\pi}{5}$; $0.3545 + \frac{n\pi}{5}, n = 0, 1, 2, \ldots$
(k) $0.3545 + \frac{n\pi}{5}, n = 0, 1, 2, \ldots$

13. 120 lb/ft; $x(t) = \frac{\sqrt{3}}{12}\sin 8\sqrt{3}\,t$

17. **(a)** above
(b) heading upward

19. **(a)** below
(b) heading upward

21. $\frac{1}{4}$ s; $\frac{1}{2}$ s, $x\left(\frac{1}{2}\right) = e^{-2}$; that is, the weight is approximately 0.14 ft below the equilibrium position.

23. **(a)** $x(t) = \frac{4}{3}e^{-2t} - \frac{1}{3}e^{-8t}$
(b) $x(t) = -\frac{2}{3}e^{-2t} + \frac{5}{3}e^{-8t}$

25. **(a)** $x(t) = e^{-2t}\left(-\cos 4t - \frac{1}{2}\sin 4t\right)$
(b) $x(t) = \frac{\sqrt{5}}{2}e^{-2t}\sin\left(4t + 4.249\right)$
(c) $t = 1.294$ s

27. **(a)** $\beta > \frac{5}{2}$ **(b)** $\beta = \frac{5}{2}$ **(c)** $0 < \beta < \frac{5}{2}$

29. $x(t) = e^{-t/2}\left(-\frac{4}{3}\cos\frac{\sqrt{47}}{2}t - \frac{64}{3\sqrt{47}}\sin\frac{\sqrt{47}}{2}t\right)$
$+ \frac{10}{3}(\cos 3t + \sin 3t)$

31. $x(t) = \frac{1}{4}e^{-4t} + te^{-4t} - \frac{1}{4}\cos 4t$

33. $x(t) = -\frac{1}{2}\cos 4t + \frac{9}{4}\sin 4t + \frac{1}{2}e^{-2t}\cos 4t$
$- 2e^{-2t}\sin 4t$

35. **(a)** $m\frac{d^2x}{dt^2} = -k(x - h) - \beta\frac{dx}{dt}$ or
$\frac{d^2x}{dt^2} + 2\lambda\frac{dx}{dt} + \omega^2 x = \omega^2 h(t)$,
where $2\lambda = \beta/m$ and $\omega^2 = k/m$
(b) $x(t) = e^{-2t}\left(-\frac{56}{13}\cos 2t - \frac{72}{13}\sin 2t\right) + \frac{56}{13}\cos t$
$+ \frac{32}{13}\sin t$

37. $x(t) = -\cos 2t - \frac{1}{8}\sin 2t + \frac{3}{4}t\sin 2t + \frac{5}{4}t\cos 2t$

39. **(b)** $\frac{F_0}{2\omega}t\sin\omega t$

45. 4.568 C; 0.0509 s

47. $q(t) = 10 - 10e^{-3t}(\cos 3t + \sin 3t)$
$i(t) = 60e^{-3t}\sin 3t$; 10.432 C

49. $q_p = \frac{100}{13}\sin t + \frac{150}{13}\cos t$
$i_p = \frac{100}{13}\cos t - \frac{150}{13}\sin t$

53. $q(t) = -\frac{1}{2}e^{-10t}(\cos 10t + \sin 10t) + \frac{3}{2}$; $\frac{3}{2}$ C

57. $q(t) = \left(q_0 - \frac{E_0C}{1 - \gamma^2 LC}\right)\cos\frac{t}{\sqrt{LC}}$
$+ \sqrt{LC}\,i_0\sin\frac{t}{\sqrt{LC}} + \frac{E_0C}{1 - \gamma^2 LC}\cos\gamma t$
$i(t) = i_0\cos\frac{t}{\sqrt{LC}} - \frac{1}{\sqrt{LC}}\left(q_0 - \frac{E_0C}{1 - \gamma^2 LC}\right)\sin\frac{t}{\sqrt{LC}}$
$- \frac{E_0C\gamma}{1 - \gamma^2 LC}\sin\gamma t$

EXERCISES 5.2 (PAGE 215)

1. **(a)** $y(x) = \frac{w_0}{24EI}(6L^2x^2 - 4Lx^3 + x^4)$

3. **(a)** $y(x) = \frac{w_0}{48EI}(3L^2x^2 - 5Lx^3 + 2x^4)$

5. **(a)** $y(x) = \frac{w_0}{360EI}(7L^4x - 10L^2x^3 + 3x^5)$
(c) $x \approx 0.51933$, $y_{\max} \approx 0.234799$

7. $y(x) = -\frac{w_0EI}{P^2}\cosh\sqrt{\frac{P}{EI}}x$
$+ \left(\frac{w_0EI}{P^2}\sinh\sqrt{\frac{P}{EI}}L - \frac{w_0L\sqrt{EI}}{P\sqrt{P}}\right)\frac{\sinh\sqrt{\frac{P}{EI}}x}{\cosh\sqrt{\frac{P}{EI}}L}$
$+ \frac{w_0}{2P}x^2 + \frac{w_0EI}{P^2}$

9. $\lambda_n = n^2, n = 1, 2, 3, \ldots$; $y = \sin nx$

11. $\lambda_n = \frac{(2n-1)^2\pi^2}{4L^2}, n = 1, 2, 3, \ldots$;
$y = \cos\frac{(2n-1)\pi x}{2L}$

13. $\lambda_n = n^2, n = 0, 1, 2, \ldots; \quad y = \cos nx$

15. $\lambda_n = \dfrac{n^2\pi^2}{25}, n = 1, 2, 3, \ldots; \quad y = e^{-x}\sin\dfrac{n\pi x}{5}$

17. $\lambda_n = n^2, n = 1, 2, 3, \ldots; \quad y = \sin(n \ln x)$

19. $\lambda_n = n^4\pi^4, \quad n = 1, 2, 3, \ldots; \quad y = \sin n\pi x$

21. $x = L/4, x = L/2, x = 3L/4$

25. $\omega_n = \dfrac{n\pi\sqrt{T}}{L\sqrt{\rho}}, n = 1, 2, 3, \ldots; \quad y = \sin\dfrac{n\pi x}{L}$

27. $u(r) = \left(\dfrac{u_0 - u_1}{b - a}\right)\dfrac{ab}{r} + \dfrac{u_1 b - u_0 a}{b - a}$

EXERCISE 5.3 (PAGE 224)

7. $\dfrac{d^2x}{dt^2} + x = 0$

15. **(a)** 5 ft **(b)** $4\sqrt{10}$ ft/s **(c)** $0 \le t \le \frac{3}{8}\sqrt{10}$; 7.5 ft

17. **(a)** $xy'' = r\sqrt{1 + (y')^2}$.

When $t = 0, x = a, y = 0, dy/dx = 0$.

(b) When $r \ne 1$,

$$y(x) = \frac{a}{2}\left[\frac{1}{1 + r}\left(\frac{x}{a}\right)^{1+r} - \frac{1}{1 - r}\left(\frac{x}{a}\right)^{1-r}\right] + \frac{ar}{1 - r^2}$$

When $r = 1$,

$$y(x) = \frac{1}{2}\left[\frac{1}{2a}(x^2 - a^2) + \frac{1}{a}\ln\frac{a}{x}\right]$$

(c) The paths intersect when $r < 1$.

19. **(a)** $\theta(t) = \omega_0\sqrt{\dfrac{l}{g}}\sin\sqrt{\dfrac{g}{l}}\,t$

(b) use at θ_{max}, $\sin\sqrt{g/l}\,t = 1$

(c) use $\cos\theta_{max} \approx 1 - \frac{1}{2}\theta_{max}^2$

(d) $v_b \approx 21{,}797$ cm/s

CHAPTER 5 IN REVIEW (PAGE 228)

1. 8 ft **3.** $\frac{5}{4}$ m

5. False; there could be an impressed force driving the system.

7. overdamped

9. $y = 0$ since $\lambda = 8$ is not an eigenvalue

11. 14.4 lb

13. $x(t) = -\frac{2}{3}e^{-2t} + \frac{1}{3}e^{-4t}$

15. $0 < m \le 2$

17. $\gamma = \frac{8}{3}\sqrt{3}$

19. $x(t) = e^{-4t}\left(\frac{26}{17}\cos 2\sqrt{2}t + \frac{28}{17}\sqrt{2}\sin 2\sqrt{2}t\right) + \frac{8}{17}e^{-t}$

21. **(a)** $q(t) = -\frac{1}{150}\sin 100t + \frac{1}{75}\sin 50t$

(b) $i(t) = -\frac{2}{3}\cos 100t + \frac{2}{3}\cos 50t$

(c) $t = \dfrac{n\pi}{50}, n = 0, 1, 2, \ldots$

25. $m\dfrac{d^2x}{dt^2} + kx = 0$

27. $mx'' + f_k\,\mathrm{sgn}(x') + kx = 0$

EXERCISES 6.1 (PAGE 237)

1. $(-1, 1], R = 1$ **3.** $[-\frac{1}{2}, \frac{1}{2}), R = \frac{1}{2}$

5. $(-5, 15), R = 10$ **7.** $[0, \frac{2}{3}], R = \frac{1}{3}$

9. $(-\frac{75}{32}, \frac{75}{32}), R = \frac{75}{32}$ **11.** $\displaystyle\sum_{n=0}^{\infty}\frac{(-1)^n}{n!2^n}x^n$

13. $\displaystyle\sum_{n=0}^{\infty}\frac{(-1)^n}{2^{n+1}}x^n$ **15.** $\displaystyle\sum_{n=1}^{\infty}\frac{-1}{n}x^n$

17. $\displaystyle\sum_{n=0}^{\infty}\frac{(-1)^n}{(2n + 1)!}(x - 2\pi)^{2n+1}$

19. $x - \frac{2}{3}x^3 + \frac{2}{15}x^5 - \frac{4}{315}x^7 + \cdots$

21. $1 + \frac{1}{2}x^2 + \frac{5}{24}x^4 + \frac{61}{720}x^6 + \cdots, (-\pi/2, \pi/2)$

23. $\displaystyle\sum_{k=3}^{\infty}(k - 2)c_{k-2}x^k$ **25.** $\displaystyle\sum_{k=0}^{\infty}[(k + 1)c_{k+1} - c_k]x^k$

27. $2c_1 + \displaystyle\sum_{k=1}^{\infty}[2(k + 1)c_{k+1} + 6c_{k-1}]x^k$

29. $c_0 + 2c_2 + \displaystyle\sum_{k=1}^{\infty}[(k + 2)(k + 1)c_{k+2} - (2k - 1)c_k]x^k$

35. $y = c_0\displaystyle\sum_{k=0}^{\infty}\frac{1}{k!}(5x)^k$ **37.** $y = c_0\displaystyle\sum_{k=0}^{\infty}\frac{1}{k!}\left(\frac{x^2}{2}\right)^k$

EXERCISES 6.2 (PAGE 246)

1. 5; 4

3. $y_1(x) = c_0\left[1 - \dfrac{1}{2!}x^2 + \dfrac{1}{4!}x^4 - \dfrac{1}{6!}x^6 + \cdots\right]$

$y_2(x) = c_1\left[x - \dfrac{1}{3!}x^3 + \dfrac{1}{5!}x^5 - \dfrac{1}{7!}x^7 + \cdots\right]$

5. $y_1(x) = c_0$

$y_2(x) = c_1\left[x + \dfrac{1}{2!}x^2 + \dfrac{1}{3!}x^3 + \dfrac{1}{4!}x^4 + \cdots\right]$

7. $$y_1(x) = c_0\left[1 + \frac{1}{3\cdot 2}x^3 + \frac{1}{6\cdot 5\cdot 3\cdot 2}x^6 + \frac{1}{9\cdot 8\cdot 6\cdot 5\cdot 3\cdot 2}x^9 + \cdots\right]$$

$$y_2(x) = c_1\left[x + \frac{1}{4\cdot 3}x^4 + \frac{1}{7\cdot 6\cdot 4\cdot 3}x^7 + \frac{1}{10\cdot 9\cdot 7\cdot 6\cdot 4\cdot 3}x^{10} + \cdots\right]$$

9. $y_1(x) = c_0\left[1 - \frac{1}{2!}x^2 - \frac{3}{4!}x^4 - \frac{21}{6!}x^6 - \cdots\right]$

$y_2(x) = c_1\left[x + \frac{1}{3!}x^3 + \frac{5}{5!}x^5 + \frac{45}{7!}x^7 + \cdots\right]$

11. $y_1(x) = c_0\left[1 - \frac{1}{3!}x^3 + \frac{4^2}{6!}x^6 - \frac{7^2 \cdot 4^2}{9!}x^9 + \cdots\right]$

$y_2(x) = c_1\left[x - \frac{2^2}{4!}x^4 + \frac{5^2 \cdot 2^2}{7!}x^7 - \frac{8^2 \cdot 5^2 \cdot 2^2}{10!}x^{10} + \cdots\right]$

13. $y_1(x) = c_0$; $y_2(x) = c_1\sum_{n=1}^{\infty}\frac{1}{n}x^n$

15. $y_1(x) = c_0\left[1 + \frac{1}{2}x^2 + \frac{1}{6}x^3 + \frac{1}{6}x^4 + \cdots\right]$

$y_2(x) = c_1\left[x + \frac{1}{2}x^2 + \frac{1}{2}x^3 + \frac{1}{4}x^4 + \cdots\right]$

17. $y_1(x) = c_0\left[1 + \frac{1}{4}x^2 - \frac{7}{4 \cdot 4!}x^4 + \frac{23 \cdot 7}{8 \cdot 6!}x^6 - \cdots\right]$

$y_2(x) = c_1\left[x - \frac{1}{6}x^3 + \frac{14}{2 \cdot 5!}x^5 - \frac{34 \cdot 14}{4 \cdot 7!}x^7 - \cdots\right]$

19. $y(x) = -2\left[1 + \frac{1}{2!}x^2 + \frac{1}{3!}x^3 + \frac{1}{4!}x^4 + \cdots\right] + 6x$

$= 8x - 2e^x$

21. $y(x) = 3 - 12x^2 + 4x^4$

23. $y_1(x) = c_0\left[1 - \frac{1}{6}x^3 + \frac{1}{120}x^5 + \cdots\right]$

$y_2(x) = c_1\left[x - \frac{1}{12}x^4 + \frac{1}{180}x^6 + \cdots\right]$

EXERCISES 6.3 (PAGE 255)

1. $x = 0$, irregular singular point

3. $x = -3$, regular singular point; $x = 3$, irregular singular point

5. $x = 0, 2i, -2i$, regular singular points

7. $x = -3, 2$, regular singular points

9. $x = 0$, irregular singular point; $x = -5, 5, 2$, regular singular points

11. for $x = 1$: $p(x) = 5$, $q(x) = \frac{x(x-1)^2}{x+1}$

for $x = -1$: $p(x) = \frac{5(x+1)}{x-1}$, $q(x) = x^2 + x$

13. $r_1 = \frac{1}{3}$, $r_2 = -1$

15. $r_1 = \frac{3}{2}$, $r_2 = 0$

$y(x) = C_1x^{3/2}\left[1 - \frac{2}{5}x + \frac{2^2}{7 \cdot 5 \cdot 2}x^2 - \frac{2^3}{9 \cdot 7 \cdot 5 \cdot 3!}x^3 + \cdots\right] + C_2\left[1 + 2x - 2x^2 + \frac{2^3}{3 \cdot 3!}x^3 - \cdots\right]$

17. $r_1 = \frac{7}{8}$, $r_2 = 0$

$y(x) = c_1x^{7/8}\left[1 - \frac{2}{15}x + \frac{2^2}{23 \cdot 15 \cdot 2}x^2 - \frac{2^3}{31 \cdot 23 \cdot 15 \cdot 3!}x^3 + \cdots\right] + c_2\left[1 - 2x + \frac{2^2}{9 \cdot 2}x^2 - \frac{2^3}{17 \cdot 9 \cdot 3!}x^3 + \cdots\right]$

19. $r_1 = \frac{1}{3}$, $r_2 = 0$

$y(x) = C_1x^{1/3}\left[1 + \frac{1}{3}x + \frac{1}{3^2 \cdot 2}x^2 + \frac{1}{3^3 \cdot 3!}x^3 + \cdots\right] + C_2\left[1 + \frac{1}{2}x + \frac{1}{5 \cdot 2}x^2 + \frac{1}{8 \cdot 5 \cdot 2}x^3 + \cdots\right]$

21. $r_1 = \frac{5}{2}$, $r_2 = 0$

$y(x) = C_1x^{5/2}\left[1 + \frac{2 \cdot 2}{7}x + \frac{2^2 \cdot 3}{9 \cdot 7}x^2 + \frac{2^3 \cdot 4}{11 \cdot 9 \cdot 7}x^3 + \cdots\right] + C_2\left[1 + \frac{1}{3}x - \frac{1}{6}x^2 - \frac{1}{6}x^3 - \cdots\right]$

23. $r_1 = \frac{2}{3}$, $r_2 = \frac{1}{3}$

$y(x) = C_1x^{2/3}\left[1 - \frac{1}{2}x + \frac{5}{28}x^2 - \frac{1}{21}x^3 + \cdots\right] + C_2x^{1/3}\left[1 - \frac{1}{2}x + \frac{1}{5}x^2 - \frac{7}{120}x^3 + \cdots\right]$

25. $r_1 = 0$, $r_2 = -1$

$y(x) = C_1\sum_{n=0}^{\infty}\frac{1}{(2n+1)!}x^{2n} + C_2x^{-1}\sum_{n=0}^{\infty}\frac{1}{(2n)!}x^{2n}$

$= C_1x^{-1}\sum_{n=0}^{\infty}\frac{1}{(2n+1)!}x^{2n+1} + C_2x^{-1}\sum_{n=0}^{\infty}\frac{1}{(2n)!}x^{2n}$

$= \frac{1}{x}[C_1\sinh x + C_2\cosh x]$

27. $r_1 = 1$, $r_2 = 0$

$y(x) = C_1x + C_2\left[x\ln x - 1 + \frac{1}{2}x^2 + \frac{1}{12}x^3 + \frac{1}{72}x^4 + \cdots\right]$

29. $r_1 = r_2 = 0$

$y(x) = C_1y(x) + C_2\left[y_1(x)\ln x + y_1(x)\left(-x + \frac{1}{4}x^2 - \frac{1}{3 \cdot 3!}x^3 + \frac{1}{4 \cdot 4!}x^4 - \cdots\right)\right]$

where $y_1(x) = \sum_{n=0}^{\infty}\frac{1}{n!}x^n = e^x$

33. (b) $y_1(t) = \sum_{n=0}^{\infty} \frac{(-1)^n}{(2n+1)!}(\sqrt{\lambda}\, t)^{2n} = \frac{\sin(\sqrt{\lambda}\, t)}{\sqrt{\lambda}\, t}$

$y_2(t) = t^{-1} \sum_{n=0}^{\infty} \frac{(-1)^n}{(2n)!}(\sqrt{\lambda}\, t)^{2n} = \frac{\cos(\sqrt{\lambda}\, t)}{t}$

(c) $y = C_1 x \sin\left(\frac{\sqrt{\lambda}}{x}\right) + C_2 x \cos\left(\frac{\sqrt{\lambda}}{x}\right)$

EXERCISES 6.4 (PAGE 267)

1. $y = c_1 J_{1/3}(x) + c_2 J_{-1/3}(x)$
3. $y = c_1 J_{5/2}(x) + c_2 J_{-5/2}(x)$
5. $y = c_1 J_0(x) + c_2 Y_0(x)$
7. $y = c_1 J_2(3x) + c_2 Y_2(3x)$
9. $y = c_1 J_{2/3}(5x) + c_2 J_{-2/3}(5x)$
11. $y = c_1 x^{-1/2} J_{1/2}(\alpha x) + c_2 x^{-1/2} J_{-1/2}(\alpha x)$
13. $y = x^{-1/2}\left[c_1 J_1(4x^{1/2}) + c_2 Y_1(4x^{1/2})\right]$
15. $y = x\left[c_1 J_1(x) + c_2 Y_1(x)\right]$
17. $y = x^{1/2}\left[c_1 J_{3/2}(x) + c_2 Y_{3/2}(x)\right]$
19. $y = x^{-1}\left[c_1 J_{1/2}\left(\tfrac{1}{2}x^2\right) + c_2 J_{-1/2}\left(\tfrac{1}{2}x^2\right)\right]$
23. $y = x^{1/2}\left[c_1 J_{1/2}(x) + c_2 J_{-1/2}(x)\right]$
$= C_1 \sin x + C_2 \cos x$
25. $y = x^{-1/2}\left[c_1 J_{1/2}\left(\tfrac{1}{8}x^2\right) + c_2 J_{-1/2}\left(\tfrac{1}{8}x^2\right)\right]$
$= C_1 x^{-3/2} \sin\left(\tfrac{1}{8}x^2\right) + C_2 x^{-3/2} \cos\left(\tfrac{1}{8}x^2\right)$
35. $y = c_1 x^{1/2} J_{1/3}\left(\tfrac{2}{3}\alpha x^{3/2}\right) + c_2 x^{1/2} J_{-1/3}\left(\tfrac{2}{3}\alpha x^{3/2}\right)$
45. $P_2(x)$, $P_3(x)$, $P_4(x)$, and $P_5(x)$ are given in the text,
$P_6(x) = \tfrac{1}{16}(231x^6 - 315x^4 + 105x^2 - 5)$,
$P_7(x) = \tfrac{1}{16}(429x^7 - 693x^5 + 315x^3 - 35x)$
47. $\lambda_1 = 2$, $\lambda_2 = 12$, $\lambda_3 = 30$
53. $y = x - 4x^3 + \tfrac{16}{5}x^5$

CHAPTER 6 IN REVIEW (PAGE 271)

1. False
3. $[-\tfrac{1}{2}, \tfrac{1}{2}]$
7. $x^2(x-1)y'' + y' + y = 0$
9. $r_1 = \tfrac{1}{2}$, $r_2 = 0$
$y_1(x) = C_1 x^{1/2}\left[1 - \tfrac{1}{3}x + \tfrac{1}{30}x^2 - \tfrac{1}{630}x^3 + \cdots\right]$
$y_2(x) = C_2\left[1 - x + \tfrac{1}{6}x^2 - \tfrac{1}{90}x^3 + \cdots\right]$
11. $y_1(x) = c_0\left[1 + \tfrac{3}{2}x^2 + \tfrac{1}{2}x^3 + \tfrac{5}{8}x^4 + \cdots\right]$
$y_2(x) = c_1\left[x + \tfrac{1}{2}x^3 + \tfrac{1}{4}x^4 + \cdots\right]$
13. $r_1 = 3$, $r_2 = 0$
$y_1(x) = C_1 x^3\left[1 + \tfrac{1}{4}x + \tfrac{1}{20}x^2 + \tfrac{1}{120}x^3 + \cdots\right]$
$y_2(x) = C_2\left[1 + x + \tfrac{1}{2}x^2\right]$
15. $y(x) = 3\left[1 - x^2 + \tfrac{1}{3}x^4 - \tfrac{1}{15}x^6 + \cdots\right]$
$-2\left[x - \tfrac{1}{2}x^3 + \tfrac{1}{8}x^5 - \tfrac{1}{48}x^7 + \cdots\right]$
17. $\tfrac{1}{6}\pi$
19. $x = 0$ is an ordinary point
21. $y(x) = c_0\left[1 - \frac{1}{3}x^3 + \frac{1}{3^2 \cdot 2!}x^6 - \frac{1}{3^3 \cdot 3!}x^9 + \cdots\right]$
$+ c_1\left[x - \frac{1}{4}x^4 + \frac{1}{4 \cdot 7}x^7 - \frac{1}{4 \cdot 7 \cdot 10}x^{10} + \cdots\right] + \left[\frac{5}{2}x^2 - \frac{1}{3}x^3 + \frac{1}{3^2 \cdot 2!}x^6 - \frac{1}{3^3 \cdot 3!}x^9 + \cdots\right]$

EXERCISES 7.1 (PAGE 280)

1. $\frac{2}{s}e^{-s} - \frac{1}{s}$
3. $\frac{1}{s^2} - \frac{1}{s^2}e^{-s}$
5. $\frac{1 + e^{-\pi s}}{s^2 + 1}$
7. $\frac{1}{s}e^{-s} + \frac{1}{s^2}e^{-s}$
9. $\frac{1}{s} - \frac{1}{s^2} + \frac{1}{s^2}e^{-s}$
11. $\frac{e^7}{s - 1}$
13. $\frac{1}{(s-4)^2}$
15. $\frac{1}{s^2 + 2s + 2}$
17. $\frac{s^2 - 1}{(s^2 + 1)^2}$
19. $\frac{48}{s^5}$
21. $\frac{4}{s^2} - \frac{10}{s}$
23. $\frac{2}{s^3} + \frac{6}{s^2} - \frac{3}{s}$
25. $\frac{6}{s^4} + \frac{6}{s^3} + \frac{3}{s^2} + \frac{1}{s}$
27. $\frac{1}{s} + \frac{1}{s - 4}$
29. $\frac{1}{s} + \frac{2}{s - 2} + \frac{1}{s - 4}$
31. $\frac{8}{s^3} - \frac{15}{s^2 + 9}$
33. Use $\sinh kt = \frac{e^{kt} - e^{-kt}}{2}$ and linearity to show that
$$\mathscr{L}\{\sinh kt\} = \frac{k}{s^2 - k^2}.$$
35. $\frac{1}{2(s-2)} - \frac{1}{2s}$
37. $\frac{2}{s^2 + 16}$
39. $\frac{4\cos 5 + (\sin 5)s}{s^2 + 16}$
43. $\frac{\sqrt{\pi}}{s^{1/2}}$
45. $\frac{3\sqrt{\pi}}{4s^{5/2}}$

EXERCISES 7.2 (PAGE 288)

1. $\tfrac{1}{2}t^2$
3. $t - 2t^4$
5. $1 + 3t + \tfrac{3}{2}t^2 + \tfrac{1}{6}t^3$
7. $t - 1 + e^{2t}$
9. $\tfrac{1}{4}e^{-t/4}$
11. $\tfrac{5}{7}\sin 7t$
13. $\cos\frac{t}{2}$
15. $2\cos 3t - 2\sin 3t$
17. $\tfrac{1}{3} - \tfrac{1}{3}e^{-3t}$
19. $\tfrac{3}{4}e^{-3t} + \tfrac{1}{4}e^{t}$
21. $0.3e^{0.1t} + 0.6e^{-0.2t}$
23. $\tfrac{1}{2}e^{2t} - e^{3t} + \tfrac{1}{2}e^{6t}$
25. $\tfrac{1}{5} - \tfrac{1}{5}\cos\sqrt{5}t$
27. $-4 + 3e^{-t} + \cos t + 3\sin t$

29. $\frac{1}{3}\sin t - \frac{1}{6}\sin 2t$ **31.** $y = -1 + e^t$

33. $y = \frac{1}{10}e^{4t} + \frac{19}{10}e^{-6t}$ **35.** $y = \frac{4}{3}e^{-t} - \frac{1}{3}e^{-4t}$

37. $y = 10\cos t + 2\sin t - \sqrt{2}\sin\sqrt{2}t$

39. $y = -\frac{8}{9}e^{-t/2} + \frac{1}{9}e^{-2t} + \frac{5}{18}e^{t} + \frac{1}{2}e^{-t}$

41. $y = \frac{1}{4}e^{-t} - \frac{1}{4}e^{-3t}\cos 2t + \frac{1}{4}e^{-3t}\sin 2t$

EXERCISES 7.3 (PAGE 297)

1. $\dfrac{1}{(s-10)^2}$ **3.** $\dfrac{6}{(s+2)^4}$

5. $\dfrac{1}{(s-2)^2} + \dfrac{2}{(s-3)^2} + \dfrac{1}{(s-4)^2}$ **7.** $\dfrac{3}{(s-1)^2+9}$

9. $\dfrac{s}{s^2+25} - \dfrac{s-1}{(s-1)^2+25} + 3\dfrac{s+4}{(s+4)^2+25}$

11. $\frac{1}{2}t^2e^{-2t}$ **13.** $e^{3t}\sin t$

15. $e^{-2t}\cos t - 2e^{-2t}\sin t$ **17.** $e^{-t} - te^{-t}$

19. $5 - t - 5e^{-t} - 4te^{-t} - \frac{3}{2}t^2e^{-t}$

21. $y = te^{-4t} + 2e^{-4t}$ **23.** $y = e^{-t} + 2te^{-t}$

25. $y = \frac{1}{9}t + \frac{2}{27} - \frac{2}{27}e^{3t} + \frac{10}{9}te^{3t}$ **27.** $y = -\frac{3}{2}e^{3t}\sin 2t$

29. $y = \frac{1}{2} - \frac{1}{2}e^{t}\cos t + \frac{1}{2}e^{t}\sin t$

31. $y = (e+1)te^{-t} + (e-1)e^{-t}$

33. $x(t) = -\frac{3}{2}e^{-7t/2}\cos\frac{\sqrt{15}}{2}t - \frac{7\sqrt{15}}{10}e^{-7t/2}\sin\frac{\sqrt{15}}{2}t$

37. $\dfrac{e^{-s}}{s^2}$ **39.** $\dfrac{e^{-2s}}{s^2} + 2\dfrac{e^{-2s}}{s}$

41. $\dfrac{s}{s^2+4}e^{-\pi s}$ **43.** $\frac{1}{2}(t-2)^2\,\mathcal{U}(t-2)$

45. $-\sin t\,\mathcal{U}(t-\pi)$ **47.** $\mathcal{U}(t-1) - e^{-(t-1)}\,\mathcal{U}(t-1)$

49. (c) **51.** (f)

53. (a)

55. $f(t) = 2 - 4\mathcal{U}(t-3)$; $\mathscr{L}\{f(t)\} = \dfrac{2}{s} - \dfrac{4}{s}e^{-3s}$

57. $f(t) = t^2\,\mathcal{U}(t-1)$; $\mathscr{L}\{f(t)\} = 2\dfrac{e^{-s}}{s^3} + 2\dfrac{e^{-s}}{s^2} + \dfrac{e^{-s}}{s}$

59. $f(t) = t - t\,\mathcal{U}(t-2)$; $\mathscr{L}\{f(t)\} = \dfrac{1}{s^2} - \dfrac{e^{-2s}}{s^2} - 2\dfrac{e^{-2s}}{s}$

61. $f(t) = \mathcal{U}(t-a) - \mathcal{U}(t-b)$; $\mathscr{L}\{f(t)\} = \dfrac{e^{-as}}{s} - \dfrac{e^{-bs}}{s}$

63. $y = [5 - 5e^{-(t-1)}]\,\mathcal{U}(t-1)$

65. $y = -\frac{1}{4} + \frac{1}{2}t + \frac{1}{4}e^{-2t} - \frac{1}{4}\mathcal{U}(t-1)$
$\quad - \frac{1}{2}(t-1)\,\mathcal{U}(t-1) + \frac{1}{4}e^{-2(t-1)}\,\mathcal{U}(t-1)$

67. $y = \cos 2t - \frac{1}{6}\sin 2(t-2\pi)\,\mathcal{U}(t-2\pi)$
$\quad + \frac{1}{3}\sin(t-2\pi)\,\mathcal{U}(t-2\pi)$

69. $y = \sin t + [1 - \cos(t-\pi)]\mathcal{U}(t-\pi)$
$\quad - [1 - \cos(t-2\pi)]\,\mathcal{U}(t-2\pi)$

71. $x(t) = \frac{5}{4}t - \frac{5}{16}\sin 4t - \frac{5}{4}(t-5)\,\mathcal{U}(t-5)$
$\quad + \frac{5}{16}\sin 4(t-5)\,\mathcal{U}(t-5) - \frac{25}{4}\,\mathcal{U}(t-5)$
$\quad + \frac{25}{4}\cos 4(t-5)\,\mathcal{U}(t-5)$

73. $q(t) = \frac{2}{5}\,\mathcal{U}(t-3) - \frac{2}{5}e^{-5(t-3)}\,\mathcal{U}(t-3)$

75. **(a)** $i(t) = \dfrac{1}{101}e^{-10t} - \dfrac{1}{101}\cos t + \dfrac{10}{101}\sin t$

$$- \frac{10}{101}e^{-10(t-3\pi/2)}\,\mathcal{U}\left(t - \frac{3\pi}{2}\right)$$

$$+ \frac{10}{101}\cos\left(t - \frac{3\pi}{2}\right)\mathcal{U}\left(t - \frac{3\pi}{2}\right)$$

$$+ \frac{1}{101}\sin\left(t - \frac{3\pi}{2}\right)\mathcal{U}\left(t - \frac{3\pi}{2}\right)$$

(b) $i_{\max} \approx 0.1$ at $t \approx 1.7$, $i_{\min} \approx -0.1$ at $t \approx 4.7$

77. $y(x) = \dfrac{w_0L^2}{16EI}x^2 - \dfrac{w_0L}{12EI}x^3 + \dfrac{w_0}{24EI}x^4$

$$- \frac{w_0}{24EI}\left(x - \frac{L}{2}\right)^4\mathcal{U}\left(x - \frac{L}{2}\right)$$

79. $y(x) = \dfrac{w_0L^2}{48EI}x^2 - \dfrac{w_0L}{24EI}x^3$

$$+ \frac{w_0}{60EIL}\left[\frac{5L}{2}x^4 - x^5 + \left(x - \frac{L}{2}\right)^5\mathcal{U}\left(x - \frac{L}{2}\right)\right]$$

81. **(a)** $\dfrac{dT}{dt} = k[T - 70 - 57.5t - (230 - 57.5t)\mathcal{U}(t-4)]$

EXERCISES 7.4 (PAGE 309)

1. $\dfrac{1}{(s+10)^2}$ **3.** $\dfrac{s^2-4}{(s^2+4)^2}$

5. $\dfrac{6s^2+2}{(s^2-1)^3}$ **7.** $\dfrac{12s-24}{[(s-2)^2+36]^2}$

9. $y = -\frac{1}{2}e^{-t} + \frac{1}{2}\cos t - \frac{1}{2}t\cos t + \frac{1}{2}t\sin t$

11. $y = 2\cos 3t + \frac{5}{3}\sin 3t + \frac{1}{6}t\sin 3t$

13. $y = \frac{1}{4}\sin 4t + \frac{1}{8}t\sin 4t$
$\quad - \frac{1}{8}(t-\pi)\sin 4(t-\pi)\mathcal{U}(t-\pi)$

17. $y = \frac{2}{3}t^3 + c_1t^2$ **19.** $\dfrac{6}{s^5}$

21. $\dfrac{s-1}{(s+1)[(s-1)^2+1]}$ **23.** $\dfrac{1}{s(s-1)}$

25. $\dfrac{s+1}{s[(s+1)^2+1]}$ **27.** $\dfrac{1}{s^2(s-1)}$

29. $\dfrac{3s^2+1}{s^2(s^2+1)^2}$ **31.** $e^t - 1$

33. $e^t - \frac{1}{2}t^2 - t - 1$

37. $f(t) = \sin t$

39. $f(t) = -\frac{1}{8}e^{-t} + \frac{1}{8}e^t + \frac{3}{4}te^t + \frac{1}{4}t^2e^t$

41. $f(t) = e^{-t}$

43. $f(t) = \frac{3}{8}e^{2t} + \frac{1}{8}e^{-2t} + \frac{1}{2}\cos 2t + \frac{1}{4}\sin 2t$

45. $y(t) = \sin t - \frac{1}{2}t\sin t$

47. $i(t) = 100[e^{-10(t-1)} - e^{-20(t-1)}]\mathscr{U}(t-1) - 100[e^{-10(t-2)} - e^{-20(t-2)}]\mathscr{U}(t-2)$

49. $\dfrac{1 - e^{-as}}{s(1 + e^{-as})}$

51. $\dfrac{a}{s}\left(\dfrac{1}{bs} - \dfrac{1}{e^{bs} - 1}\right)$

53. $\dfrac{\coth(\pi s/2)}{s^2 + 1}$

55. $i(t) = \dfrac{1}{R}\left(1 - e^{-Rt/L}\right) + \dfrac{2}{R}\displaystyle\sum_{n=1}^{\infty}(-1)^n(1 - e^{-R(t-n)/L})\mathscr{U}(t-n)$

57. $x(t) = 2(1 - e^{-t}\cos 3t - \frac{1}{3}e^{-t}\sin 3t) + 4\displaystyle\sum_{n=1}^{\infty}(-1)^n\left[1 - e^{-(t-n\pi)}\cos 3(t - n\pi) - \frac{1}{3}e^{-(t-n\pi)}\sin 3(t - n\pi)\right]\mathscr{U}(t - n\pi)$

EXERCISES 7.5 (PAGE 315)

1. $y = e^{3(t-2)}\mathscr{U}(t-2)$

3. $y = \sin t + \sin t\,\mathscr{U}(t - 2\pi)$

5. $y = -\cos t\,\mathscr{U}\left(t - \frac{\pi}{2}\right) + \cos t\,\mathscr{U}\left(t - \frac{3\pi}{2}\right)$

7. $y = \frac{1}{2} - \frac{1}{2}e^{-2t} + \left[\frac{1}{2} - \frac{1}{2}e^{-2(t-1)}\right]\mathscr{U}(t-1)$

9. $y = e^{-2(t-2\pi)}\sin t\,\mathscr{U}(t - 2\pi)$

11. $y = e^{-2t}\cos 3t + \frac{2}{3}e^{-2t}\sin 3t + \frac{1}{3}e^{-2(t-\pi)}\sin 3(t-\pi)\,\mathscr{U}(t-\pi) + \frac{1}{3}e^{-2(t-3\pi)}\sin 3(t - 3\pi)\,\mathscr{U}(t - 3\pi)$

13. $y(x) = \begin{cases} \dfrac{P_0}{EI}\left(\dfrac{L}{4}x^2 - \dfrac{1}{6}x^3\right), & 0 \le x < \dfrac{L}{2} \\ \dfrac{P_0L^2}{4EI}\left(\dfrac{1}{2}x - \dfrac{L}{12}\right), & \dfrac{L}{2} \le x \le L \end{cases}$

EXERCISES 7.6 (PAGE 319)

1. $x = -\frac{1}{3}e^{-2t} + \frac{1}{3}e^t$
$y = \frac{1}{3}e^{-2t} + \frac{2}{3}e^t$

3. $x = -\cos 3t - \frac{5}{3}\sin 3t$
$y = 2\cos 3t - \frac{7}{3}\sin 3t$

5. $x = -2e^{3t} + \frac{5}{2}e^{2t} - \frac{1}{2}$
$y = \frac{8}{3}e^{3t} - \frac{5}{2}e^{2t} - \frac{1}{6}$

7. $x = -\frac{1}{2}t - \frac{3}{4}\sqrt{2}\sin\sqrt{2}t$
$y = -\frac{1}{2}t + \frac{3}{4}\sqrt{2}\sin\sqrt{2}t$

9. $x = 8 + \dfrac{2}{3!}t^3 + \dfrac{1}{4!}t^4$
$y = -\dfrac{2}{3!}t^3 + \dfrac{1}{4!}t^4$

11. $x = \frac{1}{2}t^2 + t + 1 - e^{-t}$
$y = -\frac{1}{3} + \frac{1}{3}e^{-t} + \frac{1}{3}te^{-t}$

13. $x_1 = \dfrac{1}{5}\sin t + \dfrac{2\sqrt{6}}{15}\sin\sqrt{6}t + \dfrac{2}{5}\cos t - \dfrac{2}{5}\cos\sqrt{6}t$
$x_2 = \dfrac{2}{5}\sin t - \dfrac{\sqrt{6}}{15}\sin\sqrt{6}t + \dfrac{4}{5}\cos t + \dfrac{1}{5}\cos\sqrt{6}t$

15. **(b)** $i_2 = \frac{100}{9} - \frac{100}{9}e^{-900t}$
$i_3 = \frac{80}{9} - \frac{80}{9}e^{-900t}$
(c) $i_1 = 20 - 20e^{-900t}$

17. $i_2 = -\frac{20}{13}e^{-2t} + \frac{375}{1469}e^{-15t} + \frac{145}{113}\cos t + \frac{85}{113}\sin t$
$i_3 = \frac{30}{13}e^{-2t} + \frac{250}{1469}e^{-15t} - \frac{280}{113}\cos t + \frac{810}{113}\sin t$

19. $i_1 = \dfrac{6}{5} - \dfrac{6}{5}e^{-100t}\cosh 50\sqrt{2}t - \dfrac{9\sqrt{2}}{10}e^{-100t}\sinh 50\sqrt{2}t$
$i_2 = \dfrac{6}{5} - \dfrac{6}{5}e^{-100t}\cosh 50\sqrt{2}t - \dfrac{6\sqrt{2}}{5}e^{-100t}\sinh 50\sqrt{2}t$

CHAPTER 7 IN REVIEW (PAGE 320)

1. $\dfrac{1}{s^2} - \dfrac{2}{s^2}e^{-s}$

3. false

5. true

7. $\dfrac{1}{s+7}$

9. $\dfrac{2}{s^2 + 4}$

11. $\dfrac{4s}{(s^2 + 4)^2}$

13. $\frac{1}{6}t^5$

15. $\frac{1}{2}t^2e^{5t}$

17. $e^{5t}\cos 2t + \frac{5}{2}e^{5t}\sin 2t$

19. $\cos\pi(t-1)\mathscr{U}(t-1) + \sin\pi(t-1)\mathscr{U}(t-1)$

21. -5

23. $e^{-k(s-a)}F(s-a)$

25. $f(t)\mathscr{U}(t - t_0)$

27. $f(t - t_0)\mathscr{U}(t - t_0)$

29. $f(t) = t - (t-1)\mathscr{U}(t-1) - \mathscr{U}(t-4)$;
$\mathscr{L}\{f(t)\} = \dfrac{1}{s^2} - \dfrac{1}{s^2}e^{-s} - \dfrac{1}{s}e^{-4s}$;
$\mathscr{L}\{e^tf(t)\} = \dfrac{1}{(s-1)^2} - \dfrac{1}{(s-1)^2}e^{-(s-1)} - \dfrac{1}{s-1}e^{-4(s-1)}$

31. $f(t) = 2 + (t-2)\,\mathscr{U}(t-2)$;
$\mathscr{L}\{f(t)\} = \dfrac{2}{s} + \dfrac{1}{s^2}e^{-2s}$;
$\mathscr{L}\{e^tf(t)\} = \dfrac{2}{s-1} + \dfrac{1}{(s-1)^2}e^{-2(s-1)}$

33. $y = 5te^t + \frac{1}{2}t^2e^t$

35. $y = -\frac{6}{25} + \frac{1}{5}t + \frac{3}{2}e^{-t} - \frac{13}{50}e^{-5t} - \frac{4}{25}\mathscr{U}(t-2) - \frac{1}{5}(t-2)\,\mathscr{U}(t-2) + \frac{1}{4}e^{-(t-2)}\,\mathscr{U}(t-2) - \frac{9}{100}e^{-5(t-2)}\,\mathscr{U}(t-2)$

37. $y(t) = e^{-2t} + \left[-\frac{1}{4} + \frac{1}{2}(t-1) + \frac{1}{4}e^{-2(t-1)}\right]\mathcal{U}(t-1)$
$\quad - 2\left[-\frac{1}{4} + \frac{1}{2}(t-2) + \frac{1}{4}e^{-2(t-2)}\right]\mathcal{U}(t-2)$
$\quad + \left[-\frac{1}{4} + \frac{1}{2}(t-3) + \frac{1}{4}e^{-2(t-3)}\right]\mathcal{U}(t-3)$

39. $y = 1 + t + \frac{1}{2}t^2$

41. $x = -\frac{1}{4} + \frac{9}{8}e^{-2t} + \frac{1}{8}e^{2t}$
$y = t + \frac{9}{4}e^{-2t} - \frac{1}{4}e^{2t}$

43. $i(t) = -9 + 2t + 9e^{-t/5}$

45. $$y(x) = \frac{w_0}{12EIL}\left[-\frac{1}{5}x^5 + \frac{L}{2}x^4 - \frac{L^2}{2}x^3 + \frac{L^3}{4}x^2 + \frac{1}{5}\left(x - \frac{L}{2}\right)^5 \mathcal{U}\left(x - \frac{L}{2}\right)\right]$$

47. **(a)** $$\theta_1(t) = \frac{\theta_0 + \psi_0}{2}\cos \omega t + \frac{\theta_0 - \psi_0}{2}\cos\sqrt{\omega^2 + 2K}t$$
$$\theta_2(t) = \frac{\theta_0 + \psi_0}{2}\cos \omega t - \frac{\theta_0 - \psi_0}{2}\cos\sqrt{\omega^2 + 2K}t$$

49. **(a)** $x(t) = (v_0 \cos\theta)\,t, \quad y(t) = -\frac{1}{2}gt^2 + (v_0\sin\theta)t$
(b) $y(x) = -\dfrac{g}{2v_0^2\cos^2\theta}x^2 + \dfrac{\sin\theta}{\cos\theta}x$; solve $y(x) = 0$ and use the double-angle formula for $\sin 2\theta$
(d) approximately 2729 ft; approximately 11.54 s

EXERCISES 8.1 (PAGE 332)

1. $\mathbf{X}' = \begin{pmatrix} 3 & -5 \\ 4 & 8 \end{pmatrix}\mathbf{X}$, where $\mathbf{X} = \begin{pmatrix} x \\ y \end{pmatrix}$

3. $\mathbf{X}' = \begin{pmatrix} -3 & 4 & -9 \\ 6 & -1 & 0 \\ 10 & 4 & 3 \end{pmatrix}\mathbf{X}$, where $\mathbf{X} = \begin{pmatrix} x \\ y \\ z \end{pmatrix}$

5. $\mathbf{X}' = \begin{pmatrix} 1 & -1 & 1 \\ 2 & 1 & -1 \\ 1 & 1 & 1 \end{pmatrix}\mathbf{X} + \begin{pmatrix} 0 \\ -3t^2 \\ t^2 \end{pmatrix} + \begin{pmatrix} t \\ 0 \\ -t \end{pmatrix} + \begin{pmatrix} -1 \\ 0 \\ 2 \end{pmatrix}$,
where $\mathbf{X} = \begin{pmatrix} x \\ y \\ z \end{pmatrix}$

7. $\dfrac{dx}{dt} = 4x + 2y + e^t$
$\dfrac{dy}{dt} = -x + 3y - e^t$

9. $\dfrac{dx}{dt} = x - y + 2z + e^{-t} - 3t$
$\dfrac{dy}{dt} = 3x - 4y + z + 2e^{-t} + t$
$\dfrac{dz}{dt} = -2x + 5y + 6z + 2e^{-t} - t$

17. Yes; $W(\mathbf{X}_1, \mathbf{X}_2) = -2e^{-8t} \neq 0$ implies that $\mathbf{X}_1$ and $\mathbf{X}_2$ are linearly independent on $(-\infty, \infty)$.

19. No; $W(\mathbf{X}_1, \mathbf{X}_2, \mathbf{X}_3) = 0$ for every t. The solution vectors are linearly dependent on $(-\infty, \infty)$. Note that $\mathbf{X}_3 = 2\mathbf{X}_1 + \mathbf{X}_2$.

EXERCISES 8.2 (PAGE 346)

1. $\mathbf{X} = c_1\begin{pmatrix} 1 \\ 2 \end{pmatrix}e^{5t} + c_2\begin{pmatrix} -1 \\ 1 \end{pmatrix}e^{-t}$

3. $\mathbf{X} = c_1\begin{pmatrix} 2 \\ 1 \end{pmatrix}e^{-3t} + c_2\begin{pmatrix} 2 \\ 5 \end{pmatrix}e^{t}$

5. $\mathbf{X} = c_1\begin{pmatrix} 5 \\ 2 \end{pmatrix}e^{8t} + c_2\begin{pmatrix} 1 \\ 4 \end{pmatrix}e^{-10t}$

7. $\mathbf{X} = c_1\begin{pmatrix} 1 \\ 0 \\ 0 \end{pmatrix}e^{t} + c_2\begin{pmatrix} 2 \\ 3 \\ 1 \end{pmatrix}e^{2t} + c_3\begin{pmatrix} 1 \\ 0 \\ 2 \end{pmatrix}e^{-t}$

9. $\mathbf{X} = c_1\begin{pmatrix} -1 \\ 0 \\ 1 \end{pmatrix}e^{-t} + c_2\begin{pmatrix} 1 \\ 4 \\ 3 \end{pmatrix}e^{3t} + c_3\begin{pmatrix} 1 \\ -1 \\ 3 \end{pmatrix}e^{-2t}$

11. $\mathbf{X} = c_1\begin{pmatrix} 4 \\ 0 \\ -1 \end{pmatrix}e^{-t} + c_2\begin{pmatrix} -12 \\ 6 \\ 5 \end{pmatrix}e^{-t/2} + c_3\begin{pmatrix} 4 \\ 2 \\ -1 \end{pmatrix}e^{-3t/2}$

13. $\mathbf{X} = 3\begin{pmatrix} 1 \\ 1 \end{pmatrix}e^{t/2} + 2\begin{pmatrix} 0 \\ 1 \end{pmatrix}e^{-t/2}$

19. $\mathbf{X} = c_1\begin{pmatrix} 1 \\ 3 \end{pmatrix} + c_2\left[\begin{pmatrix} 1 \\ 3 \end{pmatrix}t + \begin{pmatrix} \frac{1}{4} \\ -\frac{1}{4} \end{pmatrix}\right]$

21. $\mathbf{X} = c_1\begin{pmatrix} 1 \\ 1 \end{pmatrix}e^{2t} + c_2\left[\begin{pmatrix} 1 \\ 1 \end{pmatrix}te^{2t} + \begin{pmatrix} -\frac{1}{3} \\ 0 \end{pmatrix}e^{2t}\right]$

23. $\mathbf{X} = c_1\begin{pmatrix} 1 \\ 1 \\ 1 \end{pmatrix}e^{t} + c_2\begin{pmatrix} 1 \\ 1 \\ 0 \end{pmatrix}e^{2t} + c_3\begin{pmatrix} 1 \\ 0 \\ 1 \end{pmatrix}e^{2t}$

25. $\mathbf{X} = c_1\begin{pmatrix} -4 \\ -5 \\ 2 \end{pmatrix} + c_2\begin{pmatrix} 2 \\ 0 \\ -1 \end{pmatrix}e^{5t}$
$\quad + c_3\left[\begin{pmatrix} 2 \\ 0 \\ -1 \end{pmatrix}te^{5t} + \begin{pmatrix} -\frac{1}{2} \\ -\frac{1}{2} \\ -1 \end{pmatrix}e^{5t}\right]$

27. $\mathbf{X} = c_1\begin{pmatrix} 0 \\ 1 \\ 1 \end{pmatrix}e^{t} + c_2\left[\begin{pmatrix} 0 \\ 1 \\ 1 \end{pmatrix}te^{t} + \begin{pmatrix} 0 \\ 1 \\ 0 \end{pmatrix}e^{t}\right]$
$\quad + c_3\left[\begin{pmatrix} 0 \\ 1 \\ 1 \end{pmatrix}\frac{t^2}{2}e^{t} + \begin{pmatrix} 0 \\ 1 \\ 0 \end{pmatrix}te^{t} + \begin{pmatrix} \frac{1}{2} \\ 0 \\ 0 \end{pmatrix}e^{t}\right]$

29. $\mathbf{X} = -7\begin{pmatrix} 2 \\ 1 \end{pmatrix} e^{4t} + 13\begin{pmatrix} 2t+1 \\ t+1 \end{pmatrix} e^{4t}$

31. Corresponding to the eigenvalue $\lambda_1 = 2$ of multiplicity five, the eigenvectors are

$$\mathbf{K}_1 = \begin{pmatrix} 1 \\ 0 \\ 0 \\ 0 \\ 0 \end{pmatrix}, \quad \mathbf{K}_2 = \begin{pmatrix} 0 \\ 0 \\ 1 \\ 0 \\ 0 \end{pmatrix}, \quad \mathbf{K}_3 = \begin{pmatrix} 0 \\ 0 \\ 0 \\ 1 \\ 0 \end{pmatrix}.$$

33. $\mathbf{X} = c_1\begin{pmatrix} \cos t \\ 2\cos t + \sin t \end{pmatrix} e^{4t} + c_2\begin{pmatrix} \sin t \\ 2\sin t - \cos t \end{pmatrix} e^{4t}$

35. $\mathbf{X} = c_1\begin{pmatrix} \cos t \\ -\cos t - \sin t \end{pmatrix} e^{4t} + c_2\begin{pmatrix} \sin t \\ -\sin t + \cos t \end{pmatrix} e^{4t}$

37. $\mathbf{X} = c_1\begin{pmatrix} 5\cos 3t \\ 4\cos 3t + 3\sin 3t \end{pmatrix} + c_2\begin{pmatrix} 5\sin 3t \\ 4\sin 3t - 3\cos 3t \end{pmatrix}$

39. $\mathbf{X} = c_1\begin{pmatrix} 1 \\ 0 \\ 0 \end{pmatrix} + c_2\begin{pmatrix} -\cos t \\ \cos t \\ \sin t \end{pmatrix} + c_3\begin{pmatrix} \sin t \\ -\sin t \\ \cos t \end{pmatrix}$

41. $\mathbf{X} = c_1\begin{pmatrix} 0 \\ 2 \\ 1 \end{pmatrix} e^{t} + c_2\begin{pmatrix} \sin t \\ \cos t \\ \cos t \end{pmatrix} e^{t} + c_3\begin{pmatrix} \cos t \\ -\sin t \\ -\sin t \end{pmatrix} e^{t}$

43. $\mathbf{X} = c_1\begin{pmatrix} 28 \\ -5 \\ 25 \end{pmatrix} e^{2t} + c_2\begin{pmatrix} 4\cos 3t - 3\sin 3t \\ -5\cos 3t \\ 0 \end{pmatrix} e^{-2t}$

$+ c_3\begin{pmatrix} 3\cos 3t + 4\sin 3t \\ -5\sin 3t \\ 0 \end{pmatrix} e^{-2t}$

45. $\mathbf{X} = -\begin{pmatrix} 25 \\ -7 \\ 6 \end{pmatrix} e^{t} - \begin{pmatrix} \cos 5t - 5\sin 5t \\ \cos 5t \\ \cos 5t \end{pmatrix}$

$+ 6\begin{pmatrix} 5\cos 5t + \sin 5t \\ \sin 5t \\ \sin 5t \end{pmatrix}$

EXERCISES 8.3 (PAGE 354)

1. $\mathbf{X} = c_1\begin{pmatrix} -1 \\ 1 \end{pmatrix} e^{-t} + c_2\begin{pmatrix} -3 \\ 1 \end{pmatrix} e^{t} + \begin{pmatrix} -1 \\ 3 \end{pmatrix}$

3. $\mathbf{X} = c_1\begin{pmatrix} 1 \\ -1 \end{pmatrix} e^{-2t} + c_2\begin{pmatrix} 1 \\ 1 \end{pmatrix} e^{4t} + \begin{pmatrix} -\frac{1}{4} \\ \frac{3}{4} \end{pmatrix} t^2$

$+ \begin{pmatrix} \frac{1}{4} \\ -\frac{1}{4} \end{pmatrix} t + \begin{pmatrix} -2 \\ \frac{3}{4} \end{pmatrix}$

5. $\mathbf{X} = c_1\begin{pmatrix} 1 \\ -3 \end{pmatrix} e^{3t} + c_2\begin{pmatrix} 1 \\ 9 \end{pmatrix} e^{7t} + \begin{pmatrix} \frac{55}{36} \\ -\frac{19}{4} \end{pmatrix} e^{t}$

7. $\mathbf{X} = c_1\begin{pmatrix} 1 \\ 0 \\ 0 \end{pmatrix} e^{t} + c_2\begin{pmatrix} 1 \\ 1 \\ 0 \end{pmatrix} e^{2t} + c_3\begin{pmatrix} 1 \\ 2 \\ 2 \end{pmatrix} e^{5t} - \begin{pmatrix} \frac{3}{2} \\ \frac{7}{2} \\ 2 \end{pmatrix} e^{4t}$

9. $\mathbf{X} = 13\begin{pmatrix} 1 \\ -1 \end{pmatrix} e^{t} + 2\begin{pmatrix} -4 \\ 6 \end{pmatrix} e^{2t} + \begin{pmatrix} -9 \\ 6 \end{pmatrix}$

11. $\mathbf{X} = c_1\begin{pmatrix} 1 \\ 1 \end{pmatrix} + c_2\begin{pmatrix} 3 \\ 2 \end{pmatrix} e^{t} - \begin{pmatrix} 11 \\ 11 \end{pmatrix} t - \begin{pmatrix} 15 \\ 10 \end{pmatrix}$

13. $\mathbf{X} = c_1\begin{pmatrix} 2 \\ 1 \end{pmatrix} e^{t/2} + c_2\begin{pmatrix} 10 \\ 3 \end{pmatrix} e^{3t/2} - \begin{pmatrix} \frac{13}{2} \\ \frac{13}{4} \end{pmatrix} te^{t/2} - \begin{pmatrix} \frac{15}{2} \\ \frac{9}{4} \end{pmatrix} e^{t/2}$

15. $\mathbf{X} = c_1\begin{pmatrix} 2 \\ 1 \end{pmatrix} e^{t} + c_2\begin{pmatrix} 1 \\ 1 \end{pmatrix} e^{2t} + \begin{pmatrix} 3 \\ 3 \end{pmatrix} e^{t} + \begin{pmatrix} 4 \\ 2 \end{pmatrix} te^{t}$

17. $\mathbf{X} = c_1\begin{pmatrix} 4 \\ 1 \end{pmatrix} e^{3t} + c_2\begin{pmatrix} -2 \\ 1 \end{pmatrix} e^{-3t} + \begin{pmatrix} -12 \\ 0 \end{pmatrix} t - \begin{pmatrix} \frac{4}{3} \\ \frac{4}{3} \end{pmatrix}$

19. $\mathbf{X} = c_1\begin{pmatrix} 1 \\ -1 \end{pmatrix} e^{t} + c_2\begin{pmatrix} t \\ \frac{1}{2} - t \end{pmatrix} e^{t} + \begin{pmatrix} \frac{1}{2} \\ -2 \end{pmatrix} e^{-t}$

21. $\mathbf{X} = c_1\begin{pmatrix} \cos t \\ \sin t \end{pmatrix} + c_2\begin{pmatrix} \sin t \\ -\cos t \end{pmatrix} + \begin{pmatrix} \cos t \\ \sin t \end{pmatrix} t$

$+ \begin{pmatrix} -\sin t \\ \cos t \end{pmatrix} \ln|\cos t|$

23. $\mathbf{X} = c_1\begin{pmatrix} \cos t \\ \sin t \end{pmatrix} e^{t} + c_2\begin{pmatrix} \sin t \\ -\cos t \end{pmatrix} e^{t} + \begin{pmatrix} \cos t \\ \sin t \end{pmatrix} te^{t}$

25. $\mathbf{X} = c_1\begin{pmatrix} \cos t \\ -\sin t \end{pmatrix} + c_2\begin{pmatrix} \sin t \\ \cos t \end{pmatrix} + \begin{pmatrix} \cos t \\ -\sin t \end{pmatrix} t$

$+ \begin{pmatrix} -\sin t \\ \sin t \tan t \end{pmatrix} - \begin{pmatrix} \sin t \\ \cos t \end{pmatrix} \ln|\cos t|$

27. $\mathbf{X} = c_1\begin{pmatrix} 2\sin t \\ \cos t \end{pmatrix} e^{t} + c_2\begin{pmatrix} 2\cos t \\ -\sin t \end{pmatrix} e^{t} + \begin{pmatrix} 3\sin t \\ \frac{3}{2}\cos t \end{pmatrix} te^{t}$

$+ \begin{pmatrix} \cos t \\ -\frac{1}{2}\sin t \end{pmatrix} e^{t} \ln|\sin t| + \begin{pmatrix} 2\cos t \\ -\sin t \end{pmatrix} e^{t} \ln|\cos t|$

29. $\mathbf{X} = c_1\begin{pmatrix} 1 \\ -1 \\ 0 \end{pmatrix} + c_2\begin{pmatrix} 1 \\ 1 \\ 0 \end{pmatrix} e^{2t} + c_3\begin{pmatrix} 0 \\ 0 \\ 1 \end{pmatrix} e^{3t}$

$+ \begin{pmatrix} -\frac{1}{4}e^{2t} + \frac{1}{2}te^{2t} \\ -e^{t} + \frac{1}{4}e^{2t} + \frac{1}{2}te^{2t} \\ \frac{1}{2}t^2e^{3t} \end{pmatrix}$

31. $\mathbf{X} = \begin{pmatrix} 2 \\ 2 \end{pmatrix} te^{2t} + \begin{pmatrix} -1 \\ 1 \end{pmatrix} e^{2t} + \begin{pmatrix} -2 \\ 2 \end{pmatrix} te^{4t} + \begin{pmatrix} 2 \\ 0 \end{pmatrix} e^{4t}$

33. $\begin{pmatrix} i_1 \\ i_2 \end{pmatrix} = 2\begin{pmatrix} 1 \\ 3 \end{pmatrix} e^{-2t} + \frac{6}{29}\begin{pmatrix} 3 \\ -1 \end{pmatrix} e^{-12t} - \frac{4}{29}\begin{pmatrix} 19 \\ 42 \end{pmatrix} \cos t$

$+ \frac{4}{29}\begin{pmatrix} 83 \\ 69 \end{pmatrix} \sin t$

EXERCISES 8.4 (PAGE 359)

1. $e^{\mathbf{A}t} = \begin{pmatrix} e^t & 0 \\ 0 & e^{2t} \end{pmatrix}; \quad e^{-\mathbf{A}t} = \begin{pmatrix} e^{-t} & 0 \\ 0 & e^{-2t} \end{pmatrix}$

3. $e^{\mathbf{A}t} = \begin{pmatrix} t+1 & t & t \\ t & t+1 & t \\ -2t & -2t & -2t+1 \end{pmatrix}$

5. $\mathbf{X} = c_1\begin{pmatrix} 1 \\ 0 \end{pmatrix}e^t + c_2\begin{pmatrix} 0 \\ 1 \end{pmatrix}e^{2t}$

7. $\mathbf{X} = c_1\begin{pmatrix} t+1 \\ t \\ -2t \end{pmatrix} + c_2\begin{pmatrix} t \\ t+1 \\ -2t \end{pmatrix} + c_3\begin{pmatrix} t \\ t \\ -2t+1 \end{pmatrix}$

9. $\mathbf{X} = c_3\begin{pmatrix} 1 \\ 0 \end{pmatrix}e^t + c_4\begin{pmatrix} 0 \\ 1 \end{pmatrix}e^{2t} + \begin{pmatrix} -3 \\ \frac{1}{2} \end{pmatrix}$

11. $\mathbf{X} = c_1\begin{pmatrix} \cosh t \\ \sinh t \end{pmatrix} + c_2\begin{pmatrix} \sinh t \\ \cosh t \end{pmatrix} - \begin{pmatrix} 1 \\ 1 \end{pmatrix}$

13. $\mathbf{X} = \begin{pmatrix} t+1 \\ t \\ -2t \end{pmatrix} - 4\begin{pmatrix} t \\ t+1 \\ -2t \end{pmatrix} + 6\begin{pmatrix} t \\ t \\ -2t+1 \end{pmatrix}$

15. $e^{\mathbf{A}t} = \begin{pmatrix} \frac{3}{2}e^{2t} - \frac{1}{2}e^{-2t} & \frac{3}{4}e^{2t} - \frac{3}{4}e^{-2t} \\ -e^{2t} + e^{-2t} & -\frac{1}{2}e^{2t} + \frac{3}{2}e^{-2t} \end{pmatrix};$

$\mathbf{X} = c_1\begin{pmatrix} \frac{3}{2}e^{2t} - \frac{1}{2}e^{-2t} \\ -e^{2t} + e^{-2t} \end{pmatrix} + c_2\begin{pmatrix} \frac{3}{4}e^{2t} - \frac{3}{4}e^{-2t} \\ -\frac{1}{2}e^{2t} + \frac{3}{2}e^{-2t} \end{pmatrix}$ or

$\mathbf{X} = c_3\begin{pmatrix} 3 \\ -2 \end{pmatrix}e^{2t} + c_4\begin{pmatrix} 1 \\ -2 \end{pmatrix}e^{-2t}$

17. $e^{\mathbf{A}t} = \begin{pmatrix} e^{2t} + 3te^{2t} & -9te^{2t} \\ te^{2t} & e^{2t} - 3te^{2t} \end{pmatrix};$

$\mathbf{X} = c_1\begin{pmatrix} 1+3t \\ t \end{pmatrix}e^{2t} + c_2\begin{pmatrix} -9t \\ 1-3t \end{pmatrix}e^{2t}$

23. $\mathbf{X} = c_1\begin{pmatrix} \frac{3}{2}e^{3t} - \frac{1}{2}e^{5t} \\ \frac{3}{2}e^{3t} - \frac{3}{2}e^{5t} \end{pmatrix} + c_2\begin{pmatrix} -\frac{1}{2}e^{3t} + \frac{1}{2}e^{5t} \\ -\frac{1}{2}e^{3t} + \frac{3}{2}e^{5t} \end{pmatrix}$ or

$\mathbf{X} = c_3\begin{pmatrix} 1 \\ 1 \end{pmatrix}e^{3t} + c_4\begin{pmatrix} 1 \\ 3 \end{pmatrix}e^{5t}$

CHAPTER 8 IN REVIEW (PAGE 360)

1. $k = \frac{1}{3}$

5. $\mathbf{X} = c_1\begin{pmatrix} 1 \\ -1 \end{pmatrix}e^t + c_2\left[\begin{pmatrix} 1 \\ -1 \end{pmatrix}te^t + \begin{pmatrix} 0 \\ 1 \end{pmatrix}e^t\right]$

7. $\mathbf{X} = c_1\begin{pmatrix} \cos 2t \\ -\sin 2t \end{pmatrix}e^t + c_2\begin{pmatrix} \sin 2t \\ \cos 2t \end{pmatrix}e^t$

9. $\mathbf{X} = c_1\begin{pmatrix} -2 \\ 3 \\ 1 \end{pmatrix}e^{2t} + c_2\begin{pmatrix} 0 \\ 1 \\ 1 \end{pmatrix}e^{4t} + c_3\begin{pmatrix} 7 \\ 12 \\ -16 \end{pmatrix}e^{-3t}$

11. $\mathbf{X} = c_1\begin{pmatrix} 1 \\ 0 \end{pmatrix}e^{2t} + c_2\begin{pmatrix} 4 \\ 1 \end{pmatrix}e^{4t} + \begin{pmatrix} 16 \\ -4 \end{pmatrix}t + \begin{pmatrix} 11 \\ -1 \end{pmatrix}$

13. $\mathbf{X} = c_1\begin{pmatrix} \cos t \\ \cos t - \sin t \end{pmatrix} + c_2\begin{pmatrix} \sin t \\ \sin t + \cos t \end{pmatrix} - \begin{pmatrix} 1 \\ 1 \end{pmatrix}$

$+ \begin{pmatrix} \sin t \\ \sin t + \cos t \end{pmatrix}\ln|\csc t - \cot t|$

15. (b) $\mathbf{X} = c_1\begin{pmatrix} -1 \\ 1 \\ 0 \end{pmatrix} + c_2\begin{pmatrix} -1 \\ 0 \\ 1 \end{pmatrix} + c_3\begin{pmatrix} 1 \\ 1 \\ 1 \end{pmatrix}e^{3t}$

EXERCISES 9.1 (PAGE 367)

1. for $h = 0.1$, $y_5 = 2.0801$; for $h = 0.05$, $y_{10} = 2.0592$
3. for $h = 0.1$, $y_5 = 0.5470$; for $h = 0.05$, $y_{10} = 0.5465$
5. for $h = 0.1$, $y_5 = 0.4053$; for $h = 0.05$, $y_{10} = 0.4054$
7. for $h = 0.1$, $y_5 = 0.5503$; for $h = 0.05$, $y_{10} = 0.5495$
9. for $h = 0.1$, $y_5 = 1.3260$; for $h = 0.05$, $y_{10} = 1.3315$
11. for $h = 0.1$, $y_5 = 3.8254$; for $h = 0.05$, $y_{10} = 3.8840$; at $x = 0.5$ the actual value is $y(0.5) = 3.9082$

13. (a) $y_1 = 1.2$

(b) $y''(c)\dfrac{h^2}{2} = 4e^{2c}\dfrac{(0.1)^2}{2} = 0.02e^{2c} \le 0.02e^{0.2}$

$= 0.0244$

(c) Actual value is $y(0.1) = 1.2214$. Error is 0.0214.

(d) If $h = 0.05$, $y_2 = 1.21$.

(e) Error with $h = 0.1$ is 0.0214. Error with $h = 0.05$ is 0.0114.

15. (a) $y_1 = 0.8$

(b) $y''(c)\dfrac{h^2}{2} = 5e^{-2c}\dfrac{(0.1)^2}{2} = 0.025e^{-2c} \le 0.025$

for $0 \le c \le 0.1$.

(c) Actual value is $y(0.1) = 0.8234$. Error is 0.0234.

(d) If $h = 0.05$, $y_2 = 0.8125$.

(e) Error with $h = 0.1$ is 0.0234. Error with $h = 0.05$ is 0.0109.

17. (a) Error is $19h^2e^{-3(c-1)}$.

(b) $y''(c)\dfrac{h^2}{2} \le 19(0.1)^2(1) = 0.19$

(c) If $h = 0.1$, $y_5 = 1.8207$.
If $h = 0.05$, $y_{10} = 1.9424$.

(d) Error with $h = 0.1$ is 0.2325. Error with $h = 0.05$ is 0.1109.

19. (a) Error is $\frac{1}{(c+1)^2}\frac{h^2}{2}$.

(b) $\left|y''(c)\frac{h^2}{2}\right| \le (1)\frac{(0.1)^2}{2} = 0.005$

(c) If $h = 0.1$, $y_5 = 0.4198$. If $h = 0.05$, $y_{10} = 0.4124$.

(d) Error with $h = 0.1$ is 0.0143. Error with $h = 0.05$ is 0.0069.

EXERCISES 9.2 (PAGE 371)

1. $y_5 = 3.9078$; actual value is $y(0.5) = 3.9082$

3. $y_5 = 2.0533$

5. $y_5 = 0.5463$

7. $y_5 = 0.4055$

9. $y_5 = 0.5493$

11. $y_5 = 1.3333$

13. (a) 35.7130

(c) $v(t) = \sqrt{\frac{mg}{k}}\tanh\sqrt{\frac{kg}{m}}\,t;\quad v(5) = 35.7678$

15. (a) for $h = 0.1$, $y_4 = 903.0282$;
for $h = 0.05$, $y_8 = 1.1 \times 10^{15}$

17. (a) $y_1 = 0.82341667$

(b) $y^{(5)}(c)\frac{h^5}{5!} = 40e^{-2c}\frac{h^5}{5!} \le 40e^{2(0)}\frac{(0.1)^5}{5!}$

$= 3.333 \times 10^{-6}$

(c) Actual value is $y(0.1) = 0.8234134413$. Error is $3.225 \times 10^{-6} \le 3.333 \times 10^{-6}$.

(d) If $h = 0.05$, $y_2 = 0.82341363$.

(e) Error with $h = 0.1$ is 3.225×10^{-6}. Error with $h = 0.05$ is 1.854×10^{-7}.

19. (a) $y^{(5)}(c)\frac{h^5}{5!} = \frac{24}{(c+1)^5}\frac{h^5}{5!}$

(b) $\frac{24}{(c+1)^5}\frac{h^5}{5!} \le 24\frac{(0.1)^5}{5!} = 2.0000 \times 10^{-6}$

(c) From calculation with $h = 0.1$, $y_5 = 0.40546517$.
From calculation with $h = 0.05$, $y_{10} = 0.40546511$.

EXERCISES 9.3 (PAGE 375)

1. $y(x) = -x + e^x$; actual values are $y(0.2) = 1.0214$, $y(0.4) = 1.0918$, $y(0.6) = 1.2221$, $y(0.8) = 1.4255$; approximations are given in Example 1.

3. $y_4 = 0.7232$

5. for $h = 0.2$, $y_5 = 1.5569$; for $h = 0.1$, $y_{10} = 1.5576$

7. for $h = 0.2$, $y_5 = 0.2385$; for $h = 0.1$, $y_{10} = 0.2384$

EXERCISES 9.4 (PAGE 379)

1. $y(x) = -2e^{2x} + 5xe^{2x}$; $y(0.2) = -1.4918$, $y_2 = -1.6800$

3. $y_1 = -1.4928$, $y_2 = -1.4919$

5. $y_1 = 1.4640$, $y_2 = 1.4640$

7. $x_1 = 8.3055$, $y_1 = 3.4199$;
$x_2 = 8.3055$, $y_2 = 3.4199$

9. $x_1 = -3.9123$, $y_1 = 4.2857$;
$x_2 = -3.9123$, $y_2 = 4.2857$

11. $x_1 = 0.4179$, $y_1 = -2.1824$;
$x_2 = 0.4173$, $y_2 = -2.1821$

EXERCISES 9.5 (PAGE 383)

1. $y_1 = -5.6774$, $y_2 = -2.5807$, $y_3 = 6.3226$

3. $y_1 = -0.2259$, $y_2 = -0.3356$, $y_3 = -0.3308$, $y_4 = -0.2167$

5. $y_1 = 3.3751$, $y_2 = 3.6306$, $y_3 = 3.6448$, $y_4 = 3.2355$, $y_5 = 2.1411$

7. $y_1 = 3.8842$, $y_2 = 2.9640$, $y_3 = 2.2064$, $y_4 = 1.5826$, $y_5 = 1.0681$, $y_6 = 0.6430$, $y_7 = 0.2913$

9. $y_1 = 0.2660$, $y_2 = 0.5097$, $y_3 = 0.7357$, $y_4 = 0.9471$, $y_5 = 1.1465$, $y_6 = 1.3353$, $y_7 = 1.5149$, $y_8 = 1.6855$, $y_9 = 1.8474$

11. $y_1 = 0.3492$, $y_2 = 0.7202$, $y_3 = 1.1363$, $y_4 = 1.6233$, $y_5 = 2.2118$, $y_6 = 2.9386$, $y_7 = 3.8490$

13. (c) $y_0 = -2.2755$, $y_1 = -2.0755$, $y_2 = -1.8589$, $y_3 = -1.6126$, $y_4 = -1.3275$

CHAPTER 9 IN REVIEW (PAGE 384)

1. Comparison of numerical methods with $h = 0.1$:

x_n	Euler	Improved Euler	RK4
1.10	2.1386	2.1549	2.1556
1.20	2.3097	2.3439	2.3454
1.30	2.5136	2.5672	2.5695
1.40	2.7504	2.8246	2.8278
1.50	3.0201	3.1157	3.1197

Comparison of numerical methods with $h = 0.05$:

x_n	Euler	Improved Euler	RK4
1.10	2.1469	2.1554	2.1556
1.20	2.3272	2.3450	2.3454
1.30	2.5409	2.5689	2.5695
1.40	2.7883	2.8269	2.8278
1.50	3.0690	3.1187	3.1197

3. Comparison of numerical methods with $h = 0.1$:

x_n	Euler	Improved Euler	RK4
0.60	0.6000	0.6048	0.6049
0.70	0.7095	0.7191	0.7194
0.80	0.8283	0.8427	0.8431
0.90	0.9559	0.9752	0.9757
1.00	1.0921	1.1163	1.1169

Comparison of numerical methods with $h = 0.05$:

x_n	Euler	Improved Euler	RK4
0.60	0.6024	0.6049	0.6049
0.70	0.7144	0.7193	0.7194
0.80	0.8356	0.8430	0.8431
0.90	0.9657	0.9755	0.9757
1.00	1.1044	1.1168	1.1169

5. $h = 0.2$: $y(0.2) \approx 3.2$; $h = 0.1$: $y(0.2) \approx 3.23$

7. $x(0.2) \approx 1.62$, $y(0.2) \approx 1.84$

EXERCISES FOR APPENDIX I (PAGE APP-2)

1. (a) 24 **(b)** 720 **(c)** $\dfrac{4\sqrt{\pi}}{3}$ **(d)** $-\dfrac{8\sqrt{\pi}}{15}$

3. 0.297

EXERCISES FOR APPENDIX II (PAGE APP-18)

1. (a) $\begin{pmatrix} 2 & 11 \\ 2 & -1 \end{pmatrix}$ **(b)** $\begin{pmatrix} -6 & 1 \\ 14 & -19 \end{pmatrix}$
(c) $\begin{pmatrix} 2 & 28 \\ 12 & -12 \end{pmatrix}$

3. (a) $\begin{pmatrix} -11 & 6 \\ 17 & -22 \end{pmatrix}$ **(b)** $\begin{pmatrix} -32 & 27 \\ -4 & -1 \end{pmatrix}$
(c) $\begin{pmatrix} 19 & -18 \\ -30 & 31 \end{pmatrix}$ **(d)** $\begin{pmatrix} 19 & 6 \\ 3 & 22 \end{pmatrix}$

5. (a) $\begin{pmatrix} 9 & 24 \\ 3 & 8 \end{pmatrix}$ **(b)** $\begin{pmatrix} 3 & 8 \\ -6 & -16 \end{pmatrix}$
(c) $\begin{pmatrix} 0 & 0 \\ 0 & 0 \end{pmatrix}$ **(d)** $\begin{pmatrix} -4 & -5 \\ 8 & 10 \end{pmatrix}$.

7. (a) 180 **(b)** $\begin{pmatrix} 4 & 8 & 10 \\ 8 & 16 & 20 \\ 10 & 20 & 25 \end{pmatrix}$ **(c)** $\begin{pmatrix} 6 \\ 12 \\ -5 \end{pmatrix}$

9. (a) $\begin{pmatrix} 7 & 38 \\ 10 & 75 \end{pmatrix}$ **(b)** $\begin{pmatrix} 7 & 38 \\ 10 & 75 \end{pmatrix}$

11. $\begin{pmatrix} -14 \\ 1 \end{pmatrix}$

13. $\begin{pmatrix} -38 \\ -2 \end{pmatrix}$

15. singular

17. nonsingular; $\mathbf{A}^{-1} = \dfrac{1}{4}\begin{pmatrix} -5 & -8 \\ 3 & 4 \end{pmatrix}$

19. nonsingular; $\mathbf{A}^{-1} = \dfrac{1}{2}\begin{pmatrix} 0 & -1 & 1 \\ 2 & 2 & -2 \\ -4 & -3 & 5 \end{pmatrix}$

21. nonsingular; $\mathbf{A}^{-1} = -\dfrac{1}{9}\begin{pmatrix} -2 & -2 & -1 \\ -13 & 5 & 7 \\ 8 & -1 & -5 \end{pmatrix}$

23. $\mathbf{A}^{-1}(t) = \dfrac{1}{2e^{3t}}\begin{pmatrix} 3e^{4t} & -e^{4t} \\ -4e^{-t} & 2e^{-t} \end{pmatrix}$

25. $\dfrac{d\mathbf{X}}{dt} = \begin{pmatrix} -5e^{-t} \\ -2e^{-t} \\ 7e^{-t} \end{pmatrix}$

27. $\dfrac{d\mathbf{X}}{dt} = 4\begin{pmatrix} 1 \\ -1 \end{pmatrix}e^{2t} - 12\begin{pmatrix} 2 \\ 1 \end{pmatrix}e^{-3t}$

29. (a) $\begin{pmatrix} 4e^{4t} & -\pi \sin \pi t \\ 2 & 6t \end{pmatrix}$ **(b)** $\begin{pmatrix} \frac{1}{4}e^{8} - \frac{1}{4} & 0 \\ 4 & 6 \end{pmatrix}$
(c) $\begin{pmatrix} \frac{1}{4}e^{4t} - \frac{1}{4} & (1/\pi) \sin \pi t \\ t^2 & t^3 - t \end{pmatrix}$

31. $x = 3$, $y = 1$, $z = -5$

33. $x = 2 + 4t$, $y = -5 - t$, $z = t$

35. $x = -\frac{1}{2}$, $y = \frac{3}{2}$, $z = \frac{7}{2}$

37. $x_1 = 1$, $x_2 = 0$, $x_3 = 2$, $x_4 = 0$

41. $\mathbf{A}^{-1} = \begin{pmatrix} 0 & \frac{2}{3} & \frac{1}{3} \\ 0 & -\frac{1}{3} & -\frac{2}{3} \\ \frac{1}{3} & -\frac{2}{3} & 0 \end{pmatrix}$

43. $\mathbf{A}^{-1} = \begin{pmatrix} 5 & 6 & -3 \\ 2 & 2 & -1 \\ -1 & -1 & 1 \end{pmatrix}$

45. $\mathbf{A}^{-1} = \begin{pmatrix} -\frac{1}{2} & -\frac{2}{3} & -\frac{1}{6} & \frac{7}{6} \\ 1 & \frac{1}{3} & \frac{1}{3} & -\frac{4}{3} \\ 0 & -\frac{1}{3} & -\frac{1}{3} & \frac{1}{3} \\ -\frac{1}{2} & 1 & \frac{1}{2} & \frac{1}{2} \end{pmatrix}$

47. $\lambda_1 = 6$, $\lambda_2 = 1$, $\mathbf{K}_1 = \begin{pmatrix} 2 \\ 7 \end{pmatrix}$, $\mathbf{K}_2 = \begin{pmatrix} 1 \\ 1 \end{pmatrix}$

49. $\lambda_1 = \lambda_2 = -4$, $\mathbf{K}_1 = \begin{pmatrix} 1 \\ -4 \end{pmatrix}$

51. $\lambda_1 = 0$, $\lambda_2 = 4$, $\lambda_3 = -4$,

$\mathbf{K}_1 = \begin{pmatrix} 9 \\ 45 \\ 25 \end{pmatrix}$, $\mathbf{K}_2 = \begin{pmatrix} 1 \\ 1 \\ 1 \end{pmatrix}$, $\mathbf{K}_3 = \begin{pmatrix} 1 \\ 9 \\ 1 \end{pmatrix}$

53. $\lambda_1 = \lambda_2 = \lambda_3 = -2$,

$\mathbf{K}_1 = \begin{pmatrix} 2 \\ -1 \\ 0 \end{pmatrix}$, $\mathbf{K}_2 = \begin{pmatrix} 0 \\ 0 \\ 1 \end{pmatrix}$

55. $\lambda_1 = 3i$, $\lambda_2 = -3i$,

$\mathbf{K}_1 = \begin{pmatrix} 1 - 3i \\ 5 \end{pmatrix}$, $\mathbf{K}_2 = \begin{pmatrix} 1 + 3i \\ 5 \end{pmatrix}$

Index

D

E

F

G

H

M

T

U

V

W

Y

Z

INDEX

TABLE OF LAPLACE TRANSFORMS

$f(t)$	$\mathscr{L}\{f(t)\} = F(s)$
1. 1	$\dfrac{1}{s}$
2. t	$\dfrac{1}{s^2}$
3. t^n	$\dfrac{n!}{s^{n+1}}$, n a positive integer
4. $t^{-1/2}$	$\sqrt{\dfrac{\pi}{s}}$
5. $t^{1/2}$	$\dfrac{\sqrt{\pi}}{2s^{3/2}}$
6. t^{α}	$\dfrac{\Gamma(\alpha + 1)}{s^{\alpha+1}}$, $\alpha > -1$
7. $\sin kt$	$\dfrac{k}{s^2 + k^2}$
8. $\cos kt$	$\dfrac{s}{s^2 + k^2}$
9. $\sin^2 kt$	$\dfrac{2k^2}{s(s^2 + 4k^2)}$
10. $\cos^2 kt$	$\dfrac{s^2 + 2k^2}{s(s^2 + 4k^2)}$
11. e^{at}	$\dfrac{1}{s - a}$
12. $\sinh kt$	$\dfrac{k}{s^2 - k^2}$
13. $\cosh kt$	$\dfrac{s}{s^2 - k^2}$
14. $\sinh^2 kt$	$\dfrac{2k^2}{s(s^2 - 4k^2)}$
15. $\cosh^2 kt$	$\dfrac{s^2 - 2k^2}{s(s^2 - 4k^2)}$
16. te^{at}	$\dfrac{1}{(s - a)^2}$
17. $t^n e^{at}$	$\dfrac{n!}{(s - a)^{n+1}}$, n a positive integer
18. $e^{at} \sin kt$	$\dfrac{k}{(s - a)^2 + k^2}$
19. $e^{at} \cos kt$	$\dfrac{s - a}{(s - a)^2 + k^2}$
20. $e^{at} \sinh kt$	$\dfrac{k}{(s - a)^2 - k^2}$
21. $e^{at} \cosh kt$	$\dfrac{s - a}{(s - a)^2 - k^2}$
22. $t \sin kt$	$\dfrac{2ks}{(s^2 + k^2)^2}$
23. $t \cos kt$	$\dfrac{s^2 - k^2}{(s^2 + k^2)^2}$
24. $\sin kt + kt \cos kt$	$\dfrac{2ks^2}{(s^2 + k^2)^2}$
25. $\sin kt - kt \cos kt$	$\dfrac{2k^3}{(s^2 + k^2)^2}$
26. $t \sinh kt$	$\dfrac{2ks}{(s^2 - k^2)^2}$
27. $t \cosh kt$	$\dfrac{s^2 + k^2}{(s^2 - k^2)^2}$
28. $\dfrac{e^{at} - e^{bt}}{a - b}$	$\dfrac{1}{(s - a)(s - b)}$
29. $\dfrac{ae^{at} - be^{bt}}{a - b}$	$\dfrac{s}{(s - a)(s - b)}$
30. $1 - \cos kt$	$\dfrac{k^2}{s(s^2 + k^2)}$
31. $kt - \sin kt$	$\dfrac{k^3}{s^2(s^2 + k^2)}$
32. $\dfrac{a \sin bt - b \sin at}{ab(a^2 - b^2)}$	$\dfrac{1}{(s^2 + a^2)(s^2 + b^2)}$
33. $\dfrac{\cos bt - \cos at}{a^2 - b^2}$	$\dfrac{s}{(s^2 + a^2)(s^2 + b^2)}$
34. $\sin kt \sinh kt$	$\dfrac{2k^2 s}{s^4 + 4k^4}$
35. $\sin kt \cosh kt$	$\dfrac{k(s^2 + 2k^2)}{s^4 + 4k^4}$
36. $\cos kt \sinh kt$	$\dfrac{k(s^2 - 2k^2)}{s^4 + 4k^4}$
37. $\cos kt \cosh kt$	$\dfrac{s^3}{s^4 + 4k^4}$
38. $J_0(kt)$	$\dfrac{1}{\sqrt{s^2 + k^2}}$